Special Functions with Applications to Mathematical Physics

Special Functions with Applications to Mathematical Physics

Editor

Francesco Mainardi

MDPI • Basel • Beijing • Wuhan • Barcelona • Belgrade • Manchester • Tokyo • Cluj • Tianjin

Editor
Francesco Mainardi
University of Bologna
Bologna, Italy

Editorial Office
MDPI
St. Alban-Anlage 66
4052 Basel, Switzerland

This is a reprint of articles from Special Issue published online in the open access journal *Mathematics* (ISSN 2227-7390) (available at: https://www.mdpi.com/journal/mathematics/special_issues/Special_Functions_Mathematical_Physics).

For citation purposes, cite each article independently as indicated on the article page online and as indicated below:

LastName, A.A.; LastName, B.B.; LastName, C.C. Article Title. *Journal Name* **Year**, *Volume Number*, Page Range.

ISBN 978-3-0365-6990-1 (Hbk)
ISBN 978-3-0365-6991-8 (PDF)

© 2023 by the authors. Articles in this book are Open Access and distributed under the Creative Commons Attribution (CC BY) license, which allows users to download, copy and build upon published articles, as long as the author and publisher are properly credited, which ensures maximum dissemination and a wider impact of our publications.

The book as a whole is distributed by MDPI under the terms and conditions of the Creative Commons license CC BY-NC-ND.

Contents

Preface to "Special Functions with Applications to Mathematical Physics" vii

Richard Paris
Asymptotic Expansion of the Modified Exponential Integral Involving the Mittag-Leffler Function
Reprinted from: *Mathematics* **2020**, *8*, 428, doi:10.3390/math8030428 1

Arak M. Mathai and Hans J. Haubold
Mathematical Aspects of Krätzel Integral and Krätzel Transform
Reprinted from: *Mathematics* **2020**, *8*, 526, doi:10.3390/math8040526 15

Alexander Apelblat
Differentiation of the Mittag-Leffler Functions with Respect to Parameters in the Laplace Transform Approach
Reprinted from: *Mathematics* **2020**, *8*, 657, doi:10.3390/math8050657 33

Francesco Mainardi and Armando Consiglio
The Wright Functions of the Second Kind in Mathematical Physics
Reprinted from: *Mathematics* **2020**, *8*, 884, doi:10.3390/math8060884 55

Yuri Luchko
The Four-Parameters Wright Function of the Second kind and its Applications in FC
Reprinted from: *Mathematics* **2020**, *8*, 970, doi:10.3390/math8060970 81

Victor Kowalenko
Exact Values of the Gamma Function from Stirling's Formula
Reprinted from: *Mathematics* **2020**, *8*, 1058, doi:10.3390/math8071058 97

Dmitrii Karp and Elena Prilepkina
Transformations of the Hypergeometric $_4F_3$ with One Unit Shift: A Group Theoretic Study
Reprinted from: *Mathematics* **2020**, *8*, 1966, doi:10.3390/math8111966 125

Paolo Emilio Ricci
Laguerre-Type Exponentials, Laguerre Derivatives and Applications. A Survey
Reprinted from: *Mathematics* **2020**, *8*, 2054, doi:10.3390/math8112054 147

Hyun Soo Chung
Some Relationships for the Generalized Integral Transform on Function Space
Reprinted from: *Mathematics* **2020**, *8*, 2246, doi:10.3390/math8122246 165

Virginia Kiryakova
A Guide to Special Functions in Fractional Calculus
Reprinted from: *Mathematics* **2021**, *9*, 106, doi:10.3390/math9010106 181

Árpád Baricz, Dragana Jankov Maširević and Tibor K. Pogány
Approximation of CDF of Non-Central Chi-Square Distribution by Mean-Value Theorems for Integrals
Reprinted from: *Mathematics* **2021**, *9*, 129, doi:10.3390/math9020129 221

Yuriy Povstenko
Some Applications of the Wright Function in Continuum Physics: A Survey
Reprinted from: *Mathematics* **2021**, *9*, 198, doi:10.3390/math9020198 233

Lotfi Boudabsa and Thomas Simon
Some Properties of the Kilbas-Saigo Function
Reprinted from: *Mathematics* 2021, 9, 217, doi:10.3390/math9030217 247

Noufe H. Aljahdaly and S. A. El-Tantawy
On the Multistage Differential Transformation Method for AnalyzingDamping Duffing Oscillator and Its Applications to Plasma Physics
Reprinted from: *Mathematics* 2021, 9, 432, doi:10.3390/math9040432 271

Alexander Apelblat, Armando Consiglio and Francesco Mainardi
The Bateman Functions Revisited after 90 Years—A Survey of Old and New Results
Reprinted from: *Mathematics* 2021, 9, 1273, doi:10.3390/math9111273 283

Jordanka Paneva-Konovska
Series in Le Roy Type Functions: A Set of Results in the Complex Plane—A Survey
Reprinted from: *Mathematics* 2021, 9, 1361, doi:10.3390/math9121361 311

Richard Paris
The Asymptotic Expansion of a Function Introduced by L.L. Karasheva
Reprinted from: *Mathematics* 2021, 9, 1454, doi:10.3390/math9121454 327

Sergei Rogosin and Maryna Dubatovskaya
Fractional Calculus in Russia at the End of XIX Century
Reprinted from: *Mathematics* 2021, 9, 1736, doi:10.3390/math9151736 337

Yuri Luchko
Special Functions of Fractional Calculus in the Form of Convolution Series and Their Applications
Reprinted from: *Mathematics* 2021, 9, 2132, doi:10.3390/math9172132 353

Hyun Soo Chung
Basic Fundamental Formulas for Wiener Transforms Associated with a Pair of Operators on Hilbert Space
Reprinted from: *Mathematics* 2021, 9, 2738, doi:10.3390/math9212738 369

Artem Bykov, Anastasia Grecheneva, Oleg Kuzichkin, Dmitry Surzhik, Gleb Vasilyev and Yerbol Yerbayev
Mathematical Description and Laboratory Study of Electrophysical Methods of Localization of Geodeformational Changes during the Control of the Railway Roadbed
Reprinted from: *Mathematics* 2021, 9, 3164, doi:10.3390/math9243164 381

Alexander Apelblat and Juan Luis González-Santander
The Integral Mittag-Leffler, Whittaker and Wright Functions
Reprinted from: *Mathematics* 2021, 9, 3255, doi:10.3390/math9243255 395

Preface to "Special Functions with Applications to Mathematical Physics"

This MDPI booklet lists the articles published in the Special Issue of the journal Mathematics devoted to special functions with applications in mathematical physics in the years 2020–2021.

The call for papers considered theories and applications of high transcendental functions, including topics found mainly in the list of keywords:

- Mittag-Leffler and related functions, and their applications in mathematical physics;
- Wright and related functions and their applications in mathematical physics;
- Exponential integrals and their extensions with applications in mathematical physics;
- Generalized hypergeometric functions and their extensions with applications.

However, the Special Issue were not limited to the above list, for example, when the content of a paper was clearly related to some high transcendental functions and their applications.

Special attention was reserved for distinct functions exhibiting some relevance in the framework of the theories and applications of the fractional calculus and in their visualization through illuminating plots.

Both research and survey articles were included in this booklet, according to the content list.

Francesco Mainardi
Editor

Article

Asymptotic Expansion of the Modified Exponential Integral Involving the Mittag-Leffler Function

Richard Paris

Division of Computing and Mathematics, Abertay University, Dundee DD1 1HG, UK; r.paris@abertay.ac.uk

Received: 21 February 2020; Accepted: 10 March 2020; Published: 16 March 2020

Abstract: We consider the asymptotic expansion of the generalised exponential integral involving the Mittag-Leffler function introduced recently by Mainardi and Masina [*Fract. Calc. Appl. Anal.* **21** (2018) 1156–1169]. We extend the definition of this function using the two-parameter Mittag-Leffler function. The expansions of the similarly extended sine and cosine integrals are also discussed. Numerical examples are presented to illustrate the accuracy of each type of expansion obtained.

Keywords: asymptotic expansions; exponential integral; Mittag-Leffler function; sine and cosine integrals

MSC: 26A33; 33E12; 34A08; 34C26

1. Introduction

The complementary exponential integral $\mathrm{Ein}(z)$ is defined by

$$\mathrm{Ein}(z) = \int_0^z \frac{1-e^{-t}}{t}\,dt = \sum_{n=1}^\infty \frac{(-)^{n-1} z^n}{n n!} \qquad (z \in \mathbf{C}) \tag{1}$$

and is an entire function. Its connection with the classical exponential integral $\mathcal{E}_1(z) = \int_z^\infty t^{-1} e^{-t}\,dt$, valid in the cut plane $|\arg z| < \pi$, is [1], p. 150.

$$\mathrm{Ein}(z) = \log z + \gamma + \mathcal{E}_1(z), \tag{2}$$

where $\gamma = 0.5772156\ldots$ is the Euler-Mascheroni constant.

In a recent paper, Mainardi and Masina [2] proposed an extension of $\mathrm{Ein}(z)$ by replacing the exponential function in (1) by the one-parameter Mittag-Leffler function

$$E_\alpha(z) = \sum_{n=0}^\infty \frac{z^n}{\Gamma(\alpha n+1)} \qquad (z \in \mathbf{C},\ \alpha > 0),$$

which generalises the exponential function e^z. They introduced the function for any $\alpha > 0$ in the cut plane $|\arg z| < \pi$

$$\mathrm{Ein}_\alpha(z) = \int_0^z \frac{1 - E_\alpha(-t^\alpha)}{t^\alpha}\,dt = \sum_{n=0}^\infty \frac{(-)^n z^{\alpha n+1}}{(\alpha n+1)\Gamma(\alpha n + \alpha + 1)}, \tag{3}$$

which when $\alpha = 1$ reduces to the function $\mathrm{Ein}(z)$. A physical application of this function for $0 \leq \alpha \leq 1$ arises in the study of the creep features of a linear viscoelastic model; see Reference [3] for details. An analogous extension of the generalised sine and cosine integrals was also considered in Reference [2]. Plots of all these functions for $\alpha \in [0,1]$ were given.

Here we consider a slightly more general version of (3) based on the two-parameter Mittag-Leffler function given by

$$E_{\alpha,\beta}(z) = \sum_{n=0}^{\infty} \frac{z^n}{\Gamma(\alpha n + \beta)} \qquad (z \in \mathbf{C}, \ \alpha > 0),$$

where β will be taken to be real. Then the extended complementary exponential integral we shall consider is

$$\mathrm{Ein}_{\alpha,\beta}(z) = \int_0^z \frac{1 - E_{\alpha,\beta}(-t^\alpha)}{t^\alpha} dt = \sum_{n=1}^{\infty} \frac{(-)^{n-1}}{\Gamma(\alpha n + \beta)} \int_0^z t^{\alpha n - \alpha} dt$$

$$= z \sum_{n=0}^{\infty} \frac{(-)^n z^{\alpha n}}{(\alpha n + 1)\Gamma(\alpha n + \alpha + \beta)}, \tag{4}$$

upon replacement of $n - 1$ by n in the last summation. When $\beta = 1$ this reduces to (3) so that $\mathrm{Ein}_{\alpha,1}(z) = \mathrm{Ein}_\alpha(z)$.

The asymptotic expansion of this function will be obtained for large complex z with the parameters α, β held fixed. We achieve this by consideration of the asymptotics of a related function using the theory developed for integral functions of hypergeometric type as discussed, for example, in Reference [4], §2.3. An interesting feature of the expansion of $\mathrm{Ein}_{\alpha,\beta}(x)$ for $x \to +\infty$ when $\alpha \in (0,1]$ is the appearance of a logarithmic term whenever $\alpha = 1, \frac{1}{2}, \frac{1}{3}, \ldots$. Similar expansions are obtained for the extended sine and cosine integrals in Section 4. The paper concludes with the presentation of some numerical results that demonstrate the accuracy of the different expansions obtained.

2. The Asymptotic Expansion of a Related Function for $|z| \to \infty$

To determine the asymptotic expansion of $\mathrm{Ein}_{\alpha,\beta}(z)$ for large complex z with the parameters α and β held fixed, we shall find it convenient to consider the related function defined by

$$F(\chi) := \sum_{n=0}^{\infty} \frac{\chi^n}{(\alpha n + \gamma)\Gamma(\alpha n + \alpha + \beta)} = \sum_{n=0}^{\infty} g(n)\frac{\chi^n}{n!} \qquad (\chi \in \mathbf{C}), \tag{5}$$

where

$$g(n) = \frac{\Gamma(n+1)}{(\alpha n + \gamma)\Gamma(\alpha n + \alpha + \beta)} = \frac{\Gamma(\alpha n + \gamma)\Gamma(n+1)}{\Gamma(\alpha n + \gamma + 1)\Gamma(\alpha n + \alpha + \beta)}.$$

It is readily seen that, when $\gamma = 1$,

$$\mathrm{Ein}_{\alpha,\beta}(z) = z\,F(-z^\alpha).$$

The parameter $\gamma > 0$, but will be chosen to have two specific values in Sections 3 and 4; namely, $\gamma = 1$ and $\gamma = 1 + \alpha$. It will be shown that the asymptotic expansion of $F(\chi)$ consists of an algebraic and an exponential expansion valid in different sectors of the complex χ-plane.

The function $F(\chi)$ in (5) is a case of the Fox-Wright function

$$_p\Psi_q(\chi) = \sum_{n=0}^{\infty} \frac{\prod_{r=1}^{p} \Gamma(\alpha_r n + a_r)}{\prod_{r=1}^{q} \Gamma(\beta_r n + b_r)} \frac{\chi^n}{n!}, \tag{6}$$

corresponding to $p = q = 2$. In (6) the parameters α_r and β_r are real and positive and a_r and b_r are arbitrary complex numbers. We also assume that the α_r and a_r are subject to the restriction

$$\alpha_r n + a_r \neq 0, -1, -2, \ldots \qquad (n = 0, 1, 2, \ldots\,;\, 1 \leq r \leq p)$$

so that no gamma function in the numerator in (6) is singular. We introduce the following parameters associated (empty sums and products are to be interpreted as zero and unity, respectively) with ${}_p\Psi_q(\chi)$ which play a key role in the analysis of its asymptotic behaviour. are given by

$$\kappa = 1 + \sum_{r=1}^{q} \beta_r - \sum_{r=1}^{p} \alpha_r, \qquad h = \prod_{r=1}^{p} \alpha_r^{\alpha_r} \prod_{r=1}^{q} \beta_r^{-\beta_r},$$

$$\vartheta = \sum_{r=1}^{p} a_r - \sum_{r=1}^{q} b_r + \tfrac{1}{2}(q-p), \qquad \vartheta' = 1 - \vartheta. \tag{7}$$

The asymptotic expansion of $F(\chi)$ is discussed in detail in Reference [5] Section 12, and is summarised in [4,6]. The algebraic expansion of $F(\chi)$ is obtained from the Mellin-Barnes integral representation [4], p. 56.

$$F(\chi) = \frac{1}{2\pi i} \int_{c-\infty i}^{c+\infty i} \frac{\Gamma(-s)\Gamma(1+s)(\chi e^{\mp \pi i})^s}{(\alpha s + \gamma)\Gamma(\alpha s + \alpha + \beta)} ds, \qquad |\arg(-\chi)| < \pi(1 - \tfrac{1}{2}\alpha),$$

where, with $-\gamma/\alpha < c < 0$, the integration path lies to the left of the poles of $\Gamma(-s)$ at $s = 0, 1, 2, \ldots$ but to the right of the poles at $s = -\gamma/\alpha$ and $s = -k-1, k = 0, 1, 2, \ldots$. The upper or lower sign is taken according as $\arg \chi > 0$ or $\arg \chi < 0$, respectively. It is seen that when $\alpha = \gamma/m$, $m = 1, 2, \ldots$ the pole at $s = -m$ is double and its residue must be evaluated accordingly. Displacement of the integration path to the left when $0 < \alpha < 2$ and evaluation of the residues then produces the algebraic expansion $H(\chi e^{\mp \pi i})$, where

$$H(\chi) = \begin{cases} \dfrac{\pi/\alpha}{\sin \gamma \pi/\alpha} \dfrac{\chi^{-\gamma/\alpha}}{\Gamma(\alpha+\beta-\gamma)} + \sum_{k=0}^{\infty} \dfrac{(-)^k \chi^{-k-1}}{(\gamma - \alpha(k+1))\Gamma(\beta - \alpha k)} & (\alpha \neq \tfrac{\gamma}{m}) \\[1em] \dfrac{(-)^{m-1} \chi^{-m}}{\Gamma(\alpha+\beta-\gamma)} \left\{ \dfrac{m}{\gamma} \log \chi - \psi(\alpha+\beta-\gamma) \right\} + \sum_{\substack{k=0 \\ k \neq m-1}}^{\infty} \dfrac{(-)^k \chi^{-k-1}}{(\gamma - \alpha(k+1))\Gamma(\beta - \alpha k)} & (\alpha = \tfrac{\gamma}{m}), \end{cases} \tag{8}$$

and ψ denotes the logarithmic derivative of the gamma function.

The exponential expansion associated with ${}_p\Psi_q(\chi)$ is given by [6] p. 299, [4] p. 57.

$$\mathcal{E}(\chi) := X^{\vartheta} e^X \sum_{j=0}^{\infty} A_j X^{-j}, \qquad X = \kappa(h\chi)^{1/\kappa}, \tag{9}$$

where the coefficients A_j are those appearing in the inverse factorial expansion

$$\frac{1}{\Gamma(1+s)} \frac{\prod_{r=1}^{p} \Gamma(\alpha_r n + a_r)}{\prod_{r=1}^{q} \Gamma(\beta_r n + b_r)} = \kappa A_0 (h\kappa^\kappa)^s \left\{ \sum_{j=0}^{M-1} \frac{c_j}{\Gamma(\kappa s + \vartheta' + j)} + \frac{\rho_M(s)}{\Gamma(\kappa s + \vartheta' + M)} \right\} \tag{10}$$

with $c_0 = 1$. Here M is a positive integer and $\rho_M(s) = O(1)$ for $|s| \to \infty$ in $|\arg s| < \pi$. The constant A_0 is specified by

$$A_0 = (2\pi)^{\frac{1}{2}(p-q)} \kappa^{-\frac{1}{2}-\vartheta} \prod_{r=1}^{p} \alpha_r^{a_r - \frac{1}{2}} \prod_{r=1}^{q} \beta_r^{\frac{1}{2} - b_r}.$$

The coefficients c_j are independent of s and depend only on the parameters $p, q, \alpha_r, \beta_r, a_r$ and b_r.

For the function $F(\chi)$, we have

$$\kappa = \alpha, \quad h = \alpha^{-\alpha}, \quad \vartheta = -\alpha - \beta, \quad A_0 = \alpha^{-1}.$$

We are in the fortunate position that the normalised coefficients c_j in this case can be determined explicitly as $c_j = (\alpha + \beta - \gamma)_j$. This follows from the well-known (convergent) expansion given in Reference [4,7], p. 41.

$$\frac{1}{(\alpha s + \gamma)\Gamma(\alpha s + \alpha + \beta)} = \sum_{j=0}^{\infty} \frac{(\alpha + \beta - \gamma)_j}{\Gamma(\alpha s + \vartheta' + j)} \qquad (\Re(s) > -\gamma/\alpha), \tag{11}$$

to which, in the case of $F(\chi)$, the ratio of gamma functions appearing on the left-hand side of (10) reduces. Then, with $X = \chi^{1/\alpha}$ we have from (9) the exponential expansion associated with $F(\chi)$ given by

$$\mathcal{E}(\chi) = \frac{1}{\alpha}\chi^{\vartheta/\alpha} \exp[\chi^{1/\alpha}] \sum_{j=0}^{\infty} (\alpha + \beta - \gamma)_j \chi^{-j/\alpha}. \tag{12}$$

From Reference [4] pp. 57–58, we then obtain the asymptotic expansion for $|\chi| \to \infty$ when $0 < \alpha < 2$

$$F(\chi) \sim \begin{cases} \mathcal{E}(\chi) + H(\chi e^{\mp\pi i}) & |\arg \chi| < \tfrac{1}{2}\pi\alpha \\ H(\chi e^{\mp\pi i}) & |\arg(-\chi)| < \pi(1 - \tfrac{1}{2}\alpha) \end{cases} \tag{13}$$

and, when $\alpha = 2$,

$$F(\chi) \sim \mathcal{E}(\chi) + \mathcal{E}(\chi e^{\mp 2\pi i}) + H(\chi e^{\mp\pi i}) \qquad |\arg \chi| \leq \pi. \tag{14}$$

The upper and lower signs are chosen according as $\arg \chi > 0$ or $\arg \chi < 0$, respectively. It may be noted that the expansions $\mathcal{E}(\chi e^{\mp 2\pi i})$ in (14) only become significant in the neighbourhood of $\arg \chi = \pm\pi$. When $\alpha > 2$, the expansion of $F(\chi)$ is exponentially large for all values of $\arg \chi$ (see Reference [4], p. 58) and accordingly we omit this case as it is unlikely to be of physical interest.

Remark 1. *The exponential expansion $\mathcal{E}(\chi)$ in (13) continues to hold beyond the sector $|\arg \chi| < \tfrac{1}{2}\pi\alpha$, where it becomes exponentially small in the sectors $\pi\alpha \leq |\arg \chi| < \tfrac{1}{2}\pi\alpha$ when $0 < \alpha \leq 1$. The rays $\arg \chi = \pm\pi\alpha$ are Stokes lines, where $\mathcal{E}(\chi)$ is maximally subdominant relative to the algebraic expansion $H(\chi e^{\mp\pi i})$. On these rays, $\mathcal{E}(\chi)$ undergoes a Stokes phenomenon, where the exponentially small expansion "switches off" in a smooth manner as $|\arg \chi|$ increases [1], §2.11(iv), with its value to leading order given by $\tfrac{1}{2}\mathcal{E}(\chi)$; see Reference [8] for a more detailed discussion of this point in the context of the confluent hypergeometric functions. We do not consider exponentially small contributions to $F(\chi)$ here, except to briefly mention in Section 3 the situation pertaining to the case $\alpha = 1$.*

3. The Asymptotic Expansion of $\text{Ein}_{\alpha,\beta}(z)$ for $|z| \to \infty$

The asymptotic expansion of $\text{Ein}_{\alpha,\beta}(z)$ defined in (4) can now be constructed from that of $F(\chi)$ with the parameter $\gamma = 1$. It is sufficient, for real α, β, to consider $0 \leq \arg z \leq \pi$, since the expansion when $\arg z < 0$ is given by the conjugate value. With $\chi = -z^\alpha = e^{-\pi i}z^\alpha$, the exponentially large sector $|\arg \chi| < \tfrac{1}{2}\pi\alpha$ becomes $|-\pi + \alpha \arg z| < \tfrac{1}{2}\pi\alpha$; that is

$$\theta_0 < \arg z < \theta_0 + \pi, \qquad \theta_0 := \frac{\pi}{2\alpha}(2 - \alpha). \tag{15}$$

On the boundaries of this sector the exponential expansion is of an oscillatory character. When $0 < \alpha < \tfrac{2}{3}$, we note that the exponentially large sector (15) lies outside the sector of interest $0 \leq \arg z \leq \pi$.

We define the algebraic and exponential asymptotic expansions

$$H_{\alpha,\beta}(z) = \begin{cases} \dfrac{\pi/\alpha}{\sin(\pi/\alpha)\,\Gamma(\alpha+\beta-1)} + \sum_{k=0}^{\infty} \dfrac{(-)^k z^{1-\alpha(k+1)}}{(1-\alpha(k+1))\Gamma(\beta-\alpha k)} & (\alpha \neq m^{-1}) \\[2ex] \dfrac{(-)^{m-1}}{\Gamma(\alpha+\beta-1)}\{\log z - \psi(\alpha+\beta-1)\} + \sum_{\substack{k=0 \\ k\neq m-1}}^{\infty} \dfrac{(-)^k z^{1-\alpha(k+1)}}{(1-\alpha(k+1))\Gamma(\beta-\alpha k)} & (\alpha = m^{-1}), \end{cases} \qquad (16)$$

where $m = 1, 2, \ldots$, and

$$\mathcal{E}_{\alpha,\beta}(z) = \frac{(e^{-\pi i/\alpha}z)^{\vartheta}}{\alpha} \exp\left[e^{-\pi i/\alpha}z\right] \sum_{j=0}^{\infty} (\alpha+\beta-1)_j (e^{-\pi i/\alpha}z)^{-j} \qquad (17)$$

where we recall that $\vartheta = -\alpha - \beta$. Then the following result holds:

Theorem 1. *Let m be a positive integer, with $\alpha > 0$ and β real and $\theta_0 = \pi(2-\alpha)/(2\alpha)$. Then the following expansions hold for $|z| \to \infty$*

$$\mathrm{Ein}_{\alpha,\beta}(z) \sim H_{\alpha,\beta}(z) \qquad (0 \leq \arg z \leq \pi) \qquad (18)$$

when $0 < \alpha < \frac{2}{3}$, and

$$\mathrm{Ein}_{\alpha,\beta}(z) \sim \begin{cases} H_{\alpha,\beta}(z) & (0 \leq \arg z < \theta_0) \\ z\mathcal{E}_{\alpha,\beta}(z) + H_{\alpha,\beta}(z) & (\theta_0 \leq \arg z \leq \pi) \end{cases} \qquad (19)$$

when $\frac{2}{3} \leq \alpha < 2$. Finally, when $\alpha = 2$ we have $\mathrm{Ein}_{2,\beta}(-z) = -\mathrm{Ein}_{2,\beta}(z)$ and it is therefore sufficient to consider $0 \leq \arg z \leq \frac{1}{2}\pi$. Then, from (14), we obtain the expansion when $\alpha = 2$

$$\mathrm{Ein}_{2,\beta}(z) \sim z\{\mathcal{E}_{2,\beta}(z) + \mathcal{E}_{2,\beta}(ze^{\pi i})\} + H_{2,\beta}(z) \qquad (0 \leq \arg z \leq \tfrac{1}{2}\pi). \qquad (20)$$

We note from Theorem 1 that when $z \to -\infty$ the value of $\mathrm{Ein}_{\alpha,\beta}(z)$ is, in general, complex-valued.

In the case of main physical interest, when $z = x > 0$ is a real variable, we have the following expansion:

Theorem 2. *When $z = x\ (>0)$ we have from Theorem 1 the expansions*

$$\mathrm{Ein}_{\alpha,\beta}(x) \sim H_{\alpha,\beta}(x) \qquad (21)$$

for $0 < \alpha < 2$, and from (17) and (20) when $\alpha = 2$

$$\mathrm{Ein}_{2,\beta}(x) \sim H_{2,\beta}(x) - x^{-1-\beta} \sum_{j=0}^{\infty} \frac{(1+\beta)_j}{x^j} \cos\left[x - \tfrac{1}{2}\pi(\beta+j)\right] \qquad (22)$$

as $x \to +\infty$.

It is worth noting that a logarithmic term is present in the asymptotic expansion of $\mathrm{Ein}_{\alpha,\beta}(x)$ whenever $\alpha = 1, \frac{1}{2}, \frac{1}{3}, \ldots$.

The Case α = 1

The special case $\alpha = 1$ deserves further consideration. From (16) and (21) we obtain the expansion

$$\text{Ein}_{1,\beta}(x) \sim \frac{1}{\Gamma(\beta)}\{\log x - \psi(\beta)\} - \sum_{k=1}^{\infty} \frac{(-x)^{-k}}{k\Gamma(\beta - k)} \qquad (x \to +\infty). \qquad (23)$$

If $\beta = 1$, the asymptotic sum in (23) vanishes and

$$\text{Ein}_{1,1}(x) \sim \log x + \gamma \qquad (24)$$

for large x. But we have the exact evaluation (compare (2))

$$\text{Ein}_{1,1}(x) = x \sum_{n=0}^{\infty} \frac{(-x)^n}{(n+1)^2 n!} = \log x + \gamma + \mathcal{E}_1(x)$$

$$\sim \log x + \gamma + \frac{e^{-x}}{x} \sum_{j=0}^{\infty} \frac{(-)^j j!}{x^j} \qquad (x \to +\infty) \qquad (25)$$

by Reference [1], (6.12.1). The additional asymptotic sum appearing in (25) is exponentially small as $x \to +\infty$ and is consequently not accounted for in the result (24).

From Remark 1, it is seen that there are Stokes lines at $\arg z = \pm\pi(1-\alpha)$, which coalesce on the positive real axis when $\alpha = 1$. In the sense of increasing $\arg z$ in the neighbourhood of the positive real axis, the exponential expansion $\mathcal{E}_{1,\beta}(z)$ is in the process of *switching on* across $\arg z = \pi(1-\alpha)$ and $\overline{\mathcal{E}_{1,\beta}(z)}$ (where the bar denotes the complex conjugate) is in the process of *switching off* across $\arg z = -\pi(1-\alpha)$. When $\alpha = 1$, this produces the exponential contribution

$$\tfrac{1}{2}x\{\mathcal{E}_{1,\beta}(x) + \overline{\mathcal{E}_{1,\beta}(x)}\} = \frac{e^{-x}}{x^\beta} \cos \pi\beta \sum_{j=0}^{\infty} \frac{(-)^{j+1}(\beta)_j}{x^j}$$

for large x. Thus, the more accurate version of (23) should read

$$\text{Ein}_{1,\beta}(x) \sim \frac{1}{\Gamma(\beta)}\{\log x - \psi(\beta)\} - \sum_{k=1}^{\infty} \frac{(-x)^{-k}}{k\Gamma(\beta-k)} - \frac{e^{-x}}{x^\beta} \cos\pi\beta \sum_{j=0}^{\infty} \frac{(-)^j(\beta)_j}{x^j} \qquad (26)$$

as $x \to +\infty$. When $\beta = 1$, this correctly reduces to (25).

When $\beta = 2$, we have [9]

$$\text{Ein}_{1,2}(x) = x \sum_{n=0}^{\infty} \frac{(-x)^n}{(n+1)^2(n+2)n!} = \log x - \psi(2) + \frac{1}{x} + \mathcal{E}_1(x) - \frac{e^{-x}}{x}$$

$$\sim \log x - \psi(2) + \frac{1}{x} + \frac{e^{-x}}{x} \sum_{j=1}^{\infty} \frac{(-)^j j!}{x^j} \qquad (x \to +\infty).$$

This can be seen also to agree with (26) after a little rearrangement.

4. The Generalised Sine and Cosine Integrals

The sine and cosine integrals are defined by [1], §6.2,

$$\text{Si}(z) = \int_0^z \frac{\sin t}{t}\, dt, \qquad \text{Cin}(z) = \int_0^z \frac{1 - \cos t}{t}\, dt.$$

Mainardi and Masina [2] generalised these definitions by replacing the trigonometric functions by

$$\sin_\alpha(t) = t^\alpha E_{2\alpha,\alpha+\beta}(-t^{2\alpha}) = \sum_{n=0}^\infty \frac{(-)^n t^{(2n+1)\alpha}}{\Gamma(2n\alpha+\alpha+\beta)}, \quad \cos_\alpha(t) = E_{2\alpha,\beta}(-t^{2\alpha}) = \sum_{n=0}^\infty \frac{(-)^n t^{2n\alpha}}{\Gamma(2n\alpha+\beta)}$$

with $\beta = 1$ to produce

$$\begin{cases} \mathrm{Sin}_\alpha(z) = \displaystyle\int_0^z \frac{\sin_\alpha(t)}{t^\alpha} dt = \sum_{n=0}^\infty \frac{(-)^n z^{2n\alpha+1}}{(2n\alpha+1)\Gamma(2n\alpha+\alpha+1)} \\[2ex] \mathrm{Cin}_\alpha(z) = \displaystyle\int_0^\infty \frac{1-\cos_\alpha(t)}{t^\alpha} dt = \sum_{n=0}^\infty \frac{(-)^n z^{2n\alpha+\alpha+1}}{(2n\alpha+\alpha+1)\Gamma(2n\alpha+2\alpha+1)} \end{cases} \tag{27}$$

Here we extend the definitions (27) by including the additional parameter $\beta \in \mathbf{R}$ in the Mittag-Leffler functions and consider the functions

$$\begin{cases} \mathrm{Sin}_{\alpha,\beta}(z) = z \sum_{n=0}^\infty \frac{(-)^n z^{2n\alpha}}{(2n\alpha+1)\Gamma(2n\alpha+\alpha+\beta)} \\[2ex] \mathrm{Cin}_{\alpha,\beta}(z) = z^{1+\alpha} \sum_{n=0}^\infty \frac{(-)^n z^{2n\alpha}}{(2n\alpha+\alpha+1)\Gamma(2n\alpha+2\alpha+\beta)} \end{cases} \tag{28}$$

The asymptotics of $\mathrm{Sin}_{\alpha,\beta}(z)$ and $\mathrm{Cin}_{\alpha,\beta}(z)$ can be deduced from the results in Section 2. However, here we restrict ourselves to determining the asymptotic expansion of these functions for large $|z|$ in a sector enclosing the positive real z-axis, where for $0 < \alpha < 1$ they only have an algebraic-type expansion. We observe in passing that

$$\mathrm{Sin}_{\alpha,\beta}(z) = \mathrm{Ein}_{2\alpha,\beta-\alpha}(z). \tag{29}$$

Comparison of the series expansion for $\mathrm{Sin}_{\alpha,\beta}(z)$ with $F(\chi)$ in Section 2, with the substitutions $\alpha \to 2\alpha$, $\beta \to \beta - \alpha$ and $\gamma = 1$ (or from the above identity combined with Theorems 1 and 2), produces the following expansion:

Theorem 3. *For $m = 1, 2, \ldots$ and $0 < \alpha < 1$ we have the algebraic expansions*

$\mathrm{Sin}_{\alpha,\beta}(z)$

$$\sim \begin{cases} \dfrac{\pi/(2\alpha)}{\sin(\pi/(2\alpha))\Gamma(\alpha+\beta-1)} + \displaystyle\sum_{k=0}^\infty \dfrac{(-)^k z^{1-2\alpha(k+1)}}{(1-2\alpha(k+1))\Gamma(\beta-(2k+1)\alpha)} & (\alpha \neq (2m)^{-1}) \\[3ex] \dfrac{(-)^{m-1}}{\Gamma(\alpha+\beta-1)}\{\log z - \psi(\alpha+\beta-1)\} + \displaystyle\sum_{\substack{k=0 \\ k \neq m-1}}^\infty \dfrac{(-)^k z^{1-2\alpha(k+1)}}{(1-2\alpha(k+1))\Gamma(\beta-(2k+1)\alpha)} & (\alpha = (2m)^{-1}) \end{cases} \tag{30}$$

as $|z| \to \infty$ in the sector $|\arg z| < \pi(1-\alpha)/(2\alpha)$.

A similar treatment for $\mathrm{Cin}_{\alpha,\beta}(z)$ shows that with the substitutions $\alpha \to 2\alpha$, $\beta \to \beta$ and $\gamma = 1+\alpha$ we obtain the following expansion:

Theorem 4. *For $m = 1, 2, \ldots$ and $0 < \alpha < 1$ we have the algebraic expansions*

$\mathrm{Cin}_{\alpha,\beta}(z)$

$$\sim \begin{cases} \dfrac{\pi/(2\alpha)}{\cos(\pi/(2\alpha))\Gamma(\alpha+\beta-1)} + \displaystyle\sum_{k=0}^{\infty} \dfrac{(-)^k z^{1-(2k+1)\alpha}}{(1-(2k+1)\alpha)\Gamma(\beta-2\alpha k)} & (\alpha \neq (2m-1)^{-1}) \\[2ex] \dfrac{(-)^{m-1}}{\Gamma(\alpha+\beta-1)}\{\log z - \psi(\alpha+\beta-1)\} + \displaystyle\sum_{\substack{k=0 \\ k\neq m-1}}^{\infty} \dfrac{(-)^k z^{1-(2k+1)\alpha}}{(1-(2k+1)\alpha)\Gamma(\beta-2\alpha k)} & (\alpha = (2m-1)^{-1}) \end{cases} \quad (31)$$

as $|z| \to \infty$ in the sector $|\arg z| < \pi(1-\alpha)/(2\alpha)$.

The expansions of $\text{Sin}_{\alpha,\beta}(x)$ and $\text{Cin}_{\alpha,\beta}(x)$ as $x \to +\infty$ when $0 < \alpha < 1$ follow immediately from Theorems 3 and 4.

As $x \to +\infty$ when $\alpha = 1$, the exponentially oscillatory contribution to $\text{Sin}_{1,\beta}(x)$ can be obtained directly from (22) together with (29). In the case of $\text{Cin}_{1,\beta}(x)$, we obtain from (9) with $\kappa = 2$, $h = \frac{1}{4}$, $\vartheta = -2 - \beta$, $X = \chi^{1/2}$ and $A_0 = \frac{1}{2}$ the exponential expansion

$$\mathcal{E}(\chi) = \frac{1}{2}\chi^{\vartheta/2} \exp[\chi^{1/2}] \sum_{j=0}^{\infty} c_j \chi^{-j/2}, \qquad \chi = e^{-\pi i} x^2,$$

with the coefficients $c_j = (\beta)_j$. Then the exponential contribution to $\text{Cin}_{1,\beta}(x)$ is

$$x^2\{\mathcal{E}(\chi) + \mathcal{E}(\chi e^{\pi i})\} = -x^{-\beta} \sum_{j=0}^{\infty} \frac{(\beta)_j}{x^j} \cos[x - \tfrac{1}{2}\pi(\beta+j)] \qquad (x \to +\infty).$$

Collecting together these results we finally obtain the following theorem.

Theorem 5. *When $\alpha = 1$ and β is real the following expansions hold:*

$$\text{Sin}_{1,\beta}(x) \sim \frac{\pi}{2\Gamma(\beta)} - \sum_{k=0}^{\infty} \frac{(-)^k x^{-2k-1}}{(2k+1)\Gamma(\beta-1-2k)}$$

$$+ x^{-\beta} \sum_{j=0}^{\infty} \frac{(\beta)_j}{x^j} \sin[x - \tfrac{1}{2}\pi(\beta+j)] \quad (32)$$

and

$$\text{Cin}_{1,\beta}(x) \sim \frac{1}{\Gamma(\beta)}\{\log x - \psi(\beta)\} - \sum_{k=1}^{\infty} \frac{(-)^k x^{-2k}}{2k\Gamma(\beta-2k)}$$

$$- x^{-\beta} \sum_{j=0}^{\infty} \frac{(\beta)_j}{x^j} \cos[x - \tfrac{1}{2}\pi(\beta+j)] \quad (33)$$

as $x \to +\infty$.

When $\beta > 0$, it is seen that $\text{Sin}_{1,\beta}(x)$ approaches the constant value $\pi/(2\Gamma(\beta))$ whereas $\text{Cin}_{1,\beta}(x)$ grows logarithmically like $\log(x)/\Gamma(\beta)$ as $x \to +\infty$.

5. Numerical Results

In this section we present numerical results confirming the accuracy of the various expansions obtained in this paper. In all cases we have employed optimal truncation (that is truncation at, or near, the least term in modulus) of the algebraic and (when appropriate) the exponential expansions. The numerical values of $\text{Ein}_{\alpha,\beta}(x)$ were computed from (4) using high-precision evaluation of the terms in the suitably truncated sum.

We first present results in the physically interesting case of $0 < \alpha \leq 1$ and $\beta = 1$ considered in Reference [2]. Table 1 shows the values (In the tables we write the values as $x(y)$ instead of $x \times 10^y$.) of the absolute relative error in the computation of $\text{Ein}_{\alpha,1}(x)$ from the asymptotic expansions in

Theorem 2 for several values of x and different α in the extended range $0 < \alpha \leq 2$. The expansion for $0 < \alpha < 2$ is given by the algebraic expansion in (21); this contains a logarithmic term for the values $\alpha = \frac{1}{4}, \frac{1}{2}, 1$. The progressive loss of accuracy when $\alpha > 1$ can be attributed to the presence of the approaching exponentially large sector, whose lower boundary is, from (15), given by $\theta_0 = \pi(2-\alpha)/(2\alpha)$. In the final case $\alpha = 2$, the accuracy is seen to suddenly increase considerably. This is due to the inclusion of the (oscillatory) exponential contribution, which from (22), takes the form

$$\text{Ein}_{2,1}(x) \sim \frac{1}{2}\pi - \frac{1}{x} - \sum_{j=1}^{\infty} \frac{j!}{x^{j+1}} \cos(x - \tfrac{1}{2}\pi j) \qquad (x \to +\infty).$$

In Figure 1 we show some plots of $\text{Ein}_{\alpha,1}(x)$ for values of α in the range $0 < \alpha \leq 1$. In Figure 2 the asymptotic approximations for two values of α are shown compared with the corresponding curves of $\text{Ein}_{\alpha,1}(x)$.

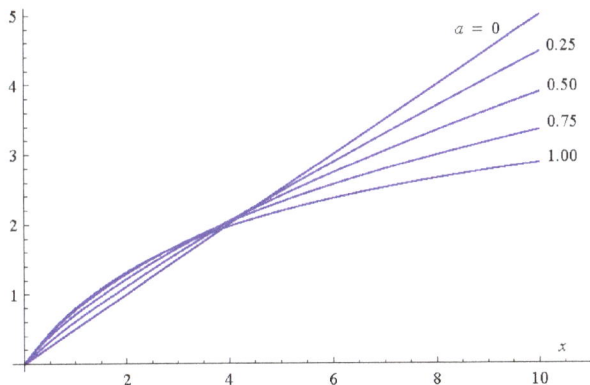

Figure 1. Plots of $\text{Ein}_{\alpha,1}(x)$ for different values of α.

Table 1. The absolute relative error in the computation of $\text{Ein}_{\alpha,\beta}(x)$ from Theorem 2 for different values of α and x when $\beta = 1$.

x	$\alpha = 0.25$	$\alpha = 0.40$	$\alpha = 0.50$	$\alpha = 0.75$	$\alpha = 1.00$
5	1.602×10^{-4}	1.678×10^{-5}	2.012×10^{-4}	2.115×10^{-4}	5.249×10^{-4}
10	5.733×10^{-7}	1.735×10^{-7}	4.413×10^{-7}	2.339×10^{-7}	1.442×10^{-6}
20	3.680×10^{-11}	3.031×10^{-11}	6.526×10^{-12}	9.362×10^{-12}	2.753×10^{-11}
30	1.808×10^{-16}	9.384×10^{-16}	1.543×10^{-16}	1.337×10^{-16}	7.595×10^{-16}
x	$\alpha = 1.20$	$\alpha = 1.40$	$\alpha = 1.60$	$\alpha = 1.80$	$\alpha = 2.00$
5	1.121×10^{-3}	1.301×10^{-4}	5.279×10^{-3}	1.407×10^{-2}	1.550×10^{-3}
10	4.345×10^{-6}	3.168×10^{-5}	2.103×10^{-4}	1.536×10^{-4}	2.849×10^{-6}
20	2.147×10^{-10}	2.277×10^{-8}	1.671×10^{-6}	4.751×10^{-5}	4.926×10^{-10}
30	4.388×10^{-14}	2.363×10^{-11}	2.125×10^{-8}	6.216×10^{-6}	1.613×10^{-14}

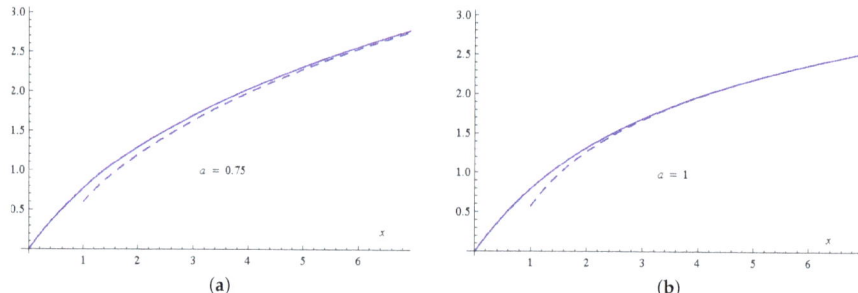

Figure 2. Plots of $\text{Ein}_{\alpha,1}(x)$ (solid curves) and the leading asymptotic approximation (dashed curves) for (**a**) $\alpha = 0.75$ and (**b**) $\alpha = 1$.

Table 2 shows the values of the absolute relative error in the computation of $\text{Ein}_{\alpha,\beta}(z)$ from the asymptotic expansions in Theorem 1 for complex z for values of α in the range $0 < \alpha \leq 2$. It will noticed that there is a sudden reduction in the error when $\alpha = 1$ and $\theta = \pi/4$. In this case, the value of $\theta_0 = \frac{1}{2}\pi$ and a more accurate treatment would include the exponentially small contribution $z\mathcal{E}_{\alpha,\beta}(z)$. When this term is included we find the absolute relative error equal to 6.935×10^{-11}.

Table 2. The absolute relative error in the computation of $\text{Ein}_{\alpha,\beta}(z)$ from Theorem 1 for different α and θ when $z = 20e^{i\theta}$ and $\beta = 1/3$.

θ	$\alpha = 0.40$	$\alpha = 0.50$	$\alpha = 1.00$	$\alpha = 1.50$	$\alpha = 2.00$
0	2.400×10^{-8}	5.494×10^{-10}	2.702×10^{-10}	1.572×10^{-6}	5.119×10^{-10}
$\pi/4$	2.553×10^{-8}	1.820×10^{-9}	1.142×10^{-7}	1.202×10^{-8}	8.204×10^{-8}
$\pi/2$	3.026×10^{-8}	4.057×10^{-9}	1.756×10^{-10}	2.021×10^{-8}	3.684×10^{-7}
$3\pi/4$	3.897×10^{-8}	8.028×10^{-9}	1.423×10^{-9}	2.320×10^{-7}	8.204×10^{-8}
π	5.398×10^{-8}	1.617×10^{-8}	6.457×10^{-9}	3.005×10^{-3}	5.119×10^{-10}

Finally, in Table 3 we present the error associated with the expansions of the generalised sine and cosine integrals $\text{Sin}_{\alpha,\beta}(x)$ and $\text{Cin}_{\alpha,\beta}(x)$ as $x \to +\infty$ given in Theorems 3–5. For $\text{Sin}_{\alpha,\beta}(x)$, the logarithmic expansion in (30) arises for $\alpha = \frac{1}{4}$ and $\alpha = \frac{1}{2}$; for $\text{Cin}_{\alpha,\beta}(x)$ the logarithmic expansion in (31) arises for $\alpha = \frac{1}{3}$. In Figure 3 are shown plots (We remark that the plot of $\text{Cin}_{\alpha 1}(x)$ in Figure 3b differs from that shown in Figure 4 of Reference [2].) of $\text{Sin}_{\alpha,1}(x)$ and $\text{Cin}_{\alpha,1}(x)$ for different α and in Figure 4 the leading asymptotic approximations from the expansions in Theorem 5 are compared with the corresponding plots of these functions.

In conclusion, it is worth mentioning that the function $\text{Ein}_{\alpha,\beta}(z)$, and also the generalised sine and cosine integrals, can be extended by using the three-parameter Mittag-Leffler function (or Prabhakar function) defined by

$$E^\rho_{\alpha,\beta}(z) = \sum_{n=0}^{\infty} \frac{(\rho)_n}{\Gamma(\alpha n + \beta)} \frac{z^n}{n!}.$$

A comprehensive discussion of this function and its applications can be found in Reference [10]; see also Reference [6] Section 5.1, for details of its large-z asymptotic expansion.

Table 3. The absolute relative error in the computation of $\mathrm{Sin}_{\alpha,\beta}(x)$ and $\mathrm{Cin}_{\alpha,\beta}(x)$ from Theorems 3–5 for different α and x when $\beta = 4/3$.

	$\mathrm{Sin}_{\alpha,\beta}(x)$				
x	$\alpha = 1/4$	$\alpha = 1/3$	$\alpha = 1/2$	$\alpha = 2/3$	$\alpha = 1$
10	4.396×10^{-7}	1.394×10^{-8}	1.785×10^{-6}	3.410×10^{-6}	1.012×10^{-5}
20	3.213×10^{-11}	1.171×10^{-13}	3.920×10^{-11}	2.076×10^{-8}	3.094×10^{-11}
25	2.373×10^{-13}	3.792×10^{-14}	2.098×10^{-13}	4.437×10^{-10}	3.270×10^{-12}
30	1.879×10^{-15}	5.065×10^{-15}	1.172×10^{-15}	8.197×10^{-12}	8.010×10^{-15}
	$\mathrm{Cin}_{\alpha,\beta}(x)$				
x	$\alpha = 1/4$	$\alpha = 1/3$	$\alpha = 1/2$	$\alpha = 2/3$	$\alpha = 1$
10	9.237×10^{-8}	3.787×10^{-7}	6.608×10^{-7}	2.270×10^{-5}	7.756×10^{-6}
20	1.293×10^{-12}	4.473×10^{-12}	1.090×10^{-11}	2.462×10^{-10}	2.576×10^{-10}
25	8.066×10^{-14}	2.334×10^{-16}	5.326×10^{-14}	6.881×10^{-11}	1.437×10^{-12}
30	1.160×10^{-16}	9.285×10^{-17}	2.764×10^{-16}	2.934×10^{-12}	7.716×10^{-15}

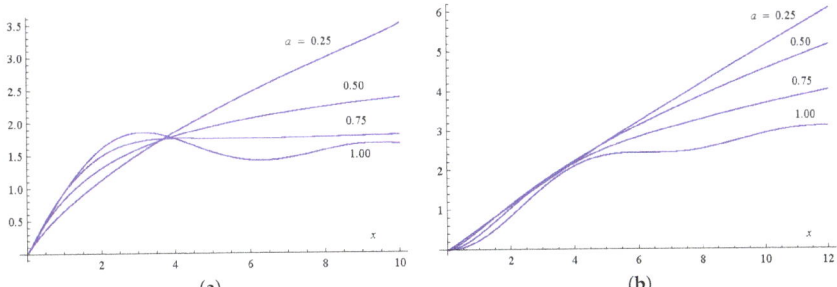

Figure 3. Plots of the generalised sine and cosine integrals (**a**) $\mathrm{Sin}_{\alpha,1}(x)$ and (**b**) $\mathrm{Cin}_{\alpha,1}(x)$ for $\alpha = 0.25, 0.50, 0.75, 1$.

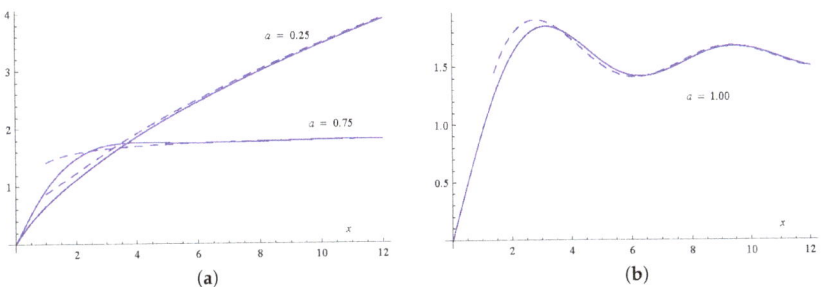

Figure 4. *Cont.*

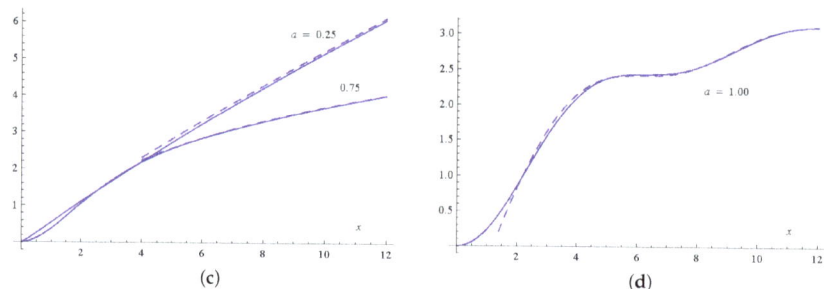

Figure 4. Plots of the generalised sine and cosine integrals (solid curves) and their leading asymptotic approximations (dashed curves) from Theorems 3, 4 and 5: (**a**) $\text{Sin}_{\alpha,1}(x)$ when $\alpha = 0.25, 0.75$, (**b**) $\text{Sin}_{\alpha,1}(x)$ when $\alpha = 1$, (**c**) $\text{Cin}_{\alpha,1}(x)$ when $\alpha = 0.25, 0.75$ and (**d**) $\text{Cin}_{\alpha,1}(x)$ when $\alpha = 1$.

6. Conclusions

The large-z asymptotic expansions of the modified exponential integral $\text{Ein}_{\alpha,\beta}(z)$ involving the two-parameter Mittag-Leffler function have been determined by exploiting the known asymptotic theory developed for integral functions of hypergeometric type, namely the Fox-Wright function. The appearance of logarithmic terms in the expansion of $\text{Ein}_{\alpha,\beta}(x)$ for $x \to +\infty$ for certain values of $\alpha \in (0,1]$ is emphasised. Similar expansions have been obtained for the extended sine and cosine integrals.

Funding: This research received no external funding.

Acknowledgments: I would like to thank Francesco Mainardi for the invitation to contribute to this special edition.

Conflicts of Interest: The author declares no conflict of interest.

References

1. Olver, F.W.J.; Lozier, D.W.; Boisvert, R.F.; Clark, C.W. (Eds.) *NIST Handbook of Mathematical Functions*; Cambridge University Press: Cambridge, UK, 2010.
2. Mainardi, F.; Masina, E. On modifications of the exponential integral with the Mittag-Leffler function. *Fract. Calc. Appl. Anal.* **2018**, *21*, 1156–1169. [CrossRef]
3. Mainardi, F.; Masina, E.; Spada, G. A generalization of the Becker model in linear viscoelasticity: Creep, relaxation and internal friction. *Mech. Time-Depend. Mater.* **2019**, *23*, 283–294. [CrossRef]
4. Paris, R.B.; Kaminski, D. *Asymptotics and Mellin-Barnes Integrals*; Cambridge University Press: Cambridge, UK, 2001.
5. Braaksma, B.L.J. Asymptotic expansions and analytic continuations for a class of Barnes integrals. *Compos. Math.* **1963**, *15*, 239–341.
6. Paris, R.B. Asymptotics of the special functions of fractional calculus. In *Handbook of Fractional Calculus with Applications*; Kochubei, A., Luchko, Y., Eds.; De Gruyter: Berlin, Germany, 2019; Volume 1, pp. 297–325.
7. Ford, W.B. *The Asymptotic Developments of Functions defined by Maclaurin Series*; University of Michigan Studies, Scientific Series; University of Michigan Press: Ann Arbor, MI, USA, 1936; p. 11.
8. Paris, R.B. Exponentially small expansions of the confluent hypergeometric functions. *Appl. Math. Sci.* **2013**, *7*, 6601–6609. [CrossRef]

9. Wolfram Research Inc. *Mathematica*, version 7; Wolfram Research Inc.: Champaign, IL, USA, 2008.
10. Guisti, A.; Colombaro, I.; Garra, R.; Garrappa, R.; Polito, F.; Popolizio, M.; Mainardi, F. A practical guide to Prabhakar fractional calculus. *Fract. Calc. Appl. Anal.* **2020**, *23*, 9–54. [CrossRef]

© 2020 by the author. Licensee MDPI, Basel, Switzerland. This article is an open access article distributed under the terms and conditions of the Creative Commons Attribution (CC BY) license (http://creativecommons.org/licenses/by/4.0/).

Article

Mathematical Aspects of Krätzel Integral and Krätzel Transform

Arak M. Mathai [1] and Hans J. Haubold [2,*]

[1] Department of Mathematics and Statistics, McGill University, Montreal, PQ H3A 2K6, Canada; a.mathai@mcgill.ca
[2] Office for Outer Space Affairs, United Nations, Vienna International Centre, A-1400 Vienna, Austria
* Correspondence: hans.haubold@gmail.com

Received: 11 February 2020; Accepted: 22 March 2020; Published: 3 April 2020

Abstract: A real scalar variable integral is known in the literature by different names in different disciplines. It is basically a Bessel integral called specifically Krätzel integral. An integral transform with this Krätzel function as kernel is known as Krätzel transform. This article examines some mathematical properties of Krätzel integral, its connection to Mellin convolutions and statistical distributions, its computable representations, and its extensions to multivariate and matrix-variate cases, in both the real and complex domains. An extension in the pathway family of functions is also explored.

Keywords: Mellin convolutions; Krätzel integrals; reaction-rate probability integral; continuous mixtures; Bayesian structures; fractional integrals; statistical distribution of products and ratios; multivariate and matrix-variate cases; real and complex domains

MSC: 26B12; 26A33; 60E10; 62E15; 33C60

1. Introduction

In this paper, real scalar mathematical or random variables are denoted by small letters $x, y, z, \ldots$ and the corresponding vector/matrix variables are denoted by capital letters $X, Y, \ldots$. Variables in the complex domain are denoted with a tilde such as $\tilde{x}, \tilde{y}, \tilde{X}, \tilde{Y} \ldots$. Constant vectors/matrices are denoted by capital letters $A, B, \ldots$ whether in the real or complex domain. Scalar constants are denoted by $a, b, \ldots$. If $X = (x_{ij})$ is a $p \times q$ matrix where the x_{ij}s are distinct real scalar variables, then the wedge product of the differentials is denoted by $dX = \wedge_{i=1}^{p} \wedge_{j=1}^{q} dx_{ij}$. If x and y are real scalar variables, then the wedge product of their differentials is defined as $dx \wedge dy = -dy \wedge dx$ so that $dx \wedge dx = 0, dy \wedge dy = 0$. If $\tilde{X}$ is in the complex domain, then $\tilde{X} = X_1 + iX_2$ where X_1, X_2 are real and $i = \sqrt{(-1)}$. Then, $d\tilde{X} = dX_1 \wedge dX_2$. The determinant of a $p \times p$ real matrix X is denoted by $|X|$ or $\det(X)$ and when in the complex domain the absolute value of the determinant is denoted by $|\det(\tilde{X})|$. The trace of a square matrix A is denoted by $\text{tr}(A)$. The integral

$$\int_A^B f(X) dX = \int_{O<A<X<B} f(X) dX$$

means a real-valued scalar function $f(X)$ of the $p \times p$ real positive definite matrix X is integrated out over $X > O$ (positive definite), $X - A > O, B - X > O, A > O, B > O$ where A and B are $p \times p$ constant positive definite matrices. The corresponding integral in the complex domain is denoted as $\int_A^B \tilde{f}(\tilde{X}) d\tilde{X}$.

1.1. Krätzel Integral

Let x be a real scalar variable. Consider the following integrals:

$$K_1 = \int_0^\infty x^{\gamma-1} e^{-ax-\frac{b}{x}} dx, a > 0, b > 0, \gamma > 0 \tag{1}$$

$$K_2 = \int_0^\infty x^{\gamma-1} e^{-ax^\delta - bx^{-\rho}} dx, a > 0, b > 0, \gamma > 0, \delta > 0, \rho > 0. \tag{2}$$

This K_2 in Equation (2) is known as the generalized Krätzel integral and Equation (1) as the basic Krätzel integral. When $\delta = 1$ in Equation (2), we have the Laplace transform of $x^{\gamma-1} e^{-bx^{-\rho}}$ with Laplace parameter a. For $\delta = 1, \rho = \frac{1}{2}$ in Equation (2), we have the basic reaction-rate probability integral in nuclear and solar neutrino astrophysics (see [1,2]). When $\delta = 1, \rho = 1$, the integrand in Equation (1) is the inverse Gaussian density for appropriate values of a, b, γ and multiplied by a normalizing constant. In addition, Equation (2) is a generalized situation of the same and Equation (1) provides the moment expression for the inverse Gaussian density, multiplied by a normalizing constant. Krätzel transform is associated with Equation (1) (see [3]). Some authors call Equation (2) as the generalized gamma, ultra gamma, Bessel integral, etc. In [4], it is shown that in the simple poles case it is a Bessel series and hence it is more appropriate to call it as a generalized Bessel integral.

The highlight of the present discussion is to point out the importance and usefulness of Krätzel function in various topics in widely different areas and to consider its extensions of various types. Krätzel integrals appear in Mellin convolution of product of two functions; in statistical distribution theory as the density of a product of two independently distributed generalized gamma random variables; in Bayesian analysis when the conditional and marginal densities belong to generalized gamma densities; in model building, especially in the pathway models where the limiting forms end up in Krätzel functions; in nuclear reaction-rate theory; and in inverse Gaussian models in stochastic processes, to mention a few topics. Krätzel function is also associated with generalized gamma and ultra gamma integrals, Kobayashi integrals and generalized special functions such as G- and H-functions. In the present discussion, we also consider extensions of Krätzel function to multivariate cases involving many scalar variables, matrix-variate cases in the real and complex domains and extensions involving multiple integrals.

1.2. Evaluation of the Integral in Equation (2)

One can evaluate Equation (2) by using different approaches. One can interpret Equation (2) as the Mellin convolution of a product and then take the inverse Mellin transform to evaluate the integral. One can draw a parallel to the statistical density of a product of two positive real scalar random variables and then evaluate the density to obtain the value of Equation (2). One can treat Equation (2) as a function $g(b)$ of b. Then, the Mellin transform of $g(b)$ with Mellin parameter s is the following for $\gamma > 0, \delta > 0, a > 0, b > 0, \eta > 0$:

$$M_g(s) = \int_0^\infty b^{s-1} \left\{ \int_0^\infty x^{\gamma-1} e^{-ax^\delta - bx^{-\rho}} dx \right\} db$$

$$= \int_0^\infty \int_0^\infty b^{s-1} x^{\gamma-1} e^{-ax^\delta - bx^{-\rho}} dx \wedge db.$$

Integrating out b first and then x, we have the following:

$$\int_0^\infty b^{s-1} e^{-bx^{-\rho}} db = \Gamma(s) x^{\rho s}, \Re(s) > 0$$

$$\int_0^\infty x^{\gamma+\rho s-1} e^{-ax^\delta} dx = \frac{1}{\delta} \Gamma\left(\frac{\gamma+\rho s}{\delta}\right) a^{-\left(\frac{\gamma+\rho s}{\delta}\right)}, \Re(\gamma+\rho s) > 0$$

where $\Re(\cdot)$ means the real part of $(\cdot)$. That is,

$$M_g(s) = \frac{1}{\delta a^{\frac{\gamma}{\delta}}} \Gamma(s) \Gamma\left(\frac{\gamma + \rho s}{\delta}\right) a^{-\frac{\rho}{\delta}s}. \tag{3}$$

Taking the inverse Mellin transform of Equation (3) we have $g(b)$ or the integral in Equation (2) as the following:

$$K_2 = \frac{1}{\delta a^{\frac{\gamma}{\delta}}} \frac{1}{2\pi i} \int_{c-i\infty}^{c+i\infty} \Gamma(s) \Gamma\left(\frac{\gamma}{\delta} + \frac{\rho}{\delta}s\right) (ba^{\frac{\rho}{\delta}})^{-s} ds, i = \sqrt{(-1)} \tag{4}$$

where the c in the contour is > 0. Note that Equation (4) can be written as a H-function.

$$K_2 = \frac{1}{\delta a^{\frac{\gamma}{\delta}}} H_{0,2}^{2,0}\left[ba^{\frac{\rho}{\delta}}\Big|_{(0,1),(\frac{\gamma}{\delta},\frac{\rho}{\delta})}\right]. \tag{5}$$

For the theory and applications of the H-function, see [5]. When $\rho = \delta$, we have Equation (5) reducing to a Meijer's G-function as the following:

$$K_2 = \frac{1}{\delta a^{\frac{\gamma}{\delta}}} G_{0,2}^{2,0}\left[ab\Big|_{0,\frac{\gamma}{\delta}}\right]. \tag{6}$$

For the theory and applications of G-function, see [6].

1.3. Computable Series form for Equation (2)

Consider the Mellin–Barnes integral representation in Equation (4). This integral can be evaluated as the sum of the residues at the poles of the gammas $\Gamma(s)$ and $\Gamma(\frac{\gamma}{\delta} + \frac{\rho}{\delta}s)$. The poles of $\Gamma(s)$ are at $s = 0, -1, -2, \ldots$. When the poles of the integrand are simple, then the sum of the residues at the poles of $\Gamma(s)$ is the following:

$$(A) \quad \delta a^{\frac{\gamma}{\delta}})^{-1} \sum_{\nu=0}^{\infty} \frac{(-1)^\nu}{\nu!} \Gamma\left(\frac{\gamma}{\delta} - \frac{\rho}{\delta}\nu\right)(ba^{\frac{\rho}{\delta}})^\nu.$$

The poles of $\Gamma(\frac{\gamma}{\delta} + \frac{\rho}{\delta}s)$ are at $\frac{\gamma}{\delta} + \frac{\rho}{\delta}s = -\nu, \nu = 0, 1, 2, \ldots$ or the poles are at $s = -\frac{\gamma}{\rho} - \frac{\delta}{\rho}\nu$ and in the simple poles case the sum of the residues is the following:

$$(B) \quad \frac{b^{\frac{\gamma}{\rho}}}{\delta} \sum_{\nu=0}^{\infty} \frac{(-1)^\nu}{\nu!} \Gamma\left(-\frac{\gamma}{\rho} - \frac{\delta}{\rho}\nu\right)(ab^{\frac{\delta}{\rho}})^\nu.$$

Hence, the sum of residues from (A) and (B) in the simple poles case is the following:

$$K_2 = (\delta a^{\frac{\gamma}{\delta}})^{-1} \sum_{\nu=0}^{\infty} \frac{(-1)^\nu}{\nu!} \Gamma\left(\frac{\gamma}{\delta} - \frac{\rho}{\delta}\nu\right)(ba^{\frac{\rho}{\delta}})^\nu$$
$$+ \frac{b^{\frac{\gamma}{\rho}}}{\delta} \sum_{\nu=0}^{\infty} \frac{(-1)^\nu}{\nu!} \Gamma\left(-\frac{\gamma}{\rho} - \frac{\delta}{\rho}\nu\right)(ab^{\frac{\delta}{\rho}})^\nu. \tag{7}$$

1.4. G-function in the Simple Poles Case

Let $\rho = \delta$ so that the H-function in Equation (5) becomes the G-function in Equation (6) and when $\frac{\gamma}{\delta}$ is not an integer then the G-function has simple poles. Consider this case and it is available from Equation (7) by putting $\delta = \rho$. Then, the gammas reduce to the following:

$$\Gamma\left(\frac{\gamma}{\rho} - \nu\right) = \frac{\Gamma(\frac{\gamma}{\rho})}{(-1)^\nu(-\frac{\gamma}{\rho}+1)_\nu} \text{ and } \Gamma\left(-\frac{\gamma}{\rho} - \nu\right) = \frac{\Gamma(-\frac{\gamma}{\rho})}{(-1)^\nu(\frac{\gamma}{\rho}+1)_\nu},$$

where, in general, the notation $(a)_m = a(a+1)...(a+m-1), a \neq 0, (a)_0 = 1$ is the Pochhammer symbol. Hence, K_2 in Equation (2) for this simple poles case and for $\delta = \rho$ is the following:

$$K_2 = \frac{\Gamma(\frac{\gamma}{\rho})}{\rho a^{\frac{\gamma}{\rho}}} \sum_{\nu=0}^{\infty} \frac{1}{(-\frac{\gamma}{\rho}+1)_\nu \nu!}(ab)^\nu + \frac{\Gamma(-\frac{\gamma}{\rho})b^{\frac{\gamma}{\rho}}}{\rho} \sum_{\nu=0}^{\infty} \frac{1}{(\frac{\gamma}{\rho}+1)_\nu \nu!}(ab)^\nu$$

$$= \frac{\Gamma(\frac{\gamma}{\rho})}{\rho a^{\frac{\gamma}{\rho}}} {}_0F_1(;-\frac{\gamma}{\rho}+1;ab) + \frac{\Gamma(-\frac{\gamma}{\rho})b^{\frac{\gamma}{\rho}}}{\rho} {}_0F_1(;\frac{\gamma}{\rho}+1;ab), \tag{8}$$

where ${}_0F_1$ is a hypergeometric series with no upper and one lower parameters. Observe that, in this simple poles case, Equation (2) or K_2 of Equation (8) is a linear function of Bessel series and hence it is appropriate to call Equation (1) as Bessel integral and Equation (2) as the generalized Bessel integral rather than calling them as ultra gamma integral or generalized gamma integral or anything connected with gamma integral.

1.5. Poles of Order Two, $\rho = \delta$, $\frac{\gamma}{\delta} = m, m = 1, 2, ...$

In this case, the poles at $s = 0, -1, -2, ..., -(m-1)$ are simple and poles at $s = -m, -m-1, ...$ are of order two each. In this case, we may write (2) as the following:

$$K_2 = \frac{1}{\rho a^{\frac{\gamma}{\rho}}} \frac{1}{2\pi i} \int_{c-i\infty}^{c+i\infty} \Gamma(s)\Gamma(m+s)(ab)^{-s} ds. \tag{9}$$

Sum of the residues at the poles $s = 0, -1, ..., -(m-1)$, coming from (9), is the following:

$$(C) \quad \frac{1}{\rho a^{\frac{\gamma}{\rho}}} \sum_{\nu=0}^{m-1} \frac{(-1)^\nu}{\nu!} \Gamma(m-\nu)(ab)^\nu.$$

For $s = -m - \nu, \nu = 0, 1, ...$ or $s = -\nu, \nu = m, m+1, ...$ the poles are of order two and the residue, denoted by R_ν, is the following: Let $h(s) = \Gamma(s)\Gamma(m+s)(ab)^{-s}$. Then,

$$R_\nu = \lim_{s \to -\nu} \frac{d}{ds}[(s+\nu)^2 \Gamma(s) \Gamma(m+s)(ab)^{-s}]$$

$$= \lim_{s \to -\nu} \frac{d}{ds}[(s+\nu)^2 \frac{(s+\nu-1)^2...(s+m)^2}{(s+\nu-1)^2...(s+m)^2} \frac{(s+m-1)...s}{(s+m-1)...s} \Gamma(s)\Gamma(m+s)(ab)^{-s}]$$

$$= \lim_{s \to -\nu} \frac{d}{ds}[\frac{\Gamma^2(s+\nu+1)}{(s+\nu-1)^2...(s+m)^2(s+m-1)...s}(ab)^{-s}].$$

Observe that $\frac{d}{ds}h(s) = h(s)\frac{d}{ds}\ln h(s)$ and $(ab)^{-s} = e^{-s\ln(ab)}$. Note that

$$\lim_{s \to -\nu} h(s) = \frac{(-1)^m (ab)^\nu}{\nu!(\nu-m)!}, \nu = m, m+1, ...$$

$$\lim_{s \to -\nu} \frac{d}{ds} \ln h(s) = \lim_{s \to -\nu}[2\psi(s+\nu+1) - \frac{2}{s+\nu-1} - ... - \frac{2}{s+m}$$

$$- \frac{1}{s+m-1} - ... - \frac{1}{s} - \ln(ab)]$$

$$= 2\psi(1) + 2[1 + \frac{1}{2} + ... + \frac{1}{\nu-m}] + [\frac{1}{\nu-m+1} + ... + \frac{1}{\nu}] - \ln(ab)$$

$$= \psi(\nu+1) + \psi(\nu-m+1) - \ln(ab).$$

Therefore,
$$R_\nu = [\psi(\nu+1) + \psi(\nu-m+1) - \ln(ab)]\left[\frac{(-1)^m (ab)^\nu}{\nu!(\nu-m)!}\right], \nu = m, m+1, \ldots.$$

Then, in this case, (2) reduces to the following:
$$K_2 = \frac{1}{\rho a^{\frac{\gamma}{\rho}}} \sum_{\nu=0}^{m-1} \frac{(-1)^\nu}{\nu!} \Gamma(m-\nu)(ab)^\nu$$
$$+ \sum_{\nu=m}^{\infty} [\psi(\nu+1) + \psi(\nu-m+1) - \ln(ab)]\left[\frac{(-1)^m}{\nu!(\nu-m)!}(ab)^\nu\right], \nu = m, m+1, \ldots$$

where $\psi(\cdot)$ is the psi function or the logarithmic derivative of the gamma function, $\psi(z) = \frac{d}{dz} \ln \Gamma(z)$.

The most general case is to consider $\Gamma(s)\Gamma(\frac{\gamma}{\delta} + \frac{\rho}{\delta}s)$ having some poles of order one and the remaining of order two. After writing this situation in a convenient way, one can use the procedure in Section 1.5 to obtain the final result. Since the expressions would take up too much space, it is not discussed here.

2. Krätzel Integral from Mellin Convolution

Let $x_1 > 0$ and $x_2 > 0$ be real scalar variables. Let $f_1(x_1)$ and $f_2(x_2)$ be real-valued scalar functions associated with x_1 and x_2, respectively. Then, the Mellin transforms of f_1 and f_2, with Mellin parameter s, are the following, whenever they exist:

$$M_{f_1}(s) = \int_0^\infty x_1^{s-1} f_1(x_1) dx_1, \quad M_{f_2}(s) = \int_0^\infty x_2^{s-1} f_2(x_2) dx_2. \tag{10}$$

Then,
$$M_{f_1}(s) M_{f_2}(s) = \int_0^\infty \int_0^\infty x_1^{s-1} x_2^{s-1} f_1(x_1) f_2(x_2) dx_1 \wedge dx_2$$
$$= \int_0^\infty \int_0^\infty u^{s-1} f_1(v) f_2\left(\frac{u}{v}\right) \frac{1}{v} du \wedge dv, u = x_1 x_1, v = x_1$$
$$= \int_0^\infty u^{s-1} g(u) du$$

where
$$g(u) = \int_0^\infty \frac{1}{v} f_1(v) f_2\left(\frac{u}{v}\right) dv$$
$$= \int_0^\infty \frac{1}{v} f_1\left(\frac{u}{v}\right) f_2(v) dv. \tag{11}$$

That is,
$$M_g(s) = M_{f_1}(s) M_{f_2}(s). \tag{12}$$

This Equation (12) is the Mellin convolution of the product involving two functions and Equation (11) is the corresponding integral representation. Let f_1 and f_2 be generalized exponential functions of the following types:

$$(D) \quad f_j(x_j) = x_j^{\gamma_j - 1} e^{-a_j x_j^{\delta_j}}, a_j > 0, \delta_j > 0, \gamma_j > 0, j = 1, 2.$$

Then, Equation (11) becomes the following:

$$(E) \quad g(u) = u^{\gamma_2-1} \int_0^\infty v^{\gamma_1-\gamma_2-1} e^{-a_1 v^{\delta_1} - a_2(\frac{u}{v})^{\delta_2}} dv$$

$$(F) \quad = u^{\gamma_1-1} \int_0^\infty v^{\gamma_2-\gamma_1-1} e^{-a_1(\frac{u}{v})^{\delta_1} - a_2 v^{\delta_2}} dv.$$

Here, (E) and (Fi) provide equivalent representations for $g(u)$. In (E), if $\delta_1 = \delta, a_1 = a, \delta_2 = \rho, a_2 u^{\delta_2} = b, \gamma_1 - \gamma_2 = \gamma$, then the integral becomes Krätzel integral of (2) in Section 1. Hence, Krätzel integral is also available as a Mellin convolution of a product involving two functions, see [7].

Instead of taking $f_j(x_j)$ of the form in (D), if we take $f_1(x_1) = \frac{1}{\Gamma(\alpha)} x_1^\gamma (1-x_1)^{\alpha-1}$ for $\Re(\gamma) > -1, \Re(\alpha) > 0$ or $\alpha > 0, \gamma > -1$ when real, and $f_2(x_2) = f(x_2)$ where $f(x_2)$ is an arbitrary function, then Equation (11) becomes the following:

$$g(u) = \int_v \frac{1}{v} f_1(\frac{u}{v}) f_2(v) dv = \int_v \frac{1}{\Gamma(\alpha)} \frac{1}{v} (\frac{u}{v})^\gamma (1 - \frac{u}{v})^{\alpha-1} f(v) dv, \Re(\alpha) > 0, \Re(\gamma) > -1$$

$$= \frac{u^\gamma}{\Gamma(\alpha)} \int_{v \geq u} v^{-\gamma-\alpha}(v-u)^{\alpha-1} f(v) dv = K_{2,\gamma}^{-\alpha} f \tag{13}$$

where $K_{2,\gamma}^{-\alpha} f$ in (13) is Erdélyi–Kober fractional integral of the second kind of order α and parameter γ, see [8]. Thus, the Mellin convolution of a product is also associated with fractional integral of the second kind. A general definition of all versions of fractional integrals in terms of Mellin convolutions of products and ratios is given in [8].

3. Krätzel Integral as the Density of a Product

Let $x_1 > 0$ and $x_2 > 0$ be two real scalar positive random variables, independently distributed with density functions $f_1(x_1)$ and $f_2(x_2)$, respectively. Due to statistical independence their joint density, denoted by $f(x_1, x_2)$, is the product, $f(x_1, x_2) = f_1(x_1) f_2(x_2)$. Let $u = x_1 x_2$ be the product and let $x_1 = v$ or $x_2 = v$. Then, $dx_1 \wedge dx_2 = \frac{1}{v} du \wedge dv$. Let $g(u,v)$ be the joint density of u and v. Then,

$$g(u,v) = \frac{1}{v} f_1(v) f_2(\frac{u}{v}) = \frac{1}{v} f_1(\frac{u}{v}) f_2(v)$$

and the marginal density of u, denoted by $g_1(u)$ is the following:

$$g_1(u) = \int_v \frac{1}{v} f_1(v) f_2(\frac{u}{v}) dv$$

$$= \int_v \frac{1}{v} f_1(\frac{u}{v}) f_2(v) dv. \tag{14}$$

Let $f_j(x_j)$ be a generalized gamma density of the form

$$f_j(x_j) = c_j x_j^{\gamma_j-1} e^{-a_j x_j^{\delta_j}}, a_j > 0, \gamma_j > 0, \delta_j > 0, j = 1, 2 \tag{15}$$

where c_j is the normalizing constant. For the $f_j(x_j)$ in Equation (15), we have Equation (14) as the following:

$$g_1(u) = c_1 c_2 u^{\gamma_2-1} \int_0^\infty v^{\gamma_1-\gamma_2-1} e^{-a_1 v^{\delta_1} - a_2(\frac{u}{v})^{\delta_2}} dv$$

$$= c_1 c_2 u^{\gamma_1-1} \int_0^\infty v^{\gamma_2-\gamma_1-1} e^{-a_1(\frac{u}{v})^{\delta_1} - a_2 v^{\delta_2}} dv. \tag{16}$$

Observe that the two expressions for $g_1(u)$ in Equation (16) are not only generalized Krätzel integrals but they are also statistical densities of a product. We can evaluate the explicit form of the density by using arbitrary moments and then inverting the expression. Consider the $(s-1)$th moments of x_1 and x_2. Then, $E[x_1 x_2]^{s-1} = E[x_1^{s-1}]E[x_2^{s-1}]$ due to statistical independence, where $E[\cdot]$ denotes the expected value of $[\cdot]$. That is,

$$E[x_j^{s-1}] = \int_0^\infty x_j^{s-1} f_j(x_j) dx_j = M_{f_j}(s), j = 1, 2$$

whenever the expected values exist, where $M_{f_j}(s)$ is the Mellin transform of the density f_j, with Mellin parameter s, when this Mellin transform exists. Evaluating $E[x_j^{s-1}]$ for the density in Equation (15), we have the following:

$$E[x_j^{s-1}] = \frac{a_j^{-\frac{(s-1)}{\delta_j}} \Gamma(\frac{\gamma_j + s - 1}{\delta_j})}{\Gamma(\frac{\gamma_j}{\delta_j})}, \Re(\gamma_j + s - 1) > 0, j = 1, 2. \quad (17)$$

Observe that in Equation (17) the explicit form of the normalizing constant c_j is used, c_j is such that $E[x_j^{s-1}] = 1$ when $s = 1$. Then, taking the product

$$E[u^{s-1}] = \{\prod_{j=1}^2 \frac{a_j^{\frac{1}{\delta_j}}}{\Gamma(\frac{\gamma_j}{\delta_j})}\}\{\prod_{j=1}^2 \Gamma(\frac{\gamma_j - 1}{\delta_j} + \frac{s}{\delta_j}) a_j^{-\frac{s}{\delta_j}}\}, \quad (18)$$

for $\Re(\gamma_j + s - 1) > 0, j = 1, 2$. Then, the density $g_1(u)$ is available from the inverse Mellin transform or by inverting Equation (18). That is,

$$g_1(u) = C \frac{1}{2\pi i} \int_{c-i\infty}^{c+i\infty} \{\prod_{j=1}^2 \Gamma(\frac{\gamma_j - 1}{\delta_j} + \frac{s}{\delta_j})\}(a_1^{\frac{1}{\delta_1}} a_2^{\frac{1}{\delta_2}} u)^{-s} ds$$

$$= C H_{0,2}^{2,0}\left[a_1^{\frac{1}{\delta_1}} a_2^{\frac{1}{\delta_2}} u \Big|_{(\frac{\gamma_1 - 1}{\delta_1}, \frac{1}{\delta_1}), (\frac{\gamma_2 - 1}{\delta_2}, \frac{1}{\delta_2})}\right], \quad (19)$$

$$C = \prod_{j=1}^2 \frac{a_j^{\frac{1}{\delta_j}}}{\Gamma(\frac{\gamma_j}{\delta_j})}.$$

Note that Equation (19) is the explicit form of the Krätzel integral as well as the statistical density $g_1(u)$. Instead of generalized gamma density for $f_j(x_j)$, suppose that the density of x_1 is a type-1 beta density with the parameters $(\gamma + 1, \alpha)$ and $f_2(x_2)$ is an arbitrary density then f_1 is of the form

$$f_1(x_1) = \frac{\Gamma(\alpha + \gamma + 1)}{\Gamma(\gamma + 1)\Gamma(\alpha)} x_1^\gamma (1 - x_1)^{\alpha - 1}, 0 \leq x_1 \leq 1, \alpha > 0, \gamma > -1.$$

Usually, the parameters in a statistical density are real. Then, $g_1(u)$ becomes the following:

$$g_1(u) = \int_v \frac{1}{v} f_1(\frac{u}{v}) f_2(v) dv$$

$$= \frac{\Gamma(\alpha + \gamma + 1)}{\Gamma(\gamma + 1)\Gamma(\alpha)} \int_{v \geq u} \frac{1}{v} (\frac{u}{v})^\gamma (1 - \frac{u}{v})^{\alpha - 1} f(v) dv$$

$$= \frac{\Gamma(\gamma + \alpha + 1)}{\Gamma(\gamma + 1)} \frac{u^\gamma}{\Gamma(\alpha)} \int_{v \geq u} v^{-\gamma - \alpha} (v - u)^{\alpha - 1} f(v) dv$$

$$= \frac{\Gamma(\alpha + \gamma + 1)}{\Gamma(\gamma + 1)} K_{2,\gamma}^{-\alpha} f, \alpha > 0, \gamma > -1 \quad (20)$$

where $K_{2,\gamma}^{-\alpha}f$ is Erdélyi–Kober fractional integral of the second kind of order α and parameter γ. From Equation (20), note that this fractional integral is a constant multiple of a statistical density of a product of positive random variables also. For generalizations of this result for the matrix-variate case, in real and complex domains, see [8]. By taking the density of a ratio of real scalar positive random variables, where the variables are independently distributed, with x_1 having a type-1 beta density with the parameters (γ, α) and x_2 having an arbitrary density we can show that the density of the ratio $u = \frac{x_2}{x_1}$ will produce a constant multiple of Erdélyi–Kober fractional integral of the first kind of order α and parameter γ, details or the generalizations of this result may be seen [8].

4. Krätzel Integral and Bayesian Structures

In a simple Bayesian structure in Bayesian statistical analysis, we have a conditional density of a random variables x, conditioned on a parameter θ, or written as $f_1(x|\theta)$ or the density of x, given θ. Then, θ has its own marginal density denoted by $f_2(\theta)$. Then, the joint density of x and θ is $f_1(x|\theta)f_2(\theta)$. When both x and θ are continuous variables, we call this situation as a continuous mixture. When one variable is discrete and the other continuous, we call it simply a mixture density. Then, the unconditional density of x, denoted by $f(x)$, is given by

$$f(x) = \int_\theta f_1(x|\theta) f_2(\theta) d\theta. \tag{21}$$

A general format of the structure in Equation (21) is of the following type:

$$f(x_1) = \int_{x_2} \cdots \int_{x_k} f_1(x_1|x_2,...,x_k) f_2(x_2|x_3,...,x_k) ... f_{k-1}(x_{k-1}|x_k) f_k(x_k) dx_2 \wedge ... \wedge dx_k. \tag{22}$$

For an application of this type of unconditional density for $k = 3$, see [9]. When all the densities involved in Equations (21) and (22) are continuous, we also call Equations (21) and (22) as continuous mixtures. Consider Equation (21), where

$$f_1(x|\theta) = \frac{\theta^{\gamma\delta}}{\Gamma(\gamma)} x^{\gamma-1} e^{-\theta^\delta x}, x \geq 0, \theta > 0, \delta > 0, \gamma > 0$$

and

$$f_2(\theta) = \frac{\rho b^{\frac{\alpha}{\rho}}}{\Gamma(\frac{\alpha}{\rho})} \theta^{-\alpha-1} e^{-b\theta^{-\rho}}, b > 0, \alpha > 0, \rho > 0, \theta > 0$$

so that

$$f_1(x|\theta) f_2(\theta) = \frac{\rho b^{\frac{\alpha}{\rho}}}{\Gamma(\gamma)\Gamma(\frac{\alpha}{\rho})} x^{\gamma-1} \theta^{\gamma\delta-\alpha-1} e^{-x\theta^\delta - b\theta^{-\rho}}.$$

Then, the unconditional density is the following, denoting $\theta = v$ in the integral and denoting the unconditional density of x, again by $f(x)$:

$$f(x) = C_1 \int_{v=0}^\infty v^{\gamma\delta-\alpha-1} e^{-xv^\delta - bv^{-\rho}} dv \tag{23}$$

where

$$C_1 = \frac{\rho b^{\frac{\alpha}{\rho}}}{\Gamma(\gamma)\Gamma(\frac{\alpha}{\rho})} x^{\gamma-1}, \alpha > 0, \rho > 0, \delta > 0, \gamma > 0, \rho > 0, x > 0.$$

Observe that Equation(23) is of the same structure of the Krätzel integral of Equation (2) of Section 1. Note that, if we use the general structure in Equation (22) and consider all densities as generalized gamma densities, then we obtain a generalization and extension of Krätzel integral to a multivariate situation. Such generalizations is considered below in this paper.

5. Pathway Extension of Krätzel Integral

The author of [10] introduced a pathway model for rectangular matrix-variate case. By using a pathway parameter there, one can go to three different families of functions. When a model is fitted to a given data, then one member from the pathway family is sure to fit the data if the data fall into one of the three wide families of functions or in the transitional stages of going from one family to another family. The pathway model for real positive scalar variable situation is the following:

$$f_3(x) = c_3 x^{\gamma-1}[1 + a(\alpha-1)x^\delta]^{-\frac{\eta}{\alpha-1}}, x > 0, \alpha > 1, \eta > 0, \delta > 0, a > 0. \tag{24}$$

When $\alpha < 1$, then we can write $\alpha - 1 = -(1-\alpha)$ so that the model in (24) switches to the model

$$f_4(x) = c_4 x^{\gamma-1}[1 - a(1-\alpha)x^\delta]^{\frac{\eta}{1-\alpha}}, \alpha < 1, \eta > 0, a > 0, \delta > 0 \tag{25}$$

and, further, $1 - a(1-\alpha)x^\delta > 0$ in order to create statistical density out of $f_4(x)$. Its support is finite or it is a finite-range density, whereas in Equation (24) it is of infinite range and $x > 0$ there. When $\alpha \to 1$, both Equations (24) and (25) go to the model

$$f_5(x) = c_5 x^{\gamma-1} e^{-a\eta x^\delta}, a > 0, x > 0, \delta > 0, \eta > 0. \tag{26}$$

Thus, through the pathway parameter α one can move among the three families of functions $f_j(x), j = 3, 4, 5$. Both Equations (24) and (25) can be taken as extensions of Equation (26). If Equation (26) is the ideal or stable situation in a physical system, then the unstable neighborhoods are given by Equations (24) and (25). The movement of α also describes the transitional stages. For the properties, generalizations and extension of the pathway model, see [11]. The model in Equation (25) for $\gamma = 1, a = 1, \eta = 1$ and for $\alpha < 1, \alpha > 1, \alpha \to 1$ is Tsallis' statistics in non-extensive statistical mechanics [12]. Some properties and other aspects of the pathway model see [11,13]. The model in Equation (24) for $a = 1, \eta = 1, \alpha > 1, \alpha \to 1$ is superstatistics (see [14]). Superstatistics considerations come from the unconditional density described in Section 4 when the conditional and marginal densities belong to the exponential and gamma families of densities. Consider the model in Equation (24) with different parameters, take f_1 and f_2 of Section 1, and consider Mellin convolutions. Let f_{31} and f_{32} be two densities belonging to Equation (24) with different parameters. That is, let

$$f_{3j}(x_j) = c_{3j} x_j^{\gamma_j - 1}[1 + a_j(\alpha_j - 1)x_j^{\delta_j}]^{-\frac{\eta_j}{\alpha_j-1}}, x_j > 0, \alpha_j > 1, a_j > 0, \gamma_j > 0, \delta_j > 0 \tag{27}$$

for $j = 1, 2$. Let $u = x_1 x_2, v = x_1$. Consider the Mellin convolution of a product or let $x_j > 0, j = 1, 2$ be independently distributed real scalar positive random variables with the densities f_{31} and f_{32} of (27) respectively. Then, the density of $u = x_1 x_2$, denoted by $g_p(u)$, where p stands for the pathway model, is the following:

$$g_p(u) = \int_v \frac{1}{v} f_{31}(v) f_{32}\left(\frac{u}{v}\right) dv$$
$$(G) \quad = c_{31} c_{32} u^{\gamma_2 - 1} \int_{v=0}^\infty v^{\gamma_1 - \gamma_2 - 1}[1 + a_1(\alpha_1 - 1)v^{\delta_1}]^{-\frac{\eta_1}{\alpha_1 - 1}}$$
$$\times [1 + a_2(\alpha_2 - 1)\left(\frac{u}{v}\right)^{\delta_2}]^{-\frac{\eta_2}{\alpha_2 - 1}} dv \tag{28}$$

for $\alpha_j > 1, a_j > 0, \delta_j > 0, \eta_j > 0, j = 1, 2$. See also the versatile integral discussed in [15]. Various types of extensions of Krätzel integrals are involved in Equation (28). When $\alpha_1 \to 1$, the first factor or the density in (G) goes to the exponential form whereas the second part in Equation (28) remains in the type-2 beta family form. This is one extension. In addition, when $\alpha_2 \to 1$, the second part density in Equation (28) goes to the exponential form whereas the first part remains in the type-2 beta family of functions. When $\alpha_1 \to 1$ and $\alpha_2 \to 1$, Equation (28) goes to the format of the Krätzel integral in Equation (2) of Section 1. A model of the form in Equation (28) for the cases $\alpha_j < 1, \alpha_j > 1, \alpha_j \to 1$, individually, is studied in detail in [15].

Connection to Kobayashi Integrals

In Equation (28), let $\alpha_1 \to 1$ and α_2 remain the same. Then, Equation (28) reduces to the following form:

$$g_p(u) = c_{31}c_{32}u^{\gamma_2-1}\int_{v=0}^{\infty} v^{\gamma_1-\gamma_2-1}e^{-a_1\eta_1 v^{\delta_1}}$$
$$\times [1 + a_2(\alpha_2-1)(\frac{u}{v})^{\delta_2}]^{-\frac{\eta_2}{\alpha_2-1}}dv. \qquad (29)$$

Observe that Equation (29) is a more general form of ultra gamma integral and Kobayashi integral. The Kobayashi form is available from the Mellin convolution of a ratio. Let $u_1 = \frac{x_2}{x_1}$ with $x_1 = v$, and let x_1 and x_2 be independently distributed pathway random variables as described in Section 5. Then, $x_1 = v, x_2 = u_1 v$ and $dx_1 \wedge dx_2 = v du_1 \wedge dv$. Then, the pathway density of u_1, denoted by $g_{p1}(u_1)$, is the following for $\alpha_1 \to 1$:

$$g_{p1}(u_1) = c_{31}c_{32}u^{\gamma_2-1}\int_{v=0}^{\infty} v^{\gamma_1+\gamma_2-1}e^{-a_1\eta_1 v^{\delta_1}}$$
$$\times [1 + a_2(\alpha_2-1)(u_1 v)^{\delta_2}]^{-\frac{\eta_2}{\alpha_2-1}} \qquad (30)$$

for $a_j > 0, \gamma_j > 0, \delta_j > 0, \eta_j > 0, j = 1, 2, \alpha_2 > 1$. Kobayashi integral is obtained from Equation (30) by putting $a_2(\alpha_2-1)u_1^{\delta_2} = \lambda$ and $\frac{\eta_2}{\alpha_2-1} = \eta$, (see [16,17]). Some people call Kobayashi form as ultra gamma integral. Observe that Equation (30) is a much more general and flexible format and for varying α_2 we have three families of functions in Equation (30) including Kobayashi format. The Mellin transform of $g_{p1}(u_1)$, with Mellin parameter s, is available from $u_1 = \frac{x_2}{x_1}$ form, namely

$$M_{g_{p1}}(s) = M_{f_1}(2-s)M_{f_2}(s) \text{ or } E[u_1^{s-1}] = E[x_1^{-s+1}]E[x_2^{s-1}]$$

and these moments are available from the pathway densities of x_1 and x_2 with $\alpha_1 \to 1$.

6. Multivariate Extensions of Krätzel Integrals

Let us start with the case of three variables. Let $x_j > 0, j = 1, 2, 3$ be three real scalar variables and let the associated functions be $f_j(x_j), j = 1, 2, 3$, respectively. If $x_j > 0, j = 1, 2, 3$ are real scalar random variables, independently distributed, then $f_j(x_j), j = 1, 2, 3$ may be the corresponding densities. Let $u = x_1 x_2 x_3$ be the product and let $v = x_2 x_3, w = x_3$. Then, $x_1 \wedge dx_2 \wedge dx_3 = \frac{1}{vw}du \wedge dv \wedge dw$. Mellin convolution of a product involving three real scalar variables is considered in [18]. Let

$$f_j(x_j) = c_j x_j^{\gamma_j-1}e^{-a_j x_j^{\delta_j}}, a_j > 0, \delta_j > 0, \gamma_j > 0, j = 1, 2, 3 \qquad (31)$$

where c_j is a constant and it may be normalizing constant if f_j in Equation (31) is a density. Then, the density of u or Mellin convolution of the product, again denoted by $g(u)$, is the following:

$$g(u) = \int_v \int_w \frac{1}{vw}f_1(\frac{u}{vw})f_2(\frac{v}{w})f_3(w)dv \wedge dw \qquad (32)$$
$$= c_1 c_2 c_3 \int_v \int_w \frac{1}{vw}(\frac{u}{v})^{\gamma_1-1}(\frac{v}{w})^{\gamma_2-1}w^{\gamma_3-1}$$
$$\times e^{-a_1(\frac{u}{v})^{\delta_1}-a_2(\frac{v}{w})^{\delta_2}-a_3 w^{\delta_3}}dv \wedge dw \qquad (33)$$

where Equation (32) is the general structure whatever be the f_js, and Equation (33) is the case when f_js belong to Equation (31). Then, Equation (33) can be taken as a bivariate version of the Krätzel integral. Observe that in the exponent we have v and w with positive and negative exponents. If we take $u = x_1 x_2 x_3, v = x_2, w = x_3$, then the exponential part in $g(u)$ is of the following form:

$$e^{-a_1(\frac{u}{vw})^{\delta_1} - a_2 v^{\delta_2} - a_3 w^{\delta_3}}.$$

In the format of Equation (33), we can take $v = x_1 x_2, w = x_2$ or $v = x_2 x_3, w = x_1$. These produce two more different forms corresponding to Equation (33). We can also take $u = x_1 x_2 x_3 = u_{12} x_3, u_{12} = x_1 x_2$. We can get the density of u_{12} first by using f_1 and f_2. Let the density of u_{12} be denoted as $g_{12}(u_{12})$. Then, by using g_{12} and f_3, we can get the density of u. This produces another bivariate extension of the Krätzel integral. Follow the same procedure by taking $u = u_{23} x_1, u_{13} x_2$ where $u_{23} = x_2 x_3, u_{13} = x_1 x_3$. In these cases, obtain the densities of u_{13} and u_{23} first and then proceed. These produce other different bivariate extensions of Krätzel integrals. For example, let $u = x_1 x_2 x_3 = u_{12} x_3, u_{12} = x_1 x_2$. Let the density of u_{12} be $g_{12}(u_{12})$. Then, from the two-variables case,

$$(H) \quad g_{12}(u_{12}) = \int_v \frac{1}{v} f_1(\frac{u_{12}}{v}) f_2(v) dv.$$

Let the density of u be $g(u)$. Then,

$$(I) \quad g(u) = \int_w \frac{1}{w} g_{12}(\frac{u}{u_{12}}) f_3(w) dw$$
$$= \int_w \frac{1}{w} [\int_v \frac{1}{v} f_1(\frac{u_{12}}{v}) f_2(v) dv] f_3(w) dw$$
$$= \int_v \int_w \frac{1}{vw} f_1(\frac{u_{12}}{v}) f_2(v) f_3(w) dv \wedge dw.$$

However, we also have

$$(J) \quad g_{12}(u_{12}) = \int_v \frac{1}{v} f_1(v) f_2(\frac{u_{12}}{v}) dv.$$

Substituting for g_{12} from (J) into (H), we have the following and other forms from the symmetry also:

$$(K) \quad \begin{aligned} g(u) &= \int_w \frac{1}{w} [\int_v \frac{1}{v} f_1(v) f_2(\frac{u}{v}) dv] f_3(w) dw \\ &= \int_v \int_w \frac{1}{vw} f_1(v) f_2(\frac{u}{v}) f_3(w) dv \wedge dw \\ &= \int_v \int_w \frac{1}{vw} f_1(v) f_2(w) f_3(\frac{u}{v}) dv \wedge dw \\ &= \int_v \int_w \frac{1}{vw} f_1(\frac{u}{v}) f_2(w) f_3(v) dv \wedge dw \\ &= \int_v \int_w \frac{1}{vw} f_1(w) f_2(v) f_3(\frac{u}{v}) dv \wedge dw \\ &= \int_v \int_w \frac{1}{vw} f_1(w) f_2(\frac{u}{v}) f_3(v) dv \wedge dw. \end{aligned}$$

A few such forms, as in (K), are described in [7] and hence these are not repeated here. From the products of four or more variables $x_j > 0, j = 4, 5, ..., k$, we can have several different extensions of Krätzel integral for bivariate, trivariate and general multivariate cases. The method is similar to what is explained above and hence further discussion is omitted. Even though hundreds of different integral representations are available for the density of $u = x_1...x_k$, the explicit evaluation of the density $g(u)$ of u is possible by inverting the corresponding Mellin transform, namely

$$M_g(s) = \prod_{j=1}^{k} M_{f_j}(s)$$

and take the inverse Mellin transform of $\prod_{j=1}^{k} M_{f_j}(s)$ to obtain the density g of $u = x_1 x_2...x_k$.

Connections to Fractional Integrals

Let $x_j > 0, j = 1,2,3$ be real scalar random variables, independently distributed with densities $f_j(x_j), j = 1,2,3$, respectively. Let $u = x_1 x_2 x_3, v = x_2, w = x_3$. Then, $dx_1 \wedge dx_2 \wedge dx_3 = \frac{1}{vw} du \wedge dv \wedge dw$. Let f_1 be a real scalar type-1 beta density with the parameters $(\gamma + 1, \alpha)$, or with the density:

$$f_1(x_1) = \frac{\Gamma(\gamma+1+\alpha)}{\Gamma(\gamma+1)\Gamma(\alpha)} x_1^{\gamma} (1 - x_1)^{\alpha-1}, 0 \leq x_1 \leq 1, \alpha > 0, \gamma > -1.$$

Let f_2 and f_3 be arbitrary densities. Then,

$$f_1(x_1) = f(\frac{u}{vw}) = \frac{\Gamma(\gamma+1+\alpha)}{\Gamma(\gamma+1)\Gamma(\alpha)} (\frac{u}{vw})^{\gamma} (1 - \frac{u}{vw})^{\alpha-1}. \tag{34}$$

Then, the density of u from (34), f_2 and f_3, denoted again by $g(u)$, is the following:

$$g(u) = \frac{\Gamma(\gamma+1+\alpha)}{\Gamma(\gamma+1)} \frac{u^{\gamma}}{\Gamma(\alpha)} \int_v \int_w (vw)^{-\gamma-\alpha} (vw - u)^{\alpha-1} f_2(v) f_3(w) dv \wedge dw$$

$$= \frac{\Gamma(\gamma+1+\alpha)}{\Gamma(\gamma+1)} K_{2,\gamma}^{-\alpha}(f_2, f_3). \tag{35}$$

If f_3 and the corresponding w are absent, then $K_{2,\gamma}^{-\alpha}(f_2, f_3) = K_{2,\gamma}^{-\alpha} f_2$ which is Erdélyi–Kober fractional integral of the second kind and of order α and parameter γ where the arbitrary function is f_2. Similarly, when f_2 and v are absent, we get Erdélyi–Kober fractional integral of the second kind of order α and parameter γ with the arbitrary function f_3. Hence, Equation (35) is a bivariate generalization of Erdélyi–Kober fractional integral of the second kind. This generalization in Equation (35) is different from the multivariate case of Mathai [8] and multi-index case of Kiryakova [19]. Other extension to bivariate case of fractional integrals are available from the various representations in (K) of Section 6 by taking one or two, out of the three functions there, as real scalar type-1 beta densities.

Let $u_1 = \frac{x_1}{x_2}$ with $x_1 = v$ so that $x_2 = \frac{v}{u_1}$ and $dx_1 \wedge dx_2 = -\frac{v}{u_1^2} du_1 \wedge dv$. Then, the density of u_1, denoted by $g_1(u_1)$, is the following:

$$g_1(u_1) = \int_v \frac{v}{u_1^2} f_1(v) f_2(\frac{v}{u_1}) dv. \tag{36}$$

Let $f_1(v) = f(v)$, be an arbitrary density and let $f_2(x_2)$ be a real scalar type-1 beta density with the parameters (γ, α). Then, from Equation (36),

$$g_1(u_1) = \frac{\Gamma(\gamma+\alpha)}{\Gamma(\gamma)\Gamma(\alpha)} \int_v \frac{v}{u_1^2} f(v) (\frac{v}{u_1})^{\gamma-1} (1 - \frac{v}{u_1})^{\alpha-1} dv$$

$$= \frac{\Gamma(\gamma+\alpha)}{\Gamma(\gamma)} \frac{u_1^{-\alpha-\gamma}}{\Gamma(\alpha)} \int_{v \leq u_1} v^{\gamma} (u - v)^{\alpha-1} f(v) dv$$

$$= \frac{\Gamma(\gamma+\alpha)}{\Gamma(\gamma)} K_{1,\gamma}^{-\alpha} f \tag{37}$$

where $K_{1,\gamma}^{-\alpha}f$ is Erdélyi–Kober fractional integral of the first kind of order α and parameter γ. Consider the generalization to three variables. Let $u_1 = \frac{x_2 x_3}{x_1}, x_2 = v, x_3 = w \Rightarrow x_1 = \frac{vw}{u_1}$. Then, $dx_1 \wedge dx_2 \wedge dx_3 = -\frac{vw}{u_1^2} du_1 \wedge dv \wedge dw$ and the marginal density of u_1, again denoted by $g_1(u_1)$, is the following:

$$g_1(u_1) = \frac{\Gamma(\gamma+\alpha)}{\Gamma(\gamma)\Gamma(\alpha)} \int_v \int_w \frac{vw}{u_1^2} \left(\frac{vw}{u_1}\right)^{\gamma-1} \left(1 - \frac{vw}{u_1}\right)^{\alpha-1} f_2(v) f_3(w) dv \wedge dw$$

$$= \frac{\Gamma(\gamma+\alpha)}{\Gamma(\gamma)} \frac{u_1^{-\gamma-\alpha}}{\Gamma(\alpha)} \int_v \int_w (vw)^\gamma (u_1 - vw)^{\alpha-1} f_2(v) f_3(w) dv \wedge dw$$

$$= \frac{\Gamma(\gamma+\alpha)}{\Gamma(\gamma)} K_{1,\gamma}^{-\alpha}(f_1, f_2) \qquad (38)$$

where $K_{1,\gamma}^{-\alpha}(f_2, f_3)$ of Equation (38) may be called Erdélyi–Kober fractional integral of the first kind of order α and parameter γ in the bivariate case or with two arbitrary functions. Here, the integrals are over $0 \le v \le 1, 0 \le w \le 1, 0 \le vw \le u_1$. This type of generalization is different from the ones available in the literature. Various definitions of fractional integrals, fractional derivatives, and fractional differentials equations and their properties may be seen in [20–22].

7. Krätzel Integral in the Real Matrix-variate Case

It is easier to interpret Krätzel integral in terms of statistical distributions. Let X_1 and X_2 be two $p \times p$ real positive definite matrix random variables with the densities $f_1(X_1)$ and $f_2(X_2)$, respectively. Density here means a real-valued scalar function $f(X)$ of the positive definite matrix $X > O$, such that $f(X) \ge 0$ for all $X > O$ and $\int_{X>O} f(X) dX = 1$. That is, for $X_j > O, j = 1, 2$ (positive definite), $f_j(X_j) \ge 0$ for all $X_j > O$ and $\int_{X_j > O} f_j(X_j) dX_j = 1, j = 1, 2$. Let $X_j > O$ have a real matrix-variate gamma density. That is,

$$f_j(X_j) = \frac{|A_j|^{\gamma_j}}{\Gamma_p(\gamma_j)} |X_j|^{\gamma_j - \frac{p+1}{2}} e^{-\text{tr}(A_j X_j)}, X_j > O, A_j > O, \Re(\gamma_j) > \frac{p-1}{2}, j = 1, 2 \qquad (39)$$

where, in Equation (39), $A_j > O$ is a $p \times p$ real positive definite constant matrix for $j = 1, 2..$ When $p = 1$, we have the corresponding scalar variable gamma density. The real matrix-variate gamma function $\Gamma_p(\gamma_j)$ is explained below. In the scalar case we have taken exponents $\delta_j > 0, j = 1, 2$ but if we take exponents in the matrix-variate case then the transformations will not produce nice forms for further derivations, see the types of difficulties from [23], and hence we have taken $\delta_1 = \delta_2 = 1$ in the matrix-variate case. Let us consider symmetric product $U = X_2^{\frac{1}{2}} X_1 X_2^{\frac{1}{2}}$ where $X_2^{\frac{1}{2}} > O$ is the positive definite square root of the positive definite matrix $X_2 > O$. We have taken the symmetric product because the transformations are on symmetric cases. Let $V = X_2$. Then, from Mathai [23], we can derive $dX_1 \wedge dX_2 = |V|^{-\frac{p+1}{2}} dU \wedge dV$ and then proceeding as in the scalar variable case, the density of U, denoted again by $g(U)$, is given by the following:

$$g(U) = \int_V |V|^{-\frac{p+1}{2}} f_1(V^{-\frac{1}{2}} U V^{-\frac{1}{2}}) f_2(V) dV \qquad (40)$$

where f_1 and f_2 in Equation (40) are some general densities. Consider the case when $f_j(X_j)$ is a real matrix-variate gamma density given by the following:

$$f_j(X_j) = \frac{|A_j|^{\gamma_j}}{\Gamma_p(\gamma_j)} |X_j|^{\gamma_j - \frac{p+1}{2}} e^{-\text{tr}(A_j X_j)}, \qquad (41)$$

for $A_j > O, X_j > O, \Re(\gamma_j) > \frac{p-1}{2}, j = 1, 2$, where $\Gamma_p(\gamma_j)$ is the real matrix-variate gamma given by

$$\Gamma_p(\alpha) = \pi^{\frac{p(p-1)}{4}} \Gamma(\alpha) \Gamma(\alpha - \frac{1}{2})...\Gamma(\alpha - \frac{p-1}{2}), \Re(\alpha) > \frac{p-1}{2}. \tag{42}$$

For the densities in Equation (41), with $\Gamma_p(\gamma_j)$ defined in Equation (42), the density of U is given by the following:

$$g(U) = C|U|^{\gamma_1 - \frac{p+1}{2}} \int_{V>O} |V|^{\gamma_2 - \gamma_1 - \frac{p+1}{2}} e^{-\text{tr}(V^{-\frac{1}{2}} A_1 V^{-\frac{1}{2}} U) - \text{tr}(A_2 V)} dV \tag{43}$$

for $A_j > O, V > O, U > O, \Re(\gamma_j) > \frac{p-1}{2}, j = 1, 2$ where

$$C = \prod_{j=1}^{2} \frac{|A_j|^{\gamma_j}}{\Gamma_p(\gamma_j)}.$$

This Equation (43) is the Krätzel integral in the real matrix-variate case. Note that, if A_1 is a positive scalar quantity, then it can be taken out of V and then V^{-1} will be obtained corresponding to the real scalar case.

The model in Equation (41) is also connected to Maxwell-Boltzmann and Raleigh densities in physics. Their matrix-variate, multivariate and rectangular matrix-variate extensions and some applications in reliability analysis are given in [24]. Their complex matrix-variate analogs can be worked out but they do not seem to be in print in the literature yet.

8. Krätzel Integral in the Complex Matrix-variate Case

Here, we consider $p \times p$ Hermitian positive definite matrices $\tilde{X}_j > O, j = 1, 2$ and Hermitian positive definite square root $\tilde{X}_2^{\frac{1}{2}}$. Consider the symmetric product $\tilde{U} = \tilde{X}_2^{\frac{1}{2}} \tilde{X}_1 \tilde{X}_2^{\frac{1}{2}}, \tilde{V} = \tilde{X}_2$. Then, from [23] we have $d\tilde{X}_1 \wedge d\tilde{X}_2 = |\det(V)|^{-p} d\tilde{U} \wedge d\tilde{V}$. Let the density of $\tilde{U}$ be denoted by $\tilde{g}(\tilde{U})$ when $\tilde{X}_j, j = 1, 2$ are independently distributed with the complex matrix-variate gamma densities given by

$$\tilde{f}_j(\tilde{X}_j) = \frac{|\det(A_j)|^{\gamma_j}}{\tilde{\Gamma}_p(\gamma_j)} |\det(\tilde{X}_j)|^{\gamma_j - p} e^{-\text{tr}(A_j \tilde{X}_j)}, \tilde{X}_j > O, \Re(\gamma_j) > p-1, j = 1, 2 \tag{44}$$

where $\tilde{\Gamma}_p(\alpha)$ is the complex matrix-variate gamma given by the following:

$$\tilde{\Gamma}_p(\alpha) = \pi^{\frac{p(p-1)}{2}} \Gamma(\alpha) \Gamma(\alpha - 1)...\Gamma(\alpha - p + 1), \Re(\alpha) > p - 1. \tag{45}$$

Then, from Equations (44) and (45), proceeding as in the real matrix-variate case the density of $\tilde{U}$, denoted by $\tilde{g}(\tilde{U})$, is the following:

$$\tilde{g}(\tilde{U}) = \tilde{C} |\det(\tilde{U})|^{\gamma_1 - p} \int_{\tilde{V}>O} |\det(\tilde{V})|^{\gamma_2 - \gamma_1 - p} e^{-\text{tr}(\tilde{V}^{-\frac{1}{2}} A_1 \tilde{V}^{-\frac{1}{2}} \tilde{U}) - \text{tr}(A_2 \tilde{V})} d\tilde{V}$$

for $\Re(\gamma_j) > p - 1, A_j > O, \tilde{V} > O, \tilde{U} > O, j = 1, 2$ where

$$\tilde{C} = \prod_{j=1}^{2} \frac{|\det(A_j)|^{\gamma_j}}{\tilde{\Gamma}_p(\gamma_j)}.$$

9. Extension to Rectangular Matrix-variate Case

Let $X = (x_{ij})$ be a $p \times q, q \geq p$ matrix of full rank p where the elements x_{ij}s are distinct real scalar variables. Let $A > O$ be $p \times p$ and $B > O$ be $q \times q$ constant real positive definite matrices. Let a prime denote the transpose, let tr$(\cdot)$ be the trace of $(\cdot)$, and let, for example, $A^{\frac{1}{2}}$ be the positive definite square root of the positive definite matrix $A > O$. Consider the model

$$f(X) = C|A^{\frac{1}{2}}XBX'A^{\frac{1}{2}}|^{\gamma}|I + a_1(q_1-1)(A^{\frac{1}{2}}XBX'A^{\frac{1}{2}})|^{-\frac{1}{q_1-1}}$$
$$\times |I + a_2(q_2-1)(A^{\frac{1}{2}}XBX'A^{\frac{1}{2}})^{-1}|^{-\frac{1}{q_2-1}} \qquad (46)$$

for $a_j > 0, q_j > 1, j = 1, 2, \gamma > -\frac{q}{2} + \frac{p-1}{2}$. Observe that

$$\lim_{q_j \to 1} |I + a_j(q_j-1)(A^{\frac{1}{2}}XBX'A^{\frac{1}{2}})|^{-\frac{1}{q_j-1}} = e^{-\text{tr}(A^{\frac{1}{2}}XBX'A^{\frac{1}{2}})} \qquad (47)$$

for $j = 1, 2$. Let

$$f_1(X) = \lim_{q_1 \to 1} f(X), f_2(X) = \lim_{q_2 \to 1} f(X), f_3(X) = \lim_{q_1 \to 1, q_2 \to 1} f(X).$$

Then,

$$f_1(X) = C_1|A^{\frac{1}{2}}XBX'A^{\frac{1}{2}}|^{\gamma} e^{-a_1(A^{\frac{1}{2}}XBX'A^{\frac{1}{2}})}$$
$$\times |I + a_2(q_2-1)(A^{\frac{1}{2}}XBX'A^{\frac{1}{2}})^{-1}|^{-\frac{1}{q_2-1}}. \qquad (48)$$

$$f_2(X) = C_2|A^{\frac{1}{2}}XBX'A^{\frac{1}{2}}|^{\gamma}|I + a_1(q_1-1)(A^{\frac{1}{2}}XBX'A^{\frac{1}{2}})|^{-\frac{1}{q_1-1}}$$
$$\times e^{-a_2\text{tr}(A^{\frac{1}{2}}XBX'A^{\frac{1}{2}})^{-1}}. \qquad (49)$$

$$f_3(X) = C_3|A^{\frac{1}{2}}XBX'A^{\frac{1}{2}}|^{\gamma} e^{-a_1(A^{\frac{1}{2}}XBX'A^{\frac{1}{2}}) - a_2(A^{\frac{1}{2}}XBX'A^{\frac{1}{2}})^{-1}}. \qquad (50)$$

Then, $f_3(X)$, coming from Equations (46) and (47), is the real rectangular matrix-variate version of Krätzel integral. In a physical model building situation, if Equation (50) is the stable or ideal situation, then Equations (46), (48) and (49) describe the unstable neighborhoods. From the discussion in Sections 2 and 3, we can see that the model in Equations (46) and (48)–(50) can also be generated by M-convolution of product or density of a product in the real matrix-variate case. In Equation (50), for simplicity, we have taken the coefficient parameters as scalar quantities. We can evaluate the normalizing constants C, C_1, C_2, C_3 by using the following steps: Let

$$(L) \quad Y = A^{\frac{1}{2}}XB^{\frac{1}{2}} \Rightarrow dX = |A|^{-\frac{p}{2}}|B|^{-\frac{q}{2}}dY$$

from the general linear transformation (see [23] for the Jacobian in (L) and other Jacobians to follow). Let the corresponding function $f(X)$ be denoted by $f_{01}(Y)$. Then,

$$f_{01}(Y) = C|A|^{-\frac{p}{2}}|B|^{-\frac{q}{2}}|YY'|^{\gamma}|I + a_1(q_1-1)(YY')|^{-\frac{1}{q_1-1}}$$
$$\times |I + a_2(q_2-1)(YY')^{-1}|^{-\frac{1}{q_2-1}}. \qquad (51)$$

Let the corresponding functions $f_1(X), f_2(X), f_3(X)$ be denoted by $f_{11}(Y), f_{21}(Y), f_{31}(Y)$, respectively. Note that Y has pq real scalar variables whereas $S = YY'$, which is a $p \times p$ real positive definite matrix, has only $p(p+1)/2$ elements. However, we can obtain a relationship between dY and dS (see [23]). It is the following:

$$(M) \quad dY = \frac{\pi^{\frac{pq}{2}}}{\Gamma_p(\frac{q}{2})} |S|^{\frac{q}{2}-\frac{p+1}{2}} dS,$$

where Y in (M) is $p \times q$, whereas S is $p \times p$. Let the corresponding functions of S be denoted by $f_{02}(S), f_{12}(S), f_{22}(S), f_{32}(S)$, respectively. Then, for example, $f_{02}(S)$ is the following:

$$f_{02}(S) = C|A|^{-\frac{p}{2}}|B|^{-\frac{q}{2}}|S|^{\gamma+\frac{q}{2}-\frac{p+1}{2}}|I+a_1(q_1-1)(S)|^{-\frac{1}{q_1-1}}$$
$$\times |I + a_2(q_2-1)(S)^{-1}|^{-\frac{1}{q_2-1}}.$$

9.1. Multivariate Situation

In Equation (46) and Equations (48)–(50), let $p = 1$ and $q > 1$; then, Y is $1 \times q$ and of the form $Y = (y_1, ..., y_q)$. Then, $YY' = y_1^2 + ... + y_q^2$. Then, for $p = 1$, the constant matrix A is 1×1 and let it be $a_3 > 0$. Then, from Equation (51),

$$f_{01} = C a_3^{-\frac{1}{2}} |B|^{-\frac{q}{2}} (y_1^2 + ... + y_q^2)^\gamma [1 + a_1(q_1-1)(y_1^2 + ... + y_q^2)]^{-\frac{1}{q_1-1}}$$
$$\times [1 + a_2(q_2-1)(y_1^2 + ... + y_q^2)^{-1}]^{-\frac{1}{q_2-1}}.$$

Then, f_{31} becomes the following:

$$f_{31}(Y) = C_3 a_3^{-\frac{1}{2}} |B|^{-\frac{q}{2}} [(y_1^2 + ... + y_q^2)]^\gamma$$
$$\times e^{-a_1(y_1^2+...+y_q^2) - a_2(y_1^2+...+y_q^2)^{-1}} \tag{52}$$

for $-\infty < y_j < \infty, j = 1, ..., q$. We may call Equation (52) as the multivariate version of the basic Krätzel integral and f_{01} for $p = 1$ as the pathway extended form of f_{31} in Equation (52).

Note that for a general $p > 1$ we do not take exponents for $(A^{\frac{1}{2}} X B X' A^{\frac{1}{2}})$ because in the general case matrix transformations create problems while computing the Jacobians. The types of problem is described in [23]. However, for the scalar cases in $f_{02}, f_{12}, f_{22}, f_{32}$, we can take arbitrary exponents. Hence, we have the general Krätzel integrals in the multivariate case as the following:

$$f_{33}(Y) = C_3 a_3^{-\frac{1}{2}} |B|^{-\frac{q}{2}} [(y_1^2 + ... + y_q^2)]^\gamma$$
$$\times e^{-a_1(y_1^2+...+y_q^2)^\delta - a_2(y_1^2+...+y_q^2)^{-\rho}} \tag{53}$$

for $\delta > 0, \rho > 0$. Corresponding exponents can be included in f_{03}, f_{13}, f_{23} as well. For evaluating the normalizing constant, we can do the following steps. Make use of the transformation and Jacobian in (M) for $p = 1$. Then, $S = s$ is a scalar variable. Then, for $p = 1$, Equation (53) becomes the following:

$$f_{34}(s) = a_3^{-\frac{1}{2}} |B|^{-\frac{q}{2}} \frac{\pi^{\frac{q}{2}}}{\Gamma(\frac{q}{2})} s^{\gamma + \frac{q}{2} - 1} e^{-a_1 s^\delta - a_2 s^{-\rho}}.$$

Since s is a real scalar variable here, one can use the scalar version of Mellin convolution of a product or density of product of Sections 2 and 3, go to the Mellin transforms to evaluate the normalizing constant. The same procedure works for all the models f_{04}, f_{14}, f_{24} also.

9.2. Evaluation of the Normalizing Constant

Let

$$\int_{s=0}^{\infty} s^{\gamma+\frac{q}{2}-1} e^{-as^\delta - bs^{-\rho}} ds = g(b) \text{ say.}$$

Let $M_g(t)$ be the Mellin transform of $g(b)$ with Mellin parameter t. Then,

$$M_g(t) = \int_0^\infty b^{t-1} \{\int_{s=0}^\infty s^{\gamma+\frac{q}{2}-1} e^{-as^\delta - bs^{-\rho}} ds\} db.$$

Evaluating the b-integral we have the following:

$$\int_0^\infty b^{t-1} e^{-bs^{-\rho}} db = \Gamma(t) s^{\rho t}, \text{ for } \Re(t) > 0.$$

Now, evaluating the s-integral, we have the following:

$$\int_0^\infty s^{\gamma+\frac{q}{2}+\rho t-1} e^{-as^\delta} ds = \frac{\Gamma(\frac{\gamma+\rho t+q/2}{\delta})}{\delta a^{\frac{\gamma+\rho t+q/2}{\delta}}}, \Re(\gamma+\rho t+q/2)>0.$$

That is,

$$M_g(t) = \frac{1}{\delta s^{\frac{\gamma+q/2}{\delta}}} \Gamma(t) \Gamma(\frac{\gamma+q/2}{\delta} + \frac{\rho}{\delta} t) a^{-\frac{\rho}{\delta} t}.$$

By taking the inverse Mellin transform, we have $g(b)$ as the following:

$$g(b) = \frac{1}{\delta a^{\frac{\gamma+q/2}{\delta}}} \frac{1}{2\pi i} \int_{c-i\infty}^{c+i\infty} \Gamma(t) \Gamma(\frac{\gamma+q/2}{\delta} + \frac{\rho}{\delta} t)(ba^{\frac{\rho}{\delta}})^{-t} dt$$

$$= \frac{1}{\delta a^{\frac{\gamma+q/2}{\delta}}} H_{0,2}^{2,0}\left[ba^{\frac{\rho}{\delta}} \Big|_{(0,1),(\frac{\gamma+q/2}{\delta}, \frac{\rho}{\delta})} \right]$$

where $H(\cdot)$ is the H-function, see [5]. Then, the normalizing constant is the following:

$$C = a_3^{\frac{1}{2}} |B|^{\frac{q}{2}} \frac{\Gamma(\frac{q}{2})}{\pi^{\frac{q}{2}}} \frac{\delta a^{\frac{\gamma+q/2}{\delta}}}{H_{0,2}^{2,0}\left[ba^{\frac{\rho}{\delta}} \Big|_{(0,1),(\frac{\gamma+q/2}{\delta}, \frac{\rho}{\delta})} \right]}.$$

Note that, when $\rho = \delta$, the H-function reduces to the G-function of the form $G_{0,2}^{2,0}\left[ab \Big|_{0, \frac{\gamma+q/2}{\delta}} \right]$. Then, replace the H-function by the G-function. Observe that, when $p = 1$, A is 1×1 and let it be $a_3 > 0$. This is the a_3 appearing above.

Author Contributions: Conceptualization, A.M.M. and H.J.H.; methodology, A.M.M. and H.J.H.; validation, A.M.M. and H.J.H.; formal analysis, A.M.M. and H.J.H.; All authors have read and agreed to the published version of the manuscript.

Funding: This research received no external funding.

Acknowledgments: The authors would like to thank Francesco Mainardi for inviting us to contribute to this Special Issue. The authors would like to thank the reviewers for making valuable comments, which helped to improve the presentation in this paper.

Conflicts of Interest: The authors declare no conflict of interest in this paper.

References

1. Mathai, A.M.; Haubold, H.J. *Erdélyi-Kober Fractional Calculus: From a Statistical Perspective, Inspired by Solar Neutrino Physics*; Springer Briefs in Mathematical Physics: Singapore, 2018.

2. Critchfield, C.L. Analytic forms of the thermonuclear function. In *Cosmology, Fusion & Other Matters: George Gamow Memorial Volume*; Reines, F., Ed.; Colorado Associated University Press: Boulder, CO, USA, 1972; pp. 186–191.
3. Krätze, E. Integral transformations of Bessel type. In *Generalized Functions of Operational Calculus, Proc. Conf. Verna, 1975*; Bulgarian Academy of Sciences: Sofia, Bulgaria, 1979; pp. 148–165.
4. Mathai, A.M. Generalized Krätzel integral and associated statistical densities. *Int. J. Math. Anal.* **2012**, *6*, 2501–2510.
5. Mathai, A.M.; Saxena, R.K.; Haubold, H.J. *The H-Function: Theory and Applications*; Springer: New York, NY, USA, 2010.
6. Coelho, C.A.; Arnold, B.C. *Finite Form Representations for Meijer G and Fox H Functions*; Lecture Notes in Statistics 223; Springer Nature Switzerland: Basel, Switzerland, 2019.
7. Mathai, A.M. Mellin convolutions, statistical distributions and fractional calculus. *Fract. Calc. Appl. Anal.* **2018**, *21*, 376–398. [CrossRef]
8. Mathai, A.M. Fractional integral operators involving many matrix variables. *Linear Algebra Appl.* **2014**, *446*, 196–215. [CrossRef]
9. Attenburger, R.; Haitz, C.; Timmer, J. Analysis of phase-resolved partial discharge patterns of voids based on a stochastic process approach, *J. Phys. D Appl. Phys.* **2002**, *35*, 1149–1163. [CrossRef]
10. Mathai, A.M. A pathway to matrix-variate gamma and normal densities. *Linear Algebra Appl.* **2005**, *396*, 317–328. [CrossRef]
11. Mathai, A.M.; Haubold, H.J. Pathway model, superstatistics, Tsallis statistics and a generalized measure of entropy. *Phys. A* **2007**, *375*, 110–122. [CrossRef]
12. Tsallis, C. *Introduction to Nonextensive Statistical Mechanics: Approaching a Complex World*; Springer: New York, NY, USA, 2009.
13. Mathai, A.M.; Haubold, H.J.; Tsallis, C. Pathway model and nonextensive statistical mechanics. *Sun Geosph.* **2015**, *10*, 157–162.
14. Beck, C. Superstatistics: Theoretical concepts and physical applications. In *Anomalous Transport: Foundations and Applications*; Klages, R., Radons, G., Sokolov, I.M., Eds.; Wiley-VCH: Weinheim, Germany, 2008; pp. 433–457.
15. Mathai, A.M.; Haubold, H.J. A versatile integral in physics and astronomy and Fox's H-function. *Axioms* **2019**, *8*, 122. [CrossRef]
16. Kobayashi, K. Plane wave diffraction by a strip: Exact and asymptotic solutions. *J. Phys. Soc. Jpn.* **1990**, *60*, 1891–1905. [CrossRef]
17. Kobayashi, K. Generalized gamma function occurring in wave scattering problem. *J. Phys. Soc. Jpn.* **1991**, *60*, 1501–1512. [CrossRef]
18. Mathai, A.M. On products and ratios of three or more generalized gamma variables. *J. Indian Soc. Probab. Stat.* **2016**, *17*, 79–94. [CrossRef]
19. Kiryakova, V.S. Multiple (multi-index) Mittag-Leffler functions and relations to generalized fractional calculus. *J. Comput. Appl. Math.* **2000**, *118*, 241–259. [CrossRef]
20. Gorenflo, R.; Kilbas, A.A.; Mainardi, F.; Rogosin, S.V. *Mittag-Leffler Functions, Related Topics and Applications*; Springer: New York, NY, USA, 2014.
21. Gorenflo, R.; Luchko, Y.; Mainardi, F. Analytic properties and applications of the Wright function. *Fract. Calc. Appl. Anal.* **1999**, *2*, 383–414.
22. Mainardi, F.; Luchko, Y.; Pagnini, G. The fundamental solution of the space-time fractional diffusion equations. *Fract. Calc. Appl. Anal.* **2001**, *4*, 153–192.
23. Mathai, A.M. *Jacobians of Matrix Transformations and Functions of Matrix Arguments*; World Scientific Publishing: New York, NY, USA, 1997.
24. Mathai, A.M.; Princy, T. Multivariate and matrix-variate Maxwell-Boltzmann and Raleigh densities. *Phys. A* **2017**, *468*, 668–676. [CrossRef]

© 2020 by the authors. Licensee MDPI, Basel, Switzerland. This article is an open access article distributed under the terms and conditions of the Creative Commons Attribution (CC BY) license (http://creativecommons.org/licenses/by/4.0/).

Article

Differentiation of the Mittag-Leffler Functions with Respect to Parameters in the Laplace Transform Approach

Alexander Apelblat

Department of Chemical Engineering, Ben Gurion University of the Negev, Beer Sheva 84105, Israel; apelblat@bgu.ac.il

Received: 5 March 2020; Accepted: 21 April 2020; Published: 26 April 2020

Abstract: In this work, properties of one- or two-parameter Mittag-Leffler functions are derived using the Laplace transform approach. It is demonstrated that manipulations with the pair direct–inverse transform makes it far more easy than previous methods to derive known and new properties of the Mittag-Leffler functions. Moreover, it is shown that sums of infinite series of the Mittag-Leffler functions can be expressed as convolution integrals, while the derivatives of the Mittag-Leffler functions with respect to their parameters are expressible as double convolution integrals. The derivatives can also be obtained from integral representations of the Mittag-Leffler functions. On the other hand, direct differentiation of the Mittag-Leffler functions with respect to parameters produces an infinite power series, whose coefficients are quotients of the digamma and gamma functions. Closed forms of these series can be derived when the parameters are set to be integers.

Keywords: derivatives with respect to parameters; Mittag-Leffler functions; Laplace transform approach; infinite power series; integral representations; convolution integrals; quotients of digamma and gamma functions

1. Introduction

At the beginning of the previous century, the exponential function was generalized by the Swedish mathematician G.M. Mittag-Leffler, who introduced a new power series that is named after him today [1]. Quite unexpectedly, enormous interest has developed regarding the Mittag-Leffler functions over the last four decades because of their ability to describe diverse physical phenomena far more easily than other approaches in a host of scientific and engineering disciplines. Consequently, the Mittag-Leffler functions have become one of the most important special functions in mathematics. Examples where they appear include kinetics of chemical reactions, time and space fractional diffusion, nonlinear waves, viscoelastic systems, neural networks, electric field relaxations, and statistical distributions [2–8]. In mathematics, the Mittag-Leffler functions play an important role in fractional calculus, solution of systems with fractional differential, and integral equations [9,10]. As a result of all this activity, there is now extensive literature on their properties and history [11–13]. A number of reviews have been produced [14–16], and of these, the monograph by Gorenflo, Kilbas, Mainardi, and Rogosin [17] occupies a special place.

The one-parameter, classical Mittag-Leffler function $E_\alpha(z)$ is defined in the whole complex plane by the following power series:

$$E_\alpha(z) = \sum_{k=0}^{\infty} \frac{z^k}{\Gamma(\alpha k + 1)}, \qquad (1)$$

where $\text{Re}\,\alpha > 0$.

Later, Wiman [18] introduced the two-parameter Mittag-Leffler function $E_{\alpha,\beta}(z)$, which is given by

$$E_{\alpha,\beta}(z) = \sum_{k=0}^{\infty} \frac{z^k}{\Gamma(\alpha k + \beta)}, \qquad (2)$$

where $\mathrm{Re}\,\alpha > 0$ and $\mathrm{Re}\,\beta > 0$. Only these two functions, not generalizations thereafter, will be studied here.

There are two main aims in this work. The first is to show that many well-known and new functional relations can be easily derived via the Laplace transform theory and the second is to consider differentiation with respect to the parameters α and β. Throughout this paper, all mathematical operations or manipulations with functions, series, integrals, integral representations, and transforms will be formal. There will be no proofs of validity of given expressions, though they are, without doubt, correct. The following sections present many results that have been derived independently by other methods, while the new results are verified by two different numerical procedures. Thus, in the framework of applied operational calculus, the reported results are only valid for real positive values of arguments and parameters.

My previous involvement with the Mittag-Leffler functions has been limited only to establishing their connections to the Volterra functions. In my monograph devoted to the Volterra functions [19], I presented in Appendix A some representations of the Mittag-Leffler functions in terms of other special functions. They can also be derived directly using the Laplace transform technique when applied to $E_\alpha(\pm t^\alpha)$ functions. Evidently, this restricts the transform–inverse pair only to the positive real axis. New results, together with some from [19], are presented below.

According to the definitions of the Mittag-Leffler functions, there is a clear distinction between the argument, z, and the parameters, α and β, as the latter appear in the coefficients. Nevertheless, $E_\alpha(z) = f(\alpha, z)$ and $E_{\alpha,\beta}(z) = f(\alpha, \beta, z)$ can be regarded as the bivariate and trivariate functions, respectively.

As this is the first investigation dealing with mathematical operations with respect to variables α and β, its scope is only limited to derivatives of the Mittag-Leffler functions. The special forms of the Laplace transforms of $E_\alpha(\pm t^\alpha)$ and $E_{\alpha,\beta}(\pm t^\alpha)$ functions will be studied extensively to establish known properties of the Mittag-Leffler functions and to derive new functional relations. As will be demonstrated, the differentiation operations will lead to power series with coefficients being quotients of psi and gamma functions. In some cases, these series can be evaluated in a closed form, i.e., in terms of elementary and special functions. Computation methods used in this investigation to obtain the Mittag-Leffler functions and their derivatives with respect to α differ from those reported in the literature. This results from the fact that the Mittag-Leffler functions are available as the build-in functions in the MATHEMATICA program.

2. Properties of the Mittag-Leffler Functions in the Laplace Transform Approach

The Laplace transform of the Mittag-Leffler function $E_\alpha(t^\rho)$ is given by

$$L\{E_\alpha(t^\rho)\} = \frac{1}{s} \sum_{k=0}^{\infty} \frac{\Gamma(\rho k + 1)}{\Gamma(\alpha k + 1)} \left(\frac{1}{s^\rho}\right)^k, \qquad (3)$$

which is not valid to all values of ρ and α as discussed in [17].

For $\rho = \alpha$, (3) becomes

$$L\{E_\alpha(t^\alpha)\} = \frac{s^{\alpha-1}}{s^\alpha - 1}, \qquad (4)$$

where $\mathrm{Re}\,\alpha > 0$ and $\mathrm{Re}\,s > 1$ and for negative t^α is

$$L\{E_\alpha(-t^\alpha)\} = \frac{s^{\alpha-1}}{s^\alpha + 1}. \qquad (5)$$

In a similar manner, the Laplace transforms of two-parameter Mittag-Leffler functions, $t^{\beta-1}E_{\alpha,\beta}(\pm\lambda t^\alpha)$, in [17] are found to be

$$L\{t^{\beta-1}E_{\alpha,\beta}(\pm\lambda t^\alpha)\} = \frac{s^{\alpha-\beta}}{s^\alpha - \lambda}, \tag{6}$$

where $\operatorname{Re}\alpha > 0$, $\operatorname{Re}\beta > 0$ and $\operatorname{Re} s > |\lambda|^{1/\alpha}$.

Not only are the inverse transforms simple to derive from them results, but one is able to identify functions for particular values of α and β. Carrying this out will require algebraic manipulations, the similarity properties of the Laplace transformation, the Heaviside expansion theorem, the convolution (product) theorem, some substitution formulas, and other techniques and rules of the operational calculus.

In the first application of the Laplace transform theory, we consider positive integer values of α from 1 to 4. Then, the Mittag-Leffler functions reduce to elementary or special functions due to the simple inverse transforms.

For $\alpha = 1$, one finds that

$$E_1(t) = L^{-1}L\{E_1(t)\} = L^{-1}\left\{\frac{1}{s-1}\right\} = e^t. \tag{7}$$

For $\alpha = 2$, one obtains

$$E_2(t^2) = L^{-1}L\{E_2(t^2)\} = L^{-1}\left\{\frac{s}{s^2-1}\right\} = L^{-1}\left\{\frac{s}{(s-1)(s+1)}\right\} = \cosh t, \tag{8}$$

where the dominator has been decomposed into partial fractions. However, the more expedient method is to evaluate the contributions from the residues at $s = \pm 1$.

Carrying out this procedure for $-t^2$ yields

$$E_2(-t^2) = L^{-1}L\{E_2(-t^2)\} = L^{-1}\left\{\frac{s}{s^2+1}\right\} = L^{-1}\left\{\frac{s}{(s-i)(s+i)}\right\} = \left.\frac{se^{it}}{s+i}\right|_{s=+i} + \left.\frac{se^{-it}}{s-i}\right|_{s=-i} = \frac{e^{it}}{2} + \frac{e^{-it}}{2} = \cos t. \tag{9}$$

For $\alpha = 3$, one finds that

$$E_3(t^3) = L^{-1}L\{E_3(t^3)\} = L^{-1}\left\{\frac{s^2}{s^3-1}\right\} = L^{-1}\left\{\frac{s^2}{(s-1)(s^2+s+1)}\right\} =$$
$$L^{-1}\left\{\frac{s^2}{(s-1)(s+\frac{1+i\sqrt{3}}{2})(s+\frac{1-i\sqrt{3}}{2})}\right\} =$$
$$\left.\frac{s^2 e^t}{(s+\frac{1+i\sqrt{3}}{2})(s+\frac{1-i\sqrt{3}}{2})}\right|_{s=1} + \left.\frac{s^2 e^{-t(1+i\sqrt{3})/2}}{(s-1)(s+\frac{1-i\sqrt{3}}{2})}\right|_{s=-\frac{1+i\sqrt{3}}{2}} \tag{10}$$
$$+ \left.\frac{s^2 e^{-t(1-i\sqrt{3})/2}}{(s-1)(s+\frac{1+i\sqrt{3}}{2})}\right|_{s=-\frac{1-i\sqrt{3}}{2}} = \frac{1}{3}[e^t + 2e^{-t/2}\cos(\frac{\sqrt{3}}{2}t)].$$

Similarly, for negative t^α, one arrives at

$$E_3(-t^3) = L^{-1}L\{E_3(-t^3)\} = L^{-1}\left\{\frac{s^2}{s^3+1}\right\} =$$
$$\frac{1}{3}[e^{-t} + 2e^{t/2}\cos(\frac{\sqrt{3}}{2}t)]. \tag{11}$$

The calculations become more tedious as α increases. However, for $\alpha = n$, an integer, we obtain in general case

$$E_n(\pm t^n) = L^{-1}L\{E_n(\pm t^n)\} = L^{-1}\left\{\frac{s^{n-1}}{s^n - 1}\right\}. \tag{12}$$

It is obvious that for integer values of α, the Mittag-Leffler functions can be expressed in terms of elementary functions, such as combination of exponential, hyperbolic, and trigonometric functions.

When α is not an integer, special functions are involved. Then, one must use a combination of tables of inverse Laplace transforms, substitution formulas, the convolution theorem, and other rules. For example, from the table of inverse transforms [20], we have

$$L^{-1}\left\{\frac{1}{\sqrt{s}}\right\} = \frac{1}{\sqrt{\pi t}},$$
$$L^{-1}\left\{\frac{1}{\sqrt{s}-1}\right\} = \frac{1}{\sqrt{\pi t}} \pm e^t erfc(-\sqrt{t}),$$
$$erfc(-t^{1/2}) = -erfc(t^{1/2}) = erf(t^{1/2}) - 1 \tag{13}$$

Hence, we find that

$$E_{1/2}(\pm \sqrt{t}) = L^{-1}L\{E_{1/2}(\pm \sqrt{t})\} = L^{-1}\left\{\frac{1}{\sqrt{s}(\sqrt{s}-1)}\right\} =$$
$$L^{-1}\left\{-\frac{1}{\sqrt{s}} \pm \frac{1}{(\sqrt{s}-1)}\right\} = e^t[1 - erf(\sqrt{t})]. \tag{14}$$

The cases with $\alpha = \pm 1/4$ are more complex. Therefore, only the final result for $\alpha = 1/4$ from [19] is presented here. This is

$$E_{1/4}(\pm t^{1/4}) = L^{-1}L\{E_{1/4}(\pm t^{1/4})\} = L^{-1}\left\{\frac{1}{s^{3/4}(s^{1/4}-1)}\right\} =$$
$$L^{-1}\left\{\frac{1}{\sqrt{s}(\sqrt{s}-1)} \pm \frac{1}{s^{1/4}(s-1)} \pm \frac{1}{s^{3/4}(s-1)}\right\} =$$
$$e^t\left\{1 + erf(\sqrt{t}) \pm \frac{\gamma(\frac{1}{4},t)}{\Gamma(\frac{1}{4})} \pm \frac{\gamma(\frac{3}{4},t)}{\Gamma(\frac{3}{4})}\right\}, \tag{15}$$
$$\gamma(a,t) = \Gamma(a) - \Gamma(a,t) = \int_0^t x^{a-1}e^{-x}dx,$$

where the last equation in (15) is the integral representation for the incomplete gamma function.

We can also determine relations between the Mittag-Leffler functions using the Laplace transformation. Putting $\beta = \alpha + 1$ in (6) yields

$$L\{t^\alpha E_{\alpha,\alpha+1}(t^\alpha)\} = \frac{1}{s(s^\alpha - 1)}. \tag{16}$$

However, noting that

$$L\{E_\alpha(t^\alpha) - 1\} = \frac{s^{\alpha-1}}{s^\alpha - 1} - \frac{1}{s} = \frac{1}{s(s^\alpha - 1)}, \tag{17}$$

we can derive the well-known relation for the Mittag-Leffler functions

$$E_\alpha(t^\alpha) - 1 = t^\alpha E_{\alpha,\alpha+1}(t^\alpha). \tag{18}$$

A similar result for the two-parameter Mittag-Leffler function can be derived from

$$L\{t^{\alpha+\beta-1}E_{\alpha,\alpha+\beta}(t^\alpha)\} = \frac{1}{s^\beta(s^\alpha - 1)}, \tag{19}$$

and

$$L\left\{t^{\beta-1}E_{\alpha,\beta}(t^\alpha) - \frac{t^{\beta-1}}{\Gamma(\beta)}\right\} = \frac{s^{\alpha-\beta}}{(s^\alpha - 1)} - \frac{1}{s^\beta} = \frac{1}{s^\beta(s^\alpha - 1)}. \tag{20}$$

Hence, we arrive at
$$E_{\alpha,\beta}(t^\alpha) = \frac{1}{\Gamma(\beta)} + t^\alpha E_{\alpha,\alpha+\beta}(t^\alpha). \tag{21}$$

For α and β integers, (21) can be written as
$$\begin{aligned} E_{1,\beta}(t) &= \tfrac{1}{\Gamma(\beta)} + t\, E_{1,\alpha+1}(t), \\ E_{n,\beta}(t^n) &= \tfrac{1}{\Gamma(\beta)} + t^n E_{n,n+\beta}(t^n), \\ E_{n,n}(t^n) &= \tfrac{1}{(n-1)!} + t^n E_{n,2n}(t^n), \\ E_{n,m}(t^n) &= \tfrac{1}{(m-1)!} + t^n E_{n,m+n}(t^n) \end{aligned} \tag{22}$$

Of the many substitution formulas in the Laplace transform theory, only three will be employed here. From [21] we have
$$\begin{aligned} L\{f(t)\} &= F(s), \\ L^{-1}\left\{\tfrac{1}{\sqrt{s}}F(\sqrt{s})\right\} &= \tfrac{1}{\sqrt{\pi t}} \int_0^\infty e^{-u^2/4t} f(u)\, du \end{aligned} \tag{23}$$

By wring the Laplace transform of $E_\alpha(t^\alpha)$ as
$$L\{E_\alpha(t^\alpha)\} = \frac{s^{\alpha-1}}{s^\alpha - 1} = \frac{1}{\sqrt{s}} \frac{(\sqrt{s})^{2\alpha-1}}{[(\sqrt{s})^{2\alpha} - 1]}, \tag{24}$$

we find that the Mittag-Leffler function can be represented by
$$E_\alpha(t^\alpha) = \frac{1}{\sqrt{\pi t}} \int_0^\infty e^{-u^2/4t} E_{2\alpha}(u^{2\alpha})\, du. \tag{25}$$

The operational rule for the Macdonald function $K_{1/3}(z)$ is
$$L^{-1}\left\{\frac{1}{s^{2/3}} F(s^{1/3})\right\} = \frac{1}{\pi} \int_0^\infty \sqrt{\frac{u}{t}}\, K_{1/3}\!\left(\frac{2u^{3/2}}{\sqrt{27t}}\right) f(u)\, du. \tag{26}$$

Writing the Laplace transform of $E_\alpha(t^\alpha)$ as
$$L\{E_\alpha(t^\alpha)\} = \frac{s^{\alpha-1}}{s^\alpha - 1} = \frac{(s^{1/3})^{3\alpha-1}}{s^{2/3}[(s^{1/3})^{3\alpha} - 1]}, \tag{27}$$

gives
$$E_\alpha(t^\alpha) = \frac{1}{\pi} \int_0^\infty \sqrt{\frac{u}{t}}\, K_{1/3}\!\left(\frac{2u^{3/2}}{\sqrt{27\, t}}\right) E_{3\alpha}(u^{3\alpha})\, du. \tag{28}$$

For specific values of α, the Mittag-Leffler functions in the integrands of (25) and (28) can be expressed as elementary or special functions. Then, the Mittag-Leffler functions on the left-hand side will be represented by definite integrals over infinity.

The third substitution formula is
$$L^{-1}\left\{\frac{1}{s^2} F\!\left(\frac{1}{s}\right)\right\} = \int_0^\infty \sqrt{\frac{t}{u}}\, J_1(2\sqrt{tu})\, f(u)\, du, \tag{29}$$

where $J_1(z)$ is the Bessel function of the first kind and of the first order

From

$$L\{1 - E_\alpha(t^\alpha)\} = \frac{1}{s} - \frac{s^{\alpha-1}}{s^\alpha - 1} = \frac{1}{s^2} \frac{(\frac{1}{s})^{\alpha-1}}{[(\frac{1}{s})^\alpha - 1]}, \qquad (30)$$

it follows that

$$\frac{E_\alpha(t^\alpha) - 1}{\sqrt{t}} = \int_0^\infty J_1(2\sqrt{tu}) E_\alpha(u^\alpha) \frac{du}{\sqrt{u}}. \qquad (31)$$

Many properties and functional relations for the Mittag-Leffler functions can be obtained from the convolution theorem. These are found by expressing the Laplace transforms of $E_\alpha(t^\alpha)$ in various forms and then evaluating the inverses via convolution integrals. For example, using

$$L\{E_\alpha(t^\alpha)\} = \frac{s^{\alpha-1}}{s^\alpha - 1} = \frac{s^{2\alpha-1}}{s^{2\alpha} - 1} + \frac{s^{2\alpha-1}}{s^{2\alpha} - 1} \cdot \frac{1}{s^\alpha}, \qquad (32)$$

immediately yields

$$E_\alpha(t^\alpha) = E_{2\alpha}(t^{2\alpha}) + E_{2\alpha}(t^{2\alpha}) * \frac{t^{\alpha-1}}{\Gamma(\alpha)} = $$
$$E_{2\alpha}(t^{2\alpha}) + \int_0^t E_{2\alpha}(u^{2\alpha}) \frac{(t-u)^{\alpha-1}}{\Gamma(\alpha)} du. \qquad (33)$$

All convolution integrals can be transformed into finite trigonometric integrals by a suitable change of variable. Therefore, putting $u = t[\cos\theta]^2$ in (33) yields

$$\frac{1}{\Gamma(\alpha)} \int_0^t E_{2\alpha}(u^{2\alpha})(t-u)^{\alpha-1} du = $$
$$\frac{t^{\alpha-1}}{\Gamma(\alpha)} \int_0^{\pi/2} \sin(2\theta) [(\sin\theta)^2]^{\alpha-1} E_{2\alpha}[t^{2\alpha}(\cos\theta)^{4\alpha}] d\theta \qquad (34)$$

Similarly, from

$$L\{t^{\beta-1} E_{\alpha,\beta}(t^\alpha)\} = \frac{s^{\alpha-\beta}}{s^\alpha - 1} = \frac{s^{2\alpha-\beta}}{s^{2\alpha} - 1} + \frac{s^{2\alpha-\beta}}{s^{2\alpha} - 1} \cdot \frac{1}{s^\alpha}, \qquad (35)$$

it follows that

$$E_{\alpha,\beta}(t^\alpha) = E_{2\alpha,\beta}(t^{2\alpha}) + \int_0^t \left(\frac{u}{t}\right)^{\beta-1} E_{2\alpha,\beta}(u^{2\alpha}) \frac{(t-u)^{\alpha-1}}{\Gamma(\alpha)} du. \qquad (36)$$

A different convolution integral can be derived from

$$\frac{1}{s^{\beta+1}} = \frac{s^{\alpha-\beta}}{s^\alpha - 1} \cdot \left[\frac{1}{s} - \frac{1}{s^{\alpha+1}}\right], \qquad (37)$$

whose inverse Laplace transform is

$$\frac{t^\beta}{\Gamma(\beta+1)} = \int_0^t u^{\beta-1} E_{\alpha,\beta}(u^\alpha) \left[1 - \frac{(t-u)^\alpha}{\Gamma(\alpha+1)}\right] du. \qquad (38)$$

Introducing the Laplace transform of $E_{\alpha,\beta}(\pm t^\alpha)$ in the form

$$L\{t^{\beta-1} E_{\alpha,\beta}(\pm t^\alpha)\} = \frac{s^{\alpha-\beta}}{s^\alpha - 1} = \frac{s^{\alpha-1}}{s^\alpha - 1} \cdot \frac{1}{s^{\beta-1}}, \qquad (39)$$

gives

$$t^{\beta-1} E_{\alpha,\beta}(\pm t^\alpha) = E_\alpha(\pm t^\alpha) * \frac{t^{\beta-2}}{\Gamma(\beta-1)} = \int_0^t E_\alpha(\pm u^\alpha) \frac{(t-u)^{\beta-2}}{\Gamma(\beta-1)} du . \quad (40)$$

For $\beta = \alpha$, this becomes

$$t^{\alpha-1} E_{\alpha,\alpha}(\pm t^\alpha) = E_\alpha(\pm t^\alpha) * \frac{t^{\alpha-2}}{\Gamma(\alpha-1)} = \int_0^t E_\alpha(\pm u^\alpha) \frac{(t-u)^{\alpha-2}}{\Gamma(\alpha-1)} du . \quad (41)$$

For α and β, positive integers, (40) reduces to

$$t^{m-1} E_{n,m}(\pm t^n) = \int_0^t E_n(\pm u^n) \frac{(t-u)^{m-2}}{(m-2)!} . \quad (42)$$

where $n = 1, 2, 3, \ldots$ and $m = 2, 3, 4, \ldots$.

These convolution integrals are easily evaluated because the Mittag-Leffler functions reduce to elementary functions. For example, for $n = 1$ and $m = 2$ and 3 and noting that $E_1(t) = e^t$, it follows that

$$\begin{aligned} t\, E_{1,2}(t) &= \int_0^t e^u \, du = e^t - 1 , \\ t^2\, E_{1,3}(t) &= \int_0^t e^u (t-u)\, du = e^t - t - 1 \end{aligned} \quad (43)$$

The Mittag-Leffler functions for $n = 1$ to 4 and $m = 2$ to 4 are presented in [19].

The operational rules of the Laplace transformation enable us to obtain representations for derivatives of the Mittag-Leffler functions $t^{\beta-1} E_{\alpha,\beta}(t^\alpha)$. It is obvious from (2) that the derivative for any order is zero at the origin. In this case, differentiation of the Mittag-Leffler function is equivalent to multiplying the Laplace transform by powers of s. Because

$$\begin{aligned} L\{f^{(n)}(t)\} &= s^n F(s) , \\ f(0) &= f'(0) = f''(0) = \ldots = f^{(n)}(0) , \\ n &= 1, 2, 3, \ldots , \end{aligned} \quad (44)$$

we find that for $\mathrm{Re}\,\alpha > 0$, $\mathrm{Re}\,\beta \geq n + 1$ and $\mathrm{Re}\,s > 1$

$$L\left\{ \frac{d^n}{dt^n} \left[t^{\beta-1} E_{\alpha,\beta}(t^\alpha) \right] \right\} = s^n \left(\frac{s^{\alpha-\beta}}{s^\alpha - 1} \right) = \frac{s^{\alpha-(\beta-n)}}{s^\alpha - 1} . \quad (45)$$

Hence, the Laplace inverse transform becomes

$$\frac{d^n}{dt^n}\left[t^{\beta-1} E_{\alpha,\beta}(t^\alpha) \right] = t^{\beta-n-1} E_{\alpha,\beta-n}(t^\alpha) . \quad (46)$$

In case of $E_\alpha(t^\alpha)$ function, its value is unity at the origin. Only the first derivative has a simple Laplace transform, which is

$$L\left\{ \frac{d}{dt} [E_\alpha(t^\alpha)] \right\} = s\left(\frac{s^{\alpha-1}}{s^\alpha - 1} \right) - 1 = \frac{1}{s^\alpha - 1} = \left(\frac{s^{\alpha-1}}{s^\alpha - 1} \right) \cdot \frac{1}{s^{\alpha-1}} , \quad (47)$$

the inverse transform of (47) is

$$\frac{d}{dt}[E_\alpha(t^\alpha)] = E_\alpha(t^\alpha) * \frac{t^{\alpha-2}}{\Gamma(\alpha-1)}, \qquad (48)$$

However, according to (41), this convolution integral is also given by

$$\frac{d}{dt}[E_\alpha(t^\alpha)] = t^{\alpha-1} E_{\alpha,\alpha}(t^\alpha). \qquad (49)$$

The n-dimensional integrals of the Mittag-Leffler functions are easily evaluated because this is equivalent to dividing the Laplace transform, $F(s)$, by s^n

$$L\left\{\int_0^t \int_0^{u_{n-1}} \cdots \int_0^{u_1} f(u_1)\, du_1\, du_2 \cdots du_n\right\} = \frac{1}{s^n} F(s), \qquad (50)$$

Then, we obtain

$$L\left\{\int_0^t \int_0^{u_{n-1}} \cdots \int_0^{u_1} u_1^{\beta-1} E_{\alpha,\beta}(u_1^\alpha)\, du_1\, du_2 \cdots du_n\right\} = \frac{1}{s^n}\left(\frac{s^{\alpha-\beta}}{s^\alpha-1}\right) = \frac{s^{\alpha-(\beta+n)}}{s^\alpha-1}, \qquad (51)$$

The inverse transform of (51) is

$$\int_0^t \int_0^{u_{n-1}} \cdots \int_0^{u_1} u_1^{\beta-1} E_{\alpha,\beta}(u_1^\alpha)\, du_1\, du_2 \cdots du_n = t^{\beta+n-1} E_{\alpha,\beta+n}(t^\alpha). \qquad (52)$$

For $n = 1$ and $\beta = 1$,

$$\int_0^t u^{\beta-1} E_{\alpha,\beta}(u^\alpha)\, du = t^\beta E_{\alpha,\beta+1}(t^\alpha),$$
$$\int_0^t E_\alpha(u^\alpha)\, du = t\, E_{\alpha,2}(t^\alpha). \qquad (53)$$

Together with the linearity property of the Laplace transformation, operational calculus is able to determine the sums of the Mittag-Leffler functions as power series. Consider the infinite and finite geometrical series, namely,

$$\begin{aligned}1 + x + x^2 + \ldots + x^k + \ldots &= \tfrac{1}{1-x},\\ 1 + x + x^2 + \ldots x^{n-1} + x^n &= \tfrac{x^{n+1}-1}{x-1}\end{aligned} \qquad (54)$$

where $0 < x < 1$.

By taking the Laplace transforms of all the terms in the power series of the corresponding Mittag-Leffler function, one obtains for $s > 1$,

$$F(s) = \frac{s^{\alpha-1}}{s^\alpha-1} + \frac{s^{\alpha-2}}{s^\alpha-1} + \frac{s^{\alpha-3}}{s^\alpha-1} + \ldots + \frac{s^{\alpha-k}}{s^\alpha-1} + \ldots, \qquad (55)$$

The inverse transform of $F(s)$ is given by the following series of the Mittag-Leffler functions:

$$\begin{aligned}L^{-1}\{F(s)\} &= E_\alpha(t^\alpha) + t\, E_{\alpha,2}(t^\alpha) + t^2 E_{\alpha,3}(t^\alpha) + \ldots \\ &+ t^k E_{\alpha,k+1}(t^\alpha) + \ldots = \sum_{k=1}^\infty t^{k-1} E_{\alpha,k}(t^\alpha).\end{aligned} \qquad (56)$$

In order to invert $F(s)$, one must express (55) as

$$F(s) = \frac{s^{\alpha-1}}{s^\alpha - 1} + \frac{s^{\alpha-1}}{s^\alpha - 1}\left\{\frac{1}{s} + \frac{1}{s^2} + \frac{1}{s^3} \ldots + \frac{1}{s^k} + \ldots\right\}, \qquad (57)$$

The series inside the brackets is merely the geometric series. Using (54) one finds that

$$F(s) = \frac{s^{\alpha-1}}{s^\alpha - 1} + \frac{s^{\alpha-1}}{s^\alpha - 1}\left[\frac{1}{1 - (1/s)} - 1\right] = \frac{s^{\alpha-1}}{s^\alpha - 1} + \frac{s^{\alpha-1}}{s^\alpha - 1}\cdot\frac{1}{s-1}, \qquad (58)$$

Finally, inverting $F(s)$ yields

$$\sum_{k=1}^{\infty} t^{k-1} E_{\alpha,k}(t^\alpha) = E_\alpha(t^\alpha) + E_\alpha(t^\alpha) * e^t =$$
$$E_\alpha(t^\alpha) + \int_0^t e^{(t-u)} E_\alpha(u^\alpha)\, du\,. \qquad (59)$$

For the case of a finite series of the Mittag-Leffler functions, one requires the second result in (54) to determine the Laplace transform $F(s)$, which is given by

$$F(s) = \frac{s^{\alpha-1}}{s^\alpha-1} + \frac{s^{\alpha-1}}{s^\alpha-1}\left[\frac{(1/s)^n - 1}{(1/s)-1} - 1\right] =$$
$$\frac{s^{\alpha-1}}{s^\alpha-1} + \left\{\frac{s^{\alpha-1}}{s^\alpha-1} - \frac{s^{\alpha-(n+1)}}{s^\alpha-1}\right\}\cdot\frac{1}{s-1}, \qquad (60)$$

According to the convolution theorem, the inverse transform of this finite sum is

$$\sum_{k=1}^{n} t^{k-1} E_{\alpha,k}(t^\alpha) = E_\alpha(t^\alpha) + e^t * \left\{E_\alpha(t^\alpha) - t^n E_{\alpha,n+1}(t^\alpha)\right\} =$$
$$E_\alpha(t^\alpha) + \int_0^t e^{(t-u)}\left\{E_\alpha(u^\alpha) - u^n E_{\alpha,n+1}(u^\alpha)\right\} du\,. \qquad (61)$$

Similarly, we can use (54) for negative value of x

$$1 - x + x^2 - \ldots + x^k - \ldots = \frac{1}{1+x}, \qquad (62)$$

Then, the corresponding Laplace transform becomes

$$F(s) = \frac{s^{\alpha-1}}{s^\alpha-1} + \frac{s^{\alpha-1}}{s^\alpha-1}\left\{-\frac{1}{s} + \frac{1}{s^2} - \frac{1}{s^3} + \ldots + \frac{1}{s^k} + \ldots\right\} =$$
$$\frac{s^{\alpha-1}}{s^\alpha-1} + \frac{s^{\alpha-1}}{s^\alpha-1}\left(\frac{s}{s+1} - 1\right) = \frac{s^{\alpha-1}}{s^\alpha-1} - \frac{s^{\alpha-1}}{s^\alpha-1}\cdot\frac{1}{s+1}, \qquad (63)$$

Inversion of this result yields

$$\sum_{k=1}^{\infty}(-1)^{k-1} t^{k-1} E_{\alpha,k}(t^\alpha) = E_\alpha(t^\alpha) - E_\alpha(t^\alpha) * e^{-t} =$$
$$E_\alpha(t^\alpha) - \int_0^t e^{-(t-u)} E_\alpha(u^\alpha)\, du\,. \qquad (64)$$

According to the binomial theorem for $x < 1$, we have

$$P(x) = 1 - 2x + 3x^2 - 4x^3 + \ldots = \sum_{k=1}^{\infty}(-1)^{k-1} k\, x^{k-1} = \frac{1}{(1+x)^2}, \qquad (65)$$

The Laplace transform corresponding to this series is

$$F(s) = \frac{s^{a-1}}{s^a-1} + \frac{s^{a-1}}{s^a-1}\left\{-\frac{2}{s} + \frac{3}{s^2} - \frac{4}{s^3}\ldots\right\} = \frac{s^{a-1}}{s^a-1} + \frac{s^{a-1}}{s^a-1}\left[\left(\frac{s}{s+1}\right)^2 - 1\right] = \frac{s^{a-1}}{s^a-1} - \frac{s^{a-1}}{s^a-1}\left[\frac{1}{(s+1)^2} + \frac{2s}{(s+1)^2}\right], \quad (66)$$

The inverse transform of the second term in (66) is

$$L^{-1}\left\{-\frac{s^{a-1}}{s^a-1}\left[\frac{1}{(s+1)^2} + \frac{2s}{(s+1)^2}\right]\right\} = -E_\alpha(t^\alpha) * \left[t\, e^{-t} + 2e^{-t}(1-t)\right] = E_\alpha(t^\alpha) * \left[(t-2)\, e^{-t}\right], \quad (67)$$

Thus, the infinite series of the Mittag-Leffler functions in (65) and (67) is

$$E_{\alpha,1}(t^\alpha) - 2t\, E_{\alpha,2}(t^\alpha) + 3t^2\, E_{\alpha,3}(t^\alpha) - 4t^3\, E_{\alpha,4}(t^\alpha) + \ldots =$$
$$\sum_{k=1}^{\infty} (-1)^{k-1} k t^{k-1} E_{\alpha,k}(t^\alpha) = \quad (68)$$
$$E_\alpha(t^\alpha) + \int_0^t \left[(t-u-2)\, e^{-(t-u)}\right] E_\alpha(u^\alpha)\, du\ .$$

From the preceding examples, it is obvious that if the function $f(t)$ is expanded into the Taylor series,

$$f(x) = \sum_{k=0}^{\infty} \frac{f^{(k)}(0)}{k!} x^k \quad (69)$$

Then, the sum of the corresponding series of the Mittag-Leffler functions can be expressed in terms of convolution integrals. This is only possible if the inverse Laplace transforms, $L^{-1}[f(1/s) - 1]$, are known.

Now, consider the binomial series with the power of 1/2. Then, we have some derivatives of the function $f(t)$, which are equal to zero at the origin

$$f(x) = \sqrt{1+x^2} = 1 + \frac{x^2}{2} - \frac{x^4}{8} + \frac{x^6}{16} - \frac{5x^8}{128} + \ldots, \quad (70)$$

The corresponding series of the Mittag-Leffler functions is

$$S(t^\alpha) = E_{\alpha,1}(t^\alpha) + \frac{t^2}{2}E_{\alpha,3}(t^\alpha) - \frac{t^4}{8}E_{\alpha,5}(t^\alpha) + \frac{t^6}{16}E_{\alpha,7}(t^\alpha) - \frac{5t^8}{128}E_{\alpha,9}(t^\alpha) - \ldots, \quad (71)$$

while the Laplace transform of $S(t^\alpha)$ after few manipulations is given by

$$F(s) = \frac{s^{a-1}}{s^a-1} + \frac{s^{a-1}}{s^a-1}\left\{\frac{1}{2s^2} - \frac{1}{8s^4} + \frac{1}{16s^6} - \frac{5}{128s^8} + \ldots\right\} =$$
$$\frac{s^{a-1}}{s^a-1} + \frac{s^{a-1}}{s^a-1}\left[\sqrt{1+\left(\frac{1}{s}\right)^2} - 1\right] = \frac{s^{a-1}}{s^a-1} + \frac{s^{a-1}}{s^a-1}\left[\frac{s}{\sqrt{s^2+1}} - 1\right] \quad (72)$$
$$= \frac{s^{a-1}}{s^a-1} - \frac{s^{a-1}}{s^a-1} \cdot \frac{1}{\sqrt{s^2+1}\,[s+\sqrt{s^2+1}]},$$

Noting that the inverse Laplace transform of the Bessel function of the first kind and of the first order is

$$L^{-1}\left\{\frac{1}{\sqrt{s^2+1}\,[s+\sqrt{s^2+1}]}\right\} = J_1(t), \quad (73)$$

one finds that the series of the Mittag-Leffler functions in (74) can be expressed as

$$E_{\alpha,1}(t^\alpha) + \tfrac{t^2}{2} E_{\alpha,3}(t^\alpha) - \tfrac{t^4}{8} E_{\alpha,5}(t^\alpha) + \tfrac{t^6}{16} E_{\alpha,7}(t^\alpha) - \tfrac{5 t^8}{128} E_{\alpha,9}(t^\alpha) - \cdots$$
$$= E_{\alpha,1}(t^\alpha) - E_{\alpha,1}(t^\alpha) * J_1(t) = E_{\alpha,1}(t^\alpha) - \int_0^t E_{\alpha,1}(u^\alpha) J_1(t-u)\, du \qquad (74)$$
$$= E_\alpha(t^\alpha) - \int_0^{\pi/2} t \sin(2\theta)\, E_\alpha[t^\alpha (\cos\theta)^{2\alpha}]\, J_1[t(\sin\theta)^2]\, d\theta\, .$$

3. Differentiation and Integration of the Mittag-Leffler Functions with Respect to Parameters in the Laplace Transform Approach

The operational rules of the Laplace transformation are also appropriate in the evaluation of derivatives of the Mittag-Leffler functions with respect to parameters. Differentiation under the integral transform sign is permissible if the function $f(t,\alpha)$ is continuous with respect to the variable t and the parameter α. Then, we have

$$\begin{aligned} L\{f(t,\alpha)\} &= F(s,\alpha)\, , \\ L\left\{\tfrac{\partial f(t,\alpha)}{\partial \alpha}\right\} &= \tfrac{\partial F(s,\alpha)}{\partial \alpha} = G(s,\alpha)\, , \\ L^{-1}\left\{\tfrac{\partial F(s,\alpha)}{\partial \alpha}\right\} &= L^{-1}\{G(s,\alpha)\} = \tfrac{\partial f(t,\alpha)}{\partial \alpha} \end{aligned} \qquad (75)$$

The Laplace transform $G(s,\alpha)$ of the derivative of the Mittag-Leffler function $E_\alpha(t^\alpha)$ is

$$\begin{aligned} G(s,\alpha) &= L\left\{\tfrac{\partial E_\alpha(t^\alpha)}{\partial \alpha}\right\} = \tfrac{\partial}{\partial \alpha}\left(\tfrac{s^{\alpha-1}}{s^\alpha-1}\right) = \left[\tfrac{s^{\alpha-1}\ln s}{s^\alpha - 1} - \tfrac{s^{2\alpha-1}\ln s}{(s^\alpha-1)^2}\right] \\ &= -\tfrac{s^{\alpha-1}}{s^\alpha-1}\cdot \tfrac{\ln s}{s^\alpha - 1} = -\tfrac{s^{\alpha-1}}{s^\alpha-1}\cdot \tfrac{s^{\alpha-1}}{s^\alpha-1}\cdot \tfrac{\ln s}{s^{\alpha-1}}\, . \end{aligned} \qquad (76)$$

In order to avoid evaluating a complex integral in the inversion process, $G(s,\alpha)$ is expressed as the product of three Laplace transforms. The convolution theorem can be applied for $G(s,\alpha)$ because inverse of the third term in (76) is given for $\operatorname{Re}\lambda > 0$ in [20]

$$L^{-1}\left\{\tfrac{\ln s}{s^\lambda}\right\} = \tfrac{t^{\lambda-1}}{\Gamma(\lambda)}[\psi(\lambda) - \ln t]\, , \qquad (77)$$

From (76) and (77) it follows that

$$\tfrac{\partial E_\alpha(t^\alpha)}{\partial \alpha} = E_\alpha(t^\alpha) * E_\alpha(t^\alpha) * \left\{\tfrac{t^{\alpha-2}}{\Gamma(\alpha-1)}[\ln t - \psi(\alpha-1)]\right\} \qquad (78)$$

where $\alpha > 1$.

Thus, due to two convolutions, the derivative with respect to α is expressed by a double convolution integral. If the Laplace transform in (76) is written as

$$G(s,\alpha) = -\tfrac{s^{\alpha-1}}{s^\alpha - 1}\cdot \tfrac{s^{\alpha-\lambda}}{s^\alpha - 1}\cdot \tfrac{\ln s}{s^{\alpha-\lambda}}\, , \qquad (79)$$

the inverse transform of (79) becomes

$$\begin{aligned} \tfrac{\partial E_\alpha(t^\alpha)}{\partial \alpha} &= \\ E_\alpha(t^\alpha) &* \left[t^{\lambda-1} E_{\alpha,\lambda}(t^\alpha)\right] * \left\{\tfrac{t^{\alpha-\lambda-1}}{\Gamma(\alpha-\lambda)}[\ln t - \psi(\alpha-\lambda)]\right\} \end{aligned} \qquad (80)$$

where $0 < \lambda < \alpha < 1$.

The case $\alpha = 1$ will be considered in the next section.

In a similar manner, the Laplace transform of derivative of the Mittag-Leffler function $t^{\beta-1}E_{\alpha,\beta}(t^\alpha)$ with respect to α is

$$G(s,\alpha,\beta) = L\left\{\frac{\partial [t^{\beta-1}E_{\alpha,\beta}(t^\alpha)]}{\partial \alpha}\right\} = \frac{\partial}{\partial \alpha}\left(\frac{s^{\alpha-\beta}}{s^\alpha-1}\right) = \\ = -\frac{s^{\alpha-\beta}}{s^\alpha-1} \cdot \frac{\ln s}{s^\alpha-1} = -\frac{s^{\alpha-1}}{s^\alpha-1} \cdot \frac{s^{\alpha-\beta}}{s^\alpha-1} \cdot \frac{\ln s}{s^{\alpha-1}},$$
(81)

This gives

$$\frac{\partial [t^{\beta-1}E_{\alpha,\beta}(t^\alpha)]}{\partial \alpha} = \\ E_\alpha(t^\alpha) * \left\{t^{\beta-1}E_{\alpha,\beta}(t^\alpha)\right\} * \left\{\frac{t^{\alpha-2}}{\Gamma(\alpha-1)}[\ln t - \psi(\alpha-1)]\right\}$$
(82)

where $\alpha > 1$.

As expected, for $\beta = 1$, (82) reduces to (78).

For $0 < \alpha < 1$, from (79), it follows that

$$G(s,\alpha,\beta) = -\frac{s^{\alpha-\lambda}}{s^\alpha-1} \cdot \frac{s^{\alpha-\beta}}{s^\alpha-1} \cdot \frac{\ln s}{s^{\alpha-\lambda}},$$
(83)

and

$$\frac{\partial [t^{\beta-1}E_{\alpha,\beta}(t^\alpha)]}{\partial \alpha} = \\ \left\{t^{\lambda-1}E_{\alpha,\lambda}(t^\alpha)\right\} * \left\{t^{\beta-1}E_{\alpha,\beta}(t^\alpha)\right\} * \left\{\frac{t^{\alpha-\lambda-1}}{\Gamma(\alpha-\lambda)}[\ln t - \psi(\alpha-\lambda)]\right\}$$
(84)

$0 < \lambda < \alpha < 1$.

For β, a variable, the Laplace transform of $t^{\beta-1}E_{\alpha,\beta}(t^\alpha)$ derivative is

$$H(s,\alpha,\beta) = L\left\{\frac{\partial t^{\beta-1}E_{\alpha,\beta}(t^\alpha)}{\partial \beta}\right\} = \frac{\partial}{\partial \beta}\left(\frac{s^{\alpha-\beta}}{s^\alpha-1}\right) = \\ = -\frac{s^{\alpha-\beta}\ln s}{s^\alpha-1} = -\frac{s^{\alpha-(\beta-\lambda)}}{s^\alpha-1} \cdot \frac{\ln s}{s^\lambda},$$
(85)

and the inverse transform is

$$\frac{\partial t^{\beta-1}E_{\alpha,\beta}(t^\alpha)}{\partial \beta} = t^{\beta-1}\ln t \, E_{\alpha,\beta}(t^\alpha) + t^{\beta-1}\frac{\partial E_{\alpha,\beta}(t^\alpha)}{\partial \beta} = \\ \left\{t^{\beta-\lambda-1}E_{\alpha,\beta-\lambda}(t^\alpha)\right\} * \left\{\frac{t^{\lambda-1}}{\Gamma(\lambda)}[\ln t - \psi(\lambda)]\right\}.$$
(86)

where $\beta > \lambda > 0$.

As in the case with differential operations, there are rules in the Laplace transformation for evaluation of integrals. The Laplace transform of the Mittag-Leffler function $t^{\beta-1}E_{\alpha,\beta}(t^\alpha)$ enables one to derive the following integral

$$I(t,\lambda) = \int_0^\lambda t^{\beta-1}E_{\alpha,\beta}(t^\alpha)\, d\beta,$$
(87)

The Laplace transform of (87) can be determined by changing the order of integration as follows:

$$\int_0^\infty e^{-st}\left\{\int_0^\lambda t^{\beta-1}E_{\alpha,\beta}(t^\alpha)\, d\beta\right\} dt = \int_0^\lambda \left\{\int_0^\infty e^{-st}t^{\beta-1}E_{\alpha,\beta}(t^\alpha)\, dt\right\} d\beta = \\ \int_0^\lambda \frac{s^{\alpha-\beta}}{s^\alpha-1}\, d\beta = \frac{s^\alpha}{s^\alpha-1}\cdot\frac{1}{\ln s} - \frac{s^{\alpha-\lambda}}{s^\alpha-1}\cdot\frac{1}{\ln s}.$$
(88)

The inverse of $(\ln s)^{-1}$ is closely related to a Volterra function [19] as

$$L^{-1}\left\{\frac{1}{\ln s}\right\} = \int_0^\infty \frac{t^{u-1}}{\Gamma(u)}\, du,$$
(89)

It follows from (47) that

$$L^{-1}\left\{\frac{s^\alpha}{s^\alpha - 1}\right\} = \delta(t) + \frac{d}{dt}[E_\alpha(t^\alpha)], \qquad (90)$$

whereas (49) gives

$$\frac{d}{dt}[E_\alpha(t^\alpha)] = t^{\alpha-1} E_{\alpha,\alpha}(t^\alpha), \qquad (91)$$

The final result in terms of convolution integrals is

$$I(t,\lambda) = \left[\delta(t) + t^{\alpha-1} E_{\alpha,\alpha}(t^\alpha) - t^{\lambda-1} E_{\alpha,\lambda}(t^\alpha)\right] * \int_0^\infty \frac{t^{u-1}}{\Gamma(u)} du. \qquad (92)$$

Two limits of integration in (87) can be altered to

$$\begin{aligned}
\int_0^\infty t^{\beta-1} E_{\alpha,\beta}(t^\alpha)\, d\beta &= \left[\delta(t) + t^{\alpha-1} E_{\alpha,\alpha}(t^\alpha)\right] * \int_0^\infty \frac{t^{u-1}}{\Gamma(u)} du, \\
\int_\lambda^\infty t^{\beta-1} E_{\alpha,\beta}(t^\alpha)\, d\beta &= \left[t^{\lambda-1} E_{\alpha,\lambda}(t^\alpha)\right] * \int_0^\infty \frac{t^{u-1}}{\Gamma(u)} du.
\end{aligned} \qquad (93)$$

The second term on the right-hand side of (88), written in a different form as inversion of the Volterra function, is as follows

$$\begin{aligned}
L^{-1}\left\{\frac{s^{\alpha-\lambda}}{s^\alpha-1} \cdot \frac{1}{\ln s}\right\} &= L^{-1}\left\{\frac{s^{\alpha-(\lambda-1)}}{s^\alpha-1} \cdot \frac{1}{s\ln s}\right\} = t^{\lambda-2} E_{\alpha,\lambda-1}(t^\alpha) * v(t), \\
v(t) &= \int_0^\infty \frac{t^u}{\Gamma(u+1)} du.
\end{aligned} \qquad (94)$$

The connection between the Mittag-Leffler functions and the Volterra functions in the Laplace transformation is discussed in detail in [19].

4. Derivatives of the Mittag-Leffler Functions with Respect to Parameters α and β Expressed as Power Series

As it has been shown in the previous section, the differentiation with respect to parameters of the Mittag-Leffler functions can be represented formally, in closed form, in terms of double convolution integrals. Unfortunately, these convolution integrals are not amenable to numerical computations. Hence, an alternative approach is required. Differentiating (1) and (2) with respect to α and β yields

$$\begin{aligned}
\frac{\partial E_\alpha(t)}{\partial \alpha} &= G(\alpha, t) = -\sum_{k=1}^{\infty} \left(\frac{\psi(\alpha k+1)}{\Gamma(\alpha k+1)}\right) k t^k, \\
\frac{\partial E_{\alpha,\beta}(t)}{\partial \alpha} &= -\sum_{k=1}^{\infty} \left(\frac{\psi(\alpha k+\beta)}{\Gamma(\alpha k+\beta)}\right) k t^k.
\end{aligned} \qquad (95)$$

and

$$\frac{\partial E_{\alpha,\beta}(t)}{\partial \beta} = -\sum_{k=0}^{\infty} \left(\frac{\psi(\alpha k+\beta)}{\Gamma(\alpha k+\beta)}\right) t^k. \qquad (96)$$

The second derivatives are

$$\frac{\partial^2 E_\alpha(t)}{\partial \alpha^2} = G'(\alpha, t) = \sum_{k=1}^{\infty} \left\{\frac{[\psi(\alpha k+1)]^2 - \psi^{(1)}(\alpha k+1)}{\Gamma(\alpha k+1)}\right\} k^2 t^k, \qquad (97)$$

and

$$\begin{aligned}\frac{\partial^2 E_{\alpha,\beta}(t)}{\partial \beta^2} &= \sum_{k=0}^{\infty} \left\{ \frac{[\psi(\alpha\, k+\beta)]^2 - \psi^{(1)}(\alpha k+\beta)}{\Gamma(\alpha k+\beta)} \right\} t^k, \\ \frac{\partial^2 E_{\alpha,\beta}(t)}{\partial \alpha \partial \beta} &= \sum_{k=1}^{\infty} \left\{ \frac{[\psi(\alpha\, k+\beta)]^2 - \psi^{(1)}(\alpha k+\beta)}{\Gamma(\alpha k+\beta)} \right\} k t^k \end{aligned} \quad (98)$$

Higher derivatives with respect to α and β yield similar summands, only differing in powers of k.

Infinite series with the digamma functions in their summands do not appear often in mathematical investigations [22,23]. This changed in 2008 with the huge collection of results in the book by Brychkov [24]. Nevertheless, in their general form, infinite series with quotients of the digamma and gamma functions in their summands are still unsolved. However, for specific values of α and β, MATHEMATICA is able to determine closed forms for them, although they are rather cumbersome with mixture of elementary and special functions. Their validity was checked by carrying out numerical calculations with (95) and (96). Only a limited number of results will appear in this section, with the remainder appearing in Tables 1 and 2.

Table 1. First derivatives of the Mittag-Leffler functions with respect to the parameter α.

α	β	$\partial E_{\alpha,\beta}(t)/\partial \alpha$
1	3	$-\sum_{k=1}^{\infty} \frac{k\, t^k\, \psi(k+3)}{\Gamma(k+3)} = \frac{1-t+\gamma(t+2)-e^t+e^t(t-2)[Chi(t)-Shi(t)-\ln t]}{t^2}$
1	4	$-\sum_{k=1}^{\infty} \frac{k\, t^k\, \psi(k+4)}{\Gamma(k+4)} = \frac{4-8t-3t^2+2\gamma\,(t^2+4z=t+6)-4e^t}{4t^3} + \frac{e^t(t-3)[Chi(t)-Shi(t)-\ln t]}{4\, t^3}$
1	5	$-\sum_{k=1}^{\infty} \frac{k\, t^k \psi(k+5)}{\Gamma(k+5)} = \frac{36 e^t(t-4)[Chi(t)-Shi(t)-\ln t]}{36 t^4} + \frac{6\gamma[t^3+6t^2+18t+24]-[11\,t^3+54t^2+108t-36]-36 e^t}{36 t^4}$
1	6	$-\sum_{k=1}^{\infty} \frac{k\, t^k\, \psi(k+6)}{\Gamma(k+6)} = \frac{e^t(t-5)[Chi(t)-Shi(t)-\ln t]-e^t}{t^5} - \frac{25\, t^4 + 176 t^3 + 648\, t^2 + 1158\, t}{288\, t^5} + \frac{12\gamma\,[t^4+8t^3+36\, t^3+96\, t^2+120\, t+24]}{288\, t^5}$
2	3	$-\sum_{k=1}^{\infty} \frac{k\, t^k\, \psi(2k+3)}{\Gamma(2\, k+3)} = \frac{2+4\gamma+\sinh(\sqrt{t})[(2Chi(\sqrt{t})-\ln t)\sqrt{t}+4Shi(\sqrt{t})]}{4t} - \frac{2\cosh(\sqrt{t})[2Chi(\sqrt{t})+\sqrt{t}Shi(\sqrt{t})-\ln t+1]}{4t}$
2	4	$-\sum_{k=1}^{\infty} \frac{k\, t^k\, \psi(2k+4)}{\Gamma(2\, k+4)} = \frac{-\sinh(\sqrt{t})[6Chi(\sqrt{t})+2\sqrt{t}Shi(\sqrt{t})-3\ln t+2]}{4t^{3/2}} + \frac{\cosh(\sqrt{t})[2\sqrt{t}Chi(\sqrt{t})-\sqrt{t}\ln t+6Shi(\sqrt{t})]+4(\gamma-1)\sqrt{t}}{4t^{3/2}}$
2	5	$-\sum_{k=1}^{\infty} \frac{k\, t^k\, \psi(2k+5)}{\Gamma(2\, k+5)} = \frac{\sinh(\sqrt{t})[2\sqrt{t}\,Chi(\sqrt{t})-\sqrt{t}\ln t+8Shi(\sqrt{t})]+(2\gamma-3)\,t}{4\,t^2} + \frac{-2\cosh(\sqrt{t})\,[4\,Chi(\sqrt{t})+\sqrt{t}\,Shi(\sqrt{t})-2\ln t+1]+8\gamma+2}{4t^2}$
2	6	$-\sum_{k=1}^{\infty} \frac{kt^k \psi(2k+6)}{\Gamma(2k+6)} = \frac{\sqrt{t}[-11t+6\gamma(t+12)-72]}{36 t^{5/2}} + \frac{-9\sinh(\sqrt{t})(10Chi(\sqrt{t})+2\sqrt{t}Shi(\sqrt{t})-5\ln t+2]}{36t^{5/2}} + \frac{9\cosh(\sqrt{t})[\sqrt{t}(2Chi(\sqrt{t})-\ln t)+10Shi(\sqrt{t})]}{36 t^{5/2}}$

Table 1. *Cont.*

α	β	$\partial E_{\alpha,\beta}(t)/\partial\alpha$
4	0	$-\sum_{k=1}^{\infty}\frac{k\,t^k\psi(4k)}{\Gamma(4k)} = \frac{t^{1/4}[\sin(t^{1/4})-\sinh(t^{1/4})]\ln t + 4\,t^{1/4}Chi(t^{1/4})\sinh(t^{1/4})}{8}$ $+\frac{t^{1/4}[-Shi(t^{1/4})\cosh(t^{1/4})+Si(t^{1/4})\cos(t^{1/4})-4\,Ci(t^{1/4})\sin(t^{1/4})]}{2}$
4	2	$-\sum_{k=1}^{\infty}\frac{k\,t^k\psi(4k+2)}{\Gamma(4k+2)} = \frac{[\sinh(t^{1/4})-t^{1/4}\cosh(t^{1/4})]\ln t - 4\sinh(t^{1/4})}{32\,t^{1/4}} +$ $\frac{4Chi(t^{1/4})[t^{1/4}\cosh(t^{1/4})-\sinh(t^{1/4})]+4\cosh(t^{1/4})Shi(t^{1/4})}{32\,t^{1/4}} +$ $\frac{t^{1/4}\cos(z^{1/4})[4Ci(z^{1/4})-\ln t]-4t^{1/4}\sinh(t^{1/4})Shi(t^{1/4})}{32z^{1/4}} +$ $\frac{\sin(t^{1/4})[-4-4Ci(t^{1/4})+\ln t + 4t^{1/4}Si(t^{1/4})]+4Si(t^{1/4})}{32\,t^{1/4}}$

Table 2. First derivatives of the Mittag-Leffler functions with respect to the parameter β.

α	β	$\partial E_{\alpha,\beta}(t)/\partial\beta$
1	3	$-\sum_{k=0}^{\infty}\frac{t^k\psi(k+3)}{\Gamma(k+3)} = \frac{t-\gamma(t+1)+e^t[Chi(t)-Shi(t)-\ln t]}{t^2}$ $-\sum_{k=1}^{\infty}\frac{(-1)^k z^k \psi(4k)}{\Gamma(4k)} = \frac{z^{1/4}[\sin(z^{1/4})-\sinh(z^{1/4})]\ln z}{8} +$ $\frac{4\,z^{1/4}[Chi(z^{1/4})\sinh(z^{1/4})-Shi(z^{1/4})\cosh(z^{1/4})+}{8}$
1	4	$-\sum_{k=0}^{\infty}\frac{t^k\psi(k+4)}{\Gamma(k+4)} = \frac{e^t[Chi(t)-Shi(t)-\ln t]}{t^3} +$ $\frac{3t^2+4t-2\gamma(t^2+2t+2)}{4\,t^3}$
1	5	$-\sum_{k=0}^{\infty}\frac{t^k\psi(k+5)}{\Gamma(k+5)} = \frac{e^t[Chi(t)-Shi(t)-\ln t]}{t^4} +$ $\frac{11t^3+27t^2+36t-6\gamma(t^3+3t^2+6t+6)}{36t^4}$
1	6	$-\sum_{k=0}^{\infty}\frac{t^k\psi(k+6)}{\Gamma(k+6)} = \frac{e^t[Chi(t)-Shi(t)-\ln t]}{t^5} +$ $\frac{25t^4+88t^3+216t^2+288t-12\gamma(t^4+4t^3+12t^2+24t+24)}{288t^5}$
4	0	$-\sum_{k=1}^{\infty}\frac{t^k\psi(4k)}{\Gamma(4k)} = \frac{t^{1/4}[4Ci(t^{1/4})\sin(t^{1/4})+\ln t][\sin(t^{1/4})-\sinh(t^{1/4})]}{8} +$ $\frac{t^{1/4}[Shi(t^{1/4})\cos(t^{1/4})-4Chi(t^{1/4})\sinh(t^{1/4})-Shi(t^{1/4})\cosh(t^{1/4})]}{8} - 1$
4	1	$-\sum_{k=1}^{\infty}\frac{t^k\psi(4k+1)}{\Gamma(4k+1)} = \frac{-Chi(t^{1/4})\cosh(t^{1/4})-Ci(t^{1/4})\cos(t^{1/4})}{2} +$ $\frac{Shi(t^{1/4})\sinh(t^{1/4})-Si(t^{1/4})\sin(t^{1/4})}{2} + \frac{\ln t[\cos(t^{1/4})+\cosh(t^{1/4})]}{8}$
4	2	$-\sum_{k=1}^{\infty}\frac{t^k\psi(4k+2)}{\Gamma(4k+2)} = \frac{-Ci(t^{1/4})\sin(t^{1/4})-Chi(t^{1/4})\sinh(t^{1/4})}{2t^{1/4}} +$ $\frac{Shi(t^{1/4})\cosh(t^{1/4})+Si(t^{1/4})\cos(t^{1/4})}{2t^{1/4}} + \frac{\ln t[\sin(t^{1/4})+\sinh(t^{1/4})]}{8t^{1/4}}$
4	3	$-\sum_{k=1}^{\infty}\frac{t^k\psi(4k+3)}{\Gamma(4k+3)} = \frac{Ci(t^{1/4})\cos(t^{1/4})-Chi(t^{1/4})\cosh(t^{1/4})}{2\sqrt{t}} +$ $\frac{Shi(t^{1/4})\sinh(t^{1/4})+Si(t^{1/4})\sin(t^{1/4})}{2\sqrt{t}} + \frac{\ln t[\cos(t^{1/4})-\cosh(t^{1/4})]}{8\sqrt{t}}$
4	4	$-\sum_{k=1}^{\infty}\frac{t^k\psi(4k+4)}{\Gamma(4k+4)} = \frac{Ci(t^{1/4})\sin(t^{1/4})-Chi(t^{1/4})\sinh(t^{1/4})}{2t^{1/4}} +$ $\frac{Si(t^{1/4})\cos(t^{1/4})-Shi(t^{1/4})\cosh(t^{1/4})}{2t^{1/4}} + \frac{\ln t[\sin(t^{1/4})-\sinh(t^{1/4})]}{8t^{1/4}}$

Convergence conditions for the power series reported in this section were not established, and therefore t values are in some cases restricted (e.g., in (99) and (100) for $|t| < 1$). These summands were obtained from MATHEMATICA, but the validity was numerically checked for only some of them.

The simplest cases occur when α and β equal zero or unity. Then, we find that

$$\frac{\partial E_{\alpha,\beta}(t)}{\partial \alpha}\bigg|_{\alpha=0,\beta=1} = -\frac{\psi(1)}{\Gamma(1)}\sum_{k=1}^{\infty} kt^k = \frac{\gamma\, t}{(t-1)^2}, \qquad (99)$$

$$\frac{\partial E_{\alpha,\beta}(t)}{\partial \alpha}\bigg|_{\alpha=0,\beta} = -\frac{\psi(\beta)}{\Gamma(\beta)}\sum_{k=1}^{\infty} kt^k = -\frac{\psi(\beta)\, t}{\Gamma(\beta)\,(t-1)^2}, \qquad (100)$$

$$\frac{\partial E_{\alpha,\beta}(t)}{\partial \alpha}\bigg|_{\alpha=1,\beta=0} = -\sum_{k=1}^{\infty}\left(\frac{\psi(k)}{\Gamma(k)}\right) kt^k = $$
$$t\left\{e^t(1+t)[Chi(t) - Shi(t) - \ln t] + 1 - e^t\right\} \qquad (101)$$

$$\frac{\partial E_{\alpha,\beta}(t)}{\partial \alpha}\bigg|_{\alpha=1,\beta=1} = -\sum_{k=1}^{\infty}\left(\frac{\psi(k+1)}{\Gamma(k+1)}\right) kt^k = $$
$$1 - e^t\{t[\ln t + \Gamma(0,t)] + 1\}\,;\ \Gamma(0,t)] = -Ei(-t)\,, \qquad (102)$$

and

$$\frac{\partial E_{\alpha,\beta}(t)}{\partial \beta}\bigg|_{\alpha=1,\beta=1} = -\sum_{k=0}^{\infty}\left(\frac{\psi(k+1)}{\Gamma(k+1)}\right) t^k = -e^t[\ln t + \Gamma(0,t)]\,, \qquad (103)$$

where $\Gamma(0,t) = -Ei(-t)$, and the hyperbolic sine and cosine integrals and the exponential integral are defined by

$$Shi(t) = \int_0^t \frac{\sinh u}{u}\, du\,,$$
$$Chi(t) = -\int_0^t \frac{1-\cosh u}{u}\, du + \gamma + \ln t\,, \qquad (104)$$
$$-Ei(-t) = \int_t^{\infty} \frac{e^{-u}}{u}\, du\,.$$

γ represents Euler's constant.

For $\alpha, \beta = 0, 1,$ and 2, the following sums of infinite series are known:

$$\frac{\partial E_{\alpha,\beta}(t)}{\partial \alpha}\bigg|_{\alpha=1,\beta=2} = -\sum_{k=1}^{\infty}\left(\frac{\psi(k+2)}{\Gamma(k+2)}\right) kt^k = $$
$$\frac{1+\gamma+e^t[(t-1)\,Chi(t)+Shi(t)-t\,(Shi(t)+\ln t)+\ln t -1]}{t} \qquad (105)$$

$$\frac{\partial E_{\alpha,\beta}(t)}{\partial \alpha}\bigg|_{\alpha=2,\beta=0} = -\sum_{k=1}^{\infty}\left(\frac{\psi(2k)}{\Gamma(2k)}\right) kt^k = $$
$$\frac{\sqrt{t}\,[2Chi(\sqrt{t})-\ln t][\sinh(\sqrt{t})+\sqrt{t}\cosh(\sqrt{t})]}{4} - $$
$$\frac{2\sqrt{t}Shi(\sqrt{t})[\sqrt{t}\sinh(\sqrt{t})+\cosh(\sqrt{t})]-2\sqrt{t}\sinh(\sqrt{t})}{4}, \qquad (106)$$

$$\frac{\partial E_{\alpha,\beta}(t)}{\partial \alpha}\bigg|_{\alpha=2,\beta=1} = -\sum_{k=1}^{\infty}\left(\frac{\psi(2k+1)}{\Gamma(2k+1)}\right) kt^k = $$
$$\frac{\sqrt{t}\sinh(\sqrt{t})[2Chi(\sqrt{t})-\ln t]-2\cosh(\sqrt{t})\,[\sqrt{t}Shi(\sqrt{t})+1]+2}{4}, \qquad (107)$$

$$\frac{\partial E_{\alpha,\beta}(t)}{\partial \alpha}\bigg|_{\alpha=2,\beta=2} = -\sum_{k=1}^{\infty}\left(\frac{\psi(2k+2)}{\Gamma(2k+2)}\right) kt^k = $$
$$\frac{[2Chi(\sqrt{t})-\ln t][\sqrt{t}\cosh(\sqrt{t})-\sinh(\sqrt{t})]-2\sinh(\sqrt{t})}{4\sqrt{t}} + $$
$$\frac{[2Shi(\sqrt{t})][\cosh(\sqrt{t})-\sqrt{t}\sinh(\sqrt{t})]}{4\sqrt{t}}, \qquad (108)$$

and

$$\frac{\partial E_{\alpha,\beta}(t)}{\partial \beta}\bigg|_{\alpha=1,\beta=2} = -\sum_{k=0}^{\infty}\left(\frac{\psi(k+2)}{\Gamma(k+2)}\right) t^k = $$
$$-\frac{\gamma+e^t[Shi(t)-Chi(t)+\ln t]}{t}, \qquad (109)$$

$$\frac{\partial E_{\alpha,\beta}(t)}{\partial \beta}\bigg|_{\alpha=2,\beta=0} = -\sum_{k=0}^{\infty}\left(\frac{\psi(2k)}{\Gamma(2k)}\right)t^k = 1+ \\ \frac{\sqrt{t}\{\sinh(\sqrt{t})[2Chi(\sqrt{t})-\ln t]-2\cosh(\sqrt{t})Shi(t)\}}{2}, \tag{110}$$

$$\frac{\partial E_{\alpha,\beta}(-t)}{\partial \beta}\bigg|_{\alpha=2,\beta=0} = -\sum_{k=0}^{\infty}\left(\frac{\psi(2k)}{\Gamma(2k)}\right)(-t)^k = \\ \frac{\sqrt{t}\{\sin(\sqrt{t})[Ci(\sqrt{t})-\ln t]-2\cos(\sqrt{t})Si(\sqrt{t})\}}{2}, \tag{111}$$

$$\frac{\partial E_{\alpha,\beta}(t)}{\partial \beta}\bigg|_{\alpha=2,\beta=1} = -\sum_{k=0}^{\infty}\left(\frac{\psi(2k+1)}{\Gamma(2k+1)}\right)t^k = \\ -\sinh(\sqrt{t})Shi(\sqrt{t}) + \frac{\cosh(\sqrt{t})[2Chi(\sqrt{t})-\ln t]}{2}, \tag{112}$$

$$\frac{dE_{\alpha,\beta}(t)}{d\beta}\bigg|_{\alpha=2,\beta=2} = -\sum_{k=0}^{\infty}\left(\frac{\psi(2k+2)}{\Gamma(2k+2)}\right)t^k = \\ \frac{-2\cosh(\sqrt{t})Shi(\sqrt{t})+\sinh(\sqrt{t})[2Chi(\sqrt{t})-\ln t]}{2\sqrt{t}}, \tag{113}$$

and

$$\frac{dE_{\alpha,\beta}(-t)}{d\beta}\bigg|_{\alpha=2,\beta=2} = -\sum_{k=0}^{\infty}\left(\frac{\psi(2k+2)}{\Gamma(2k+2)}\right)(-t)^k = \\ \frac{[\sqrt{t}\cosh(\sqrt{t})-\sinh(\sqrt{t})][2Chi(\sqrt{t})-\ln t]}{4\sqrt{t}} + \\ \frac{Shi(\sqrt{t})[\cosh(\sqrt{t})-\sqrt{t}\sinh(\sqrt{t})]-\sinh(\sqrt{t})}{2\sqrt{t}}, \tag{114}$$

where the sine and cosine integrals are defined by

$$Si(t) = \int_0^t \frac{\sin u}{u}\, du,$$
$$Ci(t) = -\int_t^\infty \frac{\cos u}{u}\, du \tag{115}$$

A number of numerical methods for evaluating the Mittag-Leffler functions and their derivatives with respect to the argument z are given in the literature [25–27]. Fortunately, the Mittag-Leffler functions are available in MATHEMATICA, which means that the first and the second derivatives with respect to α can also be evaluated. The results for $0.05 < \alpha < 5.0$ and $0 < t < 2.25$ can be obtained from the author on request. Two numerical methods were used to verify the results. In the first method, direct summation of infinite series (95) and (96) was performed in MATHEMATICA module, while in the second method, the calculations were carried out by applying the central differences to $O(h^4)$ with $h = 0.001$.

$$\frac{\partial E_\alpha(t)}{\partial \alpha} = \frac{-E_{\alpha+2h}(t)+8E_{\alpha+h}(t)-8E_{\alpha-h}(t)+E_{\alpha-2h}(t)}{12h} \tag{116}$$

and

$$\frac{\partial^2 E_\alpha(t)}{\partial \alpha^2} = \\ \frac{-E_{\alpha+2h}(t)+16E_{\alpha+h}(t)-30E_\alpha(t)+16E_{\alpha-h}(t)-E_{\alpha-2h}(t)}{12h^2} \tag{117}$$

The above results of the Mittag-Leffler functions were evaluated in MATHEMATICA.

The Mittag-Leffler functions, $f(\alpha,t) = E_\alpha(t)$, as a function of α for constant t are plotted in Figure 1. The rapid exponential behavior of these functions means that only narrow intervals of the functions can be plotted. As can be seen, they are always positive and become more divergent as t increases. For $0 < \alpha < 1$, they possess a maximum, which moves as t is increased. For large values α and t, they tend to zero.

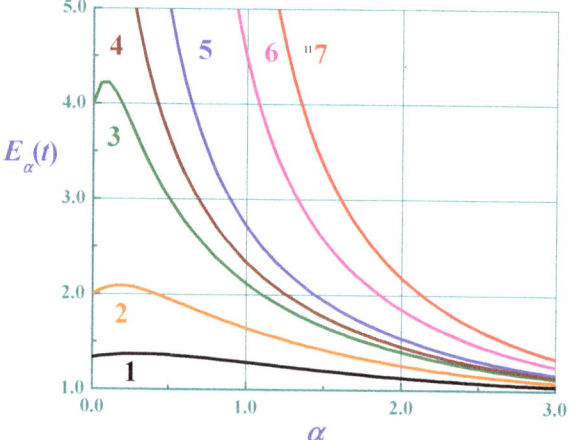

Figure 1. The Mittag-Leffler functions $E_\alpha(t)$ as a function of α at constant values of argument t. 1—0.25; 2—0.50; 3—0.75; 4—0.85; 5—1.0; 6—1.5; 7—2.0.

The first derivatives of the Mittag-Leffler with respect to α or $G(\alpha,t) = \partial E_\alpha(t)/\partial\alpha$ are plotted in Figure 2 Their behavior mirrors $E_\alpha(t)$, except that they are inverted as they are always negative.

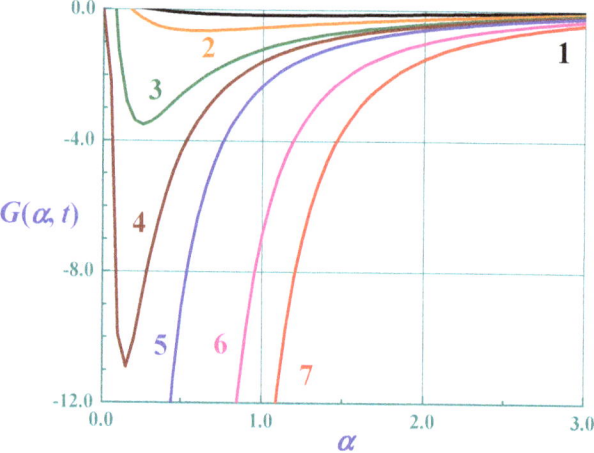

Figure 2. $G(\alpha,t)$—First derivatives of the Mittag-Leffler functions with respect to α plotted at constant values of t. 1—0.25; 2—0.50; 3—0.75; 4—0.85; 5—1.0; 6—1.5; 7—2.0.

The second derivatives with respect to α, $G'(\alpha,t) = \partial^2 E_\alpha(t)/\partial\alpha^2$ are presented in Figure 3. Their behavior resembles that of the Mittag-Leffler functions (Figure 1). However, for small values of t, they move from negative to positive values. The divergent behavior of $G'(\alpha,t)$ also applies for large values of t, but for increasing values of α and t, they tend to zero.

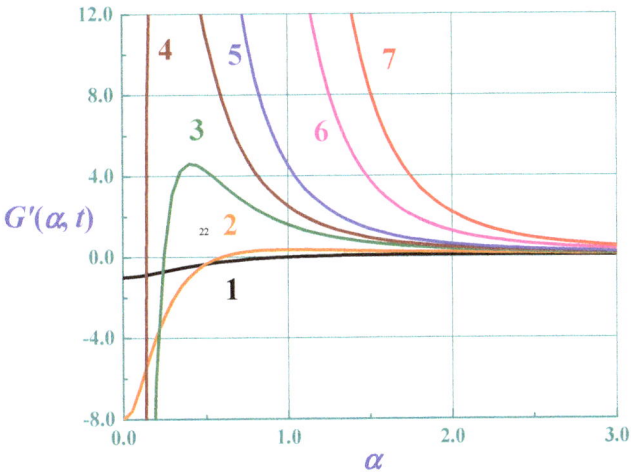

Figure 3. $G'(\alpha,t)$—Second derivatives of the Mittag-Leffler functions with respect to α plotted at constant values of t. 1—0.25; 2—0.50; 3—0.75; 4—0.85; 5—1.0; 6—1.5; 7—2.0.

5. Derivatives of the Mittag-Leffler Functions with Respect to Parameters α and β from Integral Representations

Derivatives with respect to α and β can be determined by direct differentiation of the integrands in integral representations of the Mittag-Leffler functions. Because no general expression exists for integral representations [25,27–33], it is possible to use only those that are valid for real positive and negative values of t and for restricted values of α and β.

For $0 < \alpha < 1$ and $t > 0$, these are

$$E_\alpha(t^\alpha) = \frac{e^t}{\alpha} - \frac{\sin(\pi\alpha)}{\pi} \int_0^\infty \frac{e^{-tu}\, u^{\alpha-1}}{u^{2\alpha} - 2u^\alpha \cos(\pi\alpha) + 1}\, du, \tag{118}$$

$$E_\alpha(-t^\alpha) = \frac{\sin(\pi\alpha)}{\pi} \int_0^\infty \frac{e^{-tu}\, u^{\alpha-1}}{u^{2\alpha} + 2u^\alpha \cos(\pi\alpha) + 1}\, du. \tag{119}$$

and

$$E_{\alpha,\beta}(t^\alpha) = \frac{e^t}{\alpha} - \frac{1}{\pi} \int_0^\infty \frac{e^{-tu}\, u^{\alpha-\beta}\{u^\alpha \sin(\pi\beta) + \sin[\pi(\alpha-\beta)]\}}{u^{2\alpha} - 2u^\alpha \cos(\pi\alpha) + 1}\, du, \tag{120}$$

$$E_{\alpha,\beta}(-t^\alpha) = \frac{1}{\pi} \int_0^\infty \frac{e^{-tu}\, u^{\alpha-\beta}\{u^\alpha \sin(\pi\beta) + \sin[\pi(\alpha-\beta)]\}}{u^{2\alpha} + 2u^\alpha \cos(\pi\alpha) + 1}\, du. \tag{121}$$

In (120) and (121), $0 < \beta < \alpha + 1$.

Direct differentiation of (118) and (119) with respect to α gives

$$\frac{\partial E_\alpha(t^\alpha)}{\partial \alpha} = -\frac{e^t}{\alpha^2} - \cos(\pi\alpha) \int_0^\infty \frac{e^{-ut}\, u^{\alpha-1}}{u^{2\alpha} - 2u^\alpha \cos(\pi\alpha) + 1}\, du - \frac{\sin(\pi\alpha)}{\pi} \int_0^\infty \frac{e^{-ut}\, u^{\alpha-1}\,[(1-u^{2\alpha})\ln u - 2\pi u^\alpha \sin(\pi\alpha)]}{[u^{2\alpha} - 2u^\alpha \cos(\pi\alpha) + 1]^2}\, du, \tag{122}$$

and

$$\frac{\partial E_\alpha(-t^\alpha)}{\partial \alpha} = \cos(\pi\alpha) \int_0^\infty \frac{e^{-u\,t}\,u^{\alpha-1}}{u^{2\alpha}+2u^\alpha \cos(\pi\alpha)+1}\,du +$$
$$\frac{\sin(\pi\alpha)}{\pi} \int_0^\infty \frac{e^{-u\,t}\,u^{\alpha-1}[(1-u^{2\alpha})\ln u + 2\pi u^\alpha \sin(\pi\alpha)]}{[u^{2\alpha}+2u^\alpha \cos(\pi\alpha)+1]^2}\,du \qquad (123)$$

where the first integrals in (122) and (123) can be written in terms of the Mittag-Leffler functions using (118) and (119).

In the same manner, one can obtain derivatives of the Mittag-Leffler functions $E_{\alpha,\beta}(\pm t^\alpha)$ with respect to α and β. Thus, we find that

$$\frac{\partial E_{\alpha,\beta}(t^\alpha)}{\partial \alpha} = -\frac{e^t}{\alpha^2} - \int_0^\infty \frac{e^{-tu}\,u^{\alpha-\beta}\cos[\pi(\alpha-\beta)]}{u^{2\alpha}-2u^\alpha \cos(\pi\alpha)+1}\,du -$$
$$\frac{1}{\pi}\int_0^\infty \frac{e^{-tu}\,u^{\alpha-\beta}\ln u\{2u^\alpha \sin(\pi\beta) + \sin[\pi(\alpha-\beta)]\}}{u^{2\alpha}-2u^\alpha \cos(\pi\alpha)+1}\,du +$$
$$\frac{2}{\pi}\int_0^\infty \frac{e^{-tu}\,u^{2\alpha-\beta}\ln u\{u^\alpha \sin(\pi\beta) + \sin[\pi(\alpha-\beta)]\}[u^\alpha - \cos(\pi\alpha)]}{[u^{2\alpha}-2u^\alpha \cos(\pi\alpha)+1]^2}\,du \qquad (124)$$
$$+2\int_0^\infty \frac{e^{-tu}\,u^{2\alpha-\beta}\sin(\pi\alpha)\{u^\alpha \sin(\pi\beta) + \sin[\pi(\alpha-\beta)]\}}{[u^{2\alpha}-2u^\alpha \cos(\pi\alpha)+1]^2}\,du ,$$

and

$$\frac{\partial E_{\alpha,\beta}(t^\alpha)}{\partial \beta} = \frac{1}{\pi}\int_0^\infty \frac{e^{-tu}\,u^{\alpha-\beta}\ln u\{u^\alpha \sin(\pi\beta) + \sin[\pi(\alpha-\beta)]\}}{u^{2\alpha}-2u^\alpha \cos(\pi\alpha)+1}\,du -$$
$$\int_0^\infty \frac{e^{-tu}\,u^{\alpha-\beta}\{u^\alpha \cos(\pi\beta) - \cos[\pi(\alpha-\beta)]\}}{u^{2\alpha}-2u^\alpha \cos(\pi\alpha)+1}\,du . \qquad (125)$$

For the negative real axis, one obtains

$$t^{\beta-1}\frac{\partial E_{\alpha,\beta}(-t^\alpha)}{\partial \alpha} = \int_0^\infty \frac{e^{-tu}\,u^{\alpha-\beta}\cos[\pi(\alpha-\beta)]}{u^{2\alpha}+2u^\alpha \cos(\pi\alpha)+1}\,du -$$
$$\frac{1}{\pi}\int_0^\infty \frac{e^{-tu}\,u^{\alpha-\beta}\ln u\{2u^\alpha \sin(\pi\beta) + \sin[\pi(\alpha-\beta)]\}}{u^{2\alpha}+2u^\alpha \cos(\pi\alpha)+1}\,du -$$
$$\frac{2}{\pi}\int_0^\infty \frac{e^{-tu}\,u^{2\alpha-\beta}\ln u\{u^\alpha \sin(\pi\beta) + \sin[\pi(\alpha-\beta)]\}[u^\alpha + \cos(\pi\alpha)]}{[u^{2\alpha}+2u^\alpha \cos(\pi\alpha)+1]^2}\,du \qquad (126)$$
$$+2\int_0^\infty \frac{e^{-tu}\,u^{2\alpha-\beta}\sin(\pi\alpha)\{u^\alpha \sin(\pi\beta) + \sin[\pi(\alpha-\beta)]\}}{[u^{2\alpha}+2u^\alpha \cos(\pi\alpha)+1]^2}\,du ,$$

and

$$t^{\beta-1}\ln t\, E_{\alpha,\beta}(-t^\alpha) + t^{\beta-1}\frac{\partial E_{\alpha,\beta}(-t^\alpha)}{\partial \beta} =$$
$$-\frac{1}{\pi}\int_0^\infty \frac{e^{-tu}\,u^{\alpha-\beta}\ln u\{u^\alpha \sin(\pi\beta) + \sin[\pi(\alpha-\beta)]\}}{u^{2\alpha}+2u^\alpha \cos(\pi\alpha)+1}\,du + \qquad (127)$$
$$\int_0^\infty \frac{e^{-tu}\,u^{\alpha-\beta}\{u^\alpha \cos(\pi\beta) - \cos[\pi(\alpha-\beta)]\}}{u^{2\alpha}+2u^\alpha \cos(\pi\alpha)+1}\,du .$$

The infinite integrals in (122) to (127) are valid for restricted values of α and β. As can be expected, they represent the Laplace transforms and are similar to convolution integrals in Section 3.

6. Conclusions

For the first time, the parameters of the Mittag-Leffler functions in (1) and (2) have been treated as variables, and derivatives with respect to them have consequently been determined and discussed. Thus, it has been shown that operational calculus is a powerful tool for determining the properties of the Mittag-Leffler functions. Using the Laplace transform theory, new functional relations, together with infinite and finite series of the Mittag-Leffler functions, have also been calculated. Moreover, derivatives with respect to α and β have been found to be expressible in terms of convolution integrals. Direct differentiation of (1) and (2) yields infinite power series with quotients of digamma and gamma functions in their coefficients. For small integer values of α and β, closed forms are derived in terms of elementary and special functions. The Mittag-Leffler functions, together with their first and second derivatives, are graphed as functions of α and t. On a final note, it should be mentioned that Biyajima et al. [30,31] have used (102) in their new blackbody radiation law, but not the closed form given here.

Funding: This research received no external funding.

Acknowledgments: I wish to express my profound gratitude to Francesco Mainardi, Department of Physics and Astronomy, Bologna University, Bologna, Italy, for his help, advice, and kind encouragement over the years. He was the first person to show me the importance of the Mittag-Leffler functions. I am also grateful to Yuri A. Brychkov from the Computing Center of the Russian Academy of Sciences, Moscow, Russia, and Victor Adamchik from the Computer Science Department, University of Southern California, Los Angeles, USA, for simplifying and verifying several series from MATHEMATICA in this work. I thank Juan Luis Gonzales-Santander Martinez, Department of Mathematics, Universidad de Oviedo, Oviedo, Spain, for explaining MATHEMATICA and producing more elegant and efficient programs from my original programs. Finally, I thank the referees for their constructive comments, which helped improve this work considerably.

Conflicts of Interest: The author declares no conflict of interest.

References

1. Mittag-Leffler, G.M. Sur la nouvelle function $E_\alpha(x)$. *CR Acad. Sci. Paris* **1903**, *137*, 554–558.
2. Caputo, M.; Mainardi, F. Linear models of dissipation in anelastic solids. *Riv. Nuovo Cim.* **1971**, *1*, 161–198. [CrossRef]
3. Mainardi, F. *Fractional Calculus and Waves in Linear Viscoelasticity*; Imperial College Press: London, UK, 2010.
4. Sandev, T.; Tomovski, Z.; Dubbeldam, J.L.A. Generalized Langevin equation with a three parameter Mittag-Leffler noise. *Phys. A* **2011**, *390*, 3627–3636. [CrossRef]
5. Valério, D.; Trujillo, J.J.; Rivero, M.; Machado, J.A.T.; Baleanu, D. Fractual calculus: A survey of useful formulas. *Eur. Phys. J. Spec. Top.* **2013**, *222*, 1827–1846. [CrossRef]
6. Viñales, A.D.; Paissan, G.H. Velocity autocorrelation of a free particle driven by a Mittag-Leffler noise: Fractional dynamics and temporal behaviors. *Phys. Rev. E* **2014**, *90*, 062103. [CrossRef] [PubMed]
7. Patla, S.K.; Ray, R.; Karmakar, S.; Das, S.; Tarafdar, S. Nanofiller induced ionic conductivity enhancement and relaxation property analysis of the blend polymer electrolyte using non-Debye electric field relaxation function. *J. Phys. Chem. C* **2019**, *121*, 5188–5197. [CrossRef]
8. Agahi, H.; Alipour, M. Mittag-Leffler-Gaussian distribution: Theory and applications to real data. *Math. Comput. Simul.* **2019**, *156*, 227–231. [CrossRef]
9. Samko, S.G.; Kilbas, A.A.; Marichev, O.I. *Fractional Integrals and Derivatives: Theory and Applications*; Gordon and Breach: New York, NY, USA, 1993.
10. Gorenflo, R.; Mainardi, F. *Fractional Calculus: Integral and Differential Equations of Fractional Order*; CISM Lecture Notes; International Centre for Mechanical Sciences: Udine, Italy, 1996; pp. 233–276.
11. Bansal, D.; Prajapat, J.K. Certain geometrical properties of the Mittag-Leffler functions. *Complex Var. Elliptic Equ. J.* **2016**, *61*, 338–350. [CrossRef]
12. Hanneken, J.W.; Narahari Achav, B.N.; Puzio, R.; Vaught, D.M. Properties of the Mittag-Leffler function for negative alpha. *Phys. Scr.* **2009**, *136*, 014037. [CrossRef]
13. Mainardi, F. On some properties of the Mittag-Leffler function $E_\alpha(-t^\alpha)$, completely monotonic for t > 0 with $0 < \alpha < 1$. *Discret. Contin. Dyn. Syst. B.* **2014**, *19*, 2267–2278.
14. Oberhettinger, F.; Erdélyi, A.; Magnus, W.; Tricomi, F.G. *Higher Transcendental Functions*; McGraw-Hill: New York, NY, USA, 1955; Volume 3.
15. Dzherbashyan, M.M. *Integral Transforms and Representations of Functions in the Complex Plane*; Nauka: Moscow, Russia, 1966. (In Russian)
16. Haubold, H.J.; Mathai, A.M.; Saxena, R.K. Mittag-Leffler functions and their applications. *J. App. Math.* **2011**. [CrossRef]
17. Gorenflo, R.; Kilbas, A.A.; Mainardi, F.; Rogosin, S.V. Mittag-Leffler Functions. In *Related Topics and Applications*; Springer: Berlin/Heidelberg, Germany, 2014.

18. Wiman, A. Uber den fundamental satz in der theorie der funcktionen $E_a(x)$. *Acta Math.* **1905**, *29*, 191–201. [CrossRef]
19. Apelblat, A. *Volterra Functions*; Nova Science Publishers, Inc.: New York, NY, USA, 2008.
20. Erdélyi, A.; Magnus, W.; Oberhettinger, F.; Tricomi, F.G. *Tables of Integral Transforms*; McGraw-Hill: New York, NY, USA, 1954.
21. Apelblat, A. *Laplace Transforms and Their Applications*; Nova Science Publishers, Inc.: New York, NY, USA, 2012.
22. Miller, A.R. Summations for certain series containing the digamma function. *J. Phys. A Math. Gen.* **2006**, *39*, 3011–3020. [CrossRef]
23. Cvijović, D. Closed-form summations of certain hypergeometric-type series containing the digamma function. *J. Phys. A Math. Gen.* **2008**, *41*, 455205. [CrossRef]
24. Brychkov, Y.A. Handbook of Special Functions. In *Derivatives, Integrals, Series and Other Formulas*; CRC Press: Boca Raton, FL, USA, 2008.
25. Gorenflo, R.; Loutchko, J.; Luchko, Y. Computation of the Mittag-Leffler function $E_{\alpha,\beta}(z)$ and its derivative. *Fract. Calc. Appl. Anal.* **2002**, *5*, 491–518.
26. Valério, D.; Machado, J.T. On the numerical computation of the Mittag-Leffler function. *Commun. Nonlinear Sci. Numer. Simul.* **2014**, *19*, 3419–3424. [CrossRef]
27. Ortigueira, M.D.; Lopes, A.M.; Machado, J.T. On the numerical computation of the Mittag-Leffler function. *IJNSNS* **2019**, *20*, 725–733. [CrossRef]
28. Nigmatullin, R.R.; Khamzin, A.A.; Baleanu, D. On the Laplace integral representation of multivariate Mittag-Leffler function in anomalous relaxation. *Math. Methods Appl. Sci.* **2016**, *39*, 2983–2992. [CrossRef]
29. Wu, Y.; Liu, Z. Some results about Mittag-Leffler function's integral representations. *arXiv* **2019**, arXiv:1912.0477v1.
30. Biyajima, M.; Mizoguchi, T.; Suzuki, N. A New blackbody radiation law based on fractional calculus and its application to NASA COBE data new body radiation law. *Phys. A Stat. Mech. Appl.* **2015**, *440*, 129–138. [CrossRef]
31. Biyajima, M.; Mizoguchi, T.; Suzuki, N. On effects of the chemical potential in a BE distribution and the fractional parameter in a distribution with Mittag-Leffler function. *arXiv* **2016**, arXiv:1504.01378v2.
32. Garrappa, R.; Popolizio, M. Evaluation of generalized Mittag-Leffler function on the real line. *Adv. Comput. Math.* **2013**, *39*, 205–225. [CrossRef]
33. Djrbashia, M.M. *Harmonic Analysis and Boundary Value Problems in the Complex Domain*; Springer: Berlin/Heidelberg, Germany, 1993.

© 2020 by the author. Licensee MDPI, Basel, Switzerland. This article is an open access article distributed under the terms and conditions of the Creative Commons Attribution (CC BY) license (http://creativecommons.org/licenses/by/4.0/).

Review

The Wright Functions of the Second Kind in Mathematical Physics

Francesco Mainardi [1,*] and Armando Consiglio [2]

[1] Dipartimento di Fisica e Astronomia, Università di Bologna, Via Irnerio 46, I-40126 Bologna, Italy
[2] Institut für Theoretische Physik und Astrophysik, Universität Würzburg, D-97074 Würzburg, Germany; armando.consiglio@physik.uni-wuerzburg.de
* Correspondence: francesco.mainardi@bo.infn.it

Academic Editor: Francesco Mainardi
Received: 18 April 2020; Accepted: 19 May 2020; Published: 1 June 2020

Abstract: In this review paper, we stress the importance of the higher transcendental Wright functions of the second kind in the framework of Mathematical Physics. We first start with the analytical properties of the classical Wright functions of which we distinguish two kinds. We then justify the relevance of the Wright functions of the second kind as fundamental solutions of the time-fractional diffusion-wave equations. Indeed, we think that this approach is the most accessible point of view for describing non-Gaussian stochastic processes and the transition from sub-diffusion processes to wave propagation. Through the sections of the text and suitable appendices, we plan to address the reader in this pathway towards the applications of the Wright functions of the second kind.

Keywords: fractional calculus; Wright functions; Green's functions; diffusion-wave equation; Laplace transform

MSC: 26A33; 33E12; 34A08; 34C26

1. Introduction

The special functions play a fundamental role in all fields of Applied Mathematics and Mathematical Physics because any analytical results are expressed in terms of some of these functions. Even if the topic of special functions can appear boring and their properties mainly treated in handbooks, we would promote the relevance of some of them not yet so well known. We devote our attention to the Wright functions, in particular with the class of the second kind. These functions, as we will see hereafter, are fundamental to deal with some non-standard deterministic and stochastic processes. Indeed, the Gaussian function (known as the normal probability distribution) must be generalized in a suitable way in the framework of partial differential equations of non-integer order for describing the anomalous diffusion and the transition from fractional diffusion to wave propagation.

Furthermore, their usefulness and meaningfulness also extends to other topics. For example, these functions and their Laplace Transforms can be applied in electromagnetic problems, see the 1958 paper by Ragab [1] (where the Wright functions were used without knowing their existence) and the recent 2020 paper by Stefański and Gulgowski [2]. Recently, the Wright functions have been used in the theory of coherent states by Garra, Giraldi, and Mainardi [3].

This survey article aims to discuss the relevance of the Wright Functions and also to focus on the not well-known *Four Sisters Functions* and their importance in time-fractional diffusion-wave equations.

The plan of the paper is organized as follows. In Section 2, we introduce the Wright functions, entirely in the complex plane that we distinguish in two kinds in relation to the value-range of the two parameters on which they depend. In particular, we devote our attention to two Wright functions of the second kind introduced by Mainardi with the term of auxiliary functions. One of them, known as

M-Wright function, generalizes the Gaussian function so it is expected to play a fundamental role in non-Gaussian stochastic processes.

Indeed, in Section 3, we show how the Wright functions of the second kind are relevant in the analysis of time-fractional diffusion and diffusion-wave equations being related to their fundamental solutions. This analysis leads to generalizing the known results r of the standard diffusion equation in the one-dimensional case that is recalled in Appendix A by means of auxiliary functions as particular cases of the Wright functions of the second kind known as M-Wright or Mainardi functions. For readers' convenience, in Appendix B, we will also provide an introduction to the time-derivative of fractional order in the Caputo sense We remind that nowadays, as usual, by fractional order, we mean a non-integer order, so that the term *"fractional"* is a misnomer kept only for historical reasons.

In Section 4, we consider again the Mainardi auxiliary functions functions for their role in probability theory and in particular in the framework of Lévy stable distributions whose general theory is recalled in Appendix C.

In Section 5, we show how the auxiliary functions turn out to be included in a class that we denote *the four sister functions*. On their turn, these four functions depending on a real parameter $\nu \in (0,1)$ are the natural generalization of *the three sisters functions* introduced in Appendix A devoted to the standard diffusion equation. The attribute of sisters was put in by one of us (F. M.) because of their inter-relations, in his lecture notes on Mathematical Physics, so this is only a personal reason that we hope to be shared by the readers.

Finally, in Section 6, we provide some concluding remarks paying attention to work to be done in the next future.

We point out that we have equipped our theoretical analysis with several plots hoping they will be considered illuminating for the interested readers. We also note that we have limited our review to the simplest boundary values problems of equations in one space dimension referring the readers to suitable references for more general treatments in Section 3.1.

2. The Wright Functions of the Second Kind and the Mainardi Auxiliary Functions

The classical *Wright function* that we denote by $W_{\lambda,\mu}(z)$, is defined by the series representation convergent in the whole complex plane,

$$W_{\lambda,\mu}(z) := \sum_{n=0}^{\infty} \frac{z^n}{n!\Gamma(\lambda n + \mu)}, \quad \lambda > -1, \quad \mu \in \mathbb{C}, \tag{1}$$

The *integral representation* reads as:

$$W_{\lambda,\mu}(z) = \frac{1}{2\pi i} \int_{Ha_-} e^{\sigma + z\sigma^{-\lambda}} \frac{d\sigma}{\sigma^\mu}, \quad \lambda > -1, \quad \mu \in \mathbb{C}, \tag{2}$$

where Ha_- denotes the Hankel path: this one is a loop which starts from $-\infty$ along the lower side of negative real axis, encircling it with a small circle the axes origin and ends at $-\infty$ along the upper side of the negative real axis.

$W_{\lambda,\mu}(z)$ is then an *entire function* for all $\lambda \in (-1, +\infty)$. Originally, Wright assumed $\lambda \geq 0$ in connection with his investigations on the asymptotic theory of partition [4,5] and only in 1940 he considered $-1 < \lambda < 0$, [6]. We note that, in the Vol 3, Chapter 18 of the handbook of the Bateman Project [7], presumably for a misprint, the parameter λ is restricted to be non-negative, whereas the Wright functions remained practically ignored in other handbooks. In 1993, Mainardi, being aware only of the Bateman handbook, proved that the Wright function is entire also for $-1 < \lambda < 0$ in his approaches to the time fractional diffusion equation that will be dealt with in the next section.

In view of the asymptotic representation in the complex domain and of the Laplace transform for positive argument $z = r > 0$ (r can be the time variable t or the space variable x), the Wright functions are distinguished in *first kind* ($\lambda \geq 0$) and *second kind* ($-1 < \lambda < 0$) as outlined in the Appendix F of the book by Mainardi [8]. In particular, for the asymptotic behavior, we refer the interested reader to the two papers by Wong and Zhao [9,10], and to the surveys by Luchko and by Paris in the Handbook of Fractional Calculus and Applications, see, respectively, [11,12], and references therein.

We note that the Wright functions are an entire of order $1/(1+\lambda)$; hence, only the first kind functions ($\lambda \geq 0$) are of exponential order, whereas the second kind functions ($-1 < \lambda < 0$) are not of exponential order. The case $\lambda = 0$ is trivial since $W_{0,\mu}(z) = e^z/\Gamma(\mu)$. As a consequence of the difference in the orders, we must point out the different Laplace transforms proved e.g., in [8,13], see also the recent survey on Wright functions by Luchko [11]. We have:

- for the first kind, when $\lambda \geq 0$

$$W_{\lambda,\mu}(\pm r) \div \frac{1}{s} E_{\lambda,\mu}\left(\pm\frac{1}{s}\right); \tag{3}$$

- for the second kind, when $-1 < \lambda < 0$ and putting for convenience $\nu = -\lambda$ so $0 < \nu < 1$

$$W_{-\nu,\mu}(-r) \div E_{\nu,\mu+\nu}(-s). \tag{4}$$

Above, we have introduced the Mittag–Leffler function in two parameters $\alpha > 0$, $\beta \in \mathbb{C}$ defined as its convergent series for all $z \in \mathbb{C}$

$$E_{\alpha,\beta}(z) := \sum_{n=0}^{\infty} \frac{z^n}{\Gamma(\alpha n + \beta)}. \tag{5}$$

For more details on the special functions of the Mittag–Leffler type, we refer the interested readers to the treatise by Gorenflo et al. [14], where, in the forthcoming 2nd edition, the Wright functions are also treated in some detail.

In particular, two Wright functions of the second kind, originally introduced by Mainardi and named $F_\nu(z)$ and $M_\nu(z)$ ($0 < \nu < 1$), are called *auxiliary functions* in virtue of their role in the time fractional diffusion equations considered in the next section. These functions, $F_\nu(z)$ and $M_\nu(z)$, are indeed special cases of the Wright function of the second kind $W_{\lambda,\mu}(z)$ by setting, respectively, $\lambda = -\nu$ and $\mu = 0$ or $\mu = 1 - \nu$. Hence, we have:

$$F_\nu(z) := W_{-\nu,0}(-z), \quad 0 < \nu < 1, \tag{6}$$

and

$$M_\nu(z) := W_{-\nu,1-\nu}(-z), \quad 0 < \nu < 1. \tag{7}$$

Those functions are interrelated through the following relation:

$$F_\nu(z) = \nu z M_\nu(z), \tag{8}$$

which reminds us of the second relation in (A9), seen for the standard diffusion equation.

The series representations of the auxiliary functions are derived from those of $W_{\lambda,\mu}(z)$. Then:

$$F_\nu(z) := \sum_{n=1}^{\infty} \frac{(-z)^n}{n!\Gamma(-\nu n)} = \frac{1}{\pi}\sum_{n=1}^{\infty} \frac{(-z)^{n-1}}{n!}\Gamma(\nu n + 1)\sin(\pi\nu n) \tag{9}$$

and

$$M_\nu(z) := \sum_{n=0}^{\infty} \frac{(-z)^n}{n!\Gamma[-\nu n + (1-\nu)]} = \frac{1}{\pi}\sum_{n=1}^{\infty} \frac{(-z)^{n-1}}{(n-1)!}\Gamma(\nu n)\sin(\pi\nu n), \tag{10}$$

where in both cases the *reflection formula* for the Gamma function (Equation (11)) it has been used among the first and the second step of Equations (9) and (10),

$$\Gamma(\zeta)\Gamma(1-\zeta) = \pi/\sin\pi\zeta. \tag{11}$$

In addition, the integral representations of the auxiliary functions are derived from those of $W_{\lambda,\mu}(z)$. Then:

$$F_\nu(z) := \frac{1}{2\pi i}\int_{Ha_-} e^{\sigma - z\sigma^\nu}\,d\sigma, \quad z \in \mathbb{C}, \ 0 < \nu < 1 \tag{12}$$

and

$$M_\nu(z) := \frac{1}{2\pi i}\int_{Ha_-} e^{\sigma - z\sigma^\nu}\,\frac{d\sigma}{\sigma^{1-\nu}}, \quad z \in \mathbb{C}, \ 0 < \nu < 1. \tag{13}$$

Explicit expressions of $F_\nu(z)$ and $M_\nu(z)$ in terms of known functions are expected for some particular values of ν as shown and recalled by Mainardi in the first 1990s in a series of papers [15–18] that is,

$$M_{1/2}(z) = \frac{1}{\sqrt{\pi}} e^{-z^2/4}, \tag{14}$$

$$M_{1/3}(z) = 3^{2/3}\mathrm{Ai}(z/3^{1/3}). \tag{15}$$

Liemert and Klenie [19] have added the following expression for $\nu = 2/3$

$$M_{2/3}(z) = 3^{-2/3}\left[3^{1/3}\,z\,\mathrm{Ai}\left(z^2/3^{4/3}\right) - 3\mathrm{Ai}'\left(z^2/3^{4/3}\right)\right]e^{-2z^3/27}, \tag{16}$$

where Ai and Ai$'$ denote the *Airy function* and its first derivative. Furthermore, they have suggested in the positive real field $\mathbb{R}^+$ the following remarkably integral representation

$$M_\nu(x) = \frac{1}{\pi}\frac{x^{\nu/(1-\nu)}}{1-\nu}\int_0^\pi C_\nu(\phi)\exp\left(-C_\nu(\phi)\right)x^{1/(1-\nu)}\,d\phi, \tag{17}$$

where

$$C_\nu(\phi) = \frac{\sin(1-\nu)}{\sin\phi}\left(\frac{\sin\nu\phi}{\sin\phi}\right)^{\nu/(1-\nu)} \tag{18}$$

corresponding to Equation (7) of the article written by Saa and Venegeroles [20].

The Wright function of both kinds and in particular the Mainardi auxiliary functions considerd in this paper turn out to be particular cases of more general transcendental functions as the Fox H functions, the Fox–Wright functions and the multi-index Mittag–Leffler functions. The relations with the classical Mittag–Leffler functions with two parameters have already been pointed out so; for more parameters, we refer the interested reader, e.g., to the papers by Kiryakova [21], Kilbas, Koroleva, Rogosin [22], and references therein.

We outline that for more Laplace transform pairs involving the Wright and the Mittag–Leffler functions the reader is referred to Ansari and Refahi Sheikhani [23] and to the tutorial survey by Mainardi [24].

3. The Wright Functions of the Second Kind and the Time-Fractional Diffusion Wave Equation

As we will see, the Wright functions of the second kind are relevant in the analysis of the Time-Fractional Diffusion-Wave Equation (TFDWE).

We find it convenient to show the plots of the M-Wright functions on a space symmetric interval of $\mathbb{R}$ in Figures 1 and 2, corresponding to the cases $0 \le \nu \le 1/2$ and $1/2 \le \nu \le 1$, respectively.

From these figures, we recognize the non-negativity of the M-Wright function on $\mathbb{R}$ for $1/2 \le \nu \le 1$ consistently with the analysis on distribution of zeros and asymptotics of Wright functions carried out by Luchko, see [11,25] and by Luchko and Kiryakova [26].

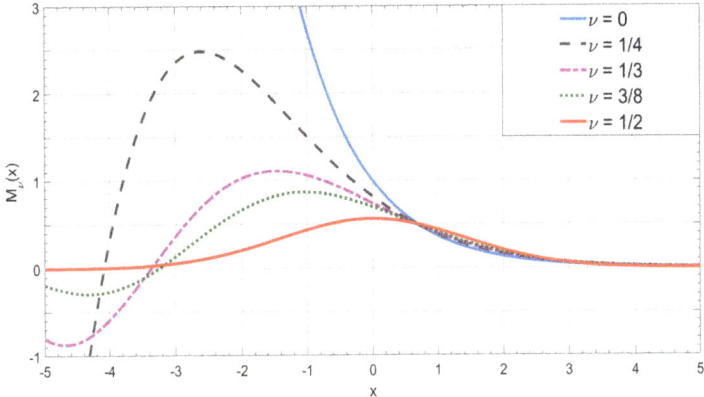

Figure 1. Plots of the M-Wright function as a function of the x variable, for $0 \leq \nu \leq 1/2$.

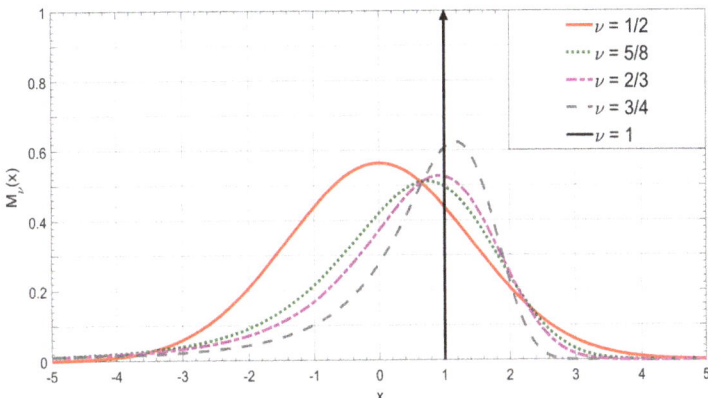

Figure 2. Plots of the M-Wright function as a function of the x variable, for $1/2 \leq \nu \leq 1$.

For this purpose, we introduce now the TFDWE as a generalization of the standard diffusion equation and we see how the two Mainardi auxiliary functions come into play. The TFDWE is thus obtained from the standard diffusion equation (or the D'Alembert wave equation) by replacing the first-order (or the second-order) time derivative by a fractional derivative (of order $0 < \beta \leq 2$) in the Caputo sense, obtaining the following Fractional PDE:

$$\frac{\partial^\beta u}{\partial t^\beta} = D \frac{\partial^2 u}{\partial x^2} \qquad 0 < \beta \leq 2, \quad D > 0, \tag{19}$$

where D is a positive constant whose dimensions are $L^2 T^{-\beta}$ and $u = u(x,t;\beta)$ is the field variable, which is assumed again to be a causal function of time. The Caputo fractional derivative is recalled in the Appendix B so that in explicit form the TFDWE (19) splits in the following integro-differential equations:

$$\frac{1}{\Gamma(1-\beta)} \int_0^t (t-\tau)^{-\beta} \left(\frac{\partial u}{\partial \tau}\right) d\tau = D \frac{\partial^2 u}{\partial x^2}, \quad 0 < \beta \leq 1; \tag{20}$$

$$\frac{1}{\Gamma(2-\beta)} \int_0^t (t-\tau)^{1-\beta} \left(\frac{\partial^2 u}{\partial \tau^2}\right) d\tau = D \frac{\partial^2 u}{\partial x^2}, \quad 1 < \beta \leq 2. \tag{21}$$

In view of our analysis, we find it convenient to put:

$$\nu = \frac{\beta}{2}, \quad 0 < \nu \leq 1. \tag{22}$$

We can then formulate the basic problems for the Time Fractional Diffusion-Wave Equation using a correspondence with the two problems for the standard diffusion equation.

Denoting by $f(x)$ and $g(t)$ two given, sufficiently well-behaved functions, we define:

(a) Cauchy problem

$$\begin{cases} u(x, 0^+; \nu) = f(x), & -\infty < x < +\infty; \\ u(\pm\infty, t; \nu) = 0, & t > 0 \end{cases} \tag{23}$$

(b) Signalling problem

$$\begin{cases} u(x, 0^+; \nu) = 0, & 0 \leq x < +\infty; \\ u(0^+, t; \nu) = g(t), \ u(+\infty, t; \nu) = 0, & t > 0 \end{cases} \tag{24}$$

If $1/2 < \nu \leq 1$ corresponding to $1 < \beta \leq 2$, we must consider also the initial value of the first time derivative of the field variable $u_t(x, 0^+; \nu)$, since, in this case, Equation (19) turns out to be akin to the wave equation and consequently two linear independent solutions are to be determined. However, to ensure the continuous dependence of the solutions to our basic problems on the parameter ν in the transition from $\nu = (1/2)^-$ to $\nu = (1/2)^+$, we agree to assume $u_t(x, 0^+; \nu) = 0$.

For the Cauchy and Signalling problems, following the approaches by Mainardi, see, e.g., [15] and related papers, we introduce now the Green functions $\mathcal{G}_c(x, t; \nu)$ and $\mathcal{G}_s(x, t; \nu)$ that for both problems can be determined by the *LT* technique, so extending the results known from the ordinary diffusion equation. We recall that the Green functions are also referred to as the fundamental solutions, corresponding respectively to $f(x) = \delta(x)$ and $g(t) = \delta(t)$ with $\delta(\cdot)$ is the Dirac delta generalized function

The expressions for the Laplace Transforms of the two Green's functions are:

$$\widetilde{\mathcal{G}}_c(x, s; \nu) = \frac{1}{2\sqrt{D} s^{1-\nu}} e^{(-|x|/\sqrt{D}) s^\nu} \tag{25}$$

and

$$\widetilde{\mathcal{G}}_s(x, s; \nu) = e^{-(x/\sqrt{D}) s^\nu} \tag{26}$$

Now, we can easily recognize the following relation:

$$\frac{d}{ds} \widetilde{\mathcal{G}}_s = -2\nu x \, \widetilde{\mathcal{G}}_c, \quad x > 0 \tag{27}$$

which implies for the original Green functions the following *reciprocity relation* for $x > 0$ and $t > 0$ and $0 < \nu < 1$:

$$2\nu x \mathcal{G}_c(x, t; \nu) = t \mathcal{G}_s(x, t; \nu) = F_\nu(z) = \nu z M_\nu(z) \quad z = \frac{x}{\sqrt{D} t^\nu} \tag{28}$$

where z is the *similarity variable* and $F_\nu(z)$ and $M_\nu(z)$ are the Mainardi auxilary functions introduced in the previous section. Indeed, Equation (28) is the generalization of Equation (A8) that we have seen for the standard diffusion equation due to the introduction of the time fractional derivative of order ν.

Then, the two Green functions of the Cauchy and Signalling problems turn out to be expressed in terms of the two auxiliary functions as follows.

For the Cauchy problem, we have

$$\mathcal{G}_C(x,t;\nu) = \frac{t^{-\nu}}{2\sqrt{D}} M_\nu\left(\frac{|x|}{\sqrt{D}t^\nu}\right) \quad -\infty < x < +\infty \quad t \geq 0 \qquad (29)$$

that generalizes Equation (A5).

For the Signalling problem, we have:

$$\mathcal{G}_S(x,t;\nu) = \frac{\nu x t^{-\nu-1}}{\sqrt{D}} M_\nu\left(\frac{x}{\sqrt{D}t^\nu}\right) \quad x \geq 0, \quad t \geq 0 \qquad (30)$$

that generalizes Equation (A7).

3.1. Complements to the Time-Fractional Diffusion-Wave Equations

The use of the Wright functions of the second kind in time fractional diffusion-wave equations has appeared in several papers for a variety of different purposes, see, e.g., Bazhlekova [27], D'Ovidio [28], Gorenflo, Luchko and Mainardi [29], Mentrelli and Pagnini [30], Mosley and Ansari [31], Pagnini [32], Povstenko [33], and references therein.

The boundary value problems dealt with previously can be considered with a source data function $f(x)$ and $g(t)$ different from the Dirac generalized functions, in particular with box-type functions as it has been carried out recently by us, see [34].

An interesting generalization of the TFDWE is obtained by considering time-fractional derivatives of distributed order. In this respect, we cite, e.g., the papers by Kochubei [35], Li, Luchko and Yamamoto [36], Mainardi, Pagnini and Gorenflo [37], and Mainardi et. al [38].

The TFDWE can also be generalized in 2D and 3D space dimensions. so consequently the Wright functions play again a fundamental role. However, we prefer to refer the interested reader to the literature, in particular to the papers by Luchko and collaborators [11,25,39–43], by Hanyga [44] and to the recent analysis by Kemppainen [45]. All of them are originated in some way from the seminal paper by Schneider and Wyss [46]. In some of these papers, the authors have considered also fractional differentiation both in time and in space, so that they have generalized to more than one dimension the former analysis by Mainardi, Luchko, and Pagnini [47] on the space-time fractional diffusion-wave equations.

4. The M-Wright Functions in Probability Theory and the Stable Distributions

We recognize that the Wright M-function with support in $\mathbb{R}^+$ can be interpreted as probability density function (pdf) because it is non negative and also it satisfies the normalization condition:

$$\int_0^\infty M_\nu(x)\,dx = 1. \qquad (31)$$

We now provide more details on these densities in the framework of the theory of probability.

Theorem 1. *Let $M_\nu(x)$ be the M-Wright function in $\mathbb{R}^+$, $0 \leq \nu < 1$ and $\delta > -1$. Then, the (finite) absolute moments of order δ are given by:*

$$\int_0^\infty x^\delta M_\nu(x)\,dx = \frac{\Gamma(\delta+1)}{\Gamma(\nu\delta+1)}. \qquad (32)$$

Proof. The proof is based on the integral representation of the M-Wright function:

$$\int_0^\infty x^\delta M_\nu(x)dx = \int_0^\infty x^\delta \left[\frac{1}{2\pi i}\int_{Ha_-} e^{\sigma-x\sigma^\nu}\frac{d\sigma}{\sigma^{1-\nu}}\right]dx$$

$$= \frac{1}{2\pi i}\int_{Ha_-} e^\sigma \left[\int_0^\infty e^{-x\sigma^\nu} x^\delta dx\right]\frac{d\sigma}{\sigma^{1-\nu}} \quad (33)$$

$$= \frac{\Gamma(\delta+1)}{2\pi i}\int_{Ha_-} \frac{e^\sigma}{\sigma^{\nu\delta+1}}d\sigma = \frac{\Gamma(\delta+1)}{\Gamma(\nu\delta+1)}$$

□

The exchange between two integrals and the following identity contributed to the final result for Equation (33):

$$\int_0^\infty e^{-x\sigma^\nu} x^\delta dx = \frac{\Gamma(\delta+1)}{(\sigma^\nu)^{\delta+1}}. \quad (34)$$

In particular, for $\delta = n \in \mathbb{N}$, the above formula provides the moments of integer order. Indeed, recalling the Mittag–Leffler function introduced in Equation (5) with $\alpha = \nu$ and $\beta = 1$:

$$E_\nu(z) := \sum_{n=0}^\infty \frac{z^n}{\Gamma(\nu n+1)}, \quad \nu > 0, \quad z \in \mathbb{C}, \quad (35)$$

the moments of integer order can also be computed from the Laplace transform pair

$$M_\nu(x) \div E_\nu(-s) \quad (36)$$

proved in the Appendix F of [8] as follows:

$$\int_0^{+\infty} x^n M_\nu(x)\,dx = \lim_{s\to 0}(-1)^n \frac{d^n}{ds^n}E_\nu(-s) = \frac{\Gamma(n+1)}{\Gamma(\nu n+1)}. \quad (37)$$

4.1. The Auxiliary Functions versus Extremal Stable Densities

We find it worthwhile to recall the relations between the Mainardi auxiliary functions and the extremal Lévy stable densities as proven in the 1997 paper by Mainardi and Tomirotti [48]. For readers' convenience, we refer to Appendix C for an essential account of the general Lévy stable distributions in probability. Indeed, from a comparison between the series expansions of stable densities in (A41) and (A42) and of the auxiliary functions in Equations (9) and (10), we recognize that the auxiliary functions are related to the extremal stable densities as follows:

$$L_\alpha^{-\alpha}(x) = \frac{1}{x}F_\alpha(x^{-\alpha}) = \frac{\alpha}{x^{\alpha+1}}M_\alpha(x^{-\alpha}) \quad 0 < \alpha < 1 \quad x \geq 0 \quad (38)$$

$$L_\alpha^{\alpha-2}(x) = \frac{1}{x}F_{1/\alpha}(x) = \frac{1}{\alpha}M_{1/\alpha}(x) \quad 1 < \alpha \leq 2 \quad -\infty < x < +\infty. \quad (39)$$

In the above equations, for $\alpha = 1$, the skewness parameter turns out to be $\theta = -1$, so we get the singular limit

$$L_1^{-1}(x) = M_1(x) = \delta(x-1). \quad (40)$$

Hereafter, we show in Figures 3 and 4 the plots the extremal stable densities according to their expressions in terms of the M-Wright functions, see Equations (38) and (39) for $\alpha = 1/2$ and $\alpha = 3/2$, respectively.

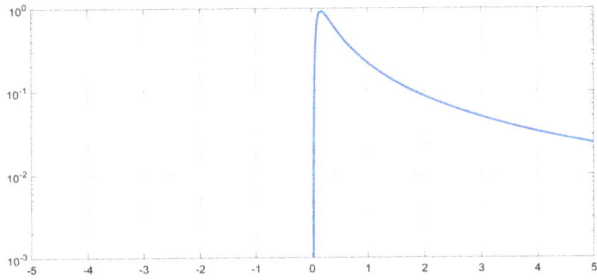

Figure 3. Plot of the unilateral extremal stable pdf for $\alpha = 1/2$.

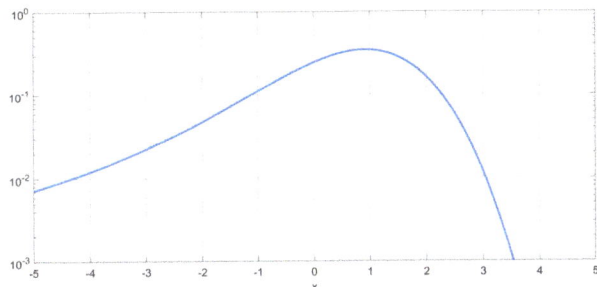

Figure 4. Plot of the bilateral extremal stable pdf for $\alpha = 3/2$.

We recognize that the above plots are consistent with the corresponding ones shown by Mainardi et al. [47] for the stable pdf's derived as fundamental solutions of a suitable space-fractional diffusion equation.

4.2. The Symmetric M-Wright Function

We easily recognize that extending the function $M_\nu(x)$ in a symmetric way to all of $\mathbb{R}$ (that is putting $x = |x|$) and dividing by 2 we have a *symmetric pdf* with support in all of $\mathbb{R}$.

As the parameter ν changes between 0 and 1, the *pdf* goes from the Laplace *pdf* to two half discrete delta *pdf*s passing for $\nu = 1/2$ through the Gaussian *pdf*.

To develop a visual intuition, also in view of the subsequent applications, we show n Figures 5 and 6 the plots of the symmetric M-Wright function on the real axis at $t = 1$ for some rational values of the parameter $\nu \in [0, 1]$

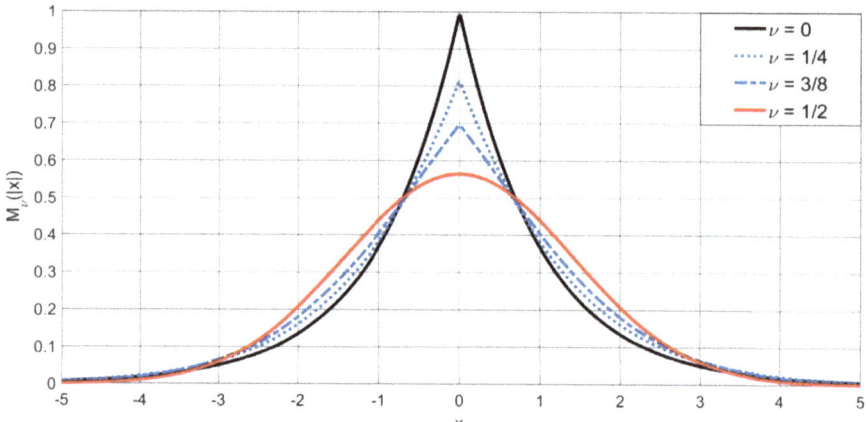

Figure 5. Plot of the symmetric M-Wright function $M_\nu(|x|)$ for $0 \leq \nu \leq 1/2$. Note that the M-Wright function becomes a Gaussian density for $\nu = 1/2$.

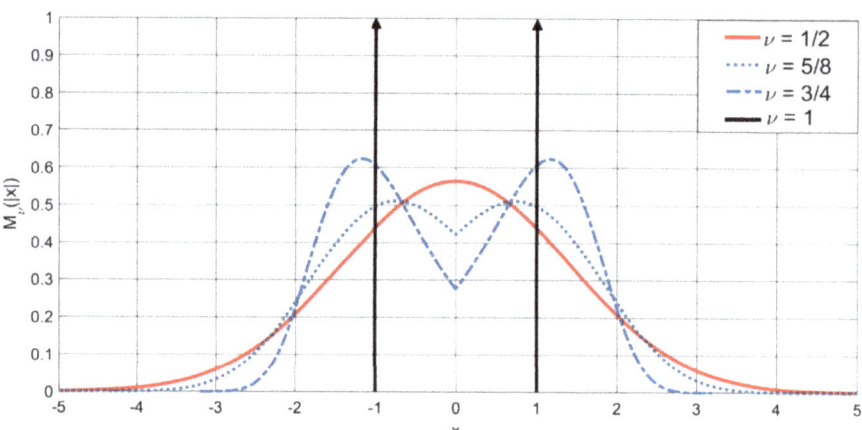

Figure 6. Plot of the symmetric M-Wright type function $M_\nu(|x|)|$ for $1/2 \leq \nu \leq 1$. Note that the M-Wright function becomes a a sum of two delta functions centered in $x = \pm 1$ for $\nu = 1$.

The readers are invited to look the YouTube video by Consiglio whose title is "Simulation of the M-Wright function", in which the author shows the evolution of this function as the parameter ν changes between 0 and 0.85 in a finite interval of $\mathbb{R}$ centered in $x = 0$.

Theorem 2. *Let $M_\nu(|x|)$ be the symmetric M-Wright function pdf. Then, its characteristic function is:*

$$\mathcal{F}\left[\frac{1}{2}M_\nu(|x|)\right] = E_{2\nu}(-\kappa^2) \tag{41}$$

Proof. The proof is based on the series development of the cosine function and on Equation (33):

$$\begin{aligned}
\mathcal{F}\left[\frac{1}{2}M_\nu(|x|)\right] &:= \frac{1}{2}\int_{-\infty}^{+\infty} e^{+i\kappa x} M_\nu(|x|)dx \\
&= \int_0^\infty \cos(\kappa x) M_\nu(x)dx \\
&= \sum_{n=0}^\infty (-1)^n \frac{\kappa^{2n}}{(2n)!}\int_0^\infty x^{2n} M_\nu(x)dx \\
&= \sum_{n=0}^\infty (-1)^n \frac{\kappa^{2n}}{\Gamma(2\nu n+1)} = E_{2\nu}(-\kappa^2)
\end{aligned} \qquad (42)$$

□

4.3. The Wright $\mathbb{M}$-Function in Two Variables

In view of time-fractional diffusion processes related to time-fractional diffusion equations, it is worthwhile to introduce the function in two variables

$$\mathbb{M}_\nu(x,t) := t^{-\nu} M_\nu(xt^{-\nu}) \quad 0<\nu<1 \quad x,t \in \mathbb{R}^+ \qquad (43)$$

which defines a spatial probability density in x evolving in time t with self-similarity exponent $H = \nu$. Of course, for $x \in \mathbb{R}$, we have to consider the symmetric version of the M-Wright function. Hereafter, we provide a list of the main properties of this function, which can be derived from the Laplace and Fourier transforms for the corresponding Wright M-function in one variable.

From Equations (39) and (43), we derive the Laplace transform of $\mathbb{M}_\nu(x,t)$ with respect to $t \in \mathbb{R}^+$,

$$\mathcal{L}\{\mathbb{M}_\nu(x,t); t \to s\} = s^{\nu-1} e^{-xs^\nu}. \qquad (44)$$

From Equation (18), we derive the Laplace transform of $\mathbb{M}_\nu(x,t)$ with respect to $x \in \mathbb{R}^+$,

$$\mathcal{L}\{\mathbb{M}_\nu(x,t); x \to s\} = E_\nu(-st^\nu). \qquad (45)$$

From Equation (55), we derive the Fourier transform of $\mathbb{M}_\nu(|x|,t)$ with respect to $x \in \mathbb{R}$,

$$\mathcal{F}\{\mathbb{M}_\nu(|x|,t); x \to \kappa\} = 2E_{2\nu}\left(-\kappa^2 t^\nu\right). \qquad (46)$$

Using the Mellin transforms, Mainardi et al. [49] derived the following interesting integral formula of composition,

$$\mathbb{M}_\nu(x,t) = \int_0^\infty \mathbb{M}_\lambda(x,\tau)\,\mathbb{M}_\mu(\tau,t)\,d\tau \quad \nu = \lambda\mu. \qquad (47)$$

Special cases of the Wright $\mathbb{M}$-function are simply derived for $\nu = 1/2$ and $\nu = 1/3$ from the corresponding ones in the complex domain, see Equations (28) and (29). We devote particular attention to the case $\nu = 1/2$ for which we get the Gaussian density in $\mathbb{R}$,

$$\mathbb{M}_{1/2}(|x|,t) = \frac{1}{2\sqrt{\pi}t^{1/2}}e^{-x^2/(4t)}. \qquad (48)$$

For the limiting case $\nu = 1$, we obtain

$$\mathbb{M}_1(|x|,t) = \frac{1}{2}[\delta(x-t) + \delta(x+t)]. \qquad (49)$$

We conclude this section pointing out that the M-Wright functions have been applied by several authors in the theory of probability and stochastic processes, see, e.g., Beghin and Orsingher [50],

Cahoy [51,52], Garra, Orsingher and Polito [53], Le Chen [54], Consiglio, Luchko and Mainardi [55], Gorenflo and Mainardi [56], Mainardi, Mura and Pagnini [57], Pagnini [58], Scalas and Viles [59], and references therein. Furthermore, these functions have been found in the first passage problem for Lévy flights dealt by the group of Prof. Metzler, see e.g., [60,61].

5. The Four Sisters

In this section, we show how some Wright functions of the second kind can provide an interesting generalization of the three sisters discussed in Appendix A. The starting point is a (not well-known) paper published in 1970 by Stankovic [62], where (in our notation) the following Laplace transform pair is proved rigorously:

$$t^{\mu-1} W_{-\nu,\mu}(x,t) \div s^{-\mu} e^{-xs^\nu} \quad 0 < \nu < 1 \quad \mu \geq 0 \tag{50}$$

where x and t are positive. We note that the Stankovic formula can be derived in a formal way by developing the exponential function in positive power of s and inverting term by term as described in the Appendix F of the book by Mainardi [8].

We recognize that the Laplace Transforms of the Three Sisters functions $\widetilde{\phi}(x,s)$, $\widetilde{\psi}(x,s)$ and $\widetilde{\chi}(x,s)$ are particular cases of the Equation (50) for $\nu = 1/2$ that is of

$$t^{\mu-1} W_{-1/2,\mu}(x,t) \div s^{-\mu} e^{-x\sqrt{s}}, \tag{51}$$

according to the following scheme:

$$\widetilde{\phi}(x,s) \text{ with } \mu = 1; \quad \widetilde{\psi}(x,s) \text{ with } \mu = 0; \quad \widetilde{\chi}(x,s) \text{ with } \mu = 1/2.$$

If ν is no longer restricted to $\nu = 1/2$, we define *Four Sisters functions* as follows:

$$\begin{aligned}
\mu = 0, & \quad e^{-xs^\nu} \div t^{-1} W_{-\nu,0}(-xt^{-\nu}), \\
\mu = 1-\nu, & \quad \frac{e^{-xs^\nu}}{s^{1-\nu}} \div t^{-\nu} W_{-\nu,1-\nu}(-xt^{-\nu}), \\
\mu = \nu, & \quad \frac{e^{-xs^\nu}}{s^\nu} \div t^{\nu-1} W_{-\nu,\nu}(-xt^{-\nu}), \\
\mu = 1, & \quad \frac{e^{-xs^\nu}}{s} \div W_{-\nu,1}(-xt^{-\nu}).
\end{aligned} \tag{52}$$

Hereafter, in Figures 7–9, we show some plots of these functions, both in the t and in the x domain for some values of ν ($\nu = 1/4, 1/2, 3/4$).

Note that for $\nu = 1/2$ we only find three functions, that is the Three Sisters functions of Appendix A.

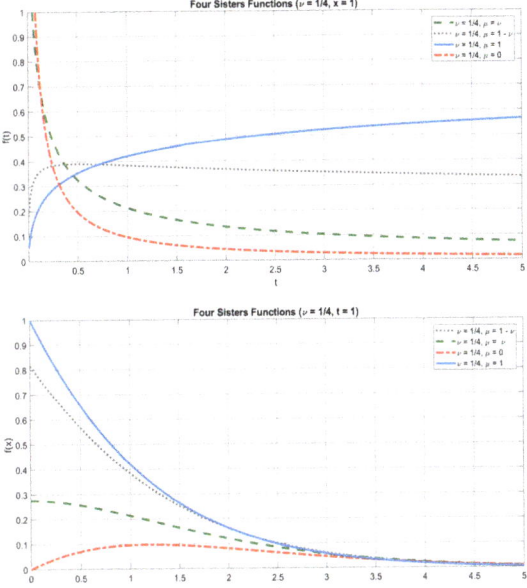

Figure 7. Plots of the four sisters functions in linear scale with $\nu = 1/4$; top: versus t ($x = 1$), bottom: versus x ($t = 1$).

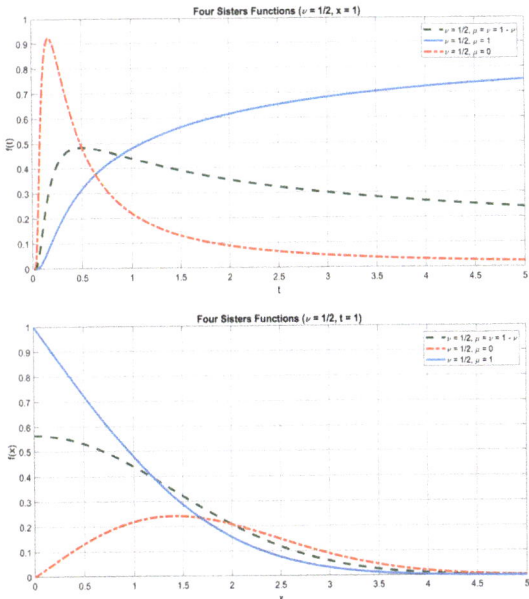

Figure 8. Plots of the three sisters functions in linear scale with $\nu = 1/2$; top: versus t ($x = 1$), bottom: versus x ($t = 1$).

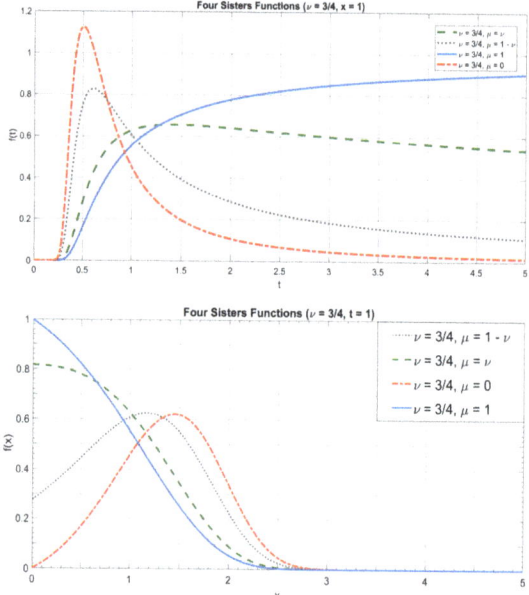

Figure 9. Plots of the four sisters functions in linear scale with $\nu = 3/4$; top: versus t ($x = 1$), bottom: versus x ($t = 1$).

6. Conclusions

In our survey on the Wright functions, we have distinguished two kinds, pointing out the particular class of the second kind. Indeed, these functions have been shown to play key roles in several processes governed by non-Gaussian processes, including sub-diffusion, transition to wave propagation, Lévy stable distributions. Furthermore, we have devoted our attention to four functions of this class that we agree to called *the Four Sisters functions*. All these items justify the relevance of the Wright functions of the second kind in Mathematical Physics.

Author Contributions: F.M.: Conceptualization of the work; A.C.: Plots. Formal analysis, review and editing have been performed by F.M. and A.C. in equal parts. All authors have read and agreed to the published version of the manuscript.

Funding: This research received no external funding.

Acknowledgments: The research activity of both the authors has been carried out (without funding) in the framework of the activities of the National Group of Mathematical Physics (GNFM, INdAM). All graphs in the figures of the present paper have been drawn using MATLAB. They have been realized mainly referring to the power series definition of the related functions by adopting a sufficiently large number of terms. The authors would like to thank the anonymous reviewers for their helpful and constructive comments.

Conflicts of Interest: The authors declare no conflict of interest.

Appendix A. The Standard Diffusion Equation and the Three Sisters

In this Appendix, let us recall the Diffusion Equation in the one-dimensional case

$$\frac{\partial u}{\partial t} = D \frac{\partial^2 u}{\partial x^2} \tag{A1}$$

where u is the field variable, the constant $D > 0$ is the diffusion coefficient, whose dimensions are $L^2 T^{-1}$, and x, t denote the space and time coordinates, respectively.

Two basic problems for Equation (A1) are the *Cauchy* and *Signalling* ones introduced hereafter In these problems, some initial values and boundary conditions are set; specify the values attained by the field variable and/or by some of its derivatives on the boundary of the space-time domain is an essential step to guarantee the existence, the uniqueness and the determination of a solution of physical interest to the problem, not only for the Diffusion Equation.

Two *data functions* $f(x)$ and $g(t)$ are then introduced to write formally these conditions; some regularities are required to be satisfied by $f(x)$ and $g(t)$, and in particular $f(x)$ must admit the Fourier transform or the Fourier series expansion if the support is finite, while $h(t)$ must admit the Laplace Transform. We also require without loss of generality that the field variable $u(x,t)$ is vanishing for $t < 0$ for every x in the spatial domain. Given these premises, we can specify the two aforementioned problems.

In the *Cauchy problem*, the medium is supposed to be unlimited ($-\infty < x < +\infty$) and to be subjected at $t = 0$ to a known disturbance provided by the data function $f(x)$. Formally:

$$\begin{cases} \lim_{t \to 0^+} u(x,t) = f(x), & -\infty < x < +\infty; \\ \lim_{x \to \pm\infty} u(x,t) = 0, & t > 0. \end{cases} \qquad (A2)$$

This is a pure *initial-value problem* (IVP) as the values are specified along the boundary $t = 0$.

In the *Signalling problem*, the medium is supposed to be semi-infinite ($0 \leq x < +\infty$) and initially undisturbed. At $x = 0$ (the accessible end) and for $t > 0$, the medium is then subjected to a known disturbance provided by the causal function $g(t)$. Formally:

$$\begin{cases} \lim_{t \to 0^+} u(x,t) = 0, & 0 \leq x < +\infty; \\ \lim_{x \to 0^+} u(x,t) = g(t), \ \lim_{x \to +\infty} u(x,t) = 0 & t > 0. \end{cases} \qquad (A3)$$

This problem is referred to as an *initial boundary value problem* (IBVP) in the quadrant $\{x, t\} > 0$.

For each problem, the solutions turn out to be expressed by a proper convolution between the data functions and the *Green functions* $\mathcal{G}$ that are the fundamental solutions of the problems.

For the Cauchy problem, we have:

$$u(x,t) = \int_{-\infty}^{+\infty} \mathcal{G}_C(\xi,t) f(x-\xi) d\xi = \mathcal{G}_C(x,t) * f(x) \qquad (A4)$$

with

$$\mathcal{G}_C(x,t) = \frac{1}{2\sqrt{\pi D t}} e^{-x^2/(4Dt)}. \qquad (A5)$$

For the Signalling problem, we have:

$$u(x,t) = \int_0^t \mathcal{G}_S(x,\tau) g(t-\tau) d\tau = \mathcal{G}_S(x,t) * g(t) \qquad -\infty < x < +\infty, \quad t \geq 0 \qquad (A6)$$

with

$$\mathcal{G}_S(x,t) = \frac{x}{2\sqrt{\pi D t^3}} e^{-x^2/(4Dt)} \qquad x \geq 0, \quad t \geq 0. \qquad (A7)$$

Following the lecture notes in Mathematical Physics by Mainardi [63], we note that the following relevant property is valid for $\{x, t\} > 0$:

$$x\mathcal{G}_C(x,t) = t\mathcal{G}_S(x,t) = F(z) \qquad (A8)$$

where

$$z = \frac{x}{\sqrt{Dt}}, \quad F(z) = \frac{z}{2} M(z), \quad M(z) = \frac{1}{\sqrt{\pi}} e^{-z^2/4}. \qquad (A9)$$

According to Mainardi's notations, Equation (A8) is known as *reciprocity relation*, $F(z)$ and $M(z)$ are called *auxiliary functions* and z is the *similarity variable*.

A particular case of the Signalling problem is obtained when $g(t) = H(t)$ (the Heaviside unit step function) and the solution $u(x,t)$ turns out to be expressed in terms of the *complementary error function*:

$$u(x,t) = \mathcal{H}_S(x,t) = \int_0^t \mathcal{G}_S(x,\tau)d\tau = \mathrm{erfc}\left(\frac{x}{2\sqrt{Dt}}\right) \quad x \geq 0, \quad t \geq 0. \tag{A10}$$

As is well known, the three above fundamental solutions can be obtained via the Fourier and Laplace transform methods. Introducing the parameter $a = |x|/\sqrt{D}$, the Laplace transforms of these functions turns out to be simply related in the Laplace domain $\mathrm{Re}(s) > 0$, as follows:

$$\phi(a,t) := \mathrm{erfc}\left(\frac{a}{2\sqrt{t}}\right) \div \frac{e^{-as^{1/2}}}{s} := \widetilde{\phi}(a,s), \tag{A11}$$

$$\psi(a,t) := \frac{a}{2\sqrt{\pi}} t^{-3/2} e^{-a^2/(4t)} \div e^{-as^{1/2}} := \widetilde{\psi}(a,s), \tag{A12}$$

$$\chi(a,t) := \frac{1}{\sqrt{\pi}} t^{-1/2} e^{-a^2/(4t)} \div \frac{e^{-as^{1/2}}}{s^{1/2}} := \widetilde{\chi}(a,s) \tag{A13}$$

where the sign $\div$ is used for the juxtaposition of a function with its Laplace transform. We easily note that Equation (A11) is related to the Step-Response problem, Equation (A12) is related to the Signalling problem and Equation (A13) is related to the Cauchy problem. Following the lecture notes by Mainardi [63], we agree to call the above functions *the three sisters functions* for their role in the standard diffusion equation. They will be discussed with details hereafter.

Everything that we have said above will be found again as a special case of the *Time Fractional Diffusion Equation* where the time derivative of the first order is replaced by a suitable time derivative of non-integer order.

It is easy to demonstrate that each of them can be expressed as a function of one of the two others *three sisters* (Table A1).

Table A1. Relations among the *three sisters* in the Laplace domain.

	$\widetilde{\phi}$	$\widetilde{\psi}$	$\widetilde{\chi}$
$\widetilde{\phi}$	$\dfrac{e^{-a\sqrt{s}}}{s}$	$\dfrac{\widetilde{\psi}}{s}$	$-\dfrac{1}{s}\dfrac{\partial \widetilde{\chi}}{\partial a}$
$\widetilde{\psi}$	$s\widetilde{\phi}$	$e^{-a\sqrt{s}}$	$-\dfrac{\partial \widetilde{\chi}}{\partial a}$
$\widetilde{\chi}$	$-\dfrac{\partial \widetilde{\phi}}{\partial a}$	$-\dfrac{2}{a}\dfrac{\partial \widetilde{\psi}}{\partial s}$	$\dfrac{e^{-a\sqrt{s}}}{\sqrt{s}}$

The *three sisters* in the t domain may be all directly calculated by making use of the *Bromwich formula* taking account of the contribution of the branch cut of $\sqrt{s}$ and of the pole of $1/s$. We obtain:

$$\widetilde{\phi}(a,s) \div \phi(a,t) = 1 - \frac{1}{\pi}\int_0^\infty e^{-rt}\sin(a\sqrt{r})\frac{dr}{r}$$

$$\widetilde{\psi}(a,s) \div \psi(a,t) = \frac{1}{\pi}\int_0^\infty e^{-rt}\sin(a\sqrt{r})\,dr$$

$$\widetilde{\chi}(a,s) \div \chi(a,t) = \frac{1}{\pi}\int_0^\infty e^{-rt}\cos(a\sqrt{r})\frac{dr}{\sqrt{r}}.$$

Then, through the substitution $\rho = \sqrt{r}$, we arrive at the Gaussian integral and, consequently, we find the previous explicit expressions of the *three sisters* that is:

$$\phi(a,t) = \mathrm{erfc}(\frac{a}{2\sqrt{t}}) = 1 - \frac{2}{\sqrt{\pi}} \int_0^{a/2\sqrt{t}} e^{-u^2} du$$

$$\psi(a,t) = \frac{a}{2\sqrt{\pi}} t^{-3/2} e^{-a^2/4t}$$

$$\chi(a,t) = \frac{1}{\sqrt{\pi}} t^{-1/2} e^{-a^2/4t},$$

reminding us of the definition of the complementary error function.

Alternatively, we can compute the *three sisters* in the t domain by using the relations among the *three sisters* in the Laplace domain listed in Table A1. However, in this case, one of the *three sisters* in the t domain must already be known. Assuming to know $\phi(a,t)$ from Equation (A11), we get:

- $\psi(a,t)$ from $\widetilde{\psi}(a,s) = s\,\widetilde{\phi}(a,s)$. Indeed, noting

$$s\,\widetilde{\phi}(a,s) \div \frac{\partial}{\partial t} \phi(a,t)$$

since $\phi(a,0^+) = 0$ we can obtain (A12), namely

$$\psi(a,t) = \frac{a}{2\sqrt{\pi}} t^{-3/2} e^{-a^2/4t};$$

- $\chi(a,t)$ from $\widetilde{\chi}(a,s) = -\frac{\partial}{\partial a}\widetilde{\phi}(a,s)$ where a is seen as a parameter. Indeed, it immediately follows Equation (A13), namely

$$\chi(a,t) = -\frac{\partial}{\partial a} \phi(a,t) = \frac{1}{\sqrt{\pi}} t^{-1/2} e^{-a^2/4t}.$$

For more details, we refer the reader again to [63].

Appendix B. Essentials of Fractional Calculus

Fractional calculus is the field of mathematical analysis which deals with the investigation and applications of integrals and derivatives of arbitrary order. The term *fractional* is a misnomer, but it is retained for historical reasons, following the prevailing use.

This appendix is based on the 1997 surveys by Gorenflo and Mainardi [64] and by Mainardi [65]. For more details on the classical treatment of fractional calculus, the reader is referred to the nice and rigorous book by Diethelm [66] published in 2010 by Springer in the series Lecture Notes in Mathematics.

According to the Riemann–Liouville approach to fractional calculus, the notion of fractional integral of order α ($\alpha > 0$) is a natural consequence of the well known formula (usually attributed to Cauchy) that reduces the calculation of the n−fold primitive of a function $f(t)$ to a single integral of convolution type. In our notation, the Cauchy formula reads

$$J^n f(t) := f_n(t) = \frac{1}{(n-1)!} \int_0^t (t-\tau)^{n-1} f(\tau)\, d\tau \quad t > 0 \quad n \in \mathbb{N} \tag{A14}$$

where $\mathbb{N}$ is the set of positive integers. From this definition, we note that $f_n(t)$ vanishes at $t = 0$ with its derivatives of order $1, 2, \ldots, n-1$. For convention, we require that $f(t)$ and henceforth $f_n(t)$ is a *causal* function, i.e., identically vanishing for $t < 0$.

In a natural way, one is led to extend the above formula from positive integer values of the index to any positive real values by using the Gamma function. Indeed, noting that $(n-1)! = \Gamma(n)$ and introducing the arbitrary *positive* real number α, one defines the *Fractional Integral of order* $\alpha > 0$:

$$J^\alpha f(t) := \frac{1}{\Gamma(\alpha)} \int_0^t (t-\tau)^{\alpha-1} f(\tau)\, d\tau \quad t > 0 \quad \alpha \in \mathbb{R}^+ \tag{A15}$$

where $\mathbb{R}^+$ is the set of positive real numbers. For complementation, we define $J^0 := I$ (Identity operator), i.e., we mean $J^0 f(t) = f(t)$. Furthermore, by $J^\alpha f(0^+)$, we mean the limit (if it exists) of $J^\alpha f(t)$ for $t \to 0^+$; this limit may be infinite.

We note the *semigroup property* $J^\alpha J^\beta = J^{\alpha+\beta}$ α $\beta \geq 0$ which implies the *commutative property* $J^\beta J^\alpha = J^\alpha J^\beta$ and the effect of our operators J^α on the power functions

$$J^\alpha t^\gamma = \frac{\Gamma(\gamma+1)}{\Gamma(\gamma+1+\alpha)} t^{\gamma+\alpha} \quad \alpha \geq 0 \quad \gamma > -1 \quad t > 0. \tag{A16}$$

These properties are of course a natural generalization of those known when the order is a positive integer.

Introducing the Laplace transform by the notation $\mathcal{L}\{f(t)\} := \int_0^\infty e^{-st} f(t)\, dt = \widetilde{f}(s)$ $s \in \mathbb{C}$ and using the sign $\div$ to denote a Laplace transform pair, i.e., $f(t) \div \widetilde{f}(s)$, we point out the following rule for the Laplace transform of the fractional integral,

$$J^\alpha f(t) \div \frac{\widetilde{f}(s)}{s^\alpha} \quad \alpha \geq 0 \tag{A17}$$

which is the generalization of the case with an n-fold repeated integral.

After the notion of fractional integral, that of fractional derivative of order α ($\alpha > 0$) becomes a natural requirement and one is attempted to substitute α with $-\alpha$ in the above formulas. However, this generalization needs some care in order to guarantee the convergence of the integrals and preserve the well known properties of the ordinary derivative of integer order.

Denoting by D^n with $n \in \mathbb{N}$ the operator of the derivative of order n, we first note that $D^n J^n = I$ $J^n D^n \neq I$ $n \in \mathbb{N}$ i.e., D^n is left-inverse (and not right-inverse) to the corresponding integral operator J^n. In fact, we easily recognize from Equation (A14) that

$$J^n D^n f(t) = f(t) - \sum_{k=0}^{n-1} f^{(k)}(0^+) \frac{t^k}{k!} \quad t > 0. \tag{A18}$$

As a consequence, we expect that D^α is defined as left-inverse to J^α. For this purpose, introducing the positive integer m such that $m - 1 < \alpha \leq m$, one defines the *Fractional Derivative of order* $\alpha > 0$ as $D^\alpha f(t) := D^m J^{m-\alpha} f(t)$ i.e.,

$$D^\alpha f(t) := \begin{cases} \dfrac{d^m}{dt^m} \left[\dfrac{1}{\Gamma(m-\alpha)} \int_0^t \dfrac{f(\tau)}{(t-\tau)^{\alpha+1-m}}\, d\tau \right], & m-1 < \alpha < m, \\ \dfrac{d^m}{dt^m} f(t) & \alpha = m. \end{cases} \tag{A19}$$

Defining for complementation $D^0 = J^0 = I$, then we easily recognize that $D^\alpha J^\alpha = I$ $\alpha \geq 0$ and

$$D^\alpha t^\gamma = \frac{\Gamma(\gamma+1)}{\Gamma(\gamma+1-\alpha)} t^{\gamma-\alpha} \quad \alpha \geq 0 \quad \gamma > -1 \quad t > 0. \tag{A20}$$

Of course, these properties are a natural generalization of those known when the order is a positive integer.

Note the remarkable fact that the fractional derivative $D^\alpha f$ is not zero for the constant function $f(t) \equiv 1$ if $\alpha \notin \mathbb{N}$. In fact, (A20) with $\gamma = 0$ teaches us that

$$D^\alpha 1 = \frac{t^{-\alpha}}{\Gamma(1-\alpha)} \quad \alpha \geq 0 \quad t > 0. \tag{A21}$$

This, of course, is $\equiv 0$ for $\alpha \in \mathbb{N}$, due to the poles of the gamma function in the points $0, -1, -2, \ldots$. We now observe that an alternative definition of fractional derivative was introduced by Caputo in 1967 [67] in a geophysical journal and in 1969 [68] in a book in Italian. Then, the Caputo definition was adopted in 1971 by Caputo and Mainardi [69,70] in the framework of the theory of *Linear Viscoelasticity*. Nowadays, it is usually referred to as the *Caputo fractional derivative* and reads $D_*^\alpha f(t) := J^{m-\alpha} D^m f(t)$ with $m - 1 < \alpha \leq m$ $m \in \mathbb{N}$ i.e.,

$$D_*^\alpha f(t) := \begin{cases} \dfrac{1}{\Gamma(m-\alpha)} \int_0^t \dfrac{f^{(m)}(\tau)}{(t-\tau)^{\alpha+1-m}} d\tau & m-1 < \alpha < m \\ \dfrac{d^m}{dt^m} f(t) & \alpha = m. \end{cases} \tag{A22}$$

We recall that there are a number of discussions on the priority of this definition that surely was formerly considered by Liouville as stated by Butzer and Westphal [71]. However, Liouville did not recognize the relevance of this representation derived by a trivial integration by part, whereas Caputo, even if unaware of the Riemann–Liouville representation, promoted his definition in several papers for all the applications where the Laplace transform plays a fundamental role. We agree to denote Equation (A22) as the *Caputo fractional derivative* to distinguish it from the standard Riemann–Liouville fractional derivative (A19).

The Caputo definition (A22) is of course more restrictive than the Riemann–Liouville definition (A19), in that it requires the absolute integrability of the derivative of order m. Whenever we use the operator D_*^α, we (tacitly) assume that this condition is met. We easily recognize that in general

$$D^\alpha f(t) := D^m J^{m-\alpha} f(t) \neq J^{m-\alpha} D^m f(t) := D_*^\alpha f(t) \tag{A23}$$

unless the function $f(t)$ along with its first $m-1$ derivatives vanishes at $t = 0^+$. In fact, assuming that the passage of the m-derivative under the integral is legitimate, one recognizes that, for $m - 1 < \alpha < m$ and $t > 0$

$$D^\alpha f(t) = D_*^\alpha f(t) + \sum_{k=0}^{m-1} \frac{t^{k-\alpha}}{\Gamma(k-\alpha+1)} f^{(k)}(0^+) \tag{A24}$$

and therefore, recalling the fractional derivative of the power functions (A20),

$$D^\alpha \left(f(t) - \sum_{k=0}^{m-1} \frac{t^k}{k!} f^{(k)}(0^+) \right) = D_*^\alpha f(t). \tag{A25}$$

The alternative definition (A22) for the fractional derivative thus incorporates the initial values of the function and of its integer derivatives of lower order. The subtraction of the Taylor polynomial of degree $m-1$ at $t = 0^+$ from $f(t)$ means a sort of regularization of the Riemann–Liouville fractional derivative. In particular, for $0 < \alpha < 1$, we get

$$D^\alpha \left(f(t) - f(0^+) \right) = D_*^\alpha f(t).$$

According to the Caputo definition, the relevant property for which the fractional derivative of a constant is still zero can be easily recognized, i.e.,

$$D_*^\alpha 1 \equiv 0 \quad \alpha > 0. \tag{A26}$$

We now explore the most relevant differences between the two fractional derivatives (A19) and (A22). We observe, again by looking at (A20), that $D^\alpha t^{\alpha-1} \equiv 0 \;\; \alpha > 0 \;\; t > 0$. From above, we thus recognize the following statements about functions which for $t > 0$ admit the same fractional derivative of order α with $m - 1 < \alpha \leq m \;\; m \in \mathbb{N}$

$$D^\alpha f(t) = D^\alpha g(t) \iff f(t) = g(t) + \sum_{j=1}^{m} c_j t^{\alpha-j} \tag{A27}$$

$$D^\alpha_* f(t) = D^\alpha_* g(t) \iff f(t) = g(t) + \sum_{j=1}^{m} c_j t^{m-j}. \tag{A28}$$

In these formulas, the coefficients c_j are arbitrary constants.

For the two definitions, we also point out a difference with respect to the *formal* limit as $\alpha \to (m-1)^+$. From (A19) and (A22) we obtain, respectively,

$$\alpha \to (m-1)^+ \implies D^\alpha f(t) \to D^m J f(t) = D^{m-1} f(t); \tag{A29}$$

$$\alpha \to (m-1)^+ \implies D^\alpha_* f(t) \to J D^m f(t) = D^{m-1} f(t) - f^{(m-1)}(0^+). \tag{A30}$$

We now consider the *Laplace transform* of the two fractional derivatives. For the standard fractional derivative D^α, the Laplace transform, assumed to exist, requires the knowledge of the (bounded) initial values of the fractional integral $J^{m-\alpha}$ and of its integer derivatives of order $k = 1, 2, \ldots, m-1$. The corresponding rule reads, in our notation,

$$D^\alpha f(t) \div s^\alpha \widetilde{f}(s) - \sum_{k=0}^{m-1} D^k J^{(m-\alpha)} f(0^+) s^{m-1-k} \quad m-1 < \alpha \leq m. \tag{A31}$$

The *Caputo fractional derivative* appears to be more suitable to be treated by the Laplace transform technique in that it requires the knowledge of the (bounded) initial values of the function and of its integer derivatives of order $k = 1, 2, \ldots, m-1$ analogous with the case when $\alpha = m$. In fact, by using Eqaution (A17) and noting that

$$J^\alpha D^\alpha_* f(t) = J^\alpha J^{m-\alpha} D^m f(t) = J^m D^m f(t) = f(t) - \sum_{k=0}^{m-1} f^{(k)}(0^+) \frac{t^k}{k!}. \tag{A32}$$

we easily prove the following rule for the Laplace transform,

$$D^\alpha_* f(t) \div s^\alpha \widetilde{f}(s) - \sum_{k=0}^{m-1} f^{(k)}(0^+) s^{\alpha-1-k} \quad m-1 < \alpha \leq m. \tag{A33}$$

Indeed, the result (A33), first stated by Caputo by using the Fubini–Tonelli theorem, appears as the most "natural" generalization of the corresponding result well known for $\alpha = m$.

In particular, Gorenflo and Mainardi have pointed out the major utility of the Caputo fractional derivative in the treatment of differential equations of fractional order for *physical applications*. In fact, in physical problems, the initial conditions are usually expressed in terms of a given number of bounded values assumed by the field variable and its derivatives of integer order, no matter if the governing evolution equation may be a generic integro-differential equation and therefore, in particular, a fractional differential equation.

Appendix C. The Lévy Stable Distributions

We now introduce the so-called *Lévy Stable Distributions*. The term stable has been assigned by the French mathematician Paul Lévy, who, in the 1920s, started a systematic research in order to generalize the celebrated *Central Limit Theorem* to probability distributions with infinite variance. For stable distributions, we can assume the following DEFINITION: *If two independent real random variables with the same shape or type of distribution are combined linearly and the distribution of the resulting random variable has the same shape, the common distribution (or its type, more precisely) is said to be stable.*

The restrictive condition of stability enabled Lévy (and then other authors) to derive the *canonic form* for the characteristic function of the densities of these distributions. Here, we follow the parameterization by Feller [72,73] revisited by Gorenflo & Mainardi in [74], see also [47]. Denoting by $L_\alpha^\theta(x)$ a generic stable density in $\mathbb{R}$, where α is the *index of stability* and and θ the asymmetry parameter, improperly called *skewness*, its characteristic function reads:

$$L_\alpha^\theta(x) \div \widehat{L}_\alpha^\theta(\kappa) = \exp\left[-\psi_\alpha^\theta(\kappa)\right] \quad \psi_\alpha^\theta(\kappa) = |\kappa|^\alpha \, e^{i(\mathrm{sign}\,\kappa)\theta\pi/2} \tag{A34}$$

$$0 < \alpha \le 2 \quad |\theta| \le \min\{\alpha, 2-\alpha\}.$$

We note that the allowed region for the parameters α and θ turns out to be a diamond in the plane $\{\alpha, \theta\}$ with vertices in the points $(0,0)$ $(1,1)$ $(1,-1)$ $(2,0)$, which we call the *Feller–Takayasu diamond*, see Figure A1. For values of θ on the border of the diamond (that is $\theta = \pm\alpha$ if $0 < \alpha < 1$, and $\theta = \pm(2-\alpha)$ if $1 < \alpha < 2$), we obtain the so-called *extremal stable densities*.

We also note the *symmetry relation* $L_\alpha^\theta(-x) = L_\alpha^{-\theta}(x)$, so that a stable density with $\theta = 0$ is symmetric.

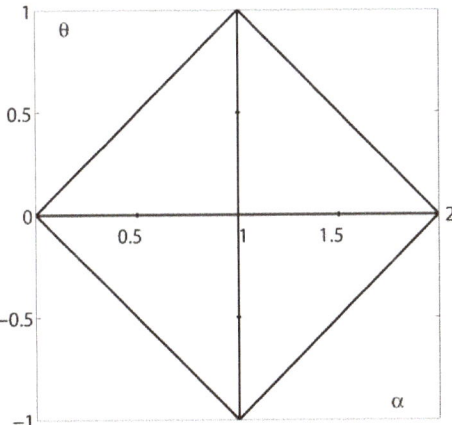

Figure A1. The Feller–Takayasu diamond for Lévy stable densities.

Stable distributions have noteworthy properties of which the interested reader can be informed from the relevant existing literature. Hereafter, we recall some peculiar PROPERTIES:

- The class of stable distributions possesses its own domain of attraction, see, e.g., [73].
- Any stable density is unimodal and indeed bell-shaped, i.e., its *n*-th derivative has exactly *n* zeros in $\mathbb{R}$, see Gawronski [75], Simon [76], and Kwaśnicki [77].
- The stable distributions are self-similar and infinitely divisible.

These properties derive from the canonic form (A34) through the scaling property of the Fourier transform.

Self-similarity means

$$L_\alpha^\theta(x,t) \div \exp\left[-t\psi_\alpha^\theta(\kappa)\right] \iff L_\alpha^\theta(x,t) = t^{-1/\alpha}\, L_\alpha^\theta(x/t^{1/\alpha})] \tag{A35}$$

where t is a positive parameter. If t is time, then $L_\alpha^\theta(x,t)$ is a spatial density evolving on time with self-similarity.

Infinite divisibility means that, for every positive integer n, the characteristic function can be expressed as the nth power of some characteristic function, so that any stable distribution can be expressed as the n-fold convolution of a stable distribution of the same type. Indeed, taking in (A34) $\theta = 0$, without loss of generality, we have

$$e^{-t|\kappa|^\alpha} = \left[e^{-(t/n)|\kappa|^\alpha}\right]^n \iff L_\alpha^0(x,t) = \left[L_\alpha^0(x,t/n)\right]^{*n} \tag{A36}$$

where

$$\left[L_\alpha^0(x,t/n)\right]^{*n} := L_\alpha^0(x,t/n) * L_\alpha^0(x,t/n) * \cdots * L_\alpha^0(x,t/n)$$

is the multiple Fourier convolution in $\mathbb{R}$ with n identical terms.

Only for a few particular cases, the inversion of the Fourier transform in (A34) can be carried out using standard tables, and well-known probability distributions are obtained.

For $\alpha = 2$ (so $\theta = 0$), we recover the *Gaussian pdf* that turns out to be the only stable density with finite variance, and more generally with finite moments of any order $\delta \geq 0$. In fact,

$$L_2^0(x) = \frac{1}{2\sqrt{\pi}} e^{-x^2/4}. \tag{A37}$$

All the other stable densities have finite absolute moments of order $\delta \in [-1, \alpha)$ as we will later show.

For $\alpha = 1$ and $|\theta| < 1$, we get

$$L_1^\theta(x) = \frac{1}{\pi} \frac{\cos(\theta\pi/2)}{[x + \sin(\theta\pi/2)]^2 + [\cos(\theta\pi/2)]^2} \tag{A38}$$

which for $\theta = 0$ includes the *Cauchy-Lorentz pdf*:

$$L_1^0(x) = \frac{1}{\pi} \frac{1}{1+x^2}. \tag{A39}$$

In the limiting cases $\theta = \pm 1$ for $\alpha = 1$, we obtain the *singular Dirac pdf's*

$$L_1^{\pm 1}(x) = \delta(x \pm 1). \tag{A40}$$

In general, we must recall the power series expansions provided in [73]. We restrict our attention to $x > 0$ since the evaluations for $x < 0$ can be obtained using the symmetry relation. The convergent expansions of $L_\alpha^\theta(x)$ ($x > 0$) turn out to be:

for $0 < \alpha < 1 \quad |\theta| \leq \alpha$:

$$L_\alpha^\theta(x) = \frac{1}{\pi x} \sum_{n=1}^\infty (-x^{-\alpha})^n \frac{\Gamma(1+n\alpha)}{n!} \sin\left[\frac{n\pi}{2}(\theta - \alpha)\right]; \tag{A41}$$

for $1 < \alpha \leq 2 \quad |\theta| \leq 2 - \alpha$:

$$L_\alpha^\theta(x) = \frac{1}{\pi x} \sum_{n=1}^\infty (-x)^n \frac{\Gamma(1+n/\alpha)}{n!} \sin\left[\frac{n\pi}{2\alpha}(\theta - \alpha)\right]. \tag{A42}$$

From the series in (A41) and the symmetry relation, we note that *the extremal stable densities for $0 < \alpha < 1$ are unilateral*, precisely vanishing for $x > 0$ if $\theta = \alpha$, vanishing for $x < 0$ if $\theta = -\alpha$.

In particular, the unilateral extremal densities $L_\alpha^{-\alpha}(x)$ with $0 < \alpha < 1$ have support in $\mathbb{R}^+$ and Laplace transform $\exp(-s^\alpha)$. For $\alpha = 1/2$, we obtain the so-called *Lévy-Smirnov pdf*:

$$L_{1/2}^{-1/2}(x) = \frac{x^{-3/2}}{2\sqrt{\pi}}\, e^{-1/(4x)} \quad x \geq 0. \tag{A43}$$

As a consequence of the convergence of the series in (A41) and (A42) and of the symmetry relation, we recognize that the stable *pdf*'s with $1 < \alpha \leq 2$ are entire functions, whereas with $0 < \alpha < 1$ have the form:

$$L_\alpha^\theta(x) = \begin{cases} (1/x)\, \Phi_1(x^{-\alpha}) & \text{for } x > 0 \\ (1/|x|)\, \Phi_2(|x|^{-\alpha}) & \text{for } x < 0 \end{cases} \tag{A44}$$

where $\Phi_1(z)$ and $\Phi_2(z)$ are distinct entire functions. The case $\alpha = 1$ ($|\theta| < 1$) must be considered in the limit for $\alpha \to 1$ of (A41) and (A42) because the corresponding series reduce to power series akin with geometric series in $1/x$ and x, respectively, with a finite radius of convergence. The corresponding stable *pdf*'s are no longer represented by entire functions, as can be noted directly from their explicit expressions (A38) and (A39).

We omit to provide the asymptotic representations of the stable densities referring the interested reader to Mainardi et al. (2001) [47]. However, based on asymptotic representations, we can state as follows: for $0 < \alpha < 2$, the stable *pdf*'s exhibit *fat tails* in such a way that their absolute moment of order δ is finite only if $-1 < \delta < \alpha$. More precisely, one can show that, for non-Gaussian, not extremal, stable densities the asymptotic decay of the tails is

$$L_\alpha^\theta(x) = O\left(|x|^{-(\alpha+1)}\right) \quad x \to \pm\infty. \tag{A45}$$

For the extremal densities with $\alpha \neq 1$, this is valid only for one tail (as $|x| \to \infty$), the other (as $|x| \to \infty$) being of exponential order. For $1 < \alpha < 2$, the extremal *pdf*'s are two-sided and exhibit an exponential left tail (as $x \to -\infty$) if $\theta = +(2-\alpha)$ or an exponential right tail (as $x \to +\infty$) if $\theta = -(2-\alpha)$. Consequently, the Gaussian *pdf* is the unique stable density with finite variance. Furthermore, when $0 < \alpha \leq 1$, the first absolute moment is infinite so we should use the median instead of the non-existent expected value in order to characterize the corresponding *pdf*.

Let us also recall a relevant identity between stable densities with index α and $1/\alpha$ (a sort of reciprocity relation) pointed out in [73], that is, assuming $x > 0$,

$$\frac{1}{x^{\alpha+1}} L_{1/\alpha}^\theta(x^{-\alpha}) = L_\alpha^{\theta^*}(x) \quad 1/2 \leq \alpha \leq 1 \quad \theta^* = \alpha(\theta+1) - 1. \tag{A46}$$

The condition $1/2 \leq \alpha \leq 1$ implies $1 \leq 1/\alpha \leq 2$. A check shows that θ^* falls within the prescribed range $|\theta^*| \leq \alpha$ if $|\theta| \leq 2 - 1/\alpha$.

We leave as an exercise for the interested reader the verification of this reciprocity relation in the limiting cases $\alpha = 1/2$ and $\alpha = 1$.

References

1. Ragab, F.M. The inverse Laplace transform of an exponential function. *Comm. Pure Appl. Math.* **1958**, *11*, 115–127. [CrossRef]
2. Stefański, T.P.; Gulgowski, J. Signal propagation in electromagnetic media described by fractional-order models. *Comm. Nonlinear Sci. Numer. Simul.* **2020**, *82*, 105020. [CrossRef]
3. Garra, R.; Giraldi, F.; Mainardi, F. Wright type generalized coherent states. *WSEAS Trans. Math.* **2019**, *18*, 428–431.
4. Wright, E.M. On the coefficients of power series having exponential singularities. *J. Lond. Math. Soc.* **1933**, *8*, 71–79. [CrossRef]

5. Wright, E.M. The asymptotic expansion of the generalized Bessel function. *Proc. Lond. Math. Soc. (Ser. II)* **1935**, *38*, 257–270. [CrossRef]
6. Wright, E.M. The generalized Bessel function of order greater than one. *Quart. J. Math. Oxf. Ser.* **1940**, *11*, 36–48. [CrossRef]
7. Erdélyi, A.; Magnus, W.; Oberhettinger, F.; Tricomi, F. *Higher Transcendental Functions*, 3rd ed.; McGraw-Hill: New York, NY, USA, 1955.
8. Mainardi, F. *Fractional Calculus and Waves in Linear Viscoelasticity*, 2nd ed.; Imperial College Press: London, UK, 2010.
9. Wong, R.; Zhao, Y.-Q. Smoothing of Stokes' discontinuity for the generalized Bessel function. *Proc. R. Soc. Lond.* **1999**, *455*, 1381–1400. [CrossRef]
10. Wong, R.; Zhao, Y.-Q. Smoothing of Stokes' discontinuity for the generalized Bessel function II. *Proc. R. Soc. Lond. A* **1999**, *455*, 3065–3084. [CrossRef]
11. Luchko, Y. The Wright function and its applications. In *Handbook of Fractional Calculus with Applications*; Machado, J.A.T., Ed.; De Gruyter GmbH: Berlin, Germany; Boston, MA, USA, 2019; Volume 1: Basic, Theory, pp. 241–268.
12. Paris, R.B. Asymptotics of the special functions of fractional calculus. In *Handbook of Fractional Calculus with Applications*; Machado, J.A.T., Ed.; De Gruyter GmbH: Berlin, Germany; Boston, MA, USA, 2019; Volume 1: Basic, Theory, pp. 297–325.
13. Gorenflo, R.; Luchko, Y.u.; Mainardi, F. Analytical properties and applications of the Wright function. *Fract. Calc. Appl. Anal.* **1999**, *2*, 383–414.
14. Gorenflo, R.; Kilbas, A.A.; Mainardi, F.; Rogosin, S. *Mittag–Leffler Functions. Related Topics and Applications*; Springer: Berlin, Germany, 2014; 2nd Edition to appear.
15. Mainardi, F. On the initial value problem for the fractional diffusion-wave equation. In *Waves and Stability in Continuous Media*; Rionero, S., Ruggeri, T., Eds.; World Scientific: Singapore, 1994; pp. 246–251. In Proceedings of the VII-th WASCOM, International Conference "Waves and Stability in Continuous Media", Bologna, Italy, 4–7 October 1993.
16. Mainardi, F. The time fractional diffusion-wave-equation. *Radiophys. Quantum Electron.* **1995**, 20–36. (English translation from the Russian of *Radiofisika*) [CrossRef]
17. Mainardi, F. The fundamental solutions for the fractional diffusion-wave equation. *Appl. Math. Lett.* **1996**, *9*, 23–28. [CrossRef]
18. Mainardi, F. Fractional relaxation-oscillation and fractional diffusion-wave phenomena. *Chaos Solitons Fractals* **1996**, *7*, 1461–1477. [CrossRef]
19. Liemert, A.; Klenie, A. Fundamental solution of the tempered fractional diffusion equation. *J. Math. Phys.* **2015**, *56*, 113504. [CrossRef]
20. Saa, A.; Venegeroles, R. Alternative numerical computation of one-sided Lévy and Mittag–Leffler distributions. *Phys. Rev.* **2011**, *84*, 026702.
21. Kiryakova, V. The multi-index Mittag–Leffler functions as an important class of special functions of fractional calculus. *Comp. Math. Appl.* **2010**, *59*, 1885–1895. [CrossRef]
22. Kilbas, A.A.; Koroleva, A.A.; Rogosin, S.V. Multi-parametric Mittag–Leffler functions and their extension. *Fract. Calc. Appl. Anal.* **2013**, *16*, 378–404. [CrossRef]
23. Ansari, A.; Refahi Sheikhani, A. New identities for the Wright and the Mittag–Leffler functions using the Laplace transform. *Asian-European J. Math.* **2014**, *7*, 1450038. [CrossRef]
24. Mainardi, F. A tutorial survey on the basic special functions of fractional calculus. *WSEAS Trans. Math.* **2020**, *19*, 74–98. [CrossRef]
25. Luchko, Y. On the asymptotics of zeros of the Wright function. *Z. für Analysis und ihre Anwendungen (ZAMP)* **2000**, *19*, 597–622. [CrossRef]
26. Luchko, Y.; Kiryakova, V. The Mellin integral transform in fractional calculus. *Fract. Calc. Appl. Anal.* **2013**, *16*, 405–430. [CrossRef]
27. Bazhlekova, E. Subordination in a class of generalized time-fractional diffusion-wave equations. *Fract. Calc. Appl. Anal.* **2018**, *21*, 869–900. [CrossRef]
28. D'Ovidio, M. Wright functions governed by fractional directional derivatives and fractional advection diffusion equations. *Methods Appl. Anal.* **2015**, *22*, 1–36. [CrossRef]

29. Gorenflo, R.; Luchko, Y.u.; Mainardi, F. Wright functions as scale-invariant solution of the diffusion-wave equation. *J. Comp. Appl. Math.* **2000**, *118*, 175–191. [CrossRef]
30. Mentrelli, A.; Pagnini, G. Front propagation in anomalous diffusive media governed by time-fractional diffusion. *J. Comput. Phys.* **2015**, *293*, 427–441 [CrossRef]
31. Moslehi, A.; Ansari, A. On M-Wright transforms and time-fractional diffusion equations. *Integral Transform. Spec. Funct.* **2017**, *28*, 113–120. [CrossRef]
32. Pagnini, G. Erdélyi-Kober fractional diffusion. *Fract. Calc. Appl. Anal.* **2012**, *15*, 117–127. [CrossRef]
33. Povstenko, Y. *Linear Partial Differential Equations for Engineers*; Springer: Heidelberg, Germany, 2015.
34. Consiglio, A.; Mainardi, F. On the Evolution of Fractional Diffusive Waves. *Ric. Mat.* **2019**. [CrossRef]
35. Kochubei, A.N. Distributed order calculus and equations of ultraslow diffusion. *J. Math. Anal. Appl.* **2008**, *340*, 252–281. [CrossRef]
36. Li, Z.; Luchko, Y.; Yamamoto, M. Asymptotic estimates of solutions to initial-boundary-value problems for distributed order time-fractional diffusion equations. *Fract. Calc. Appl. Anal.* **2014**, *17*, 1114–1136. [CrossRef]
37. Mainardi, F.; Pagnini, G.; Gorenflo, R. Some aspects of fractional diffusion equations of single and distributed order. *Comput. Appl. Math.* **2007**, *187*, 295–305. [CrossRef]
38. Mainardi, F.; Mura, A.; Pagnini, G.; Gorenflo, R. Time-fractional diffusion of distributed order. *Vib. Control.* **2008**, *14*, 1267–1290. [CrossRef]
39. Luchko, Y. Multi-dimensional fractional wave equation and some properties of its fundamental solution. *Commun. Appl. Ind. Math.* **2014**, *6*, 1–21. [CrossRef]
40. Luchko, Y. On some new properties of the fundamental solution to the multi-dimensional space- and time-fractional diffusion-wave equation. *Mathematics* **2017**, *5*, 76. [CrossRef]
41. Luchko, Y.; Mainardi, F. Cauchy and signaling problems for the time-fractional diffusion-wave equation, *ASME J. Vib. Acoust.* **2014**, *136*, 050904/1-7. [CrossRef]
42. Luchko, Y.; Mainardi, F. Fractional diffusion-wave phenomena. In *Handbook of Fractional Calculus with Applications*; Tarasov, V., Machado, J.A.T., Eds.; De Gruyter GmbH: Berlin, Germany; Boston, MA, USA, 2019; Volume 5: Applications in physics, Part B, pp. 71–98.
43. Boyadjiev, L.; Luchko, Y. Mellin integral transform approach to analyze the multidimensional diffusion-wave equations. *Chaos Solitons Fractals* **2017**, *102*, 127–134. [CrossRef]
44. Hanyga, A. Multidimensional solutions of time-fractional diffusion-wave equations. *Soc. Lond Proc. Ser. Math. Phys. Eng. Sci.* **2002**, *458*, 933–957. [CrossRef]
45. Kemppainen, J. Positivity of the fundamental solution for fractional diffusion and wave wave equations. *Math. Meth. Appl. Sci.* **2019**, 1–19. [CrossRef]
46. Schneider, W.R.; Wyss, W. Fractional diffusion and wave equations. *Math. Phys.* **1989**, *30*, 134–144. [CrossRef]
47. Mainardi, F.; Luchko, Y.; Pagnini, G.The fundamental solution of the space-time fractional diffusion equation. *Fract. Calc. Appl. Anal.* **2001**, *4*, 153–192.
48. Mainardi, F.; Tomirotti, M. Seismic pulse propagation with constant Q and stable probability distributions. *Ann. Geofis.* **1997**, *40*, 1311–1328.
49. Mainardi, F.; Pagnini, G.; Gorenflo, R. Mellin transform and subordination laws in fractional diffusion processes. *Fract. Calc. Appl. Anal.* **2003**, *6*, 441–459.
50. Beghin, L.; Orsingher, E. Poisson-type processes governed by fractional and higher-order recursive differential equations. *Electron. J. Probab.* **2010**, *15*, 684–709. [CrossRef]
51. Cahoy, D.O. On the parametrization the M-Wright Function. *Far East J. Theor. Stat.* **2011**, *34*, 155–164
52. Cahoy, D.O. Estimation and Simulation for the M-Wright Function. *Commun. Stat. Methods* **2012**, *41*, 1466–1477.
53. Garra, R.; Orsingher, E.; Polito, F. Fractional diffusions with time-varying coefficients. *J. Math. Phys.* **2015**, *56*, 093301. [CrossRef]
54. Chen, L. Nonlinear stochastic time-fractional diffusion equation on $\mathbb{R}$: Moments, Holder regularity and Intermittency. *Trans. Am. Math. Soc.* **2017**, *369*, 8497–8535. [CrossRef]
55. Consiglio, A.; Luchko, Y.; Mainardi, F. Some notes on the Wright functions in probability theory. *WSEAS Trans. Math.* **2019**, *18*, 389–393.
56. Gorenflo, R.; Mainardi, F. Parametric subordination in fractional diffusion processes. In *Fractional Dynamics. Recent Advances*; Klafter, J., Lim, S.C., Metzler, R., Eds.; World Scientific: Singapore, 2012; Chapter 10, pp. 227–261.

57. Mainardi, F.; Mura, A.; Pagnini, G. The *M* Wright function in time-fractional diffusion processes: A tutorial survey. *Int. J. Diff. Eqs.* **2010**, 104505. [CrossRef]
58. Pagnini, G. The *M*-Wright function as a generalization of the Gaussian density for fractional diffusion processes. *Fract. Calc. Appl. Anal.* **2013**, *16*, 436–453. [CrossRef]
59. Scalas, E.; Viles, N. On the convergence of quadratic variation for compound fractional Poisson processes. *Fract. Calc. Appl. Anal.* **2012**, *15*, 31–331. [CrossRef]
60. Koren, T.; Lomholt, M.A.; Chechkin, A.V.; Klafter, J.; Metzler, R. Leapover lengths and first passage time statistics for Lévy flights. *Phys. Rev. Lett.* **2007**, *99*, 160602. [CrossRef] [PubMed]
61. Padash, A.; Chechkin, A.V.; Dybiec, B.; Pavlyukevich, I.; Shokri, B.; Metzler, R. First-passage properties of asymmetric Lévy flights. *J. Phys. Math. Theor.* **2019**, *52*, 454004. [CrossRef]
62. Stankovič, B. On the function of E.M. Wright. *Publ. de l'Institut Mathématique, Beograd, Nouvelle Sér.* **1970**, *10*, 113–124.
63. Mainardi, F. *The Linear Diffusion Equation*; Lecture Notes in Mathematical Physics; University of Bologna, Department of Physics: Bologna, Italy, 1996–2006; p. 19. [Available since 2019 at Brown University]; Available online: www.dam.brown.edu/fractional_calculus/documents/THELINEARDIFFUSIONEQUATION.pdf (accessed on 1 June 2020).
64. Gorenflo, R.; Mainardi, F. Fractional Calculus: Integral and Differential Equations of Fractional Order. In *Fractals and Fractional Calculus in Continuum Mechanics*; Carpinteri, A., Mainardi, F., Eds.; Springer: Wien, Österreich, 1997; pp. 223–276. [E-print arXiv:0805.3823]
65. Mainardi, F. Fractional Calculus: Some Basic Problems in Continuum and Statistical Mechanics. In *Fractals and Fractional Calculus in Continuum Mechanics*; Carpinteri, A., Mainardi, F., Eds.; Springer Verlag: Wien, Österreich, 1997; pp. 291–348. [E-print arXiv:1201.0863]
66. Diethelm, K. *The Analysis of Fractional Differential Equations*; An Application-Oriented Exposition Using Differential Operators of Caputo Type, Lecture Notes in Mathematics No 2004; Springer: Berlin, Germany, 2010.
67. Caputo, M. Linear models of dissipation whose *Q* is almost frequency independent, Part II. *Geophys. J. R. Astr. Soc.* **1967**, *13*, 529–539. [Reprinted in *Fract. Calc. Appl. Anal.* **2008**, *11*, 4–14]
68. Caputo, M. *Elasticità e Dissipazione*; Zanichelli: Bologna, 1969. (In Italian)
69. Caputo, M.; Mainardi, F. A new dissipation model based on memory mechanism. *Pure Appl. Geophys. (PAGEOPH)* **1971**, *91*, 134–147. [Reprinted in *Fract. Calc. Appl. Anal.* **2007**, *10*, 309–324]
70. Caputo, M.; Mainardi, F. Linear Models of Dissipation in Anelastic Solids. *Riv. Nuovo C.* **1971**, *1*, 161–198. [CrossRef]
71. Butzer, P.L.; Westphal, U. Introduction to Fractional Calculus. In *Fractional Calculus, Applications in Physics*; Hilfer, H., Ed.; World Scientific: Singapore, 2000; pp. 1–85.
72. Feller, W. On a Generalization of Marcel Riesz' Potentials and the Semi-Groups generated by Them. *Meddelanden Lunds Universitets Matematiska Seminarium*; Comm. Sém. Mathém. Université de Lund, Tome suppl. dédié à M: Riesz, Lund, 1952; pp. 73–81,
73. Feller, W. *An Introduction to Probability Theory and its Applications*; Wiley: New York, NY, USA, 1971; Volume II.
74. Gorenflo, R.; Mainardi, F. Random walk models for space-fractional diffusion processes. *Fract. Calc. Appl. Anal.* **1998**, *1*, 167–191.
75. Gawronski, W. On the bell-shape of stable distributions. *Ann. Probab.* **1984**, *12*, 230–242. [CrossRef]
76. Simon, T. Positive Stable Densities and the Bell-Shape. *Proc. Am. Math. Soc.* **2015**, *143*, 885–895. [CrossRef]
77. Kwaśnicki, M. A new class of bell-shaped functions. *Trans. Am. Math. Soc.* **2020**, *373*, 2255–2280. [CrossRef]

© 2020 by the authors. Licensee MDPI, Basel, Switzerland. This article is an open access article distributed under the terms and conditions of the Creative Commons Attribution (CC BY) license (http://creativecommons.org/licenses/by/4.0/).

Review

The Four-Parameters Wright Function of the Second kind and its Applications in FC

Yuri Luchko

Department of Mathematics, Physics, and Chemistry, Beuth Technical University of Applied Sciences Berlin, Luxemburger Str. 10, 13353 Berlin, Germany; luchko@beuth-hochschule.de

Received: 20 May 2020; Accepted: 10 June 2020; Published: 12 June 2020

Abstract: In this survey paper, we present both some basic properties of the four-parameters Wright function and its applications in Fractional Calculus. For applications in Fractional Calculus, the four-parameters Wright function of the second kind is especially important. In the paper, three case studies illustrating a wide spectrum of its applications are presented. The first case study deals with the scale-invariant solutions to a one-dimensional time-fractional diffusion-wave equation that can be represented in terms of the Wright function of the second kind and the four-parameters Wright function of the second kind. In the second case study, we consider a subordination formula for the solutions to a multi-dimensional space-time-fractional diffusion equation with different orders of the fractional derivatives. The kernel of the subordination integral is a special case of the four-parameters Wright function of the second kind. Finally, in the third case study, we shortly present an application of an operational calculus for a composed Erdélyi-Kober fractional operator for solving some initial-value problems for the fractional differential equations with the left- and right-hand sided Erdélyi-Kober fractional derivatives. In particular, we present an example with an explicit solution in terms of the four-parameters Wright function of the second kind.

Keywords: four-parameters Wright function of the second kind; one-dimensional time-fractional diffusion-wave equation; scale-invariant solutions; multi-dimensional space-time-fractional diffusion equation; subordination formula; left- and right-hand sided Erdélyi-Kober fractional derivatives

MSC: 26A33; 33E20; 30C15; 30D15; 45J05; 45K05; 44A20

---

1. Introduction

In calculus, differential equations, and mathematical physics both elementary and most of the special functions can be expressed in terms of the so-called generalized hypergeometric function $_pF_q$ that is defined as the following series (in the case it converges):

$$_pF_q\left(a_1,\ldots,a_p;b_1,\ldots,b_q;z\right) := \sum_{k=0}^{\infty} \frac{\prod_{n=1}^{p}(a_n)_k}{\prod_{n=1}^{q}(b_n)_k} \frac{z^k}{k!} \tag{1}$$

with the Pochhammer symbol $(z)_k$, $k \in \mathbb{N}$ given by the formula

$$(z)_k = \frac{\Gamma(z+k)}{\Gamma(k)} = \prod_{n=0}^{k-1}(z+n).$$

In particular, all elementary functions can be represented in terms of the famous hypergeometric Gauss function $_2F_1$. Other particular cases and properties of the generalized hypergeometric function can be found in [1].

If $p \leq q$, the series at the right-hand side of the formula (1) is absolutely convergent for all values of $z \in \mathbb{C}$. For $p = q+1$, the series converges for $|z| < 1$ and for $|z| = 1$ under some additional conditions. If $p > q+1$, the series is divergent.

To overcome this restriction and to somehow define the function $_pF_q$ in the case $p > q+1$, in [2] Meijer introduced a very general special function presently known in the literature as the G-function. For definition, properties, and particular cases of the G-function we refer the readers to [1].

However, it turned out that the special functions of Fractional Calculus (FC) belong in general neither to particular cases of the generalized hypergeometric function $_pF_q$ nor to particular cases of the Meijer G-function. They are particular cases of the more general generalized Wright or Fox-Wright functions or the Fox H-function ([1,3–7]).

The probably most used and important special functions of FC are the Mittag–Leffler function and its generalizations and the Wright function and its generalizations. For the theory of the Mitag-Leffler type functions and their applications we refer the readers to the book [3] and the recent survey [8] (see also numerous references therein). As to the Wright function and its generalizations, parts of their theory and some applications were presented in [5,6,9–29].

In this paper, the focus is on the four-parameters Wright function and its applications in FC. Depending on the signs of the parameters, we distinguish between the four-parameters Wright function of the first kind and of the second kind. The four-parameters Wright function of the first kind was first considered by Fox in [30] and by Wright in [28] (more precisely, this function was a particular case of the generalized Fox-Wright function that satisfies some conditions). In [12], the four-parameters Wright function of the first kind was employed as a kernel of an integral transform. It is the first application of this function known to the author. Another useful application of the four-parameters Wright function of the first kind was presented in [31], where the authors developed an operational calculus for an integral operator with the Gauss hypergeometric function as the kernel. This operational calculus was then used for derivation of the exact solutions of some integral equations of Volterra-type with the Gauss hypergeometric function in the kernel in terms of the four-parameters Wright function of the first kind.

As to the four-parameters Wright function of the second kind, it was first introduced in Luchko and Gorenflo [18]. Luchko and Gorenflo also provided some important properties of this function including its integral representation via the Mittag–Leffler function and its asymptotic behavior. Moreover, they applied the four-parameters Wright function of the second kind for derivation of the explicit analytical scale-invariant solutions to a one-dimensional space-time fractional diffusion equation. In this paper, some important results from [18] and the subsequent publications [5,15,32,33] will be revisited.

The rest of the paper is organized as follows: In the 2nd Section, we introduce the Wright function, the four-parameters Wright function, and the generalized Wright or Fox-Wright function and provide some of their important properties with the special focus on the four-parameters Wright function of the second kind. In the 3rd Section, three examples of applications of the four-parameters Wright function of the second kind in FC are presented. The first example deals with analysis of the scale-invariant solutions to a one-dimensional time-fractional diffusion-wave equation ([14,15]). It turns out that they can be represented in terms of the Wright function of the second kind and the four-parameters Wright function of the second kind. The second example is devoted to a subordination formula for the solutions to a multi-dimensional space-time-fractional diffusion equation with different orders of the fractional derivatives ([33]). The kernel of the subordination integral is a special case of the four-parameters Wright function of the second kind that is non-negative and can be interpreted as a probability density function. In the third example, we present an application of the operational method suggested in [32] for derivation of solution to an initial-value problem for a fractional differential equation with the left- and right-hand sided Erdélyi-Kober fractional derivatives in terms of the four-parameters Wright function of the second kind.

2. The Four-Parameters Wright Function

The generalized hypergeometric function $_p\Psi_q$ presently known as the generalized Wright or Fox-Wright function was introduced and investigated by Fox in [30] and by Wright in [28]. It is defined by the convergent series

$$_p\Psi_q \left[\begin{matrix} (a_1, A_1), \ldots, (a_p, A_p) \\ (b_1, B_1) \ldots (b_q, B_q) \end{matrix} ; z \right] := \sum_{k=0}^{\infty} \frac{\prod_{i=1}^{p} \Gamma(a_i + A_i k)}{\prod_{i=1}^{q} \Gamma(b_i + B_i k)} \frac{z^k}{k!}, \quad z \in \mathbb{C} \qquad (2)$$

with $a_i \in \mathbb{R}$, $A_i > 0$, $i = 1, \ldots, p$, $b_i \in \mathbb{R}$, $B_i > 0$, $i = 1, \ldots, q$. In the case $A_i = 1$, $i = 1, \ldots, p$, $B_i = 1$, $i = 1, \ldots, q$, the generalized Wright function coincides with the generalized hypergeometric function (1) up to a constant factor. Even more, in the case of the positive rational parameters $A_i \in \mathbb{Q}$, $i = 1, \ldots, p$, $B_i \in \mathbb{Q}$, $i = 1, \ldots, q$, the generalized Wright function can be represented as a final sum of the generalized hypergeometric functions with the power functions weights. Say, in the case $p = 0$, $q = 1$, and $B_1 = \frac{n}{m} \in \mathbb{Q}$, $n, m > 0$, we have the following representation ([14]):

$$_0\Psi_1 \left[\begin{matrix} - \\ (\beta, \frac{n}{m}) \end{matrix} ; z \right] = \sum_{p=0}^{m-1} \frac{z^p}{p! \Gamma(\beta + \frac{n}{m} p)} {}_0F_{n+m-1}\left(-; \Delta(n, \frac{\beta}{n} + \frac{p}{m}), \Delta^*(m, \frac{p+1}{m}); \frac{z^m}{m^m n^n} \right), \qquad (3)$$

where $\Delta(k, a)$ and $\Delta^*(k, a)$ are defined by

$$\Delta(k, a) = \{a, a + \frac{1}{k}, \ldots, a + \frac{k-1}{k}\}, \quad \Delta^*(k, a) = \Delta(k, a) \setminus \{1\}.$$

In the case of the formula (3), the set $\Delta^*(k, a)$ is correctly defined since 1 is an element of any set $\Delta(m, \frac{p+1}{m})$, $0 \le p \le m-1$. The method employed in [14] for derivation of the formula (3) can be also applied to obtain similar but of course even more complicated representations for the function $_p\Psi_q$ with the positive rational parameters $A_i \in \mathbb{Q}$, $i = 1, \ldots, p$, $B_i \in \mathbb{Q}$, $i = 1, \ldots, q$ in terms of the generalized hypergeometric function (1).

It is worth mentioning that both in [28,30], the parameters A_i and B_i were supposed to be positive real numbers. However, in [29], Wright considered a particular case of the function $_p\Psi_q$ with $p = 0$ and $q = 1$ and the coefficient B_1 being any real number greater than -1. Presently this function is called the Wright function. Following Wright, it is denoted by $\phi(\rho, \beta; z)$:

$$\phi(\rho, \beta; z) := {}_0\Psi_1 \left[\begin{matrix} - \\ (\beta, \rho) \end{matrix} ; z \right] = \sum_{k=0}^{\infty} \frac{z^k}{k! \Gamma(\beta + \rho k)}, \quad z \in \mathbb{C}, \rho > -1, \beta \in \mathbb{C}. \qquad (4)$$

For $\rho > -1$, the series at the right-hand side of the formula (4) is convergent for all $z \in \mathbb{C}$. It is also convergent for $\rho = -1$ and $|z| < 1$ and for $\rho = -1$ and $|z| = 1$ under the condition $\Re(\beta) > 1$. However, the Wright function is an entire function only in the case $\rho > -1$ and thus this condition is usually included into its definition.

In [3,19], the function (4) with the positive parameter ρ was called the Wright function of the first kind, whereas in the case of the negative parameter ρ ($0 > \rho > -1$) it was called the Wright function of the second kind. In the case $\rho = 0$, the Wright function is reduced to the exponential function:

$$\phi(0, \beta; z) = \sum_{k=0}^{\infty} \frac{z^k}{k! \Gamma(\beta)} = \frac{e^z}{\Gamma(\beta)}. \qquad (5)$$

In analogy to the Wright function (4), the generalized Wright function (2) can be considered also in the case, some or even all of the parameters A_i and B_i are negative numbers. The well-known asymptotic behavior of the Euler Gamma-function allows determination of the convergence radius of the series at the right-hand side of (2): it is absolutely convergent for all $z \in \mathbb{C}$ under the condition

$\Delta > -1$, where Δ is determined by the parameters of the generalized Wright function as follows (see, e.g., [3,16,34]):

$$\Delta = \sum_{i=1}^{q} B_i - \sum_{i=1}^{p} A_i, \ \delta = \prod_{i=1}^{p} |A_i|^{-A_i} \prod_{i=1}^{q} |B_i|^{B_i}, \ \mu = \sum_{i=1}^{q} b_i - \sum_{i=1}^{p} a_i + \frac{p-q}{2}. \tag{6}$$

In the case $\Delta > -1$, the function (2) is an entire function. However, in the case $\Delta = -1$, the series at the right-hand side of (2) is also absolutely convergent for $|z| < \delta$ and for $|z| = \delta$ under the condition $\Re(\mu) > 1/2$ (see [34] for details).

In this paper, we mainly deal with another important particular case of the generalized Wright function (2), specifically with the so-called four-parameters Wright function:

$$W_{(\rho_1,\beta_1),(\rho_2,\beta_2)}(z) := {}_1\Psi_2 \left[\begin{matrix} (1,1) \\ (\beta_1,\rho_1) \ (\beta_2,\rho_2) \end{matrix} ; z \right]. \tag{7}$$

According to the definition of the generalized Wright function, the series representation of the four-parameters Wright function is as follows:

$$W_{(\rho_1,\beta_1),(\rho_2,\beta_2)}(z) = \sum_{k=0}^{\infty} \frac{z^k}{\Gamma(\beta_1+\rho_1 k)\Gamma(\beta_2+\rho_2 k)}, \ \rho_1, \rho_2 \in \mathbb{R}, \ \beta_1, \beta_2 \in \mathbb{C}, \ z \in \mathbb{C}. \tag{8}$$

For $\rho_1 + \rho_2 > 0$, the series at the right-hand side of (8) is absolutely convergent $\forall z \in \mathbb{C}$. For $\rho_1 + \rho_2 = 0$, the series is absolutely convergent for $|z| < 1$ and for $|z| = 1$ under the condition $\Re(\beta_1 + \beta_2) > 2$. Finally, the series is divergent for any $z \neq 0$ in the case $\rho_1 + \rho_2 < 0$.

Without any loss of generality, in what follows we always suppose that the condition $\rho_1 \geq \rho_2$ holds true in the definition of the four-parameters Wright function. This assumption will lead to simpler formulations of some results concerning the four-parameters Wright function. Moreover, we will distinguish between the four-parameters Wright function of the first kind ($\rho_2 > 0$) and of the second kind ($\rho_2 < 0$). The properties and applications of the four-parameters Wright function of the second kind are very different from those of the function of the first kind. Thus, we found it appropriate to introduce a separate notation for the four-parameters Wright function of the second kind:

$$\Phi_{(\rho_1,\beta_1),(\rho_2,\beta_2)}(z) := W_{(\rho_1,\beta_1),(\rho_2,\beta_2)}(z), \ \rho_2 < 0. \tag{9}$$

The notation $W_{(\rho_1,\beta_1),(\rho_2,\beta_2)}$ is kept for the four-parameters Wright function (including the cases of the functions of the first and of the second kinds).

In what follows, we always suppose that the condition $\rho_1 + \rho_2 > 0$ is satisfied. This condition along with the inequality $\rho_1 \geq \rho_2$ leads to the inequality $\rho_1 > 0$. Thus, the parameter ρ_1 of the four-parameters Wright function is always positive, whereas the parameter ρ_2 is positive in the case of the function of the first kind and negative in the case of the function of the second kind. In the case $\rho_2 = 0$, the four-parameters Wright function is reduced to the two-parameters Mittag–Leffler function:

$$W_{(\rho_1,\beta_1),(0,\beta_2)}(z) = \frac{1}{\Gamma(\beta_2)} E_{\rho_1,\beta_1}(z) = \frac{1}{\Gamma(\beta_2)} \sum_{k=0}^{\infty} \frac{z^k}{\Gamma(\beta_1+\rho_1 k)}. \tag{10}$$

For the theory and applications of the two-parameters Mittag–Leffler function we refer to the book [3]; in this paper we do not consider this function. Please note that in [3] the function (7) is called the generalized Mittag–Leffler function or the four-parametric Mittag–Leffler function.

Another important particular case of the four-parameters Wright function (7) is the Wright function (4):

$$W_{(1,1),(\rho,\beta)}(z) = \phi(\rho,\beta;z). \tag{11}$$

For the properties and applications of the Wright function we refer to the recent survey [5], see also the references therein.

As already mentioned, the four-parameters Wright function is an entire function provided the condition $\rho_1 + \rho_1 > 0$ holds true.

Theorem 1. *Let the condition $\rho_1 + \rho_1 > 0$ be satisfied. Then the four-parameters Wright function is an entire function of the variable z. Its order p and type σ are given by the relations*

$$p = \frac{1}{\rho_1 + \rho_2}, \quad \sigma = \frac{\rho_1 + \rho_2}{\left(\rho_1^{\rho_1} |\rho_2|^{\rho_2}\right)^{\frac{1}{\rho_1+\rho_2}}}. \tag{12}$$

The proof of the theorem is based on the Stirling formula for the asymptotic of the Gamma-function and can be found in [3,12,16].

Since it is a function of the hypergeometric type, the four-parameters Wright function possesses a very useful Mellin–Barnes integral representation ([35]):

$$W_{(\rho_1,\beta_1),(\rho_2,\beta_2)}(z) = \frac{1}{2\pi i} \int_{L_{-\infty}} \frac{\Gamma(s)\Gamma(1-s)}{\Gamma(\beta_1 - \rho_1 s)\Gamma(\beta_2 - \rho_2 s)} (-z)^{-s} ds, \tag{13}$$

where $L_{-\infty}$ is a left loop located in a horizontal strip. It goes from the point $-\infty + iy_1$ to the point $-\infty + iy_2$ with $y_1 < 0 < y_2$ and separates the poles of the Gamma-function $\Gamma(s)$ (the points $s_k = 0, -1, -2, \ldots$) from the poles of the Gamma-function $\Gamma(1-s)$ (the points $s_l = 1, 2, 3, \ldots$).

The formula (13) can be easily proved by evaluating the Mellin–Barnes integral taking into account the Jordan lemma, the formula

$$\operatorname{res}_{s=-k} \Gamma(s) = \frac{(-1)^k}{k!}, \quad k = 0, 1, 2, \ldots, \tag{14}$$

the known asymptotic of the Gamma-function, and the Cauchy residue theorem.

Depending on the sign of the parameter ρ_2 (the parameter ρ_1 is always positive), the right-hand side of the representation (13) can be interpreted as the Fox H-function:

$$W_{(\rho_1,\beta_1),(\rho_2,\beta_2)}(z) = H_{1,3}^{1,1} \left(\begin{array}{c} (0,1) \\ (0,1), (1-\beta_1, \rho_1), (1-\beta_2, \rho_2) \end{array} \bigg| -z \right), \quad \rho_2 > 0, \tag{15}$$

$$\Phi_{(\rho_1,\beta_1),(\rho_2,\beta_2)}(z) = H_{2,2}^{1,1} \left(\begin{array}{c} (0,1), (\beta_2, -\rho_2) \\ (0,1), (1-\beta_1, \rho_1) \end{array} \bigg| -z \right), \quad \rho_2 < 0. \tag{16}$$

It is worth mentioning that both the Mellin–Barnes integral representation (13) and the Fox H-function representations (15) and (16) can be used for derivation of several useful properties of the four-parameters Wright function including its particular cases for the rational values of the parameters ([1,3,5]) or its asymptotic behavior ([6,7]).

Because the focus of this paper is on applications of the four-parameters Wright function of the second kind in FC, in the rest of this section we mainly restrict ourselves to a short discussion of its important properties. For the proofs, we refer the interested readers to [18].

A very useful integral representation of the four-parameters Wright function is given in the following theorem:

Theorem 2 ([18]). *The four-parameters Wright function possesses the following integral representation in terms of the two parameters Mittag–Leffler function (10):*

$$W_{(\rho_1,\beta_1),(\rho_2,\beta_2)}(z) = \frac{1}{2\pi i} \int_{\gamma(\varepsilon;\varphi)} e^{\zeta} \zeta^{-\beta_2} E_{\rho_1,\beta_1}(z\zeta^{-\rho_2}) d\zeta, \tag{17}$$

where $\gamma(\varepsilon;\varphi)$ ($\varepsilon > 0$, $\frac{\pi}{2} < \varphi \leq \pi$) is a contour in the complex plane with the nondecreasing $\arg \zeta$ that consists of the ray $\arg \zeta = -\varphi$, $|\zeta| \geq \varepsilon$, the arc $-\varphi \leq \arg \zeta \leq \varphi$ of the circle $|\zeta| = \varepsilon$, and the ray $\arg \zeta = \varphi$, $|\zeta| \geq \varepsilon$.

In the case of the four-parameters Wright function of the second kind, the integration contour $\gamma(\varepsilon;\varphi)$ in Theorem 2 can be replaced by a simpler one:

Theorem 3 ([18]). *For any $k_0 \in \mathbb{N}$ satisfying the condition $k_0 > \max\{-1, \Re((1-\beta_2)/(-\rho_2))\}$, the four-parameters Wright function of the second kind can be represented as follows:*

$$\Phi_{(\rho_1,\beta_1),(\rho_2,\beta_2)}(z) = \sum_{k=0}^{k_0} \frac{z^k}{\Gamma(\beta_1 + \rho_1 k)\Gamma(\beta_2 + \rho_2 k)} + \qquad (18)$$

$$\frac{1}{2\pi i} \int_{L_-} \left(e^{\zeta} \zeta^{-\beta_2} E_{\rho_1,\beta_1}(z\zeta^{-\rho_2}) - \sum_{k=0}^{k_0} \frac{(z\zeta^{-\rho_2})^k}{\Gamma(\beta_1 + \rho_1 k)} \right) d\zeta,$$

where L_- is a cut in the complex ζ-plane along the negative real semi-axis.

Remark 1. *As already mentioned, for $\beta_1 = \rho_1 = 1$, the four-parameters Wright function is reduced to the Wright function (4) and the integral representation (17) with $\rho_2 = \rho > -1$ and $\beta_2 = \beta \in \mathbb{R}$ takes the well-known form*

$$\phi(\rho,\beta;z) = \frac{1}{2\pi i} \int_{\gamma(\varepsilon;\varphi)} \exp\{\zeta + z\zeta^{-\rho}\}\zeta^{-\beta} d\zeta. \qquad (19)$$

This integral representation was obtained by Wright in [27,29] and then used for derivation of the asymptotic behavior of the Wright function. In particular, he showed that the Wright function of the second kind has an algebraic asymptotic expansion on the positive real semi-axis provided the condition $1/3 < -\rho < 1$ holds true ($K = 0, 1, 2, \ldots$):

$$\phi(\rho,\beta;x) = \sum_{k=0}^{K-1} \frac{x^{(\beta-1-k)/(-\rho)}}{(-\rho)\Gamma(k+1)\Gamma(1+(\beta-1-k)/(-\rho))} + O(x^{(\beta-1-K)/(-\rho)}), \quad x \to +\infty. \qquad (20)$$

For the four-parameters Wright function of the second kind, a similar result was obtained in [18].

Theorem 4 ([18]). *Under the condition $\rho_1/3 < -\rho_2 < \rho_1 \leq 2$, the four-parameters Wright function of the second kind has the following asymptotic on the positive real semi-axis:*

$$\Phi_{(\rho_1,\beta_1),(\rho_2,\beta_2)}(x) = \sum_{k=0}^{K-1} \frac{x^{(\beta_2-1-k)/(-\rho_2)}}{(-\rho_2)\Gamma(k+1)\Gamma(\beta_1 + \rho_1(\beta_2-1-k)/(-\rho_2))} - \qquad (21)$$

$$\sum_{p=1}^{P} \frac{x^{-p}}{\Gamma(\beta_1 - \rho_1 p)\Gamma(\beta_2 - \rho_2 p)} + O(x^{(\beta_2-1-K)/(-\rho_2)}) + O(x^{-1-P}), \quad x \to +\infty$$

for any $K = 0, 1, 2, \ldots,$ and $P = 0, 1, 2, \ldots$.

For geometric properties of the four-parameters Wright function we refer the interested readers to the very recent paper [11].

3. Applications of the Four-Parameters Wright Function of the Second Kind

In this section, we consider three examples of applications of the four-parameters Wright function of the second kind in FC.

The first example concerns the well-studied one-dimensional time-fractional diffusion-wave equation with the Caputo derivative. For analytical treatment of this equation, the Wright functions of the second kind play a fundamental role ([10,14,15,19,36]). Say, the fundamental solution to this equation can be expressed in terms of some special cases of the Wright function of the second kind (so-called Mainardi auxiliary functions). However, it turns out that the formulas for the scale-invariant solutions to the one-dimensional diffusion-wave equation involve both the Wright function of the second kind and the four-parameters Wright function of the second kind.

In the second example, we deal with a subordination formula for solutions to a multi-dimensional space-time-fractional diffusion equation ([33]). This equation is obtained from the diffusion equation by replacing the first order time derivative by the Caputo fractional derivative and the Laplace operator by the fractional Laplacian. This time, it is the four-parameters Wright function of the second kind that is of importance for this equation. In particular, a special case of the four-parameters Wright function of the second kind appears in the kernel of a subordination formula that connects the solution operators of this equation with different orders of the fractional derivatives to the classical solution of the conventional diffusion equation. Moreover, this kernel function is non-negative and can be interpreted as a probability density function.

The third example deals with the ordinary fractional differential equations that contain both the left- and the right-hand sided fractional derivatives. In [32], an operational method for the so-called composed Erdélyi-Kober fractional derivatives was suggested and applied for derivation of the analytical solutions to the initial-value problems for a special class of such equations. In this section, we present an equation of this sort with an explicit solution expressed in terms of the four-parameters Wright function of the second kind.

3.1. Scale-Invariant Solutions to the One-Dimensional Time-Fractional Diffusion-Wave Equation

In this subsection, we deal with the fractional diffusion-wave equation, which is obtained from the conventional diffusion or wave equation by replacing the first- or second-order time derivative, respectively, by the Caputo fractional derivative:

$$\frac{\partial^\alpha u(x,t)}{\partial t^\alpha} = \frac{\partial^2 u(x,t)}{\partial x^2}, \quad 1 < \alpha < 2, \ t > 0, \ x > 0. \tag{22}$$

The Caputo fractional derivative of order α, $1 < \alpha < 2$, is defined as follows:

$$\frac{\partial^\alpha u(x,t)}{\partial t^\alpha} = \frac{1}{\Gamma(2-\alpha)} \int_0^t (t-\tau)^{1-\alpha} \frac{\partial^2 u(x,\tau)}{\partial \tau^2} \, d\tau. \tag{23}$$

In particular, we are interested in the scale-invariant solutions to this equation. First, we introduce some basic notions concerning the similarity method for the general equation

$$F(u) = 0, \quad u = u(x,t). \tag{24}$$

A one-parameter family of scaling transformations, denoted by T_λ, is called a transformation of the (x,t,u)-space of the form

$$\bar{x} = \lambda^a x, \quad \bar{t} = \lambda^b t, \quad \bar{u} = \lambda^c u, \tag{25}$$

where a, b, and c are some constants and λ is a real parameter restricted to an open interval I containing the value $\lambda = 1$.

The general Equation (24) is called invariant under the one-parameter family T_λ of scaling transformations (25) if and only if T_λ translates any solution u of (24) to a solution $\bar{u}$ of the same equation:

$$F(\bar{u}) = 0 \quad \text{if} \quad \bar{u} = T_\lambda u. \tag{26}$$

A real-valued function $\eta(x,t,u)$ is called an invariant of the one-parameter family T_λ of scaling transformations if it is unaffected by the transformations from T_λ:

$$\eta(T_\lambda(x,t,u)) = \eta(x,t,u) \quad \text{for all} \quad \lambda \in I.$$

The general theory ([37]) says that on the half-space $\{(x,t,u) : x > 0,\ t > 0\}$, the invariants of the scaling transformations (25) are provided by the functions

$$\eta_1(x,t,u) = xt^{-a/b},\ \eta_2(x,t,u) = t^{-c/b}u. \tag{27}$$

Say, let the Equation (24) be a second-order partial differential equation

$$G(x,\ t,\ u,\ u_x,\ u_t,\ u_{xx},\ u_{tt},\ u_{xt}) = 0. \tag{28}$$

If this equation is invariant under the family T_λ of scaling transformations (25), then the substitution

$$u(x,t) = t^{c/b}v(z),\quad z = xt^{-a/b} \tag{29}$$

reduces the Equation (28) to a second-order ordinary differential equation

$$g(z,\ v,\ v',\ v'') = 0. \tag{30}$$

In [9,14,15,18], the scale-invariant solutions for the equation of type (22) with the fractional derivatives in the Caputo and Riemann–Liouville sense and for the more general time- and space-fractional partial differential equations were obtained. In all cases, these solutions were expressed in terms of the Wright function of the second kind and the four-parameters Wright function of the second kind. In what follows, we present some of these results for the Equation (22).

The group of scaling transformations for the fractional diffusion-wave Equation (22) can be determined in explicit form.

Theorem 5 ([9]). *The group of scaling transformations of the Equation (22) has the form*

$$T_\lambda \circ (x,t,u) = (\lambda x,\ \lambda^{\frac{2}{\alpha}}t,\ \lambda^c u)$$

with an arbitrary constant $c \in \mathbb{R}$ and its invariants are given by the formulas

$$\eta_1(x,t) = xt^{-\alpha/2},\ \eta_2(x,t,u) = t^{-c\alpha/2}u. \tag{31}$$

In what follows, for the sake of convenience, we use the notation $\gamma = c\alpha/2$.

The general theory of the Lie groups ([37]) and Theorem 5 ensure that the scale-invariant solutions of the Equation (22) have the form

$$u(x,t) = t^\gamma v(y),\ y = xt^{-\alpha/2},\ \gamma = c\alpha/2. \tag{32}$$

Substitution of the function u from the formula (32) into the partial fractional differential Equation (22) transforms it into an ordinary fractional differential equation with an unknown function $v(y)$. More precisely, the following result holds true:

Theorem 6 ([9]). *The scale-invariant solutions of the Equation (22) in the form (32) satisfy the equation*

$$(_*P_{2/\alpha}^{\gamma-1,\alpha}v)(y) = v''(y),\ y > 0, \tag{33}$$

where the operator $_*P_{2/\alpha}^{\gamma-1,\alpha}$ is the Caputo type modification of the right-hand sided Erdélyi-Kober fractional derivative defined by

$$(_*P_\delta^{\tau,\alpha}g)(y) := (K_\delta^{\tau,n-\alpha}\prod_{j=0}^{n-1}(\tau+j-\frac{1}{\delta}u\frac{d}{du})g)(y), \ y > 0, \ \delta > 0, \ n-1 < \alpha \leq n \in \mathbb{N}. \quad (34)$$

The operator $K_\delta^{\tau,\alpha}$, $\alpha > 0$ is the right-hand sided Erdélyi-Kober fractional integral defined by

$$(K_\delta^{\tau,\alpha}g)(y) := \frac{1}{\Gamma(\alpha)}\int_1^\infty (u-1)^{\alpha-1}u^{-(\tau+\alpha)}g(yu^{1/\delta})\,du. \quad (35)$$

For $\alpha = 1$ and $\alpha = 2$, the fractional diffusion-wave Equation (22) is reduced to the conventional one-dimensional diffusion or wave equation, respectively. The Equation (33) for the scale-invariant solutions of (22) is an ordinary differential equation, not a fractional one. In the case $\alpha = 1$ (the diffusion equation) we have the representation

$$(_*P_2^{\gamma,1}v)(y) = (\gamma - \frac{1}{2}y\frac{d}{dy})v(y)$$

and the Equation (33) takes the well-known form

$$v''(z) + \frac{1}{2}yv'(y) - \gamma v(y) = 0.$$

In the case $\alpha = 2$ (the wave equation) we get the formula

$$(_*P_1^{\gamma-1,2}v)(y) = (\gamma - 1 - y\frac{d}{dy})(\gamma - y\frac{d}{dy})v(y) = y^2 v''(y) - 2(\gamma-1)yv'(y) + \gamma(\gamma-1)v(y)$$

and the equation (33) is transformed to the following ODE:

$$(y^2-1)v''(y) - 2(\gamma-1)yv'(y) + \gamma(\gamma-1)v(y) = 0.$$

These both cases are discussed in detail in [37].

It turns out that the Equation (34) can be solved in explicit form in terms of the Wright function of the second kind and the four-parameters Wright function of the second kind.

Theorem 7 ([14]). *The scale-invariant solutions of the fractional diffusion-wave Equation (22) are given by the formulas*

$$u(x,t) = C_1 t^\gamma \phi(-\frac{\alpha}{2}, 1+\gamma; -y) + C_2 t^\gamma \left(\frac{1}{2}\phi(-\frac{\alpha}{2}, 1+\gamma; y) - y^{2+2\frac{\gamma-1}{\alpha}}\Phi_{(2,3+2\frac{\gamma-1}{\alpha}),(-\alpha,2-\alpha)}(y^2)\right) \quad (36)$$

in the case $1 - \alpha < \gamma < 1$, $\gamma \neq 1 - \frac{\alpha}{2}$, $\gamma \neq 0$, and

$$u(x,t) = C_1\phi(-\frac{\alpha}{2}, 1; -y) + C_2\left(\frac{1}{2}\phi(-\frac{\alpha}{2}, 1; y) - y^{2-\frac{2}{\alpha}}\Phi_{(2,3-\frac{2}{\alpha}),(-\alpha,2-\alpha)}(y^2)\right) + C_3 \quad (37)$$

in the case $\gamma = 0$, where $y = xt^{-\frac{\alpha}{2}}$ is the first scale-invariant (31), ϕ is the Wright function of the second kind defined by (4), Φ is the four-parameters Wright function of the second kind defined by (7), and C_1, C_2, C_3 are arbitrary constants.

For further results regarding the scale-invariant solutions to the fractional diffusion-wave equations we refer to [9,14,15,18].

3.2. Subordination Formula for the Multi-Dimensional Space-Time-Fractional Diffusion Equations

The object of analysis in this subsection is the multi-dimensional space-time-fractional diffusion equation

$$D_t^\beta u(x,t) = -(-\Delta)^{\frac{\alpha}{2}} u(x,t), \quad x \in \mathbb{R}^n, \ t > 0, \ 0 < \alpha \le 2, \ 0 < \beta \le 1. \tag{38}$$

In the Equation (38), the time-fractional derivative D_t^β is defined in the Caputo sense:

$$D_t^\beta u(x,t) = \left(I_t^{n-\beta} \frac{\partial^n u}{\partial t^n} \right)(t), \quad n-1 < \beta \le n, \ n \in \mathbb{N} \tag{39}$$

with I_t^γ being the Riemann–Liouville fractional integral:

$$(I_t^\gamma u)(t) = \begin{cases} \frac{1}{\Gamma(\gamma)} \int_0^t (t-\tau)^{\gamma-1} u(x,\tau)\, d\tau & \text{for } \gamma > 0, \\ u(x,t) & \text{for } \gamma = 0. \end{cases}$$

The fractional Laplacian $-(-\Delta)^{\frac{\alpha}{2}}$ is understood as a pseudo-differential operator with the symbol $-|\kappa|^\alpha$ ([38,39]):

$$\left(\mathcal{F} - (-\Delta)^{\frac{\alpha}{2}} u \right)(\kappa) = -|\kappa|^\alpha (\mathcal{F} u)(\kappa), \tag{40}$$

where $(\mathcal{F} f)(\kappa)$ is the Fourier transform of a function u at the point $\kappa \in \mathbb{R}^n$ defined by

$$(\mathcal{F} u)(\kappa) = \hat{f}(\kappa) = \int_{\mathbb{R}^n} e^{i\kappa \cdot x} u(x)\, dx. \tag{41}$$

The fractional Laplacian can be also represented as a hypersingular integral ([39]):

$$-(-\Delta)^{\frac{\alpha}{2}} u(x) = -\frac{1}{d_{n,m}(\alpha)} \int_{\mathbb{R}^n} \frac{(\Delta_h^m u)(x)}{|h|^{n+\alpha}} dh, \quad 0 < \alpha < m, \ m \in \mathbb{N}, \ x \in \mathbb{R}^n \tag{42}$$

with a suitably defined finite differences operator $(\Delta_h^m f)(x)$ and a normalization constant $d_{n,m}(\alpha)$.

The representation (42) of the fractional Laplacian in form of the hypersingular integral does not depend on m, $m \in \mathbb{N}$ provided $\alpha < m$ ([39]). For other representations of the fractional Laplacian we refer the reader to [40].

In what follows, we consider the Cauchy problem for the space-time-fractional diffusion Equation (38) with the Dirichlet initial condition:

$$u(x,0) = f(x), \quad x \in \mathbb{R}^n. \tag{43}$$

Because the initial-value problem (38), (43) is linear, its solution can be represented in the form

$$u(x,t) = \int_{\mathbb{R}^n} G_{\alpha,\beta,n}(\zeta,t) f(x-\zeta)\, d\zeta. \tag{44}$$

In (44), the function f is the initial condition and $G_{\alpha,\beta,n}$ is the first fundamental solution of (38), i.e., its solution with the initial condition

$$u(x,0) = \prod_{i=1}^n \delta(x_i), \quad x = (x_1, x_2, \ldots, x_n) \in \mathbb{R}^n$$

where δ is the Dirac delta function.

In the case of the conventional diffusion equation ($\alpha = 2$ and $\beta = 1$ in the Equation (38)), the fundamental solution is well-known:

$$G_{2,1,n}(x,t) = \frac{1}{(\sqrt{4\pi t})^n} \exp\left(-\frac{|x|^2}{4t}\right). \tag{45}$$

It turned out that the fundamental solution $G_{\alpha,\beta,n}$ to the multi-dimensional space-time-fractional diffusion Equation (38) can be represented in terms of the fundamental solution $G_{2,1,n}$ of the conventional diffusion equation. The result obtained in [33] for the first time is given in the following theorem:

Theorem 8 ([33]). *For the fundamental solution $G_{\alpha,\beta,n}(x,t)$ to the multi-dimensional space-time-fractional diffusion-wave Equation (38) with $0 < \beta \leq 1, 0 < \alpha \leq 2$, and $2\beta + \alpha < 4$ the following subordination formula is valid:*

$$G_{\alpha,\beta,n}(x,t) = \int_0^\infty t^{-\frac{2\beta}{\alpha}} \Psi_{\alpha,\beta}(st^{-\frac{2\beta}{\alpha}}) G_{2,1,n}(x,s) \, ds, \tag{46}$$

where the fundamental solution $G_{2,1,n}(x,s)$ to the conventional diffusion equation is given by the formula (45) and the kernel function $\Psi_{\alpha,\beta}$ is a probability density function in s, $s \in \mathbb{R}_+$ for each value of t, $t > 0$ defined as follows:

$$\Psi_{\alpha,\beta}(\tau) = \begin{cases} \tau^{\frac{\alpha}{2}-1} \Phi_{(\frac{\alpha}{2},\frac{\alpha}{2}),(-\beta,1-\beta)}\left(-\tau^{\frac{\alpha}{2}}\right) & \text{if } \frac{\beta}{\alpha} < \frac{1}{2}, \\[6pt] -\tau^{-1-\frac{\alpha}{2}} \Phi_{(\beta,1-\beta),(-\frac{\alpha}{2},-\frac{\alpha}{2})}\left(-\tau^{-\frac{\alpha}{2}}\right) & \text{if } \frac{\beta}{\alpha} > \frac{1}{2}, \\[6pt] \begin{cases} \frac{\tau^{\frac{\alpha}{2}-1}}{\pi} \sum_{k=0}^\infty \sin\left(\frac{\pi\alpha}{2}(k+1)\right)\left(-\tau^{\frac{\alpha}{2}}\right)^k & \text{if } 0 < \tau < 1 \\[6pt] -\frac{\tau^{-1}}{\pi} \sum_{k=0}^\infty \sin\left(\frac{\pi\alpha}{2}k\right)\left(-\tau^{-\frac{\alpha}{2}}\right)^k & \text{if } \tau > 1 \end{cases} & \text{if } \frac{\beta}{\alpha} = \frac{1}{2}. \end{cases} \tag{47}$$

In the formula (47), the function Φ is the four-parameters Wright function of the second kind defined by (9).

It is worth mentioning that even if the subordination formula (46) concerns just the fundamental solution, it can be extended to the solution operator for the initial-value problem (38), (43). Indeed, let us suppose that a more general subordination formula for the fundamental solution $G_{\alpha,\beta,n}$ is valid:

$$G_{\alpha,\beta,n}(x,t) = \int_0^\infty \Psi(\alpha,\beta,s,t) G_{\hat{\alpha},\hat{\beta},n}(x,s) \, ds, \tag{48}$$

where the kernel function $\Psi = \Psi(\alpha,\beta,s,t)$ can be interpreted as a probability density function in s, $s \in \mathbb{R}_+$ for each value of t, $t > 0$ (the formula (47) is a particular case of the formula (48)). Then we have the following chain of relations:

$$S_{\alpha,\beta,n}(t) f = \int_{\mathbb{R}^n} G_{\alpha,\beta,n}(\zeta,t) f(x-\zeta) \, d\zeta = \int_{\mathbb{R}^n} \int_0^\infty \Psi(\alpha,\beta,s,t) G_{\hat{\alpha},\hat{\beta},n}(\zeta,s) \, ds \, f(x-\zeta) \, d\zeta =$$

$$\int_0^\infty \Psi(\alpha,\beta,s,t) \int_{\mathbb{R}^n} G_{\hat{\alpha},\hat{\beta},n}(\zeta,s) f(x-\zeta) \, d\zeta \, ds = \int_0^\infty \Psi(\alpha,\beta,s,t) S_{\hat{\alpha},\hat{\beta},n}(s) f \, ds.$$

Thus, the subordination formula

$$S_{\alpha,\beta,n}(t) f = \int_0^\infty \Psi(\alpha,\beta,s,t) S_{\hat{\alpha},\hat{\beta},n}(s) f \, ds \tag{49}$$

holds true for the solution operator $S_{\alpha,\beta,n}$. Vice versa, any subordination formula for the solution operator $S_{\alpha,\beta,n}$ to the initial-value problem (38), (43) in the form (49) induces a subordination formula of the type (48) for the fundamental solution $G_{\alpha,\beta,n}$ just by setting f to be the Dirac δ-function.

In the rest of this subsection, we provide some important remarks concerning the kernel $\Psi_{\alpha,\beta}$ of the subordination formula (46).

In [33], the kernel function $\Psi_{\alpha,\beta}$ given by the formula (47) was first deduced in form of the following Mellin–Barnes integral:

$$\Psi_{\alpha,\beta}(\tau) = \frac{2}{\alpha} \frac{1}{2\pi i} \int_{\gamma-i\infty}^{\gamma+i\infty} \frac{\Gamma\left(\frac{2}{\alpha} - \frac{2}{\alpha}s\right) \Gamma\left(1 - \frac{2}{\alpha} + \frac{2}{\alpha}s\right)}{\Gamma\left(1 - \frac{2\beta}{\alpha} + \frac{2\beta}{\alpha}s\right) \Gamma(1-s)} \tau^{-s} ds. \tag{50}$$

The series representation (47) was derived by evaluating the Mellin–Barnes integral (50) taking into account the Jordan lemma, the formula (14) for the residual of the Gamma-function $\Gamma(s)$ at the point $s = -k$, the asymptotic behavior of the Gamma-function, and the Cauchy residue theorem.

The kernel function $\Psi_{\alpha,\beta}$ can be also interpreted as the inverse Laplace transform of the Mittag–Leffler function $E_\beta(-\lambda^{\frac{\alpha}{2}})$:

$$E_\beta(-\lambda^{\frac{\alpha}{2}}) = \int_0^\infty \Psi_{\alpha,\beta}(\tau) e^{-\lambda \tau} d\tau, \tag{51}$$

where the Mittag–Leffler function E_β is defined as follows:

$$E_\beta(z) = E_{\beta,1}(z) = \sum_{k=0}^{\infty} \frac{z^k}{\Gamma(1+\beta k)}, \quad \beta > 0, \ z \in \mathbb{C}. \tag{52}$$

For the time-fractional diffusion equation ($\alpha = 2, 0 < \beta \le 1$ in the Equation (38)) the subordination formula (46) with the kernel function $\Phi_{\alpha,\beta}$ given by the 1st line of (47) is valid. In this case, the four-parameters Wright function of the second kind is reduced to the Wright function of the second kind and we arrive at the known formula ([41,42])

$$G_{2,\beta,n}(x,t) = \int_0^\infty t^{-\beta} \phi(-\beta, 1-\beta; -st^{-\beta}) G_{2,1,n}(x,s) ds, \quad 0 < \beta < 1. \tag{53}$$

For the space-fractional diffusion equation ($\beta = 1, 0 < \alpha \le 2$ in the Equation (38)), the subordination formula (46) with the kernel function $\Psi_{\alpha,\beta}$ given by the 2nd line of (47) is valid. It is easy to verify that the kernel function can be rewritten in the following form:

$$-\tau^{-1-\frac{\alpha}{2}} \Phi_{(\beta,1-\beta),(-\frac{\alpha}{2},-\frac{\alpha}{2})}\left(-\tau^{-\frac{\alpha}{2}}\right) = \tau^{-1} \Phi_{(\beta,1),(-\frac{\alpha}{2},0)}\left(-\tau^{-\frac{\alpha}{2}}\right).$$

Thus, also in the case of the space-fractional diffusion equation, the four-parameters Wright function of the second kind from the formula (47) is reduced to the Wright function of the second kind and we arrive at the subordination formula in the form

$$G_{\alpha,1,n}(x,t) = \int_0^\infty s^{-1} \phi(-\frac{\alpha}{2}, 0; -s^{-\frac{\alpha}{2}} t) G_{2,1,n}(x,s) ds, \quad 0 < \alpha < 2. \tag{54}$$

3.3. FDEs with the Left- and Right-Hand Sided Erdélyi-Kober Fractional Derivatives

In this part of the section, we consider an initial-value problem for an ordinary fractional differential equation with the left- and right-hand sided Erdélyi-Kober fractional derivatives defined on the positive semi-axis. The equations of this type appear in the fractional calculus of variations as the Euler-Lagrange equations. However, to the best knowledge of the author, the only method for analytical treatment of these equations defined on an infinite interval, say, on the positive real semi-axis, is the operational method recently suggested in [32]. Here we present an example of application of this method to the following sample equation ($a > b > 0$, $n - 1 < a\mu \leq n$, $n \in \mathbb{N}$):

$$(_*D_{1/a}^{-\alpha-a\mu,a\mu} y)(x) + \rho x^\mu (_*P_{1/b}^{\beta-b\mu,b\mu} y)(x) = f(x), \quad x > 0, \rho > 0 \tag{55}$$

subject to the initial conditions ($k = 0, \ldots, n-1$)

$$\lim_{x \to 0} x^{\frac{1}{a}(1-\alpha-a\mu+k)} \prod_{i=k+1}^{n-1} \left(1 - \alpha - a\mu + i + ax\frac{d}{dx}\right) y(x) = c_k. \tag{56}$$

In the Equation (55), the operator $_*P_{1/b}^{\beta-b\mu,b\mu}$ is the Caputo type modification of the right-hand sided Erdélyi-Kober fractional derivative given by the formula (34). The operator $_*D_{1/a}^{-\alpha-a\mu,a\mu}$ is the Caputo type modification of the left-hand sided Erdélyi-Kober fractional derivative defined as follows:

$$(_*D_\beta^{\gamma,\delta} f)(x) = (I_\beta^{\gamma+\delta,n-\delta} \prod_{k=0}^{n-1} \left(1 + \gamma + k + \frac{1}{\beta} t \frac{d}{dt}\right) f)(x), \tag{57}$$

where $I_\beta^{\gamma,\delta}$ stays for the left-hand sided Erdélyi-Kober fractional integral of order δ:

$$(I_\beta^{\gamma,\delta} f)(x) = \frac{1}{\Gamma(\delta)} \int_0^1 (1-t)^{\delta-1} t^\gamma f\left(xt^{\frac{1}{\beta}}\right) dt, \quad \delta, \beta > 0, \gamma \in \mathbb{R}. \tag{58}$$

It is worth mentioning that the initial conditions in form (56) are determined by the projector operator of the left-hand sided Erdélyi-Kober fractional integral $I_{1/a}^{-\alpha-a\mu,a\mu}$

$$(Py)(x) = y(x) - (I_{1/a}^{-\alpha-a\mu,a\mu} {_*D_{1/a}^{-\alpha-a\mu,a\mu}} y)(x) = \sum_{k=0}^{n-1} c_k x^{-\frac{1}{a}(1-\alpha-a\mu+k)}, \tag{59}$$

$$c_k = \lim_{x \to 0} x^{\frac{1}{a}(1-\alpha-a\mu+k)} \prod_{i=k+1}^{n-1} \left(1 - \alpha - a\mu + i + ax\frac{d}{dx}\right) y(x) \tag{60}$$

and thus, they are quite natural for the Equation (38).

In this paper, we do not repeat the derivation of the exact solution to the initial-value problem (55), (56) presented in [32] and restrict ourselves to formulation of the final result.

Theorem 9. *Let $a > b > 0$, $n - 1 < a\mu \leq n$, $n \in \mathbb{N}$, $f \in \mathfrak{D}$, and the condition*

$$\frac{\alpha - 1}{a} < \frac{\beta}{b} \tag{61}$$

be satisfied. Then the initial-value problem (38), (56) possesses a unique solution on the space $\mathfrak{D}$ in the form

$$y(x) = \sum_{k=0}^{n-1} c_k y_k(x) + y_f(x), \tag{62}$$

where the functions y_k, $k = 0, \ldots, n-1$ are defined by

$$y_k(x) = \Gamma(a\mu - k)\Gamma\left(\beta + \frac{b}{a}(1 - \alpha - a\mu + k)\right) x^{\mu - \frac{1}{a}(1-\alpha+k)} \Phi_{(a\mu,a\mu-k),(-b\mu,\beta+\frac{b}{a}(1-\alpha-a\mu+k))}(-\rho x^\mu), \tag{63}$$

and the function y_f is given by the formula

$$y_f(x) = g(x) + (g \overset{\lambda}{*} y_\Phi)(x), \tag{64}$$

with
$$g(x) = (I_{1/a}^{-\alpha-a\mu,a\mu} f)(x), \quad y_\Phi(x) = \rho\, x^{\mu-\lambda} \Phi_{(a\mu,1-\alpha+a(\mu-\lambda)),(-b\mu,\beta-b(\mu-\lambda))}(-\rho\, x^\mu)$$

and the convolution $\overset{\lambda}{*}$ defined as follows:

$$(f \overset{\lambda}{*} g)(x) = (I_{1/a}^{1-2\alpha-a\lambda,\alpha+a\lambda-1} *P_{1/b}^{\beta,\beta+b\lambda} f \circ g)(x) \tag{65}$$

with

$$(f \circ g)(x) = x^\lambda \int_0^1 \int_0^1 \tau_1^{-\alpha}(1-\tau_1)^{-\alpha}\tau_2^{\beta-1}(1-\tau_2)^{\beta-1} f\left(\frac{x\tau_1^a}{\tau_2^b}\right) g\left(\frac{x(1-\tau_1)^a}{(1-\tau_2)^b}\right) d\tau_1 d\tau_2. \tag{66}$$

The function y_f satisfies the inhomogeneous Equation (55) and homogeneous initial conditions, whereas the functions y_k, $k = 0, \ldots, n-1$ satisfy the homogeneous Equation (55) ($f(x) \equiv 0$, $x > 0$) and the initial conditions ($k = 0, \ldots, n-1$, $j = 0, \ldots, n-1$)

$$\lim_{x \to 0} x^{\frac{1}{a}(1-\alpha-a\mu+j)} \prod_{i=j+1}^{n-1} \left(1 - \alpha - a\mu + i + ax\frac{d}{dx}\right) y_k(x) = \begin{cases} 1, & j = k, \\ 0, & j \neq k. \end{cases} \tag{67}$$

In the formulation of the theorem, the space of functions denoted by $\mathfrak{D}$ consists of the functions that are continuous on the semi-axis $]0, \infty[$ and can be represented as the convergent power series with the power functions weights in some neighborhoods $U_{\epsilon_1}(0)$ and $U_{\epsilon_2}(+\infty)$ of the points $x = 0$ and $x = +\infty$, respectively, i.e., in the form

$$f(x) = x^\alpha \sum_{k=0}^\infty a_k (x^\rho)^k, \quad \rho > 0, \quad x \in U_{\epsilon_1}(0), \tag{68}$$

and

$$f(x) = x^\beta \sum_{k=0}^\infty b_k (x^{-\sigma})^k, \quad \sigma > 0, \quad x \in U_{\epsilon_2}(+\infty). \tag{69}$$

The functions from $\mathfrak{D}$ have a power law asymptotic behavior at the points 0 and $+\infty$ that appears to be an appropriate asymptotics for solutions of the fractional differential equations that contain both the left- and right-hand sided Erdélyi-Kober fractional derivatives.

Finally, we mention that the results formulated in Theorem 9 remain valid also in the case of the Equation (55) with a negative parameter ρ under the additional condition $a/3 < b$. This can be proved by the operational method presented in [32] and employing the asymptotic behavior of the four-parameters Wright function of the second kind on the positive semi-axis given in Theorem 4.

Funding: This research received no external funding.

Conflicts of Interest: The author declares no conflict of interest.

References

1. Prudnikov, A.P.; Brychkov, Y.A.; Marichev, O.I. *Integrals and Series. More Special Functions*; Gordon and Breach: New York, NY, USA, 1989; Volume 3.
2. Meijer, C.S. On the G-function. I, II, III, IV, V, VI, VII, VIII. *Proc. Nederl. Akad. Wet.* **1946**, *49*, 227–237, 344–356, 457–469, 632–641, 765–772, 936–943, 1063–1072, 1165–1175.
3. Gorenflo, R.; Kilbas, A.A.; Mainardi, F.; Rogosin, S.V. *Mittag-Leffler Functions, Related Topics and Applications*; Springer: Berlin, Germany, 2014.
4. Kiryakova, V. *Generalized Fractional Calculus and Applications*; J. Wiley: New York, NY, USA, 1994.

5. Luchko, Y. The Wright function and its applications. In *Handbook of Fractional Calculus with Applications. Vol.1: Basic Theory*; Kochubei, A., Luchko, Y., Eds.; Walter de Gruyter: Berlin, Germany; Boston, MA, USA, 2019; pp. 241–268.
6. Paris, R.B. Asymptotics of the special functions of fractional calculus. In *Handbook of Fractional Calculus with Applications. Vol. 1: Basic Theory*; Kochubei, A., Luchko, Y., Eds.; Walter de Gruyter: Berlin, Germany; Boston, MA, USA, 2019; pp. 297–326.
7. Srivastava, H.M.; Gupta, K.C.; Goyal, S.P. *The H-functions of One and Two Variables with Applications*; South Asian Publishers: New Delhi, India, 1982.
8. Gorenflo, R.; Mainardi, F.; Rogosin, S. Mittag-Leffler function: Properties and applications. In *Handbook of Fractional Calculus with Applications. Vol.1: Basic Theory*; Kochubei, A., Luchko, Y., Eds.; Walter de Gruyter: Berlin, Germany; Boston, MA, USA, 2019; pp. 269–298.
9. Buckwar, E.; Luchko, Y. Invariance of a partial differential equation of fractional order under the Lie group of scaling transformations. *J. Math. Anal. Appl.* **1998**, *227*, 81–97. [CrossRef]
10. Consiglio, A.; Mainardi, F. The Wright function of the second kind in mathematical physics. *Mathematics* **2020**, *8*, 884.
11. Das, S.; Mehrez, K. Geometric properties of the four parameters Wright function. *arXiv* **2020**, arXiv:2005.01354v1.
12. Djrbashian, M.M. On integral transforms generated by the generalized Mittag-Leffler function. *Izv. Akad. Nauk Armjan. SSR.* **1960**, *13*, 21–63. (In Russian)
13. Gajic̀, L.; Stankovic̀, B. Some properties of Wright's function. *Publ. de l'Institut Mathèmatique, Beograd, Nouvelle Sèr.* **1976**, *20*, 91–98.
14. Gorenflo, R.; Luchko, Y.; Mainardi, F. Analytical properties and applications of the Wright function. *Fract. Calc. Appl. Anal.* **1999**, *2*, 383–414.
15. Gorenflo, R.; Luchko, Y.; Mainardi, F. Wright functions as scale-invariant solutions of the diffusion-wave equation. *J. Computat. Appl. Math.* **2000**, *11*, 175–191. [CrossRef]
16. Kilbas, A.A.; Saigo, M.; Trujillo, J.J. On the generalized Wright function. *Fract. Calc. Appl. Anal.* **2002**, *5*, 437–460.
17. Luchko, Y. On the asymptotics of zeros of the Wright function. *Zeitschrift für Analysis und ihre Anwendungen* **2000**, *19*, 597–622. [CrossRef]
18. Luchko, Y.; Gorenflo, R. Scale-invariant solutions of a partial differential equation of fractional order. *Fract. Calc. Appl. Anal.* **1998**, *1*, 63–78.
19. Mainardi, F. *Fractional Calculus and Waves in Linear Viscoelasticity*; Imperial College Press: London, UK, 2010.
20. Mainardi, F.; Pagnini, G. The Wright functions as solutions of the time-fractional diffusion equations. *Appl. Math. Comp.* **2003**, *141*, 51–62. [CrossRef]
21. Mainardi, F.; Pagnini, G. The role of the Fox-Wright functions in fractional sub-diffusion of distributed order. *J. Comput. App. Math.* **2007**, *207*, 245–257. [CrossRef]
22. Mehrez, K. New Integral representations for the Fox-Wright functions and its applications. *arXiv* **2017**, arXiv:1711.08368.
23. Mehrez, K. Monotonicity patterns and functional inequalities for classical and generalized Wright functions. *arXiv* **2017**, arXiv:1708.00461.
24. Mehrez, K.; Sitnik, S.M. Functional inequalities for Fox-Wright functions. *arXiv* **2017**, arXiv:1708.06611.
25. Stankovic̀, B. On the function of E.M. Wright. *Publications De l'Institut Mathèmatique* **1970**, *10*, 113–124.
26. Wright, E.M. On the coefficients of power series having exponential singularities. *J. Lond. Math. Soc.* **1933**, *8*, 71–79. [CrossRef]
27. Wright, E.M. The asymptotic expansion of the generalized Bessel function, *Proc. London Math. Soc. (Ser. II)* **1935**, *38*, 257–270. [CrossRef]
28. Wright, E.M. The asymptotic expansion of the generalized hypergeometric function. *J. Lond. Math. Soc.* **1935**, *10*, 287–293. [CrossRef]
29. Wright, E.M. The generalized Bessel function of order greater than one. *Quart. J. Math. Oxford Ser.* **1940**, *11*, 36–48. [CrossRef]
30. Fox, C. The asymptotic expansion of generalized hypergeometric functions. *Proc. Lond. Math. Soc.* **1928**, *27*, 389–400. [CrossRef]
31. Gorenflo, R.; Luchko, Y.; Srivastava, H.M. Operational method for solving integral equations with Gauss's hypergeometric function as a kernel. *Int. J. Math. Stat. Sci.* **1997**, *6*, 179–200.

32. Hanna, L.A.-M.; Al-Kandari, M.; Luchko, Y. Operational method for solving fractional differential equations with the left-and right-hand sided Erdélyi-Kober fractional derivatives. *Fract. Calc. Appl. Anal.* **2020**, *23*, 103–125. [CrossRef]
33. Luchko, Y. Subordination principles for the multi-dimensional space-time-fractional diffusion- wave equation. *Theor. Probab. Math. Statist.* **2019**, *98*, 127–147. [CrossRef]
34. Kilbas, A.A.; Srivastava, H.M.; Trujillo, J.J. *Theory and Applications of Fractional Differential Equations*; Elsevier: Amsterdam, The Netherlands, 2006.
35. Marichev, O.I. *Handbook of Integral Transforms of Higher Transcendental Functions. Theory and Algorithmic Tables*; Ellis Horwood: Chichester, UK, 1983.
36. Mainardi, F.; The fundamental solutions for the fractional diffusion-wave equation. *Appl. Math. Lett.* **1996**, *9*, 23–28. [CrossRef]
37. Olver, P.J. *Applications of Lie Groups to Differential Equations*; Springer: New York, NY, USA, 1986.
38. Saichev, A.; Zaslavsky, G. Fractional kinetic equations: Solutions and applications. *Chaos* **1997**, *7*, 753–764. [CrossRef]
39. Samko, S.G.; Kilbas, A.A.; Marichev, O.I. *Fractional Integrals and Derivatives: Theory and Applications*; Gordon and Breach: New York, NY, USA, 1993.
40. Kwaśnicki, M. Fractional Laplace operator and its properties. In *Handbook of Fractional Calculus with Applications. Vol. 1: Basic Theory*; Kochubei, A., Luchko, Y., Eds.; Walter de Gruyter: Berlin, Germany; Boston, MA, USA, 2019; pp. 159–194.
41. Bazhlekova, E. Subordination principle for fractional evolution equations. *Fract. Calc. Appl. Anal.* **2000**, *3*, 213–230.
42. Bajlekova, E. Fractional Evolution Equations in Banach Spaces. Ph.D. Thesis, Technische Universiteit Eindhoven, Eindhoven, The Netherlands, 2001.

© 2020 by the authors. Licensee MDPI, Basel, Switzerland. This article is an open access article distributed under the terms and conditions of the Creative Commons Attribution (CC BY) license (http://creativecommons.org/licenses/by/4.0/).

Review

Exact Values of the Gamma Function from Stirling's Formula

Victor Kowalenko

School of Mathematics and Statistics, The University of Melbourne, Parkville, VIC 3010, Australia; vkowa@unimelb.edu.au

Received: 6 May 2020; Accepted: 19 June 2020; Published: 1 July 2020

Abstract: In this work the complete version of Stirling's formula, which is composed of the standard terms and an infinite asymptotic series, is used to obtain exact values of the logarithm of the gamma function over all branches of the complex plane. Exact values can only be obtained by regularization. Two methods are introduced: Borel summation and Mellin–Barnes (MB) regularization. The Borel-summed remainder is composed of an infinite convergent sum of exponential integrals and discontinuous logarithmic terms that emerge in specific sectors and on lines known as Stokes sectors and lines, while the MB-regularized remainders reduce to one complex MB integral with similar logarithmic terms. As a result that the domains of convergence overlap, two MB-regularized asymptotic forms can often be used to evaluate the logarithm of the gamma function. Though the Borel-summed remainder has to be truncated, it is found that both remainders when summed with (1) the truncated asymptotic series, (2) Stirling's formula and (3) the logarithmic terms arising from the higher branches of the complex plane yield identical values for the logarithm of the gamma function. Where possible, they also agree with results from Mathematica.

Keywords: asymptotic series; asymptotic form; Borel summation; complete asymptotic expansion; divergent series; domain of convergence; gamma function; Mellin–Barnes regularization; regularization; remainder; Stokes discontinuity; Stokes line/sector; Stokes phenomenon; Stirling's formula

MSC: 30B10; 30B30; 30E15; 30E20; 34E05; 34E15; 40A05; 40G10; 40G99; 41A60

---

1. Introduction

Discovered in the 1730s [1], Stirling's formula is a well-known result for determining approximate values of the gamma function, $\Gamma(z)$, which is so important in the definition of Mittag–Leffler functions. Mystery has lingered whether it is indeed possible to obtain exact values of the gamma function from the complete version of the formula as opposed to its more famous truncated form. Moreover, due to the function's rapid exponentiation, its logarithm or $\ln \Gamma(z)$ is studied more often. This, however, introduces multivaluedness, which makes the asymptotic analysis of the function more formidable. Consequently, no one has ever been able to obtain exact values of either function via the entire formula.

In its entirety, Stirling's formula is an asymptotic expansion and is, therefore, divergent. Here exact values of $\ln \Gamma(z)$ are determined for all values of $\arg z$ from the complete asymptotic expansion of the formula. This process known as exactification represents the ultimate goal of hyperasymptotics, whose primary aim is to obtain far more accurate values from asymptotic expansions than standard Poincaré asymptotics [2]. In such studies one not only includes all the terms in a dominant asymptotic series, but also, subdominant exponential terms, which are said to lie beyond all orders [3]. To observe their effect, hyperasymptotic calculations are generally carried out to more than 20 decimal places.

Since a complete asymptotic expansion is composed of divergent series, exactification involves obtaining meaningful values from them. This is achieved by the process of regularization, which is defined here as the removal of the infinity in the remainder of an asymptotic series so as to make the series summable. It was first demonstrated in [4] that the infinity in the remainder of an asymptotic series arises from an impropriety in the asymptotic method used to derive it. Hence regularization represents the method of correcting asymptotic methods.

Two very different techniques will be used to regularize the divergent series in this work. As discussed in [5,6], the most common method of regularizing a divergent series is Borel summation, but often, it produces results that are not amenable to fast and accurate computation. To overcome this drawback, the numerical technique of Mellin–Barnes regularization was developed in [7]. In this method, divergent series are expressed in terms of Mellin–Barnes integrals and divergent arc-contour integrals. Regularization removes the latter resulting in the Mellin–Barnes integrals yielding finite values, similar to the Hadamard finite part of a divergent integral [8]. Amazingly, the finite values obtained from applying the technique to an asymptotic expansion yield exact values of the original function with the main difference being that instead of dealing with Stokes sectors and lines, one now deals with overlapping domains of convergence over which the Mellin–Barnes integrals are valid.

2. Stirling's Formula

Stirling's formula [1] for the factorial function is often written for large integers, n, as

$$\ln n! = \ln \Gamma(n+1) = n \ln n - n + \frac{1}{2} \ln(2\pi n) + \cdots . \tag{1}$$

As this is accurate to within 1% for $n > 5$, it represents a good approximation in standard (Poincaré) asymptotics [2], but not so in hyperasymptotics. Moreover, our aim is to consider complex values, not large integers. Thus, we replace the factorial function by the more general gamma function, $\Gamma(z+1)$. The terms in (1), denoted here by $F(z)$, then become the leading terms of the complete asymptotic expansion for $\ln \Gamma(z)$. They will be treated as a separate contribution in all calculations of $\ln \Gamma(z)$, so that the reader will be able to observe just how inadequate standard asymptotics is compared with hyperasymptotics.

Occasionally, a problem arises where there is an interest in the missing terms in (1). Then Stirling's formula is expressed differently. For example, according to No. 6.1.41 in [9], for $z \to \infty$ and $|\arg z| < \pi$, $\ln \Gamma(z)$ is given by

$$\ln \Gamma(z) \sim F(z) + \frac{1}{12z} - \frac{1}{360z^3} + \frac{1}{1260z^5} - \frac{1}{1680z^7} + \cdots , \tag{2}$$

where $F(z)$ represents all the terms in Stirling's formula, namely,

$$F(z) = \left(z - \frac{1}{2}\right) \ln z - z + \frac{1}{2} \ln(2\pi). \tag{3}$$

Hence the leading terms are identical to those in (1). In other texts the dots in (2) are replaced by the Landau gauge symbol, which would be $O(z^{-9})$ here since it is the next highest order term. In [10] the power series after $\ln(2\pi)$ is truncated with the coefficients expressed in terms of the Bernoulli numbers, while the remainder term, $R_N(z)$ in No. 8.344, is given as

$$|R_N(z)| = \left| \sum_{k=N}^{\infty} \frac{B_{2k}}{2k(2k-1)z^{2k-1}} \right| < \frac{|B_{2n}|}{2n(2n-1)|z|^{2n-1} \cos^{2n-1}((\arg z)/2)} . \tag{4}$$

Although the remainder is dependent upon z and N, for $\Re z > 0$, the series diverges once N passes the optimal point of truncation, N_{OP}. Moreover, the above result is even more vague than (2) because the expansion is only valid for "large" values of $|z|$ without indicating what large means. Here, we shall evaluate exact values of $\ln \Gamma(z)$ from the complete version of Stirling's formula by following the concepts and theory in [6], but before this can be done, the following lemma is required.

Lemma 1. *Via regularization, the power/Taylor series expansion for* $\arctan u$, *namely,* $\sum_{k=0}^{\infty} u^{2k+1}/(2k+1)$, *can be expressed as*

$$\sum_{k=0}^{\infty} \frac{(-1)^k u^{2k+1}}{(2k+1)} \begin{cases} = \arctan u, & -1 < \Re(iu) < 1, \\ \equiv \arctan u, & \Re(iu) \leq -1, \text{ and } \Re(iu) \geq 1. \end{cases} \quad (5)$$

Proof. For brevity, the proof is not given here, but appears in [11]. □

It should be noted that an equivalence symbol appears in one of the results, indicating that one side possesses a divergent series, while the other side represents a finite regularized value. That is, $\arctan u$ is defined for all values of u, while the series representation for the function is divergent when u does not lie within $-1 < \Re(iu) < 1$. Since the equivalence symbol is less stringent than an equals sign, we can re-write the lemma as

$$\sum_{k=0}^{\infty} \frac{(-1)^k u^{2k+1}}{(2k+1)} \equiv \arctan u, \quad \forall u. \quad (6)$$

Therefore, if the series appears in a problem, then it can be replaced by the right-hand side (rhs). Though equivalence statements will appear throughout this paper, it does not necessarily mean that a power series is divergent for all values of the variable.

Now we derive the complete form of Stirling's formula. This will not be original, but we need to establish that it is complete. Binet's second expression for $\ln \Gamma(z)$ in [2] is

$$\ln \Gamma(z) = F(z) + 2 \int_0^\infty dt \, \frac{\arctan(t/z)}{e^{2\pi t} - 1}. \quad (7)$$

By making a change of variable, $y = 2\pi t$, and noting that z is complex, we can then introduce (5). Replacing k by $k+1$ yields

$$\ln \Gamma(z) - F(z) \equiv \frac{1}{\pi} \sum_{k=1}^{\infty} \frac{(-1)^{k+1}}{(2k-1)} \left(\frac{1}{2\pi z}\right)^{2k-1} \int_0^\infty dy \, \frac{y^{2k-1}}{e^y - 1}. \quad (8)$$

The left-hand side (lhs) of (8) is finite (convergent), while the rhs can be either divergent or convergent. From No. 3.411(1) in [10], the integral in the above equivalence is equal to $\Gamma(2k)\zeta(2k)$, where $\zeta(z)$ represents the Riemann zeta function. Thus, the above result becomes

$$\ln \Gamma(z) - F(z) \equiv 2z \sum_{k=1}^{\infty} \frac{(-1)^{k+1}}{(2k-1)} \frac{\Gamma(2k) \zeta(2k)}{(2\pi z)^{2k}}. \quad (9)$$

From here on, $S(z)$ denotes the series on the rhs. On the other hand, Paris and Kaminski [12,13], replace the terms on the lhs by $\Omega(z)$. With the aid of the reflection formula for the gamma function, the following continuation formula can be derived:

$$\Omega(z) + \Omega\left(ze^{\pm i\pi}\right) = -\ln\left(1 - e^{\mp 2i\pi z}\right). \quad (10)$$

This enables one to obtain values of $\ln \Gamma(z)$ whenever z is situated in the left-hand complex plane via the corresponding values in the right-hand complex plane. Furthermore, the rhs will play an important role when the Stokes phenomenon is discussed later.

In order to continue with this study, the following definitions are required:

Definition 1. *An asymptotic (power) series is defined here as an infinite power series with zero radius of absolute convergence.*

Definition 2. *An asymptotic form is composed of: (1) a complete asymptotic expansion, which not only possesses all terms in a dominant asymptotic power series, e.g., S(z) above, but also all the terms in each subdominant asymptotic series, should they exist, and (2) the common sector or ray in the complex plane over which the argument of the variable in each series is valid.*

By truncating $S(z)$ at N terms, we arrive at

$$S(z) = z \sum_{k=1}^{N-1} \frac{(-1)^k}{(2z)^{2k}} \Gamma(2k-1) c_k(1) - 2z \sum_{n=1}^{\infty} \sum_{k=N}^{\infty} \frac{(-1)^k}{(2\pi n z)^{2k}} \Gamma(2k-1), \qquad (11)$$

where the first term will be denoted as $TS_N(z)$, N is the truncation parameter and $c_k(1)$ represents a specific value of the cosecant polynomials [14], given by

$$c_k(1) = -2\zeta(2k)/\pi^{2k}. \qquad (12)$$

The infinite series over k in the second term is known as a generalized Type I terminant [6]. Terminants were first introduced by Dingle [15] because he found that special functions often possess asymptotic series whose late coefficients exhibit gamma function growth, viz. $\Gamma(k+\alpha)$. A Type II terminant differs in that the coefficients possess an extra phase factor of $(-1)^k$.

The notation $S^I_{p,q}(N,z^\beta)$ was introduced in [6] to denote the generalization of Dingle's Type I terminants, which are defined as

$$S^I_{p,q}\left(N, z^\beta\right) \doteq \sum_{k=N}^{\infty} (-1)^k \Gamma(pk+q) z^{\beta k}. \qquad (13)$$

Alternatively, (11) can be expressed as

$$S(z) = z \sum_{k=1}^{N-1} \frac{(-1)^k}{(2z)^{2k}} \Gamma(2k-1) c_k(1) - 2z \sum_{n=1}^{\infty} S^I_{2,-1}\left(N, (1/2n\pi z)^2\right). \qquad (14)$$

Thus, β and z in (13) are equal to 2 and $1/2n\pi z$ in (14). Although [6] states that both p and q have to be positive and real, it is $N+q/p$, which appears in the regularized value of a generalized terminant. Therefore, provided $\Re(N+q/p) > 0$, the regularized value of the series still exists. Alternatively, k can be replaced by $k+1$ in the infinite series in (14), in which case q equals unity. Since $S^I_{2,-1}(N,z^2) = -z^2 S^I_{2,1}(N-1,z^2)$, we can apply the result in [6] to $S^I_{2,1}(N-1,z^2)$ instead.

According to Rule A in ([15], Chapter 1), Stokes lines occur whenever the arguments or phases of the variable result in the terms of an asymptotic series becoming homogeneous in phase and having the same sign. In the case of the generalized terminant in (13), this means that Stokes lines occur whenever $\arg(-z^\beta) = 2l\pi$, for l, an integer. Then the terms in either $S^I_{2,-1}(N,1/z^2)$ or $S^I_{2,1}(N,1/z^2)$ are all positively real. Because l is arbitrary, we can replace -1 by $\exp(-i\pi)$. Thus, we find that the Stokes lines for $S(z)$ occur whenever $\arg z = -(l+1/2)\pi$, i.e., at half integer multiples of π.

The concept of a primary Stokes sector/line was introduced in [6] to indicate the first Stokes sector/line over which an asymptotic expansion is derived. It was also necessary to define asymptotic forms since two functions can have the same complete asymptotic expansion, but will still be different if the expansion applies over different primary Stokes sectors or lines. For example, in solving a problem for positive real values of the variable, one may obtain a generalized Type I terminant as the asymptotic solution. However, as the variable moves off the real axis, it will acquire subdominant semi-residue contributions of opposing signs in either direction as a result of the Stokes phenomenon. However, if the same asymptotic solution is obtained for positive imaginary values of the variable, then as the variable hits the positive and negative real axes, the asymptotic solution will acquire a semi-residue contribution. When the variable moves into the lower half of the complex plane, the asymptotic solution will acquire a full residue contribution. Clearly, both cases are different and will yield different values even though the same generalized Type I terminant was derived. Hence the original functions or solutions for these cases are different. In the first case the positive real axis becomes the primary Stokes line for the generalized Type I terminant, while in the second case, the upper half of the complex plane represents the primary Stokes sector. Then as more secondary Stokes sectors/lines are encountered either in a clockwise or anti-clockwise direction from the primary Stokes sector/line, more Stokes discontinuities arise at the boundaries. Although the choice of a primary Stokes sector/line is arbitrary, it will be taken here to be the Stokes sector/line situated in the principal branch of the complex plane, since most asymptotic expansions are derived under the condition that the variable lies initially in the principal branch of the complex plane.

Before we can regularize the asymptotic series, $S(z)$, we require the following lemma:

Lemma 2. *Regularization of the Taylor series for the logarithmic function yields*

$$\sum_{k=1}^{\infty} \frac{(-1)^{k+1}}{k} z^k \begin{cases} \equiv \ln(1+z) \,, & \Re z \leq -1 \,, \\ = \ln(1+z) \,, & \Re z > -1 \,. \end{cases} \tag{15}$$

Proof. There is no need for the proof to appear here as it can be found in [16]. □

As in the first lemma, we can replace the equals sign in the lemma by the less stringent equivalence symbol, which reduces the lemma to

$$\sum_{k=1}^{\infty} \frac{(-z)^k}{k} \equiv -\ln(1+z), \quad \forall z. \tag{16}$$

With this result we can now regularize $S(z)$, which will enable the asymptotic forms for $\ln \Gamma(z)$ to be derived.

Theorem 1. *As a result of the regularization of its asymptotic power series, the logarithm of the gamma function possesses the following asymptotic forms:*

$$\ln \Gamma(z) = F(z) + z \sum_{k=1}^{N-1} \frac{(-1)^k}{(2z)^{2k}} \Gamma(2k-1) c_k(1) + R_N^{SS}(z) + SD_M^{SS}(z), \tag{17}$$

where the remainder $R_N^{SS}(z)$ is given by

$$R_N^{SS}(z) = \frac{2(-1)^{N+1} z}{(2\pi z)^{2N}} \int_0^\infty dy \, y^{2N-2} e^{-y} \sum_{n=1}^{\infty} \frac{1}{n^{2N-2}((y/2\pi z)^2 + n^2)}, \tag{18}$$

and the Stokes discontinuity term $SD_M(z)$ is given by

$$SD_M^{SS}(z) = -\lfloor M/2 \rfloor \ln\left(-e^{\pm 2i\pi z}\right) - \frac{(1-(-1)^M)}{2} \ln\left(1 - e^{\pm 2i\pi z}\right). \tag{19}$$

The remainder is valid for either $(M-1/2)\pi < \theta = \arg z < (M+1/2)\pi$ or $-(M+1/2)\pi < \theta < -(M-1/2)\pi$, where M is a non-negative integer. However, the Stokes discontinuity term possesses two forms that are complex conjugates. The upper-signed version of (19) applies to $(M-1/2)\pi < \theta < (M+1/2)\pi$, while the lower-signed version is valid over $-(M+1/2)\pi < \theta < -(M-1/2)\pi$. For z lying on the Stokes lines, i.e., for $\theta = \pm(M+1/2)\pi$, $R_N^{SS}(z)$ and $SD_M^{SS}(z)$ are replaced by $R_N^{SL}(z)$ and $SD_M^{SL}(z)$, respectively. Then the remainder is given by

$$R_N^{SL}(z) = \frac{2z}{(2\pi|z|)^{2N-2}} P \int_0^\infty dy \, y^{2N-2} e^{-y} \sum_{n=1}^{\infty} \frac{1}{n^{2N-2}(y^2 - 4n^2\pi^2|z|^2)}, \tag{20}$$

while the Stokes discontinuity term becomes

$$SD_M^{SL}(z) = (-1)^M \left(\lfloor M/2 \rfloor + \frac{1-(-1)^M}{2} \right) 2\pi |z| - \frac{1}{2} \ln\left(1 - e^{-2\pi|z|}\right). \tag{21}$$

In (20), P denotes the Cauchy principal value.

Proof. For brevity, the proof is not presented here as it can be found in [11]. □

The remainder in Theorem 1 is conceptually different from the remainder term in standard Poincaré asymptotics, which is expressed in terms of the Landau gauge symbol, $\mathcal{O}()$, or as $+\ldots$ In fact, (17) would typically be written as

$$\ln \Gamma(z) = F(z) - \frac{c_1(1)}{4z} + \frac{c_2(1)}{8z^3} - \frac{3c_3(1)}{8z^5} + \mathcal{O}\left(\frac{1}{z^7}\right). \tag{22}$$

Moreover, by introducing $c_1(1) = -1/3$, $c_2(1) = -1/45$, $c_3(1) = -2/945$ and $c_4(1) = -1/4725$, into the above result, we obtain (2). For real values of z, (22) is referred to as a large z or $z \to \infty$ expansion with the limit point at infinity. For z complex, it becomes a large $|z|$ expansion. In other cases, where the Landau gauge symbol is omitted, a tilde often replaces the equals sign. Nevertheless, in all these representations it means that the later terms in the truncated power series have been neglected despite their eventual divergence past the optimal point of truncation.

3. Numerical Analysis

In the previous section the asymptotic forms for $\ln \Gamma(z)$ were derived via Borel summation. However, we still need to verify that these results yield exact values of the special function. This section aims to present such a numerical analysis. For the analysis to be effective, a large number of values of $|z|$ is not required. This is because the results change across Stokes sectors or rays, but within each sector or on each line, they behave uniformly with respect to z. Thus, a few values of $|z|$ are necessary for testing the validity of the asymptotic forms. In fact, only two values of $|z|$ are necessary: a relatively large one, where the asymptotic series in (17) can be truncated, and a small one, where truncation breaks down completely. Then a range of values for both N and $\arg z$ or θ, need to be considered across the Stokes sectors and lines. Note also that selecting extremely large/small values of $|z|$ may result in overflow or underflow problems in the numerical calculations. This would then give the misleading impression that the asymptotic forms are incorrect rather than implying a deficiency in the computing system. Since the variable in the asymptotic series is $1/(2n\pi z)^2$ with n ranging from unity to infinity, $|z| = 3$ is deemed to be sufficiently large, while for $|z| = 1/10$, there is no optimal point of truncation. The second value is, therefore, sufficiently small to demonstrate the breakdown of standard Poincaré asymptotics.

Before undertaking the numerical analysis, let us present plots of $\ln \Gamma(z)$ to help the reader understand the nature of the function. Figure 1 displays graphs of the real part of the function for several fixed values of $|z|$ used in this paper as a function of θ over $(0, \pi)$. There we see for the larger values of $|z|$, the real part of $\ln \Gamma(z)$ dips to a minimum before it begins to grow dramatically, which is the rapid exponentiation mentioned in the introduction. The smaller values of $|z|$ do not vary as much, although both are similar to the larger values of $|z|$ in that they dip to a minimum and rise afterwards. Unlike the other graphs, the graph for $|z| = 1/2$ has a positive minimum and increases rather slowly.

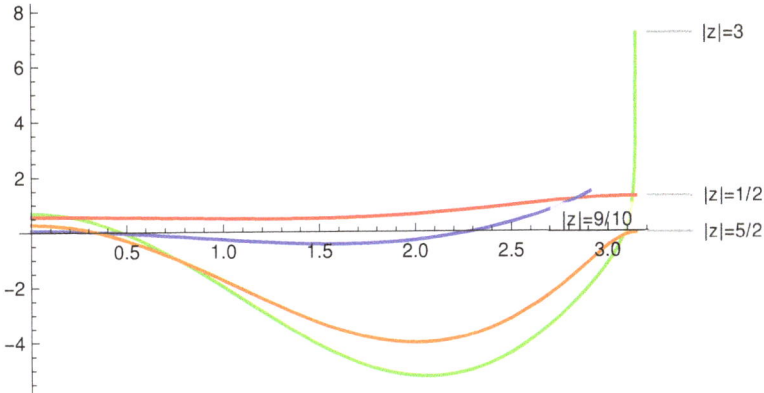

Figure 1. $\Re \ln \Gamma(z)$ as a function of θ between 0 and π for fixed values of $|z|$.

Figure 2 displays graphs of the imaginary part of $\ln \Gamma(z)$ for the same fixed values of $|z|$ as a function of θ over $(0, \pi)$. Here we see that the large values of $|z|$ rise to a positive maximum before rapidly decreasing into the negative right quadrant. The plot for $|z| = 9/10$ does not attain a positive maximum, but decreases relatively slowly from the origin into the negative right quadrant. The graph for $|z| = 1/2$ follows that for $|z| = 9/10$ until about $\theta = \pi/2$. Then it decreases faster than the $|z| = 9/10$ graph, but when θ is close to π, it rises until it meets the $|z| = 9/10$ graph at $\theta = \pi$.

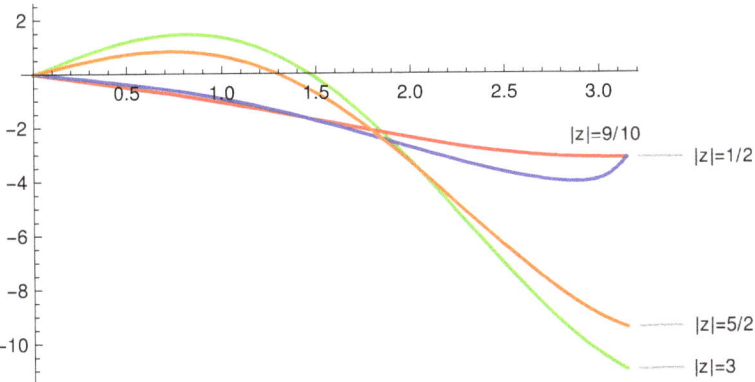

Figure 2. $\Im \ln \Gamma(z)$ as a function of θ between 0 and π for fixed values of $|z|$.

The optimal point of truncation, N_{OP}, is determined by calculating the first value of the truncation parameter, N, when successive terms in an asymptotic series begin to dominate the preceding terms. That is, it occurs at the first value of k, where the $k+1$-th term is greater than the k-th term in $S(z)$, namely,

$$\left| \frac{2k(2k-1)}{(2z)^2} \frac{c_{k+1}(1)}{c_k(1)} \right| = \left| \frac{2k(2k-1)}{(2\pi z)^2} \frac{\zeta(2k+2)}{\zeta(2k)} \right| \approx 1. \tag{23}$$

Since the ratio of the Riemann zeta functions is close to unity, we observe that N_{OP} occurs around $\pi |z|$. Therefore, for $|z| = 3$, N_{OP} will be close to 10, while for $|z| = 1/10$, it does not exist, meaning that $N_{OP} = 0$. In the latter case the first or leading term of the asymptotic series will yield the "nearest" value to $\ln \Gamma(z)$, but it will not be accurate. On the other hand, the larger N_{OP} is, the more accurate truncation of the asymptotic series becomes.

Typically, when a software package such as Mathematica [17] determines values of a special function, it only does so over the principal branch of the complex plane. Hence, the numerical analysis will be confined to $\arg z$ over $(-\pi, \pi]$, which means in turn that the numerical analysis of (17) will only be conducted over the three Stokes sectors, $-3\pi/2 < \theta < -\pi/2$, $-\pi/2 < \theta < \pi/2$ and $\pi/2 < \theta < 3\pi/2$, and the two Stokes lines at $\theta = \pm \pi/2$. In other words, only the $M = 0$ and $M = \pm 1$ results in Theorem 1 will be tested for the time being. By denoting the truncated sum in (17) by $TS_N(z)$, i.e.,

$$TS_N(z) = z \sum_{k=1}^{N-1} \frac{(-1)^k}{(2z)^{2k}} \Gamma(2k-1) c_k(1), \qquad (24)$$

we need to verify the following results:

$$\ln \Gamma(z) = \begin{cases} F(z) + TS_N(z) + R_N^{SS}(z) + SD_1^{SS,U}(z), & \pi/2 < \theta \le \pi, \\ F(z) + TS_N(z) + R_N^{SL}(z) + SD_0^{SL}(z), & \theta = \pi/2, \\ F(z) + TS_N(z) + R_N^{SS}(z), & -\pi/2 < \theta < \pi/2, \\ F(z) + TS_N(z) + R_N^{SL}(z) + SD_0^{SL}(z), & \theta = -\pi/2, \\ F(z) + TS_N(z) + R_N^{SS}(z) + SD_1^{SS,L}(z), & -\pi < \theta < -\pi/2. \end{cases} \qquad (25)$$

In the above the superscripts, U and L, have been introduced into the Stokes discontinuity terms in the Stokes sectors to indicate the upper- and lower-signed versions of (21). Although equal to zero, the Stokes discontinuity term for the third asymptotic form will be denoted as $SD_0^{SS}(z)$.

If we put $N = 4$ in the third result of (24) and neglect the final term or remainder, then we arrive at (2). However, the remaining terms in this result are now expressed as

$$R_N^{SS}(z) = \frac{2(-1)^{N+1} z}{(2\pi z)^{2N-2}} \sum_{n=1}^{\infty} \frac{1}{n^{2N-2}} \int_0^\infty dy \, \frac{y^{2N-2} e^{-y}}{(y^2 + 4\pi^2 n^2 z^2)}, \qquad (26)$$

and

$$R_N^{SL}(z) = \frac{2z}{(2\pi|z|)^{2N-2}} \sum_{n=1}^{\infty} \frac{1}{n^{2N-2}} P \int_0^\infty dy \, \frac{y^{2N-2} e^{-y}}{(y^2 - 4\pi^2 n^2 |z|^2)}, \qquad (27)$$

while the Stokes discontinuity terms are given by

$$SD_1^{SS}(z) = -\ln\left(1 - e^{\pm 2\pi z i}\right), \qquad (28)$$

and

$$SD_0^{SL}(z) = -\frac{1}{2} \ln\left(1 - e^{-2\pi|z|}\right). \qquad (29)$$

Note the connection with $\Omega(z)$ mentioned below (9).

For the numerical analysis we need to consider the results over the Stokes sectors separately from those at the Stokes lines since the latter require the evaluation of the Cauchy principal value and the Stokes discontinuity terms possess a factor of $1/2$ compared with zero when $|\theta| < \pi/2$ or unity when $|\theta| > \pi/2$. Thus, $\ln \Gamma(z)$ will be evaluated via two different Mathematica modules: one involving the standard numerical integration routine called NIntegrate, and another, where NIntegrate is adapted to evaluate only the Cauchy principal value.

When $\theta > 0$, the Stokes discontinuity terms can be combined into one expression, denoted by $SD^+(z)$. This is given by

$$SD^+(z) = -S^+ \ln\left(1 - e^{2\pi i z}\right), \tag{30}$$

where the Stokes multiplier, S^+, is written as

$$S^+ = \begin{cases} 1, & \pi/2 < \theta \leq \pi, \\ 1/2, & \theta = \pi/2, \\ 0, & -\pi/2 < \theta < \pi/2. \end{cases} \tag{31}$$

Similarly, the Stokes discontinuity terms in the lower half of the principal branch, $SD^-(z)$, can be written in terms of another Stokes multiplier, S^-, as follows:

$$SD^-(z) = S^- \ln\left(1 - e^{-2\pi i z}\right), \tag{32}$$

where S^- is given by

$$S^- = \begin{cases} 0, & -\pi/2 < \theta < \pi/2, \\ 1/2, & \theta = -\pi/2, \\ 1, & -\pi < \theta < -\pi/2. \end{cases} \tag{33}$$

From the above, we see that the Stokes multipliers are discontinuous, which is known as the conventional view of the Stokes phenomenon. However, an alternative view of the Stokes phenomenon arose in the late 1980s where they were no longer regarded as step-functions. Instead, it was proposed that they undergo a smooth, but rapid, transition from zero to unity, equalling 1/2 at the Stokes line [18]. Today, this is known as Stokes smoothing, despite the fact that Stokes never regarded the multipliers as being smooth [19]. According to this approach, first put forward by Berry and then made more "rigorous" by Olver [20], the Stokes multiplier reduces to the error function, erf(z). Later, Berry [21] and Paris and Wood [22] found an approximate form for the Stokes multipliers of $\ln \Gamma(z)$, which were given as

$$S^\pm(z) \sim \frac{1}{2} \pm \frac{1}{2} \mathrm{erf}\left((\theta \pm \pi/2)\sqrt{\pi|z|}\right). \tag{34}$$

A graph of (34) for $|z| = 3$ versus θ is displayed in Figure 3 together with the conventional view or (31). For $\theta < 1$, (34) is virtually zero, while for $\theta > 2$, it is almost equal to unity. In between, however, the rapid smoothing occurs with the greatest deviation from the step-function occurring in the vicinity of the Stokes line where both views possess a common (green) point at $(\pi/2, 1/2)$. If smoothing occurs, then Theorem 1 cannot possibly yield exact values of $\ln \Gamma(z)$, especially for θ between $13\pi/32$ and $17\pi/32$ excluding $\pi/2$.

We can establish the correct view by calculating $\ln \Gamma(z)$ for θ between $13\pi/32$ and $17\pi/32$ using (30) since smoothing implies that (30) cannot possibly yield exact values of $\ln \Gamma(z)$. However, if we obtain exact values of $\ln \Gamma(z)$, then we know that the conventional view holds and smoothing is a fallacy. The problem with testing (34) directly is that it applies to much larger values of $|z|$ than 3. The proponents of smoothing have not provided the form for smaller values of $|z|$. For very large values of $|z|$, truncating the asymptotic expansion at a few terms will yield very accurate values for $\ln \Gamma(z)$, which can obscure both views unless an extremely high precision and time-consuming analysis is undertaken. Hence much smaller values of $|z|$ will be considered in (30), so that the Stokes discontinuity term can no longer be neglected.

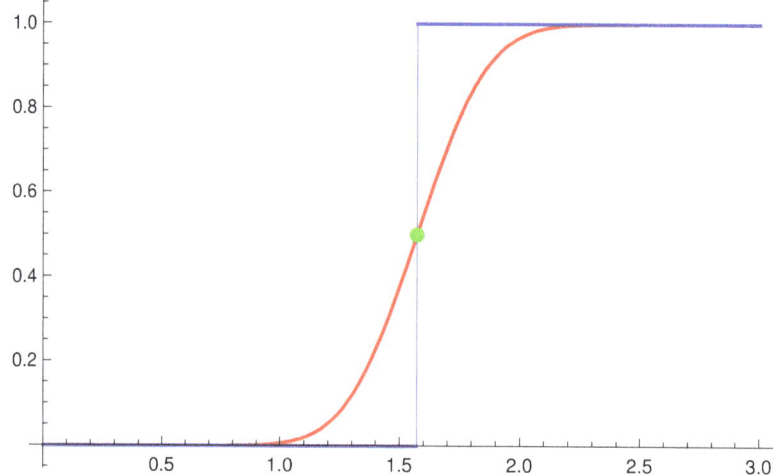

Figure 3. The conventional Stokes multiplier S^+ (blue) vs. the smoothed version (red) for $|z| = 3$ as a function of θ.

Before Stokes smoothing can be investigated, we must show that (24) behaves as a typical asymptotic expansion. That is, we must show that for large values of $|z|$, the remainder can be neglected to yield accurate, but nevertheless approximate, values of $\ln \Gamma(z)$ up to and not very far from the optimal point of truncation, while for small values of $|z|$, it is simply invalid to neglect the remainder. For this demonstration we do not require the Stokes discontinuity terms. Thus, we shall study the asymptotic series for $|\theta| < \pi/2$, in particular $\theta = 0$, because it does not require complex arithmetic.

From (26) we see that the evaluation of the remainder involves two computationally intensive tasks. The first is the infinite sum over n, which arose due to an infinite number of singularities lying on each Stokes line. The second issue is the numerical integration of the exponential integral. The latter can be avoided by decomposing the denominator into partial fractions and using No. 3.383(10) from [10]. For $|\theta| < \pi/2$, one then obtains

$$R_N^{SS}(z) = \frac{\Gamma(2N-1)}{2\pi i} \sum_{n=1}^{\infty} \frac{1}{n} \Big(e^{-2\pi n z i} \Gamma(2-2N, -2\pi n z i) - e^{2\pi n z i} \Gamma(2-2N, 2\pi n z i) \Big). \tag{35}$$

The above result can also be obtained by combining (4.3), (4.10) and (4.11) in [22].

A module was written to evaluate $\ln \Gamma(z)$ in Mathematica with the remainder given by (35) and n set to an upper limit of 10^5 to ensure 50 figure accuracy. Table 1 displays a small sample of the results obtained from the code. For more details about the code including its performance and listing as well as other results, the reader should consult [11]. Note that all the results are real, which is to be expected since $\ln \Gamma(3) = \ln 2$. In actual fact, Mathematica printed out a tiny imaginary part with each value, but it was often zero to the first 50+ decimal places and thus was discarded. The appearance of these tiny imaginary values indicates the size of the numerical error. The few cases where the errors were less than 50 decimal places will be discussed shortly.

Table 1. $\ln\Gamma(3)$ via (35) for various values of the truncation parameter, N.

N	Quantity	Value
	$F(3)$	0.66546925487494697026844282871193190148012386819465
	TS	0.02777777777777777777777777777777777777777777777777
2	$R_2^{SS}(3)$	$-0.000099852092779438597303829889692646860945357791$
	Total	0.69314718055994530944891677660001703239695628818129
	TS	0.02767489711934156378600823045267489711934156378600
5	$R_5^{SS}(3)$	$3.46840620707228089326095059290335934368930570 \times 10^{-8}$
	Total	0.69314718055994530941723212145817656807550013436025
	TS	0.02767792490305420773799675002229807193246527808017
11	$R_{11}^{SS}(3)$	$5.88900534265044578202453778763591963494785441 \times 10^{-10}$
	Total	0.69314718055994530941723212145817656807550013436025
	TS	0.02767792637739909405287985684200177174299855050138
20	$R_{20}^{SS}(3)$	$-6.04991267561310347983230543983690458667925439 \times 10^{-7}$
	Total	0.69314718055994530941723212145817656807550013436025
	TS	41.28347361380792549662137541291397749587553795756 21
30	$R_{30}^{SS}(3)$	$-41.25579568812292715747258612016773282928016169 1396$
	Total	0.69314718055994530941723212145817656807550013436025
	TS	$6.00398640887101848495574284509396382227628091 77 \times 10^{25}$
50	$R_{50}^{SS}(3)$	$-6.00398640887101848495574284232617125377644700 2 \times 10^{25}$
	Total	0.69314718055994530941723212145817656807550013436025
	$\ln\Gamma(3)$	0.69314718055994530941723212145817656807550013436025

The first column displays the values of the truncation parameter, N for each calculation. The second row in the table gives the value of Stirling's formula for $z = 3$, which only agrees with the actual value of $\ln\Gamma(3)$ at the bottom of the table to the first decimal place. For each value of N there are three rows. The first row labelled TS displays the value of the truncated sum in (25), while the row labelled $R_N^{SS}(3)$ presents the value of the remainder given by (35) with the upper limit set to 10^5. The third row labelled Sum is the sum of Stirling's formula, the truncated sum and the remainder. It yields the same value of $\ln 2$ as at the bottom of the table except for $N = 2$.

For $N = 2$, the truncated sum and remainder equal $0.027777\cdots$ and $-9.98529\cdots \times 10^{-5}$, respectively. When they are summed with $F(3)$, they yield a value that agrees with $\ln\Gamma(3)$ to 19 decimal places, which is well-below the 50 decimal figure accuracy mentioned above and nowhere near as accurate as other results such as $N = 50$. The reason this has occurred is that the factor of n^{2N-2} in the denominator of (26) affects the calculation of the remainder for the small values of N such as 1 or 2. In these cases the upper limit of 10^5 needs to be increased substantially to improve the accuracy, which does not apply for higher values of N.

The remainder is smallest in magnitude when $N = 11$, which agrees with our estimate below (23) for the optimal point of truncation, N_{OP}. For $N = N_{OP}$, the sum of the values only differs from the actual value of $\ln\Gamma(3)$ at the fifty-third decimal place. Moreover, for N close to N_{OP}, there is little deterioration in the accuracy, but for $N = 30$ and 50, well past N_{OP}, the remainder dominates, whereas in the other calculations, it is small. This is consistent with standard Poincaré asymptotics, where the remainder is neglected. Therefore, for all but the last two calculations, Stirling's formula yields the main contribution to $\ln\Gamma(3)$. For the last two values of N, the truncated sum and remainder dominate, but their divergence is cancelled out. For example, when $N = 50$, the remainder and truncated sum are $\mathcal{O}(10^{25})$. Hence the first 26 decimal places of both quantities cancel each other, thereby enabling Stirling's formula to become the main contribution. Unfortunately, losing these decimal places produces an imaginary term that is zero to a reduced number of decimal places, 23 instead of 50+ as mentioned above.

Now consider $z = 1/10$, which is unheard of in standard Poincaré asymptotics and also in the hyperasymptotic calculations of [12,18,21,23,24]. Furthermore, Paris [13] has specifically carried out a hyperasymptotic calculation of $\ln \Gamma(z)$ using Hadamard expansions for $\Omega(z)$. Depending on the number of chosen levels, his results are accurate at best to 10^{-45} for $\Re z > 8$ ($N_{OP} > 25$). Hence Table 1 displays far more accurate results, but with $z = 3$.

Table 2 presents a sample of results for $z = 1/10$ in the third asymptotic form of (25) with $R_N^{SS}(z)$ given by (35). In this case Stirling's formula is nowhere near as accurate as in Table 1. Except for $N = 2$, adding the truncated series to Stirling's formula worsens the accuracy. This has arisen because there is no optimal point of truncation. Therefore, the remainder must be evaluated. As a result that the remainder diverges far more rapidly in this case, there is a greater cancellation of decimal places than in Table 1. Thus, the total values in Table 2 are generally not as accurate, the exception being very low values of N. Despite this, these results could not have been achieved without regularization.

Table 2. $\ln \Gamma(1/10)$ via (35) for various values of the truncation parameter, N.

N	Quantity	Value
	$F(1/10)$	1.7399725704022910153875263182793633229018380690 8929
2	TS	0.83333333333333333333333333333333333333333333333333
	$R_1^{SS}(1/10)$	$-0.32059325200141753336547073402130155246488980 35500$
	Total	2.25271265173420681535538891759139510377028159887258
3	TS	$-1.94444444444444444444444444444444444444444444 44444$
	$R_3^{SS}(1/10)$	2.45718452577635938892661821233038043014208938 9324197
	Total	2.25271265173420595986970008616529930859948301 6422822
5	TS	$-5874.9603174603174603174603174603174603174603 1746031$
	$R_5^{SS}(1/10)$	5875.47305754164937526194249278840659211317401 3182747
	Total	2.25271265173420595986970164636849511861553379 1559062
9	TS	$-2.9486741947484548985872515284279990162343103 5195 \times 10^{13}$
	$R_9^{SS}(1/10)$	$2.94867419474850617259538471992244723376711926 5136 \times 10^{13}$
	Total	2.25271265173420595986970164636849511861562722 2229495
15	TS	$-3.6086855891831160967091834603534698425550101 1055 \times 10^{31}$
	$R_{15}^{SS}(1/10)$	$3.60868558918311609670918346035352116563143302 04 \times 10^{31}$
	Total	2.25271265173420595986970164636849511861562722 2294953
	$\ln \Gamma(1/10)$	2.25271265173420595986970164636849511861562638 0692264

Now we assume that the routine, Gamma[N,z], does not exist in Mathematica. Then a new program implementing the first, third and fifth asymptotic forms in (25) with the remainder given by (26) is required. As before, the upper limit in the sum will be set to 10^5. To calculate each term in the remainder, the program, which appears as the second program in the appendix of [11], employs NIntegrate inside a Do loop. Since it is a different approach for calculating $\ln \Gamma(z)$, it can be used to check the results in Table 1. The version in [11] has the precision and accuracy goals set to 30 for thirty figure accuracy, which means, in turn, that the working precision must be set to a much higher level, e.g., 60. Higher values for these options can be set, but it comes at the expense of computing time. The integrand employed in NIntegrate is called Intgrd and is basically the integrand in (26). Due to lack of space, the calculated quantities are displayed here to 25 decimal places, although they were frequently far more accurate. In addition, unlike the previous calculations, we consider complex values of z, i.e., θ takes on values within the principal branch of the complex plane except at $\pm \pi/2$. For brevity, only $|z| = 1/10$ is presented here. The results for $|z| = 3$ appear in Table 3 of [11].

Table 3 presents a very small sample of the results obtained by running the second program in the appendix of [11] with $|z| = 1/10$. Although positive values of θ were considered, only negative values are displayed here. In the table, there are six results for each value of N and θ. Stirling's formula is represented by the first value. The second value, denoted by TS, represents the value of the truncated sum in (24), while the third value is the regularized remainder, (26), as evaluated via NIntegrate. The fourth value for each calculation of $\ln \Gamma(z)$ is the Stokes discontinuity term, which according to (32) is zero for $|\theta| < \pi/2$ and is purely logarithmic for θ over $(-\pi, -\pi/2)$. The fifth value, denoted by Total, represents the sum of the four preceding values, while the final value is the actual value of $\ln \Gamma(z)$ using LogGamma[z] in Mathematica.

Since there is no optimal point of truncation, the results in Table 3 for $N > 3$ are mainly dominated by the truncated sum and its regularized remainder. In fact, both values dominate so much that many decimal places are cancelled as observed for $N = 30$ and 50 in Table 1. Once again, pressure is being put on the accuracy of the final total. For example, for $N = 9$ and $\theta = -6\pi/13$, both the truncated sum and the regularized remainder are $\mathcal{O}(10^{13})$, which means a loss of thirteen decimal places when they are summed. Since the accuracy and precision goals were set to 30, this implies that the sum of the truncated series and the regularized remainder should only be accurate to 17 decimal places. Fortunately, the total value agrees with the value of $\ln \Gamma(z)$ to 28 decimal places because the working precision was set much higher (to 60) than the precision and accuracy goals.

Table 3. $\ln \Gamma(z)$ via (25) with $|z| = 1/10$ for various values of the truncation parameter, N, and $\arg z$.

N	θ	Quantity	Value
3	$-\pi/6$	$F(z)$	$1.7580388820525170130082315272 0 + 0.3815836583429962744746012315 6\,i$
		TS	$0.7216878364870322056364359756 2 - 2.3611111111111111111111111111 1\,i$
		$R_3^{SS}(z)$	$-0.2230295240392980035338083054 + 2.5252425215223733626324734008 7\,i$
		$SD_0^{SS}(z)$	0
		Total	$2.2566971945002512151108591974 2 + 0.5457150687542585259959635213 2\,i$
		$\ln \Gamma(z)$	$2.2566971945002512151108578462 4 + 0.5457150687542585259959643014 2\,i$
9	$-6\pi/13$	$F(z)$	$1.8864834197022194013599633847 8 + 1.0353560661019478221434799816 0\,i$
		TS	$2.8756254802079423919856 1 \times 10^{13} - 7.0718880105443602759020497 \times 10^{12}\,i$
		$R_9^{SS}(z)$	$-2.8756254802079022596667 5 \times 10^{13} + 7.0718880105448298576076792 \times 10^{12}\,i$
		$SD_0^{SS}(z)$	0
		Total	$2.2878066008474191475281948431 9 + 1.5049377717315066635107508099 5\,i$
		$\ln \Gamma(z)$	$2.2878066008474191475281948431 9 + 1.5049377717315066635107508099 4\,i$
8	$-8\pi/15$	$F(z)$	$1.9381187512092596114681501910 0 + 1.1837212717094993912174218477 9\,i$
		TS	$-6.2877629092633776151775 \times 10^{10} + 1.3406164598876901506339999 \times 10^{10}\,i$
		$R_8^{SS}(z)$	$6.2877629092235027313400 9 \times 10^{10} - 1.3406164598401660891969575 \times 10^{10}\,i$
		$SD_1^{SS,L}(z)$	$0.7611055764025938317897 2540915 + 0.0752793665738315377337302724 0\,i$
		Total	$2.3004754892372078654085992631 4 + 1.7342412526537551438457543407 0\,i$
		$\ln \Gamma(z)$	$2.3004754892372078654085992631 4 + 1.7342412526537551438457543407 0\,i$
4	$-15\pi/16$	$F(z)$	$2.3366849216224335155320680197 0 + 1.8259169045164836145598810671 15\,i$
		TS	$-42.6005588915275445365792170 00 + 64.608976391113373494063197638 78\,i$
		$R_4^{SS}(z)$	$42.0897905773173511704765793260 - 64.547655655018321335102702862 63\,i$
		$SD_1^{SS,L}(z)$	$0.5412330636654141611820818172 5 + 1.0726574746608308435190398144 47\,i$
		Total	$2.3671496710776543106115121621 5 + 2.9598951152723666170394156577 12\,i$
		$\ln \Gamma(z)$	$2.3671496710776543106115121621 5 + 2.9598951152723666170394156577 12\,i$

Although $F(\exp(i\theta)/10)$ provides a substantial contribution to $\ln \Gamma(\exp(i\theta)/10)$, it is no longer accurate. The truncated sum is capable of improving the accuracy slightly for small values of the truncation parameter. For example, when the truncated sum is added to $F(z)$ for $N = 3$ and $\theta = -\pi/6$, the real part is closer to the real part of $\ln \Gamma(\exp(-i\pi/6)/10)$, but not so the imaginary part. In fact, all the results are dominated by the truncated sum and its regularized remainder, but since they act against each other, their sum is not as large as Stirling's formula. Nevertheless, one cannot neglect the remainder as in standard Poincaré asymptotics. In order to obtain the exact value of $\ln \Gamma(\exp(-i\pi/6)/10)$ via (25), the remainder must counterbalance the truncated sum, which will only occur if the regularization has been performed correctly. When the regularized remainder is included in the total, exact values of $\ln \Gamma(\exp(i\theta)/10)$ are obtained. For $\theta < -\pi/2$, however, the Stokes discontinuity term must be included. In fact, $SD_1^{SS,L}(z)$ is greater than the sum of the truncated series and the regularized remainder, which highlights its importance outside the primary Stokes sector.

So far, we have managed to verify the asymptotic forms in (25) connected with Stokes sectors. Now we consider the asymptotic forms for the two Stokes lines. As θ is fixed in both asymptotic forms, the Stokes discontinuity term will only depend upon $|z|$. In other words, it is solely real. Furthermore, since $TS_N(z)$ depends only on odd powers of z in (24), $TS_N(z)$ and $R_N^{SL,}(z)$ must be imaginary along both Stokes lines. This is consistent with Rule D in ([15], Chapter 1), which states that an asymptotic series crossing a Stokes line generates a discontinuity that is $\pi/2$ out of phase with the series on the line.

The third code in the appendix of [11] implements the second and fourth asymptotic forms of (25) in Mathematica. This program is very different from the previous program because it includes a Which statement in the Do loop. This is necessary because the singularity in the Cauchy principal value integral in (27) alters with each value of n. Moreover, the integral has been divided into several smaller intervals in order to achieve the best possible accuracy. The interval in which the singularity is situated is then determined via the Which statement. This interval is, in turn, divided into two intervals to avoid the singularity in accordance with the definition of the principal value. To ensure that the principal value is evaluated without encountering convergence problems, the option Method—>PrincipalValue must also be introduced into NIntegrate. Finally, in order to achieve the same accuracy as in Table 3, WorkingPrecision has been increased to 80. Hence the program takes much longer to execute.

Table 4 presents a sample of the results generated by running the third program in [11] with the variable modz set equal to 3. A similar set of calculations was performed for modz equal to 1/10, whose results appear in Table 6 of [11], but for brevity, they are not presented here. Although both Stokes lines were considered by putting the variable theta in the program equal to $\pm Pi/2$, only the results for positive values of theta are presented here, again for the sake of brevity. The calculations took much longer for larger values of the truncation parameter, ranging from 26 hrs for $N=1$ to 47.5 hrs for $N=50$. Because the values of $F(3\exp(i\pi/2))$ and $SD_0^{SL}(3\exp(i\pi/2))$ are independent of the truncation parameter, they only appear once at the top of the table, while their sum appears immediately below them in the row labelled Combined. As stated above, the Stokes discontinuity term is purely real, whereas the truncated sum and regularized value of the remainder are purely imaginary. Therefore, the real part of the value in the Combined row represents the real part of $\ln\Gamma(3\exp(i\pi/2))$, which can be checked by comparing it with the real part of $\ln\Gamma(3\exp(3i\pi/2))$ at the bottom of the table. Thus, the Stokes discontinuity term only corrects the real part of Stirling's formula on a Stokes line. On the other hand, the imaginary part of $\ln\Gamma(3\exp(i\pi/2))$ can only be calculated exactly if the regularization of (25) has been performed correctly. The last decimal figure of the imaginary part of $\ln\Gamma(3\exp(i\pi/2))$ was printed out as a 6 instead of a 5, because the accuracy was set to 25 decimal places in the output stage. Since more than 25 figures appear in the table, this statement should have been modified to consider a higher level of accuracy. Therefore, we should only be worried when the results agree for less than 25 decimal places. The redundant places have been introduced here to indicate that the results in the Total column have been computed via a different approach from the LogGamma routine in Mathematica at the bottom of the table. That is, we should expect differences to occur at some stage, but only outside the specified level of accuracy.

In the table we see that the regularized value of the remainder decreases steadily until the truncation parameter reaches N_{OP} around 11, before it begins to diverge. Note that the imaginary part of the Total value for $N=1$ is only accurate to 6 decimal places compared with the imaginary part of $\ln\Gamma(3\exp(i\pi/2))$. As discussed previously, this arises because the power of n in the denominator of $R_1^{SL}(z)$ is zero when N is equal to unity. Though not displayed in the table, the remainder at the optimal point of truncation, $R_{11}(3\exp(i\pi/2))$, has a minimum magnitude of $\mathcal{O}(10^{-11})$. Beyond this point, the magnitude of the regularized value of the remainder increases so that its magnitude is $\mathcal{O}(10^{-6})$ for $N=20$. By the time $N=30$, both the truncated series and regularized value of the remainder dominate the calculation, but since they act against each other, they combine to yield the extra imaginary value enabling the imaginary part in the Combined row to agree with $\ln\Gamma(3\exp(i\pi/2))$. In fact, the most surprising result in the table is the last result for $N=50$ because at least 25 decimal places cancel before we obtain the regularized value for the entire asymptotic series. As mentioned previously, the cancellation of these decimal places puts pressure on the accuracy and precision goals, which have been set to 30, as stated above. Fortunately, because WorkingPrecision was set to 80, it appears that the neglected terms in setting a limit of 10^5 in the summation are negligible. Thus, the remainder has been evaluated to a much greater accuracy than specified by the accuracy and precision goals in the program. Consequently, the Total value for $N=50$ agrees with the actual value of $\ln\Gamma(3\exp(i\pi/2))$.

Table 4. $\ln\Gamma(3\exp(i\pi/2))$ via (25) for various values of N.

N	Quantity	Value
	$F(3\exp(i\pi/2))$	$-4.3427565915140719616112579569 - 0.4895612973931192354299251350522\,i$
	$SD_0^{SL}(3\exp(i\pi/2))$	$3.2562060786428283676798164684 \times 10^{-9}$
	Combined	$-4.3427565882578658829684295892 - 0.4895612973931192354299251350522\,i$
1	TS	0
	$R_1^{SL}(3\exp(i\pi/2))$	$-0.027884089465369119932177792256\,i$
	Total	$-4.3427565882578658829684295892 - 0.5174453868584883553621029142779\,i$
6	TS	$0 - 0.0278842394252900781377131527007\,i$
	$R_6^{SL}(3\exp(i\pi/2))$	$0 - 1.8907874105339892863379255 \times 10^{-8}\,i$
	Total	$-4.3427565882578658829684295892 - 0.5174455557262834189075311513225\,i$
9	TS	$0 - 0.0278842563298976281594154202028\,i$
	$R_9^{SL}(3\exp(i\pi/2))$	$0 + 3.2562060786428283676798164 \times 10^{-9}\,i$
	Total	$-4.3427565882578658829684295892 - 0.5174455557262834189075311513225\,i$
15	TS	$0 - 0.0278842691899612112195938305035\,i$
	$R_{15}^{SL}(3\exp(i\pi/2))$	$0 + 1.0856797027741987814423624 \times 10^{-8}\,i$
	Total	$-4.3427565882578658829684295892 - 0.5174455557262834189075311513225\,i$
30	TS	$0 - 52.07235660935681329352406137393\,i$
	$R_{30}^{SL}(3\exp(i\pi/2))$	$0 + 52.04447235102364903587444012131\,4\,i$
	Total	$-4.3427565882578658829684295892 - 0.5174455557262834189075311513225\,i$
50	TS	$0 - 6.490840984334943518162045\,3 \times 10^{25}\,i$
	$R_{50}^{SL}(3\exp(i\pi/2))$	$0 + 6.490840984334943518162045\,3 \times 10^{25}\,i$
	Total	$-4.3427565882578658829684295892 - 0.5174455557262834189075311513225\,i$
	$\ln\Gamma(3\exp(i\pi/2))$	$-4.3427565882578658829684295892 - 0.5174455557226834189075311513225\,i$

So far, we have not seen any evidence of Stokes smoothing as espoused by Berry [18], Olver [20] and Paris, Kaminski and Wood [12,13,25]. As indicated earlier, smoothing implies that there is no discontinuity in the vicinity of a Stokes line, whereas we have been able to obtain exact values of $\ln\Gamma(z)$ near Stokes lines assuming the existence of a discontinuity. Because such smoothing occurs rapidly in the vicinity of Stokes lines, it could perhaps be argued that we have not investigated the asymptotic behaviour of $\ln\Gamma(z)$ sufficiently close to the Stokes lines. If a rapid transition does occur, then it means that we have still not exactified the Stokes approximation in the vicinity of the Stokes lines. From Figure 3, which represents the situation for $|z|=3$, Stokes smoothing is expected to be most pronounced for θ lying between $13\pi/32$ and $19\pi/32$. Alternatively, the Stokes multiplier is expected to be quite close to $1/2$ for small values of δ, where $\theta = \pi(1/2+\delta)$ and $|\delta| < 3/32$. On the other hand, if the conventional view of the Stokes phenomenon is valid, then the Stokes multiplier S^+ will equal unity for $0 < \delta < 1$ and zero for $-1 < \delta < 0$ according to (31). Thus, a narrow region of positive and negative values of δ exists, where one of the views can be disproved. In summary, introducing very small values of δ into the respective asymptotic forms of (24) should not yield exact values of $\ln\Gamma(z)$ if smoothing occurs since the Stokes multiplier should be close to $1/2$ and not toggle between zero and unity according to the sign of δ.

Table 5 presents a small sample of the results obtained by running the second program in [11] for $|z| = 3$ and various values of δ, where $\theta = (1/2 + \delta)\pi$. The code was run for different values of N except those close to unity for the reason given above. For each positive value of δ, there are three rows of values, while for each negative value there are only two rows because the Stokes discontinuity term is zero. The first row for each value of δ, labelled LogGamma[z] in the Method column, represents the value obtained from the LogGamma routine in Mathematica. Depending upon the sign of δ, the second row displays the Stokes discontinuity term. In general, this term was found to possess real and imaginary parts of $\mathcal{O}(10^{-8})$ or even a couple of orders lower. The next value for each value of δ is labelled either 1st AF or 3rd AF in the Method column according to whether the first or third asymptotic form in (25) was used to calculate the value of $\ln \Gamma(z)$. For brevity, the values of the truncated sum, the regularized value of the remainder and Stirling's formula do not appear in the table.

Table 5. $\ln \Gamma(3 \exp(i(1/2 + \delta)\pi))$ via (24) for various values of δ.

δ	Method	Value
1/10	LogGamma[z]	$-5.1085546405054331385771175 - 2.4350486413361823958761 3036\,i$
	$SD_1^{SS,U}(z)$	$0.0000000146924137960847328 + 0.0000000072492097873547 7097\,i$
	1st AF	$-5.1085546405054331385771175 - 2.4350486413361823958761 3036\,i$
$-1/10$	LogGamma[z]	$-3.1156770612855851062960250 + 0.7915271748617870066 3566144\,i$
	3rd AF	$-3.1156770612855851062960250 + 0.7915271748617870066 3566144\,i$
1/100	LogGamma[z]	$-4.4448078360199294879676721 - 0.6842653947061931557 9497619\,i$
	$SD_1^{SS,U}(z)$	$0.0000000054543808883397577 - 0.0000000036684566186 1183983\,i$
	1st AF	$-4.4448078360199294879676721 - 0.6842653947061931557 9497619\,i$
$-1/100$	LogGamma[z]	$-4.2360547825638102221663061 - 0.3568100346112583420 9091866\,i$
	3rd AF	$-4.2360547825638102221663061 - 0.3568100346112583420 9091866\,i$
1/1000	LogGamma[z]	$-4.3531757575591613140088085 - 0.5338516610090575526 1595669\,i$
	$SD_1^{SS,U}(z)$	$0.0000000065016016472424544 - 0.0000000003854594562 8149871\,i$
	1st AF	$-4.3531757575591613140088085 - 0.5338516610090575526 1595669\,i$
$-1/1000$	LogGamma[z]	$-4.3322909095906129602545969 - 0.5011013034712617095 1651903\,i$
	3rd AF	$-4.3322909095906129602545969 - 0.5011013034712617095 1651903\,i$
1/10,000	LogGamma[z]	$-4.3438006028809735966127763 - 0.5190833852796876654 0121412\,i$
	$SD_1^{SS,U}(z)$	$0.0000000065123040290213875 - 0.0000000000385647689 8298508\,i$
	1st AF	$-4.3438006028809735966127763 - 0.5190833852796876654 0121412\,i$
$-1/10,000$	LogGamma[z]	$-4.3417121085407199183370966 - 0.5158083447041416547 8538635\,i$
	3rd AF	$-4.3417121085407199183370966 - 0.5158083447041416547 8538635\,i$
1/20,000	LogGamma[z]	$-4.3438006028809735966127763 - 0.5190833852796876654 0121412\,i$
	$SD_1^{SS,U}(z)$	$0.0000000065123851251757157 - 0.0000000001928245580 002624\,i$
	1st AF	$-4.3438006028809735966127763 - 0.5190833852796876654 0121412\,i$
$-1/20,000$	LogGamma[z]	$-4.3422344065179726897501879 - 0.5166268728896735213 9359494\,i$
	3rd AF	$-4.3422344065179726897501879 - 0.5166268728896735213 9359494\,i$

It should be noted that when $|\delta|$ is extremely small, e.g., $\mathcal{O}(10^{-5})$, NIntegrate experiences convergence problems because the integration is now too close to the singularities on the Stokes line. For example, when $\delta = 10^{-5}$, the program printed out a value that agreed with the actual value to 25 decimal places for the real part, but the imaginary part only agreed to 18 decimal places. Although this calculation is not presented in the table, it does represent a degree of success since the imaginary part of the Stokes discontinuity term is $\mathcal{O}(10^{-12})$. That is, the Stokes discontinuity term still had to be correct to the first six decimal places for the agreement to occur at 18-th decimal place.

For $\delta > 0$ in the table, the first asymptotic form in (25) yields the exact value of $\ln \Gamma(z)$ even though the Stokes discontinuity term is very small. Nevertheless, in the case of Stokes smoothing, this term should be almost half the values appearing in the table. For $\delta < 0$, if smoothing occurs, then the third asymptotic form in (25) should also not yield exact values of $\ln \Gamma(z)$ because it is missing almost half the Stokes discontinuity term. Yet, we observe the opposite; the third asymptotic form yields exact values of $\ln \Gamma(z)$ for $\delta < 0$. Therefore, Stokes smoothing does not occur. These results are discussed in more detail in [26].

An explanation why Stokes smoothing is fallacious appears in ([6], Section 6.1), where it is shown that the form for the Stokes multiplier given by Berry and Olver is based on applying standard asymptotic techniques. Olver's "rigorous proof" [20] involves truncating an asymptotic series via Laplace's method. Since only the lowest order terms are retained in this approach, Olver arrives at the error function result for the Stokes multiplier. The neglected higher order terms are not only divergent, but are also extremely difficult to regularize. If they could be regularized, then they would produce the necessary corrections to turn the smooth function in Figure 3 into a step-function, thereby confirming the conventional view of the Stokes phenomenon.

4. Mellin–Barnes Regularization

In the preceding section we were able to exactify Stirling's formula by carrying out hyperasymptotic calculations of the asymptotic forms in (25). However, there were two drawbacks with the numerical analysis. The first is that an upper limit was applied to the infinite sums appearing in the expressions for the regularized value of the remainder. Despite this, the regularized values were extremely accurate for an upper limit of 10^5 in (26) and (27). This results in the second drawback, the considerable effort required to calculate the remainder. Ideally, we do not want to truncate any result here so that we can dispel any doubt that we are evaluating an approximation. If the infinite sum over n can be replaced by a single result, then there will be a huge reduction in the execution time since there would be only one call to NIntegrate. Such an expression emerges when we consider Mellin–Barnes regularization of $\ln \Gamma(z)$ in the following theorem.

Theorem 2. *Via the Mellin–Barnes (MB) regularization of the asymptotic series $S(z)$ given by either (9) or (11), the logarithm of the gamma function can be expressed as*

$$\ln \Gamma(z) = \left(z - \frac{1}{2}\right) \ln z - z + \frac{1}{2} \ln(2\pi) + z \sum_{k=1}^{N-1} \frac{(-1)^k}{(2z)^{2k}} \Gamma(2k-1) c_k(1)$$

$$- 2z \int_{\substack{c-i\infty \\ \text{Max}[N-1,1/2] < c = \Re s < N}}^{c+i\infty} ds \left(\frac{1}{2\pi z}\right)^{2s} \frac{e^{\pm 2Mi\pi s}}{e^{-i\pi s} - e^{i\pi s}} \zeta(2s) \Gamma(2s-1) + S_{MB}(M,z), \quad (36)$$

where

$$S_{MB}(M,z) = \pm \lfloor M/2 \rfloor \ln\left(-e^{-2i\pi z}\right) - \left(\frac{1-(-1)^M}{2}\right) \ln\left(1 - e^{\pm 2i\pi z}\right), \quad (37)$$

for $(\pm M - 1)\pi < \theta = \arg z < (\pm M + 1)\pi$, and $M \geq 0$, but excluding θ equal to half-integer values of π. The strips involving θ represent domains of convergence for the MB integral in (36) with the upper-signed forms applying to positive θ and the lower-signed ones to negative θ. For $\theta = \pm(M - 1/2)\pi$, $S_{MB}(M,z)$ reduces to

$$S_{MB}(M,z) = \left[\left(\frac{(-1)^M - 1}{2}\right) \ln\left(1 - e^{-2\pi |z|}\right) + 2(-1)^{M+1} \left\lfloor \frac{M}{2} \right\rfloor \pi |z|\right], \quad (38)$$

while for $\theta = \pm(M+1/2)\pi$, it is given by

$$S_{MB}(M,z) = \left[\left(\frac{(-1)^M - 1}{2}\right) \ln\left(1 - e^{-2\pi |z|}\right) + 2\pi |z| \left((-1)^M \left\lfloor \frac{M}{2} \right\rfloor + \frac{(-1)^M - 1}{2}\right)\right]. \quad (39)$$

Proof. For the sake of brevity, the proof is omitted as it appears in ([11], Section 4). □

Comparing the above results with those in Theorem 1, we see that not only is the remainder of the asymptotic series in (11) expressed as an MB integral, but there are also no discontinuities from crossing Stokes lines. Instead, the MB integral is valid over a strip or domain of convergence with the Stokes lines situated inside the domains of convergence. Although (38) and (39) apply at half integer values of π, they no longer represent Stokes lines as in Theorem 1. They have been isolated here as a result of the MB regularization of $S(z)$ since $\ln \Gamma(z)$ itself possesses jump discontinuities at $\theta = (l+1/2)\pi$, for l, an integer not equal to 0 or -1. Thus, MB regularization produces a totally different representation of the original function from its asymptotic forms, and relies on the continuity of the function. If the original function possesses discontinuities as $\ln \Gamma(z)$ does, then the MB-regularized value will not yield the value of the original function unless the analysis is adapted as explained in the proof.

Since Stokes multipliers do not appear in the MB regularization of $\ln \Gamma(z)$ for $\theta = \pm \pi/2$, this implies that the Stokes discontinuities obtained by Borel summation can be fictitious. That is, although we observed jumps in the Stokes multipliers at $\theta = \pm \pi/2$, it does not mean that $\ln \Gamma(z)$ is necessarily discontinuous there. In fact, discontinuities will only occur at Stokes lines if the original function possesses singularities on them. In the case of $\ln \Gamma(z)$ singularities only occur when $\theta = \pm(l+1/2)\pi$ and $l > 0$.

Another feature of the above results is that the sum over n has vanished. It has effectively been replaced by the Riemann zeta function. As a consequence, we now only have one integral to evaluate the remainder in (11). This will save much computational effort provided that the software package is able to evaluate the zeta function extremely accurately. Fortunately, this is accomplished using the Zeta routine in Mathematica [17].

Although the results in Theorem 2 have been proven, as in the case of Theorem 1, we cannot be certain that they are indeed valid because we have observed in the case of "Stokes smoothing" that proofs in asymptotics are not reliable unless they are verified by numerical analysis. Since the results in Theorem 1 have already been validated, we can use them to establish the validity of the MB-regularized forms in Theorem 2. Therefore, the next section presents a numerical analysis where the MB-regularized forms for $\ln \Gamma(z^3)$ are matched with the corresponding Borel-summed forms in Theorem 1.

5. Further Numerical Analysis

According to the definition of the regularized value [4–7], it must be invariant irrespective of how it is obtained. Therefore, we need to demonstrate that the MB-regularized forms in Theorem 2 yield identical values to the Borel-summed forms in Theorem 1, especially for the higher Stokes sectors and lines not studied previously. To access the higher/lower sectors or lines, higher powers of the variable z need to considered such as z^3 in $\ln \Gamma(z)$. This is tantamount to finding an asymptotic solution to a problem, which happens to yield the asymptotic forms of $\ln \Gamma(z^3)$. In this case the principal branch is still $(-\pi, \pi]$, but Mathematica is only able to evaluate $\ln \Gamma(z^3)$ for θ over $(-\pi/3, \pi/3]$.

From Theorem 2, two different representations exist for the regularized value of $\ln \Gamma(z)$ since replacing M by either $M-1$ or $M+1$ in (36) produces a different asymptotic form, where each is valid over one half of the domain of convergence for $M = M$. For example, the upper-signed version of (36) is valid for $\pi < \theta < 3\pi$ when $M = 2$, while for $M = 1$ and $M = 3$, it is only valid over $0 < \theta < 2\pi$ and $2\pi < \theta < 4\pi$, respectively. Thus, the $M = 1$ result is valid for the bottom half of the domain of convergence for $M = 2$, while the $M = 3$ result applies to the top half of the domain of convergence for $M = 2$. This means that we are not only able to evaluate $\ln \Gamma(z)$ for higher/lower values of θ or $\arg z$, but we can check the results against the asymptotic forms from overlapping domains of convergence. In addition, the $M = 0$ results can be checked with the values of $\ln \Gamma(z^3)$ evaluated by Mathematica. Finally, we can check to see whether the MB-regularized forms of $\ln \Gamma(z^3)$ yield identical values to the corresponding Borel-summed asymptotic forms in Theorem 1. Previously, we had no method of checking whether the Borel-summed asymptotic forms for $\ln \Gamma(z)$ outside the principal branch were correct. Now this problem can be tackled by comparing the resulting Borel-summed asymptotic forms when z is replaced by a power of itself with the corresponding MB-regularized forms.

If z is replaced by z^3, then for $M = 0$ or $-\pi/3 < \theta < \pi/3$, (36) becomes

$$\ln \Gamma\left(z^3\right) = F\left(z^3\right) + TS_N\left(z^3\right) + \Delta \ln \Gamma\left(z^3\right), \tag{40}$$

where

$$\Delta \ln \Gamma\left(z^3\right) = -2z^3 I_U(M=0) = -2z^3 I(0), \tag{41}$$

and

$$I_U(M) = \int_{\substack{c-i\infty \\ \text{Max}[N-1,1/2] < c = \Re s < N}}^{c+i\infty} ds \, \frac{(1/2\pi z^3)^{2s} e^{\pm 2i\pi Ms}}{e^{-i\pi s} - e^{i\pi s}} \, \zeta(2s) \, \Gamma(2s-1). \tag{42}$$

In (40), $TS_N(z)$ represents the truncated part of the asymptotic series, $S(z)$ at N, as in (25), while the subscript U or L in (42) denotes whether the upper-signed or lower-signed version has been used. For $M = 0$, the subscript is dropped. Thus, $\ln \Gamma(z^3)$ is composed of Stirling's formula, the truncated series and an MB integral as the regularized value of the remainder. On the other hand, for $M = 1$, the upper-signed version of (36) yields

$$\Delta \ln \Gamma\left(z^3\right) = -2z^3 I_U(1) - \ln\left(1 - e^{2i\pi z^3}\right). \tag{43}$$

The domain of convergence for this integral is $0 < \theta < 2\pi/3$, but it is not valid when $\theta = \pi/2$ since $S_{MB}(M, z^3)$ is discontinuous whenever $\theta = \pm(M \pm 1/2)\pi/3$ excluding $M = 0$. For $\theta = \pi/6$, (38) can be used, but all that happens is the logarithmic term on the right-hand side of (43) is replaced by $\ln\left(1 - e^{-2\pi |z|^3}\right)$, which indicates that there is no discontinuity in $\ln \Gamma(z^3)$ at $\theta = \pi/6$.

For $M = 1$, when $\theta = \pm(M + 1/2)\pi/3$, $\theta = \pm \pi/2$. The upper value of θ lies in the domain of convergence for (43). In (36), we substitute (39) with z equal to z^3 for $S_{MB}(M, z)$. Then we arrive at

$$\Delta \ln \Gamma\left(z^3\right) = -2z^3 I_U(1) - 2\pi |z|^3 - \ln\left(1 - e^{-2\pi |z|^3}\right). \tag{44}$$

As a result of the penultimate term, we expect a discontinuity when (44) is evaluated later. In addition, we can replace $F(z^3)$ and $TS_N(z^3)$ in (40) by $F(-i|z|^3)$ and $TS_N(-i|z|^3)$, respectively, while z^3 in (44) can be replaced by $-i|z|^3$.

When compared with (40), we see that (43) and (44) possess extra terms, which are similar to the Stokes discontinuity term in the Borel-summed asymptotic forms of Theorem 1. The difference here is that the lines of discontinuity are located in the domains of convergence. Thus, the asymptotic form is only different on the lines, whereas with Stokes lines, the regularized value is different before, on and after them. Moreover, we expect both forms for $\ln \Gamma(z^3)$ to yield identical values when the domains of convergence overlap, i.e., over $(0, \pi/3)$. This does not occur with the Stokes phenomenon, indicating again that MB regularization is different from Borel summation.

For $M = 2$ and 3, the upper-signed version of (36) with z replaced by z^3 yields

$$\Delta \ln \Gamma\left(z^3\right) = \begin{cases} -2z^3 I_U(2) + \ln\left(-e^{-2i\pi z^3}\right), & \pi/3 < \theta < \pi, \\ -2z^3 I_U(3) + \ln\left(-e^{-2i\pi z^3}\right) - \ln\left(1 - e^{2i\pi z^3}\right), & 2\pi/3 < \theta < 4\pi/3. \end{cases} \tag{45}$$

These results, which are similar to (43) except for the logarithmic terms, are not valid for $\theta = \pi/2$ and $\theta = 5\pi/6$.

For $M = 1$ and $\theta = \pm(M + 1/2)\pi/3$, (39) was used to derive the asymptotic form of $\ln \Gamma(z^3)$. However, when $\theta = \pm(M - 1/2)\pi/3$, θ can also equal $\pi/2$, but now the upper-signed version of (42) with $M = 2$ applies. Moreover, $S_{MB}(M, z)$ is determined by putting $M = 2$ and replacing z by z^3 in (38). Hence for $M = 2$ and $\theta = \pi/2$, we find that

$$\ln \Gamma\left(z^3\right) = F\left(z^3\right) + TS_N\left(z^3\right) - 2z^3 I_U(2) - 2\pi|z^3|. \tag{46}$$

For $\theta = 5\pi/6$, we have either $M = 3$ when $\theta = (M - 1/2)\pi/3$ or $M = 2$ when $\theta = (M + 1/2)\pi/3$. In the first case $S_{MB}(M, z)$ is given by (38) with $z = z^3$ and $M = 3$. In the second case (39) applies with $z = z^3$ and $M = 2$. Thus, we obtain

$$\Delta \ln \Gamma\left(z^3\right) = \begin{cases} -2z^3 I_U(3) - \ln\left(1 - e^{-2\pi|z^3|}\right) + 2\pi|z^3|, & M = 3, \\ -2z^3 I_U(2) + 2\pi|z^3|, & M = 2. \end{cases} \tag{47}$$

The corresponding lower-signed results from (36) with z replaced by z^3 are simply complex conjugates of the above results. For brevity, they are not presented here. However, the interested reader will find them in ([11], Section 5).

Two separate numerical analyses will be presented here: the first aims to show the agreement between the MB-regularized asymptotic forms for $\ln \Gamma(z^3)$ and their Borel-summed counterparts, and the second deals with the behaviour of $\ln \Gamma(z^3)$ at the Stokes lines/rays. The first one includes an explanation of how to evaluate $\ln \Gamma(z^3)$ from the MB-regularized asymptotic forms. Then the results are compared with the Borel-summed asymptotic forms in Section 3 with z replaced by z^3. We shall observe that although both MB-regularized asymptotic forms are defined at each Stokes line, they give incorrect values of $\ln \Gamma(z^3)$ with the difference being discontinuous jumps of $2\pi i$. The second study at the Stokes lines/rays will be concerned with obtaining the correct values of $\ln \Gamma(z^3)$ via both the Borel-summed and MB-regularized asymptotic forms by applying the Zwaan–Dingle principle [6,15], which states that an initially real function cannot suddenly become imaginary.

Since there are no Stokes lines of discontinuity in the above results, there are always two MB-regularized asymptotic forms that yield the values of $\ln \Gamma(z^3)$ for all values of θ or arg z, except when $\theta = k\pi/3$ and k is an integer. Thus, the values from two different asymptotic forms for the regularized value of $\ln \Gamma(z^3)$ can be checked against each other, which is simply not possible with Borel-summed results.

Because it represents a value where standard Poincaré asymptotics breaks down, we shall carry out the numerical study of the above results with $|z|$ set equal to $1/10$ as before. Note that the actual variable in the above asymptotic forms is $2\pi z^3$. Therefore, we are dealing with a very small value, which means that both the truncated series, $TS_N(z)$, and the MB integral in the above results begin to diverge very rapidly for relatively small values of the truncation parameter, e.g., $N = 4$. Consequently, a cancellation of many decimal places will occur when adding $TS_N(z)$ to the MB integral. Despite the accuracy and precision goals being set to 30, one may not necessarily obtain a final value that is accurate at this level even though WorkingPrecision is now set higher to 80, not 60 as in Section 3. As stated earlier, the problem can be overcome by specifying much larger values of AccuracyGoal, PrecisionGoal and WorkingPrecision in NIntegrate, but it will come at the expense of computing time.

Table 6 presents a very small sample of the results from the fourth program in the appendix of [11] for various values of N and θ or arg z. There are five sets of results, four with θ positive, and one where it is negative. The first row of each calculation gives the value of Stirling's formula, while the next row displays the value of $TS_N(z^3)$. Then the remainder denoted by MB Int. appears. As mentioned earlier, because the domains of convergence of the MB integrals overlap one another, two different MB integrals are computed for the remainder. The first MB integral is represented by M1, while the second is represented by M2. The second MB integral is not evaluated when $\theta = l\pi/3$ and l, an integer, as demonstrated by the third calculation. The values of N and θ appear together with the value of the first MB integral in each set. The values of $S_{MB}(M,z^3)$ are displayed in the rows immediately after the MB integrals. Then the results for the entire asymptotic form appear, which can be compared with the value of LogGamma from Mathematica.

Table 6. Values of $\ln\Gamma(z^3)$ with $|z| = 1/10$ and varying N and θ in the Mellin–Barnes (MB)-regularized forms.

θ	N	Quantity	Value
$-\pi/12$	3	$F(z^3)$	$4.366669184946739483968199 3920 + 0.3977354871318634708519397906\,i$
		$TS_3(z^3)$	$1.964244428861064224518267 \times 10^6 - 1.9641265777308664665975342 \times 10^6\,i$
		MB Int (M1 = 0)	$-1.964241888183123016922580 \times 10^6 + 1.9641269658010120780 12431 \times 10^6\,i$
		$S_{MB}(0,z^3)$	0
		Total via M1	$6.907347126154335171351562 3993 + 0.7858054943246012439632804975\,i$
		MB Int (M2 = −1)	$-1.964246960282777058252170 \times 10^6 + 1.9641277489793789404 48405 \times 10^6\,i$
		$S_{MB}(1,z^3)$	$5.072099654041329589999967 5136 - 0.7831783668624359743794750 23745\,i$
		Total via M2	$6.907347126154335171351562 3992 + 0.7858054943246012439632804 9761\,i$
		LogGamma[zcube]	$6.907347126154335171351562 3992 + 0.7858054943246012439632804 9761\,i$
$7\pi/24$	4	$F(z^3)$	$4.379070029918803338594425 0366 - 1.3800125993861205862081694 4744\,i$
		$TS_4(z^3)$	$3.037180727461076973245 84 \times 10^{11} - 7.332351579151624817641834 \times 10^{11}\,i$
		MB Int (M1 = 0)	$-3.037180727435784782 16056 \times 10^{11} + 7.332351579137933793 18864 \times 10^{11}\,i$
		$S_{MB}(0,z^3)$	0
		Total via M1	$6.908289138446136783194635 3384 - 2.7491150447054011628383556 15704\,i$
		MB Int (M2 = 1)	$-3.037180727486495598272 33 \times 10^{11} + 7.332351579146857525 73952 \times 10^{11}\,i$
		$S_{MB}(1,z^3)$	$5.071081611176593533945 417904 - 1.175195955088607468412668 235518\,i$
		Total via M2	$6.908289138446136782740881 0781 - 2.7491150447054011654092630 40026\,i$
		LogGamma[zcube]	$6.908289138446136782740881 0777 - 2.7491150447054011654092630 38133\,i$
$\pi/3$	2	$F(z^3)$	$4.380723927974723404859370 8927 - 1.5739379194484864124697843 3502\,i$
		$TS_2(z^3)$	$-83.333333333333333333333 33333 + 0\,i$
		MB Int (M1 = 1)	$80.791062865366238781576 251609 + 0\,i$
		Log. Term (M1 = 1)	$5.069879857507399578675721 5377 - 1.5676547341413068259 9285904825\,i$
		Total via M1	$6.908333317515028431778010 7065 - 3.1415926535897932384626433 83279\,i$
		LogGamma[zcube]	$6.908333317515028431778010 7065 - 3.1415926535897932384626433 83279\,i$
$4\pi/7$	5	$F(z^3)$	$4.367183997626082261177386 0371 + 0.4544218394081292974701 9906926\,i$
		$TS_5(z^3)$	$-5.952382718395083331 82790 \times 10^{17} - 7.737535164228715974 701668 \times 10^{11}\,i$
		MB Int (M1 = 1)	$5.952382718395083306 50668 \times 10^{17} + 7.737535164239864652 939712 \times 10^{11}\,i$
		$S_{MB}(1,z^3)$	$5.067421650498372367 63320 01993 + 2.466433873147543609 54113189663\,i$
		Total via M1	$6.902482816162893396835374 9191 + 4.035723536360127523 53352062801\,i$
		MB Int (M2 = 2)	$5.952382718395083357 23002 \times 10^{17} + 7.737535164233152 240154912 \times 10^{11}\,i$
		$S_{MB}^U(2,z^3)$	$5.071081611176593533945417 904 - 1.1751959550886074684126 68235518\,i$
		Total via M2	$6.902482816162893380035877 3203 + 4.035723536360127857 7347290435\,i$
$8\pi/9$	6	$F(z^3)$	$4.374956250970998118482749 8273 - 1.0550930657063033754 2065838646\,i$
		$TS_6(z^3)$	$8.417511393694925417 14725 \times 10^{23} + 5.154919990982005 385545807 \times 10^{17}\,i$
		MB Int (M1 = 2)	$-8.41751139369492541 714722 \times 10^{23} - 5.154919990820053 95943833 \times 10^{17}\,i$
		$S_{MB}^U(2,z^3)$	$0.00544139809270265 35517822347 - 3.138451060936203 44522418073989\,i$
		Total via M1	$6.913484873208586468930721 6827 - 5.233346759057508587 30776717724\,i$
		MB Int (M2 = 3)	$-8.41751139369492541 714727 \times 10^{23} - 5.154919990820053 90723539 \times 10^{17}\,i$
		$S_{MB}^U(3,z^3)$	$5.078039487244541544 2384564511 - 3.660480464761928 02988399777482\,i$
		Total via M2	$6.913484873272954124753 1300647 - 5.233346759035781054 140026255124\,i$

The first calculation in Table 6 lists the results for $\theta = -\pi/12$ and $N = 3$. Then (40) and the complex conjugate of (43) corresponding to M1 = 0 and M2 = -1, respectively, yield the value of $\ln\Gamma(\exp(-i\pi/4)/1000)$. Stirling's formula on the first row is substantial, but not accurate, compared with the actual value from the LogGamma routine in the bottom row of the calculation. The second row of the calculation displays the value of $TS_3(\exp(-i\pi/4)/1000)$, which is $\mathcal{O}(10^6)$. Thus, at least six decimal figures need to be cancelled by the remainder or MB integral, which occurs when the value on the next row is included in (40). The value of $S_{MB}(0,\exp(-i\pi/4)/1000)$ (zero since M1 = 0) appears on the fourth row of the calculation, while the sum of all the preceding quantities appears in the fifth row labelled as 'Total via M1'. The total value agrees with the actual value of $\ln\Gamma(\exp(-i\pi/4)/1000)$ to 30 decimal places, well within the accuracy and precision limits despite the cancellation of six decimal figures. The sixth row of the first calculation displays the value of the MB integral for M2 = -1. As expected, it agrees with the first six decimal figures of the values for both the truncated sum and the MB integral in (40). However, $S_{MB}(1, z^3)$, which is now non-vanishing, appears on the seventh row. There it can be seen that the real and imaginary parts of this value are much greater in magnitude than those from Stirling's formula. If this value is summed only with Stirling's formula, then the resulting value deviates from the value of $\ln \Gamma(\exp(-i\pi/4)/1000)$ far more than either value on its own, but when it is summed with the truncated sum and MB-regularized remainder, it yields $\ln \Gamma(\exp(-i\pi/4)/1000)$ to 29 decimal places despite the cancellation of six decimal figures.

The other calculations in Table 6 are similar to the first set of results except the MB-integrals and $S_{MB}(M, z^3)$ are evaluated according to the relevant domain of convergence. The third calculation presents less results because it has already been stated that there is only one MB-regularized form, viz. (44), which is applicable. Nevertheless, the final result agrees with the value obtained from Mathematica. An interesting result in this calculation is that $\Im(\ln \Gamma(z^3)) = -\pi$ for $\theta = \pi/3$ because the asymptotic series is composed of purely real terms when $\theta = k\pi/3$ and k is an integer. Hence the imaginary part of $TS_N(z^3)$ vanishes for all these values. In addition, the imaginary part of the MB integral can be shown to vanish by splitting the integral into two integrals and making the substitutions, $s = c + it$ in the upper half of the complex plane and $s = c - it$ in the lower half. Then all the terms become complex conjugates of each other. Expanding out all the terms, one is left with a real integral, while the imaginary part reduces to

$$\Im \ln \Gamma\left(|z|^3 \exp(i\pi)\right) = \Im F\left(|z|^3 e^{i\pi}\right) - \Im \ln\left(1 - e^{-2i\pi|z|^3}\right). \tag{48}$$

From Stirling's formula we obtain

$$\Im F\left(|z|^3 e^{i\pi}\right) = -\left(|z|^3 + 1/2\right)\pi, \tag{49}$$

while the second term in (48) becomes

$$\Im \ln\left(1 - e^{-2i\pi|z|^3}\right) = \Im \ln\left(e^{-i\pi|z|^3}\right) + \Im \ln\left(2i \sin(\pi|z|^3)\right). \tag{50}$$

Introducing these results into (48) yields

$$\Im \ln \Gamma\left(z^3\right)\Big|_{\theta=\pi/3} = -\pi. \tag{51}$$

In the last two calculations of Table 6, $\theta > \pi/3$, which means that the LogGamma routine can no longer be used. We are now on our own, a new frontier in mathematics, where only the totals via the M1 and M2 asymptotic forms can yield the value of $\ln \Gamma(z^3)$. Moreover, when θ equals $2\pi/3$ or π, there will only be one MB-regularized form that yields the regularized value. For these cases we require the Borel-summed regularized values as a check.

In the fifth calculation, N is set equal to 5, which yields a value of $\mathcal{O}(10^{17})$ for the truncated sum. Hence at least 16 decimal figures need to be cancelled in order to obtain $\ln \Gamma(e^{12i\pi/7}/1000)$. Since $\theta = 4\pi/7$, the domains of convergence are $(0, 2\pi/3)$ and $(\pi/3, \pi)$ corresponding to M1 = 1 and M2 = 2. Thus, (43) and (45) apply, which is interesting because $S_{MB}(M, z^3)$ is very different in these asymptotic forms, particularly the imaginary parts. As expected, the MB integrals for both asymptotic forms yield the 17 decimal figures in the real parts needed to cancel the real part of the truncated sum, $TS_5(e^{12i\pi/7}/1000)$. On the other hand, only 11 decimal figures are cancelled in the imaginary parts. As a result of the cancellation, the real parts in the totals only agree to 17 decimal figures. The same applies to the imaginary parts, which is surprising since there were less cancelled figures.

Because $4\pi/7$ is closer to the upper limit of $2\pi/3$ of the domain of convergence for (43), one expects the total obtained via M1 in the table to be the less accurate of the two forms. In actual fact, it turns out that this value is more accurate than the total via (45) by a few extra decimal places.

Nevertheless, if WorkingPrecision is set to 100 and AccuracyGoal and PrecisionGoal to 40, then both totals are found to agree to 32 decimal places, although the computation time is more than doubled. Another method of avoiding long computation times is to keep N as low as possible.

The final calculation in Table 6 is similar to the previous one except that (45) is introduced into (40) to yield the MB-regularized asymptotic forms for $\theta = 8\pi/9$. For $N = 6$, the truncated sum is $\mathcal{O}(10^{23})$. Since the highest degree of cancellation between the truncated sum and remainder occurs here, we find that this calculation yields the least accurate results of all those in the table. Despite this fact, the final results still agree with each other to 10 decimal places. Hence the results in Table 6 confirm the validity of the MB-regularized asymptotic forms for $\ln \Gamma(z^3)$.

We now consider the MB-regularized asymptotic forms near Stokes lines. Although the code should not be run when θ corresponds directly to a Stokes line, one can do so since the MB integrals are defined. Table 7 displays some of the results obtained by running the fourth program in [11] near the Stokes lines at $\theta = \pi/2$, $\theta = 5\pi/6$ and $\theta = -\pi/6$ with $|z| = 1/10$ and $N = 5$. When $\theta = \pi/2$, the code evaluates $\ln \Gamma(z^3)$ via (43) and (45) with M1 = 1 and M2 = 2, respectively. The first two results in the table display the values of (43) and (45) near the discontinuity at $\pi/2$ with $\theta = 19\pi/40$. As expected, both forms of $\ln \Gamma(z^3)$ yield identical values. At $\theta = \pi/2$, however, both forms yield different results, but only for the imaginary parts. In fact, there is a jump discontinuity of $2\pi i$ between the results with the first form yielding $-i\pi/2$ and the second, $3i\pi/2$. Note, however, that the discontinuities arise only from taking the logarithm of the gamma function. The gamma function itself is not discontinuous. As expected, neither result for $\theta = \pi/2$ is correct. The correct result is the midway between $-\pi/2$ and $3\pi/2$. That is, $\Im \ln \Gamma(z^3)\|_{\theta=\pi/2} = \pi/2$.

Table 7. $\ln \Gamma(z^3)$ via the MB-regularized forms in the vicinity of the lines of discontinuity given by $\theta = -\pi/6$, $\theta = \pi/2$ and $\theta = 5\pi/6$ with $|z| = 1/10$ and $N = 5$.

θ	Quantity	Value
$19\pi/40$	Total via M1(=1)	6.9017797138225092740511474835 − 1.3331484570580039616320161702 i
	Total via M2(=2)	6.9017797138225092740511474835 − 1.3331484570580039616320161702 i
$\pi/2$	Total via M1(=1)	6.9014712712081946221027015741 − 1.5702191115306805133718396291 i
	Total via M2(=2)	6.9014712712081946221027015741 + 4.7129661956489059635534471374 i
$21\pi/40$	Total via M1(=1)	6.9015102177011253639269574406 + 4.4758636443386950563445853873 i
	Total via M2(=2)	6.9015102177011253639269574406 + 4.4758636443386950563445853873 i
$62\pi/75$	Total via M1(=2)	6.9139890062805982640446908483 + 1.6326576822112902043838680277 i
	Total via M2(=3)	6.9139890062805982640446908483 + 1.6326576822112902043838680277 i
$5\pi/6$	Total via M1(=2)	6.9140376418225537950565521476 + 1.5702191115306805133718396291 i
	Total via M2(=3)	6.9140376418225537950565521476 + 1.5702191115306805133718396291 i
$63\pi/75$	Total via M1(=2)	6.9140614934733956410110131643 − 4.7754024883371855095758498194 i
	Total via M2(=3)	6.9140614934733956410110131643 − 4.7754024883371855095758498194 i
$-12\pi/75$	Total via M1(=0)	6.9077182194043652368591902822 + 1.5085404469087662793902283694 i
	Total via M2(=−1)	6.9077182194043652368591902822 + 1.5085404469087662793902283694 i
	$\ln \Gamma(z^3)$	6.9077182194043652368591902822 + 1.5085404469087662793902283694 i
$-\pi/6$	Total via M1(=0)	6.9077544565153742085796268609 + 1.5713735420591127250908037541 i
	Total via M2(=−1)	6.9077544565153742085796268609 + 1.5713735420591127250908037541 i
	$\ln \Gamma(z^3)$	6.9077544565153742085796268609 + 1.5713735420591127250908037541 i
$-13\pi/75$	Total via M1(=0)	6.9077907065971626138255125982 + 1.6342043592171290258017534223 i
	Total via M2(=−1)	6.9077907065971626138255125982 + 1.6342043592171290258017534223 i
	$\ln \Gamma(z^3)$	6.9077907065971626138255125982 + 1.6342043592171290258017534223 i

The next set of six results display the case where θ is very close to $5\pi/6$, viz. $\pm \pi/150$ away. In this case both forms in (45) are used to calculate $\ln \Gamma(z^3)$. Once again, both forms yield identical results below and above $\theta = 5\pi/6$. However, for $\theta = 5\pi/6$, they yield identical values for both the real and imaginary parts. In fact, although the imaginary parts have the same value of $i\pi/2$, it is incorrect because $\ln \Gamma(z^3)$ experiences a jump discontinuity of $-2\pi i$. Mathematica has simply chosen the wrong value for the logarithmic terms in (45), as explained on p. 564 of [17]. Noting that there is a jump of -2π means that $\Im \ln \Gamma(z^3)\|_{\theta=5\pi/6} = -\pi/2$, which corresponds to midway for the results before and after the Stokes line.

The final set of results in Table 7 have been obtained for θ very close to $-\pi/6$. Because $|\theta| < \pi/3$, we can evaluate $\ln \Gamma(z^3)$ via the LogGamma[z] routine in Mathematica. Hence there are more results for this calculation. This calculation employs (40) (M1 = 0) and the complex conjugate of (43) (M2 = 1). As before, the two versions of $\ln \Gamma(z^3)$ give identical values above and below the Stokes line at $\theta = -\pi/6$. Moreover, they agree with the values obtained via LogGamma[z]. The interesting point about this calculation, however, is that all three values at the Stokes line $\theta = -\pi/6$ also agree. This is expected since there is no extra logarithmic term in (42), while in the other asymptotic form the logarithmic term is purely real for $\theta = \pm\pi/6$. Thus, there is no discontinuity in $\ln \Gamma(z^3)$ at $\theta = \pm\pi/6$, which shows that Stokes discontinuities in Borel-summed regularized values can be fictitious.

The final calculation is the verification of the Borel-summed asymptotic forms for $\ln \Gamma(z^3)$. It was not possible to check these results previously because their regions of validity do not overlap. Now we use the MB-regularized asymptotic forms to verify their Borel-summed counterparts. In addition, we can confirm that the MB-regularized asymptotic forms for $\theta = k\pi/3$, where k equals ± 1, ± 2 and 3, are correct, since only one MB-regularized asymptotic form is valid for these values.

To carry out the first task, we need to replace z by z^3 in Theorem 1. Hence the Borel-summed regularized values for $\ln \Gamma(z^3)$ become

$$\ln \Gamma\left(z^3\right) = F\left(z^3\right) + TS_N\left(z^3\right) + R_N^{\pm}\left(z^3\right) + SD_M^{\pm}\left(z^3\right), \tag{52}$$

where, as before, $TS_N(z^3)$ is the truncated form of $S_N(z^3)$ at N,

$$R_N^+\left(z^3\right) = \frac{2(-1)^{N+1}z^3}{(2\pi z^3)^{2N-2}} \int_0^\infty dy\, y^{2N-2} e^{-y} \sum_{n=1}^\infty \frac{1}{n^{2N-2}(y^2+(2n\pi z^3)^2)}, \tag{53}$$

$$R_N^-\left(z^3\right) = \frac{2z^3}{(2\pi |z^3|)^{2N-2}} P\int_0^\infty dy\, y^{2N-2} e^{-y} \sum_{n=1}^\infty \frac{1}{n^{2N-2}(y^2-4n^2\pi^2|z^3|^2)}, \tag{54}$$

$$SD_M^+\left(z^3\right) = -\lfloor M/2 \rfloor \ln\left(-e^{\pm 2i\pi z^3}\right) - \frac{(1-(-1)^M)}{2} \ln\left(1-e^{\pm 2i\pi z^3}\right), \tag{55}$$

and

$$SD_M^-\left(z^3\right) = (-1)^M \left(\lfloor M/2 \rfloor + \frac{1-(-1)^M}{2}\right) 2\pi |z^3| - \frac{1}{2}\ln\left(1-e^{-2\pi|z^3|}\right). \tag{56}$$

The upper- and lower-signed versions of (55) are valid for $(M-1/2)\pi/3 < \theta < (M+1/2)\pi/3$ and $-(M+1/2)\pi/3 < \theta < -(M-1/2)\pi/3$, respectively, while (56) is only valid for $\theta = \pm(M+1/2)\pi/3$. Therefore, for $-\pi < \theta \leq \pi$ or the principal branch for z, Stokes lines occur at $\pm\pi/6$, $\pm\pi/2$, and $\pm 5\pi/6$. We shall investigate these cases after we have considered the results for the Stokes sectors first.

The Borel-summed asymptotic forms that are valid for the Stokes sectors can be expressed as

$$\ln \Gamma\left(z^{3}\right)=\begin{cases} F\left(z^{3}\right)+TS_{N}(z^{3})+R_{N}^{+}(z^{3})-\ln\left(-e^{2i\pi z^{3}}\right)-\ln\left(1-e^{2i\pi z^{3}}\right), & 5\pi/6 < \theta \leq \pi, \\ F\left(z^{3}\right)+TS_{N}(z^{3})+R_{N}^{+}(z^{3})-\ln\left(-e^{2i\pi z^{3}}\right), & \pi/2 < \theta < 5\pi/6, \\ F\left(z^{3}\right)+TS_{N}(z^{3})+R_{N}^{+}(z^{3})-\ln\left(1-e^{2i\pi z^{3}}\right), & \pi/6 < \theta \leq \pi/2, \\ F\left(z^{3}\right)+TS_{N}(z^{3})+R_{N}^{+}(z^{3}), & -\pi/6 < \theta < \pi/6, \\ F\left(z^{3}\right)+TS_{N}(z^{3})+R_{N}^{+}(z^{3})-\ln\left(1-e^{-2i\pi z^{3}}\right), & -\pi/2 < \theta < -\pi/6, \\ F\left(z^{3}\right)+TS_{N}(z^{3})+R_{N}^{+}(z^{3})-\ln\left(-e^{-2i\pi z^{3}}\right), & -5\pi/6 < \theta < -\pi/2, \\ F\left(z^{3}\right)+TS_{N}(z^{3})+R_{N}^{+}(z^{3})-\ln\left(-e^{-2i\pi z^{3}}\right)-\ln\left(1-e^{-2i\pi z^{3}}\right), & -\pi < \theta < -5\pi/6. \end{cases} \quad (57)$$

The main difference between these results and the earlier MB-regularized values is that though they possess similar logarithmic terms, they emerge in different sectors within the principal branch.

Table 8 presents a very small sample of the results obtained by running the fifth program in the appendix of [11] with $|z| = 5/2$ and the upper limit in $R_N^+(z^3)$ and $R_N^-(z^3)$ set to 10^5 as in Section 3. The first calculation displays the results obtained for $\theta = -\pi/7$ and $N = 4$. Since $N_{OP} = 8$ according to (23), Stirling's formula or $F(z^3)$ yields a reasonable approximation to the actual value of $\ln \Gamma((5/2)^3 e^{-3i\pi/7})$, which appears at the bottom of the calculation. Consequently, the truncated sum is small, only contributing at the third decimal place. Two MB-regularized asymptotic forms apply: (1) (40) denoted by M1 = 0, and (2) the complex conjugate of (43) denoted by M2 = −1. The MB-Integrals in the remainder are $\mathcal{O}(10^{-12})$. For M1 = 0, there is no logarithmic term, while for M2 = −1, there is a contribution, but it is almost negligible, $\mathcal{O}(10^{-42})$. That is, the M1 = 0 and M2 = −1 calculations are virtually identical to one another, well within the accuracy and precision goals set in NIntegrate. Therefore, the totals representing the sum of $F(z^3)$, the MB integrals and the logarithmic terms, are not only identical to one another, but they also agree with the value obtained from the LogGamma routine in Mathematica.

The result labelled Borel Rem represents the Borel-summed remainder or $R_N^+(z^3)$ in the fourth asymptotic form of (57), where the upper limit in the sum has been set to 10^5. Despite this truncation, it is identical to the values obtained from the MB regularized asymptotic forms. In actual fact, the Borel Rem value is identical to the first 34 decimal figures of the MB integrals, well within the accuracy and precision goals. Bearing in mind that the remainder is very small, this means that only the first 13 or so decimal figures of each remainder calculation will contribute to the totals. That is, the remainder is truly subdominant.

The second calculation displays the results for $\theta = 2\pi/3$ and $N = 2$, which has only one valid MB-regularized asymptotic form. In addition, Stirling's formula, the truncated sum and the MB integral are all real, while the logarithmic term yields the imaginary contribution of $-\pi/4$. Hence we see that $\Im \ln \Gamma\left(|z|^3 e^{2i\pi}\right) = -\pi/4$. As expected, the value of the MB integral is very small, $\mathcal{O}(10^{-7})$, while Stirling's formula provides an accurate value for $\Re \ln \Gamma\left(125 \exp(2i\pi)/8\right)$. Appearing below the Total is the remainder of the Borel-summed version for $\ln \Gamma\left(|z|^3 \exp(2i\pi)\right)$ or $R_N^+(z^3)$ in the second result of (57), which in turn has the same logarithmic term as (45). Hence both calculations are expected to be identical. However, a closer inspection reveals that they agree to 23 decimal figures, but not the expected 30 specified by the accuracy and precision goals. This discrepancy, which arises from the truncation of the Borel-summed remainder, is an example where the upper limit of the sum over n has to be set much higher in order to achieve the desired accuracy.

The final set of values have been obtained by setting θ equal to $11\pi/12$ and N to 4. Once again, there are two MB-regularized asymptotic forms, both obtained from (45). The Borel-summed asymptotic form in this case is given by the first form in (57). All the remainder terms are tiny, $\mathcal{O}(10^{-12})$. If only the logarithmic term for all three forms is added to Stirling's formula, then a good approximation is obtained. In this instance the logarithmic term for the Borel-summed form is identical to the second form in (45), but by comparing it with the value obtained via the M1=2 form, the extra term is very small indeed, only differing at the 20-th decimal place. Nevertheless, all three totals agree with each other as in the other cases in the table.

Table 8. $\ln \Gamma(z^3)$ for $|z| = 5/2$ and various values of θ and N.

θ	N	Quantity	Value
$-\pi/7$	4	$F(z^3)$	$-14.884860019269882183168909670 - 30.649097918823731704253309660\,5\,i$
		$TS_4(z^3)$	$0.00118723309363615315954697\,0 + 0.00520018521369322415111941337\,i$
		MB Int (M1 = 0)	$2.63163543021301667181503 \times 10^{-12} - 6.69180007011641181905\,94 \times 10^{-15}\,i$
		Log. Term (M1)	0
		Total via M1	$-14.88367278617361439457914968 - 30.64387956450513719935320401\,27\,i$
		MB Int (M2 = -1)	$2.63163543021301667181503 \times 10^{-12} - 6.69180007011641181905\,94 \times 10^{-15}\,i$
		Log. Term (M2)	$-2.67692877187577721034710 \times 10^{-42} - 3.91463936595211044303\,65 \times 10^{-43}\,i$
		Total via M2	$-14.88367278617361439457914968 - 30.64387956450513719935320401\,27\,i$
		Borel Rem	$2.63163543021301667181503 \times 10^{-12} - 6.69180007011641181905\,94 \times 10^{-15}\,i$
		Borel Log Term	0
		Borel Total	$-14.88367278617361439457914968 - 30.64387956450513719935320401\,27\,i$
		LogGamma[zcube]	$-14.88367278617361439457914968 - 30.64387956450513719935320401\,27\,i$
$2\pi/3$	2	$F(z^3)$	$26.87063049199445882448287697\,03 + 0\,i$
		$TS_2(z^3)$	$0.00533333333333333333333333333\,3 + 0\,i$
		MB Int (M1 = 2)	$-7.27328204587763887038193 \times 10^{-7} + 0\,i$
		Log. Term (M1 = 2)	$- 0.78539816339744830961566084581\,9\,i$
		Total via M1	$26.87596309799958757005254752\,5 - 0.78539816339744830961566084581\,9\,i$
		Borel Rem	$-7.27328204587763662777752 \times 10^{-7} + 0\,i$
		Borel Log Term	$- 0.78539816339744830961566084581\,9\,i$
		Borel Total	$26.87596309799958757005232326\,5 - 0.78539816339744830961566084581\,9\,i$
$11\pi/12$	4	$F(z^3)$	$-45.81050523277279074549656662 - 7.88812400847329603213924107797\,i$
		$TS_4(z^3)$	$-0.00377175046319336520004847\,5 - 0.03770720664304213164969383\,43\,i$
		MB Int (M1 = 2)	$1.84030315270991051371\,19 \times 10^{-12} - 1.86174503255187256069\,532 \times 10^{-12}\,i$
		Log Term (M1 = 2)	$69.42004590872447260962314046 - 2.83658512384077187501765734573\,i$
		Total via M1	$23.60576892549032880207923528 - 10.72847985298023386553441967\,97\,i$
		MB Int (M2 = 3)	$1.84030315270991051301\,46 \times 10^{-12} - 1.86174503255187256048\,211 \times 10^{-12}\,i$
		Log Term (M2 = 3)	$69.42004590872447260962314046 - 2.83658512384077187501765734573\,i$
		Total via M2	$23.60576892549032880207923528 - 10.72847985298023386553441967\,97\,i$
		Borel Rem	$1.84030315270991051301\,46 \times 10^{-12} - 1.86174503255187256048\,211 \times 10^{-12}\,i$
		Borel Log Term	$69.42004590872447260962314046 - 2.83658512384077187501765734573\,i$
		Borel Total	$23.60576892549032880207923528 - 10.72847985298023386553441967\,97\,i$

Before we can be assured that there is complete agreement between both sets of asymptotic forms, we need to carry out a final numerical analysis at the Stokes lines. The Borel-summed asymptotic forms at these lines are given by (52), but now with $R_N^-(z^3)$ or (54) and $SD_M^-(z^3)$ or (56). Putting M equal to 0, 1 and 2, yields the specific forms at the Stokes lines, which are

$$\ln \Gamma\left(z^3\right) = \begin{cases} F\left(z^3\right) + TS_N(z^3) + R_N^-\left(z^3\right) - \tfrac{1}{2}\ln\left(1 - e^{-2\pi|z^3|}\right), & \theta = \pm\pi/6, \\ F\left(z^3\right) + TS_N(z^3) + R_N^-\left(z^3\right) - \tfrac{1}{2}\ln\left(1 - e^{-2\pi|z^3|}\right) \\ \quad -2\pi|z^3|, & \theta = \pm\pi/2, \\ F\left(z^3\right) + TS_N(z^3) + R_N^-\left(z^3\right) - \tfrac{1}{2}\ln\left(1 - e^{-2\pi|z^3|}\right) \\ \quad +2\pi|z^3|, & \theta = \pm 5\pi/6. \end{cases} \tag{58}$$

Note the similarity of the Stokes discontinuity terms with the corresponding terms or $S_{MB}(z^3)$ in the MB-regularized asymptotic forms. The major difference occurs with the logarithmic term, which is represented by either a zero or full residue contribution in the MB-regularized asymptotic forms, while it is always represented by a semi-residue or half the contribution in the Borel-summed asymptotic forms.

Table 9 presents a small sample of the results obtained by running the final program in the appendix of [11]. Since the MB integrals yielded values of $\mathcal{O}(10^{-3})$, there was no significant cancellation of decimal figures as in Table 8. The first column of Table 9 displays the value of θ for the respective Stokes line. Here they are presented for the Stokes lines at: (1) $\theta = \pi/6$, (2) $\theta = -\pi/2$ and (3) $\theta = 5\pi/6$. As stated before, $\ln \Gamma(z^3)$ cannot be evaluated by Mathematica for the last two cases. Thus, LogGamma[z] appears as an extra result for $\theta = \pi/6$. The second column of Table 9 displays the equation that was used to calculate the value of $\ln \Gamma(z^3)$. The label 'c.c.' denotes that the complex conjugate of the equation was used, which applies here because θ is negative. The third column displays the actual values to 27 decimal places. We see that not only do the two different MB-regularized asymptotic forms agree with one another at each Stokes line, they also agree with the results obtained from the the Borel-summed asymptotic forms in (58) and where possible, with the LogGamma routine in Mathematica.

Table 9. $\ln \Gamma(z^3)$ for $|z| = 9/10$ at the Stokes lines within the principal branch.

θ	Method	Value
$\pi/6$	(40)	$-0.062979585299600601912 6614 - 1.867819809970580480394 34088\,i$
	(44)	$-0.062979585299600601912 6614 - 1.867819809970580480394 34088\,i$
	Top, (58)	$-0.062979585299600601912 6614 - 1.867819809970580480394 34088\,i$
	LogGamma[z]	$-0.062979585299600601912 6614 - 1.867819809970580480394 34088\,i$
$-\pi/2$	(44), c.c.	$-4.643421674233519143591 1954 - 1.867819809970580480394 34088\,i$
	Bottom, (47), c.c.	$-4.643421674233519143591 1954 - 1.867819809970580480394 34088\,i$
	Middle, (58)	$-4.643421674233519143591 1954 - 1.867819809970580480394 34088\,i$
$5\pi/6$	Top, (47)	$4.517462503634317939765 8872 - 1.867819809970580480394 34088\,i$
	Bottom, (47)	$4.517462503634317939765 8872 - 1.867819809970580480394 34088\,i$
	Bottom, (58)	$4.517462503634317939765 8872 - 1.867819809970580480394 34088\,i$

6. Conclusions

In [16] it was stated that a fully-fledged theory of divergent series could only be realized if more complicated problems were studied than those presented in [6]. Amongst these was the extension of the asymptotics of the gamma function to the entire complex plane since the Stokes lines possess an infinite number of singularities rather than one as studied in [6]. This has been achieved here, which leaves the development of the complete asymptotic expansion for the confluent hypergeometric function over the entire complex plane as the next problem. In this instance it will be necessary to develop and regularize infinite subdominant series throughout the complex plane.

Funding: This research received no external funding.

Acknowledgments: The author thanks Professor Mainardi for the invitation to contribute to this special issue.

Conflicts of Interest: The author declares no conflict of interest.

References

1. Wikipedia, the Free Encyclopedia, Stirling's Approximation. Available online: http://en.wikipe\protect\discretionary{\char\hyphenchar\font}{}{}dia.org/wiki/Stir\protect\discretionary{\char\hyphenchar\font}{}{}lings_approxi\protect\discretionary{\char\hyphenchar\font}{}{}mation (accessed on 8 June 2020).
2. Whittaker, E.T.; Watson, G.N. *A Course of Modern Analysis*, 4th ed.; Cambridge University Press: Cambridge, UK, 1973; p. 252.
3. Segur, H.; Tanveer, S.; Levine, H. (Eds.) *Asymptotics beyond all Orders*; Plenum Press: New York, NY, USA, 1991.
4. Kowalenko, V. Towards a theory of divergent series and its importance to asymptotics. In *Recent Research Developments in Physics*; Transworld Research Network: Trivandrum, India, 2001; Volume 2, pp. 17–68.
5. Kowalenko, V. Exactification of the asymptotics for Bessel and Hankel functions. *Appl. Math. Comput.* **2002**, *133*, 487–518. [CrossRef]
6. Kowalenko, V. *The Stokes Phenomenon, Borel Summation and Mellin-Barnes Regularisation*; Bentham Ebooks: Sharjah, UAE, 2009; Available online: http://www.bentham.org (accessed on 18 June 2020).

7. Kowalenko, V.; Frankel, N.E.; Glasser, M.L.; Taucher, T. *Generalised Euler-Jacobi Inversion Formula and Asymptotics beyond All Orders*; London Mathematical Society Lecture Note 214; Cambridge University Press: Cambridge, UK, 1995.
8. Wikipedia, the Free Encyclopedia, Hadamard Regularization. Available online: https://en.wikipedia.org/wiki/Hadamard_regularization (accessed on 11 April 2020).
9. Abramowitz, M.; Stegun, I.A. *Handbook of Mathematical Functions*; Dover: New York, NY, USA, 1965.
10. Gradshteyn, I.S.; Ryzhik, I.M.; Jeffrey, A. (Eds.) *Table of Integrals, Series and Products*, 5th ed.; Academic Press: London, OH, USA, 1994.
11. Kowalenko, V. Exactification of Stirling's approximation for the logarithm of the gamma function. *arXiv* **2014**, arXiv:1404.2705.
12. Paris, R.B.; Kaminski, D. *Asymptotics and Mellin-Barnes Integrals*; Cambridge University Press: Cambridge, UK, 2001.
13. Paris, R.B. *Hadamard Expansions and Hyperasymptotic Evaluation—An Extension of the Method of Steepest Descents*; Cambridge University Press: Cambridge, UK, 2011.
14. Kowalenko, V. Applications of the cosecant and related numbers. *Acta Appl. Math.* **2011**, *114*, 15–134. [CrossRef]
15. Dingle, R.B. *Asymptotic Expansions: Their Derivation and Interpretation*; Academic Press: London, UK, 1973.
16. Kowalenko, V. Euler and Divergent Series. *Eur. J. Pure Appl. Math.* **2011**, *4*, 370–423.
17. Wolfram, S. *Mathematica—A System for Doing Mathematics by Computer*; Addison-Wesley: Reading, MA, USA, 1992.
18. Berry, M.V. Uniform asymptotic smoothing of Stokes's discontinuities. *Proc. R. Soc. Lond. A* **1989**, *422*, 7–21.
19. Stokes, G.G. On the discontinuity of arbitrary constants which appear in divergent developments. In *Collected Mathematical and Physical Papers*; Cambridge University Press: Cambridge, UK, 1904; Volume 4, pp. 77–109.
20. Olver, F.W.J. On Stokes' phenomenon and converging factors. In *Asymptotic and Computational Analysis*; Wong, R., Ed.; Marcel-Dekker: New York, NY, USA, 1990; pp. 329–355.
21. Berry, M.V.; Howls, C.J. Hyperasymptotics for integrals with saddles. *Proc. R. Soc. Lond. A* **1991**, *434*, 657–675.
22. Paris, R.B.; Wood, A.D. Exponentially-improved asymptotics for the gamma function. *J. Comp. Appl. Math.* **1992**, *41*, 135–143. [CrossRef]
23. Berry, M.V.; Howls, C.J. Hyperasymptotics. *Proc. R. Soc. Lond. A* **1990**, *430*, 653–658.
24. Berry, M.V. Asymptotics, superasymptotics, hyperasymptotics. In *Asymptotics beyond all Orders*; Segur, H., Tanveer, S., Levine, H., Eds.; Plenum Press: New York, NY, USA, 1991; pp. 1–9.
25. Paris, R.B.; Wood, A.D. Stokes phenomenon demystified. *IMA Bull.* **1995**, *31*, 21–28.
26. Kowalenko, V. Reply to Paris's comments on exacitification for the logarithm of the gamma function. *arXiv* **2014**, arXiv:1408.1881.

© 2020 by the author. Licensee MDPI, Basel, Switzerland. This article is an open access article distributed under the terms and conditions of the Creative Commons Attribution (CC BY) license (http://creativecommons.org/licenses/by/4.0/).

Article

Transformations of the Hypergeometric $_4F_3$ with One Unit Shift: A Group Theoretic Study

Dmitrii Karp [1,*] **and Elena Prilepkina** [2,3]

1. Faculty of Mathematics and Statistics, Ton Duc Thang University, Ho Chi Minh City 700000, Vietnam
2. School of Economics and Management, Far Eastern Federal University, Vladivostok 690950, Russia; pril-elena@yandex.ru
3. Institute of Applied Mathematics, Far Eastern Branch of the Russian Academy of Sciences, Vladivostok 690041, Russia
* Correspondence: dmitriibkarp@tdtu.edu.vn

Received: 2 October 2020; Accepted: 2 November 2020; Published: 5 November 2020

Abstract: We study the group of transformations of $_4F_3$ hypergeometric functions evaluated at unity with one unit shift in parameters. We reveal the general form of this family of transformations and its group property. Next, we use explicitly known transformations to generate a subgroup whose structure is then thoroughly studied. Using some known results for $_3F_2$ transformation groups, we show that this subgroup is isomorphic to the direct product of the symmetric group of degree 5 and 5-dimensional integer lattice. We investigate the relation between two-term $_4F_3$ transformations from our group and three-term $_3F_2$ transformations and present a method for computing the coefficients of the contiguous relations for $_3F_2$ functions evaluated at unity. We further furnish a class of summation formulas associated with the elements of our group. In the appendix to this paper, we give a collection of *Wolfram Mathematica*® routines facilitating the group calculations.

Keywords: generalized hypergeometric function; hypergeometric transformations; transformation groups; symmetric group

1. Introduction and Preliminaries

Groups comprising transformation of the generalized hypergeometric functions that preserve their value at unity can be traced back to Kummer's formula ([1], Corollary 3.3.5), see (2) below. These groups play an important role in mathematical physics. In particular, the group theoretic properties of hypergeometric transformations constitute the key ingredient of a succinct description of the symmetries of Clebsh-Gordon's and Wigner's $3-j$, $6-j$ and $9-j$ coefficients from the angular momentum theory [2–5]. The Karlsson-Minton summation formula for the generalized hypergeometric function with integral parameter differences (IPD) was largely motivated by a computation of a Feymann's path integral. Furthermore, IPD hypergeometric functions appear in calculation of several integrals in high energy field theories and statistical physics [6]. See also introduction and references in [7] for further applications in mathematical physics and relation to Coxeter groups.

The generalized hypergeometric function ([1], Section 2.1.2), ([8], Chapter 16) is defined by the series

$$_{p+1}F_p\left(\begin{array}{c}a_1,\ldots,a_{p+1}\\b_1,\ldots,b_p\end{array}\bigg|z\right)=\sum_{n=0}^{\infty}\frac{(a_1)_n\cdots(a_{p+1})_n}{n!(b_1)_n\cdots(b_p)_n}z^n \qquad (1)$$

whenever it converges. When evaluated at the unit argument, $z=1$, it represents a function of $2p+1$ complex parameters with obvious symmetry with respect to separate permutation of the $p+1$ top and the p bottom parameters. As the above series diverges at $z=1$ if the parametric excess satisfies $\Re\left(\sum_{k=1}^{p}(b_k-a_k)-a_{p+1}\right)<0$, the first problem that arises is to construct an analytic continuation to the values of parameters in this domain. For $_3F_2$ function this problem is partially solved by the transformation ([1], Corollary 3.3.5)

$$_3F_2\left(\begin{array}{c}a,b,c\\d,e\end{array}\right)=\frac{\Gamma(e)\Gamma(d+e-a-b-c)}{\Gamma(e-c)\Gamma(d+e-b-a)}{}_3F_2\left(\begin{array}{c}d-b,d-a,c\\d,d+e-b-a\end{array}\right) \qquad (2)$$

discovered by Kummer in 1836. In the above formula we have omitted the argument 1 from the notation of the hypergeometric series and this convention will be adopted throughout the paper. The series the right hand side of (2) converges when $\Re(e-c)>0$ so that we get the analytic continuation to this domain. An important aspect of the above formula is that it can be applied to itself directly or after permuting some of the top and/or bottom parameters. This leads to a family of transformations which can can be studied by group theoretic methods. A notable member of this family is Thomae's (1879) transformation ([1], Corollary 3.3.6)

$$_3F_2\left(\begin{array}{c}a,b,c\\d,e\end{array}\right)=\frac{\Gamma(d)\Gamma(e)\Gamma(s)}{\Gamma(c)\Gamma(s+b)\Gamma(s+a)}{}_3F_2\left(\begin{array}{c}d-c,e-c,s\\s+a,s+b\end{array}\right), \qquad (3)$$

where $s=d+e-a-b-c$, which gave the name to the whole family of $_3F_2$ transformations generated by the algorithm described above. In an important work [9] the authors undertook a detailed group theoretic study of Thomae's transformations as well as transformations for the terminating $_4F_3$ series and Bailey's three-term relations for $_3F_2$. In particular, they have shown ([9], Theorem 3.2) that the function

$$f(x,y,z,u,v)=\frac{{}_3F_2\left(\begin{array}{c}x+u+v,y+u+v,z+u+v\\x+y+z+2u+v,x+y+z+u+2v\end{array}\right)}{\Gamma(x+y+z+2u+v)\Gamma(x+y+z+u+2v)\Gamma(x+y+z)}, \qquad (4)$$

is invariant with respect to the entire symmetric group P_5 acting on its 5 arguments (note that another, simpler version of this symmetry is given by ([2], Equation (7)). This symmetry was, in fact, first observed by Hardy in his 1940 lectures ([10], Notes on Lecture VII). The work [9] initiated the whole stream of papers on group-theoretic interpretations of hypergeometric and q-hypergeometric transformations. See, for instance, Refs. [2,4,7,11–14] and references therein.

We note in passing that the analytic continuation problem for general p was solved by Nørlund [15] and Olsson [16] with later rediscovery by Bühring [17] without resorting to group-theoretic methods. More recently, Kim, Rathie and Paris derived ([18], p. 116) the following transformation

$$_4F_3\left(\begin{array}{c}a,b,c,f+1\\d,e,f\end{array}\right)=\frac{\Gamma(e)\Gamma(\psi)}{\Gamma(e-c)\Gamma(\psi+c)}{}_4F_3\left(\begin{array}{c}d-a-1,d-b-1,c,\eta+1\\d,d+e-a-b-1,\eta\end{array}\right), \qquad (5)$$

with $\psi=d+e-a-b-c-1$ and

$$\eta=\frac{(d-a-1)(d-b-1)f}{ab+(d-a-b-1)f}.$$

This transformation can be iterated, but it is not immediately obvious what is the general form of the transformations obtained by such iterations. In our recent paper ([19], p. 14, above Theorem 2) we found another identity of a similar flavor which can be viewed as a generalization of (2):

$$
{}_4F_3\left(\begin{matrix} a,b,c,f+1 \\ d,e,f \end{matrix}\right) = \frac{(\psi f - c(d-a-b))\Gamma(e)\Gamma(\psi)}{f\Gamma(e+d-a-b)\Gamma(e-c)} {}_4F_3\left(\begin{matrix} d-a, d-b, c, \xi+1 \\ d, e+d-a-b, \xi \end{matrix}\right), \quad (6)
$$

where $\xi = f + (d-a-b)(f-c)/(e-c-1)$. The main purpose of this paper is to present a general form of the family of transformations of which the above two identities are particular cases, demonstrate that this family forms a group and analyze the structure of the subgroup generated by explicitly known transformations (5)–(8). Before we delve into this analysis let us now record two more transformations generating this subgroup. A proof will be given in Section 6.

Lemma 1. *The following identities hold*

$$
{}_4F_3\left(\begin{matrix} a,b,c,f+1 \\ d,e,f \end{matrix}\right) = \frac{(f\psi + bc)\Gamma(\psi)\Gamma(d)\Gamma(e)}{f\Gamma(a)\Gamma(\psi+b+1)\Gamma(\psi+c+1)} {}_4F_3\left(\begin{matrix} \psi, d-a, e-a, \zeta+1 \\ d+e-a-c, d+e-a-b, \zeta \end{matrix}\right), \quad (7)
$$

where $\zeta = \psi + bc/f$, $\psi = d+e-a-b-c-1$; and

$$
{}_4F_3\left(\begin{matrix} a,b,c,f+1 \\ d,e,f \end{matrix}\right) = \frac{(abc + fd\psi)\Gamma(\psi)\Gamma(e)}{fd\Gamma(e-a)\Gamma(\psi+a+1)} {}_4F_3\left(\begin{matrix} a, d-b, d-c, \nu+1 \\ d+1, \psi+a+1, \nu \end{matrix}\right), \quad (8)
$$

where $\nu = (abc + fd\psi)/(bc + f\psi)$.

Please note that each ${}_4F_3$ function containing a parameter pair $\begin{bmatrix} f+1 \\ f \end{bmatrix}$ can be decomposed into a sum of two ${}_3F_2$ functions (and we will demonstrate that there are numerous different decompositions of this type). Hence, each of the identities (5)–(8) can be written as a four-term relation for ${}_3F_2$. However, it will be seen from the subsequent considerations that, in fact, all such relations reduce to three or even two terms, and, moreover, the structure seems to be more transparent if we keep the ${}_4F_3$ function as the basic building block of our analysis. It will be revealed that the group structure of our transformations is closely related to that of the Thomae group generated by two-term transformations (2) and (3) and with contiguous three terms relations for ${}_3F_2$. We believe that our subgroup generated by (5)–(8) covers all possible two-term transformations for ${}_4F_3$ with one unit shift (more precisely all transformations of the form (10) below), but we were unable to prove this claim and leave it as a conjecture.

The paper is organized as follows. In the following section we give a general form of the transformations exemplified above and prove that they form a group. We further demonstrate that this group is isomorphic to a subgroup of $SL(\mathbb{Z})$ (integer matrices with unit determinant). In Section 3, we give a comprehensive analysis of the structure of the subgroup generated by the transformations (5)–(8) by showing that it is isomorphic to a direct product of the symmetric group P_5 and the integer lattice $\mathbb{Z}^5$. In Section 4 we explore the relation between our transformations and three-term relations for ${}_3F_2$ hypergeometric function. In particular, we show that the contiguous relations for ${}_3F_2$ functions studied recently in [20] can also be computed from the elements of our group. Section 5 provides a method of deducing summation formulas for ${}_4F_3$ with non-linearly restricted parameters while Section 6 contains the proof of Lemma 1. Finally, the Appendix A contains explicit forms of some key elements of our subgroup and several *Wolfram Mathematica*® routines facilitating the group calculations.

2. The Group Structure of the Unit Shift $_4F_3$ Transformations

Inspecting the $_4F_3$ transformations presented in Section 1 we see that they share a common structure that we will present below. To this end, let $\mathbf{r} = (a,b,c,d,e,1)^T$ be the column vector and define

$$F(\mathbf{r}, f) = {}_4F_3\left(\begin{array}{c} a,b,c,f+1 \\ d,e,f \end{array}\right). \tag{9}$$

All transformations found in Section 1 have the following general form

$$F(\mathbf{r}, f) = C(\mathbf{r}, f) F(D\mathbf{r}, \eta), \tag{10}$$

where D is a unit determinant 6×6 matrix with integer entries and the bottom row $(0,0,0,0,0,1)$;

$$\eta = \frac{\varepsilon f + \lambda(\mathbf{r})}{\alpha(\mathbf{r}) f + \beta(\mathbf{r})}, \tag{11}$$

where $\varepsilon \in \{0,1\}$, $\lambda(\mathbf{r})$, $\alpha(\mathbf{r})$ and $\beta(\mathbf{r})$ are rational functions of the arguments a,b,c,d,e (some of them may vanish identically, but $\lambda = 1$ if $\varepsilon = 0$). The coefficient $C(\mathbf{r}, f)$ has the form

$$C(\mathbf{r}, f) = \frac{N(\mathbf{r}) f + P(\mathbf{r})}{K(\mathbf{r}) f + L(\mathbf{r})} \tag{12}$$

where $N(\mathbf{r})$, $P(\mathbf{r})$, $K(\mathbf{r})$, $L(\mathbf{r})$ are (possibly vanishing) functions of Γ-type by which we mean ratios of products of gamma functions whose arguments are integer linear combinations of the components of $(a,b,c,d,e,1)$. When $N(\mathbf{r}) \neq 0$ we will additionally require that the ratio $P(\mathbf{r})/N(\mathbf{r})$ be a rational function of parameters. In fact, this last requirements is redundant, but in order to avoid it the following claim is needed: the ratio $F_2(\mathbf{r})/F_1(\mathbf{r})$ with F_i, $i=1,2$, defined in (14), is not a function of gamma type for general parameters. We were unable to find a proof of this claim in the literature although it seems to be generally accepted to be true.

Formula (10) defines a transformation T characterized by the matrix D and the functions $C(\mathbf{r}, f)$, $\eta = \eta(\mathbf{r}, f)$. Two such transformations T_1, T_2 will be considered equal if $D_1 = D_2$, $C_1(\mathbf{r}, f) \equiv C_2(\mathbf{r}, f)$ and $\eta_1(\mathbf{r}, f) \equiv \eta_2(\mathbf{r}, f)$.

According to the elementary relation $(f+1)_n = (f)_n (1 + n/f)$, we have

$$F(\mathbf{r}, f) = F_1(\mathbf{r}) + \frac{1}{f} F_2(\mathbf{r}), \tag{13}$$

where

$$F_1(\mathbf{r}) = {}_3F_2\left(\begin{array}{c} a,b,c \\ d,e \end{array}\right), \quad F_2(\mathbf{r}) = \frac{abc}{de} {}_3F_2\left(\begin{array}{c} a+1,b+1,c+1 \\ d+1,e+1 \end{array}\right). \tag{14}$$

It is not immediately obvious if the composition of two transformations (10) with η and C having the forms (11) and (12), respectively, should have the same form. The following theorem shows that it is indeed the case and these transformations form a group.

Theorem 1. *Each transformation (10) necessarily has the form*

$$F(\mathbf{r}, f) = M(\mathbf{r}) \frac{\varepsilon f + \lambda(\mathbf{r})}{f} F(D\mathbf{r}, \eta), \text{ where } \eta = \frac{\varepsilon f + \lambda(\mathbf{r})}{\alpha(\mathbf{r}) f + \beta(\mathbf{r})}, \tag{15}$$

$M(\mathbf{r})$ is a function of Γ-type, $\varepsilon \in \{0,1\}$, $\lambda(\mathbf{r})$, $\alpha(\mathbf{r})$, $\beta(\mathbf{r})$ are rational functions of the arguments a,b,c,d,e (possibly vanishing but with $\lambda = 1$ if $\varepsilon = 0$).

The collection $\mathcal{T}$ of transformations (15) forms a group with respect to composition. More explicitly, if $T_1, T_2 \in \mathcal{T}$ with parameters indexed correspondingly, then $T = T_2 \circ T_1$ is given by

(I) If $\varepsilon_1\varepsilon_2 + \alpha_1(\mathbf{r})\lambda_2(D_1\mathbf{r}) \neq 0$, then $\varepsilon = 1$, $M(\mathbf{r}) = M_1(\mathbf{r})M_2(D_1\mathbf{r})(\varepsilon_1\varepsilon_2 + \alpha_1(\mathbf{r})\lambda_2(D_1\mathbf{r}))$,

$$\lambda(\mathbf{r}) = \frac{\varepsilon_2\lambda_1(\mathbf{r}) + \lambda_2(D_1\mathbf{r})\beta_1(\mathbf{r})}{\varepsilon_1\varepsilon_2 + \alpha_1(\mathbf{r})\lambda_2(D_1\mathbf{r})}, \quad \alpha(\mathbf{r}) = \frac{\varepsilon_1\alpha_2(D_1\mathbf{r}) + \alpha_1(\mathbf{r})\beta_2(D_1\mathbf{r})}{\varepsilon_1\varepsilon_2 + \alpha_1(\mathbf{r})\lambda_2(D_1\mathbf{r})},$$

$$\beta(\mathbf{r}) = \frac{\lambda_1(\mathbf{r})\alpha_2(D_1\mathbf{r}) + \beta_1(\mathbf{r})\beta_2(D_1\mathbf{r})}{\varepsilon_1\varepsilon_2 + \alpha_1(\mathbf{r})\lambda_2(D_1\mathbf{r})}, \quad D = D_2 D_1.$$

(II) If $\varepsilon_1\varepsilon_2 + \alpha_1(\mathbf{r})\lambda_2(D_1\mathbf{r}) = 0$, then $\varepsilon = 0$, $M(\mathbf{r}) = M_1(\mathbf{r})M_2(D_1\mathbf{r})(\varepsilon_2\lambda_1(\mathbf{r}) + \lambda_2(D_1\mathbf{r})\beta_1(\mathbf{r}))$,

$$\lambda(\mathbf{r}) = 1, \quad \alpha(\mathbf{r}) = \frac{\varepsilon_1\alpha_2(D_1\mathbf{r}) + \alpha_1(\mathbf{r})\beta_2(D_1\mathbf{r})}{\varepsilon_2\lambda_1(\mathbf{r}) + \lambda_2(D_1\mathbf{r})\beta_1(\mathbf{r})},$$

$$\beta(\mathbf{r}) = \frac{\lambda_1(\mathbf{r})\alpha_2(D_1\mathbf{r}) + \beta_1(\mathbf{r})\beta_2(D_1\mathbf{r})}{\varepsilon_2\lambda_1(\mathbf{r}) + \lambda_2(D_1\mathbf{r})\beta_1(\mathbf{r})}, \quad D = D_2 D_1.$$

Each $T \in \mathcal{T}$ of the form (15) has an inverse T^{-1} determined by the parameters $\hat{\varepsilon}$, $\hat{M}(\mathbf{r})$, $\hat{\lambda}(\mathbf{r})$, $\hat{\alpha}(\mathbf{r})$, $\hat{\beta}(\mathbf{r})$, $\hat{D}$ given by:

(III) If $\beta(\mathbf{r}) \neq 0$, then $\hat{\varepsilon} = 1$ and

$$\hat{M}(\mathbf{r}) = \frac{\beta(D^{-1}\mathbf{r})}{M(D^{-1}\mathbf{r})(\varepsilon\beta(D^{-1}\mathbf{r}) - \alpha(D^{-1}\mathbf{r})\lambda(D^{-1}\mathbf{r}))}, \quad \hat{\lambda}(\mathbf{r}) = -\frac{\lambda(D^{-1}\mathbf{r})}{\beta(D^{-1}\mathbf{r})},$$

$$\hat{\alpha}(\mathbf{r}) = -\frac{\alpha(D^{-1}\mathbf{r})}{\beta(D^{-1}\mathbf{r})}, \quad \hat{\beta}(\mathbf{r}) = \frac{\varepsilon}{\beta(D^{-1}\mathbf{r})}, \quad \hat{D} = D^{-1}.$$

(IV) If $\beta(\mathbf{r}) = 0$, then $\hat{\varepsilon} = 0$ and

$$\hat{M}(\mathbf{r}) = \frac{1}{M(D^{-1}\mathbf{r})\alpha(D^{-1}\mathbf{r})}, \quad \hat{\lambda}(\mathbf{r}) = 1, \quad \hat{\alpha}(\mathbf{r}) = \frac{\alpha(D^{-1}\mathbf{r})}{\lambda(D^{-1}\mathbf{r})}, \quad \hat{\beta}(\mathbf{r}) = -\frac{\varepsilon}{\lambda(D^{-1}\mathbf{r})}, \quad \hat{D} = D^{-1}.$$

Proof of Theorem 1. We start by showing that the form of the coefficient $C(\mathbf{r},f) = (Nf+P)/(Kf+L)$ defined in (12) is restricted to

$$C(\mathbf{r},f) = M + W/f, \tag{16}$$

where $M = M(\mathbf{r})$, $W = W(\mathbf{r})$ are some functions of Γ-type, possibly one of them vanishing. It follows from (12) and (13) that transformation (10) is equivalent to

$$\frac{F_1(\mathbf{r})f + F_2(\mathbf{r})}{f} = \frac{(Nf+P)(F_1(D\mathbf{r})\eta + F_2(D\mathbf{r}))}{(Kf+L)\eta}, \tag{17}$$

where $N = N(\mathbf{r})$, $P = P(\mathbf{r})$, $K = K(\mathbf{r})$, $L = L(\mathbf{r})$. Solving this equation we get

$$\eta = \frac{f(fN+P)F_2(D\mathbf{r})}{LF_2(\mathbf{r}) + fKF_2(\mathbf{r}) - f^2NF_1(D\mathbf{r}) - fPF_1(D\mathbf{r}) + f^2KF_1(\mathbf{r}) + fLF_1(\mathbf{r})}.$$

In order that η had the form (11) the following identity must hold

$$f(fN+P)F_2(D\mathbf{r})(\alpha f+\beta)$$
$$=(\varepsilon f+\lambda)(LF_2(\mathbf{r})+fKF_2(\mathbf{r})-f^2NF_1(D\mathbf{r})-fPF_1(D\mathbf{r})+f^2KF_1(\mathbf{r})+fLF_1(\mathbf{r})). \quad (18)$$

The free term of the cubic on the right hand side equals $\lambda LF_2(\mathbf{r})$ while it vanishes on the left hand side, so that $\lambda L=0$. If $L=0$ we obtain (16). Otherwise, if $\lambda=0$ identity (18) takes the form

$$(fN+P)F_2(D\mathbf{r})(\alpha f+\beta)=LF_2(\mathbf{r})+fKF_2(\mathbf{r})-f^2NF_1(D\mathbf{r})-fPF_1(D\mathbf{r})+f^2KF_1(\mathbf{r})+fLF_1(\mathbf{r}). \quad (19)$$

If $N=0$, then $K=0$ and we again arrive at (16). If $N\neq 0$ the value $f=-P/N$ must be a root of the quadratic on the right hand side of (19). In other words, we must have

$$LF_2(\mathbf{r})-\frac{P}{N}KF_2(\mathbf{r})-\frac{P^2}{N^2}NF_1(D\mathbf{r})+\frac{P}{N}PF_1(D\mathbf{r})+\frac{P^2}{N^2}KF_1(\mathbf{r})-\frac{P}{N}LF_1(\mathbf{r})=0$$

or

$$\left(L-\frac{P}{N}K\right)\left(F_2(\mathbf{r})-\frac{P}{N}F_1(\mathbf{r})\right)=0.$$

Equality $L=PK/N$ again leads to (16). The equality $F_2(\mathbf{r})=PF_1(\mathbf{r})/N$ is impossible for rational P/N, as demonstrated by Ebisu and Iwasaki in ([20], Theorem 1.1) which proves our claim (16). If P/N is a function of gamma type then so is $F_2(\mathbf{r})/F_1(\mathbf{r})$ which would contradict the claim made before the theorem, but as we could not find a proof of this claim we explicitly prohibit this situation in the definition of $C(\mathbf{r},f)$.

Substituting $(Nf+P)/(Kf+L)$ by $M+W/f$ in (17) we can now express η as follows:

$$\eta=-\frac{(Mf+W)F_2(D\mathbf{r})}{(MF_1(D\mathbf{r})-F_1(\mathbf{r}))f+F_1(D\mathbf{r})W-F_2(\mathbf{r})}. \quad (20)$$

Next suppose $M\neq 0$. Then $C(\mathbf{r},f)=M(\varepsilon f+W/M)/f$ with $\varepsilon=1$. Comparison of (20) with (11) yields $W/M=\lambda$ which proves that the transformation (10) must have the form (15). Moreover,

$$\alpha=-\frac{M\varepsilon F_1(D\mathbf{r})-F_1(\mathbf{r})}{MF_2(D\mathbf{r})},\quad \beta=-\frac{F_1(D\mathbf{r})\lambda M-F_2(\mathbf{r})}{MF_2(D\mathbf{r})}.$$

These equalities can be rewritten as the system

$$\begin{cases} F_1(\mathbf{r})=M(\varepsilon F_1(D\mathbf{r})+\alpha F_2(D\mathbf{r})),\\ F_2(\mathbf{r})=M(\lambda F_1(D\mathbf{r})+\beta F_2(D\mathbf{r})). \end{cases} \quad (21)$$

Suppose now that $M=0$, $W\neq 0$. Then $C(\mathbf{r},f)=W(\varepsilon f+\lambda)/f$ with $\varepsilon=0$, $\lambda=1$. From (20) we have

$$\eta=\frac{\varepsilon f+\lambda}{\alpha f+\beta},$$

where again $\varepsilon=0$, $\lambda=1$, and $\alpha=F_1(\mathbf{r})/(WF_2(D\mathbf{r}))$, $\beta=-(WF_1(D\mathbf{r})-F_2(\mathbf{r}))/(WF_2(D\mathbf{r}))$ or

$$\begin{cases} F_1(\mathbf{r})=W\alpha F_2(D\mathbf{r})=W(\varepsilon F_1(D\mathbf{r})+\alpha F_2(D\mathbf{r})),\\ F_2(\mathbf{r})=W(F_1(D\mathbf{r})+\beta F_2(D\mathbf{r}))=W(\lambda F_1(D\mathbf{r})+\beta F_2(D\mathbf{r})). \end{cases} \quad (22)$$

Renaming W into M we have thus proved that the transformation again has the form (15) and the system (21) is satisfied.

The computation of composition is straightforward:

$$\begin{aligned}
T = T_2 \circ T_1 &\iff T : F(\mathbf{r}, f) = M_1(\mathbf{r})\frac{\varepsilon_1 f + \lambda_1(\mathbf{r})}{f} F(D_1\mathbf{r}, \eta_1) \\
&= M_1(\mathbf{r}) M_2(D_1\mathbf{r}) \frac{\varepsilon_1 f + \lambda_1(\mathbf{r})}{f} \frac{\varepsilon_2(\varepsilon_1 f + \lambda_1(\mathbf{r}))/(\alpha_1(\mathbf{r})f + \beta_1(\mathbf{r})) + \lambda_2(D_1\mathbf{r})}{(\varepsilon_1 f + \lambda_1(\mathbf{r}))/(\alpha_1(\mathbf{r})f + \beta_1(\mathbf{r}))} F(D_2 D_1 \mathbf{r}, \eta_2) \\
&= M_1(\mathbf{r}) M_2(D_1\mathbf{r}) \frac{\varepsilon_2(\varepsilon_1 f + \lambda_1(\mathbf{r})) + \lambda_2(D_1\mathbf{r})(\alpha_1(\mathbf{r})f + \beta_1(\mathbf{r}))}{f} F(D_2 D_1 \mathbf{r}, \eta_2) \\
&= M_1(\mathbf{r}) M_2(D_1\mathbf{r}) \frac{[\varepsilon_1 \varepsilon_2 + \alpha_1(\mathbf{r})\lambda_2(D_1\mathbf{r})]f + \varepsilon_2\lambda_1(\mathbf{r}) + \beta_1(\mathbf{r})\lambda_2(D_1\mathbf{r})}{f} F(D_2 D_1 \mathbf{r}, \eta).
\end{aligned}$$

If $\varepsilon_1\varepsilon_2 + \alpha_1(\mathbf{r})\lambda_2(D_1\mathbf{r}) \neq 0$, we can divide by this quantity leading to case (I). If it vanishes we get case (II). Given a transformation $T \in \mathcal{T}$ of the from (15) it is rather straightforward to compute its inverse. We omit the details. □

Remark 1. *Theorem 1 implies that each transformation $t \in \mathcal{T}$ is uniquely characterized by the collection $\{\varepsilon, M(\mathbf{r}), \lambda(\mathbf{r}), \alpha(\mathbf{r}), \beta(\mathbf{r}), D\}$, where $\varepsilon \in \{0, 1\}$, $M(\mathbf{r})$ is a function of gamma type, $\lambda(\mathbf{r})$, $\alpha(\mathbf{r})$ and $\beta(\mathbf{r})$ are rational functions of parameters a, b, c, d, e and D is 6×6 unit determinant integer matrix with bottom row $(0, \ldots, 0, 1)$. We will express this fact by writing $T \sim \{\varepsilon, M(\mathbf{r}), \lambda(\mathbf{r}), \alpha(\mathbf{r}), \beta(\mathbf{r}), D\}$. Occasionally, we will omit the dependence on $\mathbf{r}$ in the notation of the functions $M(\mathbf{r}), \lambda(\mathbf{r}), \alpha(\mathbf{r}), \beta(\mathbf{r})$ for brevity.*

Please note that for $\varepsilon = 1$ and non-vanishing α, β and λ the system (21) takes the form of $_4F_3 \to {}_3F_2$ reduction formulas

$$\begin{cases} F(D\mathbf{r}, \alpha(\mathbf{r})^{-1}) = M(\mathbf{r})^{-1} F_1(\mathbf{r}), \\ F(D\mathbf{r}, \lambda(\mathbf{r})/\beta(\mathbf{r})) = (M(\mathbf{r})\lambda(\mathbf{r}))^{-1} F_2(\mathbf{r}). \end{cases}$$

Next, we clarify the structure of the group $\mathcal{T}$ further. The composition rule involves all the parameters $M(\mathbf{r}), \lambda(\mathbf{r}), \alpha(\mathbf{r}), \beta(\mathbf{r})$ and D. The following theorem implies that the matrix D determines all other parameters uniquely. Denote by $\widehat{SL}(n, \mathbb{Z})$ the subgroup of the special linear group $SL(n, \mathbb{Z})$ of $n \times n$ integer matrices with unit determinant comprising matrices whose last row has the form $(0, \ldots, 0, 1)$.

Theorem 2. *The mapping $T \sim \{\varepsilon, M(\mathbf{r}), \lambda(\mathbf{r}), \alpha(\mathbf{r}), \beta(\mathbf{r}), D_T\} \to D_T$ is isomorphism, so that the group $(\mathcal{T}, \circ)$ is isomorphic to a subgroup of $\widehat{SL}(6, \mathbb{Z})$ which we denote by $(\mathcal{D}_{\mathcal{T}}, \cdot)$.*

Proof of Theorem 2. One direction is clear: each transformation $T \in \mathcal{T}$ by construction defines a matrix $D_T \in \widehat{SL}(6, \mathbb{Z})$ and the composition rule (I), (II) in Theorem 1 involves the product of matrices. Hence, to establish our claim it remains to prove that the kernel of the homomorphism $\mathcal{T} \to D_T$ is trivial. Assume the opposite: there exists a transformation $T \in \mathcal{T}$ with the identity matrix $D = I$ and non-trivial parameters $\varepsilon, M, \lambda, \alpha, \beta$. The system (21) then takes the form

$$\begin{cases} (1 - M\varepsilon)F_1(\mathbf{r}) = M\alpha F_2(\mathbf{r}), \\ M\lambda F_1(\mathbf{r}) = (1 - M\beta)F_2(\mathbf{r}). \end{cases} \tag{23}$$

If $\alpha = \lambda = 0$ we get $M = \varepsilon = 1$ from the first equation and $\beta = 1$ from the second equation, which amounts to the trivial identity transformation. We will show that all other cases are impossible. Indeed, Ebisu and Iwasaki demonstrated in ([20], Theorem 1.1) that the functions $F_1(\mathbf{r})$ and $F_2(\mathbf{r})$ are linearly independent over the field of rational functions of parameters. If $\alpha = 0$ and $\lambda \ne 0$, then $M = \varepsilon = 1$ from the first equation and $F_1(\mathbf{r})/F_2(\mathbf{r}) = (1-\beta)/\lambda$ from the second equation contradicting linear independence. Similarly, if $\alpha \ne 0$ and $\lambda = 0$, then $M = 1/\beta$ from the second equation, so that $F_2(\mathbf{r})/F_1(\mathbf{r}) = (1-\varepsilon/\beta)/(\alpha/\beta)$ is rational from the first equation leading again to contradiction. Finally, if both $\alpha \ne 0$ and $\lambda \ne 0$ we arrive at the identities

$$\frac{F_2(\mathbf{r})}{F_1(\mathbf{r})} = \frac{1-M\varepsilon}{\alpha M} = \frac{\lambda M}{1-M\beta} \;\Rightarrow\; (1-M\beta)(1-M\varepsilon) = \alpha\lambda M^2.$$

Linear independence of the functions $F_1(\mathbf{r})$, $F_2(\mathbf{r})$ over rational functions implies that the function $M = M(\mathbf{r})$ must be a ratio of products of gamma functions irreducible to a rational function. On the other hand, by the ultimate equality $M(\mathbf{r})$ solves the quadratic equation with rational coefficients:

$$M = M(\mathbf{r}) = \mu(\mathbf{r}) \pm \sqrt{\nu(\mathbf{r})}$$

with rational μ, ν. It is easy to see that this is not possible as Γ is meromorphic with infinite number of poles and no branch points, while $\mu(\mathbf{r}) \pm \sqrt{\nu(\mathbf{r})}$ may only have a finite number of poles and zeros and has branch points. $\square$

3. The Subgroup of $\mathcal{T}$ Generated by Known Transformations

We can now rewrite the transformations (5)–(8) in the standard form (15). Denote by $\psi = d + e - a - b - c - 1$ the parametric excess of the function on the left hand side of (15). Identity (7) is determined by the following set of parameters

$$M_1 = \frac{\Gamma(\psi+1)\Gamma(d)\Gamma(e)}{\Gamma(a)\Gamma(d+e-a-c)\Gamma(d+e-a-b)}, \quad \varepsilon_1 = 1, \quad \lambda_1 = \frac{bc}{\psi}, \tag{24a}$$

$$D_1 = \begin{bmatrix} -1 & -1 & -1 & 1 & 1 & -1 \\ -1 & 0 & 0 & 1 & 0 & 0 \\ -1 & 0 & 0 & 0 & 1 & 0 \\ -1 & 0 & -1 & 1 & 1 & 0 \\ -1 & -1 & 0 & 1 & 1 & 0 \\ 0 & 0 & 0 & 0 & 0 & 1 \end{bmatrix}, \quad \alpha_1 = \frac{1}{\psi}, \quad \beta_1 = 0. \tag{24b}$$

We will call this transformation T_1.

The standard form (15) of identity (6) is characterized by the following parameters:

$$M_2 = \frac{\Gamma(e)\Gamma(\psi+1)}{\Gamma(e+d-a-b)\Gamma(e-c)}, \quad \varepsilon_2 = 1, \quad \lambda_2 = \frac{c(-d+a+b)}{\psi}, \tag{25a}$$

$$D_2 = \begin{bmatrix} -1 & 0 & 0 & 1 & 0 & 0 \\ 0 & -1 & 0 & 1 & 0 & 0 \\ 0 & 0 & 1 & 0 & 0 & 0 \\ 0 & 0 & 0 & 1 & 0 & 0 \\ -1 & -1 & 0 & 1 & 1 & 0 \\ 0 & 0 & 0 & 0 & 0 & 1 \end{bmatrix}, \quad \alpha_2 = 0, \quad \beta_2 = \frac{e-c-1}{\psi}. \tag{25b}$$

We will call this transformation T_2.

The standard parameters of transformation (8) are given by

$$M_3 = \frac{\Gamma(\psi+1)\Gamma(e)}{\Gamma(e-a)\Gamma(e+d-b-c)}, \quad \varepsilon_3 = 1, \quad \lambda_3 = \frac{abc}{d\psi}, \tag{26a}$$

$$D_3 = \begin{bmatrix} 1 & 0 & 0 & 0 & 0 & 0 \\ 0 & -1 & 0 & 1 & 0 & 0 \\ 0 & 0 & -1 & 1 & 0 & 0 \\ 0 & 0 & 0 & 1 & 0 & 1 \\ 0 & -1 & -1 & 1 & 1 & 0 \\ 0 & 0 & 0 & 0 & 0 & 1 \end{bmatrix}, \quad \alpha_3 = \frac{1}{d}, \quad \beta_3 = \frac{bc}{d\psi}. \tag{26b}$$

We will call this transformation T_3.

Finally, transformation (5) in the standard form (15) is parameterized by

$$M_4 = \frac{\Gamma(e)\Gamma(\psi)}{\Gamma(e-c)\Gamma(\psi+c)}, \quad \varepsilon_4 = 1, \quad \lambda_4 = 0, \quad \alpha_4 = \frac{d-a-b-1}{(d-a-1)(d-b-1)}, \tag{27a}$$

$$D_4 = \begin{bmatrix} -1 & 0 & 0 & 1 & 0 & -1 \\ 0 & -1 & 0 & 1 & 0 & -1 \\ 0 & 0 & 1 & 0 & 0 & 0 \\ 0 & 0 & 0 & 1 & 0 & 0 \\ -1 & -1 & 0 & 1 & 1 & -1 \\ 0 & 0 & 0 & 0 & 0 & 1 \end{bmatrix}, \quad \beta_4 = \frac{ab}{(d-a-1)(d-b-1)}. \tag{27b}$$

We will call this transformation T_4. It is easy to see that it is of order 2, i.e., $T_4^2 = I$.

The four transformations T_1, T_2, T_3, T_4 (or, equivalently, (5)–(8)) combined with permutations of the upper and lower parameters generate a subgroup of $\mathcal{T}$ which we will call $\hat{\mathcal{T}}$. Isomorphism established in Theorem 2 induces an isomorphism between $\hat{\mathcal{T}}$ and a subgroup of $\widehat{SL}(6,\mathbb{Z})$ which we denote by $\mathcal{D}_{\hat{\mathcal{T}}}$.

A complete characterization of $\hat{\mathcal{T}}$ and $\mathcal{D}_{\hat{\mathcal{T}}}$ will follow. Before we turn to it, we remark that to our belief, the complete group $\mathcal{T}$ contains no elements other than those in $\hat{\mathcal{T}}$. We were unable, however, to prove this claim. Let us thus state it as a conjecture.

Conjecture. The subgroup $\hat{\mathcal{T}}$ generated by the transformations (24)–(27) coincides with the entire group $\mathcal{T}$ of all transformations of the form (10) or, equivalently, of the form (15).

Denote by S_j, $j = 1, \ldots, 5$, the transformation shifting the j-th component of the parameter vector $\mathbf{r}$ by $+1$, i.e., S_j is characterized by the matrix $\hat{S}_j$ such that $\hat{S}_1 \mathbf{r} = (a+1, b, c, d, e, 1)$, $\hat{S}_2 \mathbf{r} = (a, b+1, c, d, e, 1)$, etc. It is not *a priori* obvious that such transformations should exist among the elements of $\hat{\mathcal{T}}$. The following theorem shows that it is indeed the case.

Theorem 3. *The group $\hat{\mathcal{T}}$ contains the transformations S_j, $j = 1, \ldots, 5$.*

Proof of Theorem 3. Due to permutation symmetry it is clearly sufficient to display the transformations S_1 and S_4. We will need the inverse of the transformation T_1 defined in (24). Using Theorem 1 we calculate

$$_4F_3\left(\begin{matrix} a, b, c, f+1 \\ d, e, f \end{matrix}\right) = \frac{\hat{M}_1}{\hat{f}} {}_4F_3\left(\begin{matrix} d+e-a-b-c-1, d-a-1, e-a-1, \hat{\eta}_1 + 1 \\ d+e-a-c-1, d+e-a-b-1, \hat{\eta}_1 \end{matrix}\right), \tag{28}$$

where

$$\hat{M}_1 = \frac{\Gamma(d)\Gamma(e)\Gamma(\psi)}{\Gamma(\psi+b)\Gamma(\psi+c)\Gamma(a)}, \quad \hat{\varepsilon}_1 = 0, \quad \hat{\lambda}_1 = 1, \quad \hat{\alpha}_1 = \frac{1}{(d-a-1)(e-a-1)},$$

$$\hat{\beta}_1 = \frac{-a}{(d-a-1)(e-a-1)}, \quad \text{so that} \quad \hat{\eta}_1 = \frac{(d-a-1)(e-a-1)}{f-a}.$$

Next, exchanging the roles of $d+e-a-b-1$ and d and the roles of $d-a-1$ and c in (5) or, equivalently, post-composing T_4 with permutation (13) (45) we will obtain a transformation that we call $\hat{T}_4$. Then $\hat{T}_4 \circ \hat{T}_4$ takes the form

$$_4F_3\left(\begin{matrix} a,b,c,f+1 \\ d,e,f \end{matrix}\right) = \frac{\Gamma(e)\Gamma(d)\Gamma(\psi)}{\Gamma(b+\psi)\Gamma(c+\psi)\Gamma(a+1)} {}_4F_3\left(\begin{matrix} \psi-1, e-a-1, d-a-1, \tilde{\eta}_4+1 \\ c+\psi, b+\psi, \tilde{\eta}_4 \end{matrix}\right) \quad (29)$$

with

$$\tilde{\eta}_4 = \frac{(\psi-1)(e-a-1)(d-a-1)f}{abc + (1+2a+a^2-bc-d-ad-e-ae+de)f}.$$

Applying T_1^{-1} to the right hand side of (29) we obtain the transformation S_1:

$$_4F_3\left(\begin{matrix} a,b,c,f+1 \\ d,e,f \end{matrix}\right) = M\frac{\varepsilon f + \lambda}{f} {}_4F_3\left(\begin{matrix} a+1,b,c,\eta+1 \\ d,e,\eta \end{matrix}\right), \quad (30)$$

where $\varepsilon = 1$, and

$$M = 1 - \frac{bc}{(d-a-1)(e-a-1)}, \quad \lambda = \frac{abc}{a^2 - bc + (d-1)(e-1) - a(d+e-2)},$$

$$\alpha = \frac{d+e-a-b-c-2}{a^2 - bc + (d-1)(e-1) - a(d+e-2)}, \quad \beta = -\frac{a(d+e-a-b-c-2)}{a^2 - bc + (d-1)(e-1) - a(d+e-2)}.$$

According to (15) we thus obtain the following expression for η

$$\eta = \frac{abc + (1+2a+a^2-bc-d-ad-e-ae+de)f}{a(2+a+b+c-d-e) - (2+a+b+c-d-e)f}.$$

Application of the transformation T_3 given by (26) to itself yields $T_3 \circ T_3$ in the form:

$$_4F_3\left(\begin{matrix} a,b,c,f+1 \\ d,e,f \end{matrix}\right) = \frac{a(d-b)(d-c)(bc+f\psi) + (d+1)(e-a)(abc+fd\psi)}{fd(d+1)e\psi} {}_4F_3\left(\begin{matrix} a,b+1,c+1,\tilde{\eta}_3+1 \\ d+2, e+1, \tilde{\eta}_3 \end{matrix}\right),$$

where

$$\tilde{\eta}_3 = \frac{a(d-b)(d-c)(bc+f\psi) + (d+1)(e-a)(abc+fd\psi)}{(d-b)(d-c)(bc+f\psi) + (e-a)(abc+fd\psi)}.$$

On the other hand, using (28) we compute T_1^{-2} as follows:

$$_4F_3\left(\begin{matrix} a,b,c,f+1 \\ d,e,f \end{matrix}\right) = \frac{(d-1)(e-1)(f-a)}{f(d-a-1)(e-a-1)} {}_4F_3\left(\begin{matrix} a, b-1, c-1, \hat{\eta}'_1+1 \\ d-1, e-1, \hat{\eta}'_1 \end{matrix}\right)$$

with

$$\hat{\eta}_1' = \frac{(b-1)(c-1)(f-a)}{(d-a-1)(e-a-1) - \psi(f-a)}.$$

Comparing these formulas we see that the composition $T_1^{-2} \circ T_3^2$ gives the transformation S_4 shifting $d \to d+1$ while a, b, c, e remain intact:

$$_4F_3\begin{pmatrix} a,b,c,f+1 \\ d,e,f \end{pmatrix} = \frac{f+\lambda}{f} {}_4F_3\begin{pmatrix} a,b,c,\eta+1 \\ d+1,e,\eta \end{pmatrix}, \tag{31}$$

so that $\varepsilon = 1$, $M = 1$,

$$\lambda = \frac{abc}{d(d+e-a-b-c-1)}, \quad \alpha = \frac{1}{d}, \quad \beta = \frac{(b-d)(c-d) + a(b+c-d)}{d(d+e-a-b-c-1)}, \quad \eta = \frac{\varepsilon f + \lambda}{\alpha f + \beta}.$$

□

Each transformation S_j, $j = 1, \ldots, 5$, obviously generates a subgroup of $\hat{\mathcal{T}}$ isomorphic to $\mathbb{Z}$—the additive group of integers. Hence, in the parlance of group theory, the above theorem can be restated and enhanced as follows.

Corollary 1. *The group $\hat{\mathcal{T}}$ contains a subgroup $\mathcal{S}$ isomorphic to the 5-dimensional integer lattice $\mathbb{Z}^5$. Furthermore, this subgroup is normal.*

Proof of Corollary 1. By the previous theorem we only need to prove normality. Denote by $\mathcal{S}$ the subgroup of the matrix group $\mathcal{D}_{\hat{\mathcal{T}}}$ generated by the shift matrices $\hat{S}_j$, $j = 1, \ldots, 5$. Clearly, $\mathcal{S}$ comprises 6×6 matrices whose principal 5×5 sub-matrix equals the identity matrix I_5, the 6-th row is $(0, \ldots, 0, 1)$ and the 6-th column is $(k_1, \ldots, k_5, 1)$ for some $k_i \in \mathbb{Z}$. As all elements of $\mathcal{D}_{\hat{\mathcal{T}}}$ have integer entries and the bottom row $(0, \ldots, 0, 1)$ it is easy to see that for any shift matrix $S \in \mathcal{S}$ and any matrix $D \in \mathcal{D}_{\hat{\mathcal{T}}}$ both products DS and SD have the principal 5×5 sub-matrix equal to that of D and the last column of the form $(k_1, \ldots, k_5, 1)$ for some $k_i \in \mathbb{Z}$. Running over all elements of $\mathcal{S}$ while keeping D fixed we see that the left and right conjugacy classes of the element D with respect to $\mathcal{S}$ coincide. □

The above corollary implies that we can take the factor group $\mathcal{D}_{\hat{\mathcal{T}}}/\mathcal{S}$. Each element in $\mathcal{D}_{\hat{\mathcal{T}}}/\mathcal{S}$ is a conjugacy class containing a representative with the last column $(0, \ldots, 0, 1)^T$. Next, we note that the principal 5×5 sub-matrix of the matrix D_2 from (25b) of the transformation (6) is equal to that of the Kummer's transformation (2). This transformation together with the permutation group $P_3 \times P_2$ representing the obvious invariance with respect to separate permutations of the upper and lower parameters generate the entire group of Thomae transformations [9]. Next, comparing the principal 5×5 sub-matrices of the further generators D_1, D_3, D_4 with the matrices of the Thomae transformations found, for instance in ([4], Appendix 1), we see that all of them occur among the elements of the group of the Thomae transformations. Hence, it remains to apply Theorem 3.2 from [9] asserting that the group of the Thomae transformations is isomorphic the 120-element symmetric group P_5 of permutations on five symbols. Isomorphism is given by a linear change of variables seen in (4). Hence, our final result is the following theorem.

Theorem 4. *The group $\hat{\mathcal{T}}$ is isomorphic to $P_5 \times \mathbb{Z}^5$.*

As the entire group of the Thomae transformations for $_3F_2$ can be generated by the identity (2) and the permutation group $P_3 \times P_2$, the above theorem implies that our entire group $\hat{\mathcal{T}}$ can be generated by the identity (6) (transformation T_2) and the top parameter shift transformation S_1 together with the obvious symmetries $P_3 \times P_2$. For example, the bottom parameter shift transformation can be obtained as follows:

$$(d-c, e-c, \psi, \psi+a, \psi+b) \xmapsto{T_2^2} (a,b,c,d,e) \xmapsto{S_1 S_3^{-1}} (a+1, b, c-1, d, e)$$
$$\xmapsto{T_2^{-2}} (d-c+1, e-c+1, \psi, \psi+a+1, \psi+b) \xmapsto{S_1 S_2} (d-c, e-c, \psi, \psi+a+1, \psi+b).$$

Comparing the first and the last terms in this chain we see that we got the bottom parameter shift transformation S_4 using only T_2 and top shift transformations S_1, S_2, S_3 obtained from S_1 by permuting top parameters.

Theorem 4 further implies that there is a straightforward algorithm for computing any transformation from the group $\hat{\mathcal{T}}$. Details are given in the Appendix A to this paper.

4. Related $_3F_2$ Transformation

The proof of Theorem 1 shows that each transformation $T \in \mathcal{T}$ is associated with the system (21) of two $_3F_2$ transformations. This system leads immediately to the following corollary.

Proposition 1. *Each transformation $T \sim \{\varepsilon, M(\mathbf{r}), \lambda(\mathbf{r}), \alpha(\mathbf{r}), \beta(\mathbf{r})\} \in \mathcal{T}$ induces a transformation for the ratio*

$$\Psi(\mathbf{r}) := \frac{F_2(\mathbf{r})}{F_1(\mathbf{r})} = \frac{abc}{de} \frac{{}_3F_2\left(\begin{matrix} a+1, b+1, c+1 \\ d+1, e+1 \end{matrix}\right)}{{}_3F_2\left(\begin{matrix} a,b,c \\ d,e \end{matrix}\right)} = \frac{d}{dx} \log {}_3F_2\left(\begin{matrix} a,b,c \\ d,e \end{matrix} \middle| x\right)\bigg|_{x=1}$$

of the form

$$\Psi(\mathbf{r}) = \frac{\beta(\mathbf{r})\Psi(D\mathbf{r}) + \lambda(\mathbf{r})}{\alpha(\mathbf{r})\Psi(D\mathbf{r}) + \varepsilon}.$$

Next, we observe that any two elements of $\mathcal{T}$ generate a three-term relation for $_3F_2$.

Proposition 2. *For any two transformations from the group $\mathcal{T}$: $T_1 \sim \{\varepsilon_1, M_1(\mathbf{r}), \lambda_1(\mathbf{r}), \alpha_1(\mathbf{r}), \beta_1(\mathbf{r}), D_1\}$ and $T_2 \sim \{\varepsilon_2, M_2(\mathbf{r}), \lambda_2(\mathbf{r}), \alpha_2(\mathbf{r}), \beta_2(\mathbf{r}), D_2\}$ satisfying the condition $\alpha_2 \beta_1 - \alpha_1 \beta_2 \neq 0$, the following identities hold*

$$F_1(\mathbf{r}) = M_1 \frac{\alpha_2 \beta_1 \varepsilon_1 - \alpha_1 \alpha_2 \lambda_1}{\alpha_2 \beta_1 - \alpha_1 \beta_2} F_1(D_1 \mathbf{r}) + M_2 \frac{\alpha_1 \alpha_2 \lambda_2 - \alpha_1 \beta_2 \varepsilon_2}{\alpha_2 \beta_1 - \alpha_1 \beta_2} F_1(D_2 \mathbf{r}) \qquad (32)$$

(the dependence on $\mathbf{r}$ is omitted for brevity) and

$$F_2(\mathbf{r}) = M_1 \frac{\beta_1 \beta_2 \varepsilon_1 - \alpha_1 \beta_2 \lambda_1}{\alpha_2 \beta_1 - \alpha_1 \beta_2} F_1(D_1 \mathbf{r}) + M_2 \frac{\alpha_2 \beta_1 \lambda_2 - \beta_1 \beta_2 \varepsilon_2}{\alpha_2 \beta_1 - \alpha_1 \beta_2} F_1(D_2 \mathbf{r}), \qquad (33)$$

where as before, $F_1(\mathbf{r}) = {}_3F_2\left(\begin{matrix} a,b,c \\ d,e \end{matrix}\right)$, $F_2(\mathbf{r}) = (abc)/(de){}_3F_2\left(\begin{matrix} a+1,b+1,c+1 \\ d+1,e+1 \end{matrix}\right)$.

Proof of Proposition 2. Solving (21) for each transformation we, in particular, get the system of equations:

$$\begin{cases} F_1(D_1\mathbf{r}) = (\beta_1 F_1(\mathbf{r}) - \alpha_1 F_2(\mathbf{r}))/(M_1(\beta_1\varepsilon_1 - \alpha_1\lambda_1)), \\ F_1(D_2\mathbf{r}) = (\beta_2 F_1(\mathbf{r}) - \alpha_2 F_2(\mathbf{r}))/(M_2(\beta_2\varepsilon_2 - \alpha_2\lambda_2)). \end{cases}$$

Solving the above system for $F_1(\mathbf{r}), F_2(\mathbf{r})$ we arrive at (32) and (33). □

If the matrices D_1, D_2 contain no shifts (i.e., the last column is $(0,0,0,0,0,1)^T$), then they correspond to Thomae's relations, so that $F_1(D_1\mathbf{r}), F_1(D_2\mathbf{r})$ are equal to each other up to a factor of gamma type. In this case, identities (32) and (33) become two-term transformations. However, for non-zero shifts Proposition 2 generates genuine three-term relations for $_3F_2(a,b,c;d,e)$. For example, we obtain

$$_3F_2\begin{pmatrix} a,b,c \\ d,e \end{pmatrix} = \frac{\Gamma(d+1)\Gamma(e)\Gamma(d+e-a-b-c)}{\Gamma(a+1)\Gamma(d+e-a-b)\Gamma(d+e-a-c)} {}_3F_2\begin{pmatrix} d+e-a-b-c-1, d-a, e-a \\ d+e-a-c, e+d-a-b \end{pmatrix}$$
$$+ \frac{(a-d)(d-b)(d-c)}{d(1+d)e} {}_3F_2\begin{pmatrix} a+1,b+1,c+1 \\ d+2,e+1 \end{pmatrix}. \quad (34)$$

An important subclass of these transformations are pure shifts (the principal 5×5 submatrices of D_1, D_2 are identity matrices). This subclass comprises the so-called contiguous relations, studied recently in detail in [20]. In particular, Theorem 1.1 from [20] claims the existence of the unique rational functions $u(\mathbf{r}), v(\mathbf{r})$ such that

$$_3F_2\begin{pmatrix} a,b,c \\ d,e \end{pmatrix} = u(\mathbf{r}) {}_3F_2\begin{pmatrix} a+k_1,b+k_2,c+k_3 \\ d+k_4,e+k_5 \end{pmatrix} + v(\mathbf{r}) {}_3F_2\begin{pmatrix} a+m_1,b+m_2,c+m_3 \\ d+m_4,e+m_5 \end{pmatrix} \quad (35)$$

for any two distinct non-zero integer vectors $(k_1, k_2, k_3, k_4, k_5)$, $(m_1, m_2, m_3, m_4, m_5)$. Furthermore, Ebisu and Iwasaki presented a rather explicit algorithm in [20] for computing the functions $u(\mathbf{r}), v(\mathbf{r})$ for given shifts. Proposition 2 furnishes an alternative method for computing these functions. For its realization we provide a collection of *Mathematica* routines in the Appendix A to this paper. Our algorithm works as follows: first step is to calculate transformations $T_1, T_2 \in \hat{\mathcal{T}}$ associated with the matrices

$$D_1 = \begin{bmatrix} 1 & 0 & 0 & 0 & 0 & k_1 \\ 0 & 1 & 0 & 0 & 0 & k_2 \\ 0 & 0 & 1 & 0 & 0 & k_3 \\ 0 & 0 & 0 & 1 & 0 & k_4 \\ 0 & 0 & 0 & 0 & 1 & k_5 \\ 0 & 0 & 0 & 0 & 0 & 1 \end{bmatrix}, D_2 = \begin{bmatrix} 1 & 0 & 0 & 0 & 0 & m_1 \\ 0 & 1 & 0 & 0 & 0 & m_2 \\ 0 & 0 & 1 & 0 & 0 & m_3 \\ 0 & 0 & 0 & 1 & 0 & m_4 \\ 0 & 0 & 0 & 0 & 1 & m_5 \\ 0 & 0 & 0 & 0 & 0 & 1 \end{bmatrix}.$$

To this end we simply iterate transformations $S^\pm, S_\pm$ realizing the shifts by ± 1 of the first and forth parameters, respectively, combining them with the necessary permutations of the upper and lower parameters. To calculate the resulting λ, α and β the composition rule from Theorem 1 is used with the help of *Mathematica* routine. Then it remains to apply formula (32). For example, we get:

$$_3F_2\begin{pmatrix} a,b,c \\ d,e \end{pmatrix} = \frac{d+e-a-b-c-1}{e} {}_3F_2\begin{pmatrix} a+1,b+1,c+1 \\ d+1,e+1 \end{pmatrix}$$
$$+ \frac{(a-d)(d-b)(d-c)}{d(d+1)e} {}_3F_2\begin{pmatrix} a+1,b+1,c+1 \\ d+2,e+1 \end{pmatrix}. \quad (36)$$

Please note that identity (34) is obtained from (36) by an application of a Thomae relation to the first term on the right hand side. In a similar fashion, contiguous relations and Thomae transformations generate all three-term relations from Proposition 2, induced by the elements of the the group $\hat{\mathcal{T}}$. We note that the relations covered by Proposition 2 are different from the three-term relations for $_3F_2$ summarized by Bailey in ([21], Section 3.7) and studied from group-theoretic viewpoint in ([9], Section IV). This can be seen for example by comparing the matrices ([9], Equation (2.6c)) with the matrices D associated with $\hat{\mathcal{T}}$.

The system (21) follows from the representation (13) of $_4F_3$ with one unit shift as a linear combination of two $_3F_2$ functions. However, Formula (13) is just one example of such decomposition. The two propositions below give many more ways to expand the $_4F_3$ with unit shift into linear combination of $_3F_2$. Proposition 3 is proved directly in terms of hypergeometric series manipulations as its results will be used below in Section 6 to prove Lemma 1 used to generate the group $\hat{\mathcal{T}}$.

Proposition 3. *The following identities hold true:*

$$_3F_2\left(\begin{matrix}\alpha,b,c\\d,e\end{matrix}\right) + \gamma\, _3F_2\left(\begin{matrix}\alpha-1,b,c\\d,e\end{matrix}\right) = (\gamma+1)\, _4F_3\left(\begin{matrix}\alpha-1,b,c,\xi+1\\d,e,\xi\end{matrix}\right), \tag{37}$$

where $\xi = (\gamma+1)(\alpha-1)$;

$$_3F_2\left(\begin{matrix}\alpha,b,c\\d,e\end{matrix}\right) + \gamma\, _3F_2\left(\begin{matrix}\alpha+1,b,c\\d+1,e\end{matrix}\right) = (\gamma+1)\, _4F_3\left(\begin{matrix}\alpha,b,c,\nu+1\\d+1,e,\nu\end{matrix}\right), \tag{38}$$

where $\nu = (\gamma+1)\alpha d/(\gamma d+\alpha)$; *and*

$$_3F_2\left(\begin{matrix}\alpha,b,c\\d,e\end{matrix}\right) + \gamma\, _3F_2\left(\begin{matrix}\alpha,b+1,c+1\\d+1,e+1\end{matrix}\right) = \, _4F_3\left(\begin{matrix}\alpha-1,b,c,\lambda+1\\d,e,\lambda\end{matrix}\right), \tag{39}$$

where $\lambda = (\alpha-1)bc/(bc+\gamma de)$.

Proof of Proposition 3. We have

$$_3F_2\left(\begin{matrix}\alpha,b,c\\d,e\end{matrix}\right) + \gamma\, _3F_2\left(\begin{matrix}\alpha-1,b,c\\d,e\end{matrix}\right) = 1+\gamma+\sum_{n=1}^{\infty}\frac{(\alpha)_n(b)_n(c)_n+\gamma(\alpha-1)_n(b)_n(c)_n}{(d)_n(e)_n n!}$$

$$= (1+\gamma)\left(1+\sum_{n=1}^{\infty}\frac{(\alpha-1)_n(b)_n(c)_n}{(d)_n(e)_n n!}\left(1+\frac{n}{(\alpha-1)(\gamma+1)}\right)\right)$$

$$= (\gamma+1)\, _4F_3\left(\begin{matrix}\alpha-1,b,c,\xi+1\\d,e,\xi\end{matrix}\right),$$

where $\xi = (\gamma+1)(\alpha-1)$ and we used $(\alpha)_n = (\alpha-1)_n(1+n/(\alpha-1))$. Next,

$$_3F_2\left(\begin{matrix}\alpha,b,c\\d,e\end{matrix}\right) + \gamma\, _3F_2\left(\begin{matrix}\alpha+1,b,c\\d+1,e\end{matrix}\right) = 1+\gamma+\sum_{n=1}^{\infty}\frac{(\alpha)_n(b)_n(c)_n}{(d+1)_n(e)_n n!}\left(1+\frac{n}{d}+\gamma+\frac{\gamma n}{\alpha}\right)$$

$$= (\gamma+1)\, _4F_3\left(\begin{matrix}\alpha,b,c,\nu+1\\d+1,e,\nu\end{matrix}\right),$$

where $\nu = (\gamma+1)\alpha d/(\gamma d+\alpha)$ and we used $(\alpha+1)_n = (\alpha)_n(1+n/\alpha)$.

Finally, using the obvious identities $(b)_n = b(b+1)_{n-1}$ and $(\alpha)_n = (\alpha-1)_{n+1}/(\alpha-1)$ we get

$$
{}_3F_2\!\left(\begin{array}{c}a,b,c\\d,e\end{array}\right)+\gamma_3 F_2\!\left(\begin{array}{c}a,b+1,c+1\\d+1,e+1\end{array}\right)=1+\sum_{n=1}^{\infty}\frac{bc(a)_{n-1}(b+1)_{n-1}(c+1)_{n-1}(\alpha+n-1)}{de(d+1)_{n-1}(e+1)_{n-1}n!}
$$

$$
+\gamma_3 F_2\!\left(\begin{array}{c}\alpha,b+1,c+1\\d+1,e+1\end{array}\right)=1+\sum_{n=0}^{\infty}\frac{(\alpha)_n(b+1)_n(c+1)_n}{(d+1)_n(e+1)_n n!}\left(\frac{bc(\alpha+n)}{de(n+1)}+\gamma\right)
$$

$$
=1+\sum_{n=0}^{\infty}\frac{(\alpha-1)_{n+1}(b)_{n+1}(c)_{n+1}}{(d)_{n+1}(e)_{n+1}(n+1)!}\frac{de}{bc(\alpha-1)}\left(\frac{bc(\alpha+n)}{de}+\gamma(n+1)\right)=
$$

$$
1+\sum_{n=1}^{\infty}\frac{(\alpha-1)_n(b)_n(c)_n}{(d)_n(e)_n n!}\left(1+n\frac{bc+\gamma de}{(\alpha-1)bc}\right)={}_4F_3\!\left(\begin{array}{c}\alpha-1,b,c,\lambda+1\\d,e,\lambda\end{array}\right),
$$

where $\lambda = (\alpha-1)bc/(bc+\gamma de)$. □

Other ways to represent ${}_4F_3$ with one unit shift as a linear combination of ${}_3F_2$ are found by substituting (32) and (33) into (13). This is done in the following proposition.

Proposition 4. *Any two transformations from the group* $\mathcal{T}$: $T_1 \sim \{\varepsilon_1, M_1(\mathbf{r}), \lambda_1(\mathbf{r}), \alpha_1(\mathbf{r}), \beta_1(\mathbf{r}), D_1\}$ *and* $T_2 \sim \{\varepsilon_2, M_2(\mathbf{r}), \lambda_2(\mathbf{r}), \alpha_2(\mathbf{r}), \beta_2(\mathbf{r}), D_2\}$ *satisfying the condition* $\alpha_2\beta_1 - \alpha_1\beta_2 \neq 0$ *(for brevity we omit the dependence on* $\mathbf{r}$ *in the parameters) induce the decomposition*

$$
{}_4F_3\!\left(\begin{array}{c}a,b,c,f+1\\d,e,f\end{array}\right)=M_1\frac{\beta_1\varepsilon_1-\alpha_1\lambda_1}{\alpha_2\beta_1-\alpha_1\beta_2}\left(\alpha_2+\frac{\beta_2}{f}\right)F_1(D_1\mathbf{r})+M_2\frac{\alpha_2\lambda_2-\beta_2\varepsilon_2}{\alpha_2\beta_1-\alpha_1\beta_2}\left(\alpha_1+\frac{\beta_1}{f}\right)F_1(D_2\mathbf{r}),\quad(40)
$$

where $F_1(\mathbf{r}) = {}_3F_2\!\left(\begin{array}{c}a,b,c\\d,e\end{array}\right)$.

Let us exemplify (40) with the following two decompositions:

$$
{}_4F_3\!\left(\begin{array}{c}a,b,c,f+1\\d,e,f\end{array}\right)=\left(\frac{d+e-a-b-c-1}{e}+\frac{abc}{def}\right){}_3F_2\!\left(\begin{array}{c}a+1,b+1,c+1\\d+1,e+1\end{array}\right)
$$

$$
+\frac{(a-d)(d-b)(d-c)}{ed(1+d)}{}_3F_2\!\left(\begin{array}{c}a+1,b+1,c+1\\d+2,e+1\end{array}\right)
$$

and

$$
{}_4F_3\!\left(\begin{array}{c}a,b,c,f+1\\d,e,f\end{array}\right)=A{}_3F_2\!\left(\begin{array}{c}a+1,b,c\\d,e\end{array}\right)+B{}_3F_2\!\left(\begin{array}{c}a+1,b+1,c+1\\d+2,e+1\end{array}\right),
$$

where

$$
A=1+\frac{bc(f-a)}{f(b(d-c)-d(d+e-a-c-1))},\quad B=\frac{bc(a-d)(b-d)(c-d)(f-a)}{def(1+d)(b(c-d)+d(d+e-a-c-1))}.
$$

5. Summation Formulas

In ([22], Equation (45)) we established the following summation formula

$$_4F_3\left(\begin{array}{c} a,b,c,f+1 \\ d,e,f \end{array}\right) = \frac{\Gamma(d)\Gamma(e)}{\Gamma(a+1)\Gamma(b+1)\Gamma(c+1)}, \tag{41a}$$

valid if

$$e_1(d,e) - e_1(a,b,c) = 2 \text{ and } f = \frac{e_3(a,b,c)}{e_2(a,b,c) - e_2(1-d, 1-e)}, \tag{41b}$$

where $e_k(\cdot)$ denotes the k-th elementary symmetric polynomial. Now, if we apply any transformation of the form (15) and impose the above restrictions on the parameters on the right hand side, we obtain

$$_4F_3\left(\begin{array}{c} a,b,c,f+1 \\ d,e,f \end{array}\right) = M(\mathbf{r})\frac{\varepsilon f + \lambda(\mathbf{r})}{f}F(\mathbf{q}, \eta) = \frac{M(\mathbf{r})(\varepsilon f + \lambda(\mathbf{r}))\Gamma(q_4)\Gamma(q_5)}{f\Gamma(q_1+1)\Gamma(q_2+1)\Gamma(q_3+1)}, \tag{42a}$$

where $(q_1, q_2, q_3, q_4, q_5, 1) = D\mathbf{r}$, and the conditions $e_1(q_4, q_5) - e_1(q_1, q_2, q_3) = 2$ and

$$\eta = \frac{\varepsilon f + \lambda(\mathbf{r})}{\alpha(\mathbf{r})f + \beta(\mathbf{r})} = \frac{e_3(q_1, q_2, q_3)}{e_2(q_1, q_2, q_3) - e_2(1-q_4, 1-q_5)}$$

must hold. Expressing f they are equivalent to

$$e_1(q_4, q_5) - e_1(q_1, q_2, q_3) = 2 \text{ and } f = \frac{\lambda(\mathbf{r})(e_2(q_1, q_2, q_3) - e_2(1-q_4, 1-q_5)) - \beta(\mathbf{r})e_3(q_1, q_2, q_3)}{\alpha(\mathbf{r})e_3(q_1, q_2, q_3) - \varepsilon(e_2(q_1, q_2, q_3) - e_2(1-q_4, 1-q_5))}. \tag{42b}$$

As $q_i = q_i(a,b,c,d,e)$, $i = 1, \ldots, 5$, are linear functions we arrive at the following proposition:

Proposition 5. *Each transformation $T \in \mathcal{T}$ as characterized by the collection $\{\varepsilon, M(\mathbf{r}), \lambda(\mathbf{r}), \alpha(\mathbf{r}), \beta(\mathbf{r}), D\}$ corresponds to a summation formula (42a) valid under restrictions (42b) with $(q_1, \ldots, q_5, 1) = D\mathbf{r}$.*

We will illustrate Proposition 5 by applying it to transformation (25). First condition in (42b) becomes $e = c + 2$. In view of this condition formula (42a) takes the form

$$_4F_3\left(\begin{array}{c} a,b,c,f+1 \\ d,c+2,f \end{array}\right) = \frac{(c+1)\Gamma(d)\Gamma(d-a-b+2)(f\psi + c(a+b-d))}{\Gamma(d-a+1)\Gamma(d-b+1)f\psi},$$

where $\psi = d - a - b + 1$ and, by the second condition in (42b),

$$f = -\frac{c(a+b-d)}{\psi} + \frac{(d-a)(d-b)c}{\psi((d-a)(d-b+c) + (d-b)c + (d-1)(a+b-d-c-1))}.$$

Further examples will be given in [23].

6. Proof of Lemma 1

Write identity (13) in expanded form

$$_4F_3\left(\begin{array}{c} a,b,c,f+1 \\ d,e,f \end{array}\right) = {}_3F_2\left(\begin{array}{c} a,b,c \\ d,e \end{array}\right) + \frac{abc}{fde} {}_3F_2\left(\begin{array}{c} a+1, b+1, c+1 \\ d+1, e+1 \end{array}\right). \tag{43}$$

Applying Thomae's transformation (3) to both $_3F_2$ functions on the right hand side, we get ($\psi = d + e - a - b - c - 1$):

$$_4F_3\left(\begin{matrix} a,b,c,f+1 \\ d,e,f \end{matrix}\right) = \frac{\Gamma(\psi+1)\Gamma(d)\Gamma(e)}{\Gamma(a)\Gamma(\psi+b+1)\Gamma(\psi+c+1)} \times$$

$$\left[_3F_2\left(\begin{matrix} \psi+1, d-a, e-a \\ \psi+b+1, \psi+c+1 \end{matrix}\right) + \frac{bc}{f\psi} {}_3F_2\left(\begin{matrix} \psi, d-a, e-a \\ \psi+b+1, \psi+c+1 \end{matrix}\right) \right].$$

Now we employ Proposition 3. Application of Formula (37) to the linear combination in brackets yields

$$_4F_3\left(\begin{matrix} a,b,c,f+1 \\ d,e,f \end{matrix}\right) = \frac{(f\psi+bc)\Gamma(\psi)\Gamma(d)\Gamma(e)}{f\Gamma(a)\Gamma(\psi+b+1)\Gamma(\psi+c+1)} {}_4F_3\left(\begin{matrix} \psi, d-a, e-a, \eta+1 \\ \psi+b+1, \psi+c+1, \eta \end{matrix}\right),$$

where $\eta = \psi + bc/f$. This proves transformation given by (7).

In a similar fashion, if we apply the Kummer transformation (2) to $_3F_2$ on the right hand side of (43) we get:

$$_4F_3\left(\begin{matrix} a,b,c,f+1 \\ d,e,f \end{matrix}\right) = \frac{\Gamma(\psi+1)\Gamma(d)}{\Gamma(d-a)\Gamma(\psi+a+1)} \left[_3F_2\left(\begin{matrix} a, e-b, e-c \\ e, \psi+a+1 \end{matrix}\right) + \frac{abc}{fe\psi} {}_3F_2\left(\begin{matrix} a+1, e-b, e-c \\ e+1, \psi+a+1 \end{matrix}\right) \right].$$

Applying the relation (38) to the linear combination in brackets we then obtain

$$_4F_3\left(\begin{matrix} a,b,c,f+1 \\ d,e,f \end{matrix}\right) = \frac{(abc+fe\psi)\Gamma(\psi)\Gamma(d)}{fe\Gamma(d-a)\Gamma(\psi+a+1)} {}_4F_3\left(\begin{matrix} a, e-b, e-c, \lambda+1 \\ e+1, \psi+a+1, \lambda \end{matrix}\right),$$

where

$$\lambda = \frac{abc+fe\psi}{bc+f\psi}.$$

This proves transformation (8).

Author Contributions: The authors contributed equally to this work. Both authors have read and agreed to the published version of the manuscript.

Funding: The second author was funded by the Ministry of Science and Higher Education of the Russian Federation (supplementary agreement No. 075-02-2020-1482-1 of 21 April 2020).

Conflicts of Interest: The authors declare no conflict of interest.

Appendix A

In this appendix we will display the explicit form of the main building blocks needed for calculating the elements of the group $\hat{\mathcal{T}}$. Just as it stands for Thomae's transformations ([4], Appendix 1), we have ten different identities with zero shifts. They are obtained as follows: permuting $a \leftrightarrow b$ and $a \leftrightarrow c$ in Formula (7) we get three transformations, while $a \leftrightarrow b$, $a \leftrightarrow c$ and $d \leftrightarrow e$ in (6) leads to six more transformations. Adding the identity transformation we arrive at ten "Thomae-like" zero-shift transformations for $_4F_3$ containing the parameter pair $\begin{bmatrix} f+1 \\ f \end{bmatrix}$. The entire 120 element subgroup of

"Thomae-like" zero-shift transformations is obtained by the obvious 12 permutations of three top and two bottom parameters on the right hand side of each of the ten transformations described above.

All further transformations are obtained by consecutive application of the four shifting transformations $S^{\pm}$, $S_{\pm}$ and permutations of top and bottom parameters to the 120 transformations described above. Transformation S^+ shifting the top parameter a by $+1$ (denoted by S_1 in Section 3) is given by (30). Combining parameters it can be written as:

$$_4F_3\left(\begin{matrix} a,b,c,f+1 \\ d,e,f \end{matrix}\right) = \left(1 - \frac{bc}{(d-a-1)(e-a-1)}\right)\left(1 + \frac{\lambda}{f}\right){}_4F_3\left(\begin{matrix} a+1,b,c,\eta+1 \\ d,e,\eta \end{matrix}\right), \tag{A1}$$

where

$$\lambda = \frac{abc}{a(2+a-d-e)-bc+(d-1)(e-1)}, \quad \eta = \frac{abc+((a+1)(a+1-d-e)-bc+de)f}{(a-f)(2+a+b+c-d-e)}.$$

Its inverse S^- is given by:

$$_4F_3\left(\begin{matrix} a,b,c,f+1 \\ d,e,f \end{matrix}\right) = \left(1 + \frac{bc}{\psi f}\right){}_4F_3\left(\begin{matrix} a-1,b,c,\eta+1 \\ d,e,\eta \end{matrix}\right), \tag{A2}$$

where

$$\eta = \frac{(a-1)(bc+\psi f)}{a(d+e-a)+bc-de+\psi f}.$$

The transformation S_+ shifting the bottom parameter d by $+1$ (denoted by S_4 in Section 3) is given by (31). It can be written more compactly as

$$_4F_3\left(\begin{matrix} a,b,c,f+1 \\ d,e,f \end{matrix}\right) = \frac{abc+\psi df}{\psi df}{}_4F_3\left(\begin{matrix} a,b,c,\eta+1 \\ d+1,e,\eta \end{matrix}\right), \tag{A3}$$

where $\psi = e+d-a-b-c-1$ and

$$\eta = \frac{abc+\psi df}{d(d-a-b-c)+ab+ac+bc+\psi f}.$$

Finally, its inverse transformation S_- shifting a bottom parameter by -1 has the form

$$_4F_3\left(\begin{matrix} a,b,c,f+1 \\ d,e,f \end{matrix}\right) = \frac{[((d-b-1)(d-c-1)-a(d-b-c-1))f-abc](d-1)}{(d-a-1)(d-b-1)(d-c-1)f}{}_4F_3\left(\begin{matrix} a,b,c,\eta+1 \\ d-1,e,\eta \end{matrix}\right), \tag{A4}$$

where

$$\eta = \frac{abc+[(1-d)(d-a-b-c-1)-ab-ac-bc]f}{(d+e-a-b-c-2)(f-d+1)}.$$

In the remaining part of the Appendix we present several *Wolfram Mathematica®* routines intended for dealing with the group $\mathcal{T}$ together with an example of their use. Listing A1 contains the function CMPS[T_1, T_2] that takes as input two transformations T_1, T_2 and computes their composition $T_2 \circ T_1$. The form in which the parameters ε_i, M_i, λ_i, α_i, β_i and D_i, $i = 1, 2$, should be supplied can be seen from the example in Listing A5. Similarly, Listing A2 contains the function INV[T] that computes the inverse of a given transformation T. The output provided by CMPS and INV can be printed in an easily readable form using the function PRN[T] given in Listing A3. The same Listing A3 contains the function INPT[T] that converts the output form of the functions CMPS and INV into the input form of the same functions, so that further compositions or inverses could be computed from such output. For numerical verification of the outputs of CMPS and INV the function RHS[T] presented in Listing A4 converts these outputs into an expression that can be evaluated by the *Mathematica* function N[...] after the parameters have been assigned some numerical values, see an example at the end of Listing A5.

Listing A1. Composition.

```
CMPS[T1_, T2_]:=Module[{eps1=T1[[1]], M1=T1[[2]], lam1=T1[[3]], alpha1=T1[[4]],
  beta1=T1[[5]], D1=T1[[6]], eps2=T2[[1]], M2=T2[[2]], lam2=T2[[3]], alpha2=T2[[4]], beta2=T2[[5]],
  D2=T2[[6]], R={{a},{b},{c},{d},{e},{1}}}, RR=Flatten[Drop[R,{6}]];
 If [ Simplify [eps1*eps2+alpha1@@RR*lam2@@Flatten[Drop[D1.R,{6}]]]===0,
  {0, FullSimplify [M1@@RR*M2@@Flatten[Drop[D1.R,{6}]]*(eps2*lam1@@RR+lam2@@Flatten[Drop[D1.R, {6}]]*beta1@@RR)], 1,
  Simplify [(eps1*alpha2 @@ Flatten[Drop[D1.R,{6}]]+alpha1@@RR*beta2@@Flatten[Drop[D1.R, {6}]])/
  (eps2*lam1@@RR+lam2@@Flatten[Drop[D1.R,{6}]]*beta1@@RR)], Simplify[(lam1@@RR*alpha2@@Flatten[Drop[D1.R,{6}]]
  +beta1@@RR*beta2@@Flatten[Drop[D1.R, {6}]])/(eps2*lam1@@RR+lam2@@Flatten[Drop[D1.R,{6}]]*beta1@@RR)], D2.D1},
  {1, FullSimplify [M1@@RR*M2@@Flatten[Drop[D1.R,{6}]]*(eps1*eps2+alpha1@@RR*lam2@@Flatten[Drop[D1.R, {6}]])],
  Simplify [(eps2*lam1@@RR+lam2@@Flatten[Drop[D1.R,{6}]]*beta1@@RR)/
  (eps1*eps2+alpha1@@RR*lam2@@Flatten[Drop[D1.R,{6}]])], Simplify[(eps1*alpha2@@Flatten[Drop[D1.R, {6}]]
  +alpha1@@RR*beta2@@Flatten[Drop[D1.R, {6}]])/(eps1*eps2+alpha1@@RR*lam2@@Flatten[Drop[D1.R,{6}]])],
  Simplify [(lam1@@RR*alpha2@@Flatten[Drop[D1.R,{6}]]+beta1@@RR*beta2 @@Flatten[Drop[D1.R, {6}]])/
  (eps1*eps2+alpha1@@RR*lam2@@Flatten[Drop[D1.R,{6}]])], D2.D1}]]
```

Listing A2. Inversion.

```
INV[TT_]:=Module[{eps=TT[[1]], M=TT[[2]], lam=TT[[3]], alpha=TT[[4]], beta=TT[[5]], D=TT[[6]],
  R={{a}, {b}, {c}, {d}, {e}, {1}}}, RR=Flatten[Drop[R, {6}]];
 If [ Simplify [beta@@RR]===0, {0,FullSimplify[1/M@@Flatten[Drop[Inverse[D].R,{6}]]/
  alpha@@Flatten[Drop[Inverse[D].R, {6}]]], 1, Simplify [alpha@@Flatten[Drop[Inverse[D].R,{6}]]/
  lam@@Flatten[Drop[Inverse[D].R, {6}]]], -eps/lam@@Flatten[Drop[Inverse[D].R, {6}]], Inverse [D]},
  {1, FullSimplify [beta@@Flatten[Drop[Inverse[D].R, {6}]]/(M@@Flatten[Drop[Inverse[D].R,{6}]]*
  (eps*beta@@Flatten[Drop[Inverse[D].R, {6}]]-lam@@Flatten[Drop[Inverse[D].R, {6}]]*
  alpha@@Flatten[Drop[Inverse[D].R, {6}]]) )], Simplify [-lam@@Flatten[Drop[Inverse[D].R,{6}]]/
  beta@@Flatten[Drop[Inverse[D].R, {6}]]], Simplify [-alpha@@Flatten[Drop[Inverse[D].R,{6}]]/
  beta@@Flatten[Drop[Inverse[D].R, {6}]]], Simplify [eps/beta @@ Flatten[Drop[Inverse[D].R ,{6}]]], Inverse [D]}]]
```

Listing A3. Conversion into input form and printing.

```
exprToFunction[expr_, vars_]:=ToExpression[ToString[FullForm[expr]/. MapIndexed[#1 -> Slot @@ #2 &, vars]]<>"&"];

INPT[TT_]:=List[TT[[1]], exprToFunction[TT[[2]], {a, b, c, d, e }],
  exprToFunction[TT [[3]], {a, b, c, d, e }], exprToFunction[TT [[4]], {a, b, c, d, e }],
  exprToFunction[TT [[5]], {a, b, c, d, e }], TT [[6]]]

ETA[TT_]:=Collect[Numerator[Together[(TT[[1]]*f+TT[[3]])/(TT[[4]]*f+TT[[5]])]], f]/
  Collect [Denominator[Together[(TT[[1]]*f + TT[[3]])/(TT[[4]]*f + TT[[5]])]], f]

PRN[TT_]:=Module[{}, Print["epsilon=", TT[[1]]];
  Print ["M=", FullSimplify[TT [[2]]]];
  Print ["Lambda=", FullSimplify[TT [[3]]]];  Print ["alpha=", TT [[4]]];
  Print ["Beta=", TT[[5]]];
  Print ["Parameters=", Flatten[Drop[TT[[6]].{{a}, {b}, {c}, {d}, {e}, {1}}, {6}]]];
  Print ["eta=", ETA[TT]]];
```

Listing A4. Conversion into computable form.

```
RHS[TT_]:=Simplify[TT[[2]]*(TT[[1]]*f + TT[[3]])/f]*
  HypergeometricPFQ[Join[Flatten[Drop[TT[[6]].{{a},{b},{c},{d},{e },{1}},{6}]][[1;;3]],  {ETA[TT]+1}],
  Join [ Flatten [Drop[TT[[6]].{{a},{b},{c},{d},{e },{1}},{6}]][[4;;5]],  {ETA[TT]}],1]
```

Listing A5. Example of use.

```
(* Definition of the first transformation *)
eps1=1; M1[a_,b_,c_,d_,e_]:=Gamma[d+e-a-b-c]*Gamma[d]*Gamma[e]/Gamma[a]/Gamma[d+e-a-c]/Gamma[d+e-a-b];
lam1[a_,b_,c_,d_,e_]:=b*c/(d+e-a-b-c-1); alpha1[a_,b_,c_,d_,e_]:=1/(d+e-a-b-c-1);
beta1[a_, b_, c_, d_,e_]:=0; D1={{-1,-1,-1,1,1,-1}, {-1,0,0,1,0,0}, {-1,0,0,0,1,0}, {-1,0,-1,1,1,0},
    {-1,-1,0,1,1,0},   {0,0,0,0,0,1}};

(* Definition of the second transformation *)
eps2=1; M2[a_,b_,c_,d_,e_]:=Gamma[d+e-a-b-c]*Gamma[e]/Gamma[d+e-a-b]/Gamma[e-c];
lam2[a_,b_,c_,d_,e_]:=(a+b-d)*c/(d+e-a-b-c-1); alpha2[a_,b_,c_,d_,e_]:=0;
beta2[a_,b_,c_,d_,e_]:=(e-c-1)/(d+e-a-b-c-1); D2={{-1,0,0,1,0,0}, {0,-1,0,1,0,0}, {0,0,1,0,0,0},
    {0,0,0,1,0,0},  {-1,-1,0,1,1,0},  {0,0,0,0,0,1}};

(*composition T2T1*)
T1T2=CMPS[{eps1, M1, lam1, alpha1, beta1, D1}, {eps2, M2, lam2, alpha2, beta2, D2}];

(*Inverse of T1*)
T1INV = INV[{eps1, M1, lam1, alpha1, beta1, D1}];

(*Printing the parameters of T2T1*)
PRN[T1T2]
epsilon=1
M=-(((a c+(1+b-d) e) Gamma[d] Gamma[-1-a-b-c+d+e])/(e Gamma[-b+d] Gamma[-a-c+d+e]))
Lambda=-((a b c)/(a c+(1+b-d)e))
alpha=(1+b-d)/(a c+e+b e-d e)
Beta=0
Parameters={1+b,-c+e,-a+e,-a-c+d+e,1+e}
eta=(-abc+(ac+e+be-de)f)/((1+b-d)f)

(*Computing composition of the results of previous operations *)
NEW=CMPS[INPT[T1T2], INPT[T1INV]];

(*Numerical verification of the transformation NEW using RHS[...]*)
a=1+2/3; b=-13/17+2; c=3/7; d=5/11; e=5+44/17; f=12/13;
In[51]:= N[HypergeometricPFQ[{a, b, c, f + 1}, {d, e, f}, 1], 15]
Out[51]= 2.22268615827388
In[52]:= N[RHS[NEW], 15]
Out[52]= 2.22268615827388
```

References

1. Andrews, G.E.; Askey, R.; Roy, R. *Special Functions*; Cambridge University Press: Cambridge, UK, 1999.
2. Krattenthaler, C.; Srinivasa Rao, K. On group theoretical aspects, hypergeometric transformations and symmetries of angular momentum coefficients. In *Symmetries in Science XI*; Kluwer Acad. Publ.: Dordrecht, The Netherlands, 2005; pp. 355–375.
3. Rao, K.S. Hypergeometric series and Quantum Theory of Angular Momentum. In *Selected Topics in Special Functions*; Agarwal, R.P., Manocha, H.L., Srinivasa Rao, K., Eds.; Allied Publishers Ltd.: New Delhi, India, 2001; pp. 93–134.
4. Rao, K.S.; Doebner, H.D.; Natterman, P. Generalized hypergeometric series and the symmetries of $3-j$ and $6-j$ coefficients. In *Number Theoretic Methods. Developments in Mathematics*; Kanemitsu, S., Jia, C., Eds.; Springer: Boston, MA, USA, 2002; Volume 8, pp. 381–403
5. Rao, K.S.; Lakshminarayanan, V. *Generalized Hypergeometric Functions, Transformations and Group Theoretical Aspects*; IOP Science: Bristol, UK, 2018.

6. Shpot, M.A.; Srivastava, H.M. The Clausenian hypergeometric function $_3F_2$ with unit argument and negative integral parameter differences. *Appl. Math. Comput.* **2015**, *259*, 819–827.
7. Formichella, M.; Green, R.M.; Stade, E. Coxeter group actions on $_4F_3(1)$ hypergeometric series. *Ramanujan J.* **2011**, *24*, 93–128. [CrossRef]
8. Olver, F.W.J.; Lozier, D.W.; Boisvert, R.F.; Clark, C.W. (Eds.) *NIST Handbook of Mathematical Functions*; Cambridge University Press: Cambridge, UK, 2010.
9. Beyer, W.A.; Louck, J.D.; Stein, P.R. Group theoretical basis of some identities for thegeneralized hypergeometric series. *J. Math. Phys.* **1987**, *28*, 497–508. [CrossRef]
10. Hardy, G.H.*Ramanujan: Twelve Lectures on Subjects Suggested by His Life and Work*; AMS Chelsea Pub.: Providence, RI, USA, 1999; p. 111.
11. Green, R.M.; Mishev, I.D.; Stade, E. Coxeter group actions and limits of hypergeometric series. *Ramanujan J.* **2020**. [CrossRef]
12. Mishev, I.D. Coxeter group actions on Saalschützian $_4F_3(1)$ series and very-well-poised $_7F_6(1)$ series. *J. Math. Anal. Appl.* **2012**, *385*, 1119–1133. [CrossRef]
13. Rao, K.S.; Van der Jeugt, J.; Raynal, J.; Jagannathan, R.; Rajeswari, V. Group theoretical basis for the terminating $3F2(1)$ series. *J. Phys. A Math. Gen.* **1992**, *25*, 861–876. [CrossRef]
14. Van der Jeugt, J.; Rao, K.S. Invariance groups of transformations of basic hypergeometric series. *J. Math. Phys.* **1999**, *40*, 6692–6700. [CrossRef]
15. Nørlund, N.E. Hypergeometric functions. *Acta Math.* **1955**, *94*, 289–349. [CrossRef]
16. Olsson, P.O.M. Analytic continuation of higher-order hypergeometric functions. *J. Math. Phys.* **1966**, *7*, 702–710. [CrossRef]
17. Bühring, W. Generalized hypergeometric functions at unit argument. *Proc. Am. Math. Soc.* **1992**, *114*, 145–153. [CrossRef]
18. Kim, Y.S.; Rathie, A.K.; Paris, R.B. On two Thomae-type transformations for hypergeometric series with integral parameter differences. *Math. Commun.* **2014**, *19*, 111–118.
19. Karp, D.B.; Prilepkina, E.G. Beyond the beta integral method: transformation formulas for hypergeometric functions via Meijer's G function. *arXiv* **2019**, arXiv:1912.11266.
20. Ebisu, A.; Iwasaki, K. Three-term relations for $_3F_2(1)$. *J. Math. Anal. Appl.* **2018**, *463*, 593–610. [CrossRef]
21. Bailey, W.N. *Generalized Hypergeometric Series*; Stecherthafner Service Agency: New York, NY, USA; London, UK, 1964; Reprinted from: Cambridge Tracts in Mathematics and Mathematical Physics, 1935, Volume 32.
22. Karp, D.B.; Prilepkina, E.G. Degenerate Miller-Paris transformations. *Results Math.* **2019**, *74*. [CrossRef]
23. Çetinkaya, A.; Karp, D. Summation formulas for some hypergeometric and some digamma series. *Commun. Korean Math. Soc.* **2020**, in preparation.

Publisher's Note: MDPI stays neutral with regard to jurisdictional claims in published maps and institutional affiliations.

© 2020 by the authors. Licensee MDPI, Basel, Switzerland. This article is an open access article distributed under the terms and conditions of the Creative Commons Attribution (CC BY) license (http://creativecommons.org/licenses/by/4.0/).

Article

Laguerre-Type Exponentials, Laguerre Derivatives and Applications. A Survey

Paolo Emilio Ricci

Department of Mathematics, International Telematic University UniNettuno, Corso Vittorio Emanuele II, 39, 00186 Roma, Italy; paoloemilioricci@gmail.com or p.ricci@uninettunouniversity.net

Received: 19 October 2020; Accepted: 13 November 2020; Published: 18 November 2020

Abstract: Laguerrian derivatives and related autofunctions are presented that allow building new special functions determined by the action of a differential isomorphism within the space of analytical functions. Such isomorphism can be iterated every time, so that the resulting construction can be re-submitted endlessly in a cyclic way. Some applications of this theory are made in the field of population dynamics and in the solution of Cauchy's problems for particular linear dynamical systems.

Keywords: Laguerre-type derivative; Laguerre-type exponentials; Laguerre-type special functions; multivariable and multi-index Laguerre polynomials; population dynamics models; Laguerre-type linear dynamical systems

MSC: 33C45; 33C99; 30D05; 33B10; 33C10; 92D25; 34A30

1. Introduction

This survey article is dedicated to a topic that has received little attention in the past, and therefore seems not to be very well known by the mathematical community.

Recently the role of the Laguerre derivative was considered in a few papers.

In [1], the authors introduce an interesting application of Wright functions of the first kind to solve fractional ordinary differential equations, with variable coefficients, generalizing the Bessel-type equations.

In [2], the authors use the same tool in Combinatorics, a completely different area, and in [3] an operational approach to the subject has been examined, in the framework of Clifford algebras.

Actually, in past time, the Laguerre-type exponentials and the related Laguerre derivative were introduced and studied in several articles (see [4–15]) and applications to Special functions, have been obtained. In particular, Laguerre-type functions of Bessel, Appell, Bell and multivariate functions were defined.

The operator $DxD = D + xD^2$ determines a linear differential isomorphism, acting onto the space of analytic functions of the x variable. By using this isomorphism, a sort of parallel structure is created within this space, in such a way that the differentiation properties have their counterpart, which can be immediately derived.

Furthermore, iterations of the Laguerre derivative can be defined, so that this parallelism with the space of analytic functions can be iterated too, in an endless way.

Therefore, a cyclic construction is created within the space that repeats the same structure at a higher level of differentiation order. It is one of the great cycles that sometimes occur within mathematical theories: for example, in Number theory the Fibonacci numbers F_n with Fibonacci indexes constitute a higher sequence of Fibonacci numbers which still satisfies the same recursion, i.e., $F_{F_{n+2}} = F_{F_{n+1}} + F_{F_n}$, and this property can be iterated at infinity.

However the operators $D_L = DxD$ and its iterates as $D_{nL} = DxDxDx \cdots DxD$ are not completely new, since they can be considered to be particular cases of the hyper-Bessel differential operators when $\alpha_0 = \alpha_1 = \cdots = \alpha_n = 1$ (the special case considered in operational calculus by Ditkin and Prudnikov [16]). In general, the *Bessel-type differential operators of arbitrary order n* were introduced by Dimovski, in 1966 [17] and later called by Kiryakova *hyper-Bessel operators*, because are closely related to their eigenfunctions, called hyper-Bessel by Delerue [18], in 1953. These operators were studied in 1994 by Kiryakova in her book [19] (Ch. 3).

Since the Laguerrian exponentials on the positive semi-axis of the abscissas are convex increasing functions, with a growth lower than the exponential one, in Section 7 a natural application was made in the context of population dynamics.

Laguerre-type linear dynamical systems were also considered in Section 8.

2. The Laguerre Derivative and the Relevant Exponentials

The *Laguerre derivative*, is defined by

$$D_L := DxD = D + xD^2, \qquad (1)$$

where $D = D_x = d/dx$.

It is an interesting operator. In fact, as the exponential function e^{ax} (a constat) is an eigenfunction of the derivative operator $D = D_x$, i.e.,

$$De^{ax} = ae^{ax}, \qquad (2)$$

equally the function

$$e_1(x) := \sum_{k=0}^{\infty} \frac{x^k}{(k!)^2} = C_0(-x), \qquad (3)$$

where $C_0(x)$ is the Tricomi function of order zero, is an eigenfunction of the Laguerre derivative D_L, since:

$$D_L \, e_1(ax) = a e_1(ax). \qquad (4)$$

The proof easily follows, by noting that:

$$D_L \, e_1(ax) = (D + xD^2) \sum_{k=0}^{\infty} a^k \frac{x^k}{(k!)^2} =$$

$$= \sum_{k=1}^{\infty} (k + k(k-1)) \, a^k \frac{x^{k-1}}{(k!)^2} = \sum_{k=1}^{\infty} k^2 a^k \frac{x^{k-1}}{(k!)^2} = \qquad (5)$$

$$= a \sum_{k=0}^{\infty} a^k \frac{x^k}{(k!)^2} = a e_1(ax).$$

For this reason, *the function $e_1(x)$ is called the Laguerre-type exponential* (of order 1).

In preceding articles, the role of the Laguerre derivative, in connection with the *monomiality principle*—an important technique introduced by G. Dattoli [20]—and its application to the multidimensional Hermite (Hermite-Kampé de Fériet or Gould-Hopper polynomials, see [21–23]) or Laguerre polynomials [14,24], has been shown.

The above technique can be iterated, producing Laguerre classes of exponential-type functions, of higher order, called *L-exponentials*, and the relevant *L-circular, L-hyperbolic, L-Gaussian functions* (see [4]).

Similar generalized hypergeometric functions, called trigonometric/Bessel type, exponential/confluent type and Gauss/Beta-distribution, can be found in a book by Kiryakova [19] and also in [25].

Before going on, we notice that the Laguerre derivative verifies [26]:

$$(DxD)^n = D^n x^n D^n, \qquad (6)$$

an equation which can be easily proven by recursion.

2.1. L-Exponentials of Higher Order

We consider the operator:

$$D_{2L} := DxDxD = D\left(xD + x^2D^2\right) = D + 3xD^2 + x^2D^3, \qquad (7)$$

and the function:

$$e_2(x) := \sum_{k=0}^{\infty} \frac{x^k}{(k!)^3}. \qquad (8)$$

The following theorem holds:

Theorem 1. *The function $e_2(ax)$ is an eigenfunction of the operator D_{2L}, i.e.,*

$$D_{2L}\, e_2(ax) = a e_2(ax) \qquad (9)$$

The proof (see [4]) depends on the identity: $k + 3k(k-1) + k(k-1)(k-2) = k^3$, so that, it can be recognized that the coefficients of the combination in Equation (7) are the *Stirling numbers of the second kind*, $S(3,1), S(3,2), S(3,3)$, (see [27], and [28] (p. 835 for an extended table)).

In general, we can state the following theorem:

Theorem 2. *The function*

$$e_n(x) := \sum_{k=0}^{\infty} \frac{x^k}{(k!)^{n+1}}. \qquad (10)$$

is an eigenfunction of the operator

$$D_{nL} := Dx \cdots DxDxD = D\left(xD + x^2D^2 + \cdots + x^n D^n\right) =$$
$$= S(n+1,1)D + S(n+1,2)xD^2 + \cdots + S(n+1,n+1)x^n D^{n+1}, \qquad (11)$$

i.e., for every constant a it results:

$$D_{nL}\, e_n(ax) = a e_n(ax). \qquad (12)$$

Remark 1. *The above results show that, for every positive integer n, we can define a Laguerre-exponential function, satisfying an eigenfunction property, which is an analog of the elementary property (2) of the exponential. The function $e_n(x)$ reduces to the exponential function when $n = 0$, so that we put by definition:*

$$e_0(x) := e^x, \qquad D_{0L} := D.$$

Obviously, $D_{1L} := D_L$.

Examples of the L-exponential functions are given in Figure 1.

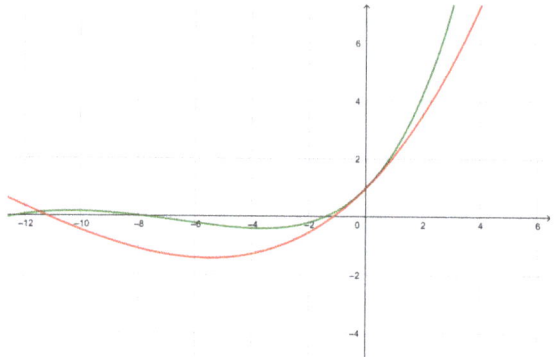

Figure 1. $e_1(x)$, (green) and $e_2(x)$, (red).

2.2. L-Circular and L-Hyperbolic Functions

Starting from the equation

$$e_1(ix) = \sum_{h=0}^{\infty}(-1)^h \frac{x^{2h}}{((2h)!)^2} + i\sum_{h=0}^{\infty}(-1)^h \frac{x^{2h+1}}{((2h+1)!)^2}, \tag{13}$$

we can define the $1L$-*circular functions* as follows

$$\cos_1(x) := \Re(e_1(ix)) = \sum_{h=0}^{\infty}(-1)^h \frac{x^{2h}}{((2h)!)^2}, \tag{14}$$

$$\sin_1(x) := \Im(e_1(ix)) = \sum_{h=0}^{\infty}(-1)^h \frac{x^{2h+1}}{((2h+1)!)^2}, \tag{15}$$

so that we find the Euler-type formulas

$$\cos_1(x) = \frac{e_1(ix) + e_1(-ix)}{2}, \quad \sin_1(x) = \frac{e_1(ix) - e_1(-ix)}{2i}, \tag{16}$$

Recalling Equation (6), we find the result:

Theorem 3. *The $1L$-circular functions (14) and (15) are solutions of the differential equation*

$$D_L^2 v + v = \left(D^2 x^2 D^2\right) v + v = 0. \tag{17}$$

The above results hold even for the generalized case.
Write the nL-exponential in the form:

$$e_n(ix) = \sum_{h=0}^{\infty}(-1)^h \frac{x^{2h}}{((2h)!)^{n+1}} + i\sum_{h=0}^{\infty}(-1)^h \frac{x^{2h+1}}{((2h+1)!)^{n+1}}. \tag{18}$$

Then we can define the nL-*circular functions* by putting

Definition 1.

$$\cos_n(x) := \Re(e_n(ix)) = \sum_{h=0}^{\infty}(-1)^h \frac{x^{2h}}{((2h)!)^{n+1}}, \tag{19}$$

$$\sin_n(x) := \Im(e_n(ix)) = \sum_{h=0}^{\infty} (-1)^h \frac{x^{2h+1}}{((2h+1)!)^{n+1}}, \qquad (20)$$

and we find again the Euler-type formulas:

$$\cos_n(x) = \frac{e_n(ix) + e_n(-ix)}{2}, \qquad \sin_n(x) = \frac{e_n(ix) - e_n(-ix)}{2i}. \qquad (21)$$

Theorem 3 becomes, in general:

Theorem 4. *The nL-circular functions (18) and (19) are solutions of the differential equation*

$$D_{nL}^2 \, v + v = 0.$$

and satisfy the conditions:

$$\cos_n(0) = 1, \qquad \sin_n(0) = 0.$$

Furthermore, we find:

Theorem 5. *The nL-circular functions satisfy*

$$D_{nL} \cos_n(x) = -\sin_n(x), \qquad D_{nL} \sin_n(x) = \cos_n(x). \qquad (22)$$

Examples of the L-circular functions are given in Figures 2 and 3.

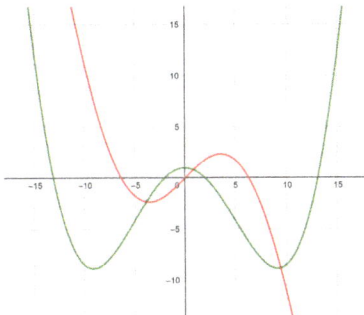

Figure 2. $\cos_1(x)$, (green) and $\sin_1(x)$, (red).

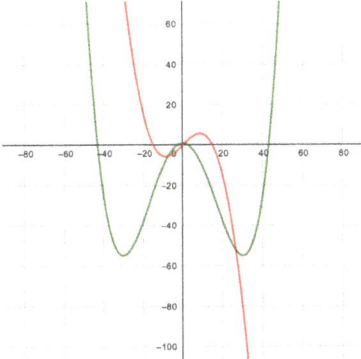

Figure 3. $\cos_2(x)$ (green) and $\sin_2(x)$ (red).

In a similar way we can define the nL-hyperbolic functions, putting

$$\cosh_n(x) := \sum_{h=0}^{\infty} \frac{x^{2h}}{((2h)!)^{n+1}},$$

$$\sinh_n(x) := \sum_{h=0}^{\infty} \frac{x^{2h+1}}{((2h+1)!)^{n+1}},$$

and the formulas analogues of that of the circular functions are easily derived (see [4]).

All the eigenfunctions $e_1(x), e_2(x), \ldots, e_n(x)$ can be expressed as generalized hypergeometric functions $_pF_q$, [29], namely: $e_1(x) = {}_0F_1(-x)$, $e_2(x) = {}_0F_2(x), \ldots, e_n(x) = {}_0F_n(x)$. In practice, starting from the Bessel function $e_1(x)$, all these eigenfunctions are special cases of the hyper-Bessel functions of Delerue [18], which are shown to be eigenfunctions of Dimovski's operators mentioned above.

Naturally, the $\cos_n$, $\sin_n$ functions, in Equations (19) and (20), and their hyperbolic variants, are special cases of the *trigonometric type generalized hypergeometric functions* considered in the Kiryakova book [19].

3. The Isomorphism $\mathcal{T}_x$ and Its Iterations

It was previously noted (see e.g., [14]) that, in the space $\mathcal{A}_x$ of analytic functions, it is possible to define an isomorphism $\mathcal{T}_x$ that preserves the differentiation properties, by means of correspondence:

$$D \to D_L, \qquad x \cdot \to D_x^{-1}, \tag{23}$$

where

$$D_x^{-1} f(x) = \int_0^x f(\xi) \, d\xi, \qquad D_x^{-n} f(x) = \frac{1}{(n-1)!} \int_0^x (x-\xi)^{n-1} f(\xi) \, d\xi, \tag{24}$$

so that

$$\mathcal{T}_x(x^n) = D_x^{-n}(1) = \frac{1}{(n-1)!} \int_0^x (x-\xi)^{n-1} \, d\xi = \frac{x^n}{n!}. \tag{25}$$

It is worth noting that this kind of isomorphism is widely used in operational calculus and differential equations also under the name of *Transmutation or Similarity operator*, since it transforms one operator into another, and eigenfunctions into each other.

In fact, in such an isomorphism we have the correspondences:

- The exponential function is transformed into the function $e_1(x)$, since

$$\mathcal{T}_x(e^x) = \sum_{k=0}^{\infty} \frac{\mathcal{T}_x(x^k)}{k!} = \sum_{k=0}^{\infty} \frac{x^k}{(k!)^2} = e_1(x).$$

- The Hermite polynomial $H_n^{(1)}(x,y) := (x-y)^n$ becomes the Laguerre polynomial

$$\mathcal{L}_n(x,y) := n! \sum_{r=0}^{n} \frac{(-1)^r y^{n-r} x^r}{(n-r)!(r!)^2}$$

and by using the *monomiality principle* we can prove thate all the relations valid in the polynomial space still hold after the substitutions stated in Equation (23).

Furthermore, an iterative application of Equation (23) gives in sequence the functions $e_1(x), e_2(x), e_3(x), \ldots$.

We have, for example:

$$T_x^2(e^x) = \sum_{k=0}^{\infty} \frac{T_x(x^k)}{(k!)^2} = \sum_{k=0}^{\infty} \frac{x^k}{(k!)^3} = e_2(x),$$

and so on.

We already noticed that the isomorphism connected with the Laguerre derivative can be iterated as many times as we wish.

Correspondently, the derivative operator is transformed into

$$\begin{aligned} D_L &= DxD, & D_{2L} &= D_L D_x^{-1} D_L = DxDxD, \\ D_{3L} &= D_L D_x^{-1} D_L D_x^{-1} D_L = DxDxDxD, \ldots, \end{aligned} \tag{26}$$

and so on.

We can conclude that the L-exponentials (and the relevant L-circular and L-hyperbolic functions) are determined by an iterative application of the considered differential isomorphism.

4. Examples of Laguerre-Type Problems

4.1. L-Diffusion Equations

Theorem 6. *For any fixed integer n, consider the problem (see [4] (Theorem 5.1)):*

$$\begin{cases} D_{nL} S(x,t) = \frac{\partial}{\partial t} S(x,t), & \text{in the half plane } t > 0, \\ S(0,t) = s(t), \end{cases} \tag{27}$$

with analytic boundary condition $s(t)$.

The operational solution of problem (27) is given by:

$$S(x,t) = e_n\left(x\frac{\partial}{\partial t}\right) s(t) = \sum_{k=0}^{\infty} \frac{x^k}{(k!)^{n+1}} \frac{d^k}{dt^k} s(t) \tag{28}$$

Representing $s(t) = \sum_{k=0}^{\infty} a_k t^k$, from Equation (28) we find, in particular:

$$S(x,0) = \sum_{k=0}^{\infty} a_k \frac{x^k}{(k!)^n}. \tag{29}$$

Please note that the operational solution becomes an effective solution whenever the series in Equation (28) is convergent. The validity of this condition depends on the growth of the coefficients a_k of the boundary data $s(t)$, but it is usually satisfied in physical problems.

More general problems are shown in [4,10], where evolution problems related to an operator of the type

$$D^{p_1} x^{q_1} D^{p_2} x^{q_2} \cdots D^{p_r} x^{q_r} D^{p_{r+1}}, \tag{30}$$

where $p_1, p_2, \ldots, p_{r+1}; q_1, q_2, \ldots q_r$ are fixed integers, have been considered.

An operational solution of the problem

$$D^{p_1} x^{q_1} D^{p_2} x^{q_2} \cdots D^{p_r} x^{q_r} D^{p_{r+1}} S(x,t) = D_t S(x,t), \quad \text{in the half plane } t > 0,$$

with suitable initial conditions have been determined, in terms of the eigenfunctions of the same operator.

Remark 2. Please note that the above operators generalize the subsequent Laguerre-type derivatives, since they are written as:

$$D_{nL}^r = \underbrace{(DxDx \cdots DxD)^r}_{(n+1)\ Derivatives} = D^r x^r D^r x^r \cdots D^r x^r D^r, \qquad (31)$$

which is an equation extending (6).

The operator (30) closely recalls the general case of hyper-Bessel B operators, in [17], since integers $q_1, q_2, \ldots q_r$ could be replaced by arbitrary real numbers, as are parameters $\alpha_0, \alpha_1, \ldots, \alpha_n$, considered in [17,30]. The solutions of the general differential equation $By(x) + \lambda y(x) = f(x)$ are given by Kiryakova et al. in [31].

4.2. L-Hyperbolic-Type Problems

Theorem 7. Let $\hat{\Omega}_x$ be a 2nd order differential operator with respect to the x variable, $D_{nL} := (D_{nL})_t$ the nL-derivative with respect to the t variable, and denote by $\psi(t)$ and $\chi(t)$ two functions such that:

$$D_{nL}\ \psi(t) = \chi(t), \qquad D_{nL}\ \chi(t) = \psi(t)$$
$$\psi(0) = 1, \qquad \chi(0) = 0 \qquad (32)$$

then the abstract L-hyperbolic-type problem:

$$\begin{cases} \hat{\Omega}_x^2\ S(x,t) = D_{nL}^2\ S(x,t), & \text{in the half plane } t > 0, \\ S(x,0) = q(x), \\ D_{nL}\ S(x,t)|_{t=0} = v(x) \end{cases} \qquad (33)$$

with analytic initial condition $q(x), v(x)$, admits the operational solution (see [4], Theorem 5.3):

$$S(x,t) = \psi\left(t\hat{\Omega}_x\right) q(x) + \chi\left(t\hat{\Omega}_x\right) w(x), \qquad (34)$$

where $w(x) := \hat{\Omega}_x^{-1} v(x)$.

Please note that conditions in (32) are satisfied, for any fixed integer n, assuming:

$$\psi(x) := \cosh_{nL}(x), \qquad \chi(x) := \sinh_{nL}(x).$$

4.3. L-Elliptic-Type Problems

Theorem 8. Let $\hat{\Omega}_x$ be a 2nd order differential operator with respect to the x variable, $D_{nL} := (D_{nL})_y$ the nL-derivative with respect to the y variable, and denote by $\varphi(y)$ and $\tau(y)$ two functions such that:

$$D_{nL}\ \varphi(y) = -\tau(y), \qquad D_{nL}\ \tau(y) = \varphi(y)$$
$$\varphi(0) = 1, \qquad \tau(0) = 0 \qquad (35)$$

then the abstract L-elliptic-type problem:

$$\begin{cases} \hat{\Omega}_x^2\ S(x,y) + D_{nL}^2\ S(x,y) = 0, & \text{in the half plane } t > 0, \\ S(x,0) = q(x), \end{cases} \qquad (36)$$

with analytic boundary condition $q(x)$, admits the operational solution (see [4], Theorem 5.4):

$$S(x,y) = \varphi\left(y\hat{\Omega}_x\right) q(x). \qquad (37)$$

Please note that conditions in (35) are satisfied, for any fixed integer n, assuming:

$$\varphi(x) := \cos_{nL}(x), \qquad \tau(x) := \sin_{nL}(x).$$

Further examples of PDE's problems involving the Laguerre derivatives can be found in [10,11].

5. Laguerre-Type Special Functions

5.1. Laguerre-Type Bessel Functions

The Laguerre-type Bessel functions, of order 1, (shortly L-Bessel functions), denoted by $_L J_n(x)$, are obtained substituting the exponential with the L-exponential $e_1(x)$ in the classic generating function, i.e., by putting

$$e_1\left[\frac{x}{2}\left(t - \frac{1}{t}\right)\right] = \sum_{n=-\infty}^{+\infty} {_L J_n}(x)\, t^n.$$

We can derive the explicit expression by applying the isomorphism $\mathcal{T}_x$ to both sides of the explicit expression of the Bessel functions, so that we find:

$$_L J_n(x) := \sum_{n=0}^{\infty} \frac{(-1)^h x^{n+2h}}{2^{n+2h}\, h!(n+h)!(n+2h)!}.$$

We proved the results:

Theorem 9. *The L-Bessel functions $_L J_n(x)$ satisfy the recurrence relation (see [8], Theorem 2.3):*

$$\begin{cases} \hat{D}_x^{-1}\left[{_L J_{n-1}}(x) + {_L J_{n+1}}(x)\right] = 2n\, {_L J_n}(x), \\ {_L J_{n-1}}(x) - {_L J_{n+1}}(x) = 2\hat{D}_L\, {_L J_n}(x). \end{cases}$$

Theorem 10. *The differential equation satisfied by the L-Bessel functions $_L J_n(x)$ is (see [8], Theorem 2.5):*

$$\left(\hat{D}_L^2 + \hat{D}_x \hat{D}_L - n^2 \hat{D}_x^2 + \hat{I}\right) {_L J_n}(x) = 0,$$

where $\hat{I}$ denotes the identity operator. This equation can be derived by applying the isomorphism $\mathcal{T}_x$ to both sides of the differential equation of the ordinary first kind Bessel functions.

5.2. Laguerre-Type Hypergeometric Functions

By using the isomorphism technique it is possible to define in general Laguerre-type special functions, and in particular, the 1st order Laguerre-type hypergeometric functions.

In fact, starting from the Gauss' hypergeometric equation:

$$x(1-x)y'' + [c - (a+b+1)x]y' - aby = 0$$

and applying the isomorphism $\mathcal{T}_x$, we find the equation

$$x(1-x)D_L^2 y + [c - (a+b+1)x]D_L y - aby = 0, \tag{38}$$

that is:

$$[x(1-x)](x^2 y^{iv} + 4xy''' + 2y'') + [c - (a+b+1)x](y' + xy'') - aby = 0. \tag{39}$$

The solution of Equation (38), corresponding to the Gauss' hypergeometric equation $F(a,b,c;x)$, is given by

$$_L F(a,b,c;x) = 1 + \sum_{n=1}^{\infty} \frac{a^{(n)} b^{(n)}}{c^{(n)}} \frac{x^n}{(n!)^2}, \tag{40}$$

where the symbol $a^{(n)}$ denotes the rising factorial.

Of course the rth order Laguerre-type hypergeometric functions are obtained by applying to both sides of the hypergeometric equation the iterated isomorphism of order r, but the corresponding differential equation becomes more and more complicated as r increases.

The generalized hypergeometric functions have their 1st order Laguerre-type counterpart, which are given by:

$$_L{}_pF_q(a_1,\ldots,a_p;b_1,\ldots,b_q;x) = \sum_{n=0}^{\infty} \frac{a_1^{(n)} \cdots a_p^{(n)}}{b_1^{(n)} \cdots b_q^{(n)}} \frac{x^n}{(n!)^2}, \tag{41}$$

and those of higher order immediately follow.

Please note that the function in (41) can be viewed as a generalized hypergeometric function of the form $_pF_{q+1}$, by moving one of the $n!$ in the first fraction under the sum and considering $\Gamma(n+1)/\Gamma(1) = 1^{(n)} = n! = b_{q+1}^{(n)}$ as the $(q+1)$-th term.

5.3. Laguerre-Type Bell Polynomials

We first note that for the Laguerre derivative, the chain rule

$$\frac{d}{dt} = \frac{d}{dx}\frac{dx}{dt}$$

becomes:

$$\frac{d}{dt} t \frac{d}{dt} = \frac{d}{dx}\frac{d}{dt} t \frac{dx}{dt}, \quad \text{that is}: \quad (D_L)_t = \frac{d}{dx}(D_L)_t\, x \tag{42}$$

and in general:

$$(D_{nL})_t = \frac{d}{dx}(D_{nL})_t\, x.$$

The problem of constructing Bell polynomials can be extended in the natural way to the case of the Laguerre-type derivatives.

To this aim, we introduce the definition:

Definition 2. *The nth Laguerre-type Bell polynomial, denoted by $_{rL}Y_n(x;[f,g]_n)$, represents the nth rLaguerre-type derivative of the composite function $f(g(t))$.*

In [12] we showed that $_{rL}Y_n$ can be expressed as a polynomial in the independent variable x, depending on $f_1, g_1; f_2, g_2; \ldots; f_n, g_n$, in terms of the classical Bell polynomials.

According to Equation (6), the Leibniz rule, gives:

$$(DxD)^n = D^n(x^n D^n) = \sum_{k=0}^{n} \binom{n}{k} D^{n-k} x^n D^{n+k} =$$

$$= \sum_{k=0}^{n} \left[\binom{n}{k}\right]^2 (n-k)!\, x^k D^{n+k} = \sum_{k=0}^{n} \frac{n!}{k!} \binom{n}{k} x^k D^{n+k}. \tag{43}$$

Therefore, the following representation formula for the Laguerre-type Bell polynomials, denoted by $_LY_n$, holds:

Theorem 11. *The $_LY_n$ polynomials are expressed in terms of the ordinary Bell polynomials according to the equation (see [12], Theorem 4.1):*

$$_LY_n(x;[f,g]_n) = \sum_{k=0}^{n} \frac{n!}{k!} \binom{n}{k} x^k Y_{n+k}([f,g]_{n+k}) . \tag{44}$$

The above results can be easily generalized, since

$$(D_{2L})^n = (DxDxD)^n = D^n(x^n D^n x^n D^n) = \\ = \sum_{k_1=0}^{n} \sum_{k_2=0}^{n} \frac{n!}{k_1!} \frac{(n+k_1)!}{(k_1+k_2)!} \binom{n}{k_1}\binom{n}{k_2} x^{k_1+k_2} D^{n+k_1+k_2} . \tag{45}$$

In [12] even the general case of polynomials $_{rL}Y_n$ Bell is considered, but we do not report here the equation which is a little more complicated.

6. The Multivariate Case

6.1. Laguerre-Type Appell Polynomials

In a preceding article [32] multivariate extensions of the Appell polynomials (including the Bernoulli and Euler cases) have been introduced, by means of the generating function [23]:

$$A(t)\exp(xt+y^j) = \sum_{n=0}^{\infty} R_n^{(j)}(x,y)\frac{t^n}{n!},$$

where j is a fixed integer.

The application of the isomorphism $\mathcal{T}_x$, and its iterations allows defining new classes of multivariate special polynomials, the Laguerre-type Appell polynomials, and to build their main properties (recurrence relations, shift operators, differential equations, etc), in an easy and uniform way.

This has been achieved in [6] starting from generating functions of the type

$$A(t)e_s(xt)e_\sigma(yt^j) = \sum_{n=0}^{\infty} R_n^{(j)}(\mathcal{T}_x^s(x), \mathcal{T}_y^\sigma(y))\frac{t^n}{n!}$$

where $e_s(\cdot)$ and $e_\sigma(\cdot)$ are Laguerre-type exponentials. Many properties of these functions have been derived, including recursions and differential equations.

The results obtained in this case are easily extended to the functions of r variables, since the technique works regardless of the number r.

6.2. Laguerre-Type Appell Series

We limit ourselves to the case of series in two variables, but the equations trivially extend to the general case. For $|x|<1$, $|y|<1$ the double series

$$_LF_1(a,b_1,b_2;c;x,y) = \sum_{m,n=0}^{\infty} \frac{a^{(m+n)}b_1^{(m)}b_2^{(n)}}{c^{(m+n)}} \frac{x^m}{(m!)^2}\frac{y^n}{(n!)^2} \tag{46}$$

is the Laguerre-type Appell series, obtained by the classical one acting on it with the two isomorphisms $\mathcal{T}_x$ and $\mathcal{T}_y$.

We avoid to consider further extension to the case of multivariate functions with several parameters, since they are trivially obtained.

7. Applications to Population Dynamics

7.1. Exponential and L-Exponential Models

In this section a possible application of the Laguerre derivative is recalled [9,13]. Since the L-exponentials for every $x \geq 0$ are convex increasing functions, with a graph lower with respect to $\exp(x)$, it is possible to use these function in the framework of population dynamics, as it seems that in some cases the growth of the exponential is too fast.

Consider the number $N(t)$ of population individuals at time t and let $N(0) = N_0$ the initial number at time $t = 0$.

In the Malthus model, the variation is assumed to be proportional to $N(t)$, i.e.,

$$\begin{cases} \dfrac{d}{dt} N(t) = rN(t), \\ N(0) = N_0, \end{cases}$$

where the *growth rate* r is a suitable constant.

The solution is given by the exponential function

$$N(t) = N_0 \, e^{rt}.$$

Using the Laguerre derivative, the Laguerre-type Malthus reads:

$$\frac{d}{dt} t \frac{dN}{dt} = rN(t) \quad \text{i.e.} \quad \frac{dN}{dt} + t \frac{d^2 N}{dt^2} = rN(t),$$

where r is a positive constant. Assuming the initial conditions

$$\begin{cases} N(0) = N_0, \\ N'(0) = N_1 = N_0 r, \end{cases}$$

we find the solution

$$N(t) = N_0 e_1(rt) = N_0 \sum_{k=0}^{+\infty} r^k \frac{t^k}{(k!)^2}.$$

In this case the population growth increases according to the Laguerre exponential function $e_1(x)$, so that the relevant increasing is slower with respect to the classical Malthus model.

In [9] it has been shown, with tables of data taken from real population dynamics, that the Laguerre-type Malthus model produces data closer to real population growth.

7.2. Logistic vs. L-logistic Model

Taking into account that the growth rate cannot be constant, since it depends on the environmental resources, Pierre Verhulst considered the so-called *logistic model*

$$\begin{cases} \dfrac{dN}{dt} = r \left[1 - \dfrac{1}{K} N(t) \right] N(t), \\ N(0) = N_0, \end{cases}$$

where r is called the *intrinsic growth rate*, and K denotes the *environmental capacity*.

The exact solution of this problem is given by

$$N(t) = \frac{N_0 K}{N_0 + (K - N_0)e^{-rt}},$$

so that, if $N_0 < K$ the solution is a function monotonically increasing to K, whereas, if $N_0 > K$, the solution is monotonically decreasing to K. In any case,

$$\lim_{t \to \infty} N(t) = K,$$

and the value $N(t) = K$ is a stable equilibrium point for the logistic equation.

The Laguerre-logistic (shortly L-logistic) model is expressed by

$$\begin{cases} N'(t) + tN''(t) = rN(t)\left(1 - \dfrac{N(t)}{K}\right), \\ N(0) = N_0, \\ N'(0) = N_1. \end{cases} \qquad (47)$$

Please note that if in the above equation N is small with respect to K, then N/K is close to 0 and consequently $D_t t D_t N \approx rN(t)$.

If $N \to K$, then $N/K \to 1$, and $D_t t D_t N \to 0$.

The L-logistic equation cannot be solved explicitly, but numerically, using a Runge-Kutta method.

The behavior of the approximate solutions for the L-logistic model is shown in Figure 4. It is worth noting that the solution tends to the environmental capacity K by an oscillating behavior.

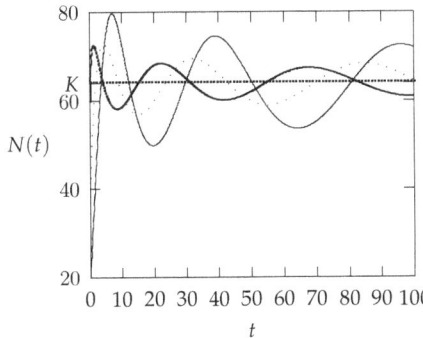
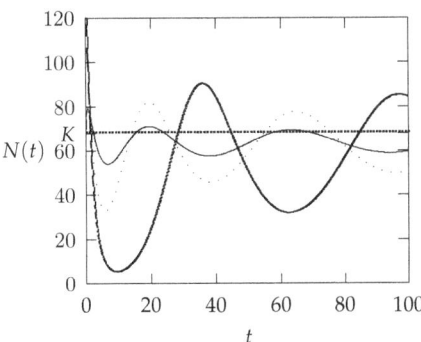

Figure 4. Solutions to the L-logistic model with $N(0) = N'(0) < K$ (on the left), $N(0) = N'(0) > K$ (on the right), $K = 64$, $r = 0.8$, $T = 100$, $\Delta t = 0.1$.

This is the main difference with respect to the ordinary logistic model, since in that case the solution was monotonically increasing or decreasing to K.

Similar results could be obtained by using the nL-derivatives, introducing suitable initial conditions which can be easily derived from the initial observations data.

Please note that as the order n increases, for $x > 0$, the Laguerrian exponential attenuates its growth and for $n \to \infty$ it tends to assume the linear value $1 + x$, so it can be used to model a growth as slow as it is needed.

Remark 3. *We recall that the oscillating asymptotic trend of solutions occurs in reality. For example, the classical experiment of G.F. Gause, relative to the protozoon paramecium shows such a typical behavior, represented in Figure 5. In this figure the true values, represented by a dotted line, are compared with the exponential trend of Malthus and with the logistic curve.*

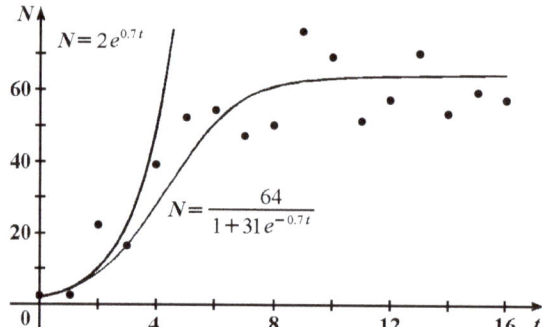

Figure 5. The behavior of growth in the Gause experiment.

7.3. Modified L-Logistic Models

Many different models modifying the basic logistic model appeared in the literature: the Bernoulli, the modified logistic, the Gompertz, the Alee, and the Beverton-Holt models.

In [13] we considered the Laguerre-type version of all of them, showing that in all cases the oscillating asymptotic behavior of solutions takes the place of the monotonic one.

Instead, it was found that the model of Volterra-Lotka model is invariant under the action of the isomorphism $\mathcal{T}_x$, since the Laguerre derivative satisfies again the chain rule, according to Equation (42).

8. Laguerre-Type Linear Dynamical Systems

Let $\mathcal{A}$ be a $r \times r$ matrix and denote by u_k, $(k = 1, 2, \ldots, r)$ the invariants of $\mathcal{A}$, i.e., the sum of principal minors (i.e., the elementary symmetric functions of the eigenvalues). The invariants of the matrix $t\mathcal{A}$ are given by $u_k(t) = t^k u_k$, $(k = 1, 2, \ldots, r)$.

Consider the vectors

$$\begin{cases} Z(t) = (Z_1(t), \ldots, Z_r(t))^T \\ Z_0 = (Z_1(0), \ldots, Z_r(0))^T. \end{cases}$$

Then the solution of the linear dynamical system

$$\begin{cases} Z'(t) = \mathcal{A} Z(t), \\ Z(0) = Z_0 \end{cases}$$

writes [33]:

$$Z(t) = e^{t\mathcal{A}} Z_0 = \sum_{h=0}^{r-1} \left[\frac{1}{2\pi i} \sum_{j=0}^{r-h-1} (-1)^j u_j(t) \oint_\gamma \frac{e^\lambda \lambda^{r-h-j-1}}{P(\lambda, t)} d\lambda \right] \cdot t^h Z_0^h,$$

where $P(\lambda, t)$ is the characteristic polynomial of the matrix $t\mathcal{A}$ and γ denotes a simple Jordan curve encircling all the eigenvalues of $\mathcal{A}$. The choice of γ, without computing the eigenvalues, can be done by using the Gershgorin theorem.

In [15], a Laguerre-type version of the above classic result has been shown.
Consider the above $r \times r$ matrix, and the vectors

$$\begin{cases} Z(t) = (Z_1(t), \ldots, Z_r(t))^T \\ Z_0 = (Z_1(0), \ldots, Z_r(0))^T \\ Z'_0 = (Z'_1(0), \ldots, Z'_r(0))^T = \mathcal{A} \cdot Z_0 \\ \vdots \\ Z_0^{r-1} = (Z_1^{r-1}(0), \ldots, Z_r^{r-1}(0))^T = \mathcal{A} \cdot Z_0^{r-2}. \end{cases}$$

The following result holds (see [15] (Theorem 10)):

Theorem 12. *The Laguerre-type Cauchy problem for a homogeneous linear differential system*

$$\begin{cases} D_L Z(t) = Z'(t) + t Z''(t) = \mathcal{A} \cdot Z(t) \\ Z(0) = Z_0 \\ Z'_0 = \mathcal{A} \cdot Z_0, \end{cases}$$

has the solution:

$$Z(t) = e_1(t\mathcal{A}) Z_0 = \sum_{h=0}^{r-1} \left[\frac{1}{2\pi i} \sum_{j=0}^{r-h-1} (-1)^j u_j(t) \oint_\gamma \frac{e_1(\lambda) \lambda^{r-h-j-1}}{P(\lambda, t)} d\lambda \right] \cdot t^h Z_0^h,$$

where $P(\lambda, t)$ and γ have been defined above.

The proof of this result is a straightforward application of the isomorphism $\mathcal{T}_t$. In [15] worked examples are reported.

9. Conclusions

The Laguerre derivative and the relevant Laguerre-type exponentials allow to associate, to any given integer n, a new class of special functions. This fact is obtained by exploiting the properties of an isomorphism, within the space of analytic functions, which acts in such a way as to preserve the differentiation properties. The successive iterations of this isomorphism produce a cyclic construction within the space that repeats the same structure at a higher level of the order of derivation.

Infinite many special functions can be defined in this way. A few of them have been presented explicitly, and the general technique to produce the others has been indicated.

This Survey has shown even possible applications of the Laguerrian derivative in the context of population dynamics and in the solution of Cauchy problems related to particular linear dynamical systems.

Funding: This research received no external funding.

Acknowledgments: The author is grateful to the referees, especially to the one who reported ties with the topics covered in the article and the hyper-Bessel operators, helping to expand the bibliography.

Conflicts of Interest: The author declares no conflict of interest.

References

1. Garra, R.; Mainardi, F. Some applications of Wright functions in fractional differential equations. *Rep. Math. Phys.* **2020**, in press.
2. Penson, K.A.; Blasiak, P.; Horzela, A.; Solomon, A.I.; Duchamp, G.H.E. Laguerre-type derivatives: Dobiński relations and combinatorial identities. *J. Math. Phys.* **2009**, *50*, 3512. doi:10.1063/1.3155380 [CrossRef]
3. Cação, I.; Falcão M.I.; Malonek, H.R. Laguerre derivative and monogenic Laguerre polynomials: An operational approach. *Math. Comput. Model.* **2011**, *53*, 1084–1094. [CrossRef]

4. Dattoli, G.; Ricci, P.E. Laguerre-type exponentials, and the relevant L-circular and L-hyperbolic functions. *Georgian Math. J.* **2003**, *10*, 481–494.
5. Bernardini, A.; Dattoli, G.; Ricci, P.E. *L*-exponentials and higher order Laguerre polynomials. In Proceedings of the Fourth International Conference of the Society for Special Functions and their Applications (SSFA), Rajasthan University, Jaipur, India, 4–6 March 2003; pp. 13–26.
6. Bretti, G.; Cesarano, C.; Ricci, P.E. Laguerre-type exponentials and generalized Appell polynomials. *Comput. Math. Appl.* **2004**, *48*, 833–839. [CrossRef]
7. Bernardini, A.; Bretti, G.; Ricci, P.E. Laguerre-type exponentials, multidimensional special polynomials and applications. *Lect. Notes TICMI* **2004**, *5*, 1–28.
8. Cesarano, C.; Germano, B.; Ricci, P.E. Laguerre-type Bessel functions. *Integral Transforms Spec. Funct.* **2005**, *16*, 315–322. [CrossRef]
9. De Andreis, S.; Ricci, P.E. Modelling population growth via Laguerre-type exponentials. *Math. Comput. Model.* **2005**, *42*, 1421–1428. [CrossRef]
10. Dattoli, G.; He, M.X.; Ricci, P.E. Eigenfunctions of Laguerre-type operators and generalized evolution problems. *Math. Comput. Model.* **2005**, *42*, 1263–1268. [CrossRef]
11. Maroscia, G.; Ricci, P.E. Laguerre-type BVP and generalized Laguerre polynomials. *Integral Transforms Spec. Funct.* **2006**, *17*, 577–590. [CrossRef]
12. Natalini, P.; Ricci, P.E. Laguerre-type Bell polynomials. *Int. J. Math. Math. Sci.* **2006**, *2006*, 45423. [CrossRef]
13. Bretti, G.; Ricci, P.E. Laguerre-type Special functions and population dynamics. *Appl. Math. Comput.* **2007**, *187*, 89–100. [CrossRef]
14. Ricci, P.E.; Tavkhelidze, I. An introduction to operational techniques and special polynomials. *J. Math. Sci.* **2009**, *157*, 161–189. [CrossRef]
15. Natalini, P.; Ricci, P.E. Laguerre-type linear dynamical systems. *Ilir. J. Math.* **2015**, *4*, 24–40.
16. Ditkin, A.P.; Prudnikov, V.A. *Transforms and Operational Calculus*; Pergamon Press: Oxford, UK, 1965.
17. Dimovski, I. Operational calculus for a class of differential operators. *C. R. Acad. Bulg. Sci.* **1966**, *19*, 1111–1114.
18. Delerue, P. Sur le calcul symbolique à *n* variables et fonctions hyper-besseliennes (II). *Ann. Soc. Sci. Brux.* **1953**, *3*, 229–274.
19. Kiryakova, V. *Generalized Fractional Calculus and Applications*; Pitman Research Notes in Mathematics Series 301; Longman: Harlow, UK, 1994.
20. Dattoli, G. Hermite-Bessel and Laguerre-Bessel functions: A by-product ot the monomiality principle. In Proceedings of the Melfi School on Advanced Topics in Mathematics and Physics, Melfi, Italy, 9–12 May 1999; Cocolicchio, D., Dattoli, G., Srivastava, H.M., Eds.; Advanced Special Functions and Applications; Aracne Editrice: Rome, Italy, 2000; pp. 147–164.
21. Appell, P.; Kampé de Fériet, J. *Fonctions Hypergéométriques et Hypersphériques. Polynômes d'Hermite*; Gauthier-Villars: Paris, France, 1926.
22. Gould, H.W.; Hopper, A.T. Operational formulas connected with two generalizations of hermite polynomials. *Duke Math. J.* **1962**, *29*, 51–62. [CrossRef]
23. Srivastava, H.M.; Manocha, H.L. *A Treatise on Generating Functions*; J. Wiley & Sons: New York, NY, USA, 1984.
24. Dattoli, G.; Ricci, P.E.; Cesarano, C. On a class of polynomials generalizing the Laguerre family. *J. Comput. Appl. Math.* **2005**, *3*, 405–412.
25. Kiryakova, V. All the special functions are fractional differintegrals of elementary functions. *J. Phys. A Math. Gen.* **1997**, *30*, 5085–5103. [CrossRef]
26. Viskov, O.V. A commutative-like noncommutation identity. *Acta Sci. Math.* **1994**, *59*, 585–590.
27. Riordan, J. *An Introduction to Combinatorial Analysis*; J. Wiley & Sons: Chichester, UK, 1958.
28. Abramowitz, M.; Stegun, I.A. *Handbook of Mathematical Functions, with Formulas, Graphs, and Mathematical Tables*; Dover: New York, NY, USA, 1965.
29. Erdélyi, A.; Magnus, W.; Oberhettinger, F.; Tricomi, F.G. *Higher Transcendental Functions*; McGraw-Hill: New York, NY, USA, 1953; Volume 1.
30. Kiryakova, V. From the hyper-Bessel opertors of Dimovski to the generalized fractional calculus. *Fract. Calc. Appl. Anal.* **2014**, *17*, 977–1000. [CrossRef]
31. Kiryakova, V.S.; McBride, A.C. Explicit solution of the nonhomogeneous hyper-Bessel differential equation. *C. R. Acad. Bulgare Sci.* **1993**, *46*, 23–26.

32. Bretti, G.; Ricci, P.E. Multidimensional extensions of the Bernoulli and Appell Polynomials. *Taiwan. J. Math.* **2004**, *8*, 415–428. [CrossRef]
33. Natalini, P.; Ricci, P.E. Avoiding higher matrix powers in the solution of linear dynamical systems. *Atlantis Trans. Geom. Model. Math.* **2017**, *2*, 117–128.

Publisher's Note: MDPI stays neutral with regard to jurisdictional claims in published maps and institutional affiliations.

© 2020 by the authors. Licensee MDPI, Basel, Switzerland. This article is an open access article distributed under the terms and conditions of the Creative Commons Attribution (CC BY) license (http://creativecommons.org/licenses/by/4.0/).

Article

Some Relationships for the Generalized Integral Transform on Function Space

Hyun Soo Chung

Department of Mathematics, Dankook University, Cheonan 3116, Korea; hschung@dankook.ac.kr

Received: 25 November 2020; Accepted: 17 December 2020; Published: 19 December 2020

Abstract: In this paper, we recall a more generalized integral transform, a generalized convolution product and a generalized first variation on function space. The Gaussian process and the bounded linear operators on function space are used to define them. We then establish the existence and various relationships between the generalized integral transform and the generalized convolution product. Furthermore, we obtain some relationships between the generalized integral transform and the generalized first variation with the generalized Cameron–Storvick theorem. Finally, some applications are demonstrated as examples.

Keywords: generalized integral transform; generalized convolution product; bounded linear operator; Gaussian process; Cameron–Storvick theorem; translation theorem

MSC: 47A60; 60J65; 28C20

1. Introduction

For $T > 0$, let $C_0[0, T]$ be the one-parameter Wiener space and let $\mathcal{M}$ denote the class of all Wiener measurable subsets of $C_0[0, T]$. Let m denote Wiener measure. Then, the space $(C_0[0, T], \mathcal{M}, m)$ is complete, and we denote the Wiener integral of a Wiener integrable functional F by

$$\int_{C_0[0,T]} F(x) dm(x).$$

Let $K \equiv K_0[0, T]$ be the space of all complex-valued continuous functions defined on $[0, T]$ which vanishes at $t = 0$ and whose real and imaginary parts are elements of $C_0[0, T]$.

In [1], Lee studied an integral transform of analytic functionals on abstract Wiener spaces

$$\mathcal{F}_{\gamma,\beta}(F)(y) = \int_{C_0[0,T]} F(\gamma x + \beta y) dm(x), \quad y \in K. \tag{1}$$

For some parameters γ and β and for certain classes of functionals, the Fourier–Wiener transform, the modified Fourier–Wiener transform, the analytic Fourier–Feynman transform and the Gauss transform are popular examples of the integral transform defined by (1) above (see [1–12]). Researchers have studied some theories of integral transform for functionals on function space. Recently, the integral transform is generalized by some methods in various papers. One of them uses the concept of Gaussian process instead of the ordinary process. For a function h on $[0, T]$, the Gaussian process is defined by the formula

$$Z_h(x, t) = \int_0^t h(s) \tilde{d}x(s)$$

where $\int_0^t h(s)\tilde{d}x(s)$ the Paley–Wiener–Zygmund (PWZ) stochastic integral. Many mathematician use this process to generalize the integral. As representative examples, the generalized integral transforms

$$\mathcal{F}^h_{\gamma,\beta}(F)(y) = \int_{C_0[0,T]} F(\gamma Z_h(x,\cdot) + \beta y) dm(x) \qquad (2)$$

and

$$\mathcal{F}^{h_1,h_2}_{\gamma,\beta}(F)(y) = \int_{C_0[0,T]} F(\gamma Z_{h_1}(x,\cdot) + \beta Z_{h_2}(y,\cdot)) dm(x) \qquad (3)$$

are studied in [13–15]. In fact, if h, h_1 and h_2 are identically 1 on $[0,T]$, then Equations (2) and (3) reduce to Equation (1).

Another method is using the operators on K. Let S and R be bounded linear operators on K. In [6,16], the authors used this operators to generalize the integral transforms. A more generalized form is given by

$$\mathcal{G}_{S,R}(F)(y) = \int_{C_0[0,T]} F(Sx + Ry) dm(x). \qquad (4)$$

If R is a constant operator and $Sx = Z_h(x,\cdot)$ for some function h, then Equation (4) reduces to Equation (2), and hence it reduces to Equation (1) again. In previous studies, many relationships among the integral transform, the convolution and the first variation have been obtained. However, most of the results consist of fixed parameters.

In this paper, we use the both concepts, the Gaussian process and the operator, to define a more generalized integral transform, a generalized convolution product and a generalized first variation of functionals on function space. We then give some necessary and sufficiently conditions for holding some relationships between the generalized integral transforms and the generalized convolution products, and between the generalized integral transforms and the generalized first variations. In addition, some examples are given to illustrate usefulness for our formulas and results. By choosing the kernel functions and operators, all results and formulas in previous papers are corollaries of our results and formulas in this paper.

2. Definitions and Preliminaries

We first list some definitions and properties needed to understand this paper.

A subset B of $C_0[0,T]$ is called scale-invariant measurable if ρB is $\mathcal{M}$-measurable for all $\rho > 0$, and a scale-invariant measurable set N is called a scale-invariant null set provided $m(\rho N) = 0$ for all $\rho > 0$. A property that holds except on a scale-invariant null set is said to hold scale-invariant almost everywhere (s-a.e.) [17]. For $v \in L_2[0,T]$ and $x \in C_0[0,T]$, let $\langle v, x \rangle$ denote the Paley–Wiener–Zygmund (PWZ) stochastic integral. Then, we have the following assertions.

(i) For each $v \in L_2[0,T]$, $\langle v, x \rangle$ exists for a.e. $x \in C_0[0,T]$.
(ii) If $v \in L_2[0,T]$ is a function of bounded variation on $[0,T]$, $\langle v, x \rangle$ equals the Riemann–Stieltjes integral $\int_0^T v(t) dx(t)$ for s-a.e. $x \in C_0[0,T]$.
(iii) The PWZ stochastic integral $\langle v, x \rangle$ has the expected linearity property.
(iv) The PWZ stochastic integral $\langle v, x \rangle$ is a Gaussian process with mean 0 and variance $\|v\|_2^2$.

For a more detailed study of the PWZ stochastic integral, see [4,5,7–9,11–15,18].
Let

$$C'_0 \equiv C'_0[0,T] = \left\{ v \in C_0[0,T] : v(t) = \int_0^t z_v(s) ds, \ z_v \in L_2[0,T] \right\}.$$

Then, C'_0 is the Hilbert space with the inner product

$$(v_1, v_2)_{C'_0} = \int_0^T z_{v_1}(t) z_{v_2}(t) dt,$$

where $v_j(t) = \int_0^t z_{v_j}(s)ds$ for $j = 1, 2$. Furthermore, we note that $C_0'[0, T] \subset C_0[0, T]$ and $(C_0'[0, T], C_0[0, T], m)$ is one example of the abstract Wiener space [1,16,19,20]. For $x \in C_0[0, T]$ and $v \in C_0'[0, T]$ with $v(t) = \int_0^t z_v(s)ds, z_v \in L_2[0, T]$, $(v, x)^\sim \equiv \langle z_v, x \rangle$ is a well-defined Gaussian random variable with mean 0 and variance $\|v\|_{C_0'}^2 = \|z_v\|_2^2$, where $(\cdot, \cdot)^\sim$ is the complex bilinear form on $K^* \times K$.

The following is a well-known integration formula which is used several times in this paper. For each $v \in C_0'$ with $v(t) = \int_0^t z_v(s)ds$,

$$\int_{C_0[0,T]} \exp\{(v, x)^\sim\} dm(x) = \exp\left\{\frac{1}{2}\|v\|_{C_0'}^2\right\} = \exp\left\{\frac{1}{2}\|z_v\|_2^2\right\}. \tag{5}$$

For each $v \in C_0'[0, T]$, let

$$\Phi_v(x) = \exp\{(v, x)^\sim\}. \tag{6}$$

These functionals are called the exponential functionals on $C_0[0, T]$. It is a well-known fact that the class

$$\mathcal{A} \equiv \{\Phi_v : v \in C_0'[0, T]\} \tag{7}$$

is a fundamental set in $L_2(C_0[0, T])$. Thus, there is a countable dense $\mathcal{S}(C_0[0, T]) = \{\Phi_{v_n}\}_{n=1}^\infty \equiv \{\Phi_n\}_{n=1}^\infty$ which is dense in $L_2(C_0[0, T])$. Thus, we have that, for each $F \in L_2(C_0[0, T])$,

$$F(x) = \lim_{n \to \infty} \sum_{j=1}^n a_j \Phi_{v_j}(x)$$

in the L_2-sense, where $\{a_j\}_{j=1}^\infty$ is a sequence of constants.

Let $\mathcal{L} \equiv \mathcal{L}(K)$ be the class of all bounded linear operators on K. Then, for each $v \in C_0'[0, T]$ and $S \in \mathcal{L}$,

$$(v, Sx)^\sim = (S^*v, x)^\sim$$

where S^* is the adjoint operator of S, see [16,19,21]. We state the conditions for the function h to obtain mathematically consistency as follows:

(i) For each $h \in L_\infty[0, T] \subset L_2[0, T]$,

$$\langle z_v, Z_h(x, \cdot)\rangle = \langle z_v h, x\rangle$$

where $v(t) = \int_0^t z_v(s)ds$ for some $z_v \in L_2[0, T]$ because, although $z_v \in L_2[0, T]$, $z_v h$ may not be an element of $L_2[0, T]$ for $h \in L_2[0, T]$.

(ii) Let

$$h(t) = \begin{cases} 0, & 0 \leq t < T/2 \\ t + 2, & T/2 \leq t \leq T \end{cases}.$$

Then, h is in $L_\infty[0, T]$ (and hence $h \in L_2[0, T]$). However, $Z_h(x, t)$ may not be a Gaussian process. A condition for h is needed. Let h be an element of $L_\infty[0, T]$ such that $m_L(supp(h)) = m_L(\{t \in [0, T] : h(t) \neq 0\}) = T$, where m_L is the Lebesgue measure. Then, we have $\|h\|_2 > 0$ and $Z_h(x, t)$ is a Gaussian process.

(iii) For each $h \in L_\infty[0, T]$ and $x \in C_0[0, T]$, $Z_h(x, t)$ is stochastically continuous but it is not continuous, namely $Z_h(x, t)$ may not element of $C_0[0, T]$. However, if h is a function of bounded variation on $[0, T]$, the Gaussian process $Z_h(x, t)$ is continuous and hence $SZ_h(x, \cdot)$ is well-defined for all $S \in \mathcal{L}$. Since for $v \in C_0'$ with $v(t) = \int_0^t z_v(s)ds$, $(v, x)^\sim = \langle z_v, x\rangle$, we have that

$$(v, SZ_h(x, \cdot))^\sim = (S^*v, Z_h(x, \cdot))^\sim = \langle z_{S^*v}, Z_h(x, \cdot)\rangle = \langle h z_{S^*v}, x\rangle. \tag{8}$$

(iv) Let $\mathcal{H} = \{h : [0, T] \to \mathbb{R} : h \in BV[0, T], m_L(supp(h)) = T\}$.

3. Generalization of the Integral Transform with Related Topics

We start this section by giving definition of generalized integral transform, generalized convolution product and the generalized first variation of functionals on K.

Definition 1. *Let h, h_1, h_2 be an element of $\mathcal{H}$ and let F and G be functionals on K. Let $S, R, A, B, C, D, S_1, S_2 \in \mathcal{L}$. Then, the generalized integral transform $\mathcal{T}_{S,R}^h(F)$ of F, a generalized convolution product $(F * G)_{A,B,C,D}^{h_1,h_2}$ of F and G, and a generalized first variation $\delta_{S_1,S_2}^{h_1,h_2} F$ of F with respect to h_1, h_2, S_1 and S_2 are defined by the formulas*

$$\mathcal{T}_{S,R}^h(F)(y) = \int_{C_0[0,T]} F(SZ_h(x,\cdot) + Ry) dm(x), \tag{9}$$

$$(F * G)_{A,B,C,D}^{h_1,h_2}(y) = \int_{C_0[0,T]} F(AZ_{h_1}(x,\cdot) + By) G(CZ_{h_2}(x,\cdot) + Dy) dm(x) \tag{10}$$

and

$$\begin{aligned}\delta_{S_1,S_2}^{h_1,h_2} F(x|u) &\equiv \delta F(S_1 Z_{h_1}(x,\cdot)|S_2 Z_{h_2}(u,\cdot)) \\ &= \frac{\partial}{\partial \alpha} F(S_1 Z_{h_1}(x,\cdot) + \alpha S_2 Z_{h_2}(u,\cdot))\bigg|_{\alpha=0}\end{aligned} \tag{11}$$

for $x, u, y \in K$ if they exist.

Remark 1.

(1) When $h(t) \equiv 1$ on $[0, T]$, the generalized integral transform $\mathcal{T}_{S,R}^1$ is the Fourier–Gauss transform $\mathcal{G}_{S,R}$ [16].
(2) When S and R are the constant operators, the generalized integral transform $\mathcal{T}_{\gamma,\beta}^h$ is a generalized integral transform $\mathcal{F}_{\gamma,\beta}^h$ used in [14,15]. In particular, if $h(t) \equiv 1$ on $[0, T]$, then $\mathcal{T}_{\gamma,\beta}^1$ is the integral transform used in [5,6,8,10,11,13,22].
(3) When $h_1(t) \equiv 1$ and $h_2(t) \equiv 1$ on $[0,T]$, $(F * G)_{A,B,C,D}^{1,1}$ is the convolution product used in [11].

We next state some notations used in this paper. For $v \in L_2[0,T], h_1, h_2, \cdots, h_n \in \mathcal{H}$ and $R_1, \cdots, R_n \in \mathcal{L}$, let

$$M(R_1,\cdots,R_n : h_1,\cdots,h_n : v) \equiv \exp\left\{\frac{1}{2}\sum_{j=1}^n \|h_j z_{R_j^* v}\|_2^2\right\}, \tag{12}$$

where $R_j^* v(t) = \int_0^t z_{R_j^* v}(s) ds$ for each $j = 1, 2, \cdots, n$. Furthermore, we have the symmetric property for $M(\cdot : \cdot : v)$.

In Theorem 1, we obtain the existence of generalized integral transform, generalized convolution product and generalized first variation of functionals in $\mathcal{S}(C_0[0,T])$. In addition, we show that they are elements of $\mathcal{S}(C_0[0,T])$.

Theorem 1. *Let h, h_1, h_2 be elements of $\mathcal{H}$ and let $S, R, A, B, C, D, S_1, S_2 \in \mathcal{L}$. Let Φ_v and Φ_w be elements of $\mathcal{S}(C_0[0,T])$ and let $u(t) = \int_0^t z_u(s) ds \in C_0'$. In addition, let $k_{h_j}(t) = \int_0^t h_j(s) ds$ for $j = 1,2$. Then, the generalized integral transform $\mathcal{T}_{S,R}^h(\Phi_v)$ of Φ_v, the generalized convolution product $(\Phi_v * \Phi_w)_{A,B,C,D}^{h_1,h_2}$ of Φ_v and Φ_w and the generalized first variation $\delta_{S_1,S_2}^{h_1,h_2} \Phi_v(x|u)$ with respect to h_1, h_2, S_1 and S_2 exist, belong to $\mathcal{S}(C_0[0,T])$ and are given by the formulas*

$$\mathcal{T}_{S,R}^h(\Phi_v)(y) = M(S : h : v)\Phi_{R^* v}(y), \tag{13}$$

$$\begin{aligned}(\Phi_v * \Phi_w)_{A,B,C,D}^{h_1,h_2}(y) \\ = M(A : h_1 : v) M(C : h_2 : w) \exp\{(h_1 z_{A^* v}, h_2 z_{C^* w})_2\} \Phi_{B^* v + D^* w}(y)\end{aligned} \tag{14}$$

and
$$\delta^{h_1,h_2}_{S_1,S_2}\Phi_v(x|u) = (h_2 z_{S_2^*v}, z_u)_2 \Phi_{h_1 z_{S_1^*v}}(x). \tag{15}$$

for $x, y, u \in K$.

Proof. First, using Equations (5), (1) and (8), it follows that, for all $y \in K$, we have

$$\mathcal{T}^h_{S,R}(\Phi_v)(y) = \int_{C_0[0,T]} \exp\left\{(v, SZ_h(x, \cdot))^\sim + (v, Ry)^\sim\right\} dm(x)$$

$$= \int_{C_0[0,T]} \exp\left\{\langle h z_{S^*v}, x\rangle + (R^*v, y)^\sim\right\} dm(x)$$

$$= \exp\left\{\frac{1}{2}\|h z_{S^*v}\|_2^2\right\} \Phi_{R^*v}(y).$$

Finally, by using Equations (12) and (13) is obtained. We next use Equations (5), (8) and (14) to obtain the following calculation

$$(\Phi_v * \Phi_w)^{h_1,h_2}_{A,B,C,D}(y) = \int_{C_0[0,T]} \Phi_v(AZ_{h_1}(x, \cdot) + By)\Phi_w(CZ_{h_2}(x, \cdot) + Dy)dm(x)$$

$$= \int_{C_0[0,T]} \exp\left\{\langle h_1 z_{A^*v} + h_2 z_{C^*w}, x\rangle + (B^*v + D^*w, y)^\sim\right\} dm(x)$$

$$= \exp\left\{\frac{1}{2}\|h_1 z_{A^*v} + h_2 z_{C^*w}\|_2^2\right\} \Phi_{B^*v}(y)\Phi_{D^*w}(y).$$

Since $\|h\|_2^2 = (h, h)_2$ for all $h \in L_2[0, T]$, we now note that

$$\frac{1}{2}\|h_1 z_{A^*v} + h_2 z_{C^*w}\|_2^2 = \frac{1}{2}\left(h_1 z_{A^*v} + h_2 z_{C^*w}, h_1 z_{A^*v} + h_2 z_{C^*w}\right)_2$$

$$= \frac{1}{2}\left[(h_1 z_{A^*v}, h_1 z_{A^*v})_2 + (h_2 z_{C^*w}, h_2 z_{C^*w})_2 + 2(h_1 z_{A^*v}, h_2 z_{C^*w})_2\right]$$

and $\Phi_v(y) + \Phi_w(y) = \Phi_{v+w}(y)$ for all $v, w \in C_0'[0, T]$. Hence, we can obtain Equation (14) as desired. Finally, we use Equations (8) and (11) to establish Equation (15) as follows:

$$\delta^{h_1,h_2}_{S_1,S_2}\Phi_v(x|u) = \frac{\partial}{\partial \alpha}\left[\Phi_v(S_1 Z_{h_1}(x, \cdot) + \alpha S_2 Z_{h_2}(u, \cdot))\right]\Big|_{\alpha=0}$$

$$= \frac{\partial}{\partial \alpha}\left[\exp\{\langle h_1 z_{S_1^*v}, x\rangle + \alpha\langle h_2 z_{S_2^*v}, u\rangle\}\right]\Big|_{\alpha=0}$$

$$= \langle h_2 z_{S_2^*v}, u\rangle \exp\{\langle h_1 z_{S_1^*v}, x\rangle\}.$$

We now note that

$$\langle h_2 z_{S_2^*v}, u\rangle = \int_0^T h_2(t) z_{S_2^*v}(t) z_u(t) dt = (h_2 z_{S_2^*v}, z_u)_2,$$

which establishes Equation (15) as desired. □

4. Some Relationships with the Generalized Convolution Products.

In this section, we obtain some relationships between the generalized integral transform and the generalized convolution product of functionals in $\mathcal{S}(C_0[0,T])$. In the first theorem in Section 4, we give a formula for the generalized integral transforms of functionals in $\mathcal{S}(C_0[0,T])$. To establish some relationships, the following lemma is needed.

Lemma 1. Let $h_1, h_2 \in \mathcal{H}$ and let $S_1, S_2, R \in \mathcal{L}$. Then, for each $v \in C_0'$,

$$M(S_1 : h_1 : R^*v)M(S_2 : h_2 : v) = M(RS_1, S_2 : h_1, h_2 : v). \tag{16}$$

Proof. Using the following fact $S_1^* R^* = (RS_1)^*$ and Equation (12) repeatedly, we have

$$\begin{aligned} M(S_1 : h_1 : R^*v)M(S_2 : h_2 : v) &= \exp\left\{\frac{1}{2}\|h_1 z_{S_1^* R^* v}\|_2^2\right\} \exp\left\{\frac{1}{2}\|h_2 z_{S_2^* v}\|_2^2\right\} \\ &= \exp\left\{\frac{1}{2}\|h_1 z_{S_1^* R^* v}\|_2^2 + \frac{1}{2}\|h_2 z_{S_2^* v}\|_2^2\right\} \\ &= \exp\left\{\frac{1}{2}\|h_1 z_{(RS_1)^* v}\|_2^2 + \frac{1}{2}\|h_2 z_{S_2^* v}\|_2^2\right\} \\ &= M(RS_1, S_2 : h_1, h_2 : v), \end{aligned}$$

which complete the proof of Lemma 1. □

Theorem 2. Let S_1, S_2, R_1 and R_2 be elements of $\mathcal{L}$ and let h_1 and h_2 be elements of $\mathcal{H}$. In addition, let Φ_v be an element of $\mathcal{S}(C_0[0,T])$. Then,

$$\mathcal{T}^{h_1}_{S_1, R_1}(\mathcal{T}^{h_2}_{S_2, R_2}(\Phi_v))(y) = M(R_2 S_1, S_2 : h_1, h_2 : v)\Phi_{(R_1 R_2)^* v}(y) \tag{17}$$

for $y \in K$.

Proof. From Theorem 1, we have

$$\mathcal{T}^{h_2}_{S_2, R_2}(\Phi_v)(y) = M(S_2 : h_2 : v)\Phi_{R_2^* v}(y).$$

Applying Theorem 1 once more,

$$\mathcal{T}^{h_1}_{S_1, R_1}(\mathcal{T}^{h_2}_{S_2, R_2}(\Phi_v))(y) = M(S_2 : h_2 : v)M(S_1 : h_1 : R_2^* v)\Phi_{(R_1 R_2)^* v}(y).$$

Finally, using Equation (16) in Lemma 1, we complete the proof of Theorem 2 as desired. □

Equations (18) and (19) in Theorem 3 are the commutative of the generalized integral transform and the Fubini theorem with respect to the generalized integral transform, respectively.

Theorem 3. Let S_1, S_2, R_1 and R_2 be elements of $\mathcal{L}$ and let h_1 and h_2 be elements of $\mathcal{H}$. In addition, let Φ_v be an element of $\mathcal{S}(C_0[0,T])$. Then,

$$\mathcal{T}^{h_1}_{S_1, R_1}(\mathcal{T}^{h_2}_{S_2, R_2}(\Phi_v))(y) = \mathcal{T}^{h_2}_{S_2, R_2}(\mathcal{T}^{h_1}_{S_1, R_1}(\Phi_v))(y) \tag{18}$$

if and only if

$$R_1 R_2 = R_2 R_1, \quad \text{and} \quad M(R_2 S_1, S_2 : h_1, h_2 : v) = M(S_1, R_1 S_2 : h_1, h_2 : v).$$

Furthermore,

$$\mathcal{T}^{h_1}_{S_1, R_1}(\mathcal{T}^{h_2}_{S_2, R_2}(\Phi_v))(y) = \mathcal{T}^{h_3}_{S_3, R_3}(\Phi_v)(y) \tag{19}$$

if and only if

$$R_1 R_2 = R_3, \quad \text{and} \quad M(R_2 S_1, S_2 : h_1, h_2 : v) = M(S_3 : h_3 : v).$$

Proof. Using Equation (17) twice, we have

$$\mathcal{T}^{h_1}_{S_1, R_1}(\mathcal{T}^{h_2}_{S_2, R_2}(\Phi_v))(y) = M(R_2 S_1, S_2 : h_1, h_2 : v)\Phi_{(R_1 R_2)^* v}(y)$$

and
$$\mathcal{T}_{S_2,R_2}^{h_2}(\mathcal{T}_{S_1,R_1}^{h_1}(\Phi_v))(y) = M(S_1, R_1 S_2 : h_1, h_2 : v)\Phi_{(R_2 R_1)^* v}(y).$$

Using these facts and Equation (13), we can establish Equations (18) and (19). □

From Theorems 2 and 3, we can establish the n-dimensional version for the generalized integral transform.

Corollary 1. *Let $S_1, \cdots, S_n, R_1, \cdots, R_{n-1}$ and R_n be elements of $\mathcal{L}$ and let h_j be an element of $\mathcal{H}$, $j = 1, 2, \cdots$. In addition, let Φ_v be an element of $\mathcal{S}(C_0[0,T])$. Then,*

$$\mathcal{T}_{S_n,R_n}^{h_n}(\cdots(\mathcal{T}_{S_1,R_1}^{h_1}(\Phi_v))\cdots)(y)$$
$$= M(S_1, R_1 S_2, R_1 R_2, S_3, \cdots, R_1 R_2 \cdots R_{n-1} S_n : h_1, \cdots, h_n : v)\Phi_{(R_1 \cdots R_n)^* v}(y).$$

In our next theorem, we show that our generalized convolution product is commutative.

Theorem 4. *Let A, B, C and D be elements of $\mathcal{L}$ and let $h_1, h_2 \in \mathcal{H}$. Let Φ_v and Φ_w be elements of $\mathcal{S}(C_0[0,T])$. Then,*
$$(\Phi_v * \Phi_w)_{A,B,C,D}^{h_1,h_2}(y) = (\Phi_w * \Phi_v)_{A,B,C,D}^{h_1,h_2}(y) \tag{20}$$

if and only if
$$M(A : h_1 : v) = M(C : h_2 : v) \text{ and } M(A : h_1 : w) = M(C : h_2 : w).$$

Proof. The proof of Theorem 4 is a straightforward application of Theorem 1. □

In Theorem 5, we give a necessary and sufficient condition for holding a relationship between the generalized integral transform and the generalized convolution product.

Theorem 5. *For $j = 1,2,3$, let $S_j, R_j \in \mathcal{L}$, and, for $= 1,2$, let $A_i, B_i, C_i, D_i \in \mathcal{L}$. In addition, for $k = 1, 2, \cdots, 7$, let $h_k \in \mathcal{H}$. Then,*

$$\mathcal{T}_{S_1,R_1}^{h_1}(\Phi_v * \Phi_w)_{A_1,B_1,C_1,D_1}^{h_2,h_3}(y) = (\mathcal{T}_{S_2,R_2}^{h_4}\Phi_v * \mathcal{T}_{S_3,R_3}^{h_5}\Phi_w)_{A_2,B_2,C_2,D_2}^{h_6,h_7}(y) \tag{21}$$

if and only if the following equations hold

$$\begin{cases} B_1 R_1 = R_2 B_2 \text{ and } D_1 R_1 = R_3 D_2 \\ M(B_1 S_1, A_1 : h_1, h_2 : v) = M(S_2, A_2 : h_4, h_6 : v) \\ M(D_1 S_1, C_1 : h_1, h_3 : w) = M(S_3, C_2 : h_5, h_7 : w) \\ (h_2 A_1^* v, h_3 C_1^* w)_2 = (h_6 A_2^* v, h_7 C_2^* w)_2 \end{cases}.$$

Proof. To complete the proof of Theorem 5, we first calculate the left hand side of Equation (21). From Equation (14) in Theorem 1, we have

$$(\Phi_v * \Phi_w)_{A_1,B_1,C_1,D_1}^{h_2,h_3}(y)$$
$$= M(A_1 : h_2 : v) M(C_1 : h_3 : w) \exp\{(h_2 A_1^* v, h_3 C_1^* w)_2\} \Phi_{B_1^* v + D_1^* w}(y). \tag{22}$$

Using Equations (13), (12), (16) and (22), we have

$$\mathcal{T}_{S_1,R_1}^{h_1}(\Phi_v * \Phi_w)_{A_1,B_1,C_1,D_1}^{h_2,h_3}(y) = M(B_1 S_1, A_1 : h_1, h_2 : v) M(D_1 S_1, C_1 : h_1, h_3 : w)$$
$$\cdot \exp\{(h_2 A_1^* v, h_3 C_1^* w)_2\} \Phi_{R_1^* B_1^* v + R_1^* D_1^* w}(y).$$

We next calculate the left hand side of Equation (21). From Equations (12) and (13) twice, we have

$$\mathcal{T}_{S_2,R_2}^{h_4}(\Phi_v)(y) = M(S_2 : h_4 : v)\Phi_{R_2^*v}(y) \qquad (23)$$

and

$$\mathcal{T}_{S_3,R_3}^{h_5}(\Phi_w)(y) = M(S_3 : h_5 : w)\Phi_{R_3^*w}(y). \qquad (24)$$

We now use Equations (14), (16), (23) and (24) repeatedly to obtain the following calculation

$$(\mathcal{T}_{S_2,R_2}^{h_4}\Phi_v * \mathcal{T}_{S_3,R_3}^{h_5}\Phi_w)_{A_2,B_2,C_2,D_2}^{h_6,h_7}(y) = M(S_2, A_2 : h_4, h_6 : v)M(S_3, C_2 : h_5, h_7 : w)$$
$$\cdot \exp\{(h_6 A_2^*v, h_7 C_2^*w)_2\}\Phi_{B_2^*R_2^*v + D_2^*R_3^*w}(y).$$

Hence, we complete the proof of Theorem 5 as desired. □

Corollary 2. *The following results and formulas stated bellow easily from Theorem 5.*

(1) Let S and R be elements of $\mathcal{L}$, and, for $= 1,2$, let $A_i, B_i, C_i, D_i \in \mathcal{L}$. In addition, for $k = 1, 2, \cdots, 5$, let $h_k \in \mathcal{H}$. Then,

$$\mathcal{T}_{S,R}^{h_1}(\Phi_v * \Phi_w)_{A_1,B_1,C_1,D_1}^{h_2,h_3}(y) = (\mathcal{T}_{S,R}^{h_1}\Phi_v * \mathcal{T}_{S,R}^{h_1}\Phi_w)_{A_2,B_2,C_2,D_2}^{h_4,h_5}(y)$$

if and only if the following equations hold

$$\begin{cases} B_1 R = RB_2 \text{ and } D_1 R = RD_2 \\ M(B_1 S, A_1 : h_1, h_2 : v) = M(S, A_2 : h_1, h_4 : v) \\ M(D_1 S, C_1 : h_1, h_3 : w) = M(S, C_2 : h_1, h_5 : w) \\ (h_2 A_1^*v, h_3 C_1^*w)_2 = (h_4 A_2^*v, h_5 C_2^*w)_2 \end{cases}$$

(2) For $j = 1, 2, 3$, let $S_j, R_j \in \mathcal{L}$ and $A, B, C, D \in \mathcal{L}$. In addition, for $k = 1, 2, \cdots, 7$, let $h_k \in \mathcal{H}$. Then,

$$\mathcal{T}_{S_1,R_1}^{h_1}(\Phi_v * \Phi_w)_{A,B,C,D}^{h_2,h_3}(y) = (\mathcal{T}_{S_2,R_2}^{h_4}\Phi_v * \mathcal{T}_{S_3,R_3}^{h_5}\Phi_w)_{A,B,C,D}^{h_2,h_3}(y)$$

if and only if the following equations hold

$$\begin{cases} BR_1 = R_2 B \text{ and } DR_1 = R_3 D \\ M(BS_1, A : h_1, h_2 : v) = M(S_2, A : h_4, h_2 : v) \\ M(DS_1, C : h_1, h_3 : w) = M(S_3, C : h_5, h_3 : w) \\ (h_2 A^*v, h_3 C^*w)_2 = (h_6 A^*v, h_3 C^*w)_2 \end{cases}$$

5. Some Relationships with the Generalized First Variations

In this section, we establish some formulas involving the generalized first variation. We next obtain a generalized Cameron–Storvick theorem for the generalized first variation and use this to apply for the generalized integral transform.

Theorem 6. *Let $h_1, h_2, h_3 \in \mathcal{H}$ and $S_1, S_2, S_3 \in \mathcal{L}$. Let $u \in C_0'$ with $u(t) = \int_0^t z_u(s)ds$. Then,*

$$\mathcal{T}_{S_1,R}^{h_1}(\delta_{S_2,S_3}^{h_2,h_3}\Phi_v(\cdot|u))(y) = \delta_{S_2,S_3}^{h_2,h_3}\mathcal{T}_{S_1,R}^{h_1}(\Phi_v)(y|u) \qquad (25)$$

if and only if $R = I$ and $M(S_1 : h_1 : \bar{v}_{S_2,h_2}) = M(S_1 : h_1 : v)$, where $\bar{v}_{S_2,h_2} = h_2 z_{S_2^*v}$.

Proof. First, using Equations (5), (12), (13) and (29), we have

$$\mathcal{T}_{S_1,R}^{h_1}(\delta_{S_2,S_3}^{h_2,h_3}\Phi_v(\cdot|u))(y)$$
$$= (h_3 z_{S_3^* v}, z_u)_2 \int_{C_0[0,T]} \Phi_{h_2 z_{S_2^* v}}(S_1 Z_{h_1}(x,\cdot) + Ry)dm(x)$$
$$= (h_3 z_{S_3^* v}, z_u)_2 \int_{C_0[0,T]} \exp\{(\bar{v}_{S_2,h_2}, S_1 Z_{h_1}(x,\cdot))^\sim + (\bar{v}_{S_2,h_2}, Ry)^\sim\}dm(x)$$
$$= (h_3 z_{S_3^* v}, z_u)_2 \int_{C_0[0,T]} \exp\{\langle h_1 z_{S_1^* \bar{v}_{S_2,h_2}}, x\rangle + (R^* \bar{v}_{S_2,h_2}, y)^\sim\}dm(x)$$
$$= (h_3 z_{S_3^* v}, z_u)_2 M(S_1 : h_1 : \bar{v}_{S_2,h_2}) \exp\{(R^* \bar{v}_{S_2,h_2}, y)^\sim\}$$
$$= (h_3 z_{S_3^* v}, z_u)_2 M(S_1 : h_1 : \bar{v}_{S_2,h_2}) \Phi_{R^* \bar{v}_{S_2,h_2}}(y).$$

On the other hands, using Equations (11)–(13), we have

$$\delta_{S_2,S_3}^{h_2,h_3}\mathcal{T}_{S_1,R}^{h_1}(\Phi_v)(y|u)$$
$$= \frac{\partial}{\partial \alpha}\left[\mathcal{T}_{S_1,R}^{h_1}(\Phi_v)(S_2 Z_{h_2}(y,\cdot) + \alpha S_3 Z_{h_3}(u,\cdot))\right]\bigg|_{\alpha=0}$$
$$= \frac{\partial}{\partial \alpha}\left[\exp\left\{\frac{1}{2}\|h_1 z_{S_1^* v}\|_2^2\right\}\Phi_{R^* v}(S_2 Z_{h_2}(y,\cdot) + \alpha S_3 Z_{h_3}(u,\cdot))\right]\bigg|_{\alpha=0}$$
$$= \exp\left\{\frac{1}{2}\|h_1 z_{S_1^* v}\|_2^2\right\}\frac{\partial}{\partial \alpha}\left[\exp\left\{(R^* v, S_2 Z_{h_2}(y,\cdot))^\sim + \alpha(R^* v, S_3 Z_{h_3}(u,\cdot))^\sim\right\}\right]\bigg|_{\alpha=0}$$
$$= \exp\left\{\frac{1}{2}\|h_1 z_{S_1^* v}\|_2^2\right\}\frac{\partial}{\partial \alpha}\left[\exp\left\{\langle h_2 z_{S_2^* R^* v}, y\rangle + \alpha\langle h_3 z_{S_3^* R^* v}, u\rangle\right\}\right]\bigg|_{\alpha=0}$$
$$= (h_3 z_{S_3^* R^* v}, z_u)_2 M(S_1 : h_1 : v) \exp\left\{\langle h_2 z_{S_2^* R^* v}, y\rangle\right\}$$
$$= (h_3 z_{S_3^* R^* v}, z_u)_2 M(S_1 : h_1 : v) \Phi_{h_2 z_{S_2^* R^* v}}(y).$$

Hence, Equation (25) holds if and only if $R = I$ and

$$M(S_1 : h_1 : \bar{v}_{S_2,h_2}) = M(S_1 : h_1 : v).$$

□

To establish a generalized Cameron–Storvick theorem for the generalized first variation, we need two lemmas with respect to the translation theorem on Wiener space.

Lemma 2. (Translation Theorem 1) *Let F be a integrable functional on $C_0[0,T]$ and let $x_0 \in C_0'$. Then,*

$$\int_{C_0[0,T]} F(x + x_0)dm(x) = \exp\left\{-\frac{1}{2}\|x_0\|_{C_0'}^2\right\}\int_{C_0[0,T]} F(x)\exp\{(x_0,x)^\sim\}dm(x). \quad (26)$$

In [23], the authors used Equation (26) to establish Equation (28), which is a generalized translation theorem. The main key in their proof is the change of kernel for the Gaussian process, i.e.

$$Z_{h_1}(\theta_0, t) = \int_0^t h_1(s)d\left(\int_0^s h_2(\tau)z_{x_0}(\tau)d\tau\right)$$
$$= \int_0^t h_1(s)h_2(s)z_{x_0}(s)ds \quad (27)$$
$$= \int_0^t h_2(s)d\left(\int_0^s h_1(\tau)z_{x_0}(\tau)d\tau\right) = Z_{h_2}(u,t)$$

where $\theta_0(t) = \int_0^t h_2(t) z_{x_0}(t) dt$ and $u(t) = \int_0^t h_1(s) z_{x_0}(s) ds$ for given $x_0 \in C_0'$.

The following lemma is said to be the translation theorem via the Gaussian process on Wiener space.

Lemma 3 (Translation Theorem 2). *Let $h_1, h_2 \in \mathcal{H}$. Let $x_0(t) = \int_0^t z_{x_0}(s) ds$ and let $F(Z_{h_1}(x, \cdot))$ be a integrable functional on $C_0[0, T]$. Let*

$$\theta_0(t) = \int_0^t h_1(s) z_{x_0}(s) ds.$$

Then,

$$\int_{C_0[0,T]} F(Z_{h_1}(x,\cdot) + Z_{h_2}(\theta_0,\cdot)) dm(x)$$
$$= \exp\left\{-\frac{1}{2}\|z_{x_0} h_2\|_2^2\right\} \int_{C_0[0,T]} F(Z_{h_1}(x,\cdot)) \exp\{(\theta_0, Z_{h_2}(x,\cdot))^{\sim}\} dm(x). \quad (28)$$

In our next theorem, we establish the generalized Cameron–Storvick theorem for the generalized first variation.

Theorem 7. *Let $x_0 \in C_0'$ be given. Let $h_1, h_2 \in \mathcal{H}$ and $S \in \mathcal{L}$. In addition, let $u(t) = \int_0^t h_1(s) z_{x_0}(s) ds$ and $\theta_0(t) = \int_0^t h_2(s) z_{x_0}(s) ds$. Then,*

$$\int_{C_0[0,T]} \delta_{S,S}^{h_1,h_2} \Phi_v(x|u) dm(x) = \int_{C_0[0,T]} (x_0, Z_{h_2}(x,\cdot))^{\sim} \Phi_v(SZ_{h_1}(x,\cdot)) dm(x). \quad (29)$$

Proof. First, by using Equation (11) and the dominated convergence theorem, we have

$$\int_{C_0[0,T]} \delta_{S,S}^{h_1,h_2} \Phi_v(x|u) dm(x)$$
$$= \frac{\partial}{\partial \alpha}\left[\int_{C_0[0,T]} \Phi_v(SZ_{h_1}(x,\cdot) + \alpha S Z_{h_2}(u,\cdot)) dm(x)\right]\bigg|_{\alpha=0}$$
$$= \frac{\partial}{\partial \alpha}\left[\int_{C_0[0,T]} \Phi_v(SZ_{h_1}(x,\cdot) + S Z_{h_2}(\alpha u,\cdot)) dm(x)\right]\bigg|_{\alpha=0}.$$

Now, let $F_S^h(x) = \Phi_v(SZ_h(x,\cdot))$. Using the key (27) used in [23], we have

$$F_S^{h_1}(x + \alpha \theta_0) = \Phi_v(SZ_{h_1}(x,\cdot) + SZ_{h_2}(\alpha u,\cdot))$$

where $\theta_0(t) = \int_0^t h_2(s) z_{x_0}(s) ds$ and $u(t) = \int_0^t h_1(s) z_{x_0}(s) ds$. This means that

$$\int_{C_0[0,T]} \delta_{S,S}^{h_1,h_2} F(x|u) dm(x) = \frac{\partial}{\partial \alpha}\left[\int_{C_0[0,T]} F_S^{h_1}(x + \alpha \theta_0) dm(x)\right]\bigg|_{\alpha=0}.$$

We next apply the translation theorem to the functional $F_S^{h_1}$ instead of F in Lemma 2 to proceed the following formula

$$\int_{C_0[0,T]} \delta_{S,S}^{h_1,h_2} F(x|u) dm(x)$$
$$= \frac{\partial}{\partial \alpha}\left[\exp\left\{-\frac{1}{2}\|\alpha \theta_0\|_{C_0'}^2\right\} \int_{C_0[0,T]} F_S^{h_1}(x) \exp\{(\alpha \theta_0, x)^{\sim}\} dm(x)\right]\bigg|_{\alpha=0}$$
$$= \frac{\partial}{\partial \alpha}\left[\exp\left\{-\frac{\alpha^2}{2}\|z_{x_0} h_2\|_2^2\right\} \int_{C_0[0,T]} F_S^{h_1}(x) \exp\{\alpha \langle z_{x_0} h_2, x\rangle\} dm(x)\right]\bigg|_{\alpha=0}$$
$$= \int_{C_0[0,T]} \langle z_{x_0} h_2, x\rangle \Phi_v(SZ_{h_1}(x,\cdot)) dm(x).$$

Since $(\theta_0, x)^\sim = \langle z_{x_0} h_2, x \rangle = (x_0, Z_{h_2}(x, \cdot))^\sim$, we complete the proof of Theorem 7 as desired. □

In the last theorem in this paper, we use Equation (29) to give an integration formula involving the generalized first variation and the generalized integral transform. This formula tells us that we can calculate the Wiener integral of generalized first variation for generalized integral transform directly without calculations of them.

Theorem 8. *Let $h_1, h_2, h_3 \in \mathcal{H}$ and let $S_1, S_2 \in \mathcal{L}$. In addition, let u, x_0, θ_0 be as in Theorem 7. Then,*

$$\int_{C_0[0,T]} \delta^{h_2,h_3}_{S_2,S_2} \mathcal{T}^{h_1}_{S_1,R}(\Phi_v)(y|u) dm(y) = M(S_1, RS_2 : h_1, h_2 : v)(h_3 z_{x_0}, h_2 z_{S_2^* R^* v})_2. \tag{30}$$

Proof. Applying Equation (29) to the functional $\mathcal{T}^{h_1}_{S_1,R}(\Phi_v)$ instead of Φ_v, we have

$$\int_{C_0[0,T]} \delta^{h_2,h_3}_{S_2,S_2} \mathcal{T}^{h_1}_{S_1,R}(\Phi_v)(y|u) dm(y)$$
$$= \int_{C_0[0,T]} (x_0, Z_{h_3}(y, \cdot))^\sim \mathcal{T}^{h_1}_{S_1,R}(\Phi_v)(S_2 Z_{h_2}(y, \cdot)) dm(y).$$

Now, using Equations (8) and (13), it becomes that

$$\int_{C_0[0,T]} \delta^{h_2,h_3}_{S_2,S_2} \mathcal{T}^{h_1}_{S_1,R}(\Phi_v)(y|u) dm(y)$$
$$= M(S_1 : h_1 : v) \int_{C_0[0,T]} (x_0, Z_{h_3}(y, \cdot))^\sim \exp\{(R^* v, S_2 Z_{h_2}(y, \cdot))^\sim\} dm(y)$$
$$= M(S_1 : h_1 : v) \int_{C_0[0,T]} \langle h_3 z_{x_0}, y \rangle \exp\{\langle h_2 z_{S_2^* R^* v}, y \rangle\} dm(y).$$

The following integration formula

$$\int_{C_0[0,T]} \langle w, x \rangle \exp\{\langle p, x \rangle\} dm(x) = (w, p)_2 \exp\left\{\frac{1}{2} \|p\|_2^2\right\}, \quad w, p \in L_2[0, T]$$

and Equation (12) yield that

$$\int_{C_0[0,T]} \delta^{h_2,h_3}_{S_2,S_2} \mathcal{T}^{h_1}_{S_1,R}(\Phi_v)(y|u) dm(y)$$
$$= M(S_1 : h_1 : v) \int_{C_0[0,T]} \langle h_3 z_{x_0}, y \rangle \exp\{\langle h_2 z_{S_2^* R^* v}, y \rangle\} dm(y)$$
$$= M(S_1 : h_1 : v)(h_3 z_{x_0}, h_2 z_{S_2^* R^* v})_2 \exp\left\{\frac{1}{2} \|h_2 z_{S_2^* R^* v}\|_2^2\right\}$$
$$= M(S_1 : h_1 : v) M(RS_2 : h_2 : v)(h_3 z_{x_0}, h_2 z_{S_2^* R^* v})_2.$$

Finally, by using Equation (16) in Lemma 1, we establish Equation (30) as desired. □

6. Application

We finish this paper by giving some examples to illustrate the usefulness of our results and formulas.

We first give a simple example used in the stack exchange and the signal process. For $x \in C_0[0, T]$, let $K_s(x)(t) = \int_0^t x(s) ds$. Then, the adjoint is given by the formula $K_s^*(x)(t) = \int_t^T x(s) ds$.

Example 1. Let $S = K_s$ and let $v(t) = -t + \frac{T}{2}$ and $h(t) = t^2$ on $[0, T]$. Then, $h \in \mathcal{H}$. In addition, we have

$$S^*v(t) = \int_t^T v(s)ds = \frac{1}{2}t^2 - \frac{t}{2}T = \int_0^t (s - \frac{1}{2}T)ds.$$

This means that $z_{S^*v}(t) = t - \frac{1}{2}T$ on $[0, T]$ and hence $\|hz_{S^*v}\|_2^2 = \frac{1}{12}T^4$. Thus, we obtain that

$$\mathcal{T}_{S,R}^h(\Phi_v)(y) = \exp\left\{\frac{1}{24}T^4\right\}\Phi_{R^*v}(y).$$

We give two examples in the quantum mechanics. To do this, we consider useful operators used in quantum mechanics. We consider two cases. However, various cases can be applied in appropriate methods as examples.

Case 1 : Multiplication operator.

In the next examples, we consider the multiplication operator T_m, which plays a role in physics (quantum theories) (see [21]). Before do this, we introduce some observations to proceed obtaining examples. Let $R \in \mathcal{L}$ such that

$$R(xy) = xR(y) \tag{31}$$

for all $x, y \in C_0[0, T]$. In addition, for $t \in [0, T]$ on $C_0[0, T]$, we define a multiplication operator T_m by

$$(T_m(x))(t) \equiv T_m(x(t)) = tx(t). \tag{32}$$

Then, we have $T_m(xy) = tx(t)y(t)$ and $xT_m(y) = x(t)ty(t)$. Hence, Equation (31) holds. In addition, one can easily check that $T_m^*v(t) = tv(t)$ for all $v \in C_0'$. Note that the expected value or corresponding mean value is

$$E(x) \equiv \int_0^T t|x(t)|^2 dt = \int_0^T T_m(|x|^2)(t) dt,$$

where x is the state function of a particle in quantum mechanics and $\int_0^T |x(t)|^2 dt$ is the probability that the particle will be found in $[0, T]$.

In the first and second examples, we give some formula with respect to the multiplication operator T_m.

Example 2. Let $S = T_m$ and let $v(t) = \frac{1}{2}t^2$ and $h(t) = t^2$ on $[0, T]$. Then, $h \in \mathcal{H}$. In addition, we have

$$v(t) = \frac{1}{2}t^2 = \int_0^t s\, ds$$

and

$$S^*v(t) = \frac{1}{2}t^3 = \int_0^t \frac{3}{2}s^2 ds.$$

This means that $z_v(t) = t$ and $z_{S^*v}(t) = \frac{3}{2}t^2$ on $[0, T]$ and hence $\|hz_{S^*v}\|_2^2 = \frac{3}{10}T^5$. Thus, we obtain that

$$\mathcal{T}_{S,R}^h(\Phi_v)(y) = \exp\left\{\frac{3}{10}T^5\right\}\Phi_{R^*v}(y).$$

Example 3. Let $S = T_m$ and let $v(t) = e^t - 1$ and $h(t) = t$ on $[0, T]$. Then, $h \in \mathcal{H}$. In addition, we have

$$v(t) = e^t - 1 = \int_0^t e^s ds$$

and
$$S^*v(t) = te^t - t = \int_0^t (se^s + e^s - 1)ds.$$

This means that $z_v(t) = e^t$ and $z_{S^*v}(t) = te^t + e^t - 1$ on $[0, T]$ and hence
$$\|hz_{S^*v}\|_2^2 = \frac{1}{4}e^{2T}(2T^4 + 2T^2 - 2T + 1) - 2e^T(T^2 + 2T - 4) + \frac{1}{3}T^3 - \frac{33}{4}.$$

Thus, we obtain that
$$T_{S,R}^h(\Phi_v)(y) = \exp\left\{\frac{1}{8}e^{2T}(2T^4 + 2T^2 - 2T + 1) \right.$$
$$\left. - e^T(T^2 + 2T - 4) + \frac{1}{6}T^3 - \frac{33}{8}\right\}\Phi_{R^*v}(y).$$

Case 2 : Quantum mechanics operators.

In the next examples, we consider some linear operators which are used to explain the solution of the diffusion equation and the Schrôdinger equation (see [24]).

Let $S : C_0'[0, T] \to C_0'[0, T]$ be the linear operator defined by
$$Sw(t) = \int_0^t w(s)ds. \tag{33}$$

Then, the adjoint operator S^* of S is given by the formula
$$S^*w(t) = w(T)t - \int_0^t w(s)ds = \int_0^t [w(T) - w(s)]ds$$

and the linear operator $A = S^*S$ is given by the formula
$$Aw(t) = \int_0^T \min\{s, t\}w(s)ds.$$

In addition, A is self-adjoint on $C_0'[0, T]$ and so
$$(w_1, Aw_2)_{C_0'} = (Sw_1, Sw_2)_{C_0'} = \int_0^T w_1(s)w_2(s)ds$$

for all $w_1, w_2 \in C_0'[0, T]$. Hence, A is a positive definite operator, i.e., $(w, Aw)_{C_0'} \geq 0$ for all $w \in C_0'[0, T]$. This means that the orthonormal eigenfunctions $\{e_m\}$ of A are given by
$$e_m(t) = \frac{\sqrt{2T}}{(m - \frac{1}{2})\pi} \sin\left(\frac{(m - \frac{1}{2})\pi}{T}t\right) \equiv \int_0^t \alpha_m(s)ds$$

with corresponding eigenvalues $\{\beta_m\}$ given by
$$\beta_m = \left(\frac{T}{(m - \frac{1}{2})\pi}\right)^2.$$

Furthermore, it can be shown that $\{e_m\}$ is a basis of $C_0'[0, T]$ and so $\{\alpha_m\}$ is a basis of Ł2, and that A is a trace class operator and so S is a Hilbert–Schmidt operator on $C_0'[0, T]$. In fact, the trace of A is given by $TrA = \frac{1}{2}T^2 = \int_0^T tdt$. By using the concept of m-lifting on abstract Wiener space, the operators S and A can be extended on $C_0[0, T]$ (see [19,25]).

We now give formulas with respect to the operators S and A, respectively.

Example 4. Let S be given by Equation (33) and let $v(t) = \frac{1}{2}t^2$ and $h(t) = t$ on $[0, T]$. Then, $h \in \mathcal{H}$. In addition, we have

$$v(t) = \frac{1}{2}t^2 = \int_0^t s\,ds$$

$$Sv(t) = \int_0^t \frac{1}{2}s^2\,ds = \frac{1}{6}t^3$$

and

$$S^*v(t) = tv(T) - Sv(t) = \frac{1}{2}tT^2 - \frac{1}{6}t^3 = \int_0^t \left[\frac{1}{2}T - \frac{1}{2}s^2\right]ds.$$

This means that $z_v(t) = t$ and $z_{S^*v}(t) = \frac{1}{2}T - \frac{1}{2}t^2$ on $[0, T]$ and hence $\|hz_{S^*v}\|_2^2 = \frac{1}{40}T^7$. Thus, we obtain that

$$\mathcal{T}_{S,R}^h(\Phi_v)(y) = \exp\left\{\frac{1}{80}T^7\right\}\Phi_{R^*v}(y).$$

Example 5. Let $S = A$ and let $v(t) = \frac{1}{2}t^2$ and $h(t) = t$ on $[0, T]$. Then, $h \in \mathcal{H}$. In addition, we have

$$v(t) = \frac{1}{2}t^2 = \int_0^t s\,ds,$$

$$Av(t) = SS^*v(t) = \int_0^t S^*v(s)\,ds = \int_0^t \int_0^s [v(T) - v(u)du]ds$$

$$= \int_0^t \left[sv(T) - \int_0^s \frac{1}{2}u^2\,du\right]ds = \int_0^t \left[\frac{1}{2}sT^2 - \frac{1}{6}s^3\right]ds$$

$$= \frac{1}{4}T^2t^2 - \frac{1}{24}t^4$$

and

$$A^*v(t) = Av(t) = \int_0^t \left[\frac{1}{2}sT^2 - \frac{1}{6}s^3\right]ds.$$

This means that $z_v(t) = t$ and $z_{S^*v}(t) = \frac{1}{2}tT^2 - \frac{1}{6}t^3$ on $[0, T]$ and hence $\|hz_{A^*v}\|_2^2 = \frac{58}{2835}T^9$. Thus, we obtain that

$$\mathcal{T}_{A,R}^h(\Phi_v)(y) = \exp\left\{\frac{29}{2835}T^9\right\}\Phi_{R^*v}(y).$$

We now give an example with respect to Theorem 8.

Example 6. Let $s_1 = T_m$ and $S_2 = S$, as used in the examples above. Let $R = I$ and let $h_1(t) = h_2(t) = t$, $h_3(t) = t^2$ on $[0, T]$. Furthermore, let $v(t) = \frac{1}{2}t^2$ on $[0, T]$ and let $x_0(t) = t = \int_0^t 1\,ds \in C_0'$. Then, we have $z_v(t) = t$, $z_{S_1^*v}(t) = \frac{3}{2}t^2$, $z_{S_2^*v}(t) = \frac{1}{2}T - \frac{1}{2}t^2$ and $z_{x_0}(t) = 1$ on $[0, T]$. Furthermore, we have

$$M(S_1, RS_2 : h_1, h_2 : v) = \exp\left\{\frac{5}{28}T^7 + \frac{1}{24}T^4 - \frac{1}{20}T^5\right\}$$

and

$$(h_3 z_{x_0}, h_2 z_{S_2^*R^*v})_2 = \frac{1}{8}T^4 - \frac{1}{12}T^6.$$

Hence, by using Equation (30) in Theorem 8, we can conclude that

$$\int_{C_0[0,T]} \delta_{S_2,S_2}^{h_2,h_3} \mathcal{T}_{S_1,R}^{h_1}(\Phi_v)(y|u)dm(y)$$
$$= \exp\left\{\frac{5}{28}T^7 + \frac{1}{24}T^4 - \frac{1}{20}T^5\right\}\left(\frac{1}{8}T^4 - \frac{1}{12}T^6\right). \tag{34}$$

7. Conclusions

In Sections 3 and 4, we establish some fundamental formulas for the generalized integral transform, the generalized convolution product and the generalized first variation involving the generalized Cameron–Storvick theorem. As shown in Examples 2, 4 and 6, various applications are established by choosing the kernel functions and operators. The results and formulas are more generalized forms than those in previous papers. From these, we can conclude that various examples can also be explained very easily.

Funding: This research was supported by the Basic Science Research Program through the National Research Foundation of Korea (NRF) funded by the Ministry of Science, ICT and Future Planning (2017R1E1A1A03070041).

Acknowledgments: The author would like to express gratitude to the referees for their valuable comments and suggestions, which have improved the original paper.

Conflicts of Interest: The author declares no conflict of interest.

References

1. Lee, Y.J. Integral transforms of analytic functions on abstract Wiener spaces. *J. Funct. Anal.* **1982**, *47*, 153–164. [CrossRef]
2. Cameron, R.H.; Martin, W.T. Transformations of Wiener integrals under translations. *Ann. Math.* **1944**, *45*, 386–396. [CrossRef]
3. Cameron, R.H. Some examples of Fourier-Wiener transforms of analytic functionals. *Duke Math. J.* **1945**, *12*, 485–488. [CrossRef]
4. Cameron, R.H.; Martin, W.T. Fourier-Wiener transforms of functionals belonging to L_2 over the space C. *Duke Math. J.* **1947**, *14*, 99–107. [CrossRef]
5. Chang, K.S.; Kim, B.S.; Yoo, I. Integral transforms and convolution of analytic functionals on abstract Wiener space. *Numer. Funct. Anal. Optim.* **2000**, *21*, 97–105.
6. Ji, U.C.; Obata, N. Quantum white noise calculus. In *Non-Commutativity, Infinite-Dimensionality and Probability at the Crossroads*; QP-PQ: Quantum Probability and White Noise Analysis; World Scientific Publishing: River Edge, NJ, USA, 2002; Volume 16, pp. 143–191,
7. Kim, B.J.; Kim, B.S.; Yoo, I. Integral transforms of functionals on a function space of two variables. *J. Chungcheong Math. Soc.* **2010**, *23*, 349–362.
8. Kim, B.J.; Kim, B.S.; Skoug, D. Integral transforms, convolution products and first variations. *Int. J. Math. Math. Soc.* **2004**, *11*, 579–598.
9. Kim, B.J.; Kim, B.S.; Skoug, D. Conditional integral transforms, conditional convolution products and first variations. *Pan Amer. Math. J.* **2004**, *14*, 27–47.
10. Kim, B.J.; Kim, B.S. Parts formulas involving conditional integral transforms on function space. *Korea J. Math.* **2014**, *22*, 57–69. [CrossRef]
11. Kim, B.S.; Yoo, I. Generalized convolution product for integral transform on Wiener space. *Turk. J. Math.* **2017**, *41*, 940–955. [CrossRef]
12. Kim, B.S.; Skoug, D. Integral transforms of functionals in $L_2(C_0[0,T])$. *Rocky Mt. J. Math.* **2003**, *33*, 1379–1393. [CrossRef]
13. Chung, H.S.; Tuan, V.K. Generalized integral transforms and convolution products on function space. *Integral Transform. Spec. Funct.* **2011**, *22*, 573–586. [CrossRef]
14. Lee, I.Y.; Chung, H.S.; Chang, S.J. Integration formulas for the conditional transform involving the first variation. *Bull. Iran. Math. Soc.* **2015**, *41*, 771–783.
15. Lee, I.Y.; Chung, H.S.; Chang, S.J. Generalized conditional transform with respect to the Gaussian process on function space. *Integ. Trans. Spec. Funct.* **2015**, *26*, 925–938.
16. Im, M.K.; Ji, U.C.; Park, Y.J. Relations among the first variation, the convolutions and the generalized Fourier-Gauss transforms. *Bull. Korean Math. Soc.* **2011**, *48*, 291–302.
17. Johnson, G.W.; Skoug, D.L. Scale-invariant measurability in Wiener space. *Pac. J. Math.* **1979**, *83*, 157–176. [CrossRef]
18. Pierce, I.; Skoug, D. Integration formulas for functionals on the function space $C_{a,b}[0,T]$ involving Paley-Wiener-Zygumund stochastic integrals. *Pan Am. Math. J.* **2008**, *18*, 101–112.

19. Kuo, H.-H. *Gaussian Measure in Banach Space*; Lecture Notes in Mathematics; Springer: Berlin, Germany, 1975; Volume 463
20. Lee, Y.J. Unitary operators on the space of L^2-functions over abstract Wiener spaces. *Soochow J. Math.* **1987**, *13*, 165–174.
21. Kreyszig, E. *Introductory Functional Analysis with Applications*; John Wiley and Sons: Hoboken, NJ, USA, 1978.
22. Chung, D.M.; Ji, U.C. Transforms on white noise functionals with their applications to Cauchy problems. *Nagoya Math. J.* **1997**, *147*, 1–23. [CrossRef]
23. Chang, S.J.; Choi, J.G. A Cameron-Storvick theorem for the analytic Feynman integral associated with Gaussian paths on Wiener space and application. *Comm. Pure Appl. Anal.* **2018**, *17*, 2225–2238. [CrossRef]
24. Chang, S.J.; Choi, J.G.; Chung, H.S. An approach to solution of the Schrödinger equation using Fourier-type functionals. *J. Korean Math. Soc.* **2013**, *53*, 259–274. [CrossRef]
25. Chang, S.J.; Choi, J.G. An analytic bilateral Laplace-Feynman transform on Hilbert space. *Int. J. Math.* **2014**, *25*, 1450118. [CrossRef]

Publisher's Note: MDPI stays neutral with regard to jurisdictional claims in published maps and institutional affiliations.

© 2020 by the author. Licensee MDPI, Basel, Switzerland. This article is an open access article distributed under the terms and conditions of the Creative Commons Attribution (CC BY) license (http://creativecommons.org/licenses/by/4.0/).

Review

A Guide to Special Functions in Fractional Calculus

Virginia Kiryakova

Institute of Mathematics and Informatics, Bulgarian Academy of Sciences, 1113 Sofia, Bulgaria; virginia@diogenes.bg

Abstract: *Dedicated to the memory of Professor Richard Askey (1933–2019) and to pay tribute to the Bateman Project.* Harry Bateman planned his "shoe-boxes" project (accomplished after his death as *Higher Transcendental Functions*, Vols. 1–3, 1953–1955, under the editorship by A. Erdélyi) as a "*Guide to the Functions*". This inspired the author to use the modified title of the present survey. Most of the standard (classical) Special Functions are representable in terms of the Meijer G-function and, specially, of the generalized hypergeometric functions ${}_pF_q$. These appeared as solutions of differential equations in mathematical physics and other applied sciences that are of integer order, usually of second order. However, recently, mathematical models of fractional order are preferred because they reflect more adequately the nature and various social events, and these needs attracted attention to "*new*" *classes* of special functions as their solutions, the so-called *Special Functions of Fractional Calculus (SF of FC)*. Generally, under this notion, we have in mind the Fox H-functions, their most widely used cases of the Wright generalized hypergeometric functions ${}_p\Psi_q$ and, in particular, the Mittag–Leffler type functions, among them the "Queen function of fractional calculus", the Mittag–Leffler function. These fractional indices/parameters extensions of the classical special functions became an unavoidable tool when fractalized models of phenomena and events are treated. Here, we try to review some of the basic results on the theory of the SF of FC, obtained in the author's works for more than 30 years, and support the wide spreading and important role of these functions by several examples.

Keywords: special functions; generalized hypergeometric functions; fractional calculus operators; integral transforms

MSC: 33C60; 33E12; 26A33; 44A20

Citation: Kiryakova, V. A Guide to Special Functions in Fractional Calculus. *Mathematics* **2021**, *9*, 106. https://doi.org/10.3390/math9010106

Received: 15 December 2020
Accepted: 24 December 2020
Published: 5 January 2021

Publisher's Note: MDPI stays neutral with regard to jurisdictional claims in published maps and institutional affiliations.

Copyright: © 2021 by the author. Licensee MDPI, Basel, Switzerland. This article is an open access article distributed under the terms and conditions of the Creative Commons Attribution (CC BY) license (https://creativecommons.org/licenses/by/4.0/).

1. Historical Introduction

Special functions are particular mathematical functions that have more or less established names and notations due to their importance in mathematical analysis, functional analysis, geometry, physics, astronomy, statistics or other applications (Wikipedia: Special Functions [1]). It might be Euler, who started to talk, since 1720, about lots of the standard special functions. He defined the Gamma-function as a continuation of the factorial, also the Bessel functions and looked after the elliptic functions. Several (theoretical and applied) scientists started to use such functions, introduced their notations and named them after famous contributors. Thus, the notions as the Bessel and cylindrical functions; the Gauss, Kummer, Tricomi, confluent and generalized hypergeometric functions; the classical orthogonal polynomials (as Laguerre, Jacobi, Gegenbauer, Legendre, Tchebisheff, Hermite, etc.); the incomplete Gamma- and Beta-functions; and the Error functions, the Airy, Whittaker, etc. functions appeared and a long list of handbooks on the so-called "*Special Functions of Mathematical Physics*" or "*Named Functions*" (we call them also "*Classical Special Functions*") were published. We mention only some of them in this survey.

As Richard Askey (*to whose memory we dedicate this survey*) confessed in his lectures [2] on orthogonal polynomials and special functions: "Now, there are relatively large number of people who know a fair amount about this topic. Nevertheless, ...most of the mathematicians are totally unaware of the power of the special functions. They react to a paper which contains a Bessel function or Legendre polynomial by turning immediately to the next paper", and continued: "Hopefully these lectures will show ... how useful hypergeometric functions can be. Very few facts about them are known, but these few facts can be very useful in many different contexts. So, my advice is to learn something about hypergeometric functions: or, if this seems too hard or dull a task, get to know someone who knows something about them. If you already know something about these functions, share your knowledge with a colleague or two, or a group of students. Every large university and research laboratory should have a person who not only finds things in the Bateman Project (i.e., [3]), but can fill in a few holes in this set of books ... In any case, I hope my point has been made; special functions are useful and those who need them and those who know them should start to talk to each other ... The mathematical community at large needs the education on the usefulness of special functions more than most other people who could use them ... ".

As a participant in the NATO International Conference on Special Functions and Applications 2000 (Arizona State University), the author had the chance to witness the late night long discussions (mainly between Richard Askey and Oleg Marichev) for the merits and competition of the two great projects on Special Functions, on which the Computer Algebra Systems packages *Mathematica* and *Maple* are based, the *Bateman Project* [3] and the *NIST Project* [4] based on the Abramowitz–Stegun handbook [5] and on a more recent one, edited by Olver–Lozier–Boisvert–Clark [6].

The author of this survey was tempted to start paying attention to Special Functions by having the handbook [3] on her desk, while working on a M.Sc. thesis. We cite some texts from the Preface of this encyclopedia book, known as the *Bateman Project*: "...During his last years he (Professor Harry Bateman) had embarked upon a project whose successful completion, he believed, would prove of great value to scientists in all fields. He planned an extensive compilation of "special functions"—solutions of a wide class of mathematically and physically relevant functional equations. He intended to investigate and to tabulate properties of such functions, inter-relations between such functions, their representations in various forms, their macro- and micro-scopic behavior, and to construct tables of important definite integrals involving such functions ... anyone who has been faced with the task of handling and discussing and understanding in detail the solution to an applied problem which is described by a differential equation is painfully familiar with the disproportionately large amount of scattered research on special functions one must wade through in the hope of extracting the desired information ... ". In the time of Bateman's death (1946) his notes amounted to a veritable mountain of paper. His card-catalogue alone filled several dozen cardboard boxes (the famous *"shoe-boxes"*). ..."Bateman planned his Project as a *'Guide to the Functions'* on a gigantic scale ... the great importance of such a work hardly needs emphasis ... (this) would have made this book as a kind of 'Greater Oxford Dictionary of Special Functions' (from the Introduction to [3])". This project resulted in publication of five important reference volumes ([3,7]), under the editorship of Arthur Erdélyi, in association with W. Magnus, F. Oberhettinger and F.G. Tricomi.

In 2007, the *Askey–Bateman Project* was announced by Mourad Ismail as a five- or six-volume encyclopedic book series on special functions, based on the works of both Harry Bateman and Richard Askey. Starting in 2020, Cambridge University Press began publishing Volumes 1 and 2 of this Encyclopedia of Special Functions with series editors Mourad Ismail and Walter Van Assche [8].

2. Preliminaries—Basic Definitions and Facts

We give here only a short background on the considered *Special Functions of Fractional Calculus (SF of FC)*. As for the standard special functions and same for the SF of FC, most of them are entire functions of the complex variable z or analytic ones in disks in $\mathbb{C}$. We skip the details on defining single-valued branches of the considered functions, functional spaces and operators' properties there (see our previous works, e.g., ref. [9] (§5.5.i)). In addition, we limit ourselves to the Fox H-functions of one complex variable, as enough general level to expose our approach and results.

Among the long list of handbooks and surveys dedicated not only to classical SF but also to the SF of FC, we mention only few of them *of the few decades*: Mathai–Saxena [10], 1973; Marichev [11], 1978; Srivastava–Gupta–Goyal [12], 1982; Srivastava–Kashyap [13], 1982; Prudnikov–Brychkov–Marichev [14], 1992; Kiryakova [9], 1994; Yakubovich–Luchko [15], 1994; Podlubny [16], 1999; Kilbas–Saigo [17], 2004; Kilbas–Srivastava–Trujillo [18], 2006; Mathai–Haubold [19], 2008; Mathai–Saxena–Haubold [20], 2010; Mainardi [21], 2010; Gorenflo–Kilbas–Mainardi–Rogosin [22], 2014–2020; the recent ones as Cohl–Ismail [23], 2020; Assche–Ismail [8], 2020; Mainardi [24], 2020; etc. (see more sources also the survey paper Machado-Kiryakova [25]). The basic classes of SF considered here are shortly discussed below.

Definitions of the Basic Special Functions

We refer to the survey by Mainardi–Pagnini [26] that points out the pioneering role of Salvatore Pincherle on developing the generalized hypergeometric functions (and, thus, later appearing G-functions) by means of Mellin–Barnes integrals, where a historical note from the Bateman Project [3] (Vol. 1, p. 49) is cited: "... Of all integrals which contain Gamma functions in their integrands the most important ones are the so-called *Mellin-Barnes integrals*. Such integrals were first introduced by S. Pincherle, in 1888; their theory has been developed in 1910 by H. Mellin ... and they were used for a complete integration of the hypergeometric differential equation by E.W. Barnes, 1908."

Definition 1. (Ch. Fox [27], 1961, see books as [9,12,14,18], and other earlier and latest ones) *The Fox H-function is a generalized hypergeometric function, defined by means of the Mellin–Barnes type contour integral*

$$H_{p,q}^{m,n}\left[z \left| \begin{array}{c} (a_i, A_i)_1^p \\ (b_j, B_j)_1^q \end{array} \right. \right] = \frac{1}{2\pi i} \int_{\mathcal{L}} \mathcal{H}_{p,q}^{m,n}(s) z^{-s} ds, \text{ with } \mathcal{H}_{p,q}^{m,n}(s) = \frac{\prod_{j=1}^{m} \Gamma(b_j + B_j s) \prod_{i=1}^{n} \Gamma(1 - a_i - A_i s)}{\prod_{j=m+1}^{q} \Gamma(1 - b_j - B_j s) \prod_{i=n+1}^{p} \Gamma(a_i + A_i s)}, \quad (1)$$

with complex variable $z \neq 0$ and a contour $\mathcal{L}$ in the complex domain; the orders (m, n, p, q) are non-negative integers so that $0 \leq m \leq q$, $0 \leq n \leq p$, the parameters $A_i > 0, B_j > 0$ are positive and $a_i, b_j, i = 1, \ldots, p; j = 1, \ldots, q$ are arbitrary complex such that $A_i(b_j + l) \neq B_j(a_i - l' - 1), l, l' = 0, 1, 2, \ldots; i = 1, \ldots, n; j = 1, \ldots, m$. Note that the integrand $\mathcal{H}_{p,q}^{m,n}(s)$ with $s \mapsto -s$ is the Mellin transform of the H-function (1).

The details on the properties of the Fox H-function and types of contour $\mathcal{L}$ can be found in many contemporary handbooks on SF as [12,14,18], where its behavior is described in terms of the following parameters:

$$\rho = \prod_{i=1}^{p} A_i^{-A_i} \prod_{j=1}^{q} B_j^{B_j} \ ; \ \Delta = \sum_{j=1}^{q} B_j - \sum_{i=1}^{p} A_i; \ \gamma = \lim_{s \to \infty, s \in \mathcal{L}_{i\infty}} \text{Re } s,$$
$$\mu = \sum_{j=1}^{q} b_j - \sum_{i=1}^{p} a_i + \frac{p-q}{2} \ ; \ a^* = \sum_{i=1}^{n} A_i - \sum_{i=n+1}^{p} A_i + \sum_{j=1}^{m} B_j - \sum_{j=m+1}^{q} B_j. \quad (2)$$

Depending on the values in (2), the H-function is a function analytic of z in disks $|z| < \rho$ or outside them, in some sectors, or in the whole complex plane. In particular, the integral (1) converges (see [14] (§8.3)), if one of the following conditions is satisfied: (1) $\mathcal{L} = \mathcal{L}_{i\infty}$: $a^* > 0$, $|\arg z| < a^*\pi/2$; (2) $\mathcal{L} = \mathcal{L}_{i\infty}$: $a^* \geq 0$, $|\arg z| = a^*\pi/2$, $\gamma\Delta < -1 - \text{Re}\,\mu$; (3) $\mathcal{L} = \mathcal{L}_{-i\infty}$: $\Delta > 0, 0 < |z| < \infty$, or $\Delta = 0, 0 < |z| < \rho$, or $\Delta = 0, a^* \geq 0, |z| = \rho, \text{Re}\,\mu < 0$; or (4) $\mathcal{L} = \mathcal{L}_{+i\infty}$: $\Delta < 0, 0 < |z| < \infty$, or $\Delta = 0, |z| > \rho$, or $\Delta = 0, a^* \geq 0, |z| = \rho, \text{Re}\,\mu < 0$. The contour $\mathcal{L}_{-i\infty}$ (respectively, $\mathcal{L}_{+i\infty}$) is a left (respectively, right) loop in some horizontal strip that begins the point $-\infty + i\varphi_1$ (respectively, $+\infty + i\varphi_1$) keeping all poles of the functions $\Gamma(b_j + B_j s), j = 1, 2, ..., m$ on the left side, and those of $\Gamma(1 - a_i - A_i s), i = 1, 2, ..., n$

on its right side, and ends at the point $-\infty + i\varphi_2$ (respectively, $+\infty + i\varphi_2$), where $\varphi_1 < \varphi_2$. The contour $\mathcal{L}_{i\infty}$ starts at the point $\gamma - i\infty$ and ends at $\gamma + i\infty$ in a way to separate the mentioned poles, same as for $\mathcal{L}_{\pm i\infty}$.

For studies on the behavior of the H-function around the singular points, one can see also the work of Karp [28], commenting and revisiting the results of Braaksma [29].

If all $A_i = B_j = 1$, $i = 1, ..., p$; $j = 1, ..., q$, the H-function $H_{p,q}^{m,n}\left[z \left| \begin{array}{c} (a_i, 1)_1^p \\ (b_j, 1)_1^q \end{array}\right.\right]$ reduces to the *Meijer G-function* (C.S. Meijer [30], 1936–1941; see details in [3] (Vol. 1) and all above-mentioned books):

$$G_{p,q}^{m,n}\left[z \left| \begin{array}{c} (a_i)_1^p \\ (b_j)_1^q \end{array}\right.\right] = \frac{1}{2\pi i}\int_{\mathcal{L}} \mathcal{G}_{p,q}^{m,n}(s) z^{-s} ds$$

$$= \frac{1}{2\pi i}\int_{\mathcal{L}} \frac{\prod_{j=1}^{m}\Gamma(b_j + s)\prod_{i=1}^{n}\Gamma(1 - a_i - s)}{\prod_{j=m+1}^{q}\Gamma(1 - b_j - s)\prod_{i=n+1}^{p}\Gamma(a_i + s)} z^{-s} ds, \quad z \neq 0. \quad (3)$$

In this case, the behavior of the function (3) depends on conditions (2) with $\rho = 1$, $\Delta = q - p$, $\delta = m + n - \frac{p+q}{2}$. Although simpler than (1), the G-function is yet enough general as it incorporates most of the Classical SF (known also as Named SF) and many elementary functions (see lists of examples in [3] (Vol. 1), [9] (Appendix C)).

The basic SF of FC that are Fox H-functions but do *not* reduce to Meijer G-functions in the general case (of *irrational* A_j, B_k) are the following generalized hypergeometric functions, extending the more popular $_pF_q$-functions.

Definition 2. (see [9,14,22]) *The Wright generalized hypergeometric function $_p\Psi_q(z)$, called also Fox–Wright function (abbrev. as F-W g.h.f. or Wright g.h.f.), is defined as:*

$$_p\Psi_q\left[\begin{array}{c} (a_1, A_1), ..., (a_p, A_p) \\ (b_1, B_1), ..., (b_q, B_q) \end{array}\bigg| z\right] = \sum_{k=0}^{\infty} \frac{\Gamma(a_1 + kA_1)...\Gamma(a_p + kA_p)}{\Gamma(b_1 + kB_1)...\Gamma(b_q + kB_q)} \frac{z^k}{k!} \quad (4)$$

$$= H_{p,q+1}^{1,p}\left[-z \left| \begin{array}{c} (1 - a_1, A_1), ..., (1 - a_p, A_p) \\ (0, 1), (1 - b_1, B_1), ..., (1 - b_q, B_q) \end{array}\right.\right]. \quad (5)$$

It was introduced and studied by Sir Edward Maitland (E.-M.) Wright in a series of his works (e.g., [31,32], pp. 1933–1940). In denotations for the parameters (2), the power series (4) defines an entire function of z if $\Delta > -1$; it is absolutely convergent in the disk $\{|z| < \rho\}$ for $\Delta = -1$; and it is the same for $|z| = \rho$ if $\text{Re}(\mu) > 1/2$, (see details, for example, in [33]).

When all $A_1 = \cdots = A_p = 1$, $B_1 = \cdots = B_q = 1$, the Wright g.h.f. reduces to the *generalized hypergeometric $_pF_q$-function* which itself is a case of the G-function (3) (for early details, see [3] (Vol. 1)):

$$_p\Psi_q\left[\begin{array}{c} (a_1, 1), ..., (a_p, 1) \\ (b_1, 1), ..., (b_q, 1) \end{array}\bigg| z\right] = c\, _pF_q(a_1, ..., a_p; b_1, ..., b_q; z) = \sum_{k=0}^{\infty} \frac{(a_1)_k ... (a_p)_k}{(b_1)_k ... (b_q)_k} \frac{z^k}{k!} \quad (6)$$

$$= G_{p,q+1}^{1,p}\left[-z \left| \begin{array}{c} 1 - a_1, ..., 1 - a_p \\ 0, 1 - b_1, ..., 1 - b_q \end{array}\right.\right];$$

where

$$c = \left[\prod_{i=1}^{p}\Gamma(a_i) \Big/ \prod_{j=1}^{q}\Gamma(b_j)\right], \quad (a)_k := \Gamma(a+k)/\Gamma(a).$$

In general (that is, except in certain integer values of parameters when the series terminates or fails to make sense), $_pF_q$ converges for all finite z if $p \leq q$, converges for

$|z| < 1$ if $p = q + 1$ and diverges for all $z \neq 0$ if $p > q + 1$. The simplest particular cases are the Gauss hypergeometric function $_2F_1$, the Kummer (confluent hypergeometric) function $_1F_1$ and the Bessel type functions $_0F_1$.

A very special and important case of SF of FC, as a H-function and also as a Wright $_p\Psi_q$-function, is the "Queen"-function of FC (see [34]), namely the *Mittag–Leffler (M-L) function*, which has recently enjoyed many extensions (along with many basic elementary and known SF as its particular cases) and wide applications in solutions of fractional order models. This is the topic of Sections 4 and 5.

3. On the Use of Some *G*- and *H*-Functions in Theory of Integral Transforms and Special Functions

The Meijer G-function includes most of elementary and special functions (the classical ones) as particular cases, one can find lists of these, say in [3] (Vol. 1), [9] (Appendix C), [11,14]. Naturally, the more general Fox H-function incorporates all cases of the G-functions, and much more the SF of FC. Here, we attract readers' attention to the use of *two basic classes of G- and H-functions* with specific orders: $G_{0,m}^{m,0}$, respectively, $H_{0,m}^{m,0}$ with $m = q$, $n = p = 0$; and $G_{m,m}^{m,0}$, respectively, $H_{m,m}^{m,0}$ with $m = p = q$, $n = 0$.

3.1. Use of G- and H-Functions as Kernels of Laplace Type Integral Transforms

The Laplace transform

$$\mathcal{L}\{f(t); s\} = \int_0^\infty \exp(-st) f(t) dt, \quad \operatorname{Re} s > \mu, \tag{7}$$

is usually considered for functions $f(t)$ of the form

$$\left\{ f(t) = t^p \widetilde{f}(t),\ p > -1, \widetilde{f} \in C[0, \infty);\ f(t) = O(\exp \mu t),\ t \to \infty, \mu \in \mathbb{R} \right\}.$$

Definition 3. *The G- and H-transforms (see, for example [35], also [15,36]) of the form*

$$\mathcal{G}\{f(t); s\} = \int_0^\infty G_{p,q}^{m,n}\left[st \,\bigg|\, \begin{matrix} (a_j)_1^p \\ (b_k)_1^q \end{matrix} \right] f(t) dt,\ \text{resp.}\ \mathcal{H}\{f(t); s\} = \int_0^\infty H_{p,q}^{m,n}\left[st \,\bigg|\, \begin{matrix} (a_j, A_j)_1^p \\ (b_k, B_k)_1^q \end{matrix} \right] f(t) dt,$$

are said to be generalized integral transforms of Laplace type when

$$\delta = m + n - \frac{p+q}{2} > 0,\ \text{resp.}\ a^* = \sum_{j=1}^n A_j - \sum_{j=n+1}^p A_j + \sum_{k=1}^m B_k - \sum_{k=m+1}^q B_k > 0,$$

and are considered in suitable functional spaces of "transformable" functions.

In 1958, Obrechkoff [37] introduced a far reaching generalization of the Laplace and Meijer transforms, particular cases of which were studied by many authors years later, mainly for the purposes of operational calculi for different classes of differential operators. His aims were to extend the theorem of S. Bernstein for absolutely monotonic functions representable by means of Laplace–Stieltjes transforms, when the conditions for nth derivatives are replaced by similar ones with more general differential operators. The *Obrechkoff transform* was defined as

$$\mathcal{F}(s) = \int_0^\infty \Phi(st) f(t) dt$$

with a kernel $\Phi(s)$ given by the integral representation

$$\Phi(s) = \int_0^\infty \cdots \int_0^\infty u_1^{\beta_1} \cdots u_p^{\beta_p} \exp\left(-u_1 - \cdots - u_p - \frac{s}{u_1 \ldots u_p} \right) du_1 \cdots du_p. \tag{8}$$

Later, in 1966, Dimovski [38] introduced a class of differential operators of Bessel type and of arbitrary integer order $m > 1$, called by the author (as for example in [9]) as *hyper-Bessel differential operators*. They have the alternative representations

$$Bf(t) = t^{\alpha_0}\frac{d}{dt}t^{\alpha_1}\frac{d}{dt}\cdots t^{\alpha_{m-1}}\frac{d}{dt}t^{\alpha_m}f(t)$$

$$= t^{-\beta}P_m\left(t\frac{d}{dt}\right)f(t) = t^{-\beta}\prod_{k=1}^{m}\left(t\frac{d}{dt} + \beta\gamma_k\right)f(t), \quad t > 0, \tag{9}$$

with arbitrary parameters $\alpha_0, \alpha_1, ..., \alpha_m$, $\beta := m - (\alpha_0 + \alpha_1 + ... + \alpha_m) > 0$, $\gamma_k := \frac{1}{\beta}(\alpha_k + \alpha_{k+1} + ... + \alpha_m)$, $k = 1, ..., m$, P_m a polynomial of degree m. Evidently for $m = \beta = 2$, $\gamma_{1,2} = \pm\frac{\nu}{2}$, one has the second-order Bessel differential operator B_ν with the Bessel function $y(t) = J_\nu(t)$ satisfying $B_\nu y(t) = -y(t)$. For other choices of parameters, many other particular differential operators appear in equations of mathematical physics, operational calculus and applied analysis. To combine the Mikusinski type algebraic approach to operational calculus for (9) with a transform method, Dimovski used a *modified Obrechkoff transform* (we shortly call it also Obrechkoff transform), defined as

$$\mathcal{O}\{f(t);s\} = \beta\int_0^\infty t^{\beta(\gamma_m+1)-1}K\left[(st)^\beta\right]f(t)dt = \beta\int_0^\infty \lambda(t,s)f(t)dt,$$

with the kernel-function

$$K(s) = \int_0^\infty\cdots\int_0^\infty \exp\left(-u_1 - ... - u_{m-1} - \frac{s}{u_1...u_{m-1}}\right)\prod_{k=1}^{m}u_k^{\gamma_m-\gamma_k-1}du_1...du_{m-1}. \tag{10}$$

In [9] (Ch.3), also in other works like [39], we proved that the kernel-functions (8) and (10) of the Obrechkoff transforms are representable as Meijer's $G_{0,m}^{m,0}$-functions, namely (for a proof see, e.g., Lemma 1 of [39]):

$$\Phi(s) = G_{0,p+1}^{p+1,0}\left[s\;\middle|\;\begin{array}{c}--\\(\beta_k+1)_1^p, 0\end{array}\right], \quad \lambda(t,s) = s^{-\beta(\gamma_m+1)+1}G_{m,m}^{m,0}\left[(st)^\beta\;\middle|\;\begin{array}{c}--\\(\gamma_k-\frac{1}{\beta}+1)_1^m\end{array}\right]. \tag{11}$$

Therefore, *the Obrechkoff transform appears to be a G-transform of Laplace type* (because $\delta = m/2 > 0$), and its theory has been further developed in whole details (convolution theorems, real and complex inversion formulas, images, examples, etc.) more easily by using the tools of the G-functions (see for example [9] (Ch.3), [39]).

Another not observed fact was that functions of the form of kernels (8) and (10) of the Obrehkoff transform were studied yet in 1937 by Erdélyi [40]. He might be the first who derived a relation between the $_0F_{m-1}$-functions (what we mention next as hyper-Bessel functions) and these kernel-functions (formula (7.4) in [40]). However, at the time of Erdélyi's work [40], 1937, the next step in introducing the G-functions had not yet been done by Meijer [30], 1946. Obrechkoff himself made no attempts to identify the kernel-function $\Phi(s)$ with some known special functions and studied its properties "ad hoc". Thus, the $G_{0,m}^{m,0}$-functions seemed to appear in use for the hyper-Bessel operators and related integral transforms in author's works since 1980 (see [9] (Ch.3), [41]).

Next, the *generalized Obrechkoff transform* (a *fractionalized* analog) was introduced and studied by Kiryakova [9] (Ch.5), Al-Mussalam–Kiryakova–Tuan [42] and Yakubovich–Luchko [15], with the Fox $H_{m,m}^{m,0}$-function as kernel:

$$\mathcal{B}(s) = \mathcal{B}_{(\rho_i),(\mu_i)}\{f(t);s\} = \int_0^\infty H_{0,m}^{m,0}\left[st\;\middle|\;\begin{array}{c}--\\(\mu_i-\frac{1}{\rho_i},\frac{1}{\rho_i})_1^m\end{array}\right]f(t)dt. \tag{12}$$

We call it as *multi-index Borel–Dzrbashjan transform*, because for $m = 1$ it is reduced to the *Borel transform*

$$\mathcal{B}_{(\rho),(\mu)}\{f(t);s\} = \int_0^\infty \exp(-s^\rho t^\rho)t^{\mu\rho-1}f(t)dt \tag{13}$$

whose kernel appears to be a $H_{0,1}^{1,0}$-function. This integral transform was shown by Dzrbashjan [43,44] to have inversion formula involving the Mittag–Leffler function $E_{1/\rho,\mu}$. The generalized Obrechkoff transform (12) is a tool in operational calculus for *fractional multi-order analogs of hyper-Bessel differential operators* (9), formally of the kind

$$D_{(\rho_i),(\mu_i)} f(t) = t^{-1} \prod_{i=1}^{m} \left(t^{1+(1-\mu_i)\rho_i} D_{t^{\rho_i}}^{1/\rho_i} t^{(\mu_i-1)\rho_i} \right) f(t), \tag{14}$$

in the same way as the Laplace transform, the Obrechkoff transform and its particular cases serve for the classical differentiation, respectively for the hyper-Bessel operators (9).

In the studies on these Laplace type G- and H-integral transforms, we used essentially the theory of the G- and H-functions, mainly of the cases of orders $(m,0;0,m)$. Note that, for example, $G_{0,m}^{m,0}(s)$ is an analytic function in the sector $|\arg s| < (m/2)\pi$ (where in this case $\delta = m/2 > 0$). Some additional necessary results on these G- and H- kernel functions were derived by Kiryakova [9] (Appendix), as Lemmas B.1–B.4, Corollaries B.5–B.7, Formula (E.21), etc.

From the known representations of some elementary and special functions in G- and H-terms, one observes many particular cases of simpler Laplace type integral transforms. Namely, the Laplace and Borel–Dzrbashjan transforms (7) and (13) are Obrechkoff transform (10) and multi-index Borel–Dzrbashjan transform (12), respectively, for $m = 1$, since

$$\exp(-s) = G_{0,1}^{1,0}\left[s \left|\begin{array}{c} -- \\ 0 \end{array}\right.\right], \quad \exp(-s^\rho t^\rho) = H_{0,1}^{1,0}\left[st \left|\begin{array}{c} -- \\ (\mu - \frac{1}{\rho}, \frac{1}{\rho}) \end{array}\right.\right].$$

For $m = 2$, we have the *classical Meijer transform* as a case of the Obrechkoff transform, related to the Bessel differential operator $B_\nu = \frac{d}{dt} t^{1-\nu} \frac{d}{dt} t^\nu$:

$$\mathcal{K}_\nu\{f(t);s\} = \int_0^\infty \sqrt{st}\, K_\nu(t)\, f(t) dt, \quad \text{since} \quad K_\nu(s) = \frac{1}{2} G_{0,2}^{2,0}\left[\frac{s^2}{4} \left|\begin{array}{c} -- \\ \frac{\nu}{2}, \frac{-\nu}{2} \end{array}\right.\right], \tag{15}$$

the kernel *Macdonald function* $K_\nu(s)$ has such a G-function representation.

In a series of papers [45,46], *Krätzel* introduced a generalization of the Meijer transform (again with $m = 2$), and further a more general one of the type of Obrechkoff transform for arbitrary integer $m > 1$,

$$\mathcal{L}_\nu^{(m)}\{f(t);s\} := \int_0^\infty \lambda_\nu^{(m)}[m(st)^{1/m}]\, f(t) dt := \int_0^\infty \Lambda(s,t)\, f(t) dt. \tag{16}$$

He used the transformation (16) for operational calculus for the following (hyper-Bessel type) operator or order $m > 1$:

$$B_\nu^{(m)} = \frac{d}{dt} t^{\frac{1}{m}-\nu} \left(t^{1-\frac{1}{m}} \frac{d}{dt} \right)^{m-1} t^{\nu+1-\frac{2}{m}} \quad \text{with} \quad \beta = 1, \gamma_1 = 0, \gamma_k = \nu + \frac{k-2}{m}, k = 2,...,m. \tag{17}$$

As expected, we can represent the Krätzel kernel in terms of the G-function corresponding to (11):

$$\Lambda(s,t) = \int_0^\infty \cdots \int_0^\infty \left[\prod_{k=1}^{m-1} u_k^{\nu-1+\frac{k-1}{m}} \right] \exp\left(-u_1 - ... - u_{m-1} - \frac{st}{u_1...u_{m-1}} \right) du_1...du_{m-1}$$

$$= s^{-\nu-1+\frac{2}{m}} G_{0,m}^{m,0}\left[st \left|\begin{array}{c} -- \\ 0, (\nu + \frac{k-2}{m})_2^m \end{array}\right.\right]. \tag{18}$$

Krätzel started from the simple case $m = 2$ with a kernel of the form $\int_0^\infty u^{\gamma-1} \exp(-u - st/u) du$ (with some variations as $t \mapsto t^\rho, \rho > 0$), close to the Macdonald function (15), which is often called as the *Krätzel function*. Then, many other authors continued to study it and established its relations to hypergeometric functions. We can refer to such works by Kilbas–Saxena–Trujillo [47], Mathai–Haubold [48], etc. In a paper by Glaeske–Kilbas–Saigo [49], a fractionalized analog of the Krätzel transform (7) was introduced, where instead of integer $m > 1$ in the transformation (16), they took a (fractional) parameter $\rho > 0$. Then, naturally, its kernel is represented as a H-function (due to some variations in the definition, it appears as $H_{1,2}^{2,0}$ instead of $H_{0,2}^{2,0}$). Relations with operators of fractional calculus are considered, but one should mention that such an integral transform is analog of the generalized Obrechkoff transform (12) for a fractional order differential operator of the form (14). In all these mentioned cases, the kernel functions have the form of (8) as also studied earlier by Erdélyi [40]. We conclude here the list of cases of the Obrechkoff transform with emphasize on the works by *Ditkin–Prudnikov* (as [50]) on operational calculi for (hyper-Bessel) operators of the form

$$B_1 = \frac{d}{dt} t \frac{d}{dt}, \quad \text{and more generally,} \quad B_m = \frac{d}{dt} t \frac{d}{dt} t \frac{d}{dt} \cdots \frac{d}{dt} = t^{-1} \left(t \frac{d}{dt} \right)^m, \quad m \geq 2. \quad (19)$$

For $m = 2$, the corresponding integral transform is a variant of the Meijer transform (with $\nu = 0$), and in the general case $m > 1$, Ditkin and Prudnikov [50] made use of an integral transform of the form

$$\mathcal{B}\{f(t); s\} = 2 \int_0^\infty E_{0m}(st) f(t) dt, \text{ where we can represent the kernel } E_{0m} \text{ as a } G_{0,m}^{m,0} - \text{function.}$$

For more details on the Obrechkoff type transforms with kernels $G_{0,m}^{m,0}$ and $H_{0,m}^{m,0}$, their properties, images and special cases, see Kiryakova [9] (Ch.3, Ch.5), [39].

3.2. Use of G- and H-Functions as Kernels in Generalized Fractional Calculus

For basic background on Fractional Calculus (FC) as theory of operators of integration and differentiation of arbitrary (fractional) order, and its closely related topics as special functions (SF) and integral transforms, we refer to the books by Samko–Kilbas–Marichev [51], Podlubny [16], Kilbas–Srivastava–Trujillo [18], and Yakubovich–Luchko [15], as well as one by the author [9], among many others. For wider list, see, for example, Machado–Kiryakova [25]. In our works, and mainly for the needs of the SF theory, we consider the Riemann–Liouville (R-L) type integrals and their corresponding derivatives of R-L and Caputo type, respectively, their generalizations involving G- and H-functions in the kernels. Note that we concentrate on the left-hand side variants and skip details (in most cases being similar) for the Weyl-type, right-hand sided operators.

The basic tools in our studies are the fractional integration operators of the form $\tilde{I} f(z) = z^{\delta_0} I_\beta^{\gamma,\delta} f(z)$, $\delta_0 \geq 0$, to which we refer as "classical fractional integrals", where

$$I_\beta^{\gamma,\delta} f(z) = \frac{1}{\Gamma(\delta)} \int_0^1 (1-\sigma)^{\delta-1} \sigma^\gamma f(z\sigma^{\frac{1}{\beta}}) d\sigma = \frac{z^{-\beta(\gamma+\delta)}}{\Gamma(\delta)} \int_0^z (z^\beta - \xi^\beta)^{\delta-1} \xi^{\beta\gamma} f(\xi) d(\xi^\beta), \quad (20)$$

is the *Erdélyi-Kober operator (E-K) of integration* of order $\delta \geq 0$, depending on two additional parameters $\gamma \in \mathbb{R}, \beta > 0$. In this general form, it is introduced in Sneddon's works [52] and considered in some books (for example, [9] (Ch.2), [15,18,51]). The earlier versions with $\beta = 1, \beta = 2$ are due to Kober and Erdélyi. The R-L operator of integration R^δ appears as a case with one parameter only, for $\gamma = 0, \beta = 1, \delta_0 = \delta \geq 0$,

$$R_{0+,z}^\delta f(z) := R^\delta f(z) = z^\delta I_1^{0,\delta} f(z); \quad \text{conversely,} \quad I_1^{\gamma,\delta} f(z) = z^{-\gamma-\delta} R^\delta z^\gamma f(z). \quad (21)$$

The *E-K fractional derivative* $D_\beta^{\gamma,\delta}$, corresponding to (20), is defined explicitly almost simultaneously in the works of Kiryakova [9] (Ch.2) and Yakubovich–Luchko [15] (Ch.3). It serves as an interpretation of the formal inversion formula $\left\{ I_\beta^{\gamma,\delta} \right\}^{-1} = I_\beta^{\gamma+\delta,-\delta}$, namely:

$$D_\beta^{\gamma,\delta} f(z) = D_n I_\beta^{\gamma+\delta,n-\delta} f(z) = \prod_{j=1}^n \left(\frac{1}{\beta} z \frac{d}{dz} + \gamma + j \right) I_\beta^{\gamma+\delta,n-\delta} f(z), \quad n-1 < \delta \leq n, n \in \mathbb{N}. \quad (22)$$

Here, the simplest integer order derivative $(d/dz)^n$ in the definition of the R-L fractional derivative $D^\delta f(z) := \left(\dfrac{d}{dz}\right)^n R^{n-\delta} f(z)$, is replaced by an auxiliary differential operator D_n of integer order, a polynomial of $(z\, d/dz)$. The *Caputo-type R-L and E-K fractional derivatives* are defined in the same way but with exchanged order of the nonnegative order integration and the integer order differentiation (see, e.g., [53]).

The notion for *generalized operators of fractional integration* was introduced by Kalla in his 1969–1979 works (see the survey [54]), who suggested their common form

$$If(z) = \int_0^1 \Phi(\sigma)\, \sigma^\gamma f(z\sigma) d\sigma = z^{-\gamma-1} \int_0^z \Phi(\tfrac{\xi}{z}) \xi^\gamma f(\xi) d\xi,$$

where $\Phi(\sigma)$ can be an arbitrary continuous (analytical) function for which the integral makes sense. The idea of such generalized fractional calculus is *to replace the elementary function in the kernel* of R-L and E-K operators (20) (and, say, the logarithmic kernel in the Hadamard integral) *by some special function*. Variants with the Gauss-, Bessel-, Whittaker-, arbitrary G- and H-functions appeared in papers of several authors (see historical details and references in [54,55]). If such a special function Φ is taken to be too general or too specific, only some formal operational rules for the corresponding fractional calculus can be derived. The lucky hint in our studies was to choose suitably the *kernel-functions* Φ to be of the form of $G_{m,m}^{m,0}$- and $H_{m,m}^{m,0}$-*functions*. Then, the operators of the generalized fractional calculus happen to be also commutative products of classical operators of FC, namely of finite number of Erdélyi-Kober operators. Thus, the tools of the special functions and the wide use of the classical FC are combined into a *Generalized Fractional Calculus (GFC)* in Kiryakova [9], with developed full theory and many illustrated applications in different areas of analysis, differential equations, special functions and integral transforms. Below, we briefly review the basic definitions and few results on this GFC.

Definition 4. (Kiryakova, [9] (Ch.5)) *We define the multiple E-K integral (of multiplicity $m > 1$), by means of the real parameters' sets $(\delta_1 \geq 0, ..., \delta_m \geq 0)$ (multi-order of integration) $(\gamma_1, ..., \gamma_m)$ (multi-weight) and $(\beta_1 > 0, ..., \beta_m > 0)$ (additional multi-parameter), as:*

$$I_{(\beta_k),m}^{(\gamma_k),(\delta_k)} f(z) := \int_0^1 H_{m,m}^{m,0}\left[\sigma \,\middle|\, \begin{array}{c} (\gamma_k + \delta_k + 1 - \frac{1}{\beta_k}, \frac{1}{\beta_k})_1^m \\ (\gamma_k + 1 - \frac{1}{\beta_k}, \frac{1}{\beta_k})_1^m \end{array}\right] f(z\sigma) d\sigma, \qquad (23)$$

if $\sum_{k=1}^m \delta_k > 0$; and as the identity operator: $I_{(\beta_k),m}^{(\gamma_k),(0,...,0)} f(z) = f(z)$, *if $\delta_1 = \delta_2 = \cdots = \delta_m = 0$.*

It is important to mention that, for the particular conditions (2), the above kernel $H_{m,m}^{m,0}$-function is analytic function in the unit disk and $H_{m,m}^{m,0}(\sigma) \equiv 0$ for $|\sigma| > 1$ (Kiryakova, ref. [9]).

In the case of all equal βs: $\beta_1 = \beta_2 = ... = \beta_m =: \beta > 0$, integral (23) has a simpler form with a *Meijer* $G_{m,m}^{m,0}$-*function* ([9] (Ch.1)), which is also analytic in unit disk and $G_{m,m}^{m,0}(\sigma) \equiv 0$ for $|\sigma| > 1$,

$$I_{(\beta,...,\beta),m}^{(\gamma_k),(\delta_k)} f(z) := I_{\beta,m}^{(\gamma_k),(\delta_k)} f(z) = \int_0^1 G_{m,m}^{m,0}\left[\sigma \,\middle|\, \begin{array}{c} (\gamma_k + \delta_k)_1^m \\ (\gamma_k)_1^m \end{array}\right] f(z\sigma^{1/\beta}) d\sigma = \left[\prod_{k=1}^m I_\beta^{\gamma_k,\delta_k}\right] f(z). \qquad (24)$$

In both cases of (23) and (24), the operators of the form

$$\widetilde{I} f(z) = z^{\delta_0} I_{(\beta_k),m}^{(\gamma_k),(\delta_k)} f(z), \quad \widetilde{I} f(z) = z^{\delta_0} I_{\beta,m}^{(\gamma_k),(\beta_k)} f(z), \quad \text{with} \quad \delta_0 \geq 0, \qquad (25)$$

are called *generalized fractional integrals* of *multi-order* $(\delta_1, ..., \delta_m)$.

The important decomposition property (for proof, see for example, [9] (Th.1.2.10, Th.5.2.1), says that the same GFC integrals (23) and (24) can be represented, instead of using the kernel H- and G-functions, by *repeated integral representations for the commutative product of classical E-K operators* (20):

$$I_{(\beta_k),m}^{(\gamma_k),(\delta_k)} f(z) := \left[\prod_{k=1}^{m} I_{\beta_k}^{\gamma_k,\delta_k}\right] f(z)$$

$$= \int_0^1 \cdots \int_0^1 \left[\prod_{k=1}^{m} \frac{(1-\sigma_k)^{\delta_k-1}\sigma_k^{\gamma_k}}{\Gamma(\delta_k)}\right] f\left(z\sigma_1^{1/\beta_1}\ldots\sigma_m^{1/\beta_m}\right) d\sigma_1 \ldots d\sigma_m. \qquad (26)$$

In the book [9] and subsequent papers, we provided a full set of operational properties of the operators (23) and (24) that justify their names as operators of GFC, as *semigroup property, formal inversion formula,* reduction to identity or to the conventional integration operators for special parameters' choice.

Analogously to the R-L and E-K fractional derivatives, we define the corresponding *generalized fractional derivatives*. The auxiliary differential operator D_η is chosen on the base of the *specific differential relations for the kernel function,* derived for the G-functions, and especially for $G_{m,m}^{m,0}$ by Kiryakova [9] (App., Lemmas B.3, B.4, Cor. B.6) and for $H_{m,m}^{m,0}$ by Kiryakova [9] (Ch.5, Lemma 5.1.7)and Kiryakova–Luchko [53] (Lemma 18).

Definition 5. (Kiryakova [9]) *Let D_η be the following polynomial of $z\left(\frac{d}{dz}\right)$ of degree $\eta_1 + \ldots + \eta_m$:*

$$D_\eta = \left[\prod_{r=1}^{m}\prod_{j=1}^{\eta_r}\left(\frac{1}{\beta_r}z\frac{d}{dz}+\gamma_r+j\right)\right], \text{ with } \eta_k := \begin{cases} [\delta_k]+1, & \text{for noninteger } \delta_k, \\ \delta_k, & \text{for integer } \delta_k, \end{cases} k=1,\ldots,m. \qquad (27)$$

The multiple (m-tuple) Erdélyi–Kober fractional derivative of R-L type of multi-order ($\delta_1 \geq 0,\ldots,\delta_m \geq 0$) is defined by means of the differ-integral operator:

$$D_{(\beta_k),m}^{(\gamma_k),(\delta_k)} f(z) := D_\eta I_{(\beta_k),m}^{(\gamma_k+\delta_k),(\eta_k-\delta_k)} f(z) = D_\eta \int_0^1 H_{m,m}^{m,0}\left[\sigma \left| \begin{array}{c} (\gamma_k+\eta_k+1-\frac{1}{\beta_k},\frac{1}{\beta_k})_1^m \\ (\gamma_k+1-\frac{1}{\beta_k},\frac{1}{\beta_k})_1^m \end{array}\right.\right] f(z\sigma)\, d\sigma. \qquad (28)$$

Similarly, the Caputo-type generalized fractional derivative was introduced by Kiryakova and Luchko [53], as

$$^*D_{(\beta_k),m}^{(\gamma_k),(\delta_k)} f(z) = I_{(\beta_k),m}^{(\gamma_k+\delta_k),(\eta_k-\delta_k)} D_\eta f(z). \qquad (29)$$

In the case $\beta_1 = \ldots = \beta_m := \beta > 0$, simpler representations involving the Meijer G-function hold for the R-L and Caputo-type "derivatives" which correspond to the generalized fractional integral (24):

$$D_{\beta,m}^{(\gamma_k),(\delta_k)} f(z) = D_\eta I_{\beta,m}^{(\gamma_k+\delta_k),(\eta_k-\delta_k)} f(z) = \left[\prod_{r=1}^{m}\prod_{j=1}^{\eta_r}\left(\frac{1}{\beta}z\frac{d}{dz}+\gamma_r+j\right)\right] I_{\beta,m}^{(\gamma_k+\delta_k),(\eta_k-\delta_k)} f(z),$$

$$^*D_{\beta,m}^{(\gamma_k),(\delta_k)} f(z) = I_{\beta,m}^{(\gamma_k+\delta_k),(\eta_k-\delta_k)} D_\eta f(z). \qquad (30)$$

More generally, the differ-integral/integro-differential operators of the form

$$\widetilde{D}f(z) = D_{(\beta_k),m}^{(\gamma_k),(\delta_k)} z^{-\delta_0} f(z) = z^{-\delta_0} D_{(\beta_k),m}^{(\gamma_k-\frac{\delta_0}{\beta}),(\delta_k)} f(z), \text{ and}$$

$$\widetilde{^*D}f(z) = {^*D}_{(\beta_k),m}^{(\gamma_k),(\delta_k)} z^{-\delta_0} f(z) \text{ with } \delta_0 \geq 0, \qquad (31)$$

are all called *generalized (multiple, multi-order) fractional derivatives* (of R-L or Caputo type).

Next, in Section 8, we often use also the notion of (generalized) fractional *differintegrals*. We have in mind either (generalized) fractional integrals or derivatives or compositions of some E-K fractional integrals and some E-K fractional derivatives. These appear as meanings of operators (26) when part of the order's δs are non-negative and the other parts are negative.

For the functional spaces (here, we mainly limit to weighted analytical functions of complex z), mapping properties, long list of operational properties, images, etc., we refer, for example, to the work of Kiryakova [9,53,56].

We use also a further extension of the generalized fractional integrals (23), based on the so-called *Wright–Erdélyi–Kober (W-E-K) operator of fractional integration* (see [57]), with parameters as in E-K integral: $\delta \geq 0$, γ real, $\beta > 0$ and additional parameter $\lambda > 0$, where the Wright–Bessel (Bessel-Maitland) function of the form J_ν^μ (see (57)) in Section 5 is used in the kernel:

$$W_{\beta,\lambda}^{\gamma,\delta} f(z) := I_{\beta,\lambda,1}^{\gamma,\delta} f(z) = \lambda \int_0^1 \sigma^{\lambda(\gamma+1)-1} J_{\gamma+\delta-\lambda(\gamma+1)/\beta}^{-\lambda/\beta}(\sigma^\lambda f(z\sigma) d\sigma. \tag{32}$$

One can show that, for $\lambda = \beta$, the above kernel-function reduces to the kernel of the E-K operator, therefore the W-E-K integration becomes the E-K one. Using *compositions of W-E-K operators* (32), Kalla and Galue [57] tried to develop a next step in the generalized fractional calculus with $H_{m,m}^{m,0}$ kernel-functions that have the same structure but different parameters β_ks and λ_ks in upper and low rows. Some revisions and properties of these operators were further provided by Kiryakova [58–60].

Definition 6. *For integer $m \geq 1$ and real parameters $\delta_k \geq 0$, γ_k, $\beta_k > 0$, $\lambda_k > 0$, $\beta_k \geq \lambda_k$, $k = 1, ..., m$, we define the multiple Wright–Erdélyi–Kober (W-E-K) fractional integrals, as follows:*

$$\widetilde{I}f(z) = I_{(\beta_k),(\lambda_k),m}^{(\gamma_k),(\delta_k)} f(z) := \int_0^1 H_{m,m}^{m,0}\left[\sigma \left|\begin{array}{c}(\gamma_i+\delta_i+1-\frac{1}{\beta_i},\frac{1}{\beta_i})_1^m\\(\gamma_i+1-\frac{1}{\lambda_i},\frac{1}{\lambda_i})_1^m\end{array}\right.\right] f(z\sigma) d\sigma = \left[\prod_{k=0}^m W_{\beta_k,\lambda_k}^{\gamma_k,\delta_k}\right] f(z), \tag{33}$$

if $\sum_{i=1}^m \delta_i > 0$; and as the identity operator: $\widetilde{I}f(z) = f(z)$, when $\delta_1 = \delta_2 = ... = \delta_m = 0$ and $\lambda_k = \beta_k, k = 1, ..., m$. For $\gamma_k > -1, k = 1, ..., m$ and the above-mentioned conditions on the other parameters, the operators (33) are shown to preserve the space of analytic functions in disks or in starlike complex domains.

If $\beta_k = \lambda_k, k = 1, ..., m$, the "new" operators of GFC (33) coincide with operators (23). The corresponding generalized fractional derivatives $D_{(\beta_k),(\lambda_k),m}^{(\gamma_k),(\delta_k)}$ are defined by means of differential-integral operators similar to those for (28).

Here, we mention some few of the numerous *special cases of the above defined GFC operators*, to emphasize the particular elementary and special functions appearing in their kernels, and thus as *cases of the kernel $H_{m,m}^{m,0}$- and $G_{m,m}^{m,0}$-functions*.

For $\underline{m = 1}$, we have the kernel-functions:

$$H_{1,1}^{1,0}\left[\sigma \left|\begin{array}{c}(\gamma+\delta,1/\beta)\\(\gamma,1/\beta)\end{array}\right.\right] = \beta \sigma^{\beta-1} G_{1,1}^{1,0}\left[\sigma^\beta \left|\begin{array}{c}\gamma+\delta\\\gamma\end{array}\right.\right] = \beta \frac{\sigma^{\beta\gamma+\beta-1}(1-\sigma^\beta)^{\delta-1}}{\Gamma(\delta)}, \tag{34}$$

thus the generalized fractional integrals and derivatives (23) and (28) reduce to the corresponding E-K (20) and (22) and R-L operators (21): $I_{\beta,1}^{\gamma,\delta} = I_\beta^{\gamma,\delta}$, $D_{\beta,1}^{\gamma,\delta} = D_\beta^{\gamma,\delta}$, R^δ and D^δ. Many other integration and differentiation operators introduced and used by different authors appear as their special cases.

In the case $\underline{m = 2}$, the kernel functions $H_{2,2}^{2,0}$ and $G_{2,2}^{2,0}$ reduce to a *Gauss hypergeometric function* or its variations, for example:

$$H_{2,2}^{2,0}\left[\sigma \left|\begin{array}{c}(\gamma_1+\delta_1+1-\frac{1}{\beta},\frac{1}{\beta}),(\gamma_2+\delta_2+1-\frac{1}{\beta},\frac{1}{\beta})\\(\gamma_1+1-\frac{1}{\beta},\frac{1}{\beta}),(\gamma_2+1-\frac{1}{\beta},\frac{1}{\beta})\end{array}\right.\right] = G_{2,2}^{2,0}\left[\sigma^\beta \left|\begin{array}{c}\gamma_1+\delta_1,\gamma_2+\delta_2\\\gamma_1,\gamma_2\end{array}\right.\right]$$

$$= \frac{\sigma^{\beta\gamma_2}(1-\sigma^\beta)^{\delta_1+\delta_2-1}}{\Gamma(\delta_1+\delta_2)} \,_2F_1(\gamma_2+\delta_2-\gamma_1,\delta_1;\delta_1+\delta_2;1-\sigma^\beta). \tag{35}$$

Therefore, the generalized fractional integrals in this case are known as *hypergeometric fractional integrals*; some of them were introduced and studied by, e.g. Love, Saxena, Saigo and Hohlov (see [54]).

For $\underline{m = 3}$, we have as special case the *Marichev–Saigo–Maeda (M-S-M) operators* of FC, the integration operators introduced and studied by Marichev (1974) and Saigo et al. (1996, 1998) (see [55]). This is because their kernel-function, the *Appel F_3 function (Horn function)*

$$F_3(a, a', b, b', c, z, \xi) = \sum_{m,n=0}^{\infty} \frac{(a)_m (a')_n (b)_m (b')_n}{(c)_{m+n}} \frac{z^m \xi^n}{m! n!}, \quad |z| < 1, |\xi| < 1 (\text{see, e.g., } [3,14]),$$

is a case of the GFC kernel-functions $H_{3,3}^{3,0}$ and $G_{3,3}^{3,0}$ (see, for example, [14], §8.4.51, (2)):

$$\frac{(1-\sigma)^{c-1}}{\Gamma(c)} F_3\left(a, a', b, b', c, 1 - \frac{1}{\sigma}, 1 - \sigma\right)$$

$$= G_{3,3}^{3,0}\left[\sigma \left| \begin{array}{c} a+b, c-a', c-b' \\ a, b, c-a'-b' \end{array} \right.\right] = H_{3,3}^{3,0}\left[\sigma \left| \begin{array}{c} (a+b, 1), (c-a', 1), (c-b', 1) \\ (a, 1), (b, 1), (c-a'-b', 1) \end{array} \right.\right], \quad \text{Re } c > 0. \quad (36)$$

Let $\underline{m \geq 1}$ be an arbitrary integer, but all δs be equal integers, say $\delta_1 = ... = \delta_m = 1$. Then, from (24), we obtain the *hyper-Bessel integral operators L* (we denote below their kernel by G_1) that correspond to the hyper-Bessel differential operators (9) of arbitrary (higher) integer order $m > 1$. In practice, these are operators of integer multi-orders $(1, 1, ..., 1)$, but their fractional powers L^λ, $\lambda > 0$ have been represented (Kiryakova [9,41]) as GFC integrals of multi-order $(\lambda, \lambda, ..., \lambda)$ with kernels G_λ, where

$$G_1(\sigma) = G_{m,m}^{m,0}\left[\sigma \left| \begin{array}{c} (\gamma_k + 1)_1^m \\ (\gamma_k)_1^m \end{array} \right.\right], \quad G_\lambda(\sigma) = G_{m,m}^{m,0}\left[\sigma \left| \begin{array}{c} (\gamma_k + \lambda)_1^m \\ (\gamma_k)_1^m \end{array} \right.\right].$$

The kernel of L^λ in the form G_λ appeared also in the work of McBride [61]. These expressions gave us the hint how to introduce our GFC, replacing $(\lambda, \lambda, ..., \lambda)$ by arbitrary fractional multi-order $(\delta_1, \delta_2, ..., \delta_m)$, explanations are in [41]. We can mention also the Gelfond–Leontiev [62] operator (47) generated by the *multi-index M-L functions* (see next section and the works by Kiryakova [63,64]), as a more general example of operators of fractional multi-order where the Fox $H_{m,m}^{m,0}$-functions serve as kernels.

The H-functions of the form $H_{p,q}^{q,0}$, of which the kernel functions of (23) are cases with $p = q = m$, were studied in series of papers by Karp. In [28], he revisited the Braaksma results [29] for the H-function's behavior in the neighborhood of the singular points and its analytical continuation. There he commented also works on applications of H-functions not only in fractional calculus, but also widely in statistics, including the book by Mathai–Saxena–Haubold [20].

In relation to the *use of the $G_{m,m}^{m,0}$-functions* (the kernel-functions of GFC integrals (24)) in applications to statistics, it is interesting to note that, in 1958, *Kabe* [65] explored them *in statistics, as density functions of a random variable*. He also distinguished the cases $m = 1$ and $m = 2$ (mentioned above) related to the kernel-functions of the E-K and of the hypergeometric fractional integrals, respectively, (34) and (35). Studies on the closely related $G_{m+1,m+1}^{m,1}$-functions as R-L integrals of $G_{m,m}^{m,0}$ can be found in the work by Karp [66].

4. Mittag–Leffler Functions and Their Extensions

The *Mittag–Leffler (M-L) function* $E_\alpha(z)$ was introduced by G. Mittag–Leffler ([67], 1902–1905), extended to 2-parameters as $E_{\alpha,\beta}(z)$ by A. Wiman [68] and studied later by P. Humbert and R.P. Agarwal [69]. It was presented in the Bateman Project [3], Vol. 3 (1954), in a chapter for "Miscellaneous Functions". However, for long time, it was ignored in the other handbooks on special functions because the applied scientists suffered from the lack of tables for its Laplace transforms. Although arising from the studies of Mittag–Leffler on a problem not related to fractional calculus, but on analytical continuation of series to maximal starlike domain (Mittag-Leffler star), nowadays, the M-L function is the most popular and most exploited SF of FC. It was titled as the *"Queen"-function of FC* by Gorenflo and Mainardi in 1997 (see also the very recent survey by Mainardi [34]). The basic theory and more details, can be found, for example, in [22,43] (see also, e.g., [9,16,70,71]).

Definition 7. *The Mittag–Leffler (M-L) functions E_α and $E_{\alpha,\beta}$, are entire functions of order $\rho = 1/\alpha$ and type 1, defined by the power series*

$$E_\alpha(z) = \sum_{k=0}^\infty \frac{z^k}{\Gamma(\alpha k + 1)}, \quad E_{\alpha,\beta}(z) = \sum_{k=0}^\infty \frac{z^k}{\Gamma(\alpha k + \beta)}, \quad \alpha > 0, \beta > 0. \tag{37}$$

As *"fractional index"* ($\alpha > 0$) *analogs* of the exponential and trigonometric functions that satisfy ODEs of first and second order ($\alpha = 1, 2$), the M-L functions serve as solutions of *fractional order differential and integral equations*. An example is the Rabotnov function, called also "fractional exponent", $y(z) = z^{\alpha-1} E_{\alpha,\alpha}(z^\alpha)$ that solves the simplest fractional order differential equation $D^\alpha y(z) = y(z)$. Let us refer also to the pioneering work by Hille–Tamarkin [72], where the solution of the Abel integral equation of the second kind was provided in terms of a M-L function. As far as the Laplace transform images are mentioned, one can find these for the M-L type functions and their kth derivatives in the work of Podlubny ([16] (S.1.2.2)):

$$\mathcal{L}\left\{z^{\alpha k + \beta - 1} E_{\alpha,\beta}^{(k)}(\pm \lambda z^\alpha)); s\right\} = \frac{k! \, s^{\alpha - \beta}}{(s^\alpha \mp \lambda)^{k+1}}, \quad \operatorname{Re} s > |\lambda|^{1/\alpha}.$$

A Mittag–Leffler type function with three indices, known as the *Prabhakar function* [73], is also often studied and used (for details, see [22,70,71,74,75] and other contemporary books and surveys on M-L type functions):

$$E_{\alpha,\beta}^\gamma(z) = \sum_{k=0}^\infty \frac{(\gamma)_k}{\Gamma(\alpha k + \beta)} \frac{z^k}{k!}, \quad \alpha, \beta, \gamma \in \mathbb{C}, \; \operatorname{Re} \alpha > 0; \tag{38}$$

where $(\gamma)_0 = 1, (\gamma)_k = \Gamma(\gamma + k)/\Gamma(\gamma)$ denotes the Pochhammer symbol. Its Laplace transform has the form

$$\mathcal{L}\left\{E_{\alpha,\beta}^\gamma(\lambda z^\alpha); s\right\} = \frac{s^{-\beta}}{(1 - \lambda s^{-\alpha})^\gamma}.$$

For $\gamma = 1$, we get the M-L function $E_{\alpha,\beta}$, and, if additionally $\beta = 1$, then it is E_α.

These M-L type functions are simple cases of the Wright g.h.f. and of the H-function, namely:

$$E_{\alpha,\beta}(z) = {}_1\Psi_1\left[\begin{matrix}(1,1)\\(\beta,\alpha)\end{matrix}\bigg| z\right] = H_{1,2}^{1,1}\left[-z\bigg|\begin{matrix}(0,1)\\(0,1),(1-\beta,\alpha)\end{matrix}\right],$$

$$E_{\alpha,\beta}^\gamma(z) = \frac{1}{\Gamma(\gamma)} {}_1\Psi_1\left[\begin{matrix}(\gamma,1)\\(\beta,\alpha)\end{matrix}\bigg| z\right] = H_{1,2}^{1,1}\left[-z\bigg|\begin{matrix}(1-\gamma,1)\\(0,1),(1-\beta,\alpha)\end{matrix}\right].$$

Another generalization of M-L function (37) with additional parameters, for example $l \in \mathbb{C}, \mu \in \mathbb{R}$, was considered by Gorenflo–Kilbas–Rogosin [76], and its relations to FC operators:

$$E_{\alpha,\mu,l}(z) = \sum_{k=0}^\infty c_k z^k, \quad \text{with} \quad c_k = \prod_{j=0}^{k-1} \frac{\Gamma[\alpha(j\mu + l) + 1]}{\Gamma(\alpha(j\mu + l + 1) + 1]}.$$

A *vector index extension* of (37) appeared in the works by Luchko et al. (e.g., [15,77,78]) on operational calculus' methods for some fractional order PDE and multi-term FO differential equations. Under the name *multi-index (multiple) M-L function*, it was introduced by Kiryakova [63,79] using a different approach, as to be the generating function of Gelfond–Leontiev generalized integration and differentiation operators (47) (see Definition 9) and inspired from the paper by Dzrbashjan [44] on M-L type function with 2×2 indices. Further, this class of functions were studied in details by Kiryakova [59,80], Kilbas–Koroleva–Rogosin [81], Paneva–Konovska [74] and many other followers. Luchko et al. also considered multivariate analogs of the so-called vector index M-L functions [78].

Definition 8. (Kiryakova [59,80]) *Let $m > 1$ be an integer, $(\alpha_1 > 0, \alpha_2 > 0, \ldots, \alpha_m > 0)$ and $(\beta_1, \beta_2, \ldots, \beta_m)$ be arbitrary real parameters. By means of these sets of "multi-indices", the multi-index Mittag–Leffler function (abbrev. as multi-M-L f.) is defined as:*

$$E_{(\alpha_i),(\beta_i)}(z) := E_{(\alpha_i),(\beta_i)}^{(m)}(z) = \sum_{k=0}^{\infty} \frac{z^k}{\Gamma(\alpha_1 k + \beta_1) \ldots \Gamma(\alpha_m k + \beta_m)}. \tag{39}$$

Under weakened restrictions on αs (or their real parts) not to be obligatory all non-negative, the study was extended by Kilbas et al. [81].

As a further extension of both Prabhakar function (38) and of the $(2m)$ multi-index M-L functions (39), Paneva–Konovska [74,82] introduced and studied the so-called $(3m)$-*parametric (multi-index) Mittag–Leffler functions*, similar to (39) but with additional set of parameters $(\gamma_1, \ldots, \gamma_m)$:

$$E_{(\alpha_i),(\beta_i)}^{(\gamma_i),m}(z) = \sum_{k=0}^{\infty} \frac{(\gamma_1)_k \ldots (\gamma_m)_k}{\Gamma(\alpha_1 k + \beta_1) \ldots \Gamma(\alpha_m k + \beta_m)} \frac{z^k}{(k!)^m}. \tag{40}$$

For $m = 1$, one has the Prabhakar function, and, for $\gamma_1 = \ldots = \gamma_m = 1$, these are (39). The Mellin transforms of (39), (40) and their particular cases can be found in [83].

The so-called *Le Roy type function* has been an object of several recent studies, e.g., by Gerhold [84], Garra–Polito [85], Garrappa–Rogosin–Mainardi [86], Garrappa–Orsingher–Polito [87], as a new special function

$$F_{\alpha,\beta}^{(\gamma)}(z) = \sum_{k=0}^{\infty} \frac{z^k}{[\Gamma(\alpha k + \beta)]^\gamma}, \tag{41}$$

which is an entire function of $z \in \mathbb{C}$ for parameters $\operatorname{Re}(\alpha) > 0$, $\beta \in \mathbb{R}$ and $\gamma > 0$. This resembles to the M-L function (for $\gamma = 1$) and to the multi-index M-L function (39) (for integer $\gamma = m$, all $\alpha_i = \alpha$, $\beta_i = \beta$, $i = 1, \ldots, m$). The function (41) appeared as extension of the function $R_\gamma(z) = \sum_{k=0}^{\infty} z^k / [(k+1)!]^\gamma$, introduced by E. Le Roy [88] (1899), similarly to the purposes of G. Mittag–Leffler [67] (1903) to study analytical continuations of the sums of power series, and it seems they were working in competition on such ideas. Similar to the M-L type functions, (41) is involved in solutions of various problems, including a Convey–Maxwell–Poison distribution for different degrees of over- and under-dispersion.

Some Basic Properties of the Multi-Index Mittag-Leffler Functions

The basic properties and results for the functions (39) and long lists of their examples, all of them having wide applications in solutions of integer- and fractional-order models, are provided in our previous papers (e.g., [59,60,79,80]). Some of them are reminded here.

Theorem 1. *The multi-index M-L functions (39) are entire functions with the following order ρ and type σ:*

$$\frac{1}{\rho} = \alpha_1 + \cdots + \alpha_m, \quad \frac{1}{\sigma} = (\rho \alpha_1)^{\rho \alpha_1} \cdots (\rho \alpha_m)^{\rho \alpha_m}, \tag{42}$$

respectively with α_is replaced by $\operatorname{Re}(\alpha_i)$s. Note that the type $\sigma > 1$ for $m > 1$ and only for $m = 1$ (classical case (37)): $\sigma = 1$. The following asymptotic estimate holds:

$$|E_{(\alpha_i),(\beta_i)}(z)| \leq C|z|^{\rho((1/2) + \mu - (m/2))} \exp(\sigma |z|^\rho), \quad \mu := \beta_1 + \cdots + \beta_m, \text{ for } |z| \to \infty.$$

The $(3m)$-parameters M-L type functions (40) are also entire functions with the same order and type as in (42), see [74,82].

Lemma 1. *The multi-index M-L functions* (39) *are important examples of the Wright generalized hypergeometric functions* $_p\Psi_q$ *and of the Fox H-functions:*

$$E_{(\alpha_i),(\beta_i)}(z) = E^{(m)}_{(\alpha_i),(\beta_i)}(z) = {}_1\Psi_m\left[\begin{array}{c}(1,1)\\(\beta_i,\alpha_i)_1^m\end{array}\bigg|z\right] = H^{1,1}_{1,m+1}\left[-z\bigg|\begin{array}{c}(0,1)\\(0,1),(1-\beta_i,\alpha_i)_1^m\end{array}\right]. \quad (43)$$

Thus, the following Mellin–Barnes type integral representation holds (cf. with (1)*):*

$$E_{(\alpha_i),(\beta_i)}(z) = \frac{1}{2\pi i}\int_{\mathcal{L}} \frac{\Gamma(s)\Gamma(1-s)}{\prod_{i=1}^{m}\Gamma(\beta_i - s\alpha_i)}(-z)^{-s}ds, \quad z \neq 0,$$

based on the Mellin transform (see [59,83]*; also* [18] *(p. 48)):*

$$\mathcal{M}\left\{E_{(\alpha_i),(\beta_i)}(-z);s\right\} = \frac{\Gamma(s)\Gamma(1-s)}{\prod_{i=1}^{m}\Gamma(\beta_i - s\alpha_i)}, \quad 0 < \mathrm{Re}(s) < 1. \quad (44)$$

Additionally, as shown by Paneva–Konovska [74,82], the $(3m)$-parametric functions (40) can be represented as

$$E^{(\gamma_i),m}_{(\alpha_i),(\beta_i)}(z) = A\,{}_m\Psi_{2m-1}\left[\begin{array}{c}(\gamma_1,1),...,(\gamma_m,1)\\(\beta_1,\alpha_1),...,(\beta_m,\alpha_m),(1,1),...,(1,1)\end{array}\bigg|z\right]$$

$$= A\,H^{1,m}_{m,2m}\left[-z\bigg|\begin{array}{c}(1-\gamma_1,1),...,(1-\gamma_m,1)\\ [(0,1),(1-\beta_i,\alpha_i)]_1^m\end{array}\right], \text{ with } A = \left[\prod_{i=1}^{m}\Gamma(\gamma_i)\right]^{-1}, \quad (45)$$

which is in agreement with (43) for $\gamma_1 = ... = \gamma_m = 1$.

As an analog of the Laplace transform ($\mathcal{L}$), relationship between the classical M-L function (37) and the classical Wright function: $\mathcal{L}\{\phi(\alpha,\beta;z);s\} = \frac{1}{s}E_{\alpha,\beta}(\frac{1}{s})$ (see in the books [16,18]), we derive the following *new relation*.

Lemma 2.

$$\mathcal{L}\left\{{}_0\Psi_m\left[\begin{array}{c}-\\(\beta_1,\alpha_1),...,(\beta_m,\alpha_m)\end{array}\bigg|z\right];s\right\} = \frac{1}{s}E_{(\alpha_i),(\beta_i)}(\frac{1}{s}), \quad \mathrm{Re}(s) > 0. \quad (46)$$

Note that we can consider the $_0\Psi_m$-functions on the left-hand side as *"fractional indices"* analogs of the $_0F_m$-functions, that is *of the hyper-Bessel functions* $J^{(m)}_{\nu_1,...,\nu_m}$ of Delerue [89], related to the hyper-Bessel operators (9) as their eigenfunctions, and discussed further as special cases of (39). For details on these special functions, see Kiryakova [9] (Ch.3).

Various relations for the multi-M-L functions in terms of the operators of classical FC and GFC have been derived in our previous works (e.g., [59,80]). First, let us consider the so-called *Gelfond–Leontiev (G-L) operators of generalized integration and differentiation*, generated by the coefficients of an entire function $\varphi(\sigma)$. For the theory of the G-L operators in general, see Gelfond and Leontiev's paper [62]) of 1951, and for details in the case when the mentioned entire function is taken to be the M-L function or multi-index Mittag–Leffler function, we refer to Kiryakova [9] (Ch.1), [59,63,79]. Here, we only remind the definition of the G-L operators related to $\varphi(\sigma) = E_{(\alpha_i),(\beta_i)}(\sigma) := \sum_{k=0}^{\infty} b_k z^k$ whose coefficients $b_k = 1/(\Gamma(\alpha_1 k + \beta_1)...\Gamma(\alpha_m k + \beta_k))$ are taken as multipliers' sequences below.

Definition 9. (Kiryakova [63,64]) *For functions* $f(z) = \sum_{k=0}^{\infty} a_k z^k$ *analytic in a disk* $\{|z| < R\}$, *we consider the operators*

$$\tilde{D}f(z) := D_{(\alpha_i),(\beta_i)}f(z) = \sum_{k=1}^{\infty} a_k \frac{b_{k-1}}{b_k} z^{k-1}, \quad \tilde{L}f(z) := L_{(\alpha_i),(\beta_i)}f(z) = \sum_{k=0}^{\infty} a_k \frac{b_{k+1}}{b_k} z^{k+1}, \quad (47)$$

and call them multiple (multi-index) *Dzrbashjan–Gelfond–Leointiev (D-G-L) differentiations and integrations*, respectively. These are generated by the multi-index M-L functions and the name of Dzrbashjan is used in addition to Gelfond–Leontiev to honor his contribution to one of the first deep studies on M-L type functions, the book [43].

Evidently, $D_{(\alpha_i),(\beta_i)} L_{(\alpha_i),(\beta_i)} f(z) = f(z)$, and it is proven that the radii of convergence (and analyticity) of resulting analytical functions in (47) are the same R as for $f(z)$. According to Theorem 3 in [79], operators (47) can be analytically extended outside the disks to starlike domains and represented as operators of GFC, as follows:

$$\tilde{D}f(z) = z^{-1} D^{(\gamma_i-1-\alpha_i),(\alpha_i)}_{(1/\alpha_i),m} f(z) - \left[\prod_{i=1}^{m} \frac{\Gamma(\gamma_i)}{\Gamma(\gamma_i-\alpha_i)}\right] \frac{f(0)}{z}, \quad \tilde{L}f(z) = z\, I^{(\gamma_i-1),(\alpha_i)}_{(1/\alpha_i),m} f(z). \quad (48)$$

To start with the classical FC operators for the multi-index M-L functions, we state the following

Lemma 3. (Kiryakova [80] (Lemma 3.2)) *For any fixed l, $1 \leq l \leq m$ and integration order $\delta_l > 0$, we have for the E-K fractional integral the relation*

$$I^{\gamma_l-1,\delta_l}_{1/\alpha_l} E_{(\alpha_i),(\gamma_1,\ldots,\gamma_l,\ldots,\gamma_m)}(\lambda z) = E_{(\alpha_i),(\gamma_1,\ldots,\gamma_l+\delta_l,\ldots,\gamma_m)}(\lambda z), \quad \lambda \neq 0,$$

that is, a fractional integration can transform a multi-M-L function into another one with same α_is and corresponding parameter γ_l increased by the order of integration to $\gamma_l + \delta_l$.

Applying E-K fractional integrals of the form $I^{\gamma_i-1,\delta_i}_{1/\alpha_i}$ successively m-times ($i = 1, \ldots, m$) to (39) and using the composition (decomposition) property (26), we obtain for the generalized fractional integrals (23) the image:

$$I^{(\gamma_i-1),(\delta_i)}_{(1/\alpha_i),m} E_{(\alpha_i),(\gamma_i)}(\lambda z) = E_{(\alpha_i),(\gamma_i+\delta_i)}(\lambda z). \quad (49)$$

Then, for $\delta_i := \alpha_i$, $i = 1, \ldots, m$, and applying the operational rules for the operators $I^{(\gamma_i),(\delta_i)}_{(\beta_i),m}$ and $D^{(\gamma_i),(\delta_i)}_{(\beta_i),m}$ of GFC, the following generalized fractional integration and differentiation relations follow:

$$(\lambda z)\, I^{(\gamma_i-1),(\alpha_i)}_{(1/\alpha_i),m} E_{(\alpha_i),(\gamma_i)}(\lambda z) = E_{(\alpha_i),(\gamma_i)}(\lambda z) - \frac{1}{\Gamma(\gamma_1)\ldots\Gamma(\gamma_m)},$$

$$D^{(\gamma_i-1-\alpha_i),(\alpha_i)}_{(1/\alpha_i),m} E_{(\alpha_i),(\gamma_i)}(\lambda z) = (\lambda z)\, E_{(\alpha_i),(\gamma_i)}(\lambda z) + \frac{1}{\Gamma(\gamma_1-\alpha_1)\ldots\Gamma(\gamma_m-\alpha_m)}, \quad (50)$$

as analogs of the classical relation $z^\alpha D^\alpha E_\alpha(\lambda z) = \lambda z^\alpha E_\alpha(\lambda z) + \frac{1}{\Gamma(1-\alpha)}$ for the R-L derivative $D^\alpha = z^{-\alpha} D^{-\alpha,\alpha}_1$.

It remains to combine the results (48) and (50) to verify the fact that *the multi-index M-L functions that generate the G-L operators (47) appear as their eigenfunctions*:

Theorem 2. *The multi-index Mittag–Leffler function (39) satisfies the differential equation of fractional multi-order $(\alpha_1, \ldots, \alpha_m)$:*

$$\tilde{D}\, E_{(\alpha_i),(\beta_i)}(\lambda z) = D_{\alpha_i,\beta_i} E_{(\alpha_i),(\beta_i)}(\lambda z) = \lambda\, E_{(\alpha_i),(\beta_i)}(\lambda z), \quad \lambda \neq 0. \quad (51)$$

The classical *Poisson integral formula*, representing the Bessel function via the cosine-function ([3] (Vol. 2)), can be written in terms of an E-K fractional integral, as

$$J_\nu(z) = \frac{2}{\sqrt{\pi}\,\Gamma(\nu+1/2)} \left(\frac{z}{2}\right)^\nu \int_0^1 (1-t^2)^{\nu-1/2} \cos(zt)\, dt = \frac{1}{\sqrt{\pi}} \left(\frac{z}{2}\right)^\nu I^{-1/2,\nu+1/2}_{1/2} \{\cos z\}. \quad (52)$$

This representation has been extended in our works [9] (Ch.4), [90] for the *hyper-Bessel functions* (58), $m \geq 2$, that is for the $_0F_{m-1}$-*functions*, via generalized fractional integrals (24) of the function $\cos_m$. The details follow in Section 8. For the multi-index M-L functions, a Poisson type integral representation of the kind of (52) has to explore the more general fractional calculus operators from Definition 6. This is a part of the general results discussed in Section 8, but we expose it here as to close (at least partly) the topic with some properties of the multi-index Mittag–Leffler functions.

Theorem 3. (Kiryakova [59]) *Let $\alpha_k > 1$, $\beta_k \geq \frac{k}{m}$, $k = 1, \ldots, m$. Then, we have the following Poisson-type integral representation of the multi-index M-L functions my means of multiple W-E-K fractional integrals (33) of the cosine function (54) of order m (from the next section):*

$$E_{(\alpha_k),(\beta_k)}(-z) = c^* I_{(1/\alpha_k)_1^m,(1)_1^m,m}^{(\frac{k}{m}-1)_1^m,(\beta_k-\frac{k}{m})_1^m} \left\{ \cos_m(mz^{1/m}) \right\}$$

$$= c^* \int_0^1 H_{m,m}^{m,0} \left[\sigma \, \middle| \, \begin{matrix} (\beta_k - \alpha_k, \alpha_k)_1^m \\ (k/m - 1, 1)_1^m \end{matrix} \right] \cos_m(m(z\sigma)^{1/m}) d\sigma, \text{ with } c^* := \sqrt{m/(2\pi)^{m-1}}. \quad (53)$$

Remark 1. *The above result is parallel with* (52) *for the Bessel functions. If we take $\alpha_k = 1$, $\beta_k = \frac{k}{m}$, the above GFC operator, the multiple W-E-K fractional integral, has a multi-order $(0, \ldots, 0)$ and since also $\lambda_k = \beta_k$, it turns into identity. Then, the $E_{(\alpha_k),(\beta_k)}$-function reduces to the $\cos_m(z)$-function. It is similar in the simplest case to the Bessel function with index $\nu = -1/2$: $J_{-1/2}(z) = \sqrt{\frac{2}{\pi z}} \cos z$. More generally, it is also known that the Bessel functions of "semi-integer" indices (called also "spherical functions" for their use in theory of spherical waves) are reducible to trigonometric functions or to integer order operators of them: $J_{n-1/2}(z) = \frac{(2z)^{n+1/2}}{\sqrt{\pi}} \frac{d^n}{(dz^2)^n} \left\{ \frac{\cos z}{z} \right\}$, $n = 0, 1, 2, \ldots$. In the case of multi-index M-L functions (39), we can call multi-indices of the form $\alpha_k = 1$, $\beta_k := \nu_k - \frac{k}{m} = 0, 1, 2, \ldots;$ for $k = 1, \ldots, m$, as "semi-integer multi-indices". A corollary of Theorem 3 tells that for such multi-indices the functions $E_{(\alpha_k),(\beta_k)}$ reduce either directly to generalized trigonometric functions, or to integer order integral or differential operators of them.*

The results for the *images of the multi-index Mittag–Leffler functions* (39) and (40) under *GFC integrals and derivatives*, or under their particular cases a R-L, E-K, Saigo, Marichev–Saigo–Maeda operators, etc. can be written from the general results in Section 7 according to definition via the Wright g.h.f. $_1\Psi_m$.

Series in systems of special functions, in the general cases of $2m$- and $3m$-parameters M-L functions and their particular case (mentioned in next section) as the M-L function, Parbhakar function, multi-index and fractional analogs of the Bessel- and hyper-Bessel functions, were studied recently in details by Paneva–Konovska in a series of papers and in the book [74], especially with respect to their convergence in complex domain, including Cauchy–Hadamard, Abel, Tauber type, Hardy–Littlewood and Ostrovski type theorems.

5. Examples of M-L Type and Multi-Index M-L Functions

5.1. For *m = 1*, this is the *classical M-L function* $E_{\alpha,\beta}(z)$ with all its special cases:

- $\alpha > 0, \beta = 1$: $E_{0,1}(z) = \frac{1}{1-z}$; $E_{1,1}(z) = \exp(z)$; $E_{2,1}(z^2) = \cosh(z)$, $E_{2,1}(-z^2) = \cos(z)$; $E_{1/2,1}(z^{1/2}) = \exp(z)\left[1 + \text{erf}(z^{1/2})\right] = \exp(z)\,\text{erfc}(-z^{1/2}) = \exp(z)\left[1 + \frac{1}{\sqrt{\pi}}\gamma(\frac{1}{2},z)\right]$ (*the error functions, or incomplete gamma functions*);

- $\beta \neq 1$: $E_{1,2}(z) = \frac{e^z - 1}{z}$; $E_{1/2,2}(z) = \frac{\text{sh}\sqrt{z}}{z}$; $E_{2,2}(z) = \frac{\text{sh}\sqrt{z}}{\sqrt{z}}$; the *Miller-Ross function* $z^\nu E_{1,\nu+1}(az)$; etc.;

- $\beta = \alpha$: the α-*exponential (Rabotnov) function* $y_\alpha(z) = z^{\alpha-1} E_{\alpha,\alpha}(z^\alpha)$.

- The *trigonometric functions of order m*, and, respectively the *hyperbolic functions of order m*:

$$\cos_m(z) = \sum_{j=0}^{\infty} \frac{(-1)^j z^{mj}}{(mj)!} = E_{m,1}(-z^m), \tag{54}$$

$y(z) = \cos_m(z)$ is the solution of IVP $y^{(m)}(z) = -y(z)$, $y(0) = 1$, $y^{(j)}(0) = 0$, $j = 1, ..., m-1$;

$$k_r(z, m) = \sum_{j=0}^{\infty} \frac{(-1)^j z^{mj+r-1}}{(mj+r-1)!} = z^{r-1} E_{m,r}(-z^m), \ r = 1, 2, \ldots; \ k_1(z, m) := \cos_m(z) = E_{m,1}(-z^m),$$

$$h_r(z, m) = \sum_{j=0}^{\infty} \frac{z^{mj+r-1}}{(mj+r-1)!} = z^{r-1} E_{m,r}(z^m), \ r = 1, 2, \ldots; \ h_1(z, m) := \cosh_m(z) = E_{m,1}(z^m),$$

can also be expressed in terms of the M-L function (see in [3] (Vol. 3) and [16] (Ch.1)); and the same for their *fractionalized versions*, as by Plotnikov [91] and Tseytlin [92]:

$$Sc_\alpha(z) = \sum_{k=0}^{\infty} \frac{(-1)^k z^{(2-\alpha)m+1}}{\Gamma((2-\alpha)m+2)} = z E_{2-\alpha,2}(-z^{2-\alpha}),$$

$$Cs_\alpha(z) = \sum_{k=0}^{\infty} \frac{(-1)^k z^{(2-\alpha)m}}{\Gamma((2-\alpha)m+1)} = E_{2-\alpha,1}(-z^{2-\alpha}),$$

and by Luchko–Srivastava [77]:

$$\sin_{\lambda,\mu}(z) = \sum_{k=0}^{\infty} \frac{(-1)^k z^{2k+1}}{\Gamma(2\mu k + 2\mu - \lambda + 1)} = z E_{2\mu, 2\mu-\lambda+1}(-z^2),$$

$$\cos_{\lambda,\mu}(z) = \sum_{k=0}^{\infty} \frac{(-1)^k z^{2k}}{\Gamma(2\mu k + \mu - \lambda + 1)} = E_{2\mu, \mu-\lambda+1}(-z^2),$$

(see details again in Podlubny [16] (Ch.1)).

- Here, we mention also the so-called *Lorenzo–Hartley functions* [93], the *F-function* and its generalization the *R-function*, shown to be solutions of some linear fractional differential equations. We can represent them in terms of M-L function, namely, for $z > 0$, $c = 0$, $q \geq 0$, $\nu \leq q$:

$$F_q(a, z) = \sum_{k=0}^{\infty} \frac{a^k z^{(k+1)q-1}}{\Gamma((k+1)q)} = z^{q-1} E_{q,q}(az),$$

$$R_{q,\nu}(a, 0, z) = \sum_{k=0}^{\infty} \frac{a^k z^{(k+1)q-1-\nu}}{\Gamma((k+1)q-\nu)} = z^{q-1} E_{q,q-\nu}(az).$$

5.2. For $m = 2$: We start with the not enough popular M-L type function of Dzrbashjan [44], with 2×2 indices, which he denoted alternatively by (we need to set $1/\rho_i := \alpha_i, \mu_i := \beta_i, i = 1, 2$):

$$\Phi_{\rho_1, \rho_2}(z; \mu_1, \mu_2) = \sum_{k=0}^{\infty} \frac{z^k}{\Gamma(\mu_1 + \frac{k}{\rho_1})\Gamma(\mu_2 + \frac{k}{\rho_2})} := E_{(\frac{1}{\rho_1}, \frac{1}{\rho_2}), (\mu_1, \mu_2)}(z) = E_{(\alpha_1, \alpha_2), (\beta_1, \beta_2)}(z). \tag{55}$$

Dzrbashjan found the order and type of this entire function, claimed on few simple particular cases, and considered some integral relations between (55) and Mellin transforms on a set of axes. Then, he developed a theory of integral transforms in the class L_2, involving kernel close to functions (55) and, further, proposed approximations of entire functions in L_2 for an arbitrary finite system of axes in complex plane starting from the origin.

The 2×2-indices M-L type functions (55) were also studied in detail by Luchko in the recent paper [94]. He allowed the parameters ρ_1, ρ_2 to be also negative or zero and called them "4-parameters Wright functions of second kind", separating the cases when $\rho_1 + \rho_2 > 0$, $\rho_1 + \rho_2 = 0$ or $\rho_1 + \rho_2 < 0$.

Some of the simple cases of (55), as mentioned and denoted in Dzrbashjan [44], are:

- the M-L function itself: $E_{\frac{1}{\rho},\mu}(z) = E_{(\frac{1}{\rho},0),(\mu,1)}(z) = \Phi_{\rho,\infty}(z;\mu,1)$; $\frac{1}{1-z} = E_{(0,0),(1,1)}(z) = \Phi_{\infty,\infty}(z;1,1)$; the *Bessel function*: $J_\nu(z) = (\frac{z}{2})^\nu E_{(1,1),(\nu+1,1)}\left(-\frac{z^2}{4}\right) = (\frac{z}{2})^\nu \Phi_{1,1}\left(-\frac{z^2}{4};1,\nu+1\right)$; etc.

To these examples, we added (see, e.g., Kiraykova [59]) the following cases:

- The *Struve and Lommel functions* (see [3] (Vol. 2); and details in [9] (App.,(C.8)), [79,80]):

$$s_{\mu,\nu}(z) = \frac{1}{4}z^{\mu+1} E_{(1,1),((3-\nu+\mu)/2,(3+\nu+\mu)/2)}\left(-\frac{z^2}{4}\right), \quad H_\nu(z) = \frac{1}{\pi 2^{\nu-1}(1/2)_\nu}s_{\nu,\nu}(z).$$

- The *"classical" Wright function* that arose in the studies of Fox ([95], 1928), Wright ([31], 1933) and Humbert and Agarwal ([69], 1953) and was also referred to in Erdélyi et al. [3] (Vol. 3). Initially, Wright [31] defined this function only for $\alpha > 0$, then extended its definition for $\alpha > -1$ [32]. Now, we see this is a case of multi-index M-L function with $m = 2$:

$$\phi(\alpha,\beta;z) := W_{\alpha,\beta}(z) = \sum_{k=0}^{\infty} \frac{1}{\Gamma(\alpha k+\beta)} \frac{z^k}{k!} = {}_0\Psi_1\left[\begin{array}{c}-\\(\beta,\alpha)\end{array}\Big|z\right] = E^{(2)}_{(\alpha,1),(\beta,1)}(z), \quad (56)$$

which is entire function of order $1/(1+\alpha)$. The survey papers by Gorenflo–Luchko–Mainardi [96] and Mainardi–Consiglio [97] reflect in detail its analytical properties and applications, see also the book [22] as well as the related literature. In the case $\alpha \geq 0$, the Wright function is said to be of first kind, and for $-1 < \alpha < 0$ of second kind. The latter survey [97] concentrates on the Wright function of second kind. It is noted that the first kind Wright function is of exponential order, while the second kind is not of exponential order, and naturally they have different asymptotic behaviors, Laplace transforms, etc. (see also Luchko [94]). The function (56) plays an important role in the solutions of linear partial fractional differential equations as the *fractional diffusion-wave equation* studied by Nigmatullin (1984–1986, to describe the diffusion process in media with fractal geometry, $0 < \alpha < 1$) and by Mainardi et al. (since 1994, for propagation of mechanical diffusive waves in viscoelastic media, $1 < \alpha < 2$). In the form $M(z;\beta) = \phi(-\beta,1-\beta;-z), \beta := \alpha/2$, this function is recently called as the *Mainardi function* (see [16] (Ch.1)). In our denotations, it is a multi-index M-L function with $m = 2$ and a Dzrbashjan function (55): $M(z;\beta) = E^{(2)}_{(-\beta,1),(1-\beta)}(-z)$ and has its own particular cases, such as $M(z;1/2) = 1/\sqrt{\pi}\exp(-z^2/4)$ and the *Airy function*, $M(z;1/3) = 3^{2/3}Ai(z/3^{1/3})$. Note also that, for $\alpha = 0$, the Wright function (56) reduces to the *exponent*, since $\phi(0,\beta;z) = \sum_{k=0}^{\infty}z^k/(k!\Gamma(\beta)) = (1/\Gamma(\beta))\exp(z)$.

In alternative form and denotation, the Wright function (56) is known as the *Wright–Bessel function* or is misnamed as the *Bessel–Maitland function*:

$$J_\nu^\mu(z) = \phi(\mu,\nu+1;-z) = {}_0\Psi_1\left[\begin{array}{c}-\\(\nu+1,\mu)\end{array}\Big|-z\right] = \sum_{k=0}^{\infty}\frac{(-z)^k}{\Gamma(\nu+k\mu+1)k!} = E^{(2)}_{(1/\mu,1),(\nu+1,1)}(-z), \quad (57)$$

again as an example of the Dzrbashjan function. It is an obvious (and was introduced as such by Sir E. Maitland Wright [32]) *"fractional index"* analog of the classical Bessel function $J_\nu(z) = c(z/2){}_0F_1(z^2/4)$, more exactly, of the Bessel–Clifford function $C_\nu(z)$.

Several further *"fractional-indices"* generalizations of $J_\nu(z)$ and $J_\nu^\mu(z)$ are found in the studies of other authors (details are in [59]), and we can represent all of them as multi-index M-L functions. One of them is the so-called *generalized Wright–Bessel(–Lommel) functions*, introduced by Pathak ([98], 1966),

$$J_{\nu,\lambda}^\mu(z) = (z/2)^{\nu+2\lambda}\sum_{k=0}^{\infty}\frac{(-1)^k(z/2)^{2k}}{\Gamma(\nu+k\mu+\lambda+1)\Gamma(\lambda+k+1)}$$

$$= (z/2)^{\nu+2\lambda}E^{(2)}_{(1/\mu,1),(\nu+\lambda+1,\lambda+1)}\left(-(z/2)^2\right), \quad \mu > 0.$$

For $\mu = 1$, it includes the mentioned Lommel and Struve functions, e.g., $J^1_{\nu,\lambda}(z) = $ const $S_{2\lambda+\nu-1,\nu}(z)$. A next example is the *generalized Lommel–Wright function with four indices*, introduced by de Oteiza, Kalla and Conde ([99], 1986), with $r > 0, n \in \mathbb{N}, \nu, \lambda \in \mathbb{C}$:

$$J^{r,n}_{\nu,\lambda}(z) = (z/2)^{\nu+2\lambda} \sum_{k=0}^{\infty} \frac{(-1)^k (z/2)^k}{\Gamma(\nu+kr+\lambda+1)\Gamma(\lambda+k+1)^n}$$

$$= (z/2)^{\nu+2\lambda} E^{(n+1)}_{(1/r,1,\ldots,1),(\nu+\lambda+1,\lambda+1,\ldots,\lambda+1)} \left(-(z/2)^2\right).$$

5.3. The above is an interesting example of a multi-M-L function with $m = n+1$.

Other particular cases of multi-index ($2m$-parameters) M-L functions with greater multiplicity $m \geq 2$ are:

- For arbitrary $m \geq 2$: let $\forall \alpha_i = 0$ and $\forall \beta_i = 1, i = 1, \ldots, m$. Then, from definition (39), we get again the *geometric series*

$$E^{(m)}_{(0,0,\ldots,0),(1,1,\ldots,1)}(z) = \sum_{k=0}^{\infty} z^k = \frac{1}{1-z}.$$

- Consider the case $m \geq 2$, $\forall \alpha_i = 1, i = 1, \ldots, m$. Then, the function

$$E^{(m)}_{(1,1,\ldots,1),(\beta_1,\ldots,\beta_m)}(z) = {}_1\Psi_m\left[\begin{matrix}(1,1)\\(\beta_i,1)^m_1\end{matrix}\Big| z\right] = [\prod_{i=1}^{m} \Gamma(\mu_i)]^{-1} {}_1F_m(1;\beta_1,\beta_2,\ldots,\beta_m;z)$$

reduces to ${}_1F_m$-function and also to a *Meijer's $G^{1,1}_{1,m+1}$-function*. Denote $\beta_i = \gamma_i+1, i=1,\ldots,m$, and let additionally one of the β_i be 1, e.g., $\beta_m = 1$, i.e., $\gamma_m = 0$. Then, the multi-index M-L function becomes a ${}_0F_{m-1}$-function, that is, a *hyper-Bessel function* in the sense of Delerue [89] (see also [9] (Ch.3)):

$$J^{(m-1)}_{\gamma_i,\ldots,\gamma_{m-1}}(z) = \left(\frac{z}{m}\right)^{\sum_{i=1}^{m-1}\gamma_i} E^{(m)}_{(1,1,\ldots,1),(\gamma_1+1,\gamma_2+1,\ldots,\gamma_{m-1}+1,1)}\left(-\left(\frac{z}{m}\right)^m\right) \quad (58)$$

$$= \left[\prod_{i=1}^{m-1}\Gamma(\gamma_i+1)\right]^{-1} \left(\frac{z}{m}\right)^{\sum_{i=1}^{m-1}\gamma_i} {}_0F_{m-1}\left(\gamma_1+1,\gamma_2+1,\ldots,\gamma_{m-1}+1;-\left(\frac{z}{m}\right)^m\right)$$

$$:= \left[\prod_{i=1}^{m-1}\Gamma(\gamma_i+1)\right]^{-1} \left(\frac{z}{m}\right)^{\sum_{i=1}^{m-1}\gamma_i} j^{(m-1)}_{\gamma_1,\ldots,\gamma_{m-1}}(-z), \quad (59)$$

where $j^{(m-1)}_{\gamma_1,\ldots,\gamma_{m-1}}$ is called as *normalized hyper-Bessel function*.

This representation suggests that the multi-index M-L functions (39) with arbitrary $(\alpha_1,\ldots,\alpha_m) \neq (1,\ldots,1)$ can be interpreted as *fractional-indices analogs of the hyper-Bessel functions* (58) and (59), which themselves are *multi-index (but integer) analogs* of the Bessel function. Functions (58) and (59) are closely related to the *hyper-Bessel differential operators* (9) (see Section 3.1), and form a fundamental system of solutions of the differential equations of the form $By(z) = \lambda y(z)$; the details are found in Kiryakova [9] (Ch.3, Th.3.4.3). For example, if the hyper-Bessel operator (9) is with $\beta = m, \gamma_1 < \gamma_2 < \ldots < \gamma_m = 0 < \gamma_1+1$, the solution of the Cauchy problem $By(z) = -y(z), y(0) = 1, y^{(j)}(0) = 0, j = 1,\ldots,m-1$, is given by the *normalized hyper-Bessel function* (59): $y(z) = j^{(m-1)}_{\gamma_1,\ldots,\gamma_{m-1}}(-z)$. Closely related functions are also the *Bessel–Clifford functions of order m*:

$$C_{\nu_1,\ldots,\nu_m}(z) = \sum_{k=0}^{\infty} \frac{(-1)^k z^k}{\Gamma(\nu_1+k+1)\ldots\Gamma(\nu_m+k+1)\,k!} = E^{(m+1)}_{(1,\ldots,1),(\nu_1+1,\ldots,\nu_m+1,1)}(-z).$$

Let us mention the special functions appearing in a very recent paper by Ricci [100]. He considered the so-called *Laguerre derivative* $D_L = \frac{d}{dz} z \frac{d}{dz}$ and its iterates $D_{mL} = \frac{d}{dz} z \frac{d}{dz} z \ldots z \frac{d}{dz} z$, same as the particular hyper-Bessel differential operators (19) considered in operational calculus by Ditkin and Prudnikov [50], as mentioned in Section 3.1. Then, the L-exponentials $e_1(z), e_2(z), ..., e_m(z), ...$, which are eigenfunctions of D_{mL}, that is, $D_{mL} e_m(\lambda z) = \lambda e_m(\lambda z)$, are shown in [100] to have the form

$$e_m(z) = \sum_{k=0}^{\infty} \frac{z^k}{(k!)^{m+1}} = {}_0F_m(-;1,1,...,1;z) = {}_1\Psi_{m+1}\left[\begin{array}{c}(1,1)\\(1,1),(1,1),...,(1,1)\end{array}\bigg| z\right]. \quad (60)$$

Then, these are examples of the hyper-Bessel functions (58) and of the multi-index Mittag–Leffler functions $E_{(1,...,1),(1,...,1)}^{(m+1)}(z)$ as well. In [100], applications to population dynamics and in solutions of linear dynamical systems of these SF and of the related Laguerre-type Bell polynomials and Laguerre-type generalized hypergeometric functions are discussed.

• The *Rabotnov function* (the α-exponential function), presented in **5.1.**, appeared in Rabotnov's works on application of fractional order operators in mechanics of solids. It is interesting to consider its *multi-index analog*, that is the case with all $\beta_i = \alpha_i = \alpha > 0, i = 1,...,m$. This is the function

$$y_{\alpha}^{(m)}(z) = z^{\alpha-1} E_{(\alpha,..,\alpha),(\alpha,..,\alpha)}^{(m)}(z^{\alpha}) = z^{\alpha-1} \sum_{k=0}^{\infty} \frac{z^{\alpha k}}{[\Gamma(\alpha+\alpha k)]^m}. \quad (61)$$

Observe that, for $\alpha = 1$, we get the Ricci function (60), namely: $e_{m-1}(z) = \sum_{k=0}^{\infty} \frac{z^k}{[k!]^m}$, and also a case of the *original Le Roy function* with $\gamma = m$.

• In general, for rational values of $\forall \alpha_i, i = 1,...,m$, the functions (39) are reducible to *generalized hypergeometric functions* ${}_1F_m$ and to *Meijer's G-functions* $G_{1,m+1}^{1,1}$, that is, to classical special functions.

Remark 2. Note that all the results we derived for the multi-index M-L functions can be applied for their particular cases mentioned above.

6. Other Special Cases of the Wright Generalized Hypergeometric Functions ${}_p\Psi_q$

6.1. Virchenko and Ricci generalized hypergeometric functions. In [101] and some other papers, Virchenko studied some generalized hypergeometric functions denoted by ${}_2R_1^{\tau}(z)$ and ${}_1\Phi_1^{\tau}(z)$, as well as their integral representations, relations and applications to the generalized Legendre functions $P_k^{m,\bar{m}}(z), Q_k^{m,n}(z)$, gamma functions, Laguerre's functions, etc.

•
$$ {}_2R_1^{\omega,\mu}(a,b;c;z) = \frac{\Gamma(c)}{\Gamma(a)\Gamma(b)} \sum_{k=0}^{\infty} \frac{\Gamma(a+k)\Gamma(b+\frac{\omega}{\mu}k)}{\Gamma(c+\frac{\omega}{\mu}k)} \cdot \frac{z^k}{k!}.$$

For $\frac{\omega}{\mu} := \tau > 0$, and a, b, c - complex, $a + k \neq 0, -1, -2,...; b + \tau k \neq 0, -1, -2,..., k = 0, 1, 2,...; |z| < 1$, it is rewritten as
$$ {}_2R_1^{\tau}(a,b;c;z) = \frac{\Gamma(c)}{\Gamma(a)\Gamma(b)} \sum_{k=0}^{\infty} \frac{\Gamma(a+k)\Gamma(b+\tau k)}{\Gamma(c+\tau k)} \cdot \frac{z^k}{k!},$$

which is nothing but the Wright g.h.f. $\frac{\Gamma(c)}{\Gamma(a)\Gamma(b)} {}_2\Psi_1\left[\begin{array}{c}(a,1),(b,\tau)\\(c,\tau)\end{array}\bigg| z\right]$. Virchenko also proposed some examples of elementary functions for these special functions, e.g., $(\ln(1+z))_{\tau}$ and $(\arcsin z)_{\tau}$; some generalized incomplete B-function; the Gauss function ${}_2F_1$; etc.

•
$$ {}_1\Phi_1^{\tau}(a;c;z) = \frac{\Gamma(c)}{\Gamma(a)} \sum_{k=0}^{\infty} \frac{\Gamma(a+\tau k)}{\Gamma(c+\tau k)} \cdot \frac{z^k}{k!},$$

and, in Virchenko [101], generalizations of the gamma function, incomplete gamma function, probability integrals and Laguerre's functions are introduced by means of ${}_1\Phi_1^{\tau}(z)$, which is a Wright g.h.f. of the form $\frac{\Gamma(c)}{\Gamma(a)} {}_1\Psi_1\left[\begin{array}{c}(a,\tau)\\(c,\tau)\end{array}\bigg| z\right]$, and, according to our classifications in Section 8, a confluent type g.h.f.

- In 5.3., the recent paper by Ricci [100] is mentioned for the Laguerre-type derivatives and related special functions. Along with the functions (60), there he also considered the Laguerre-type (L-) Bessel functions, L-type Gauss hypergeometric functions and the *Laguerre-type generalized hypergeometric functions* $_{L}{_pF_q}$. They can be shown to be representable by $_pF_{q+1}$, thus also as $_p\Psi_{q+1}$, namely:

$$_L{_pF_q}(a_1,..,a_p;b_1,...,b_q;z) = \sum_{k=0}^{\infty} \frac{a_1^{(k)}...a_p^{(k)}}{b_1^{(k)}...b_q^{(k)}} \cdot \frac{z^k}{(k!)^2}$$

$$= \sum_{k=0}^{\infty} \frac{a_1^{(k)}...a_p^{(k)}}{b_1^{(k)}...b_q^{(k)}(1)^{(k)}} \cdot \frac{z^k}{k!} = {_pF_{q+1}}(a_1,...,a_p;b_1,...,b_q,1;z). \tag{62}$$

6.2. *Mainardi-Masina and Paris generalized exponential integrals.* In [102], Mainardi and Masina introduced a generalized exponential integral $\text{Ein}_\alpha(z)$ by replacing the exponential function in the complementary exponential integral $\text{Ein}(z)$ by the Mittag–Leffler function $E_\alpha(z)$ and mentioned the physical applications for $0 < \alpha < 1$ in the studies of the creep features of linear viscoelastic models. In the recent paper [103], Paris made the next step to involve the two-parameter M-L function, namely to consider the generalized exponential integral

$$\text{Ein}_{\alpha,\beta}(z) = z \sum_{k=0}^{\infty} \frac{(-1)^k z^{\alpha k}}{(\alpha k+1)\Gamma(\alpha k+\alpha+\beta)}, \quad \text{which for } \beta=1 \text{ gives } \text{Ein}_\alpha(z). \tag{63}$$

As observed, this function can be seen as a case of the Wright g.h.f. with $p = q = 2$, namely

$$\text{Ein}_{\alpha,\beta}(z) = z \sum_{k=0}^{\infty} \frac{\Gamma(\alpha k+1)\Gamma(k+1)}{\Gamma(\alpha k+2)\Gamma(\alpha k+\alpha+\beta)} \frac{(-z^\alpha)^k}{k!} = z\,{_2\Psi_2}\left[\begin{array}{c}(1,\alpha),(1,1)\\(2,\alpha),(\alpha+\beta,\alpha)\end{array}\bigg| -z^\alpha\right].$$

Paris studied in details the asymptotic expansion of (63) for $|z| \to \infty$. In [102,103], generalized Sine and Cosine integrals are also considered (of the kind mentioned in **5.1.**), for example $\text{Sin}_{\alpha,\beta}(z) = \text{Ein}_{2\alpha,\beta-\alpha}(z)$, with their asymptotics and plots for different values of parameters.

6.3. *The so-called k-analogs of special functions.* Claims on inventing and studying "new" classes of special functions in several recent papers have been based on the extended notion of the *k-Gamma function*, $k > 0$. However, in all such works, its representation in terms of the classical Gamma-function is explicitly written there, and then is ignored:

$$\Gamma_k(s) = \int_0^\infty \exp(-\frac{t^k}{k}) t^{s-1} dt = k^{\frac{s}{k}-1} \Gamma(\frac{s}{k}), \quad s \in \mathbb{C},\ \text{Re}(s) > 0, \tag{64}$$

where $\Gamma(.)$ is the classical Gamma-function.

In addition, the *k-Pochhammer symbol* is used in the next denotations:

$$(\lambda)_{\nu,\kappa} := \Gamma_k(\lambda+\nu\kappa)/\Gamma_k(\lambda),\ \lambda \in \mathbb{C}\setminus\{0\}, \nu \in \mathbb{C}, \text{ with } \Gamma_k \text{ as in (64)}. \tag{65}$$

In [104], using the above two definitions, we showed that *most of these "new" functions are in fact some known special functions, namely Wright g.h.f. and its cases.* For the details of establishing the mentioned relations, see Kiryakova [104]. In addition, in the references lists of [104,105], one can find the particular authors/sources mentioned below.

- *A generalized k-Bessel function* was introduced by Gehlot ([106], 2014), and studied by Mondal ([107], 2016) and Shaktawat et al. ([108], 2017). It is defined by

$$W_{\nu,c}^k(z) = \sum_{n=0}^{\infty} \frac{(-c)^n}{\Gamma_k(nk+\nu+k)} \cdot \frac{(z/2)^{2n+\frac{\nu}{k}}}{n!}, \quad z \in \mathbb{C},\ k > 0,\ \text{Re}(\nu) > -1,\ c \in \mathbb{C}. \tag{66}$$

However, after simple exercise, the function (66) can be represented as a Wright g.h.f. $_0\Psi_1$, and even as the simpler g.h.f. $_0F_1$ of the same type as the classical Bessel function:

$$W_{\nu,c}^k(z) = (z/2)^{\frac{\nu}{k}} \sum_{n=0}^{\infty} \frac{[-c(\frac{z}{2})^2]^n}{k^{n+1+(\frac{\nu}{k})}\Gamma(n+1+(\frac{\nu}{k}))\Gamma(n+1)} = \ldots$$

$$= \frac{(\frac{z}{2})^{\frac{\nu}{k}}}{k^{1+(\frac{\nu}{k})}} \sum_{n=0}^{\infty} \frac{[-(\frac{c}{k})(\frac{z}{2})^2]^n}{\Gamma(1+(\frac{\nu}{k})+n.1)\Gamma(1+n.1)}$$

$$= \frac{(\frac{z}{2})^{\frac{\nu}{k}}}{k^{1+(\frac{\nu}{k})}} {}_1\Psi_2 \left[\begin{array}{c} (1,1) \\ (1+\frac{\nu}{k},1),(1,1) \end{array} \bigg| -\frac{c}{k}\left(\frac{z}{2}\right)^2 \right] = \frac{(z/2)^{\nu/k}}{k^{1+(\nu/k)}} {}_0\Psi_1 \left[\begin{array}{c} -- \\ (1+\frac{\nu}{k},1) \end{array} \bigg| -\frac{c}{k}\left(\frac{z}{2}\right)^2 \right]$$

$$= \frac{(\frac{z}{2})^{\frac{\nu}{k}}}{k^{1+(\frac{\nu}{k})}\Gamma(1+\nu)} {}_0F_1\left(-;1+\frac{\nu}{k};-\frac{c}{k}\frac{z^2}{4}\right). \tag{67}$$

Indeed, if we take $k = 1$ and $c = 1$, this function reduces to the classical Bessel function: $W_{\nu,1}^1(z) = \frac{(z/2)^\nu}{\Gamma(1+\nu)} {}_0F_1\left(-;1+\nu;-\frac{z^2}{4}\right)$. For $k > 0$ and $c = 1$ Gehlot [106] used (66) as a solution of a k-Bessel differential equation. Mondal [107] studied some properties of (66) for arbitrary $c \in \mathbb{C}$. Shaktawat et al. [108] evaluated the *Marichev–Saigo–Maeda (M-S-M) operators* of FC

$$I^{a,a',b,b',c}f(z) = z^{c-a-a'} \int_0^1 \frac{(1-\sigma)^{c-1}}{\Gamma(c)} \sigma^{-a'} F_3\left(a,a',b,b';c;1-\sigma,1-\frac{1}{\sigma}\right) f(z\sigma)d\sigma \tag{68}$$

of this function. Since its kernel Appel F_3-function is a H-function (36) with $m = 3$, in view of author's result from Corollary 3 in Section 7, it is well expected that the result appears in terms of a $_3\Psi_4$-function (because the indices of $_0\Psi_1$ are increased by 3 under the 3-tuple FC integral).

- *Generalized k-Mittag–Leffler function.* It was studied by many authors, for example in its simplest case by Gupta and Parihar ([109], 2014) in the form

$$E_{k,\alpha,\beta}(z) = \sum_{n=0}^{\infty} \frac{z^n}{\Gamma_k(\alpha n + \beta)}.$$

This function has various further extensions, such as the generalized k-Mittag–Leffler function by Nisar–Eata–Dhaifalla–Choi ([110], 2016):

$$E_{\kappa,\alpha,\beta}^{\eta,\delta,p,q}(z) = \sum_{n=0}^{\infty} \frac{(\eta)_{qn,\kappa}}{\Gamma_k(\alpha n + \beta)(\delta)_{pn,\kappa}} z^n, \text{ with } \kappa,p,q \in \mathbb{R}_+; \alpha,\beta,\eta,\delta \in \mathbb{C}, \tag{69}$$

and $\min\{\text{Re}(\alpha), \text{Re}(\beta), \text{Re}(\eta), \text{Re}(\delta)\} > 0$; $q \leq \text{Re}(\alpha) + p$.

Again, by using (64) and (65), it can be transformed into a Wright g.h.f. (see [104], Case 5.2), namely:

$$E_{\kappa,\alpha,\beta}^{\eta,\delta,p,q}(z) = k^{1-\frac{\beta}{k}} \frac{\Gamma(\delta/k)}{\Gamma(\eta/k)} {}_2\Psi_2 \left[\begin{array}{c} (\frac{\eta}{k},\frac{q\kappa}{k}),(1,1) \\ (\frac{\delta}{k},\frac{p\kappa}{k}),(\frac{\beta}{k},\frac{\alpha}{k}) \end{array} \bigg| k^{\frac{(q-p)\kappa-\alpha}{k}} z \right].$$

Nisar–Eata–Dhaifalla-Choi [110] put efforts to evaluate FC operators' images of (69) by the standard techniques, and as expected in view of the general results in next Section 7 Theorem 5, Corollarys 1–3) these appear in terms of $_5\Psi_5$-functions (for the M-S-M operators (68)), in particular, as $_4\Psi_4$-functions (for the Saigo operators (78)) and $_3\Psi_3$-functions (for the R-L and E-K operators). In addition, the pathway integrals (that are related to E-K integrals) are calculated there.

- **The generalized multi-index Bessel function.** In a series of papers, Nisar et al. ([111], 2017, 2019) introduced and studied the function

$$J^{(\alpha_j)_m,\gamma,c}_{(\beta_j)_m,\kappa,b}(z) = \sum_{k=0}^{\infty} \frac{c^k (\gamma)_{\kappa k}}{\prod_{j=1}^{m} \Gamma(\alpha_j k + \beta_j + \frac{b+1}{2})} \frac{z^k}{k!}, \quad m = 1, 2, 3, \ldots, \tag{70}$$

with the Pochhammer symbol denotation (65) for $(\gamma)_{\kappa k}$; and for $\alpha_j, \beta_j, \gamma, b, c \in \mathbb{C}$, $j = 1, 2, \ldots, m$; $\sum_{j=1}^{m} \mathrm{Re}\,(\alpha_j) > \max\{0, \mathrm{Re}\,(\kappa) - 1\}$; $\kappa > 0$, $\mathrm{Re}\,(\beta_j) > 0$, $\mathrm{Re}\,(\gamma) > 0$. As shown in Kiryakova [104], this is *only a very special case of the Wright generalized hypergeometric function* ${}_1\Psi_m$, namely:

$$J^{(\alpha)_m,\gamma,c}_{(\beta_j)_m,\kappa,b}(z) = \frac{1}{\Gamma(\gamma)} \sum_{n=0}^{\infty} \frac{\Gamma(\kappa n + \gamma)}{\prod_{j=1}^{m} \Gamma\left(\alpha_j n + (\beta_j + \frac{b+1}{2})\right)} \frac{(cz)^n}{n!} \tag{71}$$

$$= \frac{1}{\Gamma(\gamma)} {}_1\Psi_m \left[\begin{array}{c} (\gamma, \kappa) \\ (\beta_j + \frac{b+1}{2}, \alpha_j)_{j=1}^{m} \end{array} ; cz \right]$$

$$= \frac{1}{\Gamma(\gamma)} H^{1,1}_{1,m+1} \left[-cz \left| \begin{array}{c} (1-\gamma, \kappa) \\ (0,1), (1-\beta_j - \frac{b+1}{2})_{j=1}^{m} \end{array} \right. \right], \text{ that is, it is also a Fox } H\text{-function.}$$

Then, the R-L fractional integral (21) can be evaluated as part of Kiryakova's general results in next Section 7 (Theorem 5, in particular Corollary 1 for $m = 1$, $\gamma = \beta = 1$), or directly from Kilbas' Theorem 2 in [33], which is a variant of Lemma 1 in Kiryakova [112]. Taking there $p = 1, q = m$, $c_1 = \gamma$, $C_1 = \kappa$, $d_j = \beta_j + \frac{b+1}{2}$, $D_j = \alpha_j$ and $\mu = 1$, one obtains the following R-L image for the multi-index Bessel function (70):

$$I^\lambda \left\{ t^{\delta-1} J^{(\alpha_j)_m,\gamma,c}_{(\beta_j)_m,\kappa,b}(z) \right\} = \frac{1}{\Gamma(\gamma)} z^{\delta+\lambda-1} {}_2\Psi_{m+1} \left[\begin{array}{c} (\gamma, \kappa), (\delta, 1) \\ (\beta_j + \frac{b+1}{2}, \alpha_j)_1^m, (\lambda + \delta, 1) \end{array} ; cz \right].$$

This was to be the result in Theorem 1, Equation (2.4) in *arXiv:1706.08039* [111], its v1: 2017, but was *written wrongly there*—similarly looking but involving a ${}_2\Psi_2$-function. The evident true result involves the Wright function ${}_2\Psi_{m+1}$ (see Kiryakova [104] (5.3.)), as later corrected in v2: 2019 of [111].

- A special case of (70) appears as a kind of *generalized multi-index Mittag–Leffler function*. It was introduced by Saxena and Nishimoto ([113], 2010). As mentioned by Agarwal–Rogosin–Trujillo ([114], 2015), it is representable also as a Wright g.h.f. ${}_1\Psi_m$, namely:

$$E^{(\gamma,\kappa)}_{(\alpha_j,\beta_j)_m}(z) = \sum_{n=0}^{\infty} \frac{(\gamma)_{\kappa n}}{\prod_{j=1}^{m} \Gamma(\beta_j + \alpha_j n)} \cdot \frac{z^n}{n!} = \frac{1}{\Gamma(\gamma)} {}_1\Psi_m \left[\begin{array}{c} (\gamma, \kappa) \\ (\beta_j, \alpha_j)_1^m \end{array} \bigg| z \right]. \tag{72}$$

Therefore, all the GFC operators (say the R-L, E-K, Saigo, M-S-M operators) of this special function can be evaluated by means of the general results in Section 7, Corollaries 1–3 there. For $m = 1, b = -1$, this is the SF considered by Srivastava and Tomovski ([115], 2009):

$$E^{\gamma,\kappa}_{\alpha,\beta}(z) = \sum_{n=0}^{\infty} \frac{(\gamma)_{\kappa n}}{\Gamma(\alpha n + \beta)} \cdot \frac{z^n}{n!}.$$

- Similar, but simpler, is the case of the *generalized Lommel-Wright function* from the paper by Agarwal–Jain–Agarwal–Baleanu ([116], 2018), which is commented in Kiryakova [117]. It has a representation as a Wright g.h.f. as follows:

$$J^{\varphi,m}_{\omega,\theta}(z) = \left(\frac{z}{2}\right)^{\omega+2\theta} \sum_{k=0}^{\infty} \frac{(-1)^k \left(\frac{z}{2}\right)^{2k}}{(\Gamma(\theta+k+1))^m \, \Gamma(\omega + k\varphi + \theta + 1)} \tag{73}$$

$$= \left(\frac{z}{2}\right)^{\omega+2\theta} {}_1\Psi_{m+1}\left[(1,1); (\theta+1,1), \ldots, (\theta+1,1), (\omega+\theta+1,\varphi); -z^2/4\right], \quad \varphi > 0.$$

Note, additionally, that (73) is an example of the multi-index Mittag–Leffler function (39), namely: $J_{\omega,\theta}^{\varphi,m}(z) = (\frac{z}{2})^{\omega+2\theta}(\frac{z}{2})^{\omega+2\theta} E_{(1,...,1,\varphi),(\theta+1,...,\theta+1,\omega+\theta+1)}^{(m+1)}(-(\frac{z}{2})^2)$. Then, all the FC images of (73) evaluated in the commented paper follow at once by our general results (see details in [117]).

6.4. *The S-function*. It was introduced by Saxena-Daiya ([118], 2015) as a "new" special function extending the M-L function ($p = q = 0, k = 1$), the Prabhakar function (38), the M-series (76) of Sharma and Jain ([119], 2009) with $\gamma = 1, k = 1$, etc., as follows:

$$S[z] := S_{(p,q)}^{\alpha,\beta,\gamma,\tau,k}(a_1,...,a_p; b_1,...,b_q; z) = \sum_{n=0}^{\infty} \frac{(a_1)_n...(a_p)_n \cdot (\gamma)_{n\tau,k}}{(b_1)_n...(b_q)_n \cdot \Gamma_k(n\alpha+\beta)} \frac{z^n}{n!}, \quad (74)$$

with $k \in \mathbb{R}$; $\alpha, \beta, \gamma, \tau \in \mathbb{C}$; $\text{Re}(\alpha) > 0$; $\text{Re}(\alpha) > k\text{Re}(\tau)$, $p < q + 1$.

For $p = q = 0$, it reduces to the generalized k-Mittag–Leffler function $E_{k,\alpha,\beta}^{\gamma,\tau}(z)$, a variant of (69). However, it can be easily seen to be special case of the generalized hypergeometric function of Wright of the form $_{p+1}\Psi_{q+1}$. Unfortunately, this fact has not been observed, neither by the authors introducing (74) nor by their numerous followers. Namely, one can write (74) as follows (see details in [104]):

$$S[z] = k^{1-\frac{\beta}{k}} \frac{\Gamma(b_1)...\Gamma(b_q)}{\Gamma(a_1)...\Gamma(a_p) \cdot \Gamma(\frac{\gamma}{k})} {}_{p+1}\Psi_{q+1}\left[\begin{array}{c} (a_1,1),...,(a_p,1),(\frac{\gamma}{k},\tau) \\ (b_1,1),...,(b_q,1),(\frac{\beta}{k},\frac{\alpha}{k}) \end{array} ; zk^{\tau-\frac{\alpha}{k}} \right].$$

That is, the "new" special function $S[z]$ is nothing *but a case of the Wright function* $_{p+1}\Psi_{q+1}\left(zk^{\tau-\frac{\alpha}{k}}\right)$. Then, all results for images of FC operators, as R-L, E-K, Saigo, M-S-M and the Euler-transform, follow from the statements in Section 7.

• Special cases of the S-function in **6.4.** are the *generalized K-series* and the *M-series*. Recently, (K.) Sharma ([120], 2012) introduced an extension of both g.h.f. $_pF_q(z)$ and Prabhakar function $E_{\alpha,\beta}^{\gamma}(z)$:

$$_p K_q^{\alpha,\beta;\gamma}(a_1,...,a_p; b_1,...,b_q; z) = \sum_{n=0}^{\infty} \frac{(a_1)_n...(a_p)_n}{(b_1)_n...(b_q)_n} \frac{(\gamma)_n z^n}{\Gamma(\alpha n+\beta)}, \quad z,\alpha,\beta \in \mathbb{C}, \text{Re}\,\alpha > 0, \quad (75)$$

with integers $p \leq q$ (and additional requirement $|z| < R = \alpha^\alpha$ if $p = q+1$). For $\gamma = 1$, this gives the M-series (76) of (M.) Sharma and Jain ([119], 2009):

$$_p M_q^{\alpha,\beta}(a_1,...,a_p; b_1,...,b_q; z) := {}_p M_q^{\alpha,\beta}(z) = \sum_{n=0}^{\infty} \frac{(a_1)_n ... (a_p)_n}{(b_1)_n ... (b_q)_n} \frac{z^n}{\Gamma(\alpha n+\beta)}$$

$$= \kappa\, {}_{p+1}\Psi_{q+1}\left[\begin{array}{c} (a_1,1),...(a_p,1),(1,1) \\ (b_1,1),...,(b_q,1),(\beta,\alpha) \end{array} \bigg| z \right], \text{ where } \kappa = \prod_{j=1}^{q}\Gamma(b_j)/\prod_{i=1}^{p}\Gamma(a_i). \quad (76)$$

We can mention its particular cases, for example: (1) for $\beta = 1$, the (simpler) M-series, introduced by M. Sharma (2008); (2) for $p = q = 0$ (no upper and lower parameters), M-L function $E_{\alpha,\beta}(z)$; (3) for $p = 0, q = 1, b_1 = 1$, the Wright function $\phi(\alpha,\beta;z)$, or the generalized Bessel-Maitland function (57); (4) for $p = q = 1, a_1 = \gamma, b_1 = 1$ in (75), the Prabhakar type function (38); and (5) for $\alpha = \beta = 1$, the g.h.f. $_pF_q(a_1,...,a_p; b_1,...,b_q; z)$.

In the recent *arXiv* preprint [121], Lavault represented (75) as a Wright g.h.f.:

$$_p K_q^{\alpha,\beta;\gamma}(a_1,...,a_p; b_1,...,b_q; z) := {}_p K_q^{\alpha,\beta;\gamma}(z)$$

$$= \frac{\prod_{j=1}^{q}\Gamma(b_j)}{\Gamma(\gamma)\prod_{i=1}^{p}\Gamma(a_i)} {}_{p+2}\Psi_{q+2}\left[\begin{array}{c} (a_1,1),...,(a_p,1),(\gamma,1),(1,1) \\ (b_1,1),...,(b_q,1),(1,1),(\beta,\alpha) \end{array} \bigg| z \right], \quad (77)$$

although this can also be reduced to: $= \dfrac{\prod_{j=1}^{q}\Gamma(b_j)}{\Gamma(\gamma)\prod_{i=1}^{p}\Gamma(a_i)}{}_{p+1}\Psi_{q+1}\left[\begin{array}{c}(a_1,1),\ldots(a_p,1),(\gamma,1)\\(b_1,1),\ldots,(b_q,1),(\beta,\alpha)\end{array}\Big|z\right],$

since the two pairs $(1,1)$ of parameters in the upper and low rows eliminate each other. In [121] some FC operators of this K-series are calculated, as the R-L, Saigo and M-S-M operators. Naturally, a R-L integral is transforming a ${}_{p}K_{q}^{\alpha,\beta;\gamma}$-function into a ${}_{p+1}K_{q+1}^{\alpha,\beta;\gamma}$-function (Theorem 4.1, there), similarly to our Example 11 in [112] for the M-series. Next, in Theorem 4.2 of [121] for the M-series and Corollary 4.3 for the K-series, the *Saigo operator* (78) (with Gauss hypergeometric function (35), GFC with $m=2$) is derived,

$$I^{\alpha,\beta,\eta}f(z) = \frac{z^{-\alpha-\beta}}{\Gamma(\alpha)}\int_0^z (z-\xi)^{\alpha-1}{}_2F_1(\alpha+\beta,-\eta;\alpha;1-\frac{\xi}{z})f(\xi)d\xi$$

$$= \frac{z^{-\beta}}{\Gamma(\alpha)}\int_0^1 (1-\sigma)^{\alpha-1}{}_2F_1(\alpha+\beta,-\eta;\alpha;1-\sigma)f(z\sigma)d\sigma. \tag{78}$$

Since the K-series (75) is a ${}_{p+1}\Psi_{q+1}$-function, from our results (and Corollary 3 [112]; see also Corollary 2 in the next section), it is expected that the result should be given as a ${}_{p+3}\Psi_{q+3}$-function (the indices are to be increased by 2), which is the result (4.10) in [121]:

$$I^{\alpha,\beta,\gamma}\left\{t^{\sigma-1}{}_{p}K_{q}^{\xi,\eta;\nu}(cz^{\mu})\right\}$$
$$= \frac{\prod_1^q \Gamma(b_j)}{\prod_1^p \Gamma(a_i)}\frac{z^{\sigma-\beta-1}}{\Gamma(\nu)}{}_{p+3}\Psi_{q+3}\left[\begin{array}{c}(a_i,1)_1^p,(\sigma,\mu),(-\beta+\gamma+\sigma,\mu),(\nu,1)\\(b_j)_1^q,(\beta+\sigma,\mu),(\alpha+\gamma+\sigma,\mu),(\eta,\xi)\end{array}\Big|cz^{\mu}\right].$$

Similarly, the M-S-M-images (68) follow as ${}_{p+4}\Psi_{q+4}$-functions, according to Corollary 3 in next section.

6.5. *k-Wright generalized hypergeometric function ${}_p\Psi_q^k$.* Purohit and Badguzer ([122], 2018) introduced the *generalized k-Wright function*, as a k-extension ($k > 0$) of the Wright g.h.f. (4), by

$${}_p\Psi_q^k(z) = {}_p\Psi_q^k\left[\begin{array}{c}(a_1,A_1),\ldots,(a_p,A_p)\\(b_1,B_1),\ldots,(b_q,B_q)\end{array}\Big|z\right] = \sum_{n=0}^{\infty}\frac{\Gamma_k(a_1+nA_1)\ldots\Gamma_k(a_p+nA_p)}{\Gamma_k(b_1+nB_1)\ldots\Gamma_k(b_q+nB_q)}\frac{z^n}{n!}. \tag{79}$$

Replacing the k-Gamma function by the classical Gamma function according to (64), it is seen that the "new" function is *again a Wright generalized hypergeometric function*, of the form

$$\text{const }{}_{p+1}\Psi_{q+1}\left[\begin{array}{c}(a_i/k,A_i/k)_{i=1}^p\\(b_j/k,B_j/k)_{j=1}^q\end{array}\Big|k^{(A_1+\ldots+A_p-B_1-\ldots-B_q)/k}\cdot z\right]. \tag{80}$$

7. Results for the FC and GFC Images of SF of FC

Recently, there have appeared too many papers that deal with evaluation of FC and GFC operators of various special functions. They use the same standard techniques—replace the particular function by its power series, then interchange the orders of integration (fractional order integrals) and summation, etc. Usually only the special functions are changed and also the FC operators—with more and more general ones (but all these happen to be cases of our GFC operators). The *great number of combinations "special function + particular operator" explains the dramatically increasing production of such works*.

Based on our older results on GFC for SF, since the work in [9], in the papers [64,104,112,117,123], and in a recent survey paper [105] in this same journal, we propose an unified approach how this job can be done at once, for all SF of FC (we mean the H- and G-functions and in particular the Wright g.h.f., multi-index M-L functions and all their particular cases) and for all operators of GFC (we mean the generalized fractional integrals and derivatives of the form (25) and (28), thus including the R-L, E-K, Saigo, Marichev–Saigo–Maeda operators, etc.). For the initiating idea, we need to pay tribute to the initial classical results of 20th century in the Bateman Project on Integral Transforms [7] and in works by Askey [2], Lavoie–Osler–Tremblay [124], etc. for the R-L images of many elementary functions and of the simplest $_pF_q$-functions, as: $_0F_1$, $_1F_1$ and $_1F_2$. We combined these with the composition/decomposition rule (26) presenting the GFC operators as compositions of weighted R-L/E-K operators. As a recent survey on FC images of elementary functions, we mention also the work of Garrappa–Kaslik–Popolizio [125].

Below, we remind only the statements of the main results from the mentioned author's papers, as surveyed in [105], in this same journal.

Theorem 4. *The $I_{(\beta_k),m}^{(\gamma_k),(\delta_k)}$-image (23) of a H-function is also a H-function whose last three components of the order are increased by m (the multiplicity in GFC operators), and with additional parameters depending on those of the generalized fractional integration. Namely,*

$$I_{(\beta_k),m}^{(\gamma_k),(\delta_k)}\left\{H_{u,v}^{s,t}\left[\lambda z \left|\begin{array}{c}(c_i,C_i)_1^u\\(d_j,D_j)_1^v\end{array}\right.\right]\right\} = H_{u+m,v+m}^{s,t+m}\left[\lambda z \left|\begin{array}{c}(c_i,C_i)_1^t,(-\gamma_k)_1^m,(c_i,C_i)_{t+1}^u\\(d_j,D_j)_1^s,(-\gamma_k-\delta_k)_1^m,(d_j,D_j)_{s+1}^v\end{array}\right.\right]. \quad (81)$$

Then, GFC images of almost all SF of FC can be evaluated from (81). This result is based on a formula for the integral of product of two arbitrary H-functions, namely for the Mellin transform of such a product ([9] (App., (E.21′)), [12] (§5.1, (5.1.1)), [14] (§2.25, (1))). A similar formula presents the GFC operators (with $G_{m,m}^{m,0}$-kernel) of arbitrary G-function, in terms of another G-function with increased orders and additional parameters (Lemma 1.2.2 in [9] and Corollary 1 in [105]).

Since *most of the considered SF of FC are Wright g.h.f.*, the main and most useful result is as follows.

Theorem 5. *The image of a Wright g.h.f. $_p\Psi_q(z)$ by a generalized fractional integral (23) (multiple, m-tuple Erdélyi-Kober integral), provided $\delta_k \geq 0, \gamma_k > -1, k = 1,\ldots,m, c > -1, \mu > 0, \lambda \neq 0$, is another Wright g.h.f. with indices p and q increased by the multiplicity m and additional parameters related to these of the GFC integral:*

$$I_{(\beta_k)_1^m,m}^{(\gamma_k)_1^m,(\delta_k)_1^m}\left\{z^c \,_p\Psi_q\left[\begin{array}{c}(a_1,A_1),\ldots,(a_p,A_p)\\(b_1,B_1),\ldots,(b_q,B_q)\end{array}\right|\lambda z^\mu\right]\right\}$$

$$= z^c \,_{p+m}\Psi_{q+m}\left[\begin{array}{c}(a_i,A_i)_1^p,(\gamma_k+1+\frac{c}{\beta_k},\frac{\mu}{\beta_k})_1^m\\(b_j,B_j)_1^q,(\gamma_k+\delta_k+1+\frac{c}{\beta_k},\frac{\mu}{\beta_k})_1^m\end{array}\right|\lambda z^\mu\right]. \quad (82)$$

Specially, for $c = 0, \mu = 1$, this result is simplified to $_p\Psi_q(\lambda z) \longmapsto \,_{p+m}\Psi_{q+m}(\lambda z)$, as above.

Similarly (Theorem 4.2 in [104]; Theorem 4 in [105]),

$$D_{(\beta_k)_1^m,m}^{(\gamma_k)_1^m,(\delta_k)_1^m}\left\{z^c \,_p\Psi_q\left[\begin{array}{c}(a_1,A_1),\ldots,(a_p,A_p)\\(b_1,B_1),\ldots,(b_q,B_q)\end{array}\right|\lambda z^\mu\right]\right\}$$

$$= z^c \,_{p+m}\Psi_{q+m}\left[\begin{array}{c}(a_i,A_i)_1^p,(\gamma_k+\delta_k+1+\frac{c}{\beta_k},\frac{\mu}{\beta_k})_1^m\\(b_j,B_j)_1^q,(\gamma_k+1+\frac{c}{\beta_k},\frac{\mu}{\beta_k})_1^m\end{array}\right|\lambda z^\mu\right]. \quad (83)$$

The *simpler results for the $_pF_q$-functions* read by analogy (Corollarys 4.1 and 4.2 in [104]), for example with $\beta = 1$, as:

$$I_{1,m}^{(\gamma_k)_1^m,(\delta_k)_1^m}\left\{z^c \,_pF_q(a_1,\ldots,a_p;b_1,\ldots,b_q;\lambda z)\right\}$$

$$= \left[\prod_{k=1}^m \frac{\Gamma(\gamma_k+c+1)}{\Gamma(\gamma_k+\delta_k+c+1)}\right] z^c \,_{p+m}F_{q+m}(a_i,\ldots,a_p,(\gamma_k+c+1)_1^m;b_1,\ldots,b_q,(\gamma_k+\delta_k+c+1)_1^m;\lambda z). \quad (84)$$

We also describe the corollaries of the results (82) and (83) *for the particular cases of most often FC operators* on which the other authors have exercised their evaluations, say for: $m = 1$ (R-L and E-K), $m = 2$ (Saigo operators) and $m = 3$ (M-S-M operators). These results for arbitrary Wright g.h.f. are mentioned below.

Corollary 1. *For the Riemann–Liouville (R-L) integrals and derivatives, the simplest results are parts of Lemmas 1 and 2 in Kiryakova [105]:*

$$R^\delta \left\{ z^c {}_p\Psi_q \left[\begin{array}{c} (a_1, A_1), \ldots, (a_p, A_p) \\ (b_1, B_1), \ldots, (b_q, B_q) \end{array} \middle| \lambda z^\mu \right] \right\} = z^{c+\delta} {}_{p+1}\Psi_{q+1} \left[\begin{array}{c} (a_i, A_i)_1^p, (c+1, \mu) \\ (b_j, B_j)_1^q, (c+\delta+1, \mu) \end{array} \middle| \lambda z^\mu \right], \quad (85)$$

$$D^\delta \left\{ z^c {}_p\Psi_q \left[\begin{array}{c} (a_1, A_1), \ldots, (a_p, A_p) \\ (b_1, B_1), \ldots, (b_q, B_q) \end{array} \middle| \lambda z^\mu \right] \right\} = z^{c-\delta} {}_{p+1}\Psi_{q+1} \left[\begin{array}{c} (a_i, A_i)_1^p, (c+1, \mu) \\ (b_j, B_j)_1^q, (c+1-\delta, \mu) \end{array} \middle| \lambda z^\mu \right]. \quad (86)$$

The results for the E-K operators have same expressions as in (82) *and* (83) *with* $m = 1$.

Corollary 2. *The images of the Wright g.h.f.* ${}_p\Psi_q$ *and, in particular, of the g.h.f.* ${}_pF_q$ *under the Saigo operators* (78) *are given by the formulas:*

$$I^{\alpha,\beta,\eta} \left\{ z^c {}_p\Psi_q \left[\begin{array}{c} (a_i, A_i)_1^p \\ (b_j, B_j)_1^q \end{array} \middle| \lambda z^\mu \right] \right\} = z^{c-\beta} {}_{p+2}\Psi_{q+2} \left[\begin{array}{c} (a_i, A_i)_1^p, (\eta - \beta + 1 + c, \mu), (1 + c, \mu) \\ (b_j, B_j)_1^q, (-\beta + 1 + c, \mu), (\alpha + \eta + 1 + c, \mu) \end{array} \middle| \lambda z^\mu \right], \quad (87)$$

(*for* $c = 0$, $\mu = 1$, *this is Corollary 3 in* [112]) *and*

$$I^{\alpha,\beta,\eta} \left\{ {}_pF_q(a_1, \ldots, a_p; b_1, \ldots, b_q; \lambda z) \right\} = z^{-\beta} {}_{p+2}F_{q+2}(a_1, \ldots, a_p, \eta - \beta + 1, 1; b_1, \ldots, b_q, -\beta + 1, \alpha + \eta + 1; \lambda z). \quad (88)$$

Corollary 3. *The Marichev–Saigo–Maeda (M-S-M) operators* (68) *transform a Wright g.h.f. function into same kind of special function but with indices increased by 3:*

$$I^{a,a',b,b',c} \left\{ {}_p\Psi_q \left[\begin{array}{c} (a_i, A_i)_1^p \\ (b_j, B_j)_1^q \end{array} \middle| \lambda z^\mu \right] \right\}$$

$$= z^{c-a-a'} {}_{p+3}\Psi_{q+3} \left[\begin{array}{c} (a_i, A_i)_1^p, (a - a' + 1, 1), (b - a' + 1, 1), (c - 2a' - b' + 1, 1) \\ (b_j, B_j)_1^q, (a - a' + b + 1, 1), (c - 2a' + 1, 1), (c - a' - b' + 1, 1) \end{array} \middle| \lambda z^\mu \right]. \quad (89)$$

We state here also the more general result for images of arbitrary Wright generalized hypergeometric function *in the case of multiple Wright–Erdélyi–Kober operators* (33).

Theorem 6. (Kiryakova, [60], Theorem 9) *The image of a Wright generalized function* ${}_p\Psi_q(z)$ *by a multiple W-E-K operator* (33) *has the form*

$$I^{(\gamma_k),(\delta_k)}_{(\beta_k),(\lambda_k),m} \left\{ {}_p\Psi_q \left[\begin{array}{c} (a_1, A_1), \ldots, (a_p, A_p) \\ (b_1, B_1), \ldots, (b_q, B_q) \end{array} \middle| z \right] \right\} = {}_{p+m}\Psi_{q+m} \left[\begin{array}{c} (a_j, A_j)_1^p; (\gamma_k + 1, 1/\lambda_k)_1^m \\ (b_k, B_k)_1^q; (\gamma_k + \delta_k + 1, 1/\beta_k)_1^m \end{array} \middle| z \right]. \quad (90)$$

Conversely, the alternatively stated result reads as: each ${}_{p+m}\Psi_{q+m}$-*function can be represented by means of a multiple (m-tuple) operator* $\widetilde{I}$ *of GFC, of a* ${}_p\Psi_q$-*function, the orders of which are reduced by m:*

$${}_{p+m}\Psi_{q+m} \left[\begin{array}{c} (a_j, A_j)_{j=1}^p; (a_{p+i}, A_{p+i})_{i=1}^m \\ (b_k, B_k)_{k=1}^q; (b_{q+i}, B_{q+i})_{i=1}^m \end{array} \middle| z \right] = \widetilde{I} \left\{ {}_p\Psi_q \left[\begin{array}{c} (a_j, A_j)_{j=1}^p \\ (b_k, B_k)_{k=1}^q \end{array} \middle| z \right] \right\}, \quad (91)$$

with

$$\widetilde{I} f(z) = I^{(a_{p+i}-1)_{i=1}^m, (b_{q+i}-a_{p+i})_{i=1}^m}_{(1/B_{q+1})_{i=1}^m, (1/A_{p+1})_{i=1}^m, m} f(z) \quad \text{of the form} \quad (33).$$

A long list of examples how these general results work at once for any of the SF of FC mentioned in previous sections is provided in author's works [104,105,112,117,123], including some of the particular cases of W.g.h.f. and of multi-index M-L f., mentioned in

8. Theory of SF of FC in View of GFC Operators

Usually, the special functions of mathematical physics are defined by means of power series representations. However, some alternative representations can be used as their definitions. Let us mention the well-known *Poisson integral* (52) for the Bessel function and the analytical continuation of the Gauss hypergeometric function via the *Euler integral formula*. The *Rodrigues differential formulas*, involving repeated or fractional differentiation are also used as definitions of the classical orthogonal polynomials and their generalizations. As to the other special functions (most of them being $_pF_q$- and $_p\Psi_q$-functions), such representations have been less popular and even unknown in the general case. There exist various integral and differential formulas, but, unfortunately, quite peculiar for each corresponding special function and scattered in the literature without any common idea to relate them.

In our works since 1985 (e.g., [9] (Ch.4), [58,60]), we showed that all the classical SF and the SF of FC (in the sense of generalized hypergeometric functions $_pF_q$ and $_p\Psi_q$) can be presented by means of generalized fractional integrals or derivatives of three basic elementary functions. On this basis, these special functions have been classified into three specific classes, and several new integral and differential representations have been proposed under a unified idea. Besides, for these three classes of SF, *we provide analogs of the mentioned Poisson and Euler integral formulas and of the Rodrigues differential formulas*, which can also be used for *alternative definitions of these special functions, their analytical extensions or for numerical algorithms*.

The idea is briefly explained as follows: (i) most of the classical SF (SF of mathematical physics) and SF of FC are nothing but modifications of the g.h.f. $_pF_q$ or $_p\Psi_q$; (ii) each $_pF_q$-function or $_p\Psi_q$-function can be represented as an E-K fractional *differintegral* (i.e., integral or derivative) of a $_{p-1}F_{q-1}$-function or $_{p-1}\Psi_{q-1}$, respectively; (iii) a finite number of steps (ii) leads to one of the basic g.h.f. ($_0F_{q-p}$ (for $q-p=1$: Bessel function); $_1F_1$ (confluent h.f.) and $_0F_0$ (exponent); and $_2F_1$ (Gauss h.f.) and $_1F_0$ (beta-distribution) to the simplest functions $_0\Psi_{q-p}, _1\Psi_1, _1\Psi_0$, respectively; (iv) the above three basic g.h.f. can be considered themselves as fractional differintegrals of the three elementary functions, depending on whether $p < q, p = q$, or $p = q+1$; and (v) the compositions of E-K operators arising in Step (iii) give generalized (q-tuple) fractional integrals or derivatives.

Thus, for the *simpler case of $_pF_q$-functions*, we have the following general proposition.

Theorem 7. (Kiryakova [58]) *All the generalized hypergeometric functions $_pF_q$ can be considered as generalized (q-tuple) fractional differintegrals* (24), (30) *(with $G^{m,0}_{m,m}$-kernels) of one of the elementary functions*

$$\cos_{q-p+1}(z) \ (\text{if } p < q), \ z^\alpha \exp z \ (\text{if } p = q), \ z^\alpha (1-z)^\beta \ (\text{if } p = q+1), \quad (92)$$

depending on whether $p < q \ \ p = q \ \ p = q+1$.

It is based on the known auxiliary result coming yet from the Bateman Project on integral transforms [7], Askey [2], Lavoie–Osler–Tremblay [124] for the R-L derivatives that we have *paraphrased in terms of E-K operators* (e.g., Equation (4.2.2′) in [9] and Lemma 3.2 in [58]) as follows:

$$\frac{\Gamma(a_p)}{\Gamma(b_q)} \, _pF_q(a_1, \ldots, a_p; b_1, \ldots, b_q; z)$$

$$= \begin{cases} I^{a_p-1, b_q-a_p}_{1,1} \left\{ _{p-1}F_{q-1}(a_1, \ldots, a_{p-1}; b_1, \ldots, b_{q-1}; z) \right\} & \text{if } b_q > a_p, \\ D^{b_q-1, a_p-b_q}_{1,1} \left\{ _{p-1}F_{q-1}(a_1, \ldots, a_{p-1}; b_1, \ldots, b_{q-1}; z) \right\} & \text{if } b_q < a_p, \end{cases} \quad (93)$$

for all complex z, and if $p = q+1$ we require additionally $|z| < 1$. Then, this basic fact is to be used repeatedly, and combined with the composition/decomposition property (26) for the operators of GFC. In each of the three separate cases, we reach to one of the basic functions (92) with smallest possible first index p, namely: $_0F_{q-p}(z) = \cos_{q-p+1}(z)$; $_1F_1(z)$ and then $_0F_0(z) = \exp z$; and $_2F_1(z)$ and then $_1F_0(\beta;-;z) = (1-z)^{-\beta}$.

For the *Wright generalized hypergeometric functions* (4), this proposition reads almost the same, only the third basic function (for $p = q+1$) is more general, namely $_1\Psi_0 = H_{1,1}^{1,1}$, and the GFC operators have as kernel the $H_{m,m}^{m,0}$-function with different parameters βs and λs in the upper and low rows.

Theorem 8. (Kiryakova [60] (Theorem 14)) *All the Wright generalized hypergeometric functions $_p\Psi_q$ can be represented as multiple (q-tuple) W-E-K fractional integrals (33), or their corresponding fractional derivatives, of one of the following three basic functions:*

$$\cos_{q-p+1}(z) \ (\text{if } p<q), \quad \exp z \ (\text{if } p=q), \quad _1\Psi_0[(a,A)\,|\,z] \ (\text{if } p=q+1). \qquad (94)$$

In this case, the basic used result is Theorem 6, following similar Steps (i)–(v) as described above.

The three cases, for both Theorems 7 and 8, are considered in detail, in separate statements.

(1) $p < q$. The Poisson integral representation (52) is extended in [9] (Ch.4) and [90] for the *hyper-Bessel functions* (58), $m \geq 2$, that is for the $_0F_{m-1}$-*functions*, via generalized fractional integrals (24) of the function $\cos_m$, (54) as follows:

$$J_{\nu_1,\ldots,\nu_{m-1}}^{(m-1)}(z) = \sqrt{\frac{m}{(2\pi)^{m-1}}} \left(\frac{z}{m}\right)^{\nu_1+\ldots+\nu_{m-1}} I_{\frac{1}{m},m-1}^{(\frac{k}{m}-1),(\nu_k-\frac{k}{m}+1)}\{\cos_m(z)\}. \qquad (95)$$

By analogy with the hyper-Bessel functions (58), we consider what we call *the Wright hyper-Bessel functions*:

$$_0\Psi_m\left[\begin{array}{c}-\\(b_1,B_1),\ldots,(b_m,B_m)\end{array}\Big|-z\right] = H_{0,m+1}^{1,0}\left[z\,\Big|\,\begin{array}{c}-\\(0,1),(1-b_1,B_1),\ldots,(1-b_m,B_m)\end{array}\right]$$

$$= \sum_{k=0}^{\infty} \frac{z^k}{\Gamma(b_1+kB_1)\ldots\Gamma(b_m+kB_m)\cdot k!} := J_{b_1-1,\ldots,b_m-1}^{B_1,\ldots,B_m}(z). \qquad (96)$$

The latter denotation is to remind of the analogy with the hyper-Bessel functions (58), when $\forall B_k = 1$. It is easy to observe that (96) appears as special case of the multi-index Mittag–Leffler functions (39), namely: $J_{b_1-1,\ldots,b_m-1}^{B_1,\ldots,B_m}(z) = E_{(1,B_1,\ldots,B_m),(1,b_1,\ldots,b_m)}^{(m+1)}(-z)$.

We have then a result, analogous to (95), and more general than (53) for the multi-M-L functions, that: *each Wright hyper-Bessel function $_0\Psi_{q-p}$, $p < q$, can be represented by means of a Poisson type integral of the $\cos_{p-q+1}$-function*, written in the form

$$_0\Psi_{q-p}\left[\begin{array}{c}-\\(b_1,B_1),\ldots,(b_{q-p})\end{array}\Big|-z\right] = J_{b_1-1,\ldots,b_m-1}^{B_1,\ldots,B_m}(z)$$

$$= I_{(\frac{1}{B_k}),(1),q-p}^{(\frac{k}{q-p+1}-1),(b_k-\frac{k}{q-p+1})}\left\{\cos_{q-p+1}\left((q-p+1)z^{\frac{1}{q-p+1}}\right)\right\}. \qquad (97)$$

Let us now apply to the function $_0\Psi_{q-p}$ above, p-times the results (90), (91) (Theorem 6) with $m=1$, combined with the composition rule for the W-E-K integrals (33). Then, we obtain the following:

Theorem 9. (Kiryakova [60] (Theorem 15)) *Each $_p\Psi_q$-function with $p < q$ is a generalized q-tuple W-E-K fractional (differ-)integral operator of $\cos_{q-p+1}(z)$,*

$$
{}_p\Psi_q\left[\begin{array}{c}(a_1,A_1),\ldots,(a_p,A_p)\\(b_1,B_1),\ldots,(b_q,B_q)\end{array}\bigg|-z\right]=I^{(\gamma_k),(\delta_k)}_{(\frac{1}{B_k}),(\lambda_k),q}\left\{\cos_{q-p+1}((q-p+1)z^{\frac{1}{q-p+1}})\right\}, \quad (98)
$$

with the following parameters:

$$
\gamma_k=\begin{cases}\frac{k}{q-p+1}-1,\\a_{k-q+p}-1,\end{cases}; \quad \delta_k=\begin{cases}b_k-\frac{k}{q-p+1},\\b_k-a_{k-q+p},\end{cases}; \quad \lambda_k=\begin{cases}1, & k=1,\ldots,q-p\\\frac{1}{A_{k-q+p}}, & k=q-p+1,\ldots,q.\end{cases}
$$

If the following conditions are satisfied:

$$
b_k>\frac{k}{q-p+1},\ k=1,\ldots,q-p;\ b_k>a_{k-q+p}>0,\ k=q-p+1,\ldots,q,
$$

$$
B_k\geq 1,\ k=1,\ldots,q-p;\ B_k\geq A_{k-q+p},\ k=q-p+1,\ldots,q,
$$

then relation (98) gives a Poisson type integral representation; otherwise, the operator in the R.H.S. should be interpreted as a multiple W-E-K derivative (see, e.g., Definition 7 in [60]), and then (98) turns into a new Rodrigues type differential formula, or a mixed differ-integral representation.

The particular case of Poisson type representation (53) for the multi-index M-L function has been already stated as Theorem 3 in Section 4.

In the other two cases, $p=q$ and $p=q+1$, the starting results for ${}_p\Psi_q$ were formulated as Lemmas 11 and 12 in Kiryakova [60]:

$$
{}_1\Psi_1\left[\begin{array}{c}(a_1,A_1)\\(b_1,B_1)\end{array}\bigg|z\right]=W^{a_1-1,b_1-a_1}_{1/B_1,1/A_1}\{\exp z\},\ \text{if}\ A_1\geq B_1, b_1\geq a_1,\ \text{for}\ |z|<\infty; \quad (99)
$$

$$
{}_2\Psi_1\left[\begin{array}{c}(a_1,A_1),(a_2,A_2)\\(b_1,B_1)\end{array}\bigg|z\right]=W^{a_1-1,b_1-a_1}_{1/B_1,1/A_1}\left\{{}_1\Psi_0\left[\begin{array}{c}(a_2,A_2)\\-\end{array}\bigg|z\right]\right\} \quad (100)
$$

$$
=W^{a_1-1,b_1-a_1}_{1/B_1,1/A_1}\left\{H^{1,1}_{1,1}\left[-z\bigg|\begin{array}{c}(1-a_2,A_2)\\(0,1)\end{array}\right]\right\},
$$

if $A_1\geq B_1, b_1\geq a_1$; and if $A_2<1$, for $|z|<\infty$; or if $A_2=1$, for $|z|<1$.

After additional $(p-1)$ steps, from ${}_p\Psi_q$ passing via ${}_1\Psi_1$ to ${}_0\Psi_0$, respectively, to ${}_1\Psi_0$, the following explicit results for the statement in Theorem 8 are provided in [60].

(2) $p=q$.

Theorem 10. *If $p=q$, each g.h.f. ${}_p\Psi_p(z)$ is an p-tuple W-E-K fractional integral of the exponential function, namely, if $B_k\geq A_k>0, b_k>a_k>0, k=1,\ldots,p$:*

$$
{}_p\Psi_q\left[\begin{array}{c}(a_1,A_1),\ldots,(a_p,A_p)\\(b_1,B_1),\ldots,(b_p,B_p)\end{array}\bigg|z\right]=I^{(a_k-1),(b_k-a_k)}_{(\frac{1}{B_k}),(\frac{1}{A_k}),p}\{\exp z\},\ \text{for}\ |z|<\infty. \quad (101)
$$

If for some indices k, the above inequalities for parameters are not satisfied, representation (101) turns into differ-integral one, or in special cases to purely differential one.

Theorem 10 suggests us to separate the g.h.f-s ${}_p\Psi_p$ with $p=q$ in a class of so-called *Wright g.h.f. of confluent type*, involving the *confluent hypergeometric function* ${}_1F_1(a;b;z)=\Phi(a;b;z)$ and $\exp z$ as the simplest cases.

(3) $p=q+1$. Analogously, we call the ${}_{q+1}\Psi_q$-functions with $p=q+1$ as *Wright g.h.f. of Gauss type*, since the simplest case of such special function is the Gauss function. We have following specific result.

Theorem 11. *Each Wright g.h.f. of Gauss type ${}_p\Psi_q$, that is with $p=q+1$, is a q-tuple Wright–Erdélyi–Kober fractional integral (or differ-integral) of the ${}_1\Psi_0$-function. Namely, for $0<A_0\leq 1$ and $b_k>a_k>0, k=1,\ldots,p$:*

$$_{q+1}\Psi_q\left[\begin{matrix}(a_0,A_0),(a_1,A_1),\ldots,(a_q,A_q)\\(b_1,B_1),\ldots,(b_q,B_q)\end{matrix}\Bigg|z\right] = I_{(\frac{1}{B_k}),(\frac{1}{A_k}),q}^{(a_k-1),(b_k-a_k)}\left\{{}_1\Psi_0\left[\begin{matrix}(a_0,A_0)\\-\end{matrix}\Bigg|z\right]\right\} \quad (102)$$

$$= I_{(\frac{1}{B_k}),(\frac{1}{A_k}),q}^{(a_k-1),(b_k-a_k)}\left\{H_{1,1}^{1,1}\left[-z\Bigg|\begin{matrix}(1-a_0,A_0)\\(0,1)\end{matrix}\right]\right\}, \text{ if } A_0 < 1 \text{ for } |z| < \infty; \text{ or if } A_0 = 1, \text{ for } |z| < 1.$$

For other arrangements between b_k and a_k, the operator in (102) is a *generalized fractional derivative*.

For particular choices of parameters b_k and a_k not satisfying the conditions $b_k > a_k > 0$, some *integer order differentiations* appear in place of the fractional integrals or derivatives and lead to *Rodrigues type differential formulas*, analogous to these for the classical orthogonal polynomials.

Note that the integral representation (102) generalizes the *Euler integral formula for the Gauss hypergeometric functions* that serves for its analytical extension outside $|z| < 1$ to the domain $|\arg(1-z)| < \pi$:

$$_2F_1(a_1,a_2;b_1;z) = \frac{\Gamma(b_1)}{\Gamma(a_2)\Gamma(b_1-a_2)}\int_0^1 \frac{(1-\sigma)^{b_1-a_2-1}\sigma^{a_2-1}}{(1-z\sigma)^{a_1}}d\sigma, \quad b_1 > a_2 > 0. \quad (103)$$

This gave us the reason to name $_p\Psi_q$ with $p = q+1$ as a *Gauss type g.h.f.*

In particular, for $A_0 = 1$, the basic function in the case $p = q+1$ reduces to the geometric series:

$$_1\Psi_0\left[(a_0,1)\Big|z\right] = H_{1,1}^{1,1}\left[-z\Bigg|\begin{matrix}(1-a_0,1)\\(0,1)\end{matrix}\right] = G_{1,1}^{1,1}\left[-z\Bigg|\begin{matrix}1-a_0\\0\end{matrix}\right] = {}_1F_0(a_0;-;z) = (1-z)^{-a_0}.$$

Therefore, based on the statements in Theorems 7–11, we suggest a *classification of the classical SF and of the SF of FC into three classes*, namely "Bessel", "confluent" and "Gauss" types, depending on whether $p < q$, $p = q$ or $p = q+1$. This approach can facilitate applied scientists and engineers, escaping a deep knowledge on SF, to think of them in a very general view as similar to a cosine- (Bessel) function, exponent or geometric series, because the fractional integrations keep in some sense the asymptotic and general behavior.

The results from Theorems 7–11 for $_p\Psi_q$, and their specifications for the $_pF_q$-functions, yield also *several new integral and differential formulas for them*, with possible hints for computational procedures.

Below, we mention some few of them, say in the *simpler cases of $_pF_q$-functions*.

The case $\underline{p = q}$: For the g.h.f. $_pF_p$, the integral representation can be written not only by means of $G_{p,p}^{p,0}$-functions in the kernel, but also avoiding SF due to decomposition property (26). Thus, we have an integral formula, as follows:

$$_pF_p(a_1,\ldots,a_p;b_1,\ldots,b_p;z) = B\, z^{1-a_1}\, I_{1,p}^{(a_k-a_1),(b_k-a_k)}\left\{z^{a_1-1}\exp z\right\}$$

$$= B\int_0^1\ldots\int_0^1\prod_{k=1}^{p}\left[\frac{(1-\sigma_k)^{b_k-a_k-1}\sigma_k^{a_k-1}}{\Gamma(b_k-a_k)}\right]\exp(z\sigma_1\ldots\sigma_p)\,d\sigma_1\ldots d\sigma_p,\ B:=\prod_{j=1}^{p}\frac{\Gamma(b_j)}{\Gamma(a_j)}, \quad (104)$$

under conditions $b_k > a_k > 0, k = 1,\ldots,p$. If the parameters do not satisfy them, the GFC operator above is interpreted as generalized fractional derivative of the form (30).

Specially, let all the differences $a_k - b_k = \eta_k, k = 1,\ldots,p$ be *non-negative integers*. In this case, we call the $_pF_p$-functions as "*spherical*" *g.h.f. of confluent type*, using the analogy with the spherical Bessel, hyper-Bessel functions and spherical multi-M-L functions $E_{(\alpha_i),(\beta_i)}(z)$ with "semi-integer" indices, mentioned in Remark 1. Then, the operator in (104) turns into a differential operator D_η of integer order $\eta = \eta_1 + \ldots + \eta_p \geq 0$ of the form (27), and we obtain a differential formula of the form

$$\frac{\Gamma(a_p)}{\Gamma(b_q)}\,_pF_q(a_1,\ldots,a_p;b_1,\ldots,b_q;z)\,_pF_p(b_1+\eta_1,\ldots,b_p+\eta_p;b_1,\ldots,b_p;z)$$

$$= {}_pF_p(b_1+\eta_1,\ldots,b_p+\eta_p;b_1,\ldots,b_p;z)$$

$$= \left[\prod_{j=1}^p \frac{\Gamma(b_j)}{\Gamma(b_j+\eta_j)}\right]\left[\prod_{k=1}^p\prod_{j=1}^{\eta_k}(z\frac{d}{dz}+b_k+j-1)\right]\{\exp z\} = Q_p(z)\{\exp z\}. \quad (105)$$

The representation (105) gives an example how differential formulas for the "spherical" g.h.f. introduced by Kiryakova [9] can be used for their explicit calculation, especially in the case $p = q$ in the form $Q_p(z)\{\exp z\}$, where $Q_p(z)$ is a p-degree polynomial. A special case of (105) with $b_k = \eta_k = 1, k=1,\ldots,p$ and $Q_p(z) = \dfrac{d}{dz}\left(z\dfrac{d}{dz}\right)^p$ was presented by Prudnikov–Brychkov–Marichev [14] (p. 593).

The case $p = q + 1$: For the Gauss type g.h.f. $_{q+1}F_q$, we have in the unit disk $|z| < 1$ an integral representation (if written by repeated integral with no use of the kernel $G_{q,q}^{q,0}$-function), for $b_k > a_{k+1}$, $k = 1,\ldots,q$:

$$_{q+1}F_q(a_1,\ldots,a_{q+1};b_1,\ldots,b_q;z)$$

$$= \left[\prod_{j=1}^q \frac{\Gamma(b_j)}{\Gamma(a_{j+1})\Gamma(b_j-a_{j+1})}\right] z^{1-a_2} I_{1,q}^{(a_{k+1}-1)_1^q,(b_k-a_{k+1})_1^q}\left\{z^{a_2-1}(1-z)^{-a_1}\right\}$$

$$= \left[\prod_{j=1}^q \frac{\Gamma(b_j)}{\Gamma(a_{j+1})\Gamma(b_j-a_{j+1})}\right] \int_0^1\ldots\int_0^1 \prod_{j=1}^q \left[(1-\sigma_k)^{b_k-a_{k+1}-1}\sigma_k^{a_{k+1}-1}\right](1-z\sigma_1\ldots\sigma_q)^{-a_1}d\sigma_1\ldots d\sigma_q. \quad (106)$$

In this form, (106) can also be found in [14] (p. 438). In the case $q = 1$, this is exactly the *Euler integral formula* (103) for the Gauss hypergeometric function. Similarly, (106) proposes a way for an analytical continuation of the functions $_{q+1}F_q(z)$ outside the unit disk to the domain $|\arg(1-z)| < \pi$.

In the case when the a_k's and b_k's do not satisfy the above conditions, the operator in (106) turns into a generalized fractional derivative, and this also provides useful corollaries. By analogy with the previous two cases ($p < q$ and $p = q$), we introduce the notion of *spherical g.h.f. of Gauss type* when all the differences $a_{k+1} - b_k = \eta_k$, $k = 1,\ldots,q$ are non-negative integers. Then, $_{q+1}F_q(z)$ is representable by a purely differential operator of a function $(1-z)^{-a_1}$, and a special case of such differential formula is presented in [14] (p.572).

Another interesting case concerns the so-called *hypergeometric polynomials*

$$_{q+1}F_q(-n,a_1,\ldots,a_p;b_1,\ldots,b_q;z) = \sum_{k=0}^n \frac{(-n)_k (a_1)_k\ldots(a_p)_k}{(b_1)_k\ldots(b_q)_k}\frac{z^k}{k!}. \quad (107)$$

Taking $a_{q+1} = -n$ with integer $n \geq 0$ and $a_k > b_k > 0, k = 1,\ldots,q$, the fractional derivative form of the operator in (106) provides the *Rodrigues type formula* ([9] (Ch.4)):

$$\left[\prod_{j=1}^q \frac{\Gamma(a_j)}{\Gamma(b_j)}\right]{}_{p+1}F_q(-n,a_1,\ldots,a_q;b_1,\ldots,b_q;z) = D_{1,q}^{(b_k-1),(a_k-b_k)}\{(1-z)^n\}$$

$$= z^{1-a_q}D_{1,q}^{(b_k-a_q),(a_k-b_k)}\left\{z^{a_1-1}(1-z)^n\right\} = z^{1-b_q}D^{a_q-b_q}z^{a_p-b_q-1}D^{a_{p-1}-b_{q-1}} \quad (108)$$

$$\times \ldots z^{a_3-b_2}D^{a_2-b_2}z^{a_2-b_1}D^{a_1-b_1}\left\{z^{a_1-1}(1-z)^n\right\}.$$

Special cases of (108) yield some *classical Rodrigues formulas*. For example, $p = q = 1$ with $a_1 = n+1, b_1 = 1$ and $z \to \dfrac{1-z}{2}$ gives the Rodrigues formula for the Legendre polynomials:

$$P_n(z) = (-1)^n {}_2F_1(-n,n+1;1;\frac{1-z}{2}) = \frac{(-1)^n}{n!}\frac{d^n}{dz^n}\left[\frac{1-z}{2}^n \cdot \frac{1+z}{2}^n\right]$$

$$= \frac{1}{2^n n!}\frac{d^n}{dz^n}\left\{(z^2-1)^n\right\},$$

and $p = q = 2$ with $a_1 = n+1, b_1 = 1, a_2 = \zeta, b_2 = p\, (\zeta > p > 0)$ gives the *Rodrigues formula for the Rice polynomials*, namely

$$R_n(z) = {}_3F_2(-n, n+1, \zeta; 1, p; z) = \frac{\Gamma(p)}{n!\,\Gamma(\zeta)} \left[\frac{d^n}{dz^n} z^{1-p} \left(\frac{d}{dz} \right)^{\zeta-p} \right] \{z^n(1-z)^n\}.$$

9. Numerical Aspects of SF of FC

In the days before the electronic computers, the necessary complement to a special function was the computation, *by hand*, of extended *tables* of its values, intended to make the function available for users, similarly to the familiar logarithm tables. After *mechanical calculators* appeared and were more widespread, several huge special-function-table projects were started. Let us mention as examples the handbooks Gradshteyn–Ryzhik [126] and Magnus–Oberhettinger [127], both initiated in 1943.

R. Askey (at the Conf. "SF 2000", ASU): "... The advent of *fast computing machines* was thought to have made special functions a subject of the past. The reality has been different. Continued development of older functions and *the introduction of new special functions has been the reality* ... and still remains to be discovered ... The classical handbooks as mentioned, although useful as references, maybe no longer the primary means of accessing the special functions of mathematical physics. A number of high level programs appeared that are better suited for this challenging purpose, to mention as *Mathematica, Maple, Matlab, Mathcad,* ..."

We like to add a citation from Stephen Wolfram [128] (Wolfram *Mathematica*), "... and special functions became a big business. Table making had become a major activity for the governments, and was thought strategically important. Particularly for things like navigation, nuclear physics, military reasons, H-bomb, etc. And there were lots of tables ... The aspects of the theory then mattered might be as two: – for numerical analysis, discovery of infinite series or other analytical expression allowing rapid calculation; and – reduction of as many functions as possible to the given (better known) function ..." (*Author's comment*: compare with the approach applied in works of Kiryakova as [9] (Ch.4), [58,60], discussed in Section 8). (S. Wolfram:): "... There gradually started to appear systematic reference works on the properties of special functions. Each one based on lot of work ...", "... I guess integrals are timeless. *They don't really bear the marks of the human creators.* So we have the tables, but we really don't quite know where they came from ...". (*Author's comment*: However, it seems Marichev knew, and we refer to his book [11]).

In the 1960s and 1970s, a lot of efforts started for developing numerical algorithms for computers. Evaluation of special functions became a favorite area. S. Wolfram: "Well, a few years passed. And in 1986, I started designing *Mathematica*. I wanted to be sure to do a definitive job, and to have good numerics for all functions, for all values of parameters, anywhere in the complex plane, to any precision ... And I remember very distinctly a phone call I had with someone at a government lab. And there was a *silence*. And then he said: "Look, you have to understand that *by the end of the 1990s* we hope to have the *integer-order* Bessel functions done to quad precision." ... (S.W., cont'd): "You know, it's actually quite a difficult thing to put a special function into *Mathematica*. You don't just have to do the numerics... So what makes a special function good? Well, we can start thinking about that question empirically. Asking what's true about the special functions we normally use. And of course, from what we have in *Mathematica* and in our *Wolfram Functions Site* [129], we should be in the best position to answer this."

Let us note that the standard SF—the hypergeometric functions (Gauss, $_pF_q$-), the Meijer G-function, etc.—are well presented there ([129]), but (it seems) *none of the M-L type, Wright and H-functions, that is cases of SF of FC, are available yet.* Meanwhile, the fractional nature of the world needs better reflection by fractional order (FO) models in whose solutions the so-called SF of FC appear. *Thus, it is yet a challenging trend to be developed.*

Here, we try to provide only a *short information on some "recent" numerical jobs done* with respect to the M-L-function, classical Wright function and only few of their extensions.

For numerical algorithms and results in the case of *the Mittag–Leffler functions* (one- and two-parameter and matrix analog), we start with reference to Caputo–Mainardi [130] (1971). We note that this is one of the first works to propose a plot of the M-L function. On those times, without possibilities to take advantage of software packages such as *Mathematica*, *Maple* and *Matlab*, this task was difficult, as it was managing series expansions convergent only in the mathematical sense but not in the numerical sense. Further, some other authors worked on similar problems either simultaneously but independently, or in years afterward: Gorenflo–Loutchko–Luchko [131], 2002; Diethelm–Ford–Freed–Luchko [132], 2005; Podlubny [133], 2005–2009–2012, (v 1.2.0.0) 2021; Hilfer–Seybold and Seybold–Hilfer [134,135], 2006–2008; Garrappa [136], 2015 and Garrappa–Popolizio [137], 2018; etc.

Numerical algorithms and results on the *(classical) Wright function* (56) *and its special cases, including the Mainardi function*, can be found in works by Luchko [138], 2008; Luchko–Trujillo–Velasco [139], 2010; Consiglio [140], 2019; Mainardi–Consiglio [97], 2020; etc.

Concerning the *Prabhakar (three-parameter) M-L type function* (38), see Garrappa [136], 2015; etc.

For the generalized exponential integrals as (63) and related generalized trigonometric functions involving M-L functions (in the sense of **6.2.**), and shown to be Wright g.h.f. $_2\Psi_2$, one can find some tables and plots for physically interesting parameters and related models, proposed by Mainardi and Masina [102] (2018) and Paris [103] (2020).

This list can surely be extended with more information.

We would like to attract readers' attention to the *challenging Open Problem*: What about possibilities for numerical and graphical interpretations, plots and tables and implementing software packages for some more general Special Functions of Fractional Calculus, such as the multi-index Mittag-Leffler functions? At least, to treat illustrative examples for few typical sets if multi-indices?

10. Conclusions

In this survey, under the notion of *Special Functions of Fractional Calculus (SF of FC)*, we have in mind the Fox H-function and the Wright generalized hypergeometric functions $_p\Psi_q$, including the Mittag–Leffler function, its multi-index extensions and all their particular cases. The standard (classical) special functions (SF) naturally come as part of this scheme, as cases of the Meijer G-function and of the $_pF_q$-functions, including so many named SF and orthogonal polynomials. Here, we try to review some of the basic results on the theory of the SF of FC, obtained in author's works over more than 30 years, and support the wide spreading and important role of these functions by several examples.

The short outline of the contents is as follows:

In Section 1, we start with a historical introduction to pay tribute to the older projects that gave life to the contemporary development of the topic. Some short definitions and facts on the considered basic special functions are given in Section 2. In Section 3, we pay attention to the use of the H- and G-functions, especially of orders $(m, 0; 0, m)$ and $(m, 0; m, m)$, as kernel-functions of generalized integral transforms of Laplace type and of the operators of the so-called generalized fractional calculus (GFC). In Section 4, we introduce the Mittag-Leffler functions and the multi-index Mittag-Leffler functions, with short information on their properties derived in author's works. Sections 5 and 6 contain long lists of examples of SF that appear as cases of the multi-index Mittag-Leffler functions and in more general setting, of the Wright generalized hypergeometric functions $_p\Psi_q$. These include also citations to many other authors who introduced and applied such functions in their works. The author's unified approach to evaluate images of classical SF and of SF of FC under operators of FC and GFC is shortly described in Section 7, because the details are presented in another survey paper in the same journal [105]. In Section 8, we collect some of our basic propositions on the representations of the SF and of SF of FC as operators of GFC of three basic and simplest elementary functions and propose a classification of the SF based on the cases $p < q$, $p = q$ and $p = q + 1$. Thus, a new sight on the theory of SF is proposed. Since the computational aspects related to the considered SF are of important interest for their applications, in Section 9, we provide some short information on the state of affairs and some recent works on this direction by other authors. A provoking challenge in this respect is mentioned.

Author Contributions: The ideas and results in this paper survey and reflect the author's sole contributions, resulting from more than 30 years research on the topic. The author has read and agreed to the published version of the manuscript.

Funding: This research received no financial funding.

Institutional Review Board Statement: Not applicable.

Informed Consent Statement: Not applicable.

Data Availability Statement: Not applicable.

Acknowledgments: This paper is done under the working programs on bilateral collaboration contracts of Bulgarian Academy of Sciences with Serbian and Macedonian Academies of Sciences and Arts, and under the COST program, COST Action CA15225 'Fractional'.

Conflicts of Interest: The author declares no conflict of interest.

References

1. Wikipedia: Special Functions. Available online: https://en.wikipedia.org/wiki/Special_functions (accessed on 11 December 2020).
2. Askey, R. *Orthogonal Polynomials and Special Functions*; SIAM: Philadelphia, PA, USA, 1975.
3. Erdélyi, A. (Ed.) *Higher Transcendental Functions*; McGraw Hill: New York, NY, USA, 1953–1955; Vols. 1, 2, 3. Available online: https://en.wikipedia.org/wiki/Bateman_Manuscript_Project (accessed on 11 December 2020).
4. NIST Digital Library of Mathematical Functions (DLMF). Project of NIST, 2010–2010. Available online: https://dlmf.nist.gov/ (accessed on 11 December 2020).
5. Abramowitz, M.; Stegun, A. Handbook of Mathematical Functions with Formulas, Graphs, and Mathematical Tables; National Bureau of Standards, 1964-Dover, 1965. Available online: https://en.wikipedia.org/wiki/Abramowitz_and_Stegun (accessed on 11 December 2020).
6. Olver, F.W.J.; Lozier, D.W.; Boisvert, R.F.; Clark, C.W. (Eds.) *NIST Handbook of Mathematical Functions*; Cambridge University Press: Cambridge, UK, 2010.
7. Erdélyi, A. (Ed.) *Tables of Integral Transforms*; McGraw Hill: New York, NY, USA, 1954; Volumes 1–2.
8. Assche, W. *Encyclopedia of Special Functions: The Askey-Bateman Project*; Ismail, M., Ed.; Cambridge University Press: Cambridge, MA, USA, 2020. [CrossRef]
9. Kiryakova, V. *Generalized Fractional Calculus and Applications*; Longman—J. Wiley, Harlow—New York, Chapman and Hall/CRC: London, UK, 1994.
10. Mathai, A.M.; Saxena, R.K. *Generalized Hypergeometric Functions with Applications in Statistics and Physical Sciences*; Lect. Notes in Math.; Springer: Heidelberg, Germany, 1973.
11. Marichev, O.I. *Handbook of Integral Transforms of Higher Transcendental Functions, Theory and Algorithmic Tables*; Ellis Horwood: Chichester, UK, 1983.
12. Srivastava, H.M.; Gupta, K.S.; Goyal, S.P. *The H-Functions of One and Two Variables with Applications*; South Asian Publs: New Delhi, India, 1982.
13. Srivastava, H.M.; Kashyap, B.R.K. *Special Functions in Queuing Theory and Related Stochastic Processes*; Academic Press: New York, NY, USA, 1982.
14. Prudnikov, A.P.; Brychkov, Y.; Marichev, O.I. *Integrals and Series, Vol. 3: More Special Functions*; Gordon and Breach Sci. Publ.: London, UK, 1992.
15. Yakubovich, S.; Luchko, Y. *The Hypergeometric Approach to Integral Transforms and Convolutions*; Ser. Mathematics and Its Applications 287; Kluwer Acad. Publ.: Dordrecht, The Netherlands; Boston, MA, USA; London, UK, 1994.
16. Podlubny, I. *Fractional Differential Equations*; Acad. Press: Cambridge, MA, USA, 1999.
17. Kilbas, A.A.; Saigo, M. *H-Transforms: Theory and Applications*; Ser. on Analytic Methods and Special Functions, 9; CRC Press: Boca Raton, FL, USA, 2004.
18. Kilbas, A.A.; Srivastava, H.M.; Trujillo, J.J. *Theory and Applications of Fractional Differential Equations*; Elsevier, Amsterdam Etc.: Amsterdam, The Netherlands, 2006.
19. Mathai, A.M.; Haubold, H.J. *Special Functions for Applied Scientists*; Springer: Berlin, Germany, 2008.
20. Mathai, A.M.; Saxena, R.K.; Haubold, H.J. *The H-function. Theory and Applications*; Springer: Berlin, Germany, 2010.
21. Mainardi, F. *Fractional Calculus and Waves in Linear Viscoelasticity: An Introduction to Mathematical Models*, 2nd ed.; Imperial College Press: London, UK, 2010.

22. Gorenflo, R.; Kilbas, A.; Mainardi, F.; Rogosin, S. *Mittag-Leffler Functions, Related Topics and Applications*, 2nd ed.; Springer: New York, NY, USA, 2020. [CrossRef]
23. Cohl, H.; Ismail, M. (Eds.) *Lectures on Orthogonal Polynomials and Special Functions*; London Math. Soc. Lecture Note Ser.; Cambridge University Press: Cambridge, MA, USA, 2020. [CrossRef]
24. Mainardi, F. A tutorial on the basic special functions of fractional calculus. *WSEAS Trans. Math.* **2020**, *19*, 74–98. [CrossRef]
25. Machado, J.A.T.; Kiryakova, V. Recent history of the fractional calculus: Data and statistics. In *Handbook of Fractional Calculus with Applications. Volume 1: Basic Theory*; Kochubei, A., Luchko, Y., Eds.; De Gruyter: Berlin, Germany, 2019; pp. 1–21. [CrossRef]
26. Mainardi, F.; Pagnini, G. Salvatore Pincherle: The pioneer of the Mellin-Barnes integrals. *J. Comput. Appl. Math.* **2003**, *153*, 331–341. [CrossRef]
27. Fox, C. The G and H-functions as symmetric Fourier kernels. *Trans. Am. Math. Soc.* **1961**, *98*, 395–429.
28. Karp, D. A note on Fox's H-function in the light of Braaksma's results. Ch.12. In *Special Functions and Analysis of Differential Equations*; Agarwal, P., Agarwal, R.P., Ruzhansky, M., Eds.; Chapman and Hall/ CRC: New York, NY, USA, 2020; p. 12.
29. Braaksma, B.L.J. Asymptotic expansions and analytic continuation for a class of Barnes integrals. *Compos. Math.* **1936**, *15*, 239–341.
30. Meijer, C.S. On the G-function. *Indag. Math.* **1946**, *8*, 124–134, 213–225, 312–324, 391–400, 468–475, 595–602, 661–670, 713–723.
31. Wright, E.M. On the coefficients of power series having exponential singularities. *J. Lond. Math. Soc.* **1933**, *8*, 71–79. [CrossRef]
32. Wright, E.M. The generalized Bessel function of order greater than one. *Q. J. Math. Oxf. Ser.* **1940**, *11*, 36–48. [CrossRef]
33. Kilbas, A.A. Fractional calculus of the generalized Wright function. *Fract. Calc. Appl. Anal.* **2005**, *8*, 113–126.
34. Mainardi, F. Why the Mittag-Leffler function can be considered the Queen function of the fractional calculus? *Entropy* **2020**, *22*, 29. [CrossRef]
35. Rooney, P.G. On integral transformatons with G-function kernels. *Proc. R. Soc. Edinb.* **1983**, *93A*, 265–297. [CrossRef]
36. Tuan, V.K.; Marichev, O.I.; Yakubovich, S. Composition structure of integral transformations. *J. Soviet Math.* **1986**, *33*, 166–169.
37. Obrechkoff, N. On certain integral representations of real functions on the real semi-axis. *Proc. Inst. Math. Acad. Bulgare Sci.* 1958, *3*, 3–28 (In Bulgarian); Translate in English in *East J. Approx.* **1997**, *3*, 89–110.
38. Dimovski, I. Operational calculus for a class of differential operators. *CR Acad. Bulg. Sci.* **1966**, *19*, 1111–1114.
39. Dimovski, I.; Kiryakova, V. The Obrechkoff integral transform: Properties and relation to a generalized fractional calculus. *Numer. Funct. Anal. Optimiz.* **2000**, *21*, 121–144. [CrossRef]
40. Erdélyi, A. Integraldarstellungen hyper-gepmetrischer funktionen. *Q. J. Math.* **1937**, *8*, 267–277. [CrossRef]
41. Kiryakova, V. From the hyper-Bessel operators of Dimovski to the generalized fractional calculus. *Fract. Calc. Appl. Anal.* **2014**, *17*, 977–1000. [CrossRef]
42. Al-Musallam, F.; Kiryakova, V.; Tuan, V.K. A multi-index Borel-Dzrbashjan transform. *Rocky Mt. J. Math.* **2002**, *32*, 409–428. [CrossRef]
43. Dzrbashjan, M.M. *Integral Transforms and Representations in the Complex Domain*; Nauka: Moscow, Russia, 1966. (In Russian)
44. Dzrbashjan, M.M. On the integral transformations generated by the generalized Mittag-Leffler function. *Izv. AN Arm. SSR* 1960, *13*, 21–63. (In Russian)
45. Krätzel, E. Differentiationssätze der $\mathcal{L}$-Transformation unde Differentiagleichungen nach dem Operator. *Math. Machrichten* **1967**, *35*, 105–114. [CrossRef]
46. Krätzel, E. Integral transformations of Bessel type. In *Generalized Functions and Operational Calculus (Proc. Conf. Varna 1975)*; Bulg. Acad. Sci.: Sofia, Bulgaria, 1979; pp. 148–155.
47. Kilbas, A.A.; Saxena, R.K.; Trujillo, J.J. Krätzel function as a function of hypergeometric type. *Fract. Calc. Appl. Anal.* **2006**, *9*, 109–131.
48. Mathai, A.M.; Haubold, H.J. Mathematical aspects of Krätzel integral and Krätzel transform. *Mathematics* **2020**, *8*, 526. [CrossRef]
49. Glaeske, H.-Y.; Kilbas, A.A.; Saigo, M. A modified Bessel-type integral transform and its compositions with fractional calculus operators on spaces $F_{p,\mu}$ and $F'_{p,\mu}$. *J. Comput. Appl. Math.* **2000**, *118*, 151–168. [CrossRef]
50. Ditkin, V.A.; Prudnikov, A.P. Theory of operational calculus, generated by the Bessel equation. *Zhournal Vych. Mat. Mat. Fiziki* 1963, *3*, 223–238. (In Russian)
51. Samko, S.; Kilbas, A.; Marichev, O. *Fractional Integrals and Derivatives: Theory and Applications*; Gordon and Breach: Yverdon, Switzerland, 1993.
52. Sneddon, I.N. The use in mathematical analysis of Erdélyi-Kober operators and some of their applications. In *Fractional Calculus and Its Applications (Proc. Internat. Conf. Held in New Haven)*; Ross, B., Ed.; Lecture Notes in Math. 457; Springer: New York, NY, USA, 1975; pp. 37–79.
53. Kiryakova, V.; Luchko, Y. Riemann-Liouville and Caputo type multiple Erdélyi-Kober operators. *Cent. Eur. J. Phys.* **2013**, *11*, 1314–1336. [CrossRef]
54. Kalla, S.L. Operators of fractional integration. *Lect. Notes Math.* **1980**, *798*, 258–280.
55. Kiryakova, V. A brief story about the operators of the generalized fractional calculus. *Fract. Calc. Appl. Anal.* **2008**, *11*, 203–220.
56. Kiryakova, V. Generalized fractional calculus operators with special functions. In *Handbook of Fractional Calculus with Applications; Basic Theory*; Kochubei, A., Luchko, Y., Eds.; De Gruyter: Berlin, Germany, 2019; pp. 87–110. [CrossRef]
57. Kalla, S.L.; Galue, L. Generalized fractional calculus based upon composition of some basic operators. In *Recent Advances in Fractional Calculus*; Kalia, R.N., Ed.; Global Publ. Co.: Jacksonville, FL, USA, 1993; pp. 145–178.

58. Kiryakova, V. All the special functions are fractional diffeintegrals of elementary functions. *J. Phys. A Math. Gen.* **1997**, *30*, 5085–5103. [CrossRef]
59. Kiryakova, V. The multi-index Mittag-Leffler functions as important class of special functions of fractional calculus. *Comput. Math. Appl.* **2010**, *59*, 1885–1895. [CrossRef]
60. Kiryakova, V. The special functions of fractional calculus as generalized fractional calculus operators of some basic functions. *Comput. Math. Appl.* **2010**, *59*, 1128–1141. [CrossRef]
61. McBride, A. Fractional powers of a class of ordinary differential operators. *Proc. Lond. Math. Soc. (III)* **1982**, *45*, 519–546. [CrossRef]
62. Gelfond, A.O.; Leontiev, A.F. On a generalization of the Fourier series. *Mat. Sb.* **1951**, *29*, 477–500. (In Russian)
63. Kiryakova, V. Multiple Dzrbashjan-Gelfond-Leontiev fractional diffeintegrals. In *Recent Advances in Appl. Mathematics'96*; Proc. Intern. Workshop; Kuwait University: Kuwait, Kuwait, 1996; pp. 281–294.
64. Kiryakova, V. Gel'fond-Leont'ev integration operators of fractional (multi-)order generated by some special functions. *AIP Conf. Proc.* **2018**, *2048*, 10. [CrossRef]
65. Kabe, D.G. Some applications of Meijer-G functions to distribution problems in statistics. *Biometrica* **1958**, *45*, 578–580. [CrossRef]
66. Karp, D.; López, J.L. On a particular class of Meijer's G functions appearing in fractional calculus. *Int. J. Appl. Math.* **2018**, *31*, 521–543. [CrossRef]
67. Mittag-Leffler, G.M. Sur la nouvelle fonction $E_\alpha(x)$. *CR de l'Acad. Sci.* **1903**, *137*, 554–558.
68. Wiman, A. Über den Fundamentalsatz der Theorie der Funkntionen $E_\alpha(x)$. *Acta Math.* **1905**, *29*, 191–201. [CrossRef]
69. Humbert, P.; Agarwal, R.P. Sur la fonction de Mittag-Leffler et quelques-unes de ses généralisations. *Bull. Sci. Math.* **1953**, *77*, 180–185.
70. Haubold, H.J.; Mathai, A.M.; Saxena, R.K. Mittag-Leffler functions and their applications. *J. Appl. Math.* **2011**, *2011*, 298628. [CrossRef]
71. Rogosin, S. The role of the Mittag-Leffler function in fractional modeling. *Mathematics* **2015**, *3*, 368–381. [CrossRef]
72. Hille, E.; Tamarkin, J.D. On the theory of linear integral equations. *Ann. Math.* **1930**, *31*, 479–528. [CrossRef]
73. Prabhakar, T.R. A singular integral equation with a generalized Mittag-Leffler function in the kernel. *Yokohama Math. J.* **1971**, *19*, 7–15.
74. Paneva-Konovska, J. *From Bessel to Multi-Index Mittag-Leffler Functions: Enumerable Families, Series in Them and Convergence*; World Scientific Publishing: London, UK, 2016.
75. Garra, R.; Garrappa, R. The Prabhakar or three parameter Mittag–Leffler function: Theory and application. *Commun. Nonlinear Sci. Numer. Sim.* **2018**, *56*, 314–319. [CrossRef]
76. Gorenflo, R.; Kilbas, A.A.; Rogosin, S. On the generalized Mittag-Leffler type function. *Integr. Transform. Spec. Funct.* **1998**, *7*, 215–224. [CrossRef]
77. Luchko, Y.F.; Srivastava, H.M. The exact solution of certain differential equations of fractional order by using operational calculus. *Comput. Math. Appl.* **1995**, *29*, 73–85. [CrossRef]
78. Luchko, Y. Operational method in fractonal calculus. *Fract. Calc. Appl. Anal.* **1999**, *2*, 463–488.
79. Kiryakova, V. Multiindex Mittag-Leffler functions, related Gelfond-Leontiev operators and Laplace type integral transforms. *Fract. Calc. Appl. Anal.* **1999**, *2*, 445–462.
80. Kiryakova, V. Multiple (multiindex) Mittag-Leffler functions and relations to generalized fractional calculus. *J. Comput. Appl. Math.* **2000**, *118*, 241–259. [CrossRef]
81. Kilbas, A.A.; Koroleva, A.A.; Rogosin, S.V. Multi-parametric Mittag-Leffler functions and their extension. *Fract. Calc. Appl. Anal.* **2013**, *16*, 378–404. [CrossRef]
82. Paneva-Konovska, J. Multi-index (3m-parametric) Mittag-Leffler functions and fractional calculus. *Compt. Rend. Acad. Bulg. Sci.* **2011**, *64*, 1089–1098.
83. Paneva-Konovska, J.; Kiryakova, V. On the multi-index Mittag-Leffler functions and their Mellin transforms. *Int. J. Appl. Math.* **2020**, *33*, 549–571. [CrossRef]
84. Gerhold, S. Asymptotics for a variant of the Mittag-Leffler function. *Integr. Trans. Spec. Func.* **2012**, *23*, 397–403. [CrossRef]
85. Garra, R.; Polito, F. On some operators involving Hadamard derivatives. *Integr. Trans. Spec. Func.* **2013**, *24*, 773–782. [CrossRef]
86. Garrappa, R.; Rogosin, S.; Mainadi, F. On a generalized three-parameter Wright function of le Roy type. *Fract. Calc. Appl. Anal.* **2017**, *206*, 1196–1215. [CrossRef]
87. Garrappa, R.; Orsingher, E.; Polito, F. A note on Hadamard fractional differential equations with varying coefficients and their applications in probability. *Mathematics* **2018**, *6*, 4. [CrossRef]
88. Le Roy, É. Sur les séries divergentes et les fonctions définies par un développement de Taylor. *Ann. Fac. Sci. Toulouse Sér.* **1990**, *2*, 385–430. [CrossRef]
89. Delerue, P. Sur le calcul symboloque à *n* variables et fonctions hyperbesseliennes (II). *Ann. Soc. Sci. Brux. Ser. 1* **1953**, *3*, 229–274.
90. Dimovski, I.; Kiryakova, V. Generalized Poisson transmutations and corresponding representations of hyper-Bessel functions. *CR Acad. Bulg. Sci.* **1986**, *39*, 29–32.
91. Plotnikov, Y.I. Steady-State Vibrations of Plane and Axisymmetric Stamps on a Viscoelastic Foundation. Ph.D. Thesis, Moscow, Russia, 1979.
92. Tseytlin, A.I. *Applied Methods of Solution of Boundary Value Problems in Civil Engineering*; Stroyizdat: Moscow, Russia, 1984. (In Russian)

93. Lorenzo, C.F.; Hartley, T.T. R-Function relationships for application in the fractional calculus. *NASA/TM–2000-210361*, August 2000; 30p. Available online: https://ntrs.nasa.gov/citations/20000091004 (accessed on 11 December 2020).
94. Luchko, Y. The four-parameters Wright function of the second kind and its applications in FC. *Mathematics* **2020**, *8*, 970. [CrossRef]
95. Fox, C. The asymptotic expansion of generalized hypergeomtric functons. *Proc. Lond. Math. Soc. Ser.* **1928**, *27*, 389–400. [CrossRef]
96. Gorenflo, R.; Luchko, Y.; Mainardi, F. Analytical properties and applications of the Wright function. *Fract. Calc. Appl. Anal.* **1999**, *2*, 383–414.
97. Mainardi, F.; Consiglio, A. The Wright functions of the second kind in mathematical physics. *Mathematics* **2020**, *8*, 884. [CrossRef]
98. Pathak, R.S. Certain convergence theorems and asymptotic properties of a generalization of Lommel and Maitland transformations. *Proc. Natl. Acad. Sci. India Sect. Phys. Sci.* **1966**, *A-36*, 81–86.
99. De Oteiza, M.M.M.; Kalla, S.L.; Conde, S. Un estudio sobre la función Lommel–Maitland. *Rev. Technol. Fac. Ingr. Univ. Zulia* **1986**, *9*, 33–40.
100. Ricci, P.E. Laguerre-type exponentials, Laguerre derivatives and applications. A survey. *Mathematics* **2020**, *8*, 2054. [CrossRef]
101. Virchenko, N.A. On some generalizations of the functions of hypergeometric type. *Fract. Calc. Appl. Anal.* **1999**, *2*, 233–244.
102. Mainardi, F.; Masina, E. On modifications of the exponential integral with the Mittag-Leffler function. *Fract. Calc. Appl. Anal.* **2018**, *21*, 1156–1169. [CrossRef]
103. Paris, R. Asymptotic expansion of the modified exponential integral involving the Mittag-Leffler function. *Mathematics* **2020**, *8*, 428. [CrossRef]
104. Kiryakova, V. Fractional calculus of some "new" but not new special functions: k-, multi-index-, and S-analogues. *AIP Conf. Proc.* **2019**, *2172*, 12. [CrossRef]
105. Kiryakova, V. Unified approach to fractional calculus images of special functions—A survey. *Mathematics* **2020**, *8*, 2260. [CrossRef]
106. Gehlot, K.S. Differential equation of k-Bessel's function and its properties. *Nonlin. Anal. and Diff. Equa.* **2014**, *2*, 61–67. [CrossRef]
107. Mondal, S.R. Representation formulae and monotonicity of the generalized k-Bessel functions. *arXiv* **2016**, arXiv:1611.07499.
108. Shaktawat, B.S.; Rawat, D.S.; Gupta, R.K. On generalized fractional calculus of the generalized k-Bessel function. *J. Rajasthan Acad. Phys. Sci.* **2017**, *16*, 9–19.
109. Gupta, A.; Parihar, C.L. k-New generalized Mittag-Leffler function. *J. Fract. Calc. Appl.* **2014**, *5*, 165–176.
110. Nisar, K.S.; Eata, A.F.; Al-Dhaifallah, M.; Choi, J. Fractional calculus of generalized k-Mittag-Leffler function and its applications to statistical distribution. *Adv. Differ. Eq.* **2016**, *2016*, 304. [CrossRef]
111. Nisar, K.S.; Purohit, S.D.; Suthar, D.L.; Singh, J. Fractional calculus and certain integrals of generalized multiindex Bessel function. *arXiv* **2017**, arXiv:1706.08039.
112. Kiryakova, V. Fractional calculus operators of special functions? – The result is well predictable!. *Chaos Solitons Fractals* **2017**, *102*, 2–15. [CrossRef]
113. Saxena, R.K.; Nishimoto, K. N-fractional calculus of generalized Mittag-Leffler functions. *J. Fract. Calc.* **2010**, *37*, 43–52.
114. Agarwal, P.; Rogosin, S.V.; Trujillo, J.J. Certain fractional integral operators and the generalized multi-index Mittag-Leffler functions. *Proc. Indian Acad. Sci. (Math. Sci.)* **2015**, *125*, 291–306. [CrossRef]
115. Srivastava, H.M.; Tomovski, Ž. Fractional calculus with an integral operator containing generalized Mittag-Leffler function in the kernel. *Appl. Math. Comput.* **2009**, *211*, 198–210. [CrossRef]
116. Agarwal, R.; Jain, S.; Agarwal, R.P.; Baleanu, D. A remark on the fractional integral operators and the image formulas of generalized Lommel-Wright function. *Front. Phys.* **2018**, *6*, 79. [CrossRef]
117. Kiryakova, V. Commentary: "A remark on the fractional integral operators and the image formulas of generalized Lommel-Wright function". *Front. Phys.* **2019**, *7*, 145. [CrossRef]
118. Saxena, R.K.; Daiya, J. Integral transforms of S-functions. *Le Mat.* **2015**, *72*, 147–159.
119. Sharma, M.; Jain, R. A note on a generalized M-series as a special function of fractional calculus. *Fract. Calc. Appl. Anal.* **2009**, *12*, 449–452.
120. Sharma, K. An introduction to the generalized fractional integration. *Bol. Soc. Paran. Math.* **2012**, *30*, 85–90. [CrossRef]
121. Lavault, C. Fractional calculus and generalized Mittag-Leffler type functions. *arXiv* **2017**, arXiv:1703.01912.
122. Purohit, M.; Badguzer, A. MSM fractional integration and differentiation operators of multi-parametric K-Mittag-Leffler function and generalized multi-index Bessel function. *Int. J. Stat. Appl. Math.* **2018**, *3*, 156–161.
123. Kiryakova, V. Use of fractional calculus to evaluate some improper integrals of special functions. *AIP Conf. Proc.* **2017**, *1910*, 12. [CrossRef]
124. Lavoie, J.L.; Osler, T.J.; Tremblay, R. Fractional derivatives and special functions. *SIAM Rev.* **1976**, *18*, 240–268. [CrossRef]
125. Garrappa, R.; Kaslik, E.; Popolizio, M. Evaluation of fractional integrals and derivatives of elementary functions: Overview and tutorial. *Mathematics* **2019**, *7*, 407. [CrossRef]
126. Gradshteyn, I.S.; Ryzhik, I.M. *Tables of Integrals, Series, and Products*, 1st ed.; Acad. Press: Norfolk, UK, 2015.
127. Magnus, W.; Oberhettinger, F. *Formulas and Theorems for the Special Functions of Mathematical Physics*, 1st ed.; Springer: Berlin, Germany, 1966. (In German)
128. Wolfram, S. The History and Future of Special Functions (in Honor of 60th Birthday of O. Marichev). Available online: https://www.stephenwolfram.com/publications/history-future-special-functions (accessed on 11 December 2020).
129. Wolfram Matematica: The Mathematical Function Site. Available online: https://functions.wolfram.com/ (accessed on 11 December 2020).

130. Caputo, M.; Mainardi, F. A new dissipation model based on memory mechanism. *Pure Appl. Geophys.* **1971**, *91*, 134–147. [CrossRef]
131. Gorenflo, R.; Loutchko, J.; Luchko, Y. Computation of the Mittag-Leffler function and its derivatives. *Fract. Calc. Appl. Anal.* **2002**, *5*, 491–518.
132. Diethelm, K.; Ford, N.; Freed, A.; Luchko, Y. Algorithms for the fractional calculus: A selection of numerical methods. *Comput. Methods Appl. Mech. Eng.* **2005**, *194*, 743–773. [CrossRef]
133. Podlubny, I. Mittag-Leffler function. Version 1.2.0.0. Calculates the Mittag-Leffler function with desired accuracy. Matlab Central File Exchange. Retrieved 1 January 2021. Available online: https://www.mathworks.com/matlabcentral/fileexchange/8738-mittag-leffler-function (accessed on 1 January 2021).
134. Hilfer, R.; Seybold, H.J. Computation of the generalized Mittag-Leffler function and its inverse in the complex plane. *Integr. Transf. Spec. Funct.* **2006**, *17*, 637–652. [CrossRef]
135. Seybold, H.; Hilfer, R. Numerical algorithm for calculating the generalized Mittag-Leffler function. *SIAM J. Numer. Anal.* **2008**, *47*, 69–88. Available online: http://www.siam.org/journals/sinum/47-1/70028.html (accessed on 11 December 2020).
136. Garrappa, R. Numerical evaluation of two and three parameter Mittag-Leffler functions. *SIAM J. Numer. Anal.* **2015**, *53*, 1350–1369; *Matlab* Code Associated with this Paper, at MATLAB Central File Exchange. Available online: https://www.mathworks.com/matlabcentral/fileexchange/48154-the-mittag-leffler-function (accessed on 4 January 2021).
137. Garrappa, R.; Popolizio, M. Computing the matrix Mittag-Leffler function with applications to fractional calculus. *J. Sci. Comput.* **2018**, *17*, 129–153; Matlab Code Associated with This Paper. Available online: https://www.mathworks.com/matlabcentral/fileexchange/66272-mittag-leffler-function-with-matrix-arguments?s_tid=prof_contriblnk (accessed on 11 December 2020).
138. Luchko, Y. Algorithms for Evaluation of the Wright Function for the Real Arguments Values. *Fract. Calc. Appl. Anal.* **2008**, *11*, 57–75. Available online: http://www.math.bas.bg/complan/fcaa/volume11/fcaa111/Luchko_111.pdf (accessed on 11 December 2020).
139. Luchko, Y.; Trujillo, J.J.; Velasco, M.P. The Wright function and its numerical evaluation. *Int. J. Pure Appl. Math.* **2010**, *64*, 567–575. Available online: https://www.researchgate.net/publication/236221356 (accessed on 11 December 2020).
140. Consiglio, A. Simulation of the M-Wright Function. At *Youtube*. 2019. Available online: https://www.youtube.com/watch?v=uf_4aB1COPg (accessed on 11 December 2020).

Article

Approximation of CDF of Non-Central Chi-Square Distribution by Mean-Value Theorems for Integrals

Árpád Baricz [1,2], Dragana Jankov Maširević [3] and Tibor K. Pogány [2,4,*]

1. Department of Economics, Babeș-Bolyai University, 400591 Cluj-Napoca, Romania; arpad.baricz@econ.ubbcluj.ro
2. Institute of Applied Mathematics, Óbuda University, Bécsi út 96/b, 1034 Budapest, Hungary
3. Department of Mathematics, University of Osijek, Trg Lj. Gaja 6, 31000 Osijek, Croatia; djankov@mathos.hr
4. Faculty of Maritime Studies, University of Rijeka, Studentska 2, 51000 Rijeka, Croatia
* Correspondence: poganj@pfri.hr

Abstract: The cumulative distribution function of the non-central chi-square distribution $\chi_n'^2(\lambda)$ of n degrees of freedom possesses an integral representation. Here we rewrite this integral in terms of a lower incomplete gamma function applying two of the second mean-value theorems for definite integrals, which are of Bonnet type and Okamura's variant of the du Bois–Reymond theorem. Related results are exposed concerning the small argument cases in cumulative distribution function (CDF) and their asymptotic behavior near the origin.

Keywords: non-central χ^2 distribution; second mean-value theorem for definite integrals; modified Bessel function of the first kind; Marcum Q–function; lower incomplete gamma function

MSC: Primary: 26A24, 62E17; Secondary: 33C10, 60E99

Citation: Baricz, Á.; Jankov Maširević, D.; Pogány, T.K. Approximation of CDF of Non-Central Chi-Square Distribution by Mean-Value Theorems for Integrals. *Mathematics* **2021**, *9*, 129. https://doi.org/10.3390/math9020129

Received: 9 December 2020
Accepted: 4 January 2021
Published: 8 January 2021

Publisher's Note: MDPI stays neutral with regard to jurisdictional claims in published maps and institutional affiliations.

Copyright: © 2021 by the authors. Licensee MDPI, Basel, Switzerland. This article is an open access article distributed under the terms and conditions of the Creative Commons Attribution (CC BY) license (https://creativecommons.org/licenses/by/4.0/).

1. Introduction with Historical Notes and Motivation

The non-central χ^2 distribution with $n \in \mathbb{N}$ degrees of freedom (in general, n can be a non-negative real number, see ([1] (p. 436), [2]) and non–centrality parameter $\lambda > 0$ is usually denoted by $\chi_n'^2(\lambda)$ (see, e.g., [1] (p. 433)) and it is one of the most applied distributions: it is important in calculating the power function of some statistical tests [3], precisely in approximating to the power of χ^2-tests applied to contingency tables (goodness of fit tests) ([1] (p. 467)); it frequently occurs in finance, estimation and decision theory and time series analysis [4,5] and can also be regarded as a generalized Rayleigh distribution ([1] (p. 435)) in which case it is used in mathematical physics; when it is used in communication theory then we call the appropriate complementary cumulative distribution function the generalized Marcum Q-function and the non-centrality parameter is interpreted as a signal-to-noise ratio [1].

The beginnings of the research that led up to the model and finally results in the χ_n^2 distribution, which is the zero non-centrality parameter case of non-central $\chi_n'^2(\lambda)$, that is $\chi_n^2 \equiv \chi_n'^2(0)$, can be located around the middle of the 19th century. More precisely there are two main opinions exposed: firstly, the influential work by Lancaster [6] who attributed certain preliminary results to Bienaymé in ([7] (p. 58)) (never mentioning normal distribution), which are in fact the same as what Karl Pearson did to earn his tables [8]. It is not surprising, Bienaymé's interest in the sum of random squares and the related distribution of errors; namely, we should have in mind his celebrated result on the linearity of variance of a sum of independent random variables called the Bienaymé formula. Lancaster proceeded then to Helmert, who in [9,10] derived that which we understand in modern notation the χ_n^2 probability density function (PDF). Finally, Lancaster joined Kruskal [11], suggesting to call the distribution by Helmert's name.

However, Sheynin [12], Plackett [13] and especially Kendall [14] have mentioned the contribution of the applied mathematician and physicists Ernst Abbe, who has published in his *venia docendi* thesis [15] in Jena, 1863, the χ^2 distribution's PDF ([12] (p. 1004)). It is also worth mentioning that Helmert himself never explicitly mentioned this distribution as Abbe's result in this manner, but several times quoted the "modified Abbe's criterion" in geodetic literature ([12] (p. 1004)). Kendall emphasizes Abbe's priority (agreeing with Sheynin) and wrote a *laudatio* to his work regarding the derivation of the PDF of χ^2 distribution (in a contemporary notation) ([14] (p. 311, Equation (11))), preceding Helmert for at least twelve years.

A random variable (rv) ξ possesses non-central χ^2 distribution, which we signify with $\xi \sim \chi_n'^2(\lambda)$ if the associated probability density function is ([4] (p. 396, Equation (1.7)))

$$f_{n,\lambda}(x) = \frac{1}{2} e^{-(x+\lambda)/2} \left(\frac{x}{\lambda}\right)^{(n-2)/4} I_{n/2-1}(\sqrt{\lambda x}), \qquad \lambda > 0, x > 0; n \in \mathbb{N}, \qquad (1)$$

where I_ν stands for the modified Bessel function of the first kind of order ν ([16] (p. 77)) and has the power series representation ([17] (p. 375, Equation (9.6.10)))

$$I_\nu(z) = \sum_{k=0}^{\infty} \frac{1}{\Gamma(\nu+k+1)\,k!} \left(\frac{z}{2}\right)^{2k+\nu}; \qquad \Re(\nu) > -1, z \in \mathbb{C}. \qquad (2)$$

As for the historical background of related PDF and the associated cumulative distribution function (CDF) we consult the monographs [1,18]. In accordance with ([18] (Chapter 1, §5)) the PDF of $\xi \sim \chi_n'^2(\lambda)$ was pioneered in 1928 by Fisher [19] by a limiting process, while the explicit derivation belongs to Tang [20] ten years later (we also draw the interested reader's attention to ([1] (Chapter 29, pp. 435 *et seq.*)). In 1949 Patnaik [21], then, among others, Pearson [22], Sankaran [23] and Temme [24] have been studied the $\chi_n'^2(\lambda)$ distribution; Temme claimed that his formulae have certain computational advantages

$$F_{n,\lambda}(x) = \begin{cases} 1 - \frac{1}{2}\left(\frac{x}{\lambda}\right)^{n/4}\left[T_{n/2-1}(\sqrt{\lambda x},\omega) - \sqrt{\frac{\lambda}{x}}\,T_{n/2}(\sqrt{\lambda x},\omega)\right], & x > \lambda \\ \frac{1}{2}\left(\frac{x}{\lambda}\right)^{n/4}\left[\sqrt{\frac{\lambda}{x}}\,T_{n/2}(\sqrt{\lambda x},\omega) - T_{n/2-1}(\sqrt{\lambda x},\omega)\right], & x < \lambda \end{cases},$$

where $\omega = \frac{1}{2}(\sqrt{x}-\sqrt{\lambda})^2/\sqrt{\lambda x}$ and

$$T_\nu(\alpha,\omega) = \int_\alpha^\infty e^{-(\omega+1)t} I_\nu(t)\,\mathrm{d}t.$$

Here we are interested in the CDF used in communication theory ([25] (p. 66, Equation (1.1)))

$$F_{n,\lambda}(x) = 1 - Q_{n/2}(\sqrt{\lambda},\sqrt{x}), \qquad x > 0, \qquad (3)$$

where [26]

$$Q_\nu(a,b) = \frac{1}{a^{\nu-1}} \int_b^\infty t^\nu e^{-(t^2+a^2)/2} I_{\nu-1}(at)\,\mathrm{d}t, \qquad a,\nu > 0; b \geq 0, \qquad (4)$$

denotes the generalized Marcum Q-function of the order ν.

Finally, it is worth mentioning that Brychkov recently published a closed expression for the generalized Marcum Q-function ([27] (p. 178, Equation (7))) in terms of the complementary error function $z \mapsto \mathrm{erfc}(z)$ ([28] (p. 160, Equation (7.2.2))), which immediately implies a new formula for CDF (3) in the case when $n \in \mathbb{N}$ is odd. In turn, in the case of an even number of the degrees of freedom, Jankov Maširević derived the following expression for the appropriate CDF for all $\lambda > 0, x > 0$ [25]

$$F_{2n,\lambda}(x) = 1 - \frac{\sqrt{\lambda x}}{2} I_1(\sqrt{\lambda x})\left[K_0(\sqrt{\lambda x}) - K_0\left(\sqrt{\lambda x}, \ln\sqrt{\frac{x}{\lambda}}\right)\right]$$

$$+ \lambda I_0(\sqrt{\lambda x}) \frac{\partial}{\partial \lambda} \left[K_0(\sqrt{\lambda x}) - K_0\left(\sqrt{\lambda x}, \ln\sqrt{\frac{x}{\lambda}}\right) \right]$$
$$- e^{-\frac{\lambda+x}{2}} \sqrt{\frac{\lambda}{x}} \sum_{m=1}^{n} \left(\sqrt{\frac{x}{\lambda}}\right)^m I_{m-1}(\sqrt{\lambda x}). \tag{5}$$

Here K_ν stands for the modified Bessel functions of the second kind and

$$K_\nu(z,w) = \frac{\sqrt{\pi}}{\Gamma\left(\nu + \frac{1}{2}\right)} \left(\frac{z}{2}\right)^\nu \int_0^w e^{-z\cosh t} \sinh^{2\nu} t \, dt, \qquad \Re(\nu) > -1/2,$$

is its incomplete variant ([29] (p. 26, Equation (1.30))), while

$$\lim_{w\to\infty} K_\nu(z,w) = K_\nu(z), \qquad \Re(z) > 0,$$

in the pointwise sense. Jankov Maširević established the computational efficiency of Expression (5) versus the formulae derived by Patnaik, and those by Temme for even $n \in \mathbb{N}$, concluding that her approach is more efficient, compare ([25] (Section 3)).

The main aim of this paper is to present new results for the CDF (3) concerning approximation formulae obtained by two variants of the second mean-value theorems for definite integrals. Throughout, the non-centrality parameter $\lambda > 0$ and the variable $x > 0$.

2. Preliminaries and Auxiliary Results

Combining the integral form of the Marcum Q-function (4) and the integral ([30] (p. 306, Equation (2.15.5.4)))

$$\int_0^\infty t^{\nu+1} e^{-pt^2} I_\nu(ct) \, dt = \frac{c^\nu e^{c^2/(4p)}}{(2p)^{\nu+1}}, \qquad \Re(p) > 0, \Re(\nu) > -1, |\arg(c)| < \pi,$$

we express the CDF (3) for all $\lambda > 0$ and $x > 0$ as

$$F_{n,\lambda}(x) = \frac{e^{-\lambda/2}}{\lambda^{n/4-1/2}} \int_0^{\sqrt{x}} t^{n/2} e^{-t^2/2} I_{n/2-1}(\sqrt{\lambda}\, t) \, dt. \tag{6}$$

This formula is the starting point for our main results, which concerns the approximate calculation of the involved integral using two different types of mean-value theorems.

Our next main tools are two mean-value theorems for integrals, of which the integrands contain products of two suitable functions f, g, say. Both theorems belong to the so-called second mean-value theorems for definite integrals. The ancestor results of the first version theorem belongs to Bonnet ([31] (p. 14)); however, for the second one we are referred to the memoir by du Bois–Reymond ([32] (p. 83)) or also to Hobson's article [31]. The case in which at least one of the input functions f, g is a constant (first mean-value theorem) we skip in our present considerations. Now, recall the Bonnet variant of second mean-value theorem by Schwind–Ji–Koditschek.

Theorem 1. ([33] (p. 559, Theorem 2)). *Suppose $f \in C(a,b]$ and $g \geq 0$ is integrable on (a,b). Let $x \in (a,b]$ be fixed. If both $\lim_{t\to a}(f(t) - K)/(t-a)^r$ and $\lim_{t\to a} g(t)/(t-a)^s$ exist and differ from zero for some constant K, a non-zero r and some $s > -1$ with $r + s > -1$, then:*

1. *There exists $c_x \in (a, x]$ so, that*

$$\int_a^x f(t) g(t) \, dt = f(c_x) \int_a^x g(t) \, dt.$$

2. Moreover, for any such choice of c_x there holds

$$\lim_{x \to a} \frac{c_x - a}{x - a} = \left(\frac{s+1}{r+s+1}\right)^{\frac{1}{r}}. \tag{7}$$

Remark 1. *We notice that often a good choice for K in Theorem 1 is to take $K = \lim_{t \to a+} f(t)$ if the limit exists or $K = 0$ otherwise, consult ([33] (p. 561)), also see [34–36] for ancestry of (7), which describes the asymptotic behavior of c_x.*

Another approach in approximating the CDF of $\chi_n^{'2}(\lambda)$ is based on the use of the Okamura's version of the du Bois–Reymond's second mean-value theorem for definite integrals [37,38].

Theorem 2. ([39] (Equation (14))). *Let $f : [a,b] \mapsto \mathbb{R}$ be monotone and $g : [a,b] \mapsto \mathbb{R}$ integrable. Then there exists a $c \in [a,b]$ such that*

$$\int_a^b f(t)g(t)\,dt = f(a+)\int_a^c g(t)\,dt + f(b-)\int_c^b g(t)\,dt.$$

We point out that both Theorems 1 and 2 hold for Riemann integrable input functions. However, stronger second mean-value theorem results for definite integrals for Lebesgue integrable functions have been presented by Wituła–Hetmaniok–Słota in ([40] (p. 1614, Theorem 3)).

3. Approximating CDF of $\chi_n^{'2}(\lambda)$ Distribution

In this section we will state and prove our main results, derived from the formula (6) and the mean-value Theorems 1 and 2.

Theorem 3. *Let $n \in \mathbb{N}, \lambda > 0$ and $x > 0$.*
1. *Then, there exists $c_x \in (0, \sqrt{x}\,]$ such that*

$$F_{n,\lambda}(x) = \left(\frac{x}{\lambda}\right)^{n/4} e^{-\frac{\lambda + c_x^2}{2}} I_{n/2}(\sqrt{\lambda x}). \tag{8}$$

2. *For c_x there holds*

$$\lim_{x \to 0} \frac{c_x^2}{x} = \frac{n}{n+2}, \quad n \in \mathbb{N}, \tag{9}$$

while

$$F_{n,\lambda}(x) = \frac{e^{-\lambda/2}}{\Gamma(n/2+1)} \left(\frac{x}{2}\right)^{n/2} \left(1 + \mathcal{O}(x)\right), \quad x \to 0. \tag{10}$$

Proof. Consider the form of CDF given in (6). Making use of Theorem 1, with $f(t) = e^{-t^2/2} \in C(\mathbb{R}_+)$, which imply $K = \lim_{t \to 0} f(t) = 1$ and by L'Hospital rule and $r = 2$ follows

$$\lim_{t \to 0} \frac{f(t) - K}{t^r} = \lim_{t \to 0} \frac{e^{-t^2/2} - 1}{t^2} = -\frac{1}{2} \neq 0;$$

then, choosing $g(t) = t^{n/2} I_{n/2-1}(t\sqrt{\lambda})$ and $s = n - 1$ we have

$$\lim_{t \to 0} \frac{g(t)}{t^s} = \lim_{t \to 0} \frac{I_{n/2-1}(t\sqrt{\lambda})}{t^{n/2-1}} = \frac{\lambda^{(n-2)/4}}{2^{n/2-1} \Gamma(n/2)} \neq 0,$$

bearing in mind the asymptotics of the modified Bessel I_ν for small $z \to 0$ which is the consequence of (2):

$$I_\nu(z) = \frac{1}{\Gamma(\nu+1)} \left(\frac{z}{2}\right)^\nu (1 + \mathcal{O}(z^2)), \qquad -\nu \notin \mathbb{N}. \tag{11}$$

Hence, for $x > 0$ fixed, according to part 1 of Theorem 1, there exists a $c_x \in (0, \sqrt{x}\,]$ for which

$$F_{n,\lambda}(x) = \frac{e^{-\frac{\lambda + c_x^2}{2}}}{\lambda^{n/4 - 1/2}} \int_0^{\sqrt{x}} t^{n/2} I_{n/2-1}(t\sqrt{\lambda})\,dt = \left(\frac{x}{\lambda}\right)^{n/4} e^{-\frac{\lambda + c_x^2}{2}} I_{n/2}(\sqrt{\lambda x}),$$

where in the last equality the formula ([41] (p. 676, Equation (6.561.7)))

$$\int_0^1 t^{\nu+1} I_\nu(at)\,dt = a^{-1} I_{\nu+1}(a), \qquad \Re(\nu) > -1,$$

was taken.

By the second part of Theorem 1, bearing in mind that $c_x \in (0, \sqrt{x}\,]$ and setting $r = 2$, $s = n - 1$, $a = 0$, we have

$$\lim_{x \to 0} \frac{c_x^2}{(\sqrt{x})^2} = \lim_{x \to 0} \frac{c_x^2}{x} = \frac{n}{n+2}, \qquad n \in \mathbb{N},$$

that is (9). Now, the asymptotic behavior of the modified Bessel Function (11) approves the Relation (10). □

Corollary 1. *Let the situation be the same as in the preamble of Theorem 3. Then there exists $c = c_x \in (0,1]$ such that*

$$F_{n,\lambda}(x) = \left(\frac{x}{\lambda}\right)^{n/4} e^{-\frac{\lambda + xc^2}{2}} I_{n/2}(\sqrt{\lambda x}). \tag{12}$$

Proof. Using the substitution $u = t/\sqrt{x}$, from (6) mutatis mutandis

$$F_{n,\lambda}(x) = \sqrt{\lambda x}\,e^{-\lambda/2} \left(\frac{x}{\lambda}\right)^{n/4} \int_0^1 u^{n/2} e^{-xu^2/2} I_{n/2-1}(u\sqrt{\lambda x})\,du, \tag{13}$$

and then applying Theorem 1 repeating the above procedure for $f(u) = e^{-xu^2/2}$, $r = 2$, $g(u) = u^{n/2} I_{n/2-1}(u\sqrt{\lambda x})$ and $s = n - 1$ we readily conclude the Formula (12). □

In what follows we propose some numerical approximations for the real number c_x given in part 2 of Theorem 3 for small values of non-centrality parameter $\lambda > 0$.

Corollary 2. *Let $n \in \mathbb{N}$ and $x > 0$. When $\lambda \to 0$, in (8) we have*

$$c_x^2 = -2 \log\left[\left(\frac{2}{x}\right)^{n/2} \gamma\left(\frac{n}{2} + 1, \frac{x}{2}\right) + e^{-x/2}\right]. \tag{14}$$

Proof. Combining the Formulae (3) and ([26] (p. 70))

$$\lim_{a \to 0} Q_\nu(a, b) = \frac{1}{\Gamma(\nu)} \Gamma(\nu, b^2/2),$$

where $\Gamma(\cdot, \cdot)$ denotes the upper incomplete gamma function ([28] (p. 174, Equation (8.2.2)))

$$\Gamma(a, z) = \int_z^\infty t^{a-1} e^{-t}\,dt, \qquad \Re(a) > 0,$$

we obtain
$$\lim_{\lambda \to 0} F_{n,\lambda}(x) = 1 - \frac{\Gamma(n/2, x/2)}{\Gamma(n/2)}. \tag{15}$$

Now, from (11) we observe
$$\lim_{\lambda \to 0} \frac{I_{n/2}(\sqrt{\lambda x})}{\lambda^{n/4}} = \frac{x^{n/4}}{2^{n/2}\Gamma(n/2+1)}$$

which, in conjunction with (8) implies that
$$\lim_{\lambda \to 0} F_{n,\lambda}(x) = \left(\frac{x}{2}\right)^{n/2} \frac{e^{-c_x^2/2}}{\Gamma(n/2+1)}, \qquad c_x \in (0, \sqrt{x}\,]. \tag{16}$$

Equating the right-hand-side expressions in (15) and (16) we arrive at
$$\Gamma(n/2+1) - \frac{n}{2}\Gamma(n/2, x/2) = \left(\frac{x}{2}\right)^{n/2} e^{-c_x^2/2}. \tag{17}$$

The identities ([28] (p. 178, Equation (8.8.2–3)))
$$\Gamma(a+1, z) = a\Gamma(a, z) + z^a e^{-z}; \qquad \gamma(a, z) + \Gamma(a, z) = \Gamma(a),$$

where $\gamma(\cdot, \cdot)$ is the lower incomplete gamma function, defined by ([28] (p. 174, Equation (8.2.1)))
$$\gamma(a, z) = \int_0^z t^{a-1} e^{-t}\, dt, \qquad \Re(a) > 0,$$

one transforms (17) into
$$\left(\frac{2}{x}\right)^{n/2} \gamma(n/2+1, x/2) + e^{-x/2} = e^{-c_x^2/2}.$$

Now, obvious steps lead to the final form of c_x^2. □

Corollary 3. *For the small enough values of the non-centrality parameter λ and the argument x the magnitude of approximation satisfies the relation*
$$\frac{c_x^2}{x} - \frac{n}{n+2} = -\frac{n\,x}{4(n+4)} + o(x), \qquad x \to 0. \tag{18}$$

Proof. Recalling the asymptotic of the lower incomplete gamma function, which we deduce from the hypergeometric form expression ([17] (p. 262, Equation (6.5.12))), written in Landau's notation
$$\gamma(\alpha, z) = \frac{z^\alpha}{\alpha}\left(1 - \frac{\alpha z}{\alpha+1} + o(z)\right), \qquad z \to 0,$$

after asymptotic expansion of both expressions inside square brackets in (14), we get
$$\frac{c_x^2}{x} = -\frac{2}{x}\log\left[\frac{x}{n+2}\left(1 - \frac{(n+2)x}{2(n+4)} + o(x)\right) + 1 - \frac{x}{2} + \frac{x^2}{8} + o(x^2)\right]$$
$$= -\frac{2}{x}\log\left[1 - \frac{nx}{2(n+2)} + \frac{nx^2}{8(n+4)} + o(x^2)\right].$$

For when n is fixed and x is small enough, it is legitimate to express the logarithm via its asymptotic expansion $\log(1+h) = h + o(h), |h| < 1$, which approves (18). □

Remark 2. The associated limit result (18) enables the approximation

$$F_{n,\lambda}(x) \simeq \left(\frac{x}{\lambda}\right)^{n/4} \exp\left\{-\frac{\lambda}{2} - \frac{nx}{2(n+2)} + \frac{nx^2}{8(n+4)}\right\} I_{n/2}(\sqrt{\lambda x}).$$

This estimate we can readily take into account in numerical calculation of CDF for the purpose of comparison with another representations like Patnaik's and Temme's, for instance.

Corollary 4. For all $\lambda > 0, x > 0$ we have

$$F_{1,\lambda}(x) = \sqrt{\frac{2}{\pi \lambda}} \sinh(\sqrt{\lambda x})\, e^{-\frac{\lambda + c_x^2}{2}}, \tag{19}$$

where

$$c_x^2 = x + 2\log\left[1 + \frac{1}{\lambda} - \sqrt{\frac{x}{\lambda}} \frac{\cosh(\sqrt{\lambda x})}{\sinh(\sqrt{\lambda x})}\right].$$

Moreover,

$$F_{2,\lambda}(x) = \sqrt{\frac{x}{\lambda}}\, e^{-\frac{\lambda + c_x^2}{2}} I_1(\sqrt{\lambda x}), \tag{20}$$

where

$$c_x^2 = x + 2\log\left[1 - \sqrt{\frac{x}{\lambda}} \frac{I_2(\sqrt{\lambda x})}{I_1(\sqrt{\lambda x})}\right].$$

Proof. Having in mind that for non-negative integer m there holds ([27] (p. 178, Equation (7)))

$$Q_{m+1/2}(a,b) = \frac{1}{2}\left[\mathrm{erfc}\left(\frac{b-a}{\sqrt{2}}\right) + \mathrm{erfc}\left(\frac{b+a}{\sqrt{2}}\right)\right]$$
$$+ e^{-(a^2+b^2)/2} \sum_{k=1}^{m} \left(\frac{b}{a}\right)^{k-1/2} I_{k-1/2}(ab),$$

the Formula (3) for $n = 1$ becomes

$$F_{1,\lambda}(x) = \frac{1}{2}\left[\mathrm{erf}\left(\frac{\sqrt{x} - \sqrt{\lambda}}{\sqrt{2}}\right) + \mathrm{erf}\left(\frac{\sqrt{x} + \sqrt{\lambda}}{\sqrt{2}}\right)\right]. \tag{21}$$

As $(\mathrm{erf}(z))' = 2e^{-z^2}/\sqrt{\pi}$, equating (21) and the Formula (8) and then deriving such equality with respect to λ we get

$$e^{-x/2}\left(e^{-\sqrt{\lambda x}} - e^{\sqrt{\lambda x}}\right) = \frac{2}{\lambda} e^{-c_x^2/2}\left(\sqrt{\lambda x}\cosh(\sqrt{\lambda x}) - (1+\lambda)\sinh(\sqrt{\lambda x})\right).$$

Finally, the definition of hyperbolic sine implies (19).

The Formula (3) for $n = 2$ becomes $F_{2,\lambda}(x) = 1 - Q(\sqrt{\lambda}, \sqrt{x})$ where $Q_1(a,b) \equiv Q(a,b)$ is the Marcum Q-function. Now, knowing that ([42] (p. 1221, Equation (5)))

$$\frac{\partial Q(a,b)}{\partial a} = b\, I_1(ab)\, e^{-(a^2+b^2)/2},$$

the first derivative of (8), with respect of λ becomes

$$-\frac{\sqrt{x}}{2\sqrt{\lambda}} e^{-(\lambda + x)/2} I_1(\sqrt{\lambda x}) = \sqrt{x}\, e^{-(\lambda + c_x^2)/2}\left[\frac{\sqrt{x}}{2\lambda} I_2(\sqrt{\lambda x}) - \frac{I_1(\sqrt{\lambda x})}{2\sqrt{\lambda}}\right],$$

that is
$$e^{-x/2} = e^{-c_x^2/2}\left[1 - \sqrt{\frac{x}{\lambda}}\frac{I_2(\sqrt{\lambda x})}{I_1(\sqrt{\lambda x})}\right],$$

giving (20). □

The second approach in approximating the CDF of $\chi_n'^2(\lambda)$ is to apply Theorem 2.

Theorem 4. *Let $\lambda > 0$, $x > 0$ and $R_\rho(n) = [(2/n - 1)_+, 2/n + 1)$, where $(a)_+ = \max\{0, a\}$. Then for all $\rho \in R_\rho(1) = [1, 3)$ there exists some $c \in [0, 1]$ for which*

$$F_{1,\lambda}(x) = \frac{e^{-\lambda/2}}{\sqrt{\pi}}\left(\frac{x}{2}\right)^{(\rho-1)/4} \cosh(\sqrt{\lambda x}\,)\left[\gamma\left(\frac{3-\rho}{4}, \frac{x}{2}\right) - \gamma\left(\frac{3-\rho}{4}, \frac{xc^2}{2}\right)\right]. \quad (22)$$

When $\rho \in R_\rho(2) = [0, 2)$ there exists certain $c \in [0, 1]$ that

$$F_{2,\lambda}(x) = e^{-\lambda/2}\left(\frac{x}{2}\right)^{\frac{\rho}{2}}\left\{I_0(\sqrt{\lambda x}\,)\gamma(1 - \rho/2, x/2)\right.$$
$$\left. + [\delta_{\rho 0} - I_0(\sqrt{\lambda x}\,)]\gamma(1 - \rho/2, xc^2/2)\right\}, \quad (23)$$

where δ_{ab} stands for the Kronecker delta.

Moreover, for all $n \in \mathbb{N}_3 = \{3, 4, \dots\}$ and $\rho \in R_\rho(n)$ there exists $c \in [0, 1]$ such that

$$F_{n,\lambda}(x) = \frac{\lambda^{(2-n)/4}}{\sqrt{2}}e^{-\lambda/2}\left(\frac{x^\rho}{2^{\rho-1}}\right)^{n/4} I_{n/2-1}(\sqrt{\lambda x}\,)$$
$$\times \left[\gamma\left(\frac{(1-\rho)n + 2}{4}, \frac{x}{2}\right) - \gamma\left(\frac{(1-\rho)n + 2}{4}, \frac{xc^2}{2}\right)\right]. \quad (24)$$

We remark that the value of c is not necessarily the same throughout.

Proof. Consider the CDF's integral representation (13) in which the integration domain is the unit interval $[0, 1]$. Our intention is to specify the appropriate input functions f, g in a simple way and by scaling only the exponent of the power term—the integrand contains a product of three functions—to prepare it for the use of Okamura's Theorem 2. Precisely, consider for some real ρ (which range will be established later):

$$f_{n,\rho}(t) = t^{\rho n/2} I_{n/2-1}(t\sqrt{\lambda x}\,); \qquad g_{n,\rho}(t) = t^{(1-\rho)n/2} e^{-xt^2/2}.$$

From Formula (2) we can conclude that the function $I_\nu(x)$ increases monotonically for $\nu > 0$, $x > 0$. Therefore, $f_{n,\rho}(t)$, as a product of monotonically increasing functions, also monotonically increases. However, to establish the interconnection between the scaling parameter ρ and the degrees of freedom n we are forced to employ a more sophisticated approach. Namely, investigating the monotone behavior of $f_{n,\rho}(t)$, $t \in (0, 1]$ we start with

$$f'_{n,\rho}(t) = t^{\rho n/2-1}\left\{[(\rho + 1)n/2 - 1] I_{n/2-1}(t\sqrt{\lambda x}\,) + t\sqrt{\lambda x} I_{n/2}(t\sqrt{\lambda x}\,)\right\}. \quad (25)$$

The function I_ν is monotone decreasing with respect to the order, viz. ([43] (p. 220, Equation (2)))

$$I_\nu(x) > I_\mu(x), \qquad \mu > \nu \geq 0, x > 0,$$

also consult [44–46] regarding this question. So, evaluating (25) we get

$$f'_{n,\rho}(t) \geq t^{\rho n/2-1}\left[(\rho + 1)n/2 - 1 + t\sqrt{\lambda x}\,\right] I_{n/2}(t\sqrt{\lambda x}\,)$$
$$\geq t^{\rho n/2-1}\left[(\rho + 1)n/2 - 1\right] I_{n/2}(t\sqrt{\lambda x}\,),$$

which is sufficient to see that $f'_{n,\rho}(t) > 0$ for $\rho > 2/n - 1$ and also follows $f'_{n,2/n-1}(t) > 0$ directly from (25). On the other side we have

$$\int_0^1 g_{n,\rho}(t)\,dt = \frac{1}{2}\left(\frac{x}{2}\right)^{[(\rho-1)n-2]/4} \gamma\big([(1-\rho)n+2]/4, x/2\big); \qquad (26)$$

this expression makes sense for $\rho < 1 + 2/n$. Thus, having in mind the finiteness of $f_{n,\rho}(0+)$ and collecting all these constraints we infer that the range of the scaling parameter ρ is the interval $R_\rho(n) = [(2/n-1)_+, 2/n+1)$.

Firstly, consider $\rho \in R_\rho(1) = [1,3)$ with the associated input functions

$$f_{1,\rho}(t) = t^{\rho/2}\, I_{-1/2}(t\sqrt{\lambda x}) = \frac{\sqrt{2/\pi}}{\sqrt[4]{\lambda x}}\, t^{(\rho-1)/2} \cosh(t\sqrt{\lambda x}) \qquad (27)$$

$$g_{1,\rho}(t) = t^{(1-\rho)/2}\, e^{-xt^2/2}.$$

Being $\rho \geq 1$, the input limits are

$$f_{1,\rho}(0+) = 0; \qquad f_{1,\rho}(1) = \frac{\sqrt{2/\pi}}{\sqrt[4]{\lambda x}}\cosh(\sqrt{\lambda x}).$$

From (13) Okamura's Theorem 2 there follows (22).

The case $n = 2$, $\rho \in R_\rho(2) = [0,2)$ works since $I_0(0) = 1$. Ergo, we have two different solutions: when $\rho = 0$ and, respectively, $\rho \in R_\rho(2) \setminus \{0\} \equiv (0,2)$. Indeed, since

$$f'_{2,0}(t) = \sqrt{\lambda x}\, I_1(t\sqrt{\lambda x}) > 0,$$

$$f'_{2,\rho>0}(t) = t^{\rho-1}\left[\rho\, I_0(t\sqrt{\lambda x}) + t\sqrt{\lambda x}\, I_1(t\sqrt{\lambda x})\right] > 0, \qquad t \in (0,1],$$

both $f_{2,0}(t)$ and $f_{2,\rho>0}(t)$ monotone increase for $t \in (0,1]$. The associated limits read

$$f_{2,\rho}(0+) = \delta_{\rho 0}, \qquad f_{2,\rho}(1) = I_0(\sqrt{\lambda x}); \qquad \rho \in R_\rho(2),$$

which leads to the master Formula (23) for the CDF $F_{2,\lambda}(x)$.

It remains to see $n \in \mathbb{N}_3, \rho \in R_\rho(n)$. Knowing that $I_\nu(0) = 0$, $\Re(\nu) > 0$, we have vanishing $f_{n,\rho}(0+) = 0$ for $\rho \geq 0$ and $f_{n,\rho}(1) = I_{n/2-1}(\sqrt{\lambda x})$. By the monotonicity of $f_{n,\rho}(t)$ and the integration result (26) of $g_{n,\rho}(t)$ we get

$$F_{n,\lambda}(x) = \sqrt{\frac{\lambda}{2e^\lambda}}\left(\frac{2}{\lambda}\right)^{n/4}\left(\frac{x}{2}\right)^{\rho n/4} I_{n/2-1}(\sqrt{\lambda x})$$

$$\times \left[\gamma\Big(\frac{(1-\rho)n+2}{4}, \frac{x}{2}\Big) - \gamma\Big(\frac{(1-\rho)n+2}{4}, \frac{xc^2}{2}\Big)\right].$$

The rest is obvious. This completes the proof of the expression (24). □

Remark 3. Let ξ_1, ξ_2 be independent random variables defined on a standard probability space $(\Omega, \mathscr{A}, \mathsf{P})$ having $\chi'^2_{n_1}(\lambda_1)$, $\chi'^2_{n_2}(\lambda_2)$ distributions, respectively. Then the rv $\xi_1 + \xi_2 \sim \chi'^2_n(\lambda)$, where $n = n_1 + n_2$ and $\lambda = \lambda_1 + \lambda_2$, see, e.g., ([18] (p. 33, Teorema 27)). According to this relation we can consider $F_{2,\lambda}(x)$ as the CDF of the sum of two $\chi'^2_1(\lambda_j)$, $j = 1, 2$ distributed random variables where the linear combination $\lambda = \theta\lambda_1 + (1-\theta)\lambda_2$, $\theta \in [0,1]$ occurs between their non-centrality parameters.

Moreover, the values $\theta = 0, 1$ correspond to the problem of obtaining the CDF using the property $\chi'^2_n(\lambda) = \chi'^2_1(\lambda) + \chi'^2_{n-1}(0) \equiv \chi'^2_1(\lambda) + \chi^2_{n-1}$, where the no–central and the central rvs on the right are mutually independent, consult ([1] (p. 436)) and the related quotations therein.

Corollary 5. Let $\lambda > 0$, $x > 0$. Then for all $n \in \mathbb{N}_2 = \{2, 3, 4, \dots\}$ there exists certain $c \in [0, 1]$ such that

$$F_{n,\lambda}(x) = \sqrt{\frac{\pi \lambda}{2}}\, e^{-\lambda/2} \left(\frac{x}{\lambda}\right)^{n/4} I_{n/2-1}(\sqrt{\lambda x}\,) \left[\operatorname{erf}(\sqrt{x/2}\,) - \operatorname{erf}(c\sqrt{x/2}\,)\right]. \qquad (28)$$

Also, for all $n \in \mathbb{N}_3 = \{3, 4, \dots\}$ there exists some $c \in [0, 1]$ for which

$$F_{n,\lambda}(x) = e^{-\lambda/2} \left(\frac{2}{\lambda}\right)^{(n-2)/4} I_{n/2-1}(\sqrt{\lambda x}\,) \left[\gamma\!\left(\frac{n+2}{4}, \frac{x}{2}\right) - \gamma\!\left(\frac{n+2}{4}, \frac{xc^2}{2}\right)\right]. \qquad (29)$$

Proof. The first case occurs when $\rho = 1$ in Theorem 4. From (27) we have

$$f_{1,1}(t) = \frac{\sqrt{2/\pi}}{\sqrt[4]{\lambda x}} \cosh(t\sqrt{\lambda x})$$

which results in $f_{1,1}(0+) = \sqrt{2/\pi}/\sqrt[4]{\lambda x}$. Hence, we consider $n \in \mathbb{N}_2$ in which case $f_{n,1}(0+) = 0$ and $f_{n,1}(1) = I_{n/2-1}(\sqrt{\lambda x}\,)$. Additionally, from (25) it follows for all $n \in \mathbb{N}_2$, $t \geq 0$ that

$$f'_{n,1}(t) = t^{n/2-1} I_{n/2-1}(t\sqrt{\lambda x}\,)(n-1) + \sqrt{\lambda x}\, t^{n/2} I_{n/2}(t\sqrt{\lambda x}\,) \geq 0.$$

So, $f_{n,1}(t)$ monotone increases on $[0, 1]$. Therefore

$$F_{n,\lambda}(x) = \sqrt{\lambda x}\, e^{-\lambda/2} \left(\frac{x}{\lambda}\right)^{n/4} I_{n/2-1}(\sqrt{\lambda x}\,) \int_c^1 e^{-xt^2/2}\, dt$$

$$= \sqrt{\frac{\pi \lambda}{2}}\, e^{-\lambda/2} \left(\frac{x}{\lambda}\right)^{n/4} I_{n/2-1}(\sqrt{\lambda x}\,) \left[\operatorname{erf}(\sqrt{x/2}) - \operatorname{erf}(c\sqrt{x/2})\right].$$

Here, the notation of the error function (or probability integral)

$$\operatorname{erf}(z) = \frac{2}{\sqrt{\pi}} \int_0^z e^{-t^2}\, dt,$$

has been used.

Taking $\rho = 0$ in Theorem 4, from (27) $f_{1,0}(t) = \frac{\sqrt{2/(\pi t)}}{\sqrt[4]{\lambda x}} \cosh(t\sqrt{\lambda x})$, no right limit exists at zero, hence *a fortiori* $n > 1$. Having in mind the observations stated in the proof of Theorem 4 for $n \in \mathbb{N}_3$ the Formula (29) follows immediately from (24), setting $\rho = 0$. □

Remark 4. *Recalling the relation ([28] (p. 176, Equation (8.4.1)))*

$$\gamma(1/2, x) = \sqrt{\pi}\, \operatorname{erf}(\sqrt{x}\,)$$

the representation Formula (28) becomes

$$F_{n,\lambda}(x) = \sqrt{\frac{\lambda}{2}}\, e^{-\lambda/2} \left(\frac{x}{\lambda}\right)^{n/4} I_{n/2-1}(\sqrt{\lambda x}\,) \left[\gamma(1/2, x/2\,) - \gamma(1/2, xc^2/2\,)\right].$$

Author Contributions: The authors contributed equally to the manuscript and typed, read and approved the final version. All authors have read and agreed to the published version of the manuscript.

Funding: This research received no external funding.

Acknowledgments: The authors are grateful to the referees for careful reading of the first version of the manuscript and for helpful comments that finally encompass the article.

Conflicts of Interest: The authors declare no conflict of interest.

References

1. Johnson, N.L.; Kotz, S.; Balakrishnan, N. *Continuous Univariate Distributions*; John Wiley & Sons, Inc.: New York, NY, USA, 1995; Volume 2.
2. Robert, C. Modified Bessel functions and their applications in probability and statistics. *Stat. Probab. Lett.* **1990**, *9*, 155–161. [CrossRef]
3. Kamel, A.S.; Abdel-Samad, A.I. On the computation of non-central Chi-square distribution function. *Commun. Stat. Simul. Comput.* **1990**, *19*, 1279–1291. [CrossRef]
4. András, S.; Baricz, Á. Properties of the probability density function of the non–central chi–squared distribution. *J. Math. Anal. Appl.* **2008**, *346*, 395–402. [CrossRef]
5. Scharf, L.L. *Statistical Signal Processing: Detection, Estimation, and Time Series Analysis*; Addison–Wesley Publishing Co.: Boston, MA, USA, 1990.
6. Lancaster, H.O. Forerunners of the Pearson χ^2. *Aust. J. Stat.* **1966**, *8*, 117–126. [CrossRef]
7. Bienaymé, I.J. Sur la probabilité des erreurs d'après la méthode des moindres carrés. *Liouville's J. Math. Pures Appl.* **1852**, *17*, 33–78.
8. Pearson, K. On a criterion that a given system of deviations from the probable in the case of a correlated system of variables is such that it can be reasonably supposed to have arisen from random sampling. *Philos. Mag.* **1900**, *50*, 157–175. [CrossRef]
9. Helmert, F.R. Über die Berechnung des wahrscheinlichen Fehlers aus einer endlichen Anzahl wehrer Beobachtungsfehler. *Z. Math. Phys.* **1875**, *20*, 300–303.
10. Helmert, F.R. Über die Wahrscheinlichkeit der Potenzsummen der Beobachtungsfehler und über einige damit im Zusammenhange stehende Fragen. *Z. Math. Phys.* **1876**, *21*, 192–218.
11. Kruskal, W.H. Helmert's distribution. *Am. Math. Mon.* **1946**, *53*, 435–438. [CrossRef]
12. Sheynin, O.B. Origin of the theory of errors. *Nature* **1966**, *211*, 1003–1004. [CrossRef]
13. Plackett, R.L. Karl Pearson and the Chi-Squared Test. *Int. Stat. Rev.* **1983**, *51*, 59–72. [CrossRef]
14. Kendall, M.G. Studies in the history of probability and statistics. XXVI. The work of Ernst Abbe. *Biometrika* **1971**, *58*, 369–373. [CrossRef]
15. Abbe, E. *Über die Gesetzmässigkeit in der Vertheilung der Fehler bei Beobachtungsreihen*; Dissertation zur Erlangung der Venia Docendi bei den Phyilosophischen Fakultät in Jena; Verlag Frommann: Jena, Germany, 1863.
16. Watson, G.N. *A Treatise on the Theory of Bessel Functions*; Cambridge University Press: London, UK, 1922.
17. Abramowitz, M.; Stegun, I.A. (Eds.) *Handbook of Mathematical Functions with Formulas, Graphs, and Mathematical Tables*; Applied Mathematics Series 55; National Bureau of Standards: Washington, DC, USA, 1964; Reprinted by Dover Publications, New York, 1972.
18. Mihoc, G.; Craiu, V. *Treatise on Mathematical Statistics, Sampling and Estimation*; With an English Table of Contents; Editura Academiei Republicii Socialiste România: Bucureşti, Romania, 1976; Volume I. (In Romanian)
19. Fisher, R.A. The general sampling distribution of the multiple correlation coefficient. *Proc. R. Soc. Lond.* **1928**, *121*, 654–673.
20. Tang, P.C. The power function of the analysis of variance tests with tables and illustrations of their use. *Stat. Res. Mem. Lond.* **1938**, *2*, 126–149.
21. Patnaik, P.B. The non-central χ^2– and the *F*–distributions and their applications. *Biometrika* **1949**, *36*, 202–234. [CrossRef]
22. Pearson, E.S. Note on an approximation to the distribution of non-central χ^2. *Biometrika* **1959**, *46*, 202–232. [CrossRef]
23. Sankaran, M. Approximations to the non-central chi-square distribution. *Biometrika* **1963**, *50*, 199–204. [CrossRef]
24. Temme, N.M. Asymptotic and numerical aspects of the non-central chi-square distribution. *Comput. Math. Appl.* **1993**, *25*, 55–63. [CrossRef]
25. Jankov Maširević, D. On new formulas for the cumulative distribution function of the non-central chi-square distribution. *Mediterr. J. Math.* **2017**, *14*, 66. [CrossRef]
26. András, S.; Baricz, Á.; Sun, Y. The generalized Marcum Q–function: An orthogonal polynomial approach. *Acta Univ. Sapientiae Math.* **2011**, *3*, 60–76.
27. Brychkov, Y.A. On some properties of the Marcum Q function. *Integral Transform. Spec. Funct.* **2012**, *23*, 177–182. [CrossRef]
28. Olver, F.W.J.; Lozier, D.W.; Boisvert, R.F.; Clark, C.W. (Eds.) *NIST Handbook of Mathematical Functions*; NIST and Cambridge University Press: Cambridge, UK, 2010.
29. Agrest, M.M.; Maksimov, M.S. *Theory of Incomplete Cylindrical Functions and Their Applications*; Springer: New York, NY, USA, 1971.
30. Prudnikov, A.P.; Brychkov, Y.A.; Marichev, O.I. *Special Functions*; Integrals and Series; Gordon and Breach Science Publishers: New York, NY, USA, 1986; Volume 2.
31. Hobson, E.W. On the second mean-value theorem of the integral calculus. *Trans. Am. Math. Soc.* **1908**, *2*, 14–23. [CrossRef]
32. du, Bois–Reymond, P. Über die allgemeinen Eigenschaften der Klasse von Doppelintegralen, zu welcher das Fouriersche Doppelintegral gehört. *J. Reine Angew. Math.* **1868**, *69*, 65–108.
33. Schwind, W.J.; Ji, J.; Koditschek, D.E. A physically motivated further note on the mean-value theorem for integrals. *Am. Math. Mon.* **1999**, *106*, 559–564. [CrossRef]
34. Bao-Lin, Z. A note on the mean-value theorem for integrals. *Am. Math. Mon.* **1997**, *104*, 561–562. [CrossRef]
35. Jacobson, B. On the mean-value theorem for integrals. *Am. Math. Mon.* **1982**, *89*, 300–301. [CrossRef]
36. Polezzi, M. On the weighted mean value theorem for integrals. *Internat. J. Math. Ed. Sci. Technol.* **2006**, *37*, 868–870. [CrossRef]
37. Matsumoto, T. Hiroshi Okamura. *Mem. Coll. Sci. Univ. Kyoto Ser. A Math.* **1950**, *26*, 1–3. [CrossRef]

38. Okamura, H. On the second mean-value theorem of integral. In *Mathematics*; Kyoto Mathematical Society: Kyoto, Japan, 1947; Volume 1. (In Japanese)
39. Baricz, Á.; Pogány, T.K. On a sum of modified Bessel functions. *Mediterr. J. Math.* **2014**, *11*, 349–360. [CrossRef]
40. Wituła, R.; Hetmaniok, E.; Słota, D. A stronger version of the second mean value theorem for integrals. *Comput. Math. Appl.* **2016**, *64*, 1612–1615. [CrossRef]
41. Gradshteyn, I.S.; Ryzhik, I.M. *Table of Integrals, Series, and Products*, 6th ed.; Jeffrey, A., Zwillinger, D., Eds.; Academic Press: New York, NY, USA, 2000.
42. Pratt, W.K. Partial differentials of Marcum's Q function. *Proc. IEEE* **1968**, *56*, 1220–1221. [CrossRef]
43. Jones, A.L. An extension of an inequality involving modified Bessel functions. *J. Math. Phys.* **1968**, *47*, 220–221. [CrossRef]
44. Cochran, J.A. The monotonicity of modified Bessel functions with respect to their order. *J. Math. Phys.* **1967**, *46*, 220–222. [CrossRef]
45. Nåsell, I. Inequalities for modified Bessel functions. *Math. Comput.* **1974**, *28*, 253–256. [CrossRef]
46. Soni, R.P. On an inequality for modified Bessel functions. *J. Math. Phys.* **1965**, *44*, 406–407. [CrossRef]

Review

Some Applications of the Wright Function in Continuum Physics: A Survey

Yuriy Povstenko

Department of Mathematics and Computer Science, Faculty of Science and Technology, Jan Długosz University in Częstochowa, al. Armii Krajowej 13/15, 42-200 Częstochowa, Poland; j.povstenko@ajd.czest.pl

Abstract: The Wright function is a generalization of the exponential function and the Bessel functions. Integral relations between the Mittag–Leffler functions and the Wright function are presented. The applications of the Wright function and the Mainardi function to description of diffusion, heat conduction, thermal and diffusive stresses, and nonlocal elasticity in the framework of fractional calculus are discussed.

Keywords: fractional calculus; Caputo derivative; Mittag–Leffler functions; Wright function; Mainardi function; Laplace transform; Fourier transform; nonperfect thermal contact; nonlocal elasticity; fractional nonlocal elasticity

MSC: 26A33; 33E12; 35Q74; 74S40

Citation: Povstenko, Y. Some Applications of the Wright Function in Continuum Physics: A Survey. *Mathematics* **2021**, *9*, 198. https://doi.org/10.3390/math9020198

Received: 23 December 2020
Accepted: 17 January 2021
Published: 19 January 2021

Publisher's Note: MDPI stays neutral with regard to jurisdictional claims in published maps and institutional affiliations.

Copyright: © 2021 by the author. Licensee MDPI, Basel, Switzerland. This article is an open access article distributed under the terms and conditions of the Creative Commons Attribution (CC BY) license (https://creativecommons.org/licenses/by/4.0/).

1. Introduction

The fractional calculus (the theory of integrals and derivatives of non-integer order) has attracted considerable interest of researchers and has many applications in physics, chemistry, rheology, geology, hydrology, medicine, engineering, finance, etc. (see, for example, West–Bologna–Grigolini [1], Magin [2], Povstenko [3], Tarasov [4], Povstenko [5], Uchaikin [6], Atanacković–Pilipović–Stanković–Zorica [7], Herrmann [8], Povstenko [9], Datsko–Gafiychuk–Podlubny [10], West [11], Skiadas [12], Tarasov [13], Kumar–Singh [14], Su [15] and references therein). The Mittag–Leffler functions and the Wright function appear in solutions of various types of equations with fractional operators. The Mittag–Leffler function in one parameter $E_\alpha(z)$ was introduced in [16,17]. The generalized Mittag–Leffler function in two parameters $E_{\alpha,\beta}(z)$ was considered in [18,19]. A comprehensive treatment of properties of the Mittag–Leffler functions can be found in Erdélyi–Magnus–Oberhettinger–Tricomi [20], Gorenflo–Mainardi [21], Podlubny [22], Kilbas–Srivastava–Trujillo [23], Gorenflo–Kilbas–Mainardi–Rogosin [24]. Numerical algorithms for calculation of the Mittag–Leffler functions were proposed in [25] and implemented in [26]. The Wright function was presented in [27,28] and later on discussed by Erdélyi–Magnus–Oberhettinger–Tricomi [20], Gorenflo–Mainardi [21], Podlubny [22], Kilbas–Srivastava–Trujillo [23], Gorenflo–Kilbas–Mainardi–Rogosin [24], Luchko [29], among others. Numerical algorithms for calculating the Wright function were suggested in [30].

In 1996, Mainardi [31,32] solved the diffusion-wave equation with the Caputo fractional derivative of the order α

$$\frac{\partial^\alpha T}{\partial t^\alpha} = a \frac{\partial^2 T}{\partial x^2}, \qquad 0 < \alpha \leq 2, \tag{1}$$

on a real line (the Cauchy problem) and a half-line (the signaling problem). The solutions were obtained in terms of the Mainardi function $M(z; \frac{\alpha}{2})$ [33], where

$$z = \frac{|x|}{\sqrt{a} t^{\alpha/2}} \tag{2}$$

is the similarity variable, a can be treated as the generalized thermal diffusivity coefficient.

Equation (1) in the limiting case $\alpha \to 0$ corresponds to the Helmholtz equation (localized diffusion); the subdiffusion regime is characterized by the values $0 < \alpha < 1$. For $1 < \alpha < 2$, the diffusion-wave Equation (1) interpolates between the diffusion equation ($\alpha = 1$) and the wave equation ($\alpha = 2$).

Applications of fractional calculus to viscoelasticity have been studied by many authors. The historical notes and the extensive bibliography on this subject can be found in the book of Mainardi [34]. According to the Scott–Blair stress-strain law, the dependence between the stress $\sigma(x,t)$ and the strain $\epsilon(x,t)$ can be written as [34,35]

$$\sigma(x,t) = \rho a \frac{\partial^\nu \epsilon(x,t)}{\partial t^\nu}, \qquad 0 \leq \nu \leq 1. \tag{3}$$

The constitutive Equation (3) characterizes a viscoelastic material intermediate between a perfectly elastic solid (the Hooke law for the value $\nu = 0$) and a perfectly viscous fluid (the Newton law when $\nu = 1$) with the corresponding interpretations of the coefficient a in terms of the elasticity constant or the kinematic viscosity. The relation (3) leads to the evolution Equation (1) with $\alpha = 2 - \nu$.

The book [36] presents a picture of the state-of-the-art for solutions of the diffusion-wave equation with one, two, and three space variables in Cartesian, cylindrical, and spherical coordinates under different kinds of boundary conditions.

In the present survey article, we briefly discuss the properties of the Mittag–Leffler functions and Wright function and present the integral relations between the Mittag–Leffler functions and the Wright function. The applications of the Wright function and the Mainardi function to the description of diffusion, heat conduction, thermal and diffusive stresses, and nonlocal elasticity in the framework of fractional calculus are reviewed.

2. Mathematical Preliminaries

2.1. Integrals and Derivatives of Fractional Order

The Riemann–Liouville integral of fractional order α is defined as [21–23]:

$$I^\alpha f(t) = \frac{1}{\Gamma(\alpha)} \int_0^t (t-\tau)^{\alpha-1} f(\tau) \, d\tau, \qquad \alpha > 0, \tag{4}$$

where $\Gamma(\alpha)$ is the gamma function.

The Riemann–Liouville derivative of fractional order α has the form

$$D_{RL}^\alpha f(t) = \frac{d^n}{dt^n} \left[\frac{1}{\Gamma(n-\alpha)} \int_0^t (t-\tau)^{n-\alpha-1} f(\tau) \, d\tau \right], \qquad n-1 < \alpha < n, \tag{5}$$

whereas the Caputo fractional derivative is written as

$$D_C^\alpha f(t) \equiv \frac{d^\alpha f(t)}{dt^\alpha} = \frac{1}{\Gamma(n-\alpha)} \int_0^t (t-\tau)^{n-\alpha-1} \frac{d^n f(\tau)}{d\tau^n} \, d\tau, \qquad n-1 < \alpha < n. \tag{6}$$

The fractional operators have the following Laplace transform rules:

$$\mathcal{L}\{I^\alpha f(t)\} = \frac{1}{s^\alpha} f^*(s), \tag{7}$$

$$\mathcal{L}\{D_{RL}^\alpha f(t)\} = s^\alpha f^*(s) - \sum_{k=0}^{n-1} D^k I^{n-\alpha} f(0^+) s^{n-1-k}, \qquad n-1 < \alpha < n, \tag{8}$$

$$\mathcal{L}\left\{\frac{d^\alpha f}{dt^\alpha}\right\} = s^\alpha f^*(s) - \sum_{k=0}^{n-1} f^{(k)}(0^+) s^{\alpha-1-k}, \qquad n-1 < \alpha < n. \tag{9}$$

Here, the asterisk denotes the transform, and s is the Laplace transform variable.

2.2. Mittag–Leffler Functions

The Mittag–Leffler function in one parameter α

$$E_\alpha(z) = \sum_{k=0}^{\infty} \frac{z^k}{\Gamma(\alpha k + 1)}, \quad \alpha > 0, \ z \in \mathbb{C}, \tag{10}$$

can be considered as the extension of the exponential function $e^z = E_1(z)$, whereas the generalized Mittag–Leffler function in two parameters α and β is defined by the series representation

$$E_{\alpha,\beta}(z) = \sum_{k=0}^{\infty} \frac{z^k}{\Gamma(\alpha k + \beta)}, \quad \alpha > 0, \ \beta > 0, \ z \in \mathbb{C}. \tag{11}$$

In the general case, the parameters α and β can be treated as complex numbers with some limitations on their real parts [24], but we restrict ourselves to positive values of α and β.

The following recurrence relations [20,24]

$$E_{\alpha,\beta}(z) = \frac{1}{\Gamma(\beta)} + z E_{\alpha,\alpha+\beta}(z). \tag{12}$$

$$E_{\alpha,\beta}(z) = \beta E_{\alpha,\beta+1}(z) + \alpha z \frac{dE_{\alpha,\beta+1}(z)}{dz} \tag{13}$$

are valid for the Mittag–Leffler functions.

For investigation of the convergence of integrals containing the Mittag–Leffler functions, their asymptotic representations for large negative values of argument are useful. For $x \to \infty$, we have

$$E_\alpha(-x) \sim \frac{1}{\Gamma(1-\alpha)x}, \tag{14}$$

$$E_{\alpha,2}(-x) \sim \frac{1}{\Gamma(2-\alpha)x}, \tag{15}$$

$$E_{\alpha,\alpha}(-x) \sim -\frac{1}{\Gamma(-\alpha)x^2}, \tag{16}$$

$$E_{\alpha,\beta}(-x) \sim \frac{1}{\Gamma(\beta-\alpha)x}. \tag{17}$$

The essential role of the Mittag–Leffler functions in fractional calculus is connected with the formula for the inverse Laplace transform (see Gorenflo–Mainardi [21], Podlubny [22], Kilbas–Srivastava–Trujillo [23], Gorenflo–Kilbas–Mainardi–Rogosin [24]):

$$\mathcal{L}^{-1}\left\{\frac{s^{\alpha-\beta}}{s^\alpha + b}\right\} = t^{\beta-1} E_{\alpha,\beta}(-bt^\alpha). \tag{18}$$

2.3. Wright Function and Mainardi Function

The Wright function is a generalization of the exponential function and the Bessel functions and is defined as [27,28] (see also refs. [20–24,31,32,37–39])

$$W(\alpha,\beta;z) = \sum_{k=0}^{\infty} \frac{z^k}{k!\,\Gamma(\alpha k + \beta)}, \quad \alpha > -1, \ \beta \in \mathbb{C}, \ z \in \mathbb{C}. \tag{19}$$

The Wright function satisfies the recurrence equations [20]

$$\alpha z W(\alpha, \alpha+\beta; z) = W(\alpha, \beta-1; z) + (1-\beta) W(\alpha,\beta;z), \tag{20}$$

$$\frac{dW(\alpha,\beta;z)}{dz} = W(\alpha,\alpha+\beta;z). \tag{21}$$

The Mainardi function $M(\alpha; z)$ [22,31–33] is a particular case of the Wright function

$$M(\alpha; z) = W(-\alpha, 1-\alpha; -z) = \sum_{k=0}^{\infty} \frac{(-1)^k z^k}{k! \Gamma[-\alpha k + (1-\alpha)]}, \quad 0 < \alpha < 1, \quad z \in \mathbb{C}. \quad (22)$$

The Wright function and the Mainardi function appear in formulae for the inverse Laplace transform (see Mainardi [31,32], Stanković [40], Gajić–Stanković [41]):

$$\mathcal{L}^{-1}\{\exp(-\lambda s^\alpha)\} = \frac{\alpha \lambda}{t^{\alpha+1}} M(\alpha; \lambda t^{-\alpha}), \quad 0 < \alpha < 1, \quad \lambda > 0, \quad (23)$$

$$\mathcal{L}^{-1}\{s^{\alpha-1} \exp(-\lambda s^\alpha)\} = \frac{1}{t^\alpha} M(\alpha; \lambda t^{-\alpha}), \quad 0 < \alpha < 1, \quad \lambda > 0, \quad (24)$$

$$\mathcal{L}^{-1}\{s^{-\beta} \exp(-\lambda s^\alpha)\} = t^{\beta-1} W(-\alpha, \beta; -\lambda t^{-\alpha}), \quad 0 < \alpha < 1, \quad \lambda > 0. \quad (25)$$

2.4. The Integral Transform Relations between the Mittag–Leffler Function and Wright Function

The Laplace transform of the Wright function is expressed in terms of the Mittag–Leffler function [20,22,23]

$$\mathcal{L}\{W(\alpha, \beta; t)\} = \frac{1}{s} E_{\alpha, \beta}\left(\frac{1}{s}\right), \quad \alpha > 0, \quad \beta > 0, \quad (26)$$

and [37]

$$\mathcal{L}\{W(\alpha, \beta; -t)\} = E_{-\alpha, \beta-\alpha}(-s), \quad -1 < \alpha < 0, \quad \beta > 0, \quad (27)$$

whereas, for the Mainardi function, the corresponding relation takes the form

$$\mathcal{L}\{M(\alpha; t)\} = E_\alpha(-s), \quad 0 < \alpha < 1. \quad (28)$$

The Mittag–Leffler functions and the Wright function are related by the Fourier cosine transform (Povstenko [36,42]):

$$\int_0^\infty E_\alpha(-\xi^2) \cos(x\xi) \, d\xi = \frac{\pi}{2} M\left(\frac{\alpha}{2}; x\right), \quad 0 < \alpha < 2, \quad x > 0, \quad (29)$$

$$\int_0^\infty E_{\alpha,2}(-\xi^2) \cos(x\xi) \, d\xi = \frac{\pi}{2} W\left(-\frac{\alpha}{2}, 2-\frac{\alpha}{2}; -x\right), \quad 0 < \alpha < 2, \quad x > 0, \quad (30)$$

$$\int_0^\infty E_{\alpha,\alpha}(-\xi^2) \cos(x\xi) \, d\xi = \frac{\pi}{2} W\left(-\frac{\alpha}{2}, \frac{\alpha}{2}; -x\right), \quad 0 < \alpha < 2, \quad x > 0, \quad (31)$$

$$\int_0^\infty E_{\alpha,\beta}(-\xi^2) \cos(x\xi) \, d\xi = \frac{\pi}{2} W\left(-\frac{\alpha}{2}, \beta-\frac{\alpha}{2}; -x\right), \quad 0 < \alpha < 2, \quad \beta > 0, \quad x > 0, \quad (32)$$

as well as by the Fourier sine transform

$$\int_0^\infty \xi E_\alpha(-\xi^2) \sin(x\xi) \, d\xi = \frac{\pi}{2} W\left(-\frac{\alpha}{2}, 1-\alpha; -x\right), \quad 0 < \alpha < 2, \quad x > 0, \quad (33)$$

$$\int_0^\infty \xi E_{\alpha,2}(-\xi^2) \sin(x\xi) \, d\xi = \frac{\pi}{2} W\left(-\frac{\alpha}{2}, 2-\alpha; -x\right), \quad 0 < \alpha < 2, \quad x > 0, \quad (34)$$

$$\int_0^\infty \xi E_{\alpha,\alpha}(-\xi^2) \sin(x\xi) \, d\xi = \frac{\pi}{2} W\left(-\frac{\alpha}{2}, 0; -x\right) = \frac{\alpha \pi}{4} x M\left(\frac{\alpha}{2}; x\right), \quad 0 < \alpha < 2, \quad x > 0, \quad (35)$$

$$\int_0^\infty \xi E_{\alpha,\beta}(-\xi^2) \sin(x\xi) \, d\xi = \frac{\pi}{2} W\left(-\frac{\alpha}{2}, \beta-\alpha; -x\right), \quad 0 < \alpha < 2, \quad \beta > 0, \quad x > 0. \quad (36)$$

Due to (16), we can also obtain for $E_{\alpha,\alpha}(-\xi^2)$

$$\int_0^\infty \xi^2 E_{\alpha,\alpha}(-\xi^2)\cos(x\xi)\,d\xi = -\frac{\pi}{2} W\left(-\frac{\alpha}{2}, -\frac{\alpha}{2}; -x\right), \quad 0<\alpha<2,\ x>0, \tag{37}$$

$$\int_0^\infty \xi^3 E_{\alpha,\alpha}(-\xi^2)\sin(x\xi)\,d\xi = -\frac{\pi}{2} W\left(-\frac{\alpha}{2}, -\alpha; -x\right), \quad 0<\alpha<2,\ x>0. \tag{38}$$

The equations presented above allow us to obtain additional integral relations between the Mittag–Leffler functions and the Wright function, which can be helpful when solving problems in polar or cylindrical coordinates using the Hankel transform of order zero. Taking into account the integral representation of the Bessel function $J_0(x)$ (Watson [43], Abramowitz-Stegun [44])

$$J_0(x) = \frac{1}{\pi}\int_0^\pi \cos(x\sin\theta)\,d\theta, \tag{39}$$

$$J_0(x) = \frac{2}{\pi}\int_0^\infty \sin(x\cosh t)\,dt, \quad x>0, \tag{40}$$

$$J_0(x) = \frac{2}{\pi}\int_1^\infty \frac{\sin(xt)}{\sqrt{t^2-1}}\,dt, \quad x>0, \tag{41}$$

we get

$$\int_0^\infty E_\alpha(-\xi^2) J_0(r\xi)\,d\xi = \frac{1}{2}\int_0^\pi M\left(\frac{\alpha}{2}; r\sin\theta\right)d\theta, \quad 0<\alpha\le 2,\ r>0, \tag{42}$$

$$\int_0^\infty E_{\alpha,2}(-\xi^2) J_0(r\xi)\,d\xi = \frac{1}{2}\int_0^\pi W\left(-\frac{\alpha}{2}, 2-\frac{\alpha}{2}; -r\sin\theta\right)d\theta, \quad 0<\alpha\le 2,\ r>0, \tag{43}$$

$$\int_0^\infty E_{\alpha,\alpha}(-\xi^2) J_0(r\xi)\,d\xi = \frac{1}{2}\int_0^\pi W\left(-\frac{\alpha}{2}, \frac{\alpha}{2}; -r\sin\theta\right)d\theta, \quad 0<\alpha\le 2,\ r>0, \tag{44}$$

$$\int_0^\infty E_{\alpha,\beta}(-\xi^2) J_0(r\xi)\,d\xi = \frac{1}{2}\int_0^\pi W\left(-\frac{\alpha}{2}, \beta-\frac{\alpha}{2}; -r\sin\theta\right)d\theta, \quad 0<\alpha\le 2,\ r>0. \tag{45}$$

Similarly,

$$\int_0^\infty E_\alpha(-\xi^2) J_0(r\xi)\xi\,d\xi = \int_0^\infty W\left(-\frac{\alpha}{2}, 1-\alpha; -r\cosh t\right)dt, \quad 0<\alpha\le 2,\ r>0, \tag{46}$$

$$\int_0^\infty E_{\alpha,2}(-\xi^2) J_0(r\xi)\xi\,d\xi = \int_0^\infty W\left(-\frac{\alpha}{2}, 2-\alpha; -r\cosh t\right)dt, \quad 0<\alpha\le 2,\ r>0, \tag{47}$$

$$\int_0^\infty E_{\alpha,\alpha}(-\xi^2) J_0(r\xi)\xi\,d\xi = \int_0^\infty W\left(-\frac{\alpha}{2}, 0; -r\cosh t\right)dt, \quad 0<\alpha\le 2,\ r>0, \tag{48}$$

$$\int_0^\infty E_{\alpha,\beta}(-\xi^2) J_0(r\xi)\xi\,d\xi = \int_0^\infty W\left(-\frac{\alpha}{2}, \beta-\alpha; -r\cosh t\right)dt, \quad 0<\alpha\le 2,\ r>0, \tag{49}$$

and

$$\int_0^\infty E_\alpha(-\xi^2) J_0(r\xi)\xi\,d\xi = \int_1^\infty W\left(-\frac{\alpha}{2}, 1-\alpha; -rt\right)\frac{1}{\sqrt{t^2-1}}\,dt, \quad 0<\alpha\le 2,\ r>0, \tag{50}$$

$$\int_0^\infty E_{\alpha,2}(-\xi^2) J_0(r\xi)\xi\,d\xi = \int_1^\infty W\left(-\frac{\alpha}{2}, 2-\alpha; -rt\right)\frac{1}{\sqrt{t^2-1}}\,dt, \quad 0<\alpha\le 2,\ r>0, \tag{51}$$

$$\int_0^\infty E_{\alpha,\alpha}\left(-\xi^2\right)J_0(r\xi)\xi\,d\xi = \int_1^\infty W\left(-\frac{\alpha}{2},0;-rt\right)\frac{1}{\sqrt{t^2-1}}\,dt \qquad (52)$$
$$= \frac{\alpha r}{2}\int_0^\infty M\left(\frac{\alpha}{2};r\sqrt{1+u^2}\right)du, \quad 0 < \alpha \leq 2,\ r > 0,$$

$$\int_0^\infty E_{\alpha,\beta}\left(-\xi^2\right)J_0(r\xi)\xi\,d\xi = \int_1^\infty W\left(-\frac{\alpha}{2},\beta-\alpha;-rt\right)\frac{1}{\sqrt{t^2-1}}\,dt, \quad 0 < \alpha \leq 2,\ r > 0. \qquad (53)$$

In addition,

$$\int_0^\infty E_{\alpha,\alpha}\left(-\xi^2\right)J_0(r\xi)\xi^2\,d\xi = -\frac{1}{2}\int_0^\pi W\left(-\frac{\alpha}{2},-\frac{\alpha}{2};-r\sin\theta\right)d\theta, \quad 0 < \alpha \leq 2,\ r > 0, \qquad (54)$$

$$\int_0^\infty E_{\alpha,\alpha}\left(-\xi^2\right)J_0(r\xi)\xi^3\,d\xi = -\int_0^\infty W\left(-\frac{\alpha}{2},-\alpha;-r\cosh t\right)dt \qquad (55)$$
$$= -\int_1^\infty W\left(-\frac{\alpha}{2},-\alpha;-rt\right)\frac{1}{\sqrt{t^2-1}}\,dt, \quad 0 < \alpha \leq 2,\ r > 0.$$

3. Applications of the Wright Function

3.1. Fractional Heat Conduction in Nonhomogeneous Media under Perfect Thermal Contact

Time-fractional heat conduction in two joint half-lines was considered by Povstenko [36,45,46]. In the general case, the heat conduction equation with the Caputo derivative of the order $0 < \alpha \leq 2$ in one half-line

$$\frac{\partial^\alpha T_1}{\partial t^\alpha} = a_1 \frac{\partial^2 T_1}{\partial x^2}, \quad x > 0, \qquad (56)$$

and the corresponding equation with the Caputo derivative of the order $0 < \beta \leq 2$ in another half-line

$$\frac{\partial^\beta T_2}{\partial t^\beta} = a_2 \frac{\partial^2 T_2}{\partial x^2}, \quad x < 0, \qquad (57)$$

were treated under the boundary conditions of perfect thermal contact which state that two bodies must have the same temperature at the contact point and the heat fluxes through the contact point must be the same:

$$T_1(x,t)\Big|_{x=0^+} = T_2(x,t)\Big|_{x=0^-}, \qquad (58)$$

$$k_1 D_{RL}^{1-\alpha}\frac{\partial T_1(x,t)}{\partial x}\bigg|_{x=0^+} = k_2 D_{RL}^{1-\beta}\frac{\partial T_2(x,t)}{\partial x}\bigg|_{x=0^-}, \quad 0 < \alpha \leq 2,\ 0 < \beta \leq 2. \qquad (59)$$

In the condition (59), k_1 and k_2 are the generalized thermal conductivities of two bodies; the Riemann–Liouville fractional derivative of the negative order $D_{RL}^{-\alpha}(f(t))$ is understood as the Riemann–Liouville fractional integral $I^\alpha(f(t))$.

Here, we present the fundamental solution to the first Cauchy problem with the initial condition

$$t = 0: \ T_1 = p_0\delta(x-\varrho), \quad x > 0,\ \varrho > 0, \qquad (60)$$

for the case $\alpha = \beta$ (for details see Povstenko [46]):

$$T_1(x,t) = \frac{p_0}{2\sqrt{a_1}t^{\alpha/2}}\left[M\left(\frac{\alpha}{2};\frac{|x-\varrho|}{\sqrt{a_1}t^{\alpha/2}}\right) + \frac{\varepsilon-1}{\varepsilon+1}M\left(\frac{\alpha}{2};\frac{x+\varrho}{\sqrt{a_1}t^{\alpha/2}}\right)\right], \quad x \geq 0, \qquad (61)$$

$$T_2(x,t) = \frac{\varepsilon p_0}{(\varepsilon+1)\sqrt{a_1}t^{\alpha/2}}M\left(\frac{\alpha}{2};\frac{|x|}{\sqrt{a_2}t^{\alpha/2}} + \frac{\varrho}{\sqrt{a_1}t^{\alpha/2}}\right), \quad x \leq 0, \qquad (62)$$

where
$$\varepsilon = \frac{k_1\sqrt{a_2}}{k_2\sqrt{a_1}}. \tag{63}$$

For the corresponding problem with uniform initial temperature T_0 in one of half-lines [45], in the particular case $\alpha = \beta$, we have:

$$T_1 = T_0 - \frac{T_0}{(1+\varepsilon)} W\left(-\frac{\alpha}{2}, 1; -\frac{x}{\sqrt{a_1}t^{\alpha/2}}\right), \quad x > 0, \tag{64}$$

$$T_2 = \frac{\varepsilon T_0}{(1+\varepsilon)} W\left(-\frac{\alpha}{2}, 1; -\frac{|x|}{\sqrt{a_2}t^{\alpha/2}}\right), \quad x < 0. \tag{65}$$

The time-fractional heat conduction equations with the Caputo derivatives in a semi-infinite medium composed of a region $0 < x < L$ and a region $L < x < \infty$ under the boundary conditions of perfect thermal contact at $x = L$ and the insulated boundary condition at $x = 0$ with uniform initial temperature in a layer were investigated in [47]. The approximate solution of the considered problem for small values of time is obtained based on Tauberian theorems for the Laplace transform. For $\alpha = \beta$, this solution reads

$$T_1 \simeq T_0 - \frac{T_0}{1+\varepsilon} W\left(-\frac{\alpha}{2}, 1; -\frac{L-x}{\sqrt{a_1}t^{\alpha/2}}\right), \quad 0 \leq x \leq L, \tag{66}$$

$$T_2 \simeq \frac{\varepsilon T_0}{1+\varepsilon} W\left(-\frac{\alpha}{2}, 1; -\frac{x-L}{\sqrt{a_2}t^{\alpha/2}}\right), \quad L \leq x < \infty. \tag{67}$$

Fractional heat conduction in an infinite medium with a spherical inclusion when a sphere $0 \leq r < R$ is at the initial uniform temperature T_0 and a matrix $R < r < \infty$ is at a zero initial temperature was considered by Povstenko [36,48]. In the case of perfect thermal contact at the boundary $r = R$,

$$r = R: \quad T_1(r,t) = T_2(r,t), \tag{68}$$

$$k_1 D_{RL}^{1-\alpha} \frac{\partial T_1(r,t)}{\partial r} = k_2 D_{RL}^{1-\beta} \frac{\partial T_2(r,t)}{\partial r}, \quad 0 < \alpha \leq 2, \quad 0 < \beta \leq 2, \tag{69}$$

the approximate solution for small values of time has the following form (we present only the solution for $\alpha = \beta$):

$$T_1(r,t) \simeq T_0 - \frac{RT_0 k_2}{(k_2 - k_1)r}\left[W\left(-\frac{\alpha}{2}, 1; -\frac{R-r}{\sqrt{a_1}t^{\alpha/2}}\right) - W\left(-\frac{\alpha}{2}, 1; -\frac{R+r}{\sqrt{a_1}t^{\alpha/2}}\right)\right]$$

$$+ \frac{cRT_0}{r}\int_0^t \frac{(t-\tau)^{\alpha/2-1}}{\tau^{\alpha/2}}\left[M\left(\frac{\alpha}{2}; \frac{R-r}{\sqrt{a_1}\tau^{\alpha/2}}\right)\right. \tag{70}$$

$$\left. - M\left(\frac{\alpha}{2}; \frac{R+r}{\sqrt{a_1}\tau^{\alpha/2}}\right)\right] E_{\alpha/2,\alpha/2}\left[-b(t-\tau)^{\alpha/2}\right] d\tau,$$

$$T_2(r,t) \simeq -\frac{RT_0 k_1}{(k_2-k_1)r} W\left(-\frac{\alpha}{2}, 1; -\frac{r-R}{\sqrt{a_2}t^{\alpha/2}}\right) + \frac{cRT_0}{r}\int_0^t \frac{(t-\tau)^{\alpha/2-1}}{\tau^{\alpha/2}}$$

$$\times M\left(\frac{\alpha}{2}; \frac{r-R}{\sqrt{a_2}\tau^{\alpha/2}}\right) E_{\alpha/2,\alpha/2}\left[-b(t-\tau)^{\alpha/2}\right] d\tau, \tag{71}$$

where
$$b = \frac{(k_2-k_1)\sqrt{a_1 a_2}}{R(k_1\sqrt{a_1}+k_2\sqrt{a_2})}, \quad c = \frac{k_1 k_2(\sqrt{a_1}+\sqrt{a_2})}{(k_2-k_1)(k_1\sqrt{a_1}+k_2\sqrt{a_2})}. \tag{72}$$

It should be mentioned that, for the classical heat conduction, the method of analysis of the solution for small values of time was described by Luikov [49] and Özişik [50]. In the case of fractional diffusion equation, the decay rate at large values of time was analyzed by Sakamoto–Yamamoto [51].

3.2. Fractional Heat Conduction in Nonhomogeneous Media under Nonperfect Thermal Contact

Near the interface between two solids, a transition region arises whose state differs from the state of contacting media owing to different conditions of material–particle interaction. The transition region has its own physical, mechanical, and chemical properties, and processes occurring in it differ from those in the bulk. Small thickness of the intermediate region between two solids allows us to reduce a three-dimensional problem to a two-dimensional one for median surface endowed with equivalent physical properties. There are several approaches to reducing three-dimensional equations to the corresponding two-dimensional equations for the median surface. For example, introducing the mixed coordinate system (ξ, η, z), where ξ and η are the curvilinear coordinates in the median surface and z is the normal coordinate, the linear or polynomial dependence of the considered functions on the normal coordinate can be assumed. This assumption is often used in the theory of elastic shells.

For the classical heat conduction equation, which is based on the conventional Fourier law, the reduction of the three-dimensional problem to the simplified two-dimensional one was pioneered by Marguerre [52,53] and later on developed by many authors. In this case, the assumption of linear or polynomial dependence of temperature on the normal coordinate or more general operator method were used. An extensive literature on this subject can be found, for example, in [9]. For time-fractional heat conduction, the reduction of the three-dimensional equation to the two-dimensional one was carried out by Povstenko [9,54,55].

A solution to the problem (56), (57) with uniform initial temperature in one of halflines under conditions of nonperfect thermal contact was obtained in [56]. In the particular case $\alpha = \beta$, the solution reads

$$T_1 = T_0 - \frac{T_0}{(1+\varepsilon)} W\left(-\frac{\alpha}{2}, 1; -\frac{x}{\sqrt{a_1} t^{\alpha/2}}\right) + \frac{T_0(1-\varepsilon)}{2(1+\varepsilon)} \int_0^t \frac{(t-\tau)^{\alpha/2-1}}{\tau^{\alpha/2}}$$

$$\times M\left(\frac{\alpha}{2}; \frac{x}{\sqrt{a_1} \tau^{\alpha/2}}\right) E_{\alpha/2,\alpha/2}\left[-b_\Sigma (t-\tau)^{\alpha/2}\right] d\tau, \quad x > 0, \tag{73}$$

$$T_2 = \frac{\varepsilon T_0}{(1+\varepsilon)} W\left(-\frac{\alpha}{2}, 1; -\frac{|x|}{\sqrt{a_2} t^{\alpha/2}}\right) + \frac{T_0(1-\varepsilon)}{2(1+\varepsilon)} \int_0^t \frac{(t-\tau)^{\alpha/2-1}}{\tau^{\alpha/2}}$$

$$\times M\left(\frac{\alpha}{2}; \frac{|x|}{\sqrt{a_2} \tau^{\alpha/2}}\right) E_{\alpha/2,\alpha/2}\left[-b_\Sigma (t-\tau)^{\alpha/2}\right] d\tau, \quad x < 0, \tag{74}$$

where ε is defined by (63),

$$b_\Sigma = \frac{k_1 \sqrt{a_2} + k_2 \sqrt{a_1}}{C_\Sigma \sqrt{a_1 a_2}}, \tag{75}$$

C_Σ is the reduced heat capacity of the median surface of the transition region. When $C_\Sigma \to 0$, the solutions (73), (74) coincide with the solutions (64), (65).

3.3. Fractional Heat Conduction under Time-Harmonic Impact

Ångström [57] was the first to investigate the standard parabolic heat conduction equation under time-harmonic impact. An extensive review of literature in this field in the case of classical diffusion equation can be found in the book by Mandelis [58].

Fractional heat conduction with a source varying harmonically in time was studied by Povstenko [59]. Equation (1) with a source term

$$\frac{\partial^\alpha T}{\partial t^\alpha} = a \frac{\partial^2 T}{\partial x^2} + Q_0 \delta(x) e^{i\omega t}, \quad 0 < \alpha \leq 2, \tag{76}$$

was solved in the domain $-\infty < x < \infty$ under zero initial conditions. Temperature is expressed as

$$T(x,t) = \frac{Q_0}{2\sqrt{a}} \int_0^t \tau^{\alpha/2-1} W\left(-\frac{\alpha}{2}, \frac{\alpha}{2}; -\frac{|x|}{\sqrt{a}\tau^{\alpha/2}}\right) e^{i\omega(t-\tau)} d\tau. \tag{77}$$

The corresponding problem in the central symmetric case

$$\frac{\partial^\alpha T}{\partial t^\alpha} = a\left(\frac{\partial^2 T}{\partial r^2} + \frac{2}{r}\frac{\partial T}{\partial r}\right) + Q_0 \frac{\delta(r)}{4\pi r^2} e^{i\omega t}, \quad 0 < r < \infty, \quad 0 < \alpha \leq 2, \tag{78}$$

has the solution

$$T(x,t) = \frac{\alpha Q_0}{8\pi a^{3/2}} \int_0^t \frac{1}{\tau^{1+\alpha/2}} M\left(\frac{\alpha}{2}; \frac{r}{\sqrt{a}\tau^{\alpha/2}}\right) e^{i\omega(t-\tau)} d\tau. \tag{79}$$

3.4. Fractional Nonlocal Elasticity

Nonlocal continuum physics assumes integral constitutive equations. In the nonlocal theory of the continuum mechanics, stresses at the reference point $\mathbf{x}$ of an elastic solid at time t depend not only on the strains at this point at this time, but also on strains at all the points $\mathbf{x}'$ of a body and all the times prior to and at time t:

$$\mathbf{t}(\mathbf{x}, t, \epsilon_L, \epsilon_T) = \int_0^t \int_V \gamma(|\mathbf{x} - \mathbf{x}'|, t - t', \epsilon_L, \epsilon_T) \sigma(\mathbf{x}', t') dv(\mathbf{x}') dt', \tag{80}$$

$$\sigma(\mathbf{x}', t') = 2\mu \mathbf{e}(\mathbf{x}', t') + \lambda \operatorname{tr} \mathbf{e}(\mathbf{x}', t') \mathbf{I}, \tag{81}$$

where $\mathbf{t}$ and σ are the nonlocal and classical stress tensors, $\mathbf{x}$ and $\mathbf{x}'$ are the reference and running points, $\mathbf{e}$ the linear strain tensor, λ and μ are Lamé constants, $\mathbf{I}$ stands for the unit tensor. The volume integral in (80) is over the region occupied by the solid. The time-non-locality describes memory effects, distributed lag (distributed time delay), and frequency dispersion; the space-non-locality deals with the long-range interaction. The weight function (the non-locality kernel) $\gamma(|\mathbf{x} - \mathbf{x}'|, t - t', \epsilon_L, \epsilon_T)$ depends on two basic non-locality parameters (see Eringen [60]): the characteristic length ratio

$$\epsilon_L = \frac{\text{Internal characteristic length}}{\text{External characteristic length}}$$

and the characteristic time ratio

$$\epsilon_T = \frac{\text{Internal characteristic time}}{\text{External characteristic time}}.$$

When $\epsilon_T \to 0$, the memory effects are eliminated; for $\epsilon_L \to 0$ the space-non-locality disappears.

In the pioneering works by Podstrigach [61,62], a new nontraditional thermodynamic pair (the chemical potential tensor φ and the concentration tensor $\mathbf{c}$) was introduced (see also [63,64]). The tensor character of the chemical potential means that, for solids, the work of bringing the substance into a point in a body depends on the direction. In this case, the diffusion equation, split into the mean and deviatoric parts, has the form

$$\rho \frac{\partial (\operatorname{tr} \mathbf{c})}{\partial t} = 3a \Delta (\operatorname{tr} \varphi), \tag{82}$$

$$\rho \frac{\partial (\operatorname{dev} \mathbf{c})}{\partial t} = 2a_1 \Delta (\operatorname{dev} \varphi), \tag{83}$$

where ρ is the mass density, and a and a_1 are the corresponding diffusion coefficients.

Starting from interrelated equations describing elasticity and diffusion, Podstrigach [65] eliminated the chemical potential tensor from the constitutive equation for the stress tensor and obtained the stress–strain relation containing spatial and time derivatives. In the infinite medium, this relation can be integrated using the Fourier and Laplace integral transforms, and the final result, written for the mean and deviatoric parts, has the nonlocal integral form:

$$\operatorname{tr}\sigma = 3K_c \operatorname{tr}\mathbf{e} + 3\frac{K_\varphi - K_c}{p} \int_0^t \int_{-\infty}^\infty \int_{-\infty}^\infty \int_{-\infty}^\infty \gamma_{(p)}(x-x', y-y', z-z', t-t')$$
$$\times \operatorname{tr}\mathbf{e}(x',y',z',t')\mathrm{d}x'\,\mathrm{d}y'\,\mathrm{d}z'\,\mathrm{d}t', \tag{84}$$

$$\operatorname{dev}\sigma = 2\mu_c \operatorname{dev}\mathbf{e} + 2\frac{\mu_\varphi - \mu_c}{q} \int_0^t \int_{-\infty}^\infty \int_{-\infty}^\infty \int_{-\infty}^\infty \gamma_{(q)}(x-x', y-y', z-z', t-t')$$
$$\times \operatorname{dev}\mathbf{e}(x',y',z',t')\mathrm{d}x'\,\mathrm{d}y'\,\mathrm{d}z'\,\mathrm{d}t'. \tag{85}$$

Here, K_c, K_φ, μ_c, μ_φ, p, and q are material constants (for details, see [42,65]). The kernel $\gamma_{(p)}(x,y,z,t)$ has the following form:

$$\gamma_{(p)}(x,y,z,t) = \left(\frac{p}{2t}\right)^{5/2}\left(3 - \frac{p r^2}{2t}\right)\exp\left(-\frac{p r^2}{4t}\right), \tag{86}$$

where $r = \sqrt{x^2 + y^2 + z^2}$; the kernel $\gamma_{(q)}(x,y,z,t)$ is obtained from the kernel $\gamma_{(p)}(x,y,z,t)$ substituting p by q.

The results of Podstrigach [65] were generalized by Povstenko [42] for the case of fractional diffusion equations

$$\rho\frac{\partial^\alpha(\operatorname{tr}\mathbf{c})}{\partial t^\alpha} = 3a\,\Delta\,(\operatorname{tr}\boldsymbol{\varphi}), \tag{87}$$

$$\rho\frac{\partial^\alpha(\operatorname{dev}\mathbf{c})}{\partial t^\alpha} = 2a_1\,\Delta\,(\operatorname{dev}\boldsymbol{\varphi}). \tag{88}$$

The kernel $\gamma_{(p)}(x,y,z,t)$ in the fractional generalization of the constitutive Equation (84) for the mean part of the stress tensor is expressed in terms of the Wright function:

$$\gamma_{(p)}(x,y,z,t) = -\frac{\sqrt{\pi}p^2}{\sqrt{2}\,t^{\alpha+1}\,r}\,W\!\left(-\frac{\alpha}{2}, -\alpha; -\sqrt{p}\,\frac{r}{t^{\alpha/2}}\right). \tag{89}$$

The kernel $\gamma_{(q)}(x,y,z,t)$ in the fractional generalization of the constitutive Equation (85) for the deviatoric part of the stress tensor is obtained by substituting p with q.

In the case of only space-non-locality, the constitutive equation for the stress tensor reads

$$\mathbf{t}(\mathbf{x},\epsilon_L) = \int_V \gamma(|\mathbf{x}-\mathbf{x}'|, \epsilon_L)\sigma(\mathbf{x}')\mathrm{d}v(\mathbf{x}'). \tag{90}$$

The space-nonlocal elasticity reduces to the classical theory of elasticity in the long wavelength limit and to the atomic lattice theory in the short wave-length limit. Several versions of nonlocal elasticity based on various assumptions were proposed by different authors (see, for example, Podstrigach [65], Eringen [66,67], Kunin [68,69] and references therein).

In the case of space-nonlocal constitutive Equation (90), the nonlocal kernel $\gamma(|\mathbf{x}-\mathbf{x}'|,\epsilon_L)$ is a delta sequence and in the classical elasticity limit $\epsilon_L \to 0$ becomes the Dirac delta function. For example, slightly changing the notation, the nonlocal kernel $\gamma(|\mathbf{x}-\mathbf{x}'|,\tau)$ can be considered as the Green function of the Cauchy problem for the diffusion operator (see Eringen [67,70]):

$$\frac{\partial \gamma(\mathbf{x},\tau)}{\partial \tau} - a\Delta\,\gamma(\mathbf{x},\tau) = 0, \tag{91}$$

$$\tau = 0: \quad \gamma(\mathbf{x},\tau) = \delta(\mathbf{x}), \tag{92}$$

which results in the kernel

$$\gamma(\mathbf{x}, \tau) = \frac{1}{(2\sqrt{\pi a \tau})^n} \exp\left(-\frac{|\mathbf{x}|^2}{4a\tau}\right) \qquad (93)$$

for $n = 1, 2, 3$ space variables. In this case, the nonlocal stress tensor is a solution of the corresponding Cauchy problem:

$$\frac{\partial \mathbf{t}(\mathbf{x}, \tau)}{\partial \tau} - a\Delta\, \mathbf{t}(\mathbf{x}, \tau) = 0, \qquad (94)$$

$$\tau = 0: \quad \mathbf{t}(\mathbf{x}, \tau) = \sigma(\mathbf{x}). \qquad (95)$$

It should be emphasized that in, the formal sense, τ in the initial-value problems (91), (92) and (94), (95) looks like time, but in fact τ is a non-locality parameter related to the space-non-locality characteristic ratio ϵ_L.

In the paper [71], the nonlocal kernel $\gamma(|\mathbf{x} - \mathbf{x}'|, \tau)$ was considered as the Green function of the Cauchy problem for the fractional diffusion operator

$$\frac{\partial^\alpha \gamma(\mathbf{x}, \tau)}{\partial \tau^\alpha} - a\Delta\, \gamma(\mathbf{x}, \tau) = 0, \quad 0 < \alpha \le 1, \qquad (96)$$

$$\tau = 0: \quad \gamma(\mathbf{x}, \tau) = \delta(\mathbf{x}). \qquad (97)$$

In the framework of this approach, instead of the Cauchy problem (94)–(95), we obtain

$$\frac{\partial^\alpha \mathbf{t}(\mathbf{x}, \tau)}{\partial \tau^\alpha} - a\Delta\, \mathbf{t}(\mathbf{x}, \tau) = 0, \quad 0 < \alpha \le 1, \qquad (98)$$

$$\tau = 0: \quad \mathbf{t}(\mathbf{x}, \tau) = \sigma(\mathbf{x}). \qquad (99)$$

In the case of one spatial coordinate, the nonlocal kernel takes the form

$$\gamma(x, \tau) = \frac{1}{2\sqrt{a\tau^{\alpha/2}}}\, M\!\left(\frac{\alpha}{2}; \frac{|x|}{\sqrt{a\tau^{\alpha/2}}}\right), \qquad 0 \le \alpha \le 1, \qquad (100)$$

and in the central symmetric case

$$\gamma(r, \tau) = \frac{1}{4\pi a \tau^\alpha r}\, W\!\left(-\frac{\alpha}{2}, 1-\alpha; -\frac{r}{\sqrt{a\tau^{\alpha/2}}}\right), \qquad 0 \le \alpha \le 1. \qquad (101)$$

4. Conclusions

In this survey, we have reviewed the main applications of the Wright function and the Mainardi function in continuum physics based essentially on the author's works. We have presented the integral relations between the Mittag–Leffler functions and the Wright function, which can be useful when solving fractional differential equations. We have restricted ourselves to the standard Mittag–Leffler functions and Wright function. The interested reader is referred to publications on further generalizations of the Mittag–Leffler functions [24,72–75] and of the Wright function [24,29,76–78].

Funding: This research received no external funding.

Institutional Review Board Statement: Not applicable.

Informed Consent Statement: Not applicable.

Acknowledgments: The author would like to thank the anonymous reviewers for their helpful comments.

Conflicts of Interest: The author declares no conflict of interest.

References

1. West, B.J.; Bologna, M.; Grigolini, P. *Physics of Fractal Operators*; Springer: New York, NY, USA, 2003.
2. Magin, R.L. *Fractional Calculus in Bioengineering*; Begell House Publishers: Redding, CT, USA, 2006.
3. Povstenko, Y. Fractional heat conduction equation and associated thermal stresses. *J. Therm. Stress.* **2005**, *28*, 83–102. [CrossRef]
4. Tarasov, V.E. *Fractional Dynamics: Applications of Fractional Calculus to Dynamics of Particles, Fields and Media*; Springer: Berlin, Germany, 2010.
5. Povstenko, Y. Fractional Cattaneo-type equations and generalized thermoelasticity. *J. Therm. Stress.* **2011**, *34*, 97–114. [CrossRef]
6. Uchaikin, V.V. *Fractional Derivatives for Physicists and Engineers*; Springer: Berlin, Germany, 2013.
7. Atanacković, T.M.; Pilipović, S.; Stanković, B.; Zorica, D. *Fractional Calculus with Applications in Mechanics: Vibrations and Diffusion Processes*; John Wiley & Sons: Hoboken, NJ, USA, 2014.
8. Herrmann, R. *Fractional Calculus: An Introduction for Physicists*, 2nd ed.; World Scientific: Singapore, 2014.
9. Povstenko, Y. *Fractional Thermoelasticity*; Springer: New York, NY, USA, 2015.
10. Datsko, B.; Gafiychuk, V.; Podlubny, I. Solitary travelling auto-waves in fractional reaction–diffusion systems. *Commun. Nonlinear Sci. Numer. Simul.* **2015**, *23*, 378–387. [CrossRef]
11. West, B.J. *Fractional Calculus View of Complexity: Tomorrow's Science*; CRC Press: Boca Raton, FL, USA, 2016.
12. Skiadas, C.H. (Ed.) *Fractional Dynamics, Anomalous Transport and Plasma Science*; Springer: Cham, Switzerland, 2018.
13. Tarasov, V.E. (Ed.) *Handbook of Fractional Calculus with Applications. Volume 4: Application in Physics. Part A*; Walter de Gruyter: Berlin, Germany, 2019.
14. Kumar, D.; Singh, J. (Eds.) *Fractional Calculus in Medical and Health Science*; CRC Press: Boca Raton, FL, USA, 2020.
15. Su, N. *Fractional Calculus for Hydrology, Soil Science and Geomechanics: An Introduction to Applications*; CRC Press: Boca Raton, FL, USA, 2020.
16. Mittag–Leffler, G.M. Sur la nouvelle fonction $E_\alpha(x)$. *C. R. Acad. Sci. Paris Ser. II* **1903**, *137*, 554–558.
17. Mittag–Leffler, G.M. Sopra la funzione $E_\alpha(x)$. *Rend. Accad. Lincei Ser. V* **1904**, *13*, 3–5.
18. Humbert, P. Quelques résultats relatifs à la fonction de Mittag-Leffler. *C. R. Acad. Sci. Paris* **1953**, *236*, 1467–1468.
19. Humbert, P.; Agarwal, R.P. Sur la fonction de Mittag-Leffler et quelques-unes de ses généralisations. *Bull. Sci. Math.* **1953**, *77*, 180–185.
20. Erdélyi, A.; Magnus, W.; Oberhettinger, F.; Tricomi, F. *Higher Transcendental Functions*; McGraw-Hill: New York, NY, USA, 1955; Volume 3.
21. Gorenflo, R.; Mainardi, F. Fractional calculus: Integral and differential equations of fractional order. In *Fractals and Fractional Calculus in Continuum Mechanics*; Carpinteri, A., Mainardi, F., Eds.; Springer: Wien, Austria, 1997; pp. 223–276.
22. Podlubny, I. *Fractional Differential Equations*; Academic Press: San Diego, CA, USA, 1999.
23. Kilbas, A.A.; Srivastava, H.M.; Trujillo, J.J. *Theory and Applications of Fractional Differential Equations*; Elsevier: Amsterdam, The Netherlands, 2006.
24. Gorenflo, R.; Kilbas, A.A.; Mainardi, F.; Rogosin, S.V. *Mittag–Leffler Functions, Related Topics and Applications*, 2nd ed.; Springer: New York, NY, USA, 2020.
25. Gorenflo, R.; Loutchko, J.; Luchko, Y. Computation of the Mittag-Leffler function and its derivatives. *Fract. Calc. Appl. Anal.* **2002**, *5*, 491–518.
26. Podlubny, I. Mittag-Leffler Function; Calculates the Mittag-Leffler Function with Desired Accuracy, MATLAB Central File Exchange, File ID 8738. Available online: www.mathworks.com/matlabcentral/fileexchange/8738 (accessed on 16 November 2020).
27. Wright, E.M. On the coefficients of power series having exponential singularities. *J. Lond. Math. Soc.* **1933**, *8*, 71–79. [CrossRef]
28. Wright, E.M. The asymptotic expansion of the generalized Bessel function. *Proc. Lond. Math. Soc. Ser. II* **1935**, *38*, 257–270. [CrossRef]
29. Luchko, Y. The Wright function and its applications. In *Handbook of Fractional Calculus with Applications. Volume 1. Basic Theory*; Kochubei, A., Luchko, Y., Eds.; Walter de Gruyter: Berlin, Germany, 2019; pp. 241–268.
30. Luchko, Y. Algorithms for evaluation of the Wright function for the real arguments' values. *Fract. Calc. Appl. Anal.* **2008**, *11*, 57–75.
31. Mainardi, F. The fundamental solutions for the fractional diffusion-wave equation. *Appl. Math. Lett.* **1996**, *9*, 23–28. [CrossRef]
32. Mainardi, F. Fractional relaxation-oscillation and fractional diffusion-wave phenomena. *Chaos Solitons Fractals* **1996**, *7*, 1461–1477. [CrossRef]
33. Mainardi, F.; Tomirotti, M. On a special function arising in the time fractional diffusion-wave equation. In *Transform Methods & Special Functions, Sofia' 94*; Rusev, P., Dimovski, I., Kiryakova, V., Eds.; Science Culture Technology Publishing: Singapore, 1995; pp. 171–183.
34. Mainardi, F. *Fractional Calculus and Waves in Linear Viscoelasticity: An Introduction to Mathematical Models*; Imperial College Press: London, UK, 2010.
35. Mainardi, F.; Tomirotti, M. Seismic pulse propagation with constant Q and stable probability distributions. *Ann. Geofis.* **1997**, *40*, 1311–1328.
36. Povstenko, Y. *Linear Fractional Diffusion-Wave Equation for Scientists and Engineers*; Birkhäuser: New York, NY, USA, 2015.
37. Gorenflo, R.; Luchko, Y.; Mainardi, F. Analytical properties and applications of the Wright function. *Fract. Calc. Appl. Anal.* **1999**, *2*, 383–414.

38. Gorenflo, R.; Luchko, Y.; Mainardi, F. Wright functions as scale-invariant solutions of the diffusion-wave equation. *J. Comput. Appl. Math.* **2000**, *118*, 175–191. [CrossRef]
39. Mainardi, F.; Pagnini, G. The Wright functions as solutions of the time-fractional diffusion equation. *Appl. Math. Comput.* **2003**, *141*, 51–62. [CrossRef]
40. Stanković, B. On the function of E. M. Wright. *Publ. Inst. Math.* **1970**, *10*, 113–124.
41. Gajić, L.; Stanković, B. Some properties of Wright's function. *Publ. Inst. Math.* **1976**, *20*, 91–98.
42. Povstenko, Y. Generalized theory of diffusive stresses associated with the time-fractional diffusion equation and nonlocal constitutive equations for the stress tensor. *Comput. Math. Appl.* **2019**, *78*, 1819–1825. [CrossRef]
43. Watson, G.N. *A Treatise on the Theory of Bessel Functions*, 2nd ed.; Cambridge University Press: Cambridge, UK, 1944.
44. Abramowitz, M.; Stegun, I.A. (Eds.) *Handbook of Mathematical Functions with Formulas, Graphics and Mathematical Tables*; Dover: New York, NY, USA, 1972.
45. Povstenko, Y. Fractional heat conduction in infinite one-dimensional composite medium. *J. Therm. Stress.* **2013**, *36*, 351–363. [CrossRef]
46. Povstenko, Y. Fundamental solutions to time-fractional heat conduction equations in two joint half-lines. *Cent. Eur. J. Phys.* **2013**, *11*, 1284–1294. [CrossRef]
47. Povstenko, Y. Fractional heat conduction in a semi-infinite composite body. *Comm. Appl. Industr. Math.* **2014**, *6*, 1–13. [CrossRef]
48. Povstenko, Y. Fractional heat conduction in an infinite medium with a spherical inclusion. *Entropy* **2013**, *15*, 4122–4133. [CrossRef]
49. Luikov, A.V. *Analytical Heat Diffusion Theory*; Academic Press: New York, NY, USA, 1968.
50. Özişik, M.N. *Heat Conduction*; John Wiley: New York, NY, USA, 1980.
51. Sakamoto, K.; Yamamoto, M. Initial value/boundary value problems for fractional diffusion-wave equations and applications to some inverse problems. *J. Math. Anal. Appl.* **2011**, *382*, 426–447. [CrossRef]
52. Marguerre, K. Thermo-elastische Platten–Gleichungen. *Z. Angew. Math. Mech.* **1935**, *15*, 369–372. [CrossRef]
53. Marguerre, K. Temperaturverlauf und Temperaturspannumgen in platten- und schalenformigen Körpern. *Ing. Arch.* **1937**, *8*, 216–228. [CrossRef]
54. Povstenko, Y. Fractional thermoelasticity of thin shells. In *Shell Structures*; Pietraszkiewicz, W., Górski, J., Eds.; CRC Press: Boca Raton, FL, USA, 2014; Volume 3, pp. 141–144.
55. Povstenko, Y. Generalized boundary conditions for the time-fractional advection diffusion equation. *Entropy* **2015**, *17*, 4028–4039. [CrossRef]
56. Povstenko, Y.; Kyrylych, T. Fractional heat conduction in solids connected by thin intermediate layer: Nonperfect thermal contact. *Contin. Mech. Thermodyn.* **2019**, *31*, 1719–1731. [CrossRef]
57. Ångström, A.J. Neue Methode, das Wärmeleitungsvermögen der Körper zu bestimmen. *Ann. Phys. Chem.* **1861**, *114*, 513–530. [CrossRef]
58. Mandelis, A. *Diffusion-Wave Fields: Mathematical Methods and Green Functions*; Springer: New York, NY, USA, 2001.
59. Povstenko, Y. Fractional heat conduction in a space with a source varying harmonically in time and associated thermal stresses. *J. Therm. Stress.* **2016**, *39*, 1442–1450. [CrossRef]
60. Eringen, A.C. Vistas of nonlocal continuum physics. *Int. J. Engng. Sci.* **1992**, *30*, 1551–1565. [CrossRef]
61. Pidstryhach, Y.S. Differential equations of the diffusion theory of deformation of a solid. *Dopovidi Ukr. Acad. Sci.* **1963**, *3*, 336–340. (In Ukrainian)
62. Podstrigach, Y.S. Diffusion theory of the anelasticity of metals. *J. Appl. Mech. Tech. Phys.* **1965**, *6*, 56–60. [CrossRef]
63. Podstrigach, Y.S.; Povstenko, Y. *Introduction to Mechanics of Surface Phenomena in Deformable Solids*; Naukova Dumka: Kiev, Ukraine, 1985. (In Russian)
64. Povstenko, Y. From the chemical potential tensor and concentration tensor to nonlocal continuum theories. *J. Math. Sci.* **2020**, *249*, 389–403. [CrossRef]
65. Podstrigach, Y.S. On a nonlocal theory of solid body deformation. *Internat. Appl. Mech.* **1967**, *3*, 44–46. [CrossRef]
66. Eringen, A.C. Linear theory of nonlocal elasticity and dispersion of plane waves. *Int. J. Engng. Sci.* **1972**, *10*, 425–435. [CrossRef]
67. Eringen, A.C. *Nonlocal Continuum Field Theories*; Springer: New York, NY, USA, 2002.
68. Kunin, I.A. *Elastic Media with Microstructure I: One-Dimensional Models*; Springer: Berlin, Germany, 1982.
69. Kunin, I.A. *Elastic Media with Microstructure II: Three-Dimensional Models*; Springer: Berlin, Germany, 1983.
70. Eringen, A.C. On differential eqations of nonlocal elasticity and solutions of screw dislocation and surface waves. *J. Appl. Phys.* **1983**, *54*, 4703–4710. [CrossRef]
71. Povstenko, Y. Fractional nonlocal elasticity and solutions for straight screw and edge dislocations. *Fiz. Mesomekhanika* **2020**, *23*, 35–44.
72. Al-Bassam, M.A.; Luchko, Y.F. On generalized fractional calculus and its application to the solution of integro-diferential equations. *J. Fract. Calc.* **1995**, *7*, 69–88.
73. Kiryakova, V. The multi-index Mittag–Leffler functions as an important class of special functions of fractional calculus. *Comp. Math. Appl.* **2010**, *59*, 1885–1895. [CrossRef]
74. Luchko, Y. Initial-boundary-value problems for the generalized multi-term time-fractional diffusion equation. *J. Math. Anal. Appl.* **2011**, *374*, 538–548. [CrossRef]

75. Li, Z.; Liu, Y.; Yamamoto, M. Initial-boundary value problems for multi-term time-fractional diffusion equations with positive constant coefficients. *Appl. Math. Comput.* **2015**, *257*, 381–397. [CrossRef]
76. Kilbas, A.A.; Saigo, M.; Trujillo, J.J. On the generalized Wright function. *Fract. Calc. Appl. Anal.* **2002**, *5*, 437–460.
77. Kilbas, A.A. Fractional calculus of the generalized Wright function. *Fract. Calc. Appl. Anal.* **2005**, *8*, 113–126.
78. Khan, N.U.; Usman, T.; Aman, M. Some properties concerning the analysis of generalized Wright function. *J. Comput. Appl. Math.* **2020**, *376*, 112840. [CrossRef]

Article

Some Properties of the Kilbas–Saigo Function

Lotfi Boudabsa [1] and Thomas Simon [2,*]

[1] Institut de Mathématiques, Ecole Polytechnique Fédérale de Lausanne, CH-1015 Lausanne, Switzerland; lotfi.boudabsa@epfl.ch
[2] Laboratoire Paul Painlevé, UMR 8524, Université de Lille, Cité Scientifique, F-59655 Villeneuve d'Ascq, France
* Correspondence: thomas.simon@univ-lille.fr

Abstract: We characterize the complete monotonicity of the Kilbas–Saigo function on the negative half-line. We also provide the exact asymptotics at $-\infty$, and uniform hyperbolic bounds are derived. The same questions are addressed for the classical Le Roy function. The main ingredient for the proof is a probabilistic representation of these functions in terms of the stable subordinator.

Keywords: complete monotonicity; convex ordering; double Gamma function; fractional extreme distribution; Kilbas–Saigo function; Le Roy function; Mittag–Leffler function; stable subordinator

---

Citation: Boudabsa, L.; Simon, T. Some Properties of the Kilbas-Saigo Function. *Mathematics* **2021**, *9*, 217. https://doi.org/10.3390/math9030217

Received: 11 December 2020
Accepted: 16 January 2021
Published: 22 January 2021

Publisher's Note: MDPI stays neutral with regard to jurisdictional claims in published maps and institutional affiliations.

Copyright: © 2021 by the authors. Licensee MDPI, Basel, Switzerland. This article is an open access article distributed under the terms and conditions of the Creative Commons Attribution (CC BY) license (https://creativecommons.org/licenses/by/4.0/).

1. Introduction

The Kilbas–Saigo function is a three-parameter entire function with the convergent series representation

$$E_{\alpha,m,l}(z) = 1 + \sum_{n \geq 1} \left(\prod_{k=1}^{n} \frac{\Gamma(1 + \alpha((k-1)m + l))}{\Gamma(1 + \alpha((k-1)m + l + 1))} \right) z^n, \quad z \in \mathbb{C},$$

where the parameters are such that $\alpha, m > 0$ and $l > -1/\alpha$. It can be viewed as a generalization of the one- or two-parameter Mittag–Leffler function since, with standard notations,

$$E_{\alpha,1,0}(z) = \sum_{n \geq 0} \frac{z^n}{\Gamma(1 + \alpha n)} = E_\alpha(z)$$

and

$$E_{\alpha,1,\frac{\beta-1}{\alpha}}(z) = \Gamma(\beta) \sum_{n \geq 0} \frac{z^n}{\Gamma(\beta + \alpha n)} = \Gamma(\beta) E_{\alpha,\beta}(z)$$

for every $\alpha, \beta > 0$ and $z \in \mathbb{C}$. This function was introduced in [1] as the solution to some integro-differential equation with Abelian kernel on the half-line, and we refer to Chapter 5.2 in [2] for a more recent account, including an extension to complex values of the parameter l. In our previous paper [3], written in collaboration with P. Vallois, it was shown that certain Kilbas–Saigo functions are moment generating functions of Riemannian integrals of the stable subordinator. This observation made it possible to define rigorously some Weibull and Fréchet distributions of fractional type via an independent exponential random variable and the stable subordinator—see [3] for details. In the present paper, we wish to take the other way round and use the probabilistic connection to deduce some non-trivial analytical properties of the Kilbas–Saigo function.

In Section 2, we tackle the problem of the complete monotonicity on the negative half-line. This problem dates to Pollard in 1948 for the one-parameter Mittag–Leffler function—see e.g., Section 3.7.2 in [2] for details and references. It was shown in [3] that for every $m > 0$ and $\alpha \in (0,1]$ the function $x \mapsto E_{\alpha,m,m-1}(-x)$ is completely monotone, extending Pollard's result and solving an open problem stated in [4]. In Theorem 1 below, we characterize the complete monotonicity of $x \mapsto E_{\alpha,m,l}(-x)$ by $\alpha \in [0,1]$ and $l \geq m - 1/\alpha$.

We also give an explicit representation, albeit complicated in general, of the underlying positive random variable. Along the way, we study an interesting family of Mellin transforms given as the quotient of four double Gamma functions.

In Section 3, we establish uniform hyperbolic bounds on the negative half-line for two families of completely monotonic Kilbas–Saigo functions, extending the bounds obtained in [5] for the classical Mittag–Leffler function. The argument in [5] relied on stochastic and convex orderings and was rather lengthy. We use here the same kind of arguments, but the proof is shorter and more transparent thanks to the connection with the stable subordinator; which also enables us to derive some monotonicity properties on $m \mapsto E_{\alpha,m,l}(x)$ for every $x \in \mathbb{R}$—see Proposition 1 below.

In Section 4, we address the question of the asymptotic behavior at $-\infty$ in the completely monotonic case $\alpha \in (0,1]$ and $l \geq m - 1/\alpha$. It is shown in Theorem 5.5 of [2] that in the general case $\alpha, m > 0$ and $l > m - 1/\alpha$, the entire function $E_{\alpha,m,l}(z)$ has order $\rho = 1/\alpha$ and type $\sigma = 1/m$. However, precise asymptotics along given directions of the complex plane do not seem to have been investigated as yet, as is the case—see e.g., Proposition 3.6 in [2] for the classical Mittag–Leffler function. For the negative half-line and $\alpha \in (0,1]$, the asymptotics are different depending on whether $l = m + 1/\alpha$ or $l > m + 1/\alpha$. In the former case, the behavior is in $c_{\alpha,m} x^{-(1+1/m)}$ with a non-trivial constant $c_{\alpha,m}$ obtained from the connection with the fractional Fréchet distribution and given in terms of the double Gamma function—see Proposition 7 and Remark 8 (c) below. In the latter case, the behavior is in $c_{\alpha,m,l} x^{-1}$ with a uniform speed and a simple constant $c_{\alpha,m,l}$ given in terms of the standard Gamma function—see Proposition 6 below. The method for the case $l > m + 1/\alpha$ relies on the computation of the Mellin transform of the positive function $E_{\alpha,m,l}(-x)$, which is obtained from the proof of its complete monotonicity, and is interesting in its own right—see Remark 2 (c) below. Along the way, we provide the exact asymptotics of the fractional Weibull and Fréchet densities at both ends of their support and we give a series of probabilistic factorizations. The latter enhance the position of the fractional Fréchet distribution, which is in one-to-one correspondence with the boundary Kilbas–Saigo function $E_{\alpha,m,m-1/\alpha}(x)$, as an irreducible factor—see Remark 8 (a) below.

In the last Section 5, we pay attention to the so-called Le Roy function with parameter $\alpha > 0$. This is a simple generalization of the exponential function defined by

$$\mathcal{L}_\alpha(z) = \sum_{n \geq 0} \frac{z^n}{(n!)^\alpha}, \qquad z \in \mathbb{C}.$$

Introduced in [6] in the context of analytic continuation, a couple of years before the Mittag–Leffler function, the Le Roy function has been much less studied. It was shown in [3] that this function encodes for $\alpha \in [0,1]$ a Gumbel distribution of fractional type, as the moment generating function of the perpetuity of the α-stable subordinator. This fact is recalled in Proposition 9 below, together with a characterization of the moment generating property. The exact asymptotic behavior at $-\infty$ is also derived for $\alpha \in [0,1]$, completing the original result of Le Roy. Finally, the non-increasing character of $\alpha \mapsto \mathcal{L}_\alpha(x)$ on $[0,1]$ for every $x \in \mathbb{R}$ is established by convex ordering. It is worth mentioning that this property is an open problem—see Conjecture 5 below-for the Mittag–Leffler function.

As in [3], an important role is played throughout the paper by Barnes' double Gamma function $G(z; \delta)$ which is the unique solution to the functional equation $G(z+1; \delta) = \Gamma(z\delta^{-1})G(z; \delta)$ with normalization $G(1; \delta) = 1$, and its associated Pochhammer type symbol

$$[a; \delta]_s = \frac{G(a+s; \delta)}{G(a; \delta)}.$$

We have gathered in Appendix A all the needed facts and formulæ on this double Gamma function, whose connection with the Kilbas-Saigo function has probably a broader focus than the content of the present paper (we leave this topic open to further research).

2. Complete Monotonicity on the Negative Half-Line

In this section, we wish to characterize the property that the function $x \mapsto E_{\alpha,m,l}(-x)$ is completely monotone (CM) on $(0,\infty)$. We begin with the following result on the above generalized Pochhammer symbols, which is reminiscent of Proposition 5.1 and Theorem 6.2 in [7] and has an independent interest.

Lemma 1. *Let a, b, c, d and δ be positive parameters. There exists a positive random variable $Z = \mathbf{Z}[a,c;b,d;\delta]$ such that*

$$\mathbb{E}[Z^s] \;=\; \frac{[a;\delta]_s[c;\delta]_s}{[b;\delta]_s[d;\delta]_s} \tag{1}$$

for every $s > 0$, if and only if $b + d \leq a + c$ and $\inf\{b,d\} \leq \inf\{a,c\}$. This random variable is absolutely continuous on $(0,\infty)$, except in the degenerate case $a = b = c = d$. Its support is $[0,1]$ if $b + d = a + c$ and $[0,\infty)$ if $b + d < a + c$.

Proof of Lemma 1. We giscard the degenerate case $a = b = c = d$, which is obvious with $Z = 1$. By (A2) and some rearrangements—see also (2.15) in [8], we first rewrite

$$\log\left(\frac{[a;\delta]_s[c;\delta]_s}{[b;\delta]_s[d;\delta]_s}\right) \;=\; \kappa s \;+\; \int_{-\infty}^{0}(e^{sx} - 1 - sx)\left(\frac{e^{-b|x|} + e^{-d|x|} - e^{-a|x|} - e^{-c|x|}}{|x|(1 - e^{-|x|})(1 - e^{-\delta|x|})}\right)dx$$

for every $s > 0$, where κ is some real constant. By convexity, it is easy to see that if $b + d \leq a + c$ and $\inf\{b,d\} \leq \inf\{a,c\}$, then the function $z \mapsto z^b + z^d - z^a - z^c$ is positive on $(0,1)$. This implies that the function

$$x \mapsto \frac{e^{-b|x|} + e^{-d|x|} - e^{-a|x|} - e^{-c|x|}}{|x|(1 - e^{-|x|})(1 - e^{-\delta|x|})}$$

is positive on $(-\infty, 0)$ and that it can be viewed as the density of some Lévy measure on $(-\infty, 0)$, since it integrates $1 \wedge x^2$. By the Lévy–Khintchine formula, there exists a real infinitely divisible random variable Y such that

$$\mathbb{E}[e^{sY}] \;=\; \frac{[a;\delta]_s[c;\delta]_s}{[b;\delta]_s[d;\delta]_s}$$

for every $s > 0$, and the positive random variable $Z = e^Y$ satisfies (1). Since we have excluded the degenerate case, the Lévy measure of Y is clearly infinite and it follows from Theorem 27.7 in [9] that Y has a density and the same is true for Z.

Assuming first $b + d = a + c$, a Taylor expansion at zero shows that the density of the Lévy measure of Y integrates $1 \wedge |x|$ and we deduce from (A2) the simpler formula

$$\log \mathbb{E}[e^{sY}] \;=\; \log\left(\frac{[a;\delta]_s[c;\delta]_s}{[b;\delta]_s[d;\delta]_s}\right) \;=\; -\int_0^\infty (1 - e^{-sx})\left(\frac{e^{-bx} + e^{-dx} - e^{-ax} - e^{-cx}}{x(1 - e^{-x})(1 - e^{-\delta x})}\right)dx.$$

By the Lévy–Khintchine formula, this shows that the ID random variable Y is negative. Moreover, its support is $(-\infty, 0]$ since its Lévy measure has full support and its drift coefficient is zero—see Theorem 24.10 (iii) in [9], so that the support of Z is $[0,1]$.

Assuming second $b + d < a + c$, the same Taylor expansion as above shows that the density of the Lévy measure of Y does not integrate $1 \wedge |x|$ and the real Lévy process associated with Y is thus of type C using the terminology of [9]—see Definition 11.9 therein. By Theorem 24.10 (i) in [9], this implies that Y has full support on $\mathbb{R}$, and so does Z on $\mathbb{R}^+$.

It remains to prove the only if part of the Lemma. Assuming $a \leq d$ and $b \leq c$ without loss of generality, we first observe that if $a < b$ then the function

$$s \mapsto \frac{[a;\delta]_s [c;\delta]_s}{[b;\delta]_s [d;\delta]_s}$$

is real-analytic on $(-b, \infty)$ and vanishes at $s = -a > -b$, an impossible property for the Mellin transform of a positive random variable. The necessity of $b + d \leq a + c$ is slightly more subtle and hinges again upon infinite divisibility. First, setting $\varphi(z) = z^b + z^d - z^a - z^c$ and $z_* = \inf\{z > 0, \varphi(z) < 0\}$, it is easy to see by convexity and a Taylor expansion at 1 that if $b + d > a + c$, then $z_* < 1$ and $\varphi(z) < 0$ on $(z_*, 1)$ with $\varphi(z) \sim (b + d - a - c)(z - 1)$ as $z \to 1$. Introducing next the ID random variable V with Laplace exponent

$$\log \mathbb{E}[e^{sV}] = -\kappa s + \int_{\log z_*}^{0} (e^{sx} - 1 - sx) \left(\frac{e^{-a|x|} + e^{-c|x|} - e^{-b|x|} - e^{-d|x|}}{|x|(1 - e^{-|x|})(1 - e^{-\delta|x|})} \right) dx,$$

we obtain the decomposition

$$\log \left(\frac{[a;\delta]_s [c;\delta]_s}{[b;\delta]_s [d;\delta]_s} \right) + \log \mathbb{E}[e^{sV}] = \int_{-\infty}^{\log z_*} (e^{sx} - 1 - sx) \left(\frac{e^{-b|x|} + e^{-d|x|} - e^{-a|x|} - e^{-c|x|}}{|x|(1 - e^{-|x|})(1 - e^{-\delta|x|})} \right) dx,$$

whose right-hand side is the Laplace exponent of some ID random variable U with an atom because its Lévy measure, whose support is bounded away from zero, is finite—see Theorem 27.4 in [9]. On the other hand, the random variable V has an absolutely continuous and infinite Lévy measure and hence it has also a density. If there existed Z such that (1) holds, then the independent decomposition $U \stackrel{d}{=} V + \log Z$ would imply by convolution that U has a density as well. This contradiction finishes the proof of the Lemma. □

Remark 1. *(a) By the Mellin inversion formula, the density of $\mathbf{Z}[a, c; b, d; \delta]$ is expressed as*

$$f(x) = \frac{1}{2i\pi x} \int_{s_0 - i\infty}^{s_0 + i\infty} x^{-s} \left(\frac{[a;\delta]_s [c;\delta]_s}{[b;\delta]_s [d;\delta]_s} \right) ds$$

over $(0, \infty)$ for any $s_0 > -\inf\{b, d\}$. From this expression, it is possible to prove that this density is real-analytic over the interior of the support. We omit details. Let us also mention by Remark 28.8 in [9] that this density is positive over the interior of its support.

(b) With the standard notation for the Pochhammer symbol, the aforementioned Proposition 5.1 and Theorem 6.2 in [7] show that

$$s \mapsto \frac{(a)_s (c)_s}{(b)_s (d)_s}$$

is the Mellin transform of a positive random variable if and only if $b + d \geq a + c$ and $\inf\{b, d\} \geq \inf\{a, c\}$. This fact can be proved exactly as above, in writing

$$\log \left(\frac{(a)_s (c)_s}{(b)_s (d)_s} \right) = -\int_0^\infty (1 - e^{-sx}) \left(\frac{e^{-ax} + e^{-cx} - e^{-bx} - e^{-dx}}{x(1 - e^{-x})} \right) dx.$$

This expression also shows that the underlying random variable has support $[0, 1]$ and that it is absolutely continuous, save for $a + c = b + d$ where it has an atom at zero. We refer to [7] for an exact expression of the density on $(0, 1)$ in terms of the classical hypergeometric function.

We can now characterize the CM property for $E_{\alpha, m, l}(-x)$ on $(0, \infty)$.

Theorem 1. *Let $\alpha, m > 0$ and $l > -1/\alpha$. The Kilbas-Saigo function*

$$x \mapsto E_{\alpha,m,l}(-x)$$

is CM on $(0, \infty)$ if and only if $\alpha \leq 1$ and $l \geq m - 1/\alpha$. Its Bernstein representation is

$$E_{\alpha,m,l}(-x) = \mathbb{E}\left[\exp -x\left\{\mathbf{X}_{\alpha,m,l} \times \int_0^\infty \left(1 + \sigma_t^{(\alpha)}\right)^{-\alpha(m+1)} dt\right\}\right] \qquad (2)$$

with $\delta = 1/\alpha m$ and $\mathbf{X}_{\alpha,m,l} = \mathbf{Z}[1 + 1/m, (\alpha l + 1)\delta; 1, 1/m + (\alpha l + 1)\delta; \delta]$.

Proof of Theorem 1. Assume first $\alpha \leq 1$ and $l \geq m - 1/\alpha$ and let

$$\mathbf{Y}_{\alpha,m,l} = \mathbf{X}_{\alpha,m,l} \times \int_0^\infty \left(1 + \sigma_t^{(\alpha)}\right)^{-\alpha(m+1)} dt.$$

By Proposition 2.4 in [8], and Lemma 1, its Mellin transform is

$$\mathbb{E}[(\mathbf{Y}_{\alpha,m,l})^s] = \delta^s \frac{[1+\delta;\delta]_s[(\alpha l + 1)\delta;\delta]_s}{[1;\delta]_s[1/m + (\alpha l + 1)\delta;\delta]_s}$$

$$= \Gamma(1+s) \times \frac{[(\alpha l + 1)\delta;\delta]_s}{[1/m + (\alpha l + 1)\delta;\delta]_s}$$

where in the second equality we have used (A9). By Fubini's theorem, the moment generating function of $\mathbf{Y}_{\alpha,m,l}$ reads

$$\mathbb{E}[e^{z\mathbf{Y}_{\alpha,m,l}}] = \sum_{n \geq 0} \mathbb{E}[(\mathbf{Y}_{\alpha,m,l})^n] \frac{z^n}{n!}$$

$$= \sum_{n \geq 0} \left(\frac{[(\alpha l + 1)\delta;\delta]_n}{[1/m + (\alpha l + 1)\delta;\delta]_n}\right) z^n$$

$$= \sum_{n \geq 0} \left(\prod_{j=0}^{n-1} \frac{\Gamma(\alpha(jm+l)+1)}{\Gamma(\alpha(jm+l+1)+1)}\right) z^n = E_{\alpha,m,l}(z)$$

for every $z \geq 0$, where in the third equality we have used (A1) repeatedly. The latter identity is extended analytically to the whole complex plane and we get, in particular,

$$E_{\alpha,m,l}(-x) = \mathbb{E}[e^{-x\mathbf{Y}_{\alpha,m,l}}], \qquad x \geq 0.$$

This shows that $E_{\alpha,m,l}(-x)$ is CM with the required Bernstein representation.

We now prove the only if part. If $E_{\alpha,m,l}(-x)$ is CM, then we see by analytic continuation that $E_{\alpha,m,l}(z)$ is the moment generating function on $\mathbb{C}$ of the underlying random variable X, whose positive integer moments read

$$\mathbb{E}[X^n] = n! \times \left(\prod_{j=0}^{n-1} \frac{\Gamma(\alpha(jm+l)+1)}{\Gamma(\alpha(jm+l+1)+1)}\right), \qquad n \geq 0.$$

If $\alpha > 1$, Stirling's formula implies $\mathbb{E}[X^n]^{\frac{1}{n}} \to 0$ as $n \to \infty$ so that $X \equiv 0$, a contradiction because $E_{\alpha,m,l}$ is not a constant. If $\alpha = 1$ and $l + 1 < m$, then

$$\mathbb{E}[X^n] = \frac{n!}{(c)_n m^n} \sim \frac{n^{1-c}}{m^n} \quad \text{as } n \to \infty,$$

with $c = (l+1)/m \in (0,1)$. In particular, the Mellin transform $s \mapsto \mathbb{E}[X^s]$ is analytic on $\{\Re(s) \geq 0\}$, bounded on $\{\Re(s) = 0\}$, and has at most exponential growth on $\{\Re(s) > 0\}$ because

$$|\mathbb{E}[X^s]| \leq \mathbb{E}\left[X^{\Re(s)}\right] = \left(\mathbb{E}\left[X^{[\Re(s)]+1}\right]\right)^{\frac{\Re(s)}{[\Re(s)]+1}}$$

by Hölder's inequality. On the other hand, the Stirling type Formula (A4) implies, after some simplifications,

$$\delta^{-s}\frac{[1+\delta;\delta]_s[c;\delta]_s}{[1;\delta]_s[c+\delta;\delta]_s} = \delta^{-s}s^{1-c}(1+o(1)) \quad \text{as } |s| \to \infty \text{ with } |\arg s| < \pi$$

and this shows that the function on the left-hand side, which is analytic on $\{\Re(s) \geq 0\}$, has at most linear growth on $\{\Re(s) = 0\}$ and at most exponential growth on $\{\Re(s) > 0\}$. Moreover, the above analysis clearly shows that

$$\mathbb{E}[X^n] = \delta^{-n}\frac{[1+\delta;\delta]_n[c;\delta]_n}{[1;\delta]_n[c+\delta;\delta]_n}$$

for all $n \geq 0$ and by Carlson's theorem—see e.g., Section 5.81 in [10], we must have

$$\mathbb{E}[X^s] = \delta^{-s}\frac{[1+\delta;\delta]_s[c;\delta]_s}{[1;\delta]_s[c+\delta;\delta]_s}$$

for every $s > 0$, a contradiction since Lemma 1 shows that the right-hand side cannot be the Mellin transform of a positive random variable if $c < 1$. The case $\alpha < 1$ and $l + 1/\alpha < m$ is analogous. It consists of identifying the bounded sequence

$$\frac{1}{n!} \times \left(\prod_{j=0}^{n-1}\frac{\Gamma(\alpha(jm+l+1)+1)}{\Gamma(\alpha(jm+l)+1)}\right)$$

as the values at non-negative integer points of the function

$$\delta^{-s} \times \frac{[1;\delta]_s[1/m+(\alpha l+1)\delta;\delta]_s}{[1+\delta;\delta]_s[(\alpha l+1)\delta;\delta]_s} = \delta^{-s}e^{-(1-\alpha)s\ln(s)+\kappa s+O(1)} \quad \text{as } |s| \to \infty \text{ with } |\arg s| < \pi,$$

where the purposeless constant κ can be evaluated from (A4). On $\{\Re(s) \geq 0\}$, we see that this function has growth at most $e^{\pi(1-\alpha)|s|/2}$ and we can again apply Carlson's theorem. We leave the details to the interested reader. □

Remark 2. (a) When $m = 1$, applying (A1) we see that the random variable $\mathbf{X}_{\alpha,1,l}$ has Mellin transform

$$\mathbb{E}[(\mathbf{X}_{\alpha,1,l})^s] = \frac{[2;\delta]_s[l+1/\alpha;\delta]_s}{[1;\delta]_s[1+l+1/\alpha;\delta]_s} = \frac{(\alpha)_{\alpha s}}{(\beta)_{\alpha s}}$$

with $\beta = 1 + \alpha l \geq \alpha$. This shows $\mathbf{X}_{\alpha,1,l} \stackrel{d}{=} \mathbf{B}_{\alpha,\beta-\alpha}^{\alpha}$ where $\mathbf{B}_{a,b}$ denotes, here and throughout, a standard Beta random variable with parameters $a, b > 0$. We hence recover the Bernstein representation of the CM function $\Gamma(\beta)E_{\alpha,\beta}(-x)$ which was discussed in Remark 3.3 (c) in [3]. Notice also the very simple expression of the Mellin transform

$$\mathbb{E}[(\mathbf{Y}_{\alpha,1,l})^s] = \frac{\Gamma(1+\alpha l)\Gamma(1+s)}{\Gamma(1+\alpha(l+s))}.$$

(b) Another simplification occurs when $l + 1/\alpha = km$ for some integer $k \geq 1$. One finds

$$\mathbb{E}[(\mathbf{X}_{\alpha,m,km-1/\alpha})^s] = \frac{[k;\delta]_s[1+1/m;\delta]_s}{[1;\delta]_s[k+1/m;\delta]_s} = \prod_{j=1}^{k-1}\left(\frac{(\alpha jm)_u}{(\alpha(jm+1))_u}\right)$$

for $u = \alpha ms \geq 0$, which implies

$$\mathbf{X}_{\alpha,m,km-1/\alpha} \stackrel{d}{=} \left(\mathbf{B}_{\alpha m,\alpha} \times \cdots \times \mathbf{B}_{\alpha m(k-1),\alpha}\right)^{\alpha m}.$$

In general, the law of the absolutely continuous random variable $\mathbf{X}_{\alpha,m,l}$ valued in $[0,1]$ seems to have a complicated expression.

(c) As seen during the proof, the random variable $\mathbf{Y}_{\alpha,m,l}$ defined by the Bernstein representation

$$E_{\alpha,m,l}(-x) = \mathbb{E}[e^{-x\mathbf{Y}_{\alpha,m,l}}]$$

has Mellin transform

$$\mathbb{E}[(\mathbf{Y}_{\alpha,m,l})^s] = \Gamma(1+s) \times \frac{[(\alpha l + 1)\delta;\delta]_s}{[1/m + (\alpha l + 1)\delta;\delta]_s} \qquad (3)$$

with $\delta = 1/\alpha m$, for every $s > -1$. By Fubini's theorem, this implies the following exact computation, which seems unnoticed in the literature on the Kilbas-Saigo function.

$$\int_0^\infty E_{\alpha,m,l}(-x)\, x^{s-1}\, dx = \Gamma(s)\, \mathbb{E}[\mathbf{Y}_{\alpha,m,l}^{-s}] = \Gamma(s)\Gamma(1-s) \times \frac{[(\alpha l + 1)\delta;\delta]_{-s}}{[1/m + (\alpha l + 1)\delta;\delta]_{-s}} \qquad (4)$$

for every $s \in (0,1)$. For $m = 1$, we recover from (A1) the formula

$$\int_0^\infty E_{\alpha,\beta}(-x)\, x^{s-1}\, dx = \frac{1}{\Gamma(\beta)} \int_0^\infty E_{\alpha,1,\frac{\beta-1}{\alpha}}(-x)\, x^{s-1}\, dx = \frac{\Gamma(s)\Gamma(1-s)}{\Gamma(\beta - \alpha s)}$$

which is given in (4.10.3) of [2], as a consequence of the Mellin-Barnes representation of $E_{\alpha,\beta}(z)$. Notice that there is no such Mellin-Barnes representation for $E_{\alpha,m,l}(z)$ in general.

3. Uniform Hyperbolic Bounds

In Theorem 4 of [5], the following uniform hyperbolic bounds are obtained for the classical Mittag–Leffler function:

$$\frac{1}{1 + \Gamma(1-\alpha)x} \leq E_\alpha(-x) \leq \frac{1}{1 + \frac{1}{\Gamma(1+\alpha)}x} \qquad (5)$$

for every $\alpha \in [0,1]$ and $x \geq 0$. The constants in these inequalities are optimal because of the asymptotic behaviors

$$E_\alpha(-x) \sim \frac{1}{\Gamma(1-\alpha)x} \quad \text{as } x \to \infty \quad \text{and} \quad 1 - E_\alpha(-x) \sim \frac{x}{\Gamma(1+\alpha)} \quad \text{as } x \to 0.$$

See [11] and the references therein for some motivations on these hyperbolic bounds. In this section, we shall obtain analogous bounds for $E_{\alpha,m,m-1}(-x)$ and $E_{\alpha,m,m-\frac{1}{\alpha}}(-x)$ with $\alpha \in [0,1], m > 0$. Those peculiar functions are associated with the fractional Weibull and Fréchet distributions defined in [3]. Specifically, we will use the following representations as a moment generating function, obtained respectively in (3.1) and (3.4) therein:

$$E_{\alpha,m,m-1}(z) = \mathbb{E}\left[\exp\left\{z \int_0^\infty \left(1 - \sigma_t^{(\alpha)}\right)_+^{\alpha(m-1)} dt\right\}\right] \qquad (6)$$

and

$$E_{\alpha,m,m-\frac{1}{\alpha}}(z) = \mathbb{E}\left[\exp\left\{z \int_0^\infty \left(1 + \sigma_t^{(\alpha)}\right)^{-\alpha(m+1)} dt\right\}\right] \qquad (7)$$

for every $z \in \mathbb{C}$, where $\{\sigma_t^{(\alpha)}\, t \geq 0\}$ is the α-stable subordinator normalized such that

$$\mathbb{E}[e^{-\lambda \sigma_t^{(\alpha)}}] = e^{-t\lambda^\alpha}, \qquad \lambda, t \geq 0.$$

Observe that these two formulæ specify the general Bernstein representation (2) in terms of the α-stable subordinator only. We begin with the following monotonicity properties, of independent interest.

Proposition 1. *Fix $\alpha \in (0,1]$ and $x \in \mathbb{R}$. The functions*

$$m \mapsto E_{\alpha,m,m-1}(x) \qquad \text{and} \qquad m \mapsto E_{\alpha,m,m-\frac{1}{\alpha}}(x)$$

are decreasing on $(0,\infty)$ if $x > 0$ and increasing on $(0,\infty)$ if $x < 0$.

Proof of Proposition 1. This follows from (6) resp. (7), and the fact that $\sigma_t^{(\alpha)} > 0$ for every $t > 0$. □

Remark 3. *It would be interesting to know if the same property holds for $m \mapsto E_{\alpha,m,m-l}(x)$ and any $l \leq 1/\alpha$. In the case $l \notin \{1, 1/\alpha\}$, this would require from (2) a monotonicity analysis of the mapping $m \mapsto X_{\alpha,m,m-l}$, which does not seem easy at first sight.*

As in [5], our analysis to obtain the uniform bounds will use some notions of stochastic ordering. Recall that if X, Y are real random variables such that $\mathbb{E}[\varphi(X)] \leq \mathbb{E}[\varphi(Y)]$ for every $\varphi : \mathbb{R} \to \mathbb{R}$ convex, then Y is said to dominate X for the convex order, a property which we denote by $X \prec_{cx} Y$. Another ingredient in the proof is the following infinite independent product

$$\mathbf{T}(a,b,c) = \prod_{n \geq 0} \left(\frac{a+nb+c}{a+nb}\right) \mathbf{B}_{a+nb,c}.$$

We refer to Section 2.1 in [8] for more details on this infinite product, including the fact that it is a.s. convergent for every $a, b, c > 0$. We also mention from Proposition 2 in [8] that its Mellin transform is

$$\mathbb{E}[\mathbf{T}(a,b,c)^s] = \left(\frac{\Gamma(ab^{-1})}{\Gamma((a+c)b^{-1})}\right)^s \times \frac{[a+c;b]_s}{[a;b]_s}$$

for every $s > -a$. The following simple result on convex orderings for the above infinite independent products has an independent interest.

Lemma 2. *For every $a, b, c > 0$ and $d \geq c$, one has*

$$\mathbf{T}(a,b,c) \prec_{cx} \mathbf{T}(a,b,d).$$

Proof of Lemma 2. By the definition of $\mathbf{T}(a,b,c)$ and the stability of the convex order by mixtures—see Corollary 3.A.22 in [12], it is enough to show

$$(a+b)\mathbf{B}_{a,b} \prec_{cx} (a+c)\mathbf{B}_{a,c}$$

for every $a, b > 0$ and $c \geq b$. Using again Corollary 3.A.22 in [12] and the standard identity $\mathbf{B}_{a,c} \stackrel{d}{=} \mathbf{B}_{a,b} \times \mathbf{B}_{a+b,c-b}$, we are reduced to show

$$\left(\frac{a+b}{a+c}\right) = \mathbb{E}[\mathbf{B}_{a+b,c-b}] \prec_{cx} \mathbf{B}_{a+b,c-b}$$

which is a consequence of Jensen's inequality. □

The following result is a generalization of the inequalities (5), which deal with the case $m = 1$ only, to all Kilbas-Saigo functions $E_{\alpha,m,m-1}(-x)$. The argument is considerably simpler than in the original proof of (5).

Theorem 2. *For every $\alpha \in [0,1], m > 0$ and $x \geq 0$, one has*

$$\frac{1}{1 + \Gamma(1-\alpha)x} \leq E_{\alpha,m,m-1}(-x) \leq \frac{1}{1 + \frac{\Gamma(1+\alpha(m-1))}{\Gamma(1+\alpha m)} x}.$$

Proof of Theorem 2. The first inequality is a consequence of Proposition 1, which implies in letting $m \to 0$

$$E_{\alpha,m,m-1}(-x) \geq \mathbb{E}\left[\exp\left\{-x \int_0^\infty \left(1 - \sigma_t^{(\alpha)}\right)_+^{-\alpha} dt\right\}\right]$$
$$= \mathbb{E}\left[e^{-x\Gamma(1-\alpha)\mathbf{L}}\right] = \frac{1}{1+\Gamma(1-\alpha)x}$$

for $x \geq 0$, where the first equality follows from Theorem 1.2 (b) (ii) in [8]. For the second inequality, we come back to the infinite product representation

$$\int_0^\infty \left(1 - \sigma_t^{(\alpha)}\right)_+^{\rho-\alpha} dt \stackrel{d}{=} \frac{\Gamma(\rho+1-\alpha)}{\Gamma(\rho+1)} \mathbf{T}(1,\rho^{-1},(1-\alpha)\rho^{-1})$$

which follows from Theorem 1.2 (b) (i) in [8], exactly as in the proof of Theorem 1.1 in [3]. Lemma 2 implies then

$$\int_0^\infty \left(1 - \sigma_t^{(\alpha)}\right)_+^{\rho-\alpha} dt \prec_{cx} \frac{\Gamma(\rho+1-\alpha)}{\Gamma(\rho+1)} \mathbf{T}(1,\rho^{-1},\rho^{-1}) \stackrel{d}{=} \frac{\Gamma(\rho+1-\alpha)}{\Gamma(\rho+1)} \mathbf{L}$$

where the identity in law follows from (2.7) in [8]. Using (6) with $\rho = \alpha m$ and the convexity of $t \mapsto e^{-xt}$, we obtain the required

$$E_{\alpha,m,m-1}(-x) \leq \frac{1}{1 + \frac{\Gamma(1+\alpha(m-1))}{\Gamma(1+\alpha m)} x}.$$

□

Remark 4. (a) As for the classical case $m = 1$, these bounds are optimal because of the asymptotic behaviors

$$1 - E_{\alpha,m,m-1}(-x) \sim \frac{\Gamma(1+\alpha(m-1))}{\Gamma(1+\alpha m)} x \quad \text{as } x \to 0$$

and

$$E_{\alpha,m,m-1}(-x) \sim \frac{1}{\Gamma(1-\alpha)x} \quad \text{as } x \to \infty.$$

The behavior at zero is plain from the definition, whereas the behavior at infinity will be given after Remark 6 below.

(b) It is easy to check that the above proof also yields the upper bound

$$E_{\alpha,m,m-1}(x) \leq \frac{1}{(1-\Gamma(1-\alpha)x)_+}$$

for every $\alpha \in [0,1], m > 0$ and $x \geq 0$, which seems unnoticed even in the classical case $m = 1$.

Our next result is a uniform hyperbolic upper bound for the Kilbas-Saigo function $E_{\alpha,m,m-\frac{1}{\alpha}}(-x)$, with a power exponent which will be shown to be optimal in Remark 8 (c) below, and also an optimal constant because

$$1 - E_{\alpha,m,m-\frac{1}{\alpha}}(-x) \sim \left(1 + \frac{1}{m}\right) \times \frac{\Gamma(1+\alpha m)\, x}{\Gamma(1+\alpha(m+1))} \quad \text{as } x \to 0.$$

Proposition 2. *For every $\alpha \in (0,1]$, $m > 0$ and $x \geq 0$, one has*

$$E_{\alpha,m,m-\frac{1}{\alpha}}(-x) \leq \frac{1}{\left(1 + \frac{\Gamma(1+\alpha m)}{\Gamma(1+\alpha(m+1))} x\right)^{1+\frac{1}{m}}}.$$

Proof of Proposition 2. The inequality is derived by convex ordering as in Theorem 2: setting, here and throughout, $\mathbf{\Gamma}_a$ for a Gamma random variable with parameter $a > 0$, one has

$$\int_0^\infty \left(1 + \sigma_t^{(\alpha)}\right)_+^{-\rho-\alpha} dt \stackrel{d}{=} \frac{\Gamma(\rho)}{\Gamma(\rho+\alpha)} \mathbf{T}(1+\alpha\rho^{-1}, \rho^{-1}, (1-\alpha)\rho^{-1})$$

$$\prec_{cx} \frac{\Gamma(\rho)}{\Gamma(\rho+\alpha)} \mathbf{T}(1+\alpha\rho^{-1}, \rho^{-1}, \rho^{-1}) \stackrel{d}{=} \frac{\Gamma(\rho+1)}{\Gamma(\rho+1+\alpha)} \mathbf{\Gamma}_{1+\frac{\alpha}{\rho}}$$

where the first identity follows from Corollary 3 in [8] as in the proof of Theorem 1.1 in [3], the convex ordering from Lemma 2 and the second identity from (2.7) in [8]. Then, using (7) with $\rho = \alpha m$, we get the required inequality. □

As in Theorem 2, we believe that there is also a uniform lower bound, with a more complicated optimal constant which can be read off from the asymptotic behavior of the density at zero obtained in Proposition 7 below:

Conjecture 3. *For every $\alpha \in (0,1]$, $m > 0$ and $x \geq 0$, one has*

$$E_{\alpha,m,m-\frac{1}{\alpha}}(-x) \geq \frac{1}{\left(1 + (\alpha m)^{-\frac{\alpha}{m+1}} (\Gamma(1+\alpha)\, G(1-\alpha;\alpha m)\, G(1+\alpha;\alpha m))^{-\frac{m}{m+1}} x\right)^{1+\frac{1}{m}}}. \quad (8)$$

Unfortunately, the proof of this general inequality still eludes us. The monotonicity property observed in Proposition 1 does not help here, giving only the trivial lower bound zero. The discrete factorizations which are used in [5] are also more difficult to handle in this context, because the Mellin transform underlying $E_{\alpha,m,m-\frac{1}{\alpha}}$ is expressed in terms of generalized Pochhammer symbols. In the case $m = 1$, we could however get a proof of (8). The argument relies on the following representation, observed in Remarks 3.1 (d) and 3.3 (c) of [3]:

$$E_{\alpha,1,1-\frac{1}{\alpha}}(z) = \Gamma(\alpha) E'_{\alpha,\alpha}(z) = \Gamma(1+\alpha) E'_\alpha(z) = \Gamma(1+\alpha) \mathbb{E}\left[T_\alpha\, e^{zT_\alpha}\right] = \mathbb{E}\left[e^{zT_\alpha^{(1)}}\right] \quad (9)$$

for every $z \in \mathbb{C}$, where $T_\alpha = \inf\{t > 0,\ \sigma_t^{(\alpha)} > 1\}$ is the first-passage time above one of the α-stable subordinator and $T_\alpha^{(1)}$ its usual size-bias of order one.

Proposition 3. *For every $\alpha \in (0,1)$ and $x \geq 0$, one has*

$$E_{\alpha,1,1-\frac{1}{\alpha}}(-x) \geq \frac{1}{\left(1 + \sqrt{\frac{\Gamma(1-\alpha)}{\Gamma(1+\alpha)}}\, x\right)^2}.$$

Proof of Proposition 3. By (9) and since

$$\mathbb{E}\left[e^{-x\Gamma_2}\right] = \frac{1}{(1+x)^2}$$

for every $x \geq 0$, it is enough to show, reasoning exactly as in the proof of Theorem 4 in [5], that

$$T_\alpha^{(1)} \prec_{st} \sqrt{\frac{\Gamma(1-\alpha)}{\Gamma(1+\alpha)}} \Gamma_2, \qquad (10)$$

where $\prec_{st}$ stands for the usual stochastic order between two real random variables. Recall that $X \prec_{st} Y$ means $\mathbb{P}[X \geq x] \leq \mathbb{P}[Y \geq x]$ for every $x \in \mathbb{R}$. Since $T_{1/2} \stackrel{d}{=} 2\sqrt{\Gamma_{1/2}}$, the case $\alpha = 1/2$ is explicit and the stochastic ordering can be obtained directly. More precisely, the densities of both random variables in (10) are respectively given by

$$\frac{x}{2} e^{-x^2/4} \quad \text{and} \quad \frac{x}{2} e^{-x/\sqrt{2}}$$

on $(0, \infty)$, where they cross only once at $x = 2\sqrt{2}$. It is a well-known and an easy result that this single intersection property yields (10)—see Theorem 1.A.12 in [12].

The argument for the case $\alpha \neq 1/2$ is somehow analogous, but the details are more elaborate because the density of $T_\alpha^{(1)}$ is not explicit anymore. We proceed as in Theorem C of [5] and first consider the case where α is rational. Setting $\alpha = p/n$ with $n > p$ positive integers and $X_\alpha = T_\alpha^{(1)}$ we have, on the one hand,

$$\begin{aligned}
\mathbb{E}[(X_\alpha)^{ns}] &= \frac{\mathbb{E}[(T_\alpha)^{1+ns}]}{\mathbb{E}[T_\alpha]} \\
&= \frac{\Gamma(2+ns)\Gamma(1+pn^{-1})}{\Gamma(1+pn^{-1}+ps)} \\
&= \frac{n^{ns}}{p^{ps}} \times \mathbb{E}\left[\left(\mathbf{B}_{\frac{2}{n}, \frac{1}{p}-\frac{1}{n}}\right)^s\right] \times \frac{\prod_{i=3}^{n+1}(in^{-1})_s}{\prod_{j=2}^{p}(jp^{-1}+n^{-1})_s}
\end{aligned}$$

for every $s > -2n^{-1}$, where we have used the well-known identity $T_\alpha \stackrel{d}{=} (\sigma_1^{(\alpha)})^{-\alpha}$ in the second equality, whereas in the third equality we have used repeatedly the Legendre-Gauss multiplication formula for the Gamma function—see e.g., Theorem 1.5.2 in [13]. The same formula implies, on the other hand,

$$\begin{aligned}
\mathbb{E}\left[\left(\sqrt{\frac{\Gamma(1-\alpha)}{\Gamma(1+\alpha)}} \Gamma_2\right)^{ns}\right] &= \frac{n^{ns} \kappa_\alpha^s}{p^{ps}} \times \mathbb{E}\left[\left(\Gamma_{\frac{2}{n}}\right)^s\right] \times \left(\prod_{i=3}^{n+1}(in^{-1})_s\right) \\
&= \frac{n^{ns}}{p^{ps}} \times \mathbb{E}\left[\left(\kappa_\alpha \times \Gamma_{\frac{2}{n}} \times \prod_{j=2}^{p} \Gamma_{\frac{j+1}{p}+\frac{1}{n}}\right)^s\right] \\
&\quad \times \frac{\prod_{i=3}^{n+1}(in^{-1})_s}{\prod_{j=2}^{p}(jp^{-1}+n^{-1})_s}
\end{aligned}$$

for every $s > -2n^{-1}$, with the notation

$$\kappa_\alpha = \left(\prod_{i=1}^{p} \frac{\Gamma(ip^{-1}-n^{-1})}{\Gamma(ip^{-1}+n^{-1})}\right)^{\frac{n}{2}}.$$

Since
$$\frac{\prod_{i=3}^{n+1}(in^{-1})_s}{\prod_{j=2}^{p}(jp^{-1}+n^{-1})_s} = \mathbb{E}\left[\left(\prod_{i=2}^{p}\mathbf{B}_{\frac{i+1}{n},\frac{i}{p}-\frac{i}{n}} \times \prod_{j=p+1}^{n}\Gamma_{\frac{j+1}{n}}\right)^s\right]$$

for every $s > -3n^{-1}$, by factorization and Theorem 1.A.3 (d) in [5] we are finally reduced to show

$$\mathbf{B}_{\frac{2}{n},\frac{1}{p}-\frac{1}{n}} \prec_{st} \left(\prod_{i=1}^{p}\frac{\Gamma(ip^{-1}-n^{-1})}{\Gamma(ip^{-1}+n^{-1})}\right)^{\frac{n}{2}} \times \Gamma_{\frac{2}{n}} \times \prod_{j=2}^{p}\Gamma_{\frac{j}{p}+\frac{1}{n}}$$

for every $n > p$ positive integers. The above inequality is equivalent to

$$(\mathbf{B}_{\frac{2}{n},\frac{1}{p}-\frac{1}{n}})^{\frac{2}{n}} \prec_{st} \left(\prod_{i=2}^{p}\frac{\Gamma(ip^{-1}-n^{-1})}{\Gamma(ip^{-1}+n^{-1})}\right) \times \left(\Gamma_{\frac{2}{n}} \times \prod_{j=2}^{p}\Gamma_{\frac{j}{p}+\frac{1}{n}}\right)^{\frac{2}{n}}$$

and this is proved via the single intersection property exactly as for (5.1) in [5]: the random variable on the left-hand side has an increasing density on $(0,1)$, whereas the random variable on the right-hand side has a decreasing density on $(0,\infty)$, both densities having the same positive finite value at zero. We omit details. This completes the proof of (10) when α is rational. The case when α is irrational follows then by a density argument. □

Remark 5. *It is easy to check from (A5) and (A6) that*

$$\frac{\Gamma(1+\alpha)}{\Gamma(1-\alpha)} = \alpha^\alpha \Gamma(1+\alpha) G(1-\alpha;\alpha) G(1+\alpha;\alpha),$$

so that Proposition 3 leads to (8) for $m = 1$, in accordance with the estimate (13). In general, the absence of a tractable complement formula for the product $G(1-\alpha;\delta) G(1+\alpha;\delta)$ makes however the constant in (8) more difficult to handle.

Our last result in this section gives optimal uniform hyperbolic bounds for the generalized Mittag–Leffler functions $E_{\alpha,\beta}(-x)$ whenever they are completely monotone, that is for $\beta \geq \alpha$—see the above Remark 2 (a). This can be viewed as another generalization of (5).

Proposition 4. *For every $\alpha \in (0,1], \beta > \alpha$ and $x \geq 0$, one has*

$$\frac{1}{\left(1+\sqrt{\frac{\Gamma(1-\alpha)}{\Gamma(1+\alpha)}}\,x\right)^2} \leq \Gamma(\alpha)\, E_{\alpha,\alpha}(-x) \leq \frac{1}{\left(1+\frac{\Gamma(1+\alpha)}{\Gamma(1+2\alpha)}\,x\right)^2}$$

and

$$\frac{1}{1+\frac{\Gamma(\beta-\alpha)}{\Gamma(\beta)}\,x} \leq \Gamma(\beta)\, E_{\alpha,\beta}(-x) \leq \frac{1}{1+\frac{\Gamma(\beta)}{\Gamma(\beta+\alpha)}\,x}.$$

Proof of Proposition 4. The bounds for $E_{\alpha,\alpha}(-x)$ are a direct consequence of (9), Proposition 2 and Proposition 3. Notice that letting $\alpha \to 1$ leads to the trivial bound $0 \leq e^{-x} \leq (2/(2+x))^2$. To handle the bounds for $\beta > \alpha$, we first recall from Remark 2 (a) that

$$\Gamma(\beta)\, E_{\alpha,\beta}(-x) = \Gamma(\beta)\, E_{\alpha,1,\frac{\beta-1}{\alpha}}(-x) = \mathbb{E}\left[e^{-x\mathbf{Y}_{\alpha,1,l}}\right]$$

with $l = (\beta-1)/\alpha > 1 - 1/\alpha$ and $\mathbf{Y}_{\alpha,1,l} \stackrel{d}{=} \mathbf{B}_{\alpha,\beta-\alpha}^\alpha \times \mathbf{T}_\alpha^{(1)}$. Moreover, one has

$$\mathbb{E}[(\mathbf{Y}_{\alpha,1,l})^s] = \frac{\Gamma(1+s)\Gamma(\beta)}{\Gamma(\beta+\alpha s)} \qquad (11)$$

for every $s > -1$, which implies the factorization $\mathbf{L} \stackrel{d}{=} \mathbf{Y}_{\alpha,1,l} \times (\mathbf{\Gamma}_\beta)^\alpha$. Since, by Jensen's inequality,

$$\frac{\Gamma(\beta+\alpha)}{\Gamma(\beta)} = \mathbb{E}\big[(\mathbf{\Gamma}_\beta)^\alpha\big] \prec_{cx} (\mathbf{\Gamma}_\beta)^\alpha,$$

we deduce from Corollary 3.A.22 in [12] the convex ordering

$$\mathbf{Y}_{\alpha,1,l} \prec_{cx} \frac{\Gamma(\beta)}{\Gamma(\beta+\alpha)} \mathbf{L}$$

which, as above, implies

$$\Gamma(\beta)\, E_{\alpha,\beta}(-x) \;\leq\; \frac{1}{1 + \frac{\Gamma(\beta)}{\Gamma(\beta+\alpha)}\, x}$$

for every $x \geq 0$.

The argument for the other inequality is analogous to that of Proposition 3. By density, we only need to consider the case $\alpha = p/n$ and $\beta = (p+q)/n$ with $p < n$ and q positive integers. By (11) and the Legendre-Gauss multiplication formula, we obtain

$$\mathbb{E}[(\mathbf{Y}_{\alpha,1,l})^{ns}] = \frac{n^{ns}}{p^{ps}} \times \mathbb{E}\left[\left(\mathbf{B}_{\frac{1}{n},\frac{q}{np}}\right)^s\right] \times \frac{\prod_{i=2}^{n}(in^{-1})_s}{\prod_{j=1}^{p-1}(jp^{-1}+(p+q)(np)^{-1})_s}$$

for every $s > -n^{-1}$. On the other hand, one has

$$\mathbb{E}\left[\left(\frac{\Gamma(\beta-\alpha)}{\Gamma(\beta)}\mathbf{L}\right)^{ns}\right] = \frac{n^{ns}}{p^{ps}} \mathbb{E}\left[\left(\kappa_{\alpha,\beta} \times \mathbf{\Gamma}_{\frac{1}{n}} \times \prod_{j=1}^{p-1} \mathbf{\Gamma}_{\frac{j}{p}+\frac{p+q}{np}}\right)^s\right] \times \frac{\prod_{i=2}^{n}(in^{-1})_s}{\prod_{j=1}^{p-1}(jp^{-1}+(p+q)(np)^{-1})_s}$$

with

$$\kappa_{\alpha,\beta} = p^p \left(\frac{\Gamma(qn^{-1})}{\Gamma((p+q)n^{-1})}\right)^n.$$

Comparing these two formulæ we are reduced to show

$$\left(\mathbf{B}_{\frac{1}{n},\frac{q}{np}}\right)^{\frac{1}{n}} \prec_{st} p^{\frac{p}{n}} \left(\frac{\Gamma(qn^{-1})}{\Gamma((p+q)n^{-1})}\right) \times \left(\mathbf{\Gamma}_{\frac{1}{n}} \times \prod_{j=1}^{p-1} \mathbf{\Gamma}_{\frac{j}{p}+\frac{p+q}{np}}\right)^{\frac{1}{n}}$$

for every $p < n$ and q positive integers. This is obtained in the same way as above via the single intersection property. We leave the details to the reader. □

4. Asymptotic Behavior of Fractional Extreme Densities

In this section, which is a complement to [3], we study the behavior of the density functions of the fractional Weibull and Fréchet distributions at both ends of their support. To this end, we also evaluate their Mellin transforms in terms of Barnes' double Gamma function. Along the way, we give the exact asymptotics of $x \mapsto E_{\alpha,m,l}(x)$ on the negative half-line, in the completely monotonic case $\alpha \in [0,1]$ and $l \geq m - 1/\alpha$.

4.1. The Fractional Weibull Case

In [3], a fractional Weibull distribution function with parameters $\alpha \in [0,1]$ and $\lambda, \rho > 0$ is defined as the unique distribution function $F^{\mathbf{W}}_{\alpha,\lambda,\rho}$ on $(0,\infty)$ solving the fractional differential equation

$$\mathbf{D}^\alpha_{0+} F(x) = \lambda\, x^{\rho-\alpha} \bar{F}(x)$$

where $\bar{F} = 1 - F$ denotes the associated survival function and D_{0+}^α a progressive Liouville fractional derivative on $(0, \infty)$. The case $\alpha = 1$ corresponds to the standard Weibull distribution. In [3], it is shown that this distribution function exists and is given by

$$F_{\alpha,\lambda,\rho}^{\mathbf{W}}(x) = 1 - E_{\alpha,\frac{\rho}{\alpha},\frac{\rho}{\alpha}-1}(-\lambda x^\rho)$$

for every $x \geq 0$—see the formula following (3.1) in [3]. In particular, the density $f_{\alpha,\lambda,\rho}^{\mathbf{W}}$ is real-analytic on $(0, \infty)$ and has the following asymptotic behavior at zero:

$$f_{\alpha,\lambda,\rho}^{\mathbf{W}}(x) \sim \left(\frac{\lambda \Gamma(\rho + 1 - \alpha)}{\Gamma(\rho)}\right) x^{\rho-1} \quad \text{as } x \to 0.$$

The behavior of $f_{\alpha,\lambda,\rho}^{\mathbf{W}}$ at infinity is however less immediate, and to this aim we will need an exact expression for the Mellin transform of the random variable $\mathbf{W}_{\alpha,\lambda,\rho}$ with distribution function $F_{\alpha,\lambda,\rho}^{\mathbf{W}}$, which has an interest in itself.

Proposition 5. *The Mellin transform of $\mathbf{W}_{\alpha,\lambda,\rho}$ is*

$$\mathbb{E}\left[\mathbf{W}_{\alpha,\lambda,\rho}^s\right] = \left(\frac{\rho^\alpha}{\lambda}\right)^{\frac{s}{\rho}} \Gamma(1 + s\rho^{-1}) \times \frac{[\rho + (1-\alpha); \rho]_{-s}}{[\rho; \rho]_{-s}}$$

for every $s \in (-\rho, \rho)$. Consequently, one has

$$f_{\alpha,\lambda,\rho}^{\mathbf{W}}(x) \sim \left(\frac{\rho}{\lambda \Gamma(1-\alpha)}\right) x^{-\rho-1} \quad \text{as } x \to \infty.$$

Proof of Proposition 5. We start with a more concise expression of (3) for $l = m - 1$, which is a direct consequence of (A9):

$$\mathbb{E}[(\mathbf{Y}_{\alpha,\frac{\rho}{\alpha},\frac{\rho}{\alpha}-1})^s] = \rho^{-s} \times \frac{[1 + (1-\alpha)\rho^{-1}; \rho^{-1}]_s}{[1; \rho^{-1}]_s}.$$

By Theorem 1.1 in [3] and using the notations therein, we deduce

$$\mathbb{E}\left[\mathbf{W}_{\alpha,\lambda,\rho}^s\right] = \mathbb{E}\left[\left(\frac{\mathbf{L}}{\lambda \mathbf{Y}_{\alpha,\frac{\rho}{\alpha},\frac{\rho}{\alpha}-1}}\right)^{\frac{s}{\rho}}\right]$$

$$= \left(\frac{\rho}{\lambda}\right)^{\frac{s}{\rho}} \Gamma(1 + s\rho^{-1}) \times \frac{[1 + (1-\alpha)\rho^{-1}; \rho^{-1}]_{-s\rho^{-1}}}{[1; \rho^{-1}]_{-s\rho^{-1}}}$$

$$= \left(\frac{\rho^\alpha}{\lambda}\right)^{\frac{s}{\rho}} \Gamma(1 + s\rho^{-1}) \times \frac{[\rho + (1-\alpha); \rho]_{-s}}{[\rho; \rho]_{-s}}$$

for every $s \in (-\rho, \rho)$ as required, where the third equality comes from (A8). The asymptotic behavior of the density at infinity is then a standard consequence of Mellin inversion. First, we observe from the above formula and (A10) that the first positive pole of $s \mapsto \mathbb{E}\left[\mathbf{W}_{\alpha,\lambda,\rho}^s\right]$ is simple and isolated in the complex plane at $s = \rho$, with

$$\mathbb{E}\left[\mathbf{W}_{\alpha,\lambda,\rho}^s\right] \sim \left(\frac{\rho^\alpha}{\lambda}\right) \times \frac{[\rho + (1-\alpha); \rho]_{-\rho}}{[\rho; \rho]_{-s}}$$

$$\sim \left(\frac{\rho^{\rho+\alpha}}{\lambda}\right) \times \frac{[\rho + (1-\alpha); \rho]_{-\rho}}{[2\rho; \rho]_{-\rho}} \times (\rho)_{-s} = \frac{\rho \Gamma(\rho - s)}{\lambda \Gamma(1-\alpha)} \sim \frac{\rho}{\lambda \Gamma(1-\alpha)(\rho - s)}$$

as $s \uparrow \rho$, where the second asymptotics comes from (A9) and the equality from (A5). Therefore, applying Theorem 4 (ii) in [14] beware the correction $(\log x)^k \to (\log x)^{k-1}$ to be made in the expansion of $f(x)$ therein, we obtain

$$f^{\mathbf{W}}_{\alpha,\lambda,\rho}(x) \sim \left(\frac{\rho}{\lambda\Gamma(1-\alpha)}\right) x^{-\rho-1} \quad \text{as } x \to \infty$$

as required. □

Remark 6. (a) *Another proof of the asymptotic behavior at infinity can be obtained from that of the so-called generalized stable densities. More precisely, using the identity in law on top of p.12 in* [3] *and the notation therein, we see by multiplicative convolution, having set $f^{\mathcal{G}}_{\alpha,\rho}$ for the density of the generalized stable random variable $\mathcal{G}(\rho + 1 - \alpha, 1 - \alpha)$, that*

$$\begin{aligned}
f^{\mathbf{W}}_{\alpha,\lambda,\rho}(x) &= \lambda\, x^{\rho-1} \int_0^\infty f^{\mathcal{G}}_{\alpha,\rho}(y)\, y^{-\rho}\, e^{-\frac{\lambda}{\rho}(\frac{x}{y})^\rho} dy \\
&= \left(\frac{\lambda}{\rho}\right)^{\frac{1}{\rho}} \int_0^\infty f^{\mathcal{G}}_{\alpha,\rho}\!\left(x(\rho\lambda^{-1}t)^{-\frac{1}{\rho}}\right) t^{-\frac{1}{\rho}} e^{-t}\, dt \\
&\sim \left(\frac{\rho}{\lambda\Gamma(1-\alpha)} \int_0^\infty t\, e^{-t}\, dt\right) x^{-\rho-1} = \left(\frac{\rho}{\lambda\Gamma(1-\alpha)}\right) x^{-\rho-1}
\end{aligned}$$

as $x \to \infty$, where for the asymptotics we have used the Proposition in [15] *and a direct integration. This argument does not make use of Mellin inversion and is overall simpler than the above. However, it does not convey to the fractional Fréchet case.*

(b) *The Mellin transform simplifies for $\alpha = 0$ and $\alpha = 1$: using (A1) and (A6) we recover*

$$\mathbb{E}[\mathbf{W}^s_{0,\lambda,\rho}] = \lambda^{-\frac{s}{\rho}} \Gamma(1+s\rho^{-1})\Gamma(1-s\rho^{-1}) \quad \text{and} \quad \mathbb{E}[\mathbf{W}^s_{1,\lambda,\rho}] = \left(\frac{\rho}{\lambda}\right)^{\frac{s}{\rho}} \Gamma(1+s\rho^{-1})$$

in accordance with the scaling property $\mathbf{W}_{\alpha,\lambda,\rho} \stackrel{d}{=} \lambda^{-1/\rho} \mathbf{W}_{\alpha,1,\rho}$ and the identities given at the bottom of p.3 in [3]. *The Mellin transform takes a simpler form in two other situations.*

- *For $\rho = \alpha$, we obtain from (3), (A1) and (A5)*

$$\mathbb{E}[(\mathbf{Y}_{\alpha,1,0})^s] = \frac{\Gamma(1+s)}{\Gamma(1+\alpha s)} = \mathbb{E}[\mathbf{Z}_\alpha^{-\alpha s}],$$

in accordance with Remark 3.1 (d) in [3]. *This yields*

$$\mathbf{W}_{\alpha,\lambda,\alpha} \stackrel{d}{=} \left(\frac{\mathbf{L}}{\lambda \mathbf{Y}_{\alpha,1,0}}\right)^{\frac{1}{\alpha}} \stackrel{d}{=} \lambda^{-\frac{1}{\alpha}}\, \mathbf{Z}_\alpha \times \mathbf{L}^{\frac{1}{\alpha}},$$

an identity which was already discussed for $\lambda = 1$ in the introduction of [3] *as the solution to (1.3) therein. The Mellin transform reads*

$$\mathbb{E}[\mathbf{W}^s_{\alpha,\lambda,\alpha}] = \lambda^{-\frac{s}{\alpha}}\, \frac{\Gamma(1+s\alpha^{-1})\Gamma(1-s\alpha^{-1})}{\Gamma(1-s)}.$$

- *For $\rho = 1 - \alpha$, where we obtain from (A5)*

$$\mathbb{E}[\mathbf{W}^s_{1-\rho,\lambda,\rho}] = \left(\frac{\rho}{\lambda}\right)^{\frac{s}{\rho}} \frac{\Gamma(1+s\rho^{-1})\Gamma(\rho-s)}{\Gamma(\rho)} \quad \text{and} \quad \mathbf{W}_{1-\rho,\lambda,\rho} \stackrel{d}{=} \left(\frac{\rho}{\lambda}\right)^{\frac{1}{\rho}} \mathbf{L}^{\frac{1}{\rho}} \times \mathbf{\Gamma}_\rho^{-1}.$$

(c) *The two cases $\rho = \alpha$ and $\rho = 1 - \alpha$ have a Mellin transform expressed as the quotient of a finite number of Gamma functions. This makes it possible to use a Mellin-Barnes representation of* the density to get its full asymptotic expansion at infinity. Using the standard notation of Definition C.1.1 in [13], one obtains

$$f^{\mathbf{W}}_{\alpha,\lambda,\alpha}(x) \sim \sum_{n\geq 1} \frac{n\alpha\, x^{-1-n\alpha}}{\lambda^n \Gamma(1-n\alpha)} \quad \text{and} \quad f^{\mathbf{W}}_{1-\alpha,\lambda,\alpha}(x) \sim \frac{\alpha x^{-\alpha-1}}{\lambda \Gamma(\alpha)} \sum_{n\geq 0} (-1)^n \frac{\Gamma(\frac{n}{\rho}+2)}{n!} \left(\frac{\lambda}{\rho}\right)^{-\frac{n}{\rho}} x^{-n}$$

which are everywhere divergent. The first expansion can also be obtained from (1.8.28) in [16] using

$$f^{\mathbf{W}}_{\alpha,\lambda,\alpha}(x) = \lambda\, x^{\alpha-1}\, E_{\alpha,\alpha}(-\lambda x^\alpha).$$

Unfortunately, the Mellin transform of $\mathbf{W}_{\alpha,\lambda,\rho}$ might have poles of variable order and it seems difficult to obtain a general formula for the full asymptotic expansion at infinity of $f^{\mathbf{W}}_{\alpha,\lambda,\rho}(x)$.

Writing

$$E_{\alpha,\frac{\rho}{\alpha},\frac{\rho}{\alpha}-1}(-\lambda x^\rho) = \mathbb{P}[\mathbf{W}_{\alpha,\lambda,\rho} > x] = \int_x^\infty f^{\mathbf{W}}_{\alpha,\lambda,\rho}(y)\,dy,$$

we obtain by integration the following asymptotic behavior at infinity, which is valid for any $\alpha \in (0,1]$ and $m > 0$:

$$E_{\alpha,m,m-1}(-x) \sim \frac{1}{\Gamma(1-\alpha)\,x} \quad \text{as } x \to \infty.$$

This behavior, which turns out to be the same as that of the classical Mittag–Leffler function $E_\alpha(-x)$—see e.g., (3.4.15) in [2], gives the reason the constant in the lower bound of Theorem 2 is optimal—see the above Remark 4 (a). It is actually possible to get the exact behavior of $E_{\alpha,m,l}(-x)$ at infinity for any $\alpha \in (0,1], m > 0$ and $l > m - 1/\alpha$. We include this result here since it seems unnoticed in the literature on Kilbas-Saigo functions.

Proposition 6. *For any $\alpha \in [0,1], m > 0$ and $l > m - 1/\alpha$, one has*

$$E_{\alpha,m,l}(-x) \sim \frac{\Gamma(1+\alpha(l+1-m))}{\Gamma(1+\alpha(l-m))\,x} \quad \text{as } x \to \infty.$$

Proof of Proposition 6. The case $\alpha = 0$ is obvious since $E_{0,m,l}(x) = 1/(1-x)$. For $\alpha \in (0,1]$, setting $\delta = 1/\alpha m$, recall from (4) that for every $s \in (0,1)$ one has

$$\int_0^\infty E_{\alpha,m,l}(-x)\, x^{s-1}\, dx = \Gamma(s)\Gamma(1-s) \times \frac{[(\alpha l+1)\delta;\delta]_{-s}}{[1/m + (\alpha l+1)\delta;\delta]_{-s}}$$

$$\sim \frac{[(\alpha l+1)\delta;\delta]_{-1}}{[1/m + (\alpha l+1)\delta;\delta]_{-1}(1-s)} = \frac{\Gamma(1+\alpha(l+1-m))}{\Gamma(1+\alpha(l-m))\,(1-s)}$$

as $s \uparrow 1$, where in the equality we have used the concatenation formula (A1). The asymptotic behavior follows then by Mellin inversion as in the proof of Proposition 5. □

Remark 7. *In the boundary case $l = m - 1/\alpha$, the behavior of $E_{\alpha,m,m-1/\alpha}(-x)$ at infinity, which has different speed and a more complicated constant, will be obtained with the help of the fractional Fréchet distribution—see Remark 8 (c) below.*

We end this paragraph with the following conjecture which is natural in view of Proposition 6. We know by Theorem 2 resp. Proposition 4 that this conjecture is true for the cases $l = m - 1$ and $m = 1$.

Conjecture 4. *For every $\alpha \in (0,1], m > 0, l > m - 1/\alpha$ and $x \geq 0$, one has*

$$\frac{1}{1 + \frac{\Gamma(1+\alpha(l-m))}{\Gamma(1+\alpha(l-m+1))}\, x} \leq E_{\alpha,m,l}(-x) \leq \frac{1}{1 + \frac{\Gamma(1+\alpha l)}{\Gamma(1+\alpha(1+l))}\, x}.$$

4.2. The Fréchet Case

In [3], a fractional Fréchet distribution function with parameters $\alpha \in [0,1]$ and $\lambda, \rho > 0$ is defined as the unique distribution function $F^{\mathbf{F}}_{\alpha,\lambda,\rho}$ on $(0,\infty)$ solving the fractional differential equation

$$D^{\alpha}_{-} F(x) = \lambda x^{-\rho-\alpha} F(x)$$

where D^{α}_{-} denotes a regressive Liouville fractional derivative on $(0,\infty)$. The case $\alpha = 1$ corresponds to the standard Fréchet distribution. In [3], it is shown that this distribution function exists and is given by

$$F^{\mathbf{F}}_{\alpha,\lambda,\rho}(x) = E_{\alpha,\frac{\rho}{\alpha},\frac{\rho-1}{\alpha}}(-\lambda x^{-\rho})$$

for every $x \geq 0$—see the formula following (3.4) in [3]. In particular, the density $f^{\mathbf{F}}_{\alpha,\lambda,\rho}$ is real-analytic on $(0,\infty)$ and has the following asymptotic behavior at infinity:

$$f^{\mathbf{F}}_{\alpha,\lambda,\rho}(x) \sim \left(\frac{\lambda \Gamma(\rho+1)}{\Gamma(\rho+\alpha)}\right) x^{-\rho-1} \quad \text{as } x \to \infty.$$

The behavior of the density at zero is less immediate and we will need, as in the above paragraph, the exact expression of the Mellin transform of the random variable $\mathbf{F}_{\alpha,\lambda,\rho}$ with distribution function $F^{\mathbf{F}}_{\alpha,\lambda,\rho}$, whose strip of analyticity is larger than that of $\mathbf{W}_{\alpha,\lambda,\rho}$.

Proposition 7. *The Mellin transform of $\mathbf{F}_{\alpha,\lambda,\rho}$ is*

$$\mathbb{E}\left[\mathbf{F}^{s}_{\alpha,\lambda,\rho}\right] = \left(\frac{\rho^{\alpha}}{\lambda}\right)^{-\frac{s}{\rho}} \Gamma(1 - s\rho^{-1}) \times \frac{[\rho+1;\rho]_s}{[\rho+\alpha;\rho]_s}$$

for every $s \in (-\rho - \alpha, \rho)$. Consequently, one has

$$f^{\mathbf{F}}_{\alpha,\lambda,\rho}(x) \sim \left(\frac{\rho^{\frac{\alpha^2}{\rho}}(\rho+\alpha)}{\lambda^{1+\frac{\alpha}{\rho}}} \Gamma(1+\alpha)\, G(1-\alpha;\rho)\, G(1+\alpha;\rho)\right) x^{\rho+\alpha-1} \quad \text{as } x \to 0.$$

Proof of Proposition 7. The evaluation of the Mellin transform is done as for the fractional Weibull distribution, starting from the expression

$$\mathbb{E}[(\mathbf{Y}_{\alpha,\frac{\rho}{\alpha},\frac{\rho-1}{\alpha}})^s] = \rho^{-s} \times \frac{[1+\rho^{-1};\rho^{-1}]_s}{[1+\alpha\rho^{-1};\rho^{-1}]_s}$$

which is a consequence of (3) and (A5). By Theorem 1.2 in [3] and (A8), we obtain the required formula

$$\mathbb{E}\left[\mathbf{F}^{s}_{\alpha,\lambda,\rho}\right] = \mathbb{E}\left[\left(\frac{\mathbf{L}}{\lambda \mathbf{Y}_{\alpha,\frac{\rho}{\alpha},\frac{\rho-1}{\alpha}}}\right)^{-\frac{s}{\rho}}\right] = \left(\frac{\rho^{\alpha}}{\lambda}\right)^{-\frac{s}{\rho}} \Gamma(1 - s\rho^{-1}) \times \frac{[\rho+1;\rho]_s}{[\rho+\alpha;\rho]_s}.$$

Then the asymptotic behavior of $f^{\mathbf{F}}_{\alpha,\lambda,\rho}(x)$ at zero follows as that of $f^{\mathbf{W}}_{\alpha,\lambda,\rho}(x)$ at infinity, in considering the residue at the first negative pole $s = -(\rho+\alpha)$ which is simple and isolated in the complex plane, applying Theorem 4 (i) in [14] with the same correction as above, and making various simplifications. We omit details. □

Remark 8. *(a) Comparing the Mellin transforms, Propositions 5 and 7 imply the factorization*

$$\mathbf{W}^{-1}_{\alpha,\lambda,\rho} \stackrel{d}{=} \mathbf{F}_{\alpha,\lambda,\rho} \times \mathbf{Z}(\rho+1-\alpha, \rho+\alpha; \rho, \rho+1; \rho). \tag{12}$$

In general, it follows from Theorem 1 that for every $\alpha \in (0,1], m, \lambda > 0$ and $l > m - 1/\alpha$, there exists a positive random variable with distribution function $E_{\alpha,m,l}(-\lambda x^{-\alpha m})$, and which is given by (3), (2) and Theorem 1.2 in [3] as the independent product

$$\mathbf{F}_{\alpha,\lambda,\alpha m} \times (\mathbf{X}_{\alpha,m,l})^{\frac{1}{\alpha m}} \stackrel{d}{=} \mathbf{F}_{\alpha,\lambda,\alpha m} \times \mathbf{Z}(\alpha l + 1, \alpha(m+1); \alpha m, \alpha(l+1) + 1; \alpha m),$$

where the identity in law follows from (A8). In this respect, the fractional Fréchet distributions can be viewed as the "ground state" distributions associated with the Kilbas-Saigo functions $E_{\alpha,m,l}$, in the boundary case $l = m - 1/\alpha$.

(b) As above, the Mellin transform simplifies for $\alpha = 0, 1$: we get

$$\mathbb{E}[\mathbf{F}_{0,\lambda,\rho}^s] = \lambda^{\frac{s}{\rho}} \Gamma(1 + s\rho^{-1})\Gamma(1 - s\rho^{-1}) \quad \text{and} \quad \mathbb{E}[\mathbf{F}_{1,\lambda,\rho}^s] = \left(\frac{\lambda}{\rho}\right)^{\frac{s}{\rho}} \Gamma(1 - s\rho^{-1}),$$

in accordance with the scaling property $\mathbf{F}_{\alpha,\lambda,\rho} \stackrel{d}{=} \lambda^{1/\rho} \mathbf{F}_{\alpha,1,\rho}$ and the identities given after the statement of Theorem 1.2 in [3]. The Mellin transform also takes a simpler form in the same other situations as above.

- For $\rho = \alpha$, with

$$\mathbb{E}[\mathbf{F}_{\alpha,\lambda,\alpha}^s] = \lambda^{\frac{s}{\alpha}} \frac{\Gamma(\alpha)\Gamma(1 + s\alpha^{-1})\Gamma(1 - s\alpha^{-1})}{\Gamma(\alpha + s)}.$$

This yields the identity $\mathbf{F}_{\alpha,\lambda,\alpha} \stackrel{d}{=} \lambda^{\frac{1}{\alpha}} (\mathbf{Z}_{\alpha}^{-1})^{(\alpha)} \times \mathbf{L}^{-\frac{1}{\alpha}}$, which was discussed for $\lambda = 1$ in the introduction of [3] as the solution to (1.4) therein. This is also in accordance with Remark 3.3 (c) in [3], since

$$(T_{\alpha}^{(1)})^{\frac{1}{\alpha}} \stackrel{d}{=} ((\mathbf{Z}_{\alpha}^{-\alpha})^{(1)})^{\frac{1}{\alpha}} \stackrel{d}{=} (\mathbf{Z}_{\alpha}^{-1})^{(\alpha)}.$$

Notice that the constant appearing in the asymptotic behavior of the density at zero is also simpler: one finds

$$f_{\alpha,\lambda,\alpha}^{\mathbf{F}}(x) \sim \left(\frac{2\alpha \, \Gamma(1+\alpha)}{\lambda^2 \, \Gamma(1-\alpha)}\right) x^{2\alpha - 1} \quad \text{as } x \to 0. \tag{13}$$

- For $\rho = 1 - \alpha$, with

$$\mathbb{E}[\mathbf{F}_{1-\rho,\lambda,\rho}^s] = \left(\frac{\lambda}{\rho}\right)^{\frac{s}{\rho}} \Gamma(1 - s\rho^{-1})\Gamma(1 + s) \quad \text{and} \quad \mathbf{F}_{1-\rho,\lambda,\rho} \stackrel{d}{=} \left(\frac{\lambda}{\rho}\right)^{\frac{1}{\rho}} \mathbf{L}^{-\frac{1}{\rho}} \times \mathbf{L}.$$

Here, the density converges at zero to a simple constant: one finds

$$f_{1-\rho,\lambda,\rho}^{\mathbf{F}}(x) \to \left(\frac{\rho}{\lambda}\right)^{\frac{1}{\rho}} \Gamma(1 + \rho^{-1}) \quad \text{as } x \to 0.$$

(c) Integrating the density and using $\mathbb{P}[\mathbf{F}_{\alpha,\lambda,\rho} \leq x] = E_{\alpha,\frac{\rho}{\alpha},\frac{\rho-1}{\alpha}}(-\lambda x^{-\rho})$, we obtain the following asymptotic behavior at infinity for any $\alpha \in (0,1]$ and $m > 0$, which is more involved than that of Proposition 6:

$$E_{\alpha,m,m-\frac{1}{\alpha}}(-x) \sim (\alpha m)^{\frac{\alpha}{m}} \Gamma(1+\alpha) \, G(1-\alpha; \alpha m) \, G(1+\alpha; \alpha m) \, x^{-1-\frac{1}{m}} \quad \text{as } x \to \infty.$$

For $m = 1$, this behavior matches the first term in the full asymptotic expansion

$$E_{\alpha,1,1-\frac{1}{\alpha}}(-x) = \Gamma(\alpha) E_{\alpha,\alpha}(-x) \sim \Gamma(\alpha) \sum_{n \geq 1} \frac{(-1)^n}{\Gamma(-\alpha n) \, x^{n+1}}.$$

As for $E_{\alpha,m,m-1}(-x)$, a full asymptotic expansion of $E_{\alpha,m,m-\frac{1}{\alpha}}(-x)$ at infinity seems difficult to obtain for all values of m.

5. Some Complements on the Le Roy Function

In this section, we show some miscellaneous results on the Le Roy function

$$\mathcal{L}_\alpha(x) = \sum_{n \geq 0} \frac{x^n}{(n!)^\alpha}, \qquad \alpha > 0, \; x \in \mathbb{R}.$$

In [3], this function played a role in the construction of a fractional Gumbel distribution —see Theorem 1.3 therein. The Le Roy function, which has been much less studied than the classical Mittag–Leffler function, can be viewed as an alternative generalization of the exponential function. See also the recent paper [17] for a further generalization related to the Mittag–Leffler function. Throughout, we discard the explicit case $\mathcal{L}_1(x) = E_1(x) = e^x$.

We begin with the asymptotic behavior at infinity. Le Roy's original result—see [6] p. 263—reads

$$\mathcal{L}_\alpha(x) \sim \frac{(2\pi)^{\frac{1-\alpha}{2}}}{\sqrt{\alpha}} x^{\frac{1-\alpha}{2}} e^{\alpha x^{\frac{1}{\alpha}}} \qquad \text{as } x \to \infty,$$

and is obtained by a variation of Laplace's method. An extension of this asymptotic behavior has been given in [18] for the so-called Mittag–Leffler functions of Le Roy type. Laplace's method can also be used to solve Exercise 8.8.4 in [19], which states

$$\mathcal{L}_\alpha(-x) = \frac{2(2\pi)^{\frac{1-\alpha}{2}}}{\sqrt{\alpha}} x^{\frac{1-\alpha}{2\alpha}} e^{\alpha \cos(\pi/\alpha) x^{\frac{1}{\alpha}}} \left(\sin\left(\pi/2\alpha + \alpha \sin(\pi/\alpha) x^{\frac{1}{\alpha}}\right) + O(x^{-\frac{1}{\alpha}}) \right) \qquad (14)$$

for $\alpha \geq 2$ and

$$\mathcal{L}_\alpha(-x) \sim \frac{1}{\alpha^\alpha \, \Gamma(1-\alpha) \, x \, (\log x)^\alpha} \qquad (15)$$

for $\alpha \in (1,2)$, as $x \to \infty$. The following estimate, which seems to have passed unnoticed in the literature, completes the picture.

Proposition 8. *For every $\alpha \in (0,1)$, one has*

$$\mathcal{L}_\alpha(-x) \sim \frac{1}{\Gamma(1-\alpha) \, x \, (\log x)^\alpha} \qquad \text{as } x \to \infty.$$

Proof of Proposition 8. In the proof of Theorem 1.3 in [3] it is shown that

$$\mathcal{L}_\alpha(-x) = \mathbb{P}[\mathbf{L} > x \mathbf{L}_\alpha] = \int_0^\infty e^{-xt} f_\alpha(t) \, dt$$

where

$$\mathbf{L}_\alpha \stackrel{d}{=} \int_0^\infty e^{-\sigma_t^{(\alpha)}} dt$$

has density f_α on $(0, \infty)$ and Mellin transform

$$\mathbb{E}[\mathbf{L}_\alpha^s] = \Gamma(1+s)^{1-\alpha}, \qquad s > -1.$$

In particular, using the notation in [20], we have $f_\alpha = e_{1-\alpha}$, and Theorem 2.4 therein implies

$$f_\alpha(x) \sim \frac{1}{\Gamma(1-\alpha) \, (-\log x)^\alpha} \qquad \text{as } x \to 0. \qquad (16)$$

Plugging this estimate into the above expression of $\mathcal{L}_\alpha(-x)$, we conclude the proof by a direct integration. □

Remark 9. (a) The estimate (16) also gives the asymptotic behavior, at the right end of the support, of the density of the fractional Gumbel random variable $\mathbf{G}_{\alpha,\lambda}$ which is defined in Theorem 1.3 of [3]. Indeed, by the definition and multiplicative convolution the density of $e^{\lambda \mathbf{G}_{\alpha,\lambda}}$ on $(0,\infty)$ writes

$$\int_0^\infty e^{-xy} y f_\alpha(y)\, dy \sim \frac{1}{\Gamma(1-\alpha)\, x^2\, (\log x)^\alpha} \qquad \text{as } x \to \infty,$$

where the estimate follows from (16) as in the proof of Proposition 8. A change of variable implies then

$$f_{\alpha,\lambda}^{\mathbf{G}}(x) \sim \left(\frac{\lambda^{1-\alpha}}{\Gamma(1-\alpha)}\right) x^{-\alpha}\, e^{-\lambda x} \qquad \text{as } x \to \infty.$$

Notice that at the left end of the support, there is a convergent series representation which is given by Corollary 3.6 in [3].

(b) In the case $\alpha = 2$, one has $\mathcal{L}_2(x) = I_0(2\sqrt{x})$ and $\mathcal{L}_2(-x) = J_0(2\sqrt{x})$ for all $x \geq 0$, where I_0 and J_0 are the classical Bessel functions with index 0. In particular, a full asymptotic expansion for $\mathcal{L}_2$ at both ends of the support is available, to be deduced e.g., from (4.8.5) and (4.12.7) in [13]. These expansions also exist when α is an integer since $\mathcal{L}_\alpha$ is then a generalized Wright function—see Chapter F.2.3 in [2] and the original articles by Wright quoted therein. The case when α is not an integer does not seem to have been investigated, and might be technical in the absence of a true Mellin-Barnes representation.

Our next result characterizes the connection between the entire function $\mathcal{L}_\alpha(z)$ and random variables. Recall that a function $f : \mathbb{C} \to \mathbb{C}$ which is holomorphic in a neighborhood Ω of the origin is a moment generating function (MGF) if there exists a real random variable X such that

$$f(z) = \mathbb{E}\left[e^{zX}\right], \qquad z \in \Omega.$$

In particular, it is clear that $\mathcal{L}_0$ is the MGF of the exponential law $\mathbf{L}$ and $\mathcal{L}_1$ that of the constant variable $\mathbf{1}$. The following provides a characterization.

Proposition 9. *The function $\mathcal{L}_\alpha(z)$ is the MGF of a real random variable if and only if $\alpha \leq 1$. In this case, one has*

$$\mathcal{L}_\alpha(z) = \mathbb{E}\left[e^{z\mathbf{L}_\alpha}\right], \qquad z \in \mathbb{C}.$$

Proof of Proposition 9. The if part is a direct consequence of the proof of Proposition 8. On the other hand, the estimates (14) and (15) show that $\mathcal{L}_\alpha(z)$ takes negative values on $\mathbb{R}^-$, so that it cannot be the moment generating function of a real random variable, when $\alpha > 1$. This completes the proof. □

Observe that since $\mathbf{L}_\alpha$ is non-negative, the above result also shows $\mathcal{L}_\alpha(-x)$ is CM on $(0,\infty)$ if and only if $\alpha \leq 1$, echoing Pollard's aforementioned classical result for the Mittag–Leffler $E_\alpha(-x)$. One can ask whether there are further complete monotonicity properties for $\mathcal{L}_\alpha$, as in [21] for E_α. Our last result for the Le Roy function is a monotonicity property which is akin to Proposition 1.

Proposition 10. *The mapping $\alpha \mapsto \mathcal{L}_\alpha(x)$ is non-increasing on $[0,1]$ for every $x \in \mathbb{R}$.*

Proof of Proposition 10. The fact that $\alpha \mapsto \mathcal{L}_\alpha(x)$ decreases on $\mathbb{R}^+$ is obvious for $x \geq 0$, by the definition of $\mathcal{L}_\alpha$. To show the property on $[0,1]$ for $x < 0$, we will use a convex ordering argument. More precisely, the Malmsten Formula (A3) and the Lévy–Khintchine formula show that for every $t \in [0,1]$, the random variable $\mathbf{G}_{1-t} = \log \mathbf{L}_{1-t}$ is the marginal at time t of a real Lévy process, since $\mathbb{E}[e^{iz\mathbf{G}_{1-t}}] = \Gamma(1+iz)^t = e^{t\psi(z)}$ for every $z \in \mathbb{R}$, with

$$\psi(z) = -\gamma i z + \int_{-\infty}^0 (e^{izx} - 1 - izx)\, \frac{dx}{|x|(e^{|x|} - 1)}.$$

This is actually well known—see Example E in [22]. By independence and stationarity of the increments of a Lévy process, we deduce that there exists a multiplicative martingale $\{M_t, t \in [0,1]\}$ such that $M_t \stackrel{d}{=} \mathbf{L}_{1-t}$ for every $t \in [0,1]$. Jensen's inequality implies

$$\mathbf{L}_\beta \prec_{cx} \mathbf{L}_\alpha$$

for every $0 \leq \alpha \leq \beta \leq 1$. Applying the definition of convex ordering to the function $\varphi(x) = e^x$, we get $\mathcal{L}_\beta(x) \leq \mathcal{L}_\alpha(x)$ for every $x < 0$ and $0 \leq \alpha \leq \beta \leq 1$, as required. □

Remark 10. *(a) In the terminology of [23], the family $\{\mathbf{L}_{1-\alpha}, \alpha \in [0,1]\}$ is a peacock, whose associated multiplicative martingale is completely explicit. We refer to [23] for numerous examples of explicit peacocks related to exponential functionals of Lévy processes. Observe from Lemma 2 that the family $\{\mathbf{T}(a,b,t), t > 0\}$ is also a peacock.*

(b) Letting $\alpha \to 0$ and $\alpha \to 1$ in Proposition 10 leads to the bounds

$$e^x \leq \mathcal{L}_\beta(x) \leq \mathcal{L}_\alpha(x) \leq \frac{1}{(1-x)_+}$$

for every $x \in \mathbb{R}$ and $0 < \alpha < \beta < 1$. The hyperbolic upper bound is optimal as in Theorem 2 and Proposition 2, because $\mathcal{L}_\alpha(x) - 1 \sim x$ as $x \to 0$. The exponential lower bound is thinner than the order given in Proposition 8. On the other hand, it does not seem that stochastic ordering arguments can help for a uniform estimate involving a logarithmic term.

It is natural to ask if the statement of Proposition 10 is also true for the classical Mittag–Leffler function, and this problem seems still open.

Conjecture 5. *The mapping $\alpha \mapsto E_\alpha(x)$ is non-increasing on $[0,1]$ for every $x \in \mathbb{R}$.*

Numerical simulations suggest a positive answer. It is clear by the definition that $\alpha \mapsto E_\alpha(x)$ is non-increasing for every $x \geq 0$ on $[\alpha_0, \infty)$, where $1 + \alpha_0 = 1.46163...$ is the location of the minimum of the Gamma function on $(0, \infty)$. A direct consequence of Theorem B in [5] is also that

$$\alpha \mapsto E_\alpha(\Gamma(1+\alpha)x)$$

is non-increasing on $[1/2, 1]$ for every $x \in \mathbb{R}$. The constant $\Gamma(1+\alpha)$ appears above because of the convex ordering argument used in [5]. It seems that other kinds of arguments are necessary to study the monotonicity of $\alpha \mapsto E_\alpha(x)$ on $[0,1]$.

We would like to finish this paper with the following related monotonicity result, which relies on a stochastic ordering argument, for the generalized Mittag–Leffler function.

Proposition 11. *For every $\alpha \in [0,1]$ and $x \in \mathbb{R}$, the mapping*

$$\beta \mapsto \Gamma(\beta) E_{\alpha,\beta}(x)$$

is non-increasing on (α, ∞) if $x > 0$ and non-decreasing on (α, ∞) if $x < 0$.

Proof of Proposition 11. By Remark 3.3 (c) in [3], we have the probabilistic representation

$$\Gamma(\beta) E_{\alpha,\beta}(x) = \mathbb{E}\left[e^{x \mathbf{B}^\alpha_{\alpha,\beta-\alpha} \times T^{(1)}_\alpha}\right]$$

for every $\alpha \in [0,1], \beta > \alpha$ and $x \in \mathbb{R}$. Reasoning as in Proposition 3, we see by factorization that it suffices to show that

$$\beta \mapsto \mathbf{B}^\alpha_{\alpha,\beta-\alpha}$$

is non-increasing on (α, ∞) for the usual stochastic order. On the other hand, the density function of the random variable $\mathbf{B}^\alpha_{\alpha,\beta-\alpha}$ is

$$\frac{\Gamma(\beta)}{\Gamma(\alpha+1)\Gamma(\beta-\alpha)}\left(1-x^{\frac{1}{\alpha}}\right)^{\beta-\alpha-1}$$

on $[0,1)$ and its value at zero is by the log-convexity of the Gamma function an increasing function of β. Moreover, the density functions of $\mathbf{B}^\alpha_{\alpha,\beta-\alpha}$ and $\mathbf{B}^\alpha_{\alpha,\beta'-\alpha}$ cross only once for $\beta \neq \beta'$, at

$$\left(1-\left(\frac{\Gamma(\beta)\Gamma(\beta'-\alpha)}{\Gamma(\beta')\Gamma(\beta-\alpha)}\right)^{\frac{1}{\beta'-\beta}}\right)^\alpha \in (0,1).$$

The single intersection property finishes then the argument, as for Proposition 3. □

Author Contributions: Writing—original draft, L.B. and T.S. All authors have read and agreed to the published version of the manuscript.

Funding: This research received no external funding.

Conflicts of Interest: The authors declare no conflict of interest.

Appendix A

In this Appendix we recall some properties of Barnes' double Gamma function $G(z;\delta)$, which are used throughout the paper. For every $\delta > 0$, this function is defined as the unique solution to the functional equation

$$G(z+1;\delta) = \Gamma(z\delta^{-1})G(z;\delta) \tag{A1}$$

with normalization $G(1;\delta) = 1$. The function is holomorphic on $\mathbb{C}$ and admits the following Malmsten type representation

$$G(z;\delta) = \exp \int_0^\infty \left(\frac{1-e^{-zx}}{(1-e^{-x})(1-e^{-\delta x})} - \frac{ze^{-\delta x}}{1-e^{-\delta x}} + (z-1)(\frac{z}{2\delta}-1)e^{-\delta x} - 1\right)\frac{dx}{x} \tag{A2}$$

which is valid for $\Re(z) > 0$—see (5.1) in [24]. Putting (A1) and (A2) together and making some simplifications, we recover the standard Malmsten formula for the Gamma function

$$\Gamma(1+z) = \exp\left\{-\gamma z + \int_{-\infty}^0 (e^{zx}-1-zx)\frac{dx}{|x|(e^{|x|}-1)}\right\} \tag{A3}$$

for every $z > -1$, where γ is Euler's constant. The following Stirling type asymptotic behavior

$$\log G(z;\delta) - \frac{1}{2\delta}(z^2\log z - (\frac{3}{2}+\log\delta)z^2) - (1+\delta)z\log z - Az - B\log z \to C \tag{A4}$$

is valid for $|z| \to \infty$ with $|\arg(z)| < \pi$, for some real constants A, B and C which are given in (4.5) of [24]. There is a second concatenation formula

$$G(z+\delta;\delta) = (2\pi)^{(\delta-1)/2}\delta^{1/2-z}\Gamma(z)G(z;\delta) \tag{A5}$$

which is valid for all $z \in \mathbb{C}$, the right-hand side being understood as an analytic extension when z is a non-positive integer—see (4.6) in [25] and the references therein. Observe that (A1) and (A5) lead readily to the closed formula

$$G(\delta;\delta) = G(1+\delta;\delta) = (2\pi)^{(\delta-1)/2}\delta^{-1/2}. \tag{A6}$$

In this paper, we make an extensive use of the following Pochhammer type symbol

$$[a;\delta]_s = \frac{G(a+s;\delta)}{G(a;\delta)} \qquad (A7)$$

which is well-defined for every $a, \delta > 0$ and $s > -a$. The following formula

$$[a\delta^{-1};\delta^{-1}]_{s\delta^{-1}} = (2\pi)^{s(1/\delta-1)/2} \delta^{s^2/2\delta - s(1+(1-2a)/\delta)/2} [a;\delta]_s \qquad (A8)$$

can be deduced from (4.10) in [25] beware the different normalization for $G(1;\delta)$ therein which becomes irrelevant when considering the Pochhammer type symbol. Notice also that (A5) yields

$$\delta^s [a+\delta;\delta]_s = (a)_s [a;\delta]_s \qquad (A9)$$

with the standard notation

$$(a)_s = \frac{\Gamma(a+s)}{\Gamma(a)}$$

for the usual Pochhammer symbol. Finally, we observe from the double product representation of $G(z,\delta)$—see e.g., (4.4) in [25] that for every $a, \delta > 0$ one has

$$\inf\{s > 0, \; [a;\delta]_{-s} = 0\} = a \qquad (A10)$$

and that this zero is simple and isolated on the complex plane.

References

1. Kilbas, A.A.; Saigo, M. Fractional integral and derivatives of Mittag-Leffler type function. *Dokl. Akad. Nauk Belarus* **1995**, *39*, 22–26.
2. Gorenflo, R.; Kilbas, A.A.; Mainardi, F.; Rogosin, S.V. *Mittag-Leffler Functions, Related Topics and Applications*; Springer: Heidelberg, Germany, 2020.
3. Boudabsa, L.; Simon, T.; Vallois, P. Fractional extreme distributions. *Electron. J. Probab.* **2020**, *25*, 1–20. [CrossRef]
4. de Oliveira, E.C.; Mainardi, F.; Vaz, J. Fractional models of anomalous relaxation based on the Kilbas and Saigo function. *Meccanica* **2014**, *49*, 2049–2060. [CrossRef]
5. Simon T. Comparing Fréchet and positive stable laws. *Electron. J. Probab.* **2014**, *19*, 1–25. [CrossRef]
6. Le Roy, E. Valeurs asymptotiques de certaines séries procédant suivant les puissances entières et positives d'une variable réelle. *Darboux Bull.* **1899**, *24*, 245–268.
7. Dufresne, D. G distributions and the beta-gamma algebra. *Electron. J. Probab.* **2010**, *15*, 2163–2199. [CrossRef]
8. Letemplier, J.; Simon, T. On the law of homogeneous stable functionals. *ESAIM Probab. Stat.* **2019**, *23*, 82–111. [CrossRef]
9. Sato, K. *Lévy Processes and Infinitely Divisible Distributions*; Cambridge University Press: Cambridge, UK, 1999.
10. Titchmarsh, E.C. *The Theory of Functions*; Oxford University Press: Oxford, UK, 1939.
11. Mainardi, F. On some properties of the Mittag-Leffler function $E_\alpha(-t^\alpha)$, completely monotone for $t > 0$ with $0 < \alpha < 1$. *Discret. Cont. Dyn. Syst. Ser. B* **2014**, *19*, 2267–2278.
12. Shaked, M.; Shanthikumar, J.G. *Stochastic Orders and Their Applications*; Springer: New York, NY, USA, 2007.
13. Andrews, G.E.; Askey, R.; Roy, R. *Special Functions*; Cambridge University Press: Cambridge, UK, 1999.
14. Flajolet, P.; Gourdon, X.; Dumas, P. Mellin transforms and asymptotics: Harmonic sums. *Theoret. Comput. Sci.* **1995**, *144*, 3–58. [CrossRef]
15. Jedidi, W.; Simon, T.; Wang, M. Density solutions to a class of integro-differential equations. *J. Math. Anal. Appl.* **2018**, *458*, 134–152. [CrossRef]
16. Kilbas, A.A.; Srivastava, H.M.; Trujillo, J.J. *Theory and Applications of Fractional Differential Equations*; North-Holland: Amsterdam, The Netherlands, 2006.
17. Garrappa, R.; Mainardi, F.; Rogosin, S. On a generalized three-parameter Wright function of Le Roy type. *Frac. Calc. Appl. Anal.* **2017**, *20*, 1196–1215.
18. Gerhold, S. Asymptotics for a variant of the Mittag-Leffler function. *Int. Transf. Spec. Funct.* **2012**, *23*, 397–403. [CrossRef]
19. Olver, F.W.J. *Asymptotics and Special Functions*; Academic Press: New York, NY, USA, 1974.
20. Berg, C.; López, J.-L. Asymptotic behaviour of the Urbanik semigroup. *J. Approx. Theory* **2015**, *195*, 109–121. [CrossRef]
21. Simon, T. Mittag-Leffler functions and complete monotonicity. *Int. Transf. Spec. Funct.* **2015**, *26*, 36–50. [CrossRef]
22. Carmona, P.; Petit, F.; Yor, M. On the distribution and asymptotic results for exponential functionals of Lévy processes. In *Exponential Functionals and Principal Values Related to Brownian Motion*; Biblioteca de la Revista Matematica Iberoamericana: Madrid, Spain, 1997; pp. 73–121.
23. Hirsch, F.; Profeta, C.; Roynette, B.; Yor, M. *Peacocks and Associated Martingales, with Explicit Constructions*; Springer: Milan, Italy, 2011.

24. Billingham, J.; King, A.C. Uniform asymptotic expansions for the Barnes double gamma function. *Ser. A Math. Phys. Eng. Sci.* **1997**, *453*, 1817–1829. [CrossRef]
25. Kuznetsov, A. On extrema of stable processes. *Ann. Probab.* **2011**, *39*, 1027–1060. [CrossRef]

Article

On the Multistage Differential Transformation Method for Analyzing Damping Duffing Oscillator and Its Applications to Plasma Physics

Noufe H. Aljahdaly [1,*,†] **and S. A. El-Tantawy** [2,3,†]

1. Department of Mathematics, Faculty of Sciences and Arts-Rabigh Campus, King Abdulaziz University, Jeddah 21589, Saudi Arabia
2. Center for Physics Research (CPR), Department of Physics, Faculty of Science and Arts, Al-Mikhwah, Al-Baha University, Al-Baha 65431, Saudi Arabia; seltantawy@bu.edu.sa
3. Department of Physics, Faculty of Science, Port Said University, Port Said 42521, Egypt
* Correspondence: nhaljahdaly@kau.edu.sa
† These authors contributed equally to this work.

Citation: Alajhdaly, N.H.; El-Tantawy, S.A. On the Multistage Differential Transformation Method for Analyzing Damping Duffing Oscillator and Its Applications to Plasma Physics. *Mathematics* **2021**, *9*, 432. https://doi.org/10.3390/math9040432

Academic Editor: Francesco Mainardi

Received: 27 December 2020
Accepted: 19 February 2021
Published: 22 February 2021

Publisher's Note: MDPI stays neutral with regard to jurisdictional claims in published maps and institutional affiliations.

Copyright: © 2021 by the authors. Licensee MDPI, Basel, Switzerland. This article is an open access article distributed under the terms and conditions of the Creative Commons Attribution (CC BY) license (https://creativecommons.org/licenses/by/4.0/).

Abstract: The multistage differential transformation method (MSDTM) is used to find an approximate solution to the forced damping Duffing equation (FDDE). In this paper, we prove that the MSDTM can predict the solution in the long domain as compared to differential transformation method (DTM) and more accurately than the modified differential transformation method (MDTM). In addition, the maximum residual errors for DTM and its modification methods (MSDTM and MDTM) are estimated. As a real application to the obtained solution, we investigate the oscillations in a complex unmagnetized plasma. To do that, the fluid govern equations of plasma species is reduced to the modified Korteweg–de Vries–Burgers (mKdVB) equation. After that, by using a suitable transformation, the mKdVB equation is transformed into the forced damping Duffing equation.

Keywords: multistage differential transformation method; Duffing equation; nonlinear damping oscillations

1. Introduction

Mathematical techniques are very important tools in mathematics. Mathematicians have developed many mathematical methods to compute linear or nonlinear differential equations which describe many important phenomena and applications in science [1–7]. The mathematical techniques are classified as algebraic methods, semi-approximate, general analytical, approximate analytical, numerical, or qualitative techniques. The basic concept of approximate analytical techniques such as Adomian decomposition method (ADM), Laplacian decomposition method (LDM), or differential transformation method (DTM) is assuming that the solution is descried by a Taylor expansion form. Indeed, some solutions of equations have well-known Taylor expansions such as exponential function or hyperbolic function. In this case, it is easy to determine the exact solution by a few terms of the Taylor expansion series. Otherwise, the approximate solution will be obtained in the form of few terms of Taylor expansion series. Since Taylor expansion is local convergent about the initial condition, the method can approximate the solution in the neighborhood of the initial point. Thus, the solution is obtained in a very short domain. This feature of ADM, LDM or DTM has been mentioned by some researchers [8–12]. DTM has been improved by dividing the domain into subdomains and modifying the initial point in each subdomain. The other modification is by using the Laplacian transformation and Padé approximate. In Section 2, we describe these modifications in details. However, it is very important to determine the optimal modification of DTM to present fast and accurate techniques.

Some of the most important and famous differential equations whose solutions are related to many natural phenomena, physical concepts, and engineering phenomena are the Duffing equation (including conservative and non-conservative cases), the Helmholtz equation (including conservative and non-conservative cases), and their families [13–24]. Given the importance of the family of Duffing equation, a great effort has been made by many researchers to solve this equation and its family, with a numerical, analytical, or semi-analytical solution according to the type of Duffing equation. Examples of these approximate methods for solving the conservative Duffing equation $(u'' + \beta u(t) + \gamma u^3 = 0)$ include the homotopy perturbation method [25], harmonic balance method [26], energy balance method [27], modified variational approach [28], and coupled homotopy–variational approach [29]. On the other hand, many researchers have tried to find a solution to the damping Duffing equation (DDE), $(u'' + \alpha u' + \beta u(t) + \gamma u^3 = 0)$ [30–35], since it is more closely related to reality than the undamping Duffing equation, which is correct only for idealized isolated systems, i.e., systems in which the frictional force and viscosity are absent. One of the most important approximate methods that has been used and developed to solve many differential equations is DTM, which has been used in solving DDE [11]. Nourazar et al. [11] used the modified DTM to get an approximate solution to the DDE. The authors compared their solution with both the fourth-order Runge–Kutta (RK4) numerical solution and the DTM solution. They found that the DTM solution is suitable only for small time intervals while the MDTM solution is suitable for the whole time domain. In our study, we solve the forced damping Duffing equation (FDDE) $(u'' + \alpha u' + \beta u(t) + \gamma u^3 = F)$ using the multistage differential transformation method (MSDTM) for arbitrary initial conditions. Moreover, we compare the approximate solutions of DTM and MDTM as well as the numerical solution using RK4 in order to determine the optimal technique. Furthermore, the oscillations in complex unmagnetized plasmas are investigated by reducing the fluid govern equations of the plasma species to an evolution equation and then transform this equation to the Duffing-type equation using a suitable transformation.

2. Methodology

This section is devoted to briefly describing DTM and its modifications. Assume the following ordinary differential equation (ODE)

$$P(u, u', u'', ...) = 0, \qquad (1)$$

where $u(t)$ is the solution of this ODE in domain $[t_0, t_N]$, P is a polynomial in terms of u and its derivative, and $u(t_0) = c$.

2.1. Differential Transformation Method (DTM)

Assume that the goal is finding the approximate solution of Equation (1). The main concept of DTM is based on applying the differential transformation $u(t) \Longrightarrow U(k)$ at $t = t_0$ as follows:

$$U(k) = \frac{1}{k!}\left[\frac{d^k u(t)}{dt^k}\right]_{t=t_0}. \qquad (2)$$

The differential inverse transformation $U(k) \Longrightarrow u(t)$ is defined as

$$u(t) = \sum_{k=0}^{\infty} U(k)(t - t_0)^k, \qquad (3)$$

Inserting Equation (2) into Equation (3), $u(t)$ can be approximated in finite number series as follows

$$u_N(t) = \sum_{k=0}^{N} \frac{(t - t_0)^k}{k!}\left[\frac{d^k u(t)}{dt^k}\right]_{t=t_0} = g_N. \qquad (4)$$

Some differential transformation rules are introduced in Table 1.

Table 1. Differential transformation rules.

Original Function	Transformed Function
$u(t) \pm v(t)$	$U(t) \pm V(t)$
$cu(t)$	$cU(t)$ (c is constant)
$\frac{du(t)}{dt}$	$(k+1)U(k+1)$
$\frac{d^n u(t)}{dt^n}$	$\frac{(k+n)!}{k!}U(k+n)$
$u(t)v(t)$	$\sum_{m=0}^{k} U(m)V(k-m)$

It is well known that, since the DTM based on Taylor expansion, the approximate solution if it is locally analytic converges to the exact solution with the following approximated error

$$|u(t) - g_N(t)| \leq \frac{M}{(N+1)!}|t-t_0|^{N+1},$$

where $|u(t_N)| \leq M$.

It is obvious that the error increases when $|t-t_0|$ incenses for fixed term N. Note that DTM gives accurate results only in a small domain around the initial point. Therefore, to obtain good results, some modifications to this method must be introduced. There are some attempts to improve this method, such as the modified differential transformation method (MDTM) and the multistage differential transformation method (MSDTM).

2.2. Modified Differential Transformation Method (MDTM)

MDTM is presented in [11]. The idea is described simply by applying the Laplacian transformation into Equation (3) $\mathcal{L}u(t)$. We obtain the polynomial in terms of $1/t^s$. Next, we use Padé approximate, [3/3] or [4/4], and then apply the Laplacian inverse transform. The method is improved and able to approximate the solution in long domain.

Definition 1. *We say the function $g(t)$ is Padé approximate of order $[m/n]$ for function $u(t)$ if*

$$g(t) = \frac{a_0 + a_1 t + a_2 t^2 + \ldots + a_m t^m}{1 + b_0 + b_1 t + b_2 t^2 + \ldots + b_n t^n},$$

where $u(0) = g(0), u'(0) = g'(0), u''(0) = g''(0), \ldots, u^{(m+n)}(0) = g^{(m+n)}(0)$. The constants $a_i, i = 1, 2, \ldots, m$ and $b_j, j = 1, 2, \ldots, n$ are uniquely determined. The Padé approximate is unique for given n and m.

2.3. Multistage Differential Transformation Method (MSDTM)

The other modification is MSDTM. The main concept is dividing the domain into subdomains $[t_i, t_{i+1}] = D_i$ and applying DTM in each subdomain with the initial condition at t_i to approximate $u(t)$ at the subdomain D_i.

2.4. Example

In this section, we apply DTM and its modifications to one of the most famous equations in dynamic systems which is called the Duffing oscillator or the Duffing equation. It is known that the Duffing equation has many formulas, and, in this paper, we restrict our attention to investigating the forced damping Duffing equation (FDDE) $(u'' + \alpha u' + \beta u(t) + \gamma u^3 = F)$. This equation is non-integrable and does not have an exact solution except under certain conditions on its coefficients (α, β, γ). Therefore, the approximate solution to the following FDDE for arbitrary values of its coefficients (α, β, γ) and for arbitrary initial conditions is obtained:

$$\begin{cases} u'' + \alpha u' + \beta u(t) + \gamma u^3 = F, \\ u(0) = u_0 \;\&\; u'= u'_0. \end{cases} \quad (5)$$

In the following analysis, we give some numerical examples to solve the initial value problem (i.v.p.) (5) using the aforementioned methods and examine the accuracy of these methods for calculating the residual error for each method compared to the RK4.

2.4.1. MDTM

Firstly, let us use the same values of $(\alpha, \beta, \gamma, F) = (0.5, 25, 25, 0)$ as mentioned by [11] with the initial conditions $u(0) = 0.1$ and $u'(0) = 0$. Note that the solution of the i.v.p. (5) for unforced ($F = 0$) using MDTM is introduced in details in [11]. In the case of using Padé approximate of [3/3], we have

$$u(t) = 0.00194 + 0.000238 e^{-0.25t}(411\cos(5.068t) + 20.273\sin(5.068t)), \tag{6}$$

In the case of using Padé approximate of [4/4], the solution is approximated as

$$u(t) = Ae^{(-0.60107-15.0816i)t} + Be^{(-0.60107+15.0816i)t} + Ce^{(-0.24894-5.0125i)t} + De^{(-0.24894+5.0125i)t}, \tag{7}$$

with

$$A = 1.6932 \times 10^{-5} - 1.3567 \times 10^{-4} i,$$
$$B = 1.6932 \times 10^{-5} + 1.3567 \times 10^{-4} i,$$
$$C = 4.9983 \times 10^{-2} - 2.5633 \times 10^{-3} i,$$
$$D = 4.9983 \times 10^{-2} + 2.5633 \times 10^{-3} i.$$

In the second example, we use the values $(\alpha, \beta, \gamma, F) = (1, 20, 2, 0)$ and with initial condition $u(0) = -0.2$ and $u'(0) = 2$ and apply Padé approximate of [3/3] and [4/4]. The solution in the case of using Padé approximate of [3/3] reads

$$u(t) = 0.003101 \exp(-6.3493t) + \exp(-0.52098t)(0.434516\sin(4.4046t) - 0.203101\cos(4.4046t)), \tag{8}$$

and for [4/4] reads

$$u(t) = Ae^{(-2.0169+12.6572i)t} + Be^{(-2.0169-12.6572i)t} + Ce^{(-0.4965+4.4826i)t} + De^{(-0.4965-4.4826i)t}, \tag{9}$$

with

$$A = 2.265 \times 10^{-4}, -4.807 \times 10^{-5} i,$$
$$B = 2.265 \times 10^{-4}, +4.807 \times 10^{-5} i,$$
$$C = -0.100226 - 0.21195i,$$
$$D = -0.100226 + 0.21195i.$$

2.4.2. MSDTM

In this work, we focus our attention to solve the i.v.p. (5) for arbitrary initial conditions using MSDTM by dividing the domain $[0, 20]$ to subdomains with time step 10^{-2} and apply DTM with $k = 3$ to find u^i as follows:

$$u^i_{k+1} = \frac{k!}{(k+1)!} y^i_k, \tag{10}$$

$$y^i_{k+1} = \frac{k!}{(k+1)!}\left[-\beta u^i_k - \alpha y^i_k - \gamma \sum_{r=0}^{k}\left(\sum_{l=0}^{r}(u^i_l u^i_{r-l})\right)u^i_{k-r} + F\right], \qquad (11)$$

where $y = u'$.

To check the accuracy of the aforementioned methods as compared the RK4 solution, we use the following error formula for the maximum residual error

$$L_D(\text{method}) = \max_{t_0 \leq t \leq t_N} |\text{RK}(t) - u(t)|.$$

Figures 1 and 2 demonstrate the approximate solutions to the i.v.p. (5) for different values of the coefficients $(\alpha, \beta, \gamma, F)$. The results show that the MDTM4 and MSDTM are better approximations than MDTM3. Moreover, the comparison of the maximum residual errors for the approximate solutions shown in Table 2 proves that the accurate method is MSDTM. Aljahdaly [10] proved that the MSDTM and RK4 techniques have the same accuracy, but MSDTM is faster than RK4. Thus, we conclude that MSDTM is a fast, accurate, and reliable method for many differential equations in physics and in different branches of science. In the next section, a new application to the damping Duffing equation in plasma physics is introduced.

Table 2. The error $L_D(\text{methods})$ is estimated for different values of the coefficients $(\alpha, \beta, \gamma, u_0, u'_0)$.

$(\alpha, \beta, \gamma, u_0, u'_0)$	Time Range	$L_D(\text{MDTM3})$	$L_D(\text{MDTM4})$	$L_D(\text{MSDTM})$
$(0.5, 25, 25, 0.1, 0)$	$0 \leq t \leq 20$	1.19631×10^{-2}	1.67636×10^{-3}	4.81974×10^{-4}
$(1, 20, 2, -0.2, 2)$	$0 \leq t \leq 6$	1.83182×10^{-2}	7.99595×10^{-3}	1.96895×10^{-5}

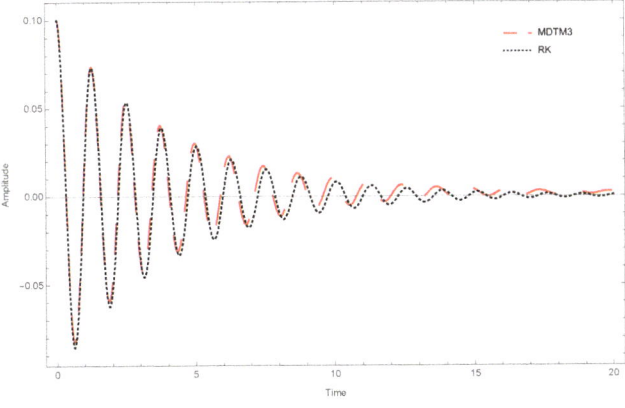

(**a**) Comparing RK method and MDTM3

Figure 1. *Cont.*

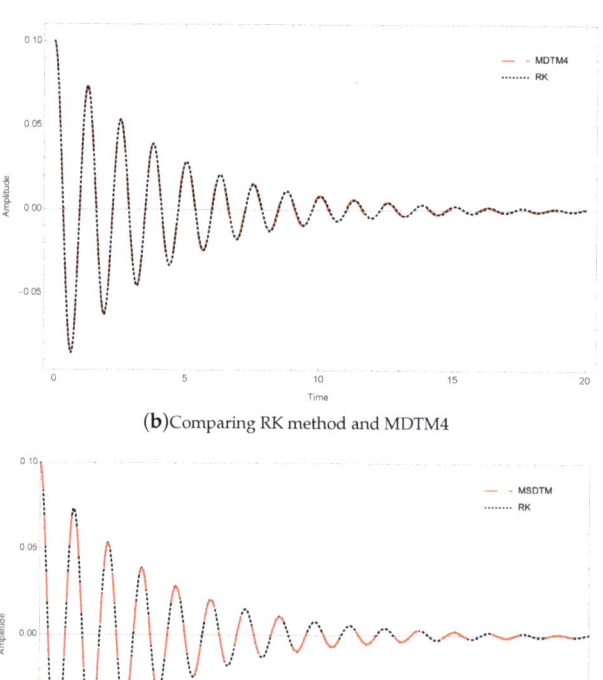

(b) Comparing RK method and MDTM4

(c) Comparing RK method and MSDTM

Figure 1. Plot the solution $u(t)$ for $\alpha = 0.5, \beta = \gamma = 25, F = 0, u(0) = 0.1, u'(0) = 0$.

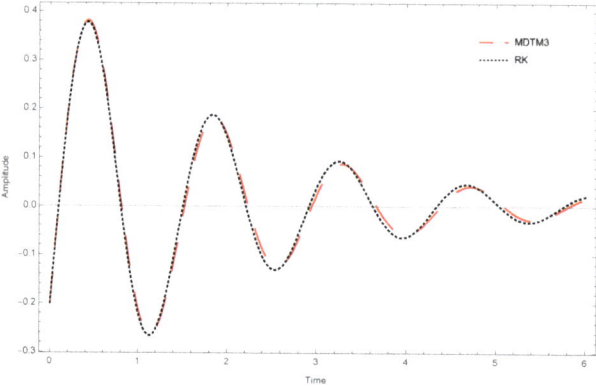

(a) Comparing RK method and MDTM3

Figure 2. *Cont.*

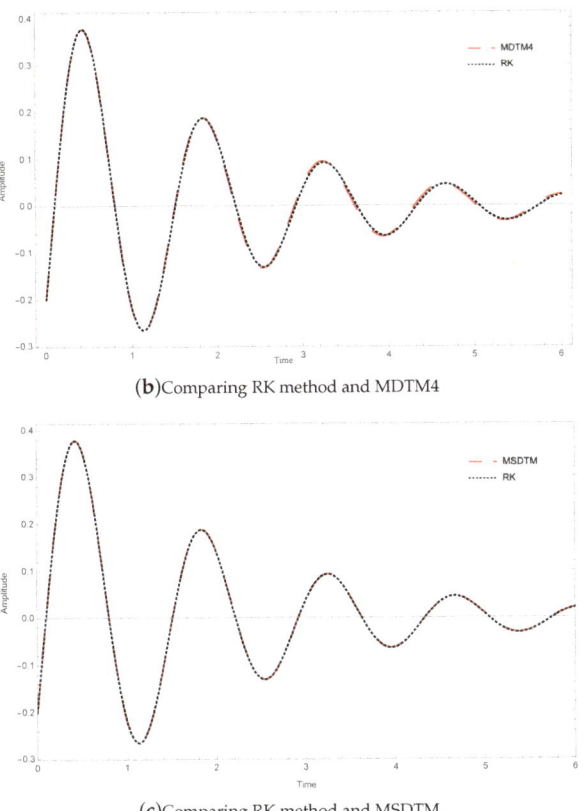

(**b**) Comparing RK method and MDTM4

(**c**) Comparing RK method and MSDTM

Figure 2. Plot the solution $u(t)$ for $\alpha = 1, \beta = 20, \gamma = 2, F = 0, u(0) = -0.2, u'(0) = 2$.

3. Application in Plasma Physics

Let us consider the propagation of nonlinear structures in a complex unmagnetized plasma composed of inertial positive ions (with subscript "i") and two different types of electrons (with subscripts "l" and "h" for the lower and higher electron temperatures, respectively) that follow the kappa distribution in addition to static dust grains with negative charge (with subscript "d") [36]. Accordingly, the neutrality condition reads $n_l^{(0)} + n_h^{(0)} + z_d n_d^{(0)} = n_i^{(0)}$, where $n_j^{(0)}$ represents the unperturbed number density of species j ($j \equiv l, h, d, i$) and z_d gives the number of electrons residing on the surface of the dust grains. The dynamics of the nonlinear structures whose phase speed is much larger than the ion thermal speed but smaller than the electron thermal speed are governed by the following dimensionless fluid continuity, momentum, and Poisson's equations, respectively,

$$\begin{cases} \partial_t n_i + \partial_x(n_i u_i) = 0, \\ \partial_t u_i + u_i \partial_x u_i + \partial_x \phi = \eta \partial_x^2 u_i, \\ \partial_x^2 \phi - n_e + n_i - \mu_d = 0, \end{cases} \quad (12)$$

where the number density of the electrons in kappa distribution is given by

$$n_e = n_l + n_h = \mu_l \left(1 - \frac{\sigma_l \phi}{R_l}\right)^{S_l} + \mu_h \left(1 - \frac{\sigma_h \phi}{R_h}\right)^{S_h}$$
$$\equiv \Gamma_0 + \Gamma_1 \phi + \Gamma_2 \phi^2 + \Gamma_3 \phi^3 + \cdots, \quad (13)$$

with

$$\Gamma_0 = \mu_l + \mu_h,$$
$$\Gamma_1 = -\left[\frac{S_l \mu_l \sigma_l}{R_l} + \frac{S_h \mu_h \sigma_h}{R_h}\right],$$
$$\Gamma_2 = \left[\frac{S_l \mu_l \sigma_l^2 (S_l - 1)}{2R_l^2} + \frac{S_h \mu_h \sigma_h^2 (S_h - 1)}{2R_h^2}\right],$$
$$\Gamma_3 = -\left[\frac{S_l \mu_l \sigma_l^3 (S_l - 1)(S_l - 2)}{6R_l^3} + \frac{S_h \mu_h \sigma_h^3 (S_h - 1)(S_h - 2)}{6R_h^3}\right],$$
$$S_l = \left(-\kappa_l + \frac{1}{2}\right) \& S_h = \left(-\kappa_h + \frac{1}{2}\right),$$
$$R_l = \left(\kappa_l - \frac{3}{2}\right) \& R_h = \left(\kappa_h - \frac{3}{2}\right).$$

where $n_i/n_l/n_h$ is the normalized number density of the positive ions/low temperature electrons/high temperature electrons, u_i refers to the normalized velocity of the positive ions, ϕ is the normalized electrostatic potential, η represents the normalized coefficient of ionic kinematic viscosity, $\sigma_{l,h} = T_{eff}/T_{l,h}$ is the electron temperature ratio, the effective electron temperature is $T_{eff} = n_e^{(0)} T_l T_h / \left(n_l^{(0)} T_h + n_h^{(0)} T_l\right)$, $n_e^{(0)} \equiv \left(n_l^{(0)} + n_h^{(0)}\right)$ is the total unperturbed electrons density, $\mu_d = z_d n_d^{(0)} / n_i^{(0)}$ is the dust concentration, $\mu_l = n_l^{(0)}/n_i^{(0)}$ is the concentration of low electron temperature, $\mu_h = n_h^{(0)}/n_i^{(0)}$ is the concentration of high electron temperature, and $\kappa_{l,h} (> 3/2)$ is the kappa index parameter [36].

To model and analyze the nonlinear structures that can propagate in the present plasma model, the reductive perturbation method (RPM) [37,38] is used to reduce the basic set of fluid Equations (12) and (13) to an evolution equation. According to this method, the independent variables (x,t,η) can be stretched as follows:

$$X = \varepsilon\left(x - v_{ph} t\right), T = \varepsilon^3 t \,\&\, \eta = \varepsilon \tilde{\eta}, \quad (14)$$

where ε is a real and small parameter ($\varepsilon \ll 1$) that measures the strength of the nonlinearity and v_{ph} represents the normalized phase velocity, which is scaled by C_i. In addition, the dependent quantities $\Pi(x,t) \equiv (n_i, u_i, \phi)$ are expanded as follows:

$$\Pi(x,t) = \Pi^{(0)} + \sum_{s=1}^{\infty} \varepsilon^s \Pi^{(s)}(X,T), \quad (15)$$

where $\Pi^{(0)} \equiv [1,0,0]^T$, $\Pi^{(s)}(X,T) \equiv \left[n_i^{(s)}, u_i^{(s)}, \phi^{(s)}\right]^T$, and T gives the matrix transpose.

Inserting both stretching (14) and expansion (15) into the basic set of fluid Equations (12) and (13), we get a system of reduced equations in different powers of ε. From the lowest-order of ε, i.e., $O(\varepsilon)$, the values of the first-order quantities $\left(n_i^{(1)}, u_i^{(1)}\right)$ and the phase velocity v_{ph} can be obtained as

$$u_i^{(1)} = v_{ph} n_i^{(1)} = \frac{1}{v_{ph}} \phi^{(1)},$$
$$v_{ph} = \frac{1}{\sqrt{\Gamma_1}}. \quad (16)$$

The solution of next-order of ε, i.e., $O(\varepsilon^2)$, gives the values of the second-order quantities $\left(n_i^{(2)}, u_i^{(2)}\right)$

$$n_i^{(2)} = \frac{1}{v_{ph}^4}\left(\frac{3}{2}\phi^{(1)2} + v_{ph}^2\phi^{(2)}\right),$$
$$u_i^{(2)} = \frac{1}{v_{ph}^3}\left(\frac{1}{2}\phi^{(1)2} + v_{ph}^2\phi^{(2)}\right), \tag{17}$$

and the Poisson's equation gives

$$A\phi^{(1)2} + A_c\phi^{(2)} = 0, \tag{18}$$

where $A = \left[3/\left(2v_{ph}^4\right) - \Gamma_2\right] = 0$ at the critical value of low electron temperature concentration $\mu_l = \mu_{lc}$ and the coefficient $A_c = \left(1/v_{ph}^2 - \Gamma_1\right)$ represents the compatibility condition, i.e., $A_c = 0$.

From the next-order of ε, i.e., $O(\varepsilon^3)$, we get

$$\partial_T n_i^{(1)} + \partial_X\left(n_i^{(1)}u_i^{(2)}\right) + \partial_X\left(n_i^{(2)}u_i^{(1)}\right) - v_{ph}\partial_X n_i^{(3)} + \partial_X u_i^{(3)} = 0, \tag{19}$$

$$\partial_T u_i^{(1)} + \partial_X\left(u_i^{(1)}u_i^{(2)}\right) + \partial_X\left(n_i^{(2)}u_i^{(1)}\right) - v_{ph}\partial_X u_i^{(3)} + \partial_X \phi^{(3)} - \tilde{\eta}\partial_X^2 u_i^{(1)} = 0, \tag{20}$$

and the Poisson's equation gives

$$\partial_X\left(n_i^{(3)} - \Gamma_3\phi^{(1)3} - 2\Gamma_2\phi^{(1)}\phi^{(2)} - \Gamma_1\phi^{(3)} + \partial_X^2\phi^{(1)}\right) = 0. \tag{21}$$

By solving Equations (19)–(21) with the help of Equations (16) and (17), we finally get the mKdVB equation

$$\partial_T \varphi + P_1\varphi^2\partial_X\varphi + P_2\partial_X^3\varphi = P_3\partial_X^2\varphi, \tag{22}$$

with

$$P_1 = \left(15 - 6\Gamma_3 v_{ph}^6\right)/\left(4v_{ph}^3\right),$$
$$P_2 = \frac{v_{ph}^3}{2} \; \& \; P_3 = \frac{\tilde{\eta}}{2},$$

where $\varphi \equiv \phi^{(1)}$.

It is known that Equation (22) supports the shock solution due to the presence of ion kinematic viscosity. However, in this paper, we want to investigate the damping oscillations in the present model. Accordingly, the transformation $\varphi(X,T) = \Phi(\zeta)$, where $\zeta = X + v_f T$, is used to transform Equation (22) into the FDDE as follows:

$$\varphi'' + \alpha\varphi' + \beta\varphi + \gamma\varphi^3 = F, \tag{23}$$

where $\alpha = -P_3/P_2$, $\beta = v_f/P_2$, $\gamma = P_1/(3P_2)$, and F is the constant of integration.

Let us now investigate the effect of typical complex plasma parameters, namely $(\kappa_l, \sigma_h, \mu_l, \mu_h) = (3, 0.1, \mu_c, 0.4)$, and different values for $(\sigma_l, \kappa_h, \tilde{\eta})$ on the profile of plasma oscillations. Some plasma data are used as an example for investigating the solution of MSDTM, as shown in Figure 3. It is clear from the results in Figure 3 that the enhancement of the viscosity parameter $\tilde{\eta}$ leads to an increase in the number of oscillations and decreasing the time of periodicity. Note that the effect σ_l has on the profile of oscillation is the same as its effect on $\tilde{\eta}$ while κ_h has the opposite effect, in which the number of oscillations decreases and the time periodicity increases with the enhancement of κ_h.

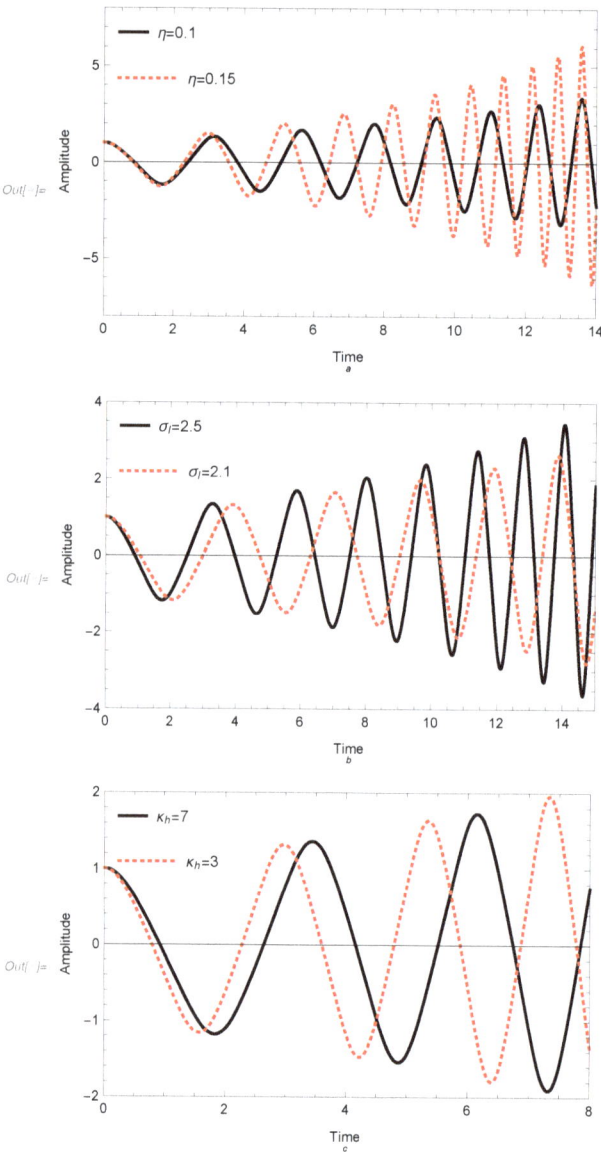

Figure 3. Plot of the initial solution $u(t)$ for $\eta_h = 0.4; \kappa_l = 3; \kappa_h = 3; \sigma_l = 2.5; \sigma_h = 0.1; u_f = 0.1; \eta = 0.3$. The plot shows the effects of: η (**a**); σ_l (**b**); and κ_h (**c**).

4. Conclusions

The forced damping Duffing equation $\left(\varphi'' + \alpha\varphi' + \beta\varphi + \gamma\varphi^3 = F\right)$ with arbitrary initial conditions is investigated numerically via the highly-accurate MSDTM. The comparison between the approximate solutions using MDTM and MSDTM with RK4 numerical solution is reported. Moreover, the maximum residual error for all approximate numerical solutions as compared to the RK4 solution is estimated. It is observed that the approximate numerical solution using MSDTM is highly accurate and better than both DTM and MDTM. Furthermore, the application of the FDDE in the practical plasma model is investigated to study the dynamics of nonlinear oscillations that occur in a complex unmagnetized plasma.

This solution might help many researchers in studying and investigating many problems in various fields of science such as plasma physics and optical fiber.

Future work: in this work, the MSDTM is devoted for solving the FDDE for constant force, but sometimes the perturbation force is not constant but periodic with time $\left(\varphi'' + \alpha\varphi' + \beta\varphi + \gamma\varphi^3 = f(t)\right)$, this is considered an important and vital problem but out of the present scope.

Author Contributions: N.H.A., conceptualization of the mathematics, methodology, software, computation, mathematical analysis, writing and editing; and S.A.E.-T., conceptualization of the physics, introducing the application, computation, physics analysis and writing. All authors have read and agreed to the published version of the manuscript.

Funding: This research was founded by the Deanship of Scientific Research (DSR) at King Abdulaziz University, Jeddah, under grant No. (D- 1441-445-662).

Institutional Review Board Statement: The study did not involve humans or animals.

Informed Consent Statement: Not applicable.

Data Availability Statement: The study did not report any data.

Acknowledgments: This Project was funded by the Deanship of Scientific Research (DSR) at King Abdulaziz University, Jeddah, under grant No. (D- 1441-445-662). The authors, therefore, acknowledge with thanks DSR technical and financial support.

Conflicts of Interest: The authors declare no conflict of interest.

References

1. Aljahdaly, N.H.; Alqudah, M.A. Analytical solutions of a modified predator-prey model through a new ecological interaction. *Comp. Math. Meth. Med.* **2019**, *2019*, 4849393
2. Salas, A.H.; El-Tantawy, S.A.; Aljahdaly, N.H. An Exact Solution to the Quadratic Damping Strong Nonlinearity Duffing Oscillator. *Math. Probl. Eng.* **2021**, *2021*, 8875589. [CrossRef]
3. Aljahdaly, N.H.; Seadawy, A.R.; Albarakati, W.A. Applications of dispersive analytical wave solutions of nonlinear seventh order Lax and Kaup-Kupershmidt dynamical wave equations. *Results Phys.* **2019**, *14*, 102372. [CrossRef]
4. Hammad, M.A.; Salas, A.H.; El-Tantawy, S.A. New method for solving strong conservative odd parity nonlinear oscillators: Applications to plasma physics and rigid rotator. *AIP Adv.* **2020**, *10*, 085001. [CrossRef]
5. Wazwaz, A.M. *Partial Differential Equations and Solitary Waves Theory*; Springer Science & Business Media: New York, NY, USA, 2010.
6. Alqudah, M.A.; Aljahdaly, N.H. Global stability and numerical simulation of a mathematical model of stem cells therapy of HIV-1 infection. *J. Comp. Sci.* **2020** *45*, 101176 [CrossRef]
7. Aljahdaly, N.H. Some applications of the modified (G'/G^2)-expansion method in mathematical physics. *Res. Phys.* **2019**, *13*, 102272. [CrossRef]
8. Do, Y.; Jang, B. Enhanced Multistage Differential Transform Method: Application to the Population Models. *Abstr. Appl. Anal.* **2012**, *2012*, 253890. [CrossRef]
9. Ashi, H.A.; Aljahdaly, N.H. Breather and solitons waves in optical fibers via exponential time differencing method. *Commun. Nonlinear Sci. Numer. Simul.* **2020**, *85*, 105237. [CrossRef]
10. Aljahdaly, N.H. New application through multistage differential transform method. *AIP Conf Proc.* **2020**, *2293*, 420025.
11. Nourazar, S.; Mirzabeigy, A. Approximate solution for nonlinear Duffing oscillator with damping effect using the modified differential transform method. *Sci. Iran. B* **2013**, *20*, 364–368
12. Gókdoğan, A.; Merdan, M.; Yildirim, A. A multistage differential transformation method for approximate solution of Hantavirus infection model. *Commun. Nonlinear Sci. Numer. Simul.* **2012**, *17*, 1–8. [CrossRef]
13. Hosen, M.A.; Chowdhury MS, H. Analytical Approximate Solutions for the Helmholtz-Duffing Oscillator. *ARPN J. Eng. App. Sci.* **2015**, *10*, 17363–17369.
14. Geng, Y. Exact solutions for the quadratic mixed-parity Helmholtz—Duffing oscillator by bifurcation theory of dynamical systems. *Chaos Solitons Fractals* **2015**, *81*, 68–77. [CrossRef]
15. Metter, E. Dynamic buckling. In *Handbook of Engineering Mechanics*; Flügge, W., Ed.; Wiley: New York, NY, USA, 1992.
16. Bikdash, M.; Balachandran, B.; Nayfeh, A. Melnikov analysis for a ship with a general roll-damping model. *Nonlinear Dyn.* **1994**, *6*, 101–124.
17. Ajjarapu, V.; Lee, B. Bifurcation theory and its application to nonlinear dynamical phenomena in an electrical power system. *Trans. Power Syst.* **1992**, *7*, 424–431. [CrossRef]
18. Kang, I.S.; Leal, L.G. Bubble dynamics in time-periodic straining flows. *J. Fluid Mech.* **1990**, *218*, 41–69. [CrossRef]

19. Wei, D.; Luo, X.; Zeng, S. Noise-triggered escapes in Helmholtz Oscillator. *Mod. Phys. Lett. B* **2014** *28*, 1450047. [CrossRef]
20. Almendral, J.A.; Seoane, J.M.; Sanjúan, M.A.F. Nonlinear dynamics of the Helmholtz oscillator. *Recent Res. Dev. Sound Vib.* **2004**, *2*, 115–150.
21. Nayfeth, N.; Mook, D.T. *Non-Linear Oscillations*; John Wiley: New York, NY, USA, 1973.
22. Almendral, J.A.; Sanjúan, M.A.F. Integrability and Symmetries for the Helmholtz Oscillator with Friction. *J. Phys. A Math. Gen.* **2003**, *36*, 695–710. [CrossRef]
23. Morfa, S.; Comte, J.C. A nonlinear oscillators network devoted to image processing. *Int. J. Bifurc. Chaos* **2009**, *14*, 1385–1394. [CrossRef]
24. Geng, Y.; Zhang, J; Li, L. Exact explicit traveling wave solutions for two nonlinear Schrodinger type equatios. *Appl. Math. Comput.* **2010**, *217*, 1509–1521.
25. Younesian, D.; Askari, H.; Saadatnia, Z.; Yazdi, M.K. Free vibration analysis of strongly nonlinear generalized Duffing oscillators using He's variational approach & homotopy perturbation method. *Nonlinear Sci. Lett. A* **2011**, *2*, 11–16.
26. Mickens, R.E. Mathematical and numerical study of the Duffing-harmonic oscillator. *J. Sound Vib.* **2001**, *244*, 563–567. [CrossRef]
27. Ganji, D.D.; Gorji, M.; Soleimani, S.; Esmaeilpour, M. Solution of nonlinear cubic-quintic Duffing oscillators using He's energy balance method. *J. Zhejiang Univ. Sci. A* **2009** *10*, 1263–1268. [CrossRef]
28. Yazdi, M.K.; Mirzabeigy, A.; Abdollahi, H. Nonlinear oscillators with non-polynomial and discontinuous elastic restoring forces. *Nonlinear Sci. Lett. A* **2012**, *3*, 48–53.
29. Khan, Y.; Akbarzade, M.; Kargar, A. Coupling of homotopy and the variational approach for a conservative oscillator with strong oddnonlinearity. *Sci. Iran.* **2012**, *19*, 417–422 . [CrossRef]
30. Wu, B.S.; Sun, W.P. Construction of approximate analytical solutions to strongly nonlinear damped oscillators. *Arch. Appl. Mech.* **2011**, *81*, 1017–1030. [CrossRef]
31. Elias-Zungia, A. Analytical solution of the damped Helmholtz–Duffing equation. *Appl. Math. Lett.* **2012**, *25*, 2349–2353. [CrossRef]
32. Elías-Zú niga, A. Exact solution of the quadratic mixed-parity Helmholtz–Duffing oscillator. *Appl. Math. Comput.* **2012**, *218*, 7590–7594.
33. Johannesen, K. The Duffing oscillator with damping. *Eur. J. Phys.* **2015**, *36*, 065020. [CrossRef]
34. Johannesen K. The Duffing Oscillator with Damping for a Softening Potential. *Int. J. Appl. Comput. Math.* **2017**, *3*, 3805 . [CrossRef]
35. Salas, A.H.; El-Tantawy, S.A. On the approximate solutions to a damped harmonic oscillator with higher-order nonlinearities and its application to plasma physics: Semi-analytical solution and moving boundary method. *Eur. Phys. J. Plus* **2020**, *135*, 1–17. [CrossRef]
36. Alama, M.S.; Masudb, M.M.; Mamun, A.A. Effects of bi-kappa distributed electrons on dust-ion-acoustic shock waves in dusty superthermal plasmas. *Chin. Phys. B* **2013**, *22*, 115202. [CrossRef]
37. El-Tantawy, S.A.; Wazwaz, A.M. Anatomy of modified Korteweg–de Vries equation for studying the modulated envelope structures in non-Maxwellian dusty plasmas: Freak waves and dark soliton collisions. *Phys. Plasmas* **2018**, *25*, 092105. [CrossRef]
38. El-Tantawy, S.A. Nonlinear dynamics of soliton collisions in electronegative plasmas: The phase shifts of the planar KdV-and mkdV-soliton collisions. *Chaos Solitons Fractals* **2016**, *93*, 162–168. [CrossRef]

Review

The Bateman Functions Revisited after 90 Years—A Survey of Old and New Results

Alexander Apelblat [1], Armando Consiglio [2] and Francesco Mainardi [3,*]

[1] Department of Chemical Engineering, Ben Gurion University of the Negev, Beer Sheva 84105, Israel; apelblat@bgu.ac.il
[2] Institut für Theoretische Physik und Astrophysik and Würzburg-Dresden Cluster of Excellence ct.qmat, Universität Würzburg, D-97074 Würzburg, Germany; armando.consiglio@physik.uni-wuerzburg.de
[3] Dipartimento di Fisica e Astronomia, Università di Bologna, & INFN, Via Irnerio 46, I-40126 Bologna, Italy
* Correspondence: francesco.mainardi@bo.infn.it

Abstract: The Bateman functions and the allied Havelock functions were introduced as solutions of some problems in hydrodynamics about ninety years ago, but after a period of one or two decades they were practically neglected. In handbooks, the Bateman function is only mentioned as a particular case of the confluent hypergeometric function. In order to revive our knowledge on these functions, their basic properties (recurrence functional and differential relations, series, integrals and the Laplace transforms) are presented. Some new results are also included. Special attention is directed to the Bateman and Havelock functions with integer orders, to generalizations of these functions and to the Bateman-integral function known in the literature.

Keywords: bateman functions; havelock functions; integral-bateman functions; confluent hypergeometric functions

Citation: Apelblat, A.; Consiglio, A.; Mainardi, F. The Bateman Functions Revisited after 90 Years—A Survey of Old and New Results. *Mathematics* 2021, 9, 1273. https://doi.org/10.3390/math9111273

Academic Editors: J. Tenreiro Machado, Manuel de León, Carsten Schneider and Abdelmejid Bayad

Received: 19 April 2021
Accepted: 26 May 2021
Published: 1 June 2021

Publisher's Note: MDPI stays neutral with regard to jurisdictional claims in published maps and institutional affiliations.

Copyright: © 2021 by the authors. Licensee MDPI, Basel, Switzerland. This article is an open access article distributed under the terms and conditions of the Creative Commons Attribution (CC BY) license (https://creativecommons.org/licenses/by/4.0/).

1. Introduction

Harry Bateman (1882–1946) has been a renowned Anglo-American applied mathematician, who made outstanding contributions to mathematical physics, namely to aero- and fluid dynamics, to electro-magnetic and optical phenomena, to thermodynamics and geophysics and to many other fields [1,2]. His main interests in mathematics were analytical solutions of partial differential and integral equations. His book published in 1932, *Partial Differential Equations of Mathematical Physics* [3] is even today, a basic textbook on this subject. Born in Manchester and educated in Trinity College, Cambridge, with a continuation in Paris and Gottingen, Bateman emigrated to USA in 1910 and starting since 1917, during nearly three decades he has been Professor of Aeronautical Research and Mathematical Physics in the California Institute of Technology (Caltech). During these years he solved a number of various applied problems and simultaneously compiled from mathematical literature a vast amount of information associated with special functions and their properties.

From an enormous scientific legacy that Bateman left behind him, it is important to mention three items which are named after him. The first is the so-called *Bateman equation* which is applied in solutions of pbharmacokinetics problems (modeling of effective therapeutic management of drugs). As usual with Bateman, the origin of this equation came from an interaction with other scientists, and this one with Ernest Rutherford. It includes the solution of a set of ordinary differential equations which describes the radioactive decay process. Mathematically, this process is similar to the behaviour of drugs in the human body and therefore is frequently used in pharmacokinetic models (see for example [4], and for prediction of the spread of COVID-19 look in [5]) listed in the fifties of the past century, and they constitute the so-called *Bateman approach*.

In mathematics, the Bateman name is mostly associated with the five red books published in the fifties of the previous century, and they constitute the so-called *Bateman*

Manuscript Project. Three volumes are devoted to the properties of special functions [1] and two volumes to tables of integral transforms [6]. This enormous collection of functions, series and integrals, together with the description of their properties is based on the material compiled largely by Bateman, and prepared for publication by four editors A. Erdélyi, R. Magnus, F. Oberhettinger and F.G. Tricomi. Even today, these five books are indispensable for everybody, mathematicians, scientists and engineers who are involved in study and use of special functions and integral transforms. They were essential as a precursor and model for later appearing in published or in modern on-line forms various compilations of mathematical reference data (for most important see for example [7–19]).

In 1931 Bateman published a paper entitled: *The k-function, a particular case of the confluent hypergeometric function*, where he presented the definite trigonometric integral (1) and derived for it many properties [20]

$$k_n(x) = \frac{2}{\pi} \int_0^{\pi/2} \cos(x \tan \theta - n\theta)\, d\theta, \quad n = 0, 1, 2, 3, \ldots \tag{1}$$

This integral represents the solution of the ordinary differential equation which appeared in Theodore von Kármán's theory of turbulent flows

$$x \frac{d^2 u(x)}{dx^2} = (x - n) u(x). \tag{2}$$

Bateman named the integral in (1) as *k*-function in tribute for the outstanding contribution of von Kármán in the field of fluid dynamics. Nowadays, denoted in the mathematical literature by small or capital *k*, this function in a more general form, is called the Bateman function of argument *x* and order (parameter) ν.

$$k_\nu(x) = \frac{2}{\pi} \int_0^{\pi/2} \cos(x \tan \theta - \nu \theta)\, d\theta. \tag{3}$$

The reason that Bateman used integer orders only, came from the fact that $k_n(x)$ functions can then be expressed by the Rodriguez type formulas and they are associated with the Laguerre polynomials. This also permitted to express sums of them in closed form and to link the Bateman functions with the confluent hypergeometric and Whittaker functions. In 1935, some new results were derived by Shastri [21], who showed that methods of operational calculus can be applied to this function.

Unfortunately, the Bateman functions found later rather limited attention in the mathematical literature. Few only topics associated with them were considered and these mainly by Indian mathematicians [22–35]. They included the generalized Bateman functions, dual, triple and multi series equations of these functions, some integral equations and recurrence relations. It is worthwhile also to mention that in mathematical textbooks and tables, the Bateman function is not considered as a some kind of minor special function, but only indicated as a particular case of the confluent hypergeometric function. Besides, no plots or tabulations of the Bateman functions are known in the literature.

One of the first attempts to enlarge a knowledge about properties of the Bateman functions, has been evidently to introduce a new function, by replacing in the integrand of integral (1) cosine function with sine function

$$T_n(x) = \frac{2}{\pi} \int_0^{\pi/2} \sin(x \tan \theta - n\theta)\, d\theta, \quad n = 0, 1, 2, 3, \ldots \tag{4}$$

In 1950 H.M. Srivastava [25] and in 1966 K.N. Srivastava [29] suggested to denote this new function as $T_n(x)$, where the capital *T* letter was adapted to honor Walter Tollmien who made pioneering works in the transition region between fully established laminar and turbulent flows. However, an unquestionably historical fact is that both trigonometric integrals as defined in (1) and (4), were already, six year earlier in 1925, considered by Havelock who investigated some problems associated with surface waves [36]. In the case of a circular cylinder immersed in a uniform flow, he needed to evaluated the following integrals which are written here in their original notation for $k > 0$

$$L_r = \int_0^{\pi/2} \cos(2r\phi - k\tan\phi)\,d\phi, \quad M_r = \int_0^{\pi/2} \sin(2r\phi - k\tan\phi)\,d\phi. \tag{5}$$

Thus, in view of that $2r = x$ and $k = n$, these integrals differ from (1) and (4) only by the normalization factor $2/\pi$ and the minus sign in the second integral. What is even more important, Havelock was able to present the first six integrals in a closed form. It is of interest also to mention that Bateman knew about the Havelock paper and of related integrals investigated by him. These integrals are included in the manuscript (later edited and published by Erdélyi) which was found among his papers [37]. Taking these facts into account, it is more fair and consistent to name the sine integral as the *Havelock function* and to use similar as in (3) notation

$$h_\nu(x) = \frac{2}{\pi} \int_0^{\pi/2} \sin(x\tan\theta - \nu\theta)\,d\theta. \tag{6}$$

In the next step, further generalizations of the Bateman function were proposed by including powers of trigonometric functions in integrands for $m, n = 0, 1, 2, 3, \ldots$,

$$\begin{aligned} k_\nu^m(x) &= \frac{2}{\pi} \int_0^{\pi/2} (\cos\theta)^m \cos(x\tan\theta - \nu\theta)\,d\theta, \\ k_\nu^{m,n}(x) &= \frac{2}{\pi} \int_0^{\pi/2} (\sin\theta)^m (\cos\theta)^n \cos(x\tan\theta - \nu\theta)\,d\theta. \end{aligned} \tag{7}$$

However, by reviewing the papers dealing with these so-called generalized Bateman functions, Erdélyi pointed out that the integrals in (7) are particular cases of confluent hypergeometric functions and the derived mathematical expressions are not new because they follow directly from manipulations with known properties of the Kummer confluent hypergeometric functions.

Probably, the most paying attention from generalized Bateman functions is that which was proposed by Chaudhuri [38]. In an analogy with the integral Bessel functions, he introduced the *Bateman-integral function*

$$ki_n(x) = -\int_x^\infty \frac{k_{2n}(u)}{u}\,du \; ; \quad x > 0, \tag{8}$$

and discussed its properties.

As already mentioned above, in the last decades, the interest in the Bateman functions was very limited, and only investigations of Koepf and Schmersau [39–41] dealing with recurrence and other relations of $F_n(x)$ functions, defined by

$$\begin{aligned} e^{-x(1+t)/(1-t)} &= \sum_{n=0}^\infty t^n F_n(x), \\ F_n(x) &= (-1)^n k_{2n}(x) = (-1)^n \frac{2}{\pi} \int_0^{\pi/2} \cos(x\tan\theta - 2n\theta)\,d\theta. \end{aligned} \tag{9}$$

should be mentioned.

Considering that at the present time, the Bateman functions are unjustly neglected and nearly entirely forgotten, we decided to prepare this survey in order to revive them and to promote them as independent functions. It seems that the Bateman functions should be treated separately, less as particular cases of the confluent hypergeometric functions or the Whittaker functions. Bearing in mind today that the literature on the subject is rather old and practically unknown, after Introduction, in the second section of this survey we collect the most important properties of the Bateman functions with integer orders $k_n(x)$. In the next section we present known results associated with the Havelock functions with integer orders $h_n(x)$. In the fourth section the generalized Bateman and Havelock functions are discussed. More general aspects related with the Bateman and Havelock functions having any order are considered in the fifth section. In these sections some new results derived by us are also included. The sixth section is dedicated to properties of the Bateman-integral functions. Concluding remarks are included in the last section.

In Appendix A we report various finite and infinite integrals of functions associated with functions considered in this survey. Differential equations and trigonometric integrals associated with the Kummer confluent hypergeometric function are discussed in Appendix B. We refer the readers to Appendix C where they can find the integral representations of known special functions recalled in the text because of their relations with the Bateman and Havelock functions.

It is expected that all results presented here in analytical and in graphical form will stimulate a new research devoted to the Bateman and Havelock functions and these functions will find a desirable and proper place in the mathematical literature.

2. The Bateman Functions with Integer Orders

The Bateman functions with integer order n and with real argument x, are defined by

$$k_n(x) = \frac{2}{\pi} \int_0^{\pi/2} \cos(x \tan\theta - n\theta)\, d\theta, \quad n = 0, 1, 2, 3, \ldots \tag{10}$$

For this integral Bateman showed that [20]

$$k_n(0) = \frac{2}{\pi n} \sin\left(\frac{\pi n}{2}\right), \quad k_{2n}(0) = 0,$$
$$\lim_{x \to \infty} k_n(x) = \lim_{x \to \infty} k'_n(x) = 0, \tag{11}$$

and

$$|k_n(x)| \leq 1$$
$$|k_n(x)| \leq \left|\frac{n}{x}\right|; \quad |k_n(x)| \leq \left|\frac{n^2+2}{x^2}\right|; \quad n > 2,$$
$$|k_{2n}(x)| \leq \left|\frac{2n}{x}\right|; \quad x > 1, \tag{12}$$
$$|k'_n(x)| \leq \left|\frac{n}{2x}\right|.$$

In the case of even integers they are associated with the Havelock integrals (5) and with $F_n(x)$ functions (9) in the following way [36,39–41]

$$k_{2n}(x) = \frac{2}{\pi} L_n(x), \quad k_{2n}(x) = (-1)^n F_n(x), \quad h_{2n}(x) = -\frac{2}{\pi} M_n(x). \tag{13}$$

The first six Bateman functions were tabulated by Havelock [36] for $x > 0$,

$$k_0(x) = e^{-x},$$
$$k_2(x) = 2xe^{-x},$$
$$k_4(x) = 2x(x-1)e^{-x},$$
$$k_6(x) = \frac{2}{3}x(2x^2 - 6x + 3)e^{-x},$$
$$k_8(x) = \frac{2}{3}x(x^3 - 6x^2 + 9x - 3)e^{-x}, \tag{14}$$
$$k_{10}(x) = \frac{2}{15}x(2x^4 - 20x^3 + 60x^2 - 60x + 15)e^{-x},$$
$$k_{12}(x) = \frac{2}{45}x(2x^5 - 30x^4 + 150x^3 - 300x^2 + 225x - 45)e^{-x}.$$

In the general case these polynomials can be derived from the Rodriguez type formula

$$k_{2n}(x) = \frac{(-1)^n x e^x}{n!} \frac{d^n}{dx^n}\left[x^{n-1} e^{-2x}\right], \tag{15}$$

which is similar to that of the generalized Laguerre polynomials $L_n^{(\alpha)}(x)$.

$$L_n^{(\alpha)}(x) = \frac{x^{-\alpha} e^x}{n!} \frac{d^n}{dx^n}\left[x^{n+\alpha} e^{-x}\right]. \tag{16}$$

Bateman showed that for his functions with even integer orders we have [20]

$$k_{2n}(x) = (-1)^n e^{-x}[L_n(2x) - L_{n-1}(2x)], \tag{17}$$

where $L_k(z)$ are the Laguerre polynomials.

It is more difficult to express the Bateman functions with odd orders in terms of other known functions. For $n = 1$, Bateman introduced a new integration variable $t = \tan\theta$ and obtained [20]

$$\begin{aligned}
k_1(x) &= \frac{2}{\pi} \int_0^{\pi/2} \cos(x\tan\theta - \theta)\, d\theta = \\
&\frac{2}{\pi} \int_0^{\pi/2} \cos(x\tan\theta)\cos\theta\, d\theta + \frac{2}{\pi}\int_0^{\pi/2} \sin(x\tan\theta)\sin\theta\, d\theta = \\
&\frac{2}{\pi} \int_0^\infty \frac{\cos(xt)}{(1+t^2)^{3/2}}\, dt + \frac{2}{\pi}\int_0^\infty \frac{t\sin(xt)}{(1+t^2)^{3/2}}\, dt = \\
&\frac{2}{\pi} \int_0^\infty \frac{\cos(xt)}{(1+t^2)^{3/2}}\, dt - \frac{2x}{\pi}\int_0^\infty \frac{\cos(xt)}{(1+t^2)^{1/2}}\, dt.
\end{aligned} \tag{18}$$

The last two integrals are the integral representations of the modified Bessel functions of the second kind of the first and zero orders [7]

$$\begin{aligned}
k_1(x) &= \frac{2x}{\pi}[K_1(x) - K_0(x)]; \quad x > 0, \\
k_1(x) &= -\frac{2x}{\pi}[K_1(-x) + K_0(-x)]; \quad x < 0.
\end{aligned} \tag{19}$$

The Bateman functions with other even and odd integer orders can also be derived by applying the recurrence relations which are in the form of difference equations and differential-difference equations

$$\begin{aligned}
(2x - 2n)k_{2n}(x) &= (n-1)k_{2n-2}(x) + (n+1)k_{2n+2}(x) \\
4x k'_n(x) &= (n-2)k_{n-2}(x) - (n+2)k_{n+2}(x) \\
k'_n(x) + k'_{n+2}(x) &= k_n(x) - k_{n+2}(x) \\
x k''_n(x) &= (x - n)k_n(x).
\end{aligned} \tag{20}$$

For example, using the second equation in (20) for $n = 1$, we have

$$\begin{aligned}
k_3(x) &= -\frac{1}{3}\left[4x\frac{dk_1(x)}{dx} + k_{-1}(x)\right] \\
\frac{dk_1(x)}{dx} &= \frac{2}{\pi}[K_1(x) - K_0(x)] + \frac{2x}{\pi}\left[\frac{dK_1(x)}{dx} - \frac{dK_0(x)}{dx}\right] \\
\frac{dK_1(x)}{dx} &= \frac{K_2(x) + K_0(x)}{2} \\
\frac{dK_0(x)}{dx} &= -K_1(x)
\end{aligned} \tag{21}$$

and $k_{-1}(x)$ can be expressed by using integrals from (18)

$$
\begin{aligned}
k_{-1}(x) &= \frac{2}{\pi} \int_0^{\pi/2} \cos(x \tan\theta + \theta) \, d\theta = \\
&\frac{2}{\pi} \int_0^{\pi/2} \cos(x \tan\theta) \cos\theta \, d\theta - \frac{2}{\pi} \int_0^{\pi/2} \sin(x \tan\theta) \sin\theta \, d\theta.
\end{aligned}
\tag{22}
$$

It is also possible to obtain the Bateman functions with odd orders in a different new procedure, for example $k_3(x)$

$$
\begin{aligned}
k_3(x) &= \frac{2}{\pi} \int_0^{\pi/2} \cos(x \tan\theta - 3\theta) \, d\theta = \\
&\frac{2}{\pi} \int_0^{\pi/2} \cos(x \tan\theta) \cos(3\theta) \, d\theta + \frac{2}{\pi} \int_0^{\pi/2} \sin(x \tan\theta) \sin(3\theta) \, d\theta,
\end{aligned}
\tag{23}
$$

but with $t = \tan\theta$

$$
\begin{aligned}
\sin(3\theta) &= 3\sin\theta - 4(\sin\theta)^3 = \sin\theta \, \frac{3 - (\tan\theta)^2}{1 + (\tan\theta)^2} = \frac{t(3 - t^2)}{(1 + t^2)^{3/2}}, \\
\cos(3\theta) &= -3\cos\theta + 4(\cos\theta)^3 = \cos\theta \, \frac{1 - 3(\tan\theta)^2}{1 + (\tan\theta)^2} = \frac{(1 - 3t^2)}{(1 + t^2)^{3/2}},
\end{aligned}
\tag{24}
$$

and therefore (23) becomes

$$
k_3(x) = \frac{2}{\pi} \int_0^\infty \frac{(1 - 3t^2) \cos(xt)}{(1 + t^2)^{5/2}} \, dt + \frac{2}{\pi} \int_0^\infty \frac{t(3 - t^2) \sin(xt)}{(1 + t^2)^{5/2}} \, dt.
\tag{25}
$$

However, this type of integrals can be evaluated by differentiating the modified Bessel functions of the second kind [14]

$$
\begin{aligned}
\int_0^\infty \frac{t^{2n+1} \sin(xt)}{(1 + t^2)^\alpha} \, dt &= (-1)^{n+1} \frac{2^{1/2 - \alpha} \sqrt{\pi}}{\Gamma(\alpha)} \frac{\partial^{2n+1}}{\partial x^{2n+1}} \left[x^{\alpha - 1/2} K_{\alpha - 1/2}(x) \right], \quad \alpha > n + 1/2, \\
\int_0^\infty \frac{t^{2n} \sin(xt)}{(1 + t^2)^\alpha} \, dt &= (-1)^n \frac{2^{1/2 - \alpha} \sqrt{\pi}}{\Gamma(\alpha)} \frac{\partial^{2n+1}}{\partial x^{2n+1}} \left[x^{\alpha - 1/2} K_{\alpha - 1/2}(x) \right], \quad \alpha > n.
\end{aligned}
\tag{26}
$$

Using known expressions for $\sin(\alpha + 2\theta)$ and $\cos(\alpha + 2\theta)$ functions with $\alpha = 2n + 1$, and taking into account that [7] with $t = \tan\theta$

$$
\begin{aligned}
\sin(2\theta) &= \frac{2 \tan\theta}{1 + (\tan\theta)^2} = \frac{2t}{(1 + t^2)}, \\
\cos(2\theta) &= \frac{1 - (\tan\theta)^2}{1 + (\tan\theta)^2} = \frac{(1 - t^2)}{(1 + t^2)},
\end{aligned}
\tag{27}
$$

the above described procedure can be extended to the Bateman functions with higher odd orders. Integrals of the type presented in (26) can be also used when derivatives with respect to the argument are considered with $m = 0, 1, 2, 3, \ldots$

$$
\begin{aligned}
\frac{\partial^{2m} k_n(x)}{\partial x^{2m}} &= (-1)^m \frac{2}{\pi} \int_0^{\pi/2} (\tan\theta)^{2m} \cos(x \tan\theta - n\theta) \, d\theta, \\
\frac{\partial^{2m+1} k_n(x)}{\partial x^{2m+1}} &= (-1)^m \frac{2}{\pi} \int_0^{\pi/2} (\tan\theta)^{2m+1} \sin(x \tan\theta - n\theta) \, d\theta.
\end{aligned}
\tag{28}
$$

In order to illustrate the behaviour of the Bateman functions as a function of argument and order, they were numerically evaluated using the MATLAB program and they are presented in Figure 1 for positive integer orders and in Figure 2 for negative integer order. As can be observed by comparing both figures, the curves are shifted with the symmetry predicted by Bateman [20]

$$
k_{-n}(x) = k_n(-x).
\tag{29}
$$

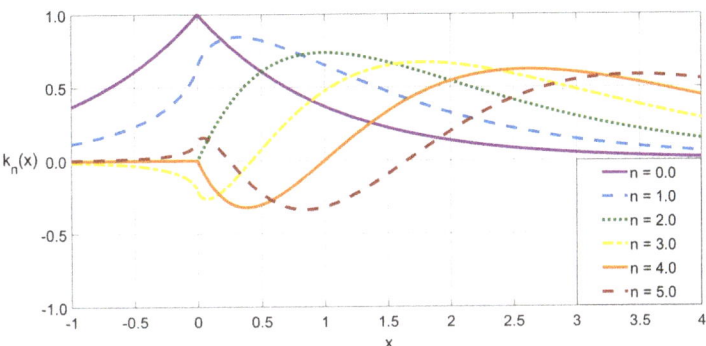

Figure 1. Bateman functions with positive integer orders as a function of argument x.

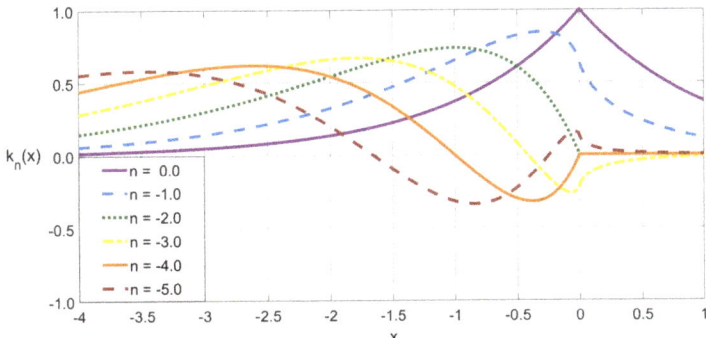

Figure 2. Bateman functions with negative integer orders as a function of argument x.

Considering similarity with the generalized Laguerre polynomials, Bateman was able to show the existence of the following expansions associated with his functions with even orders [20]

$$\sum_{n=0}^{\infty} (-1)^n t^n k_{2n}(x) = (1-t)^{\alpha+1} e^{-x} \sum_{n=0}^{\infty} t^n L_n^{(\alpha)}(2x),$$
$$\sum_{n=0}^{\infty} \frac{t^n}{2^n n!} k_{2n+2}(x) = 2e^{-(x+t/2)} \sqrt{\frac{x}{t}} I_1(2\sqrt{xt}), \qquad (30)$$
$$\sum_{n=0}^{\infty} (-1)^n k_{4n+2}(x) = \sin x, \quad \sum_{n=0}^{\infty} (-1)^n k_{4n}(x) = \cos x.$$

where I_1 denoted the modified Bessel function of order 1, see (C.8) and [7]. Shabde [22] demonstrated that

$$\sum_{n=0}^{\infty} (n+1)t^n k_{2n+2}(x) = \frac{2xe^{-x+[2xt/(1+t)]}}{(1+t)^2},$$
$$\sum_{n=0}^{\infty} (-1)^n (2n+1)t^{2n} k_{2n+2}(x) =$$
$$\frac{2xe^{-x+2xt^2/(1+t^2)}}{(1+t^2)^2} \left[(1-t^2) \cos\left(\frac{2xt}{1+t^2}\right) + 2t \sin\left(\frac{2xt}{1+t^2}\right) \right], \qquad (31)$$
$$\sum_{n=0}^{\infty} (-1)^n (2n+2)t^{2n+1} k_{4n+4}(x) =$$
$$\frac{2xe^{-x+2xt^2/(1+t^2)}}{(1+t^2)^2} \left[(1-t^2) \sin\left(\frac{2xt}{1+t^2}\right) - 2t \cos\left(\frac{2xt}{1+t^2}\right) \right],$$

and

$$\sum_{n=0}^{\infty} \frac{(-1)^n t^n}{n!} k_{2n+2}(x) = \sqrt{\frac{2x}{t}} e^{-(x+t)} J_1(2^{3/2}\sqrt{xt}),$$

$$\sum_{n=0}^{\infty} \frac{(-1)^n t^{2n}}{(2n)!} k_{2n+2}(x) = \sqrt{\frac{2x}{t}} \left[-\sin t \, \text{ber}'(2^{3/2}\sqrt{xt}) + \cos t \, \text{bei}'(2^{3/2}\sqrt{xt}) \right], \quad (32)$$

$$\sum_{n=0}^{\infty} \frac{(-1)^{n+1} t^{2n+1}}{(2n+1)!} k_{4n+4}(x) = \sqrt{\frac{2x}{t}} \left[\cos t \, \text{ber}'(2^{3/2}\sqrt{xt}) + \sin t \, \text{bei}'(2^{3/2}\sqrt{xt}) \right],$$

where $\text{ber}'(z)$ and $\text{bei}'(z)$ are the derivatives of the Kelvin functions.

Additional sums of series expansions were reported by Shastri [24]

$$\sum_{n=0}^{\infty} (-1)^n t^{2n+1} k_{4n+2}(x) = e^{x(t^2-1)/(1+t^2)} \sin\left(\frac{2xt}{1+t^2}\right); \quad |t| < 1,$$

$$\sum_{n=0}^{\infty} (-1)^n t^{2n} k_{4n}(x) = e^{x(t^2-1)/(1+t^2)} \cos\left(\frac{2xt}{1+t^2}\right); \quad |t| < 1, \quad (33)$$

$$\sum_{n=0}^{\infty} (-1)^n k_{4n+2}(x) = \sin x, \quad \sum_{n=0}^{\infty} (-1)^n k_{4n}(x) = \cos x,$$

and

$$\sum_{n=0}^{\infty} k_{2n}(x) \sin(2n\theta) = \sin(x \tan \theta),$$

$$\sum_{n=0}^{\infty} k_{2n}(x) \sin(2n\theta) = \sin(x \tan \theta), \quad (34)$$

$$\sum_{n=0}^{\infty} k_{2n}(x) = 1.$$

The orthogonal relations were established by Bateman [20]

$$\int_0^{\infty} [k_{2n}(x)]^2 \, dx = \begin{cases} 1 & ; \quad n > 0 \\ 1/2 & ; \quad n = 0 \end{cases}$$

$$\int_0^{\infty} k_{2n}(x) k_{2n+2k}(x) \, dx = \begin{cases} 0 & ; \quad k > 1 \\ 1/2 & ; \quad k = 1 \end{cases} \quad (35)$$

$$\int_0^{\infty} \frac{k_n(x) k_{2k}(x)}{x} \, dx = \frac{4 \sin\left[\frac{\pi}{2}(2k-n)\right]}{\pi n (2k-n)}; \quad k > 0,$$

and over the entire integration interval

$$\int_{-\infty}^{+\infty} k_{2k}(x) k_{2m}(x) \, dx = \frac{\sin[\pi(m-k)]}{\pi(k-m+1)(k-m)(k-m-1)},$$

$$PV \int_{-\infty}^{\infty} k_{2k+1}(x) k_{2m+1}(x) \frac{dx}{x} = \begin{cases} 0; & k \neq m, \\ \frac{2}{\pi(2k+1)}; & k = m. \end{cases} \quad (36)$$

In the literature there is a number of infinite integrals where the Bateman functions appear in integrands or in final results of integration. These integrals are collected in Appendix A, here only the Laplace transforms of the Bateman functions are presented [6,9]:

$$\int_0^{\infty} e^{-st} k_0(t) \, dt = \frac{1}{s+1}; \quad \text{Re}(s+1) > 0; \quad n = 0, 1, 2, \ldots$$

$$\int_0^{\infty} e^{-st} k_{2n+2}(t) \, dt = \frac{2(1-s)^n}{(s+1)^{n+2}} \quad (37)$$

$$\int_0^{\infty} e^{-st} k_{2\nu}(t) \, dt = \frac{\sin(\pi \nu)}{2\pi \nu (1-\nu)} \, _2F_1\left(1, 2; 2-\nu; \frac{1-s}{2}\right); \quad \text{Re} \, s > 0$$

and

$$\int_0^\infty e^{-st} e^{-t^2} k_{2n}(t^2)\, dt = \frac{(-1)^{n-1} s^{n-3/2} e^{s^2/16}}{2^{3n/2+1/4}} W_{-n/2-1/4,-n/2-1/4}\left(\frac{s^2}{8}\right)$$

$$\int_0^\infty e^{-st} k_{2m+2}\left(\frac{t}{2}\right) k_{2n+2}\left(\frac{t}{2}\right) \frac{dt}{t} = \frac{(-1)^{m+n}}{(s+1)^{m+n+2}}\, {}_2F_1\left(-m,-n;2;\frac{1}{s^2}\right)$$

$\operatorname{Re} s > -1$
(38)

$$\int_0^\infty e^{-st} \frac{e^{(\alpha+\beta)t/2}}{\alpha\beta} k_{2m+2}\left(\frac{\alpha t}{2}\right) k_{2n+2}\left(\frac{\beta t}{2}\right) \frac{dt}{t} =$$
$$\frac{(-1)^{m+n}(m+n+1)!\,(s-\alpha)^m(s-\beta)^n}{(m+1)!\,(n+1)!(s+1)^{m+n+2}}\, {}_2F_1\left(-m,-n;-m-n-1;\frac{s(s-\alpha-\beta)}{(s-\alpha)(s-\beta)}\right)$$

$m, n = 0, 1, 2, \ldots; \quad \operatorname{Re} s > 0$

where $W_{\kappa,\mu}(z)$ is the Whittaker function. Formulas in (32) and (33) are accessible in a much more general forms by applying the basic properties of the Laplace transformation

$$L\{f(t)\} = \int_0^\infty e^{-st} f(t)\, dt = F(s); \quad a > 0$$
$$L\{f(at)\} = \frac{1}{a} F\left(\frac{s}{a}\right)$$
(39)
$$L\{e^{\pm at} f(t)\} = F(s \mp a)$$
$$L\{t^n f(t)\} = (-1)^n \frac{d^n F(s)}{ds^n}$$

For example in the simple case of the function $k_2(t)$ we have from (39)

$$L\{k_2(t)\} = \frac{2}{(s+1)^2}$$
$$L\{k_2(at)\} = \frac{2a}{(s+a)^2}$$
(40)
$$L\{e^{\pm at} k_2(at)\} = \frac{2}{(s \mp a + 1)^2}$$
$$L\{t k_2(t)\} = \frac{4}{(s+1)^3}.$$

The initial and final values of the Bateman functions with even integer orders (see Figure 1) as presented in (11), can also be derived from the rules of the operational calculus

$$k_0(t \to +0) = \lim_{s \to \infty}[sF(s)] = \lim_{s \to \infty}\left[\frac{s}{s+1}\right] = 1$$
$$k_0(t \to \infty) = \lim_{s \to 0}[sF(s)] = \lim_{s \to 0}\left[\frac{s}{s+1}\right] = 0$$
(41)
$$k_{2n+2}(t \to +0) = \lim_{s \to \infty}[sF(s)] = \lim_{s \to \infty}\left[\frac{2s(1-s)^n}{(s+1)^{n+2}}\right] = 0$$
$$k_{2n+2}(t \to \infty) = \lim_{s \to 0}[sF(s)] = \lim_{s \to 0}\left[\frac{2s(1-s)^n}{(s+1)^{n+2}}\right] = 0.$$

Since the Bateman function is a particular case of the Whittaker function

$$k_{2\nu}\left(\frac{t}{2}\right) = \frac{1}{\Gamma(\nu+1)} W_{\nu,1/2}(t) \qquad (42)$$

it is possible to enlarge a number of the Laplace transforms using transforms of the Whittaker functions $W_{1/2,1/2}(t)$ and $W_{\nu,1/2}(t)$

$$L\left\{t^{1/2}e^{1/2t}k_1\left(\frac{2}{t}\right)\right\} = \frac{\sqrt{\pi}}{s}\left[H_1(2\sqrt{s}) - Y_1(2\sqrt{s})\right]$$
$$L\left\{te^{1/2t}k_1\left(\frac{2}{t}\right)\right\} = \frac{1}{2s}H_1^{(1)}(\sqrt{s})H_1^{(2)}(\sqrt{s})$$
$$L\left\{\frac{1}{t}e^{-1/2t}k_1\left(\frac{2}{t}\right)\right\} = \frac{2^{5/2}\sqrt{s}}{\pi}K_0(\sqrt{s})K_1(\sqrt{s})$$
$$L\left\{\frac{1}{t^2}e^{-1/2t}k_1\left(\frac{2}{t}\right)\right\} = \frac{4}{\pi s}\left[K_1(\sqrt{s})\right]^2$$
(43)

and

$$L\left\{t^{\alpha-1}k_{2\nu}\left(\frac{t}{2}\right)\right\} =$$
$$\frac{\Gamma(\alpha)\Gamma(\alpha+1)}{\Gamma(\nu+1)\Gamma(\alpha-\nu+1)}\left(\frac{2}{2s+1}\right)^{\alpha+1}{}_2F_1(\alpha+1,-\nu;\alpha-\nu+1;\frac{2s-1}{2s+1})$$
$$\text{Res} > -\frac{1}{2}$$
$$L\left\{t^\nu e^{1/2t}k_{2\nu}\left(\frac{2}{t}\right)\right\} = \frac{2^{1-2\nu}}{\Gamma(\nu+1)s^{\nu+1/2}}S_{2\nu,1}(2\sqrt{s}); \quad \text{Re}(\nu\pm\frac{1}{2}) > -\frac{1}{2}$$
$$L\left\{\frac{1}{t^\nu}e^{-1/2t}k_{2\nu}\left(\frac{2}{t}\right)\right\} = \frac{2s^{\nu-1/2}}{\Gamma(\nu+1)}K_1(2\sqrt{s}); \quad \text{Res} > 0,$$
(44)

where $H_\mu(t)$, $Y_\mu(t)$, $H_\mu^{(1)}(t)$, $H_\mu^{(2)}(t)$ and $S_\mu(t)$ are the Struve, Bessel, Hankel and Lommel functions, respectively.

3. The Havelock Functions with Integer Orders

As pointed out above, Havelock in solving the surface wave problem [36] encountered the following trigonometric integrals with even integer values of order (parameter) n

$$h_n(x) = \frac{2}{\pi}\int_0^{\pi/2}\sin(x\tan\theta - n\theta)\,d\theta.$$
(45)

These functions with positive and negative values of order were calculated numerically by using the MATLAB program and they are plotted in Figures 3 and 4. Comparing both figures, it is evident that the curves are shifted according to

$$h_{-n}(x) = -h_n(-x).$$
(46)

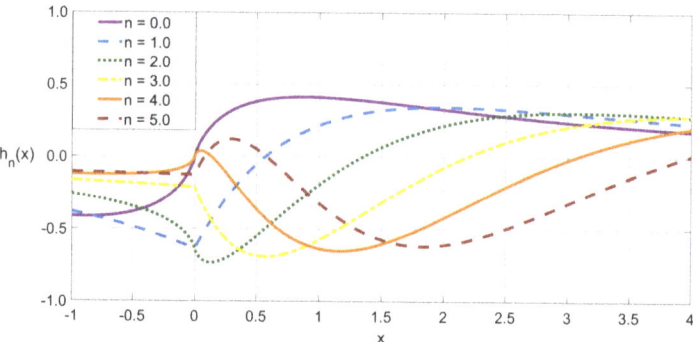

Figure 3. Havelock functions with positive integer orders as a function of argument x.

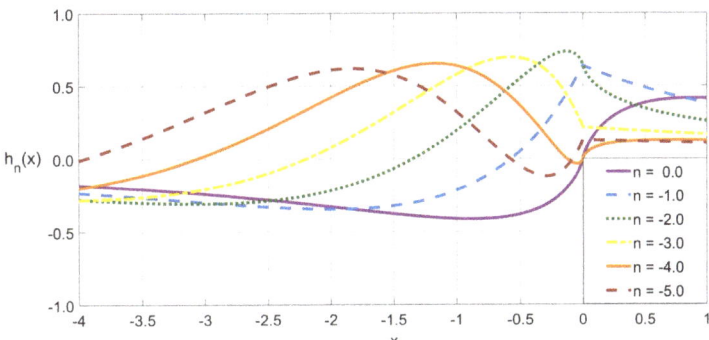

Figure 4. Havelock functions with begative integer orders as a function of argument x.

Havelock was able to present the first six integrals in terms of polynomials and the logarithmic integrals [36]

$$h_0(x) = \frac{1}{2}\left[e^x li(e^{-x}) - e^{-x} li(e^x)\right]$$
$$h_2(x) = xe^{-x} li(e^x) - 1$$
$$h_4(x) = x(x-1)e^{-x} li(e^x) - x$$
$$h_6(x) = \frac{x(2x^2 - 6x + 3)e^{-x} li(e^x) - (2x^2 - 4x + 1)}{3}$$

(47)

and

$$h_8(x) = \frac{x(x^3 - 6x^2 + 9x - 3)e^{-x} li(e^x) - x(x^2 - 5x + 5)}{3}$$
$$h_{10}(x) = \frac{x(2x^4 - 20x^3 + 60x^2 - 60x + 15)e^{-x} li(e^x)}{15} - \frac{(2x^4 - 18x^3 + 44x^2 - 28x + 3)}{15}$$
$$h_{12}(x) = \frac{x(2x^5 - 30x^4 + 150x^3 - 300x^2 + 225x - 45)e^{-x} li(e^x)}{45} - \frac{x(2x^4 - 28x^3 + 124x^2 - 198x + 93)}{45}$$

(48)

where

$$li(z) = \int_0^z \frac{dt}{\ln t} = \gamma + \ln z + \sum_{n=1}^{\infty} \frac{z^n}{n!n}, \quad z = e^x.$$

(49)

In the same way as in the Bateman paper from 1931, the properties of the Havelock functions with integer orders were studied by Srivastava in 1950 [25]. He found that

$$|h_n(x)| \leq 1$$
$$h_n(0) = \frac{2}{\pi n}\left[\cos\left(\frac{\pi n}{2}\right) - 1\right]$$
$$h_{2n}(0) = \left[\frac{1 - (-1)^n}{\pi n}\right]$$
$$h_{4n}(0) = 0$$
$$\lim_{x \to \infty} h_n(x) = \lim_{x \to \infty} h_n'(x) = 0,$$

(50)

and
$$h_0(x) = \frac{2}{\pi} \int_0^{\pi/2} \sin(x \tan \theta) \, d\theta = \frac{2}{\pi} \int_0^\infty \frac{\sin(xt)}{1+t^2} \, dt$$
$$h_1(x) = \frac{2}{\pi} \int_0^{\pi/2} \sin(x \tan \theta - \theta) \, d\theta =$$
$$\frac{2}{\pi} \int_0^{\pi/2} [\sin(x \tan \theta) \cos \theta - \cos(x \tan \theta) \sin \theta] \, d\theta =$$
$$\frac{2}{\pi} \int_0^\infty \frac{[\sin(xt) - t \cos(xt)]}{(1+t^2)^{3/2}} \, dt. \tag{51}$$

These integrals are of the type presented in (26). In 1950 Srivastava [25] showed that the infinite integral in (51) can be expressed in terms of the modified Bessel function of the first kind of zero order and the Struve function of zero order and their derivatives.

The Havelock functions satisfy the following recurrence and differential relations [25,37]

$$(2n - 4x) h_n(x) + (n - 2) h_{n-2}(x) + (n + 2) h_{n+2}(x) = -\frac{8}{\pi}$$
$$4x h'_n(x) = (n - 2) h_{n-2}(x) - (n + 2) h_{n+2}(x)$$
$$h'_{n-1}(x) + h'_{n+1}(x) = h_{n-1}(x) - h_{n+1}(x) \tag{52}$$
$$x h''_n(x) = (x - n) h_n(x) - \frac{2}{\pi}.$$

The Laplace transform of the function $h_0(x)$ can be obtained in the following way

$$L\{h_0(x)\} = \frac{2}{\pi} \int_0^\infty e^{-sx} \left(\int_0^{\pi/2} \sin(x \tan \theta) \, d\theta \right) dx =$$
$$\frac{2}{\pi} \int_0^{\pi/2} \left(\int_0^\infty e^{-sx} \sin(x \tan \theta) \, dx \right) d\theta = \frac{2}{\pi} \int_0^{\pi/2} \frac{\tan \theta}{s^2 + (\tan \theta)^2} \, d\theta = \tag{53}$$
$$\frac{2}{\pi} \int_0^\infty \frac{t}{(s^2 + t^2)(1 + t^2)} \, dt = \frac{2 \ln(s)}{\pi(s^2 - 1)}.$$

For the function $h_1(x)$ we have

$$L\{h_1(x)\} = \frac{2}{\pi} \int_0^{\pi/2} \left(\int_0^\infty e^{-sx} [\sin(x \tan \theta) \cos \theta - \cos(x \tan \theta) \sin \theta] dx \right) d\theta =$$
$$= \frac{2}{\pi} \int_0^{\pi/2} \frac{[\tan \theta \cos \theta - s \sin \theta]}{s^2 + (\tan \theta)^2} \, d\theta = \frac{2(1-s)}{\pi} \int_0^\infty \frac{t}{(s^2 + t^2)(1 + t^2)^{3/2}} \, dt = \tag{54}$$
$$\frac{2}{\pi(s+1)} \left[\frac{\sec^{-1}(s)}{\sqrt{s^2 - 1}} - 1 \right].$$

The Laplace transforms of the functions $h_0(x)$ and $h_1(x)$ were also derived by Srivastava [25] in 1950, but in the final expressions, the factor $2/\pi$ is missing.

The Havelock function $h_2(x)$ is expressed by

$$h_2(x) = \frac{2}{\pi} \int_0^{\pi/2} \sin(x \tan \theta - 2\theta) \, d\theta =$$
$$\frac{2}{\pi} \int_0^{\pi/2} [\sin(x \tan \theta) \cos(2\theta) - \cos(x \tan \theta) \sin(2\theta)] \, d\theta =$$
$$\frac{2}{\pi} \int_0^{\pi/2} \frac{\sin(x \tan \theta) [1 - (\tan \theta)^2] - 2 \tan \theta \cos(x \tan \theta)}{1 + (\tan \theta)^2} \, d\theta \tag{55}$$
$$\frac{2}{\pi} \int_0^\infty \frac{(1 - t^2) \sin(xt) - 2t \cos(xt)}{(1 + t^2)^2} \, dt$$

and its Laplace transform is therefore

$$L\{h_2(x)\} = \frac{2}{\pi} \int_0^\infty \left[\int_0^\infty e^{-sx} \frac{(1-t^2)\sin(xt) - 2t\cos(xt)}{(1+t^2)^2} dx \right] dt = -\frac{2[s+1+\ln(s)]}{\pi(s+1)^2} \tag{56}$$

where the infinite integrals in (53), (54) and (56) were verified using the MATHEMATICA program. The derived Laplace transforms allow us to obtain the initial and final values of the Havelock functions, for example for the function $h_0(x)$ we have

$$\begin{aligned} h_0(x \to +0) &= \lim_{s \to \infty} [sF(s)] = \lim_{s \to \infty} \left[\frac{2s\ln(s)}{s^2 - 1} \right] = 0 \\ h_0(x \to \infty) &= \lim_{s \to 0} [sF(s)] = \lim_{s \to 0} \left[\frac{2s\ln(s)}{s^2 - 1} \right] = 0 \end{aligned} \tag{57}$$

as it is observed in Figure 3.

There is a number of recurrence and differential expressions that include both the Bateman and the Havelock functions. They were reported by Srivastava [25] and three of them are presented here

$$\begin{aligned} (n-2)\left[k_n(x)h_{n-2}(x) - k_{n-2}(x)h_n(x)\right] + \\ (n+2)\left[k_n(x)h_{n+2}(x) - k_{n+2}(x)h_n(x)\right] &= -\frac{8}{\pi}k_n(x) \\ 4x\left[k_n(x)h'_{n-2}(x) + k'_{n-2}(x)h_n(x)\right] &= \\ (n-2)\left[k_n(x)h_{n-2}(x) + k_{n-2}(x)h_n(x)\right] + \\ (n+2)\left[k_n(x)h_{n+2}(x) + k_{n+2}(x)h_n(x)\right] \\ \left[k_n(x)h''_n(x) - k''_n(x)h_n(x)\right] &= -\frac{2}{\pi x}k_n(x), \end{aligned} \tag{58}$$

where n is an even integer.

If we consider the Havelock function in the special case

$$h_n(nx) = \frac{2}{\pi} \int_0^{\pi/2} \sin[n(x\tan\theta - \theta)] d\theta = \frac{2}{\pi} \int_0^{\pi/2} \sin[n\alpha] d\theta, \tag{59}$$

then we recognize that the sums of series of the Havelock can be expressed by finite trigonometric integrals.

For example from [42]

$$\frac{2}{\pi} \sum_{n=1}^\infty t^n \sin(n\alpha) = \frac{2}{\pi} \left[\frac{t\sin\alpha}{1 - 2t\cos\alpha + t^2} \right]; \quad t^2 < 1 \tag{60}$$

and integrating (60) with interchanging the order of summation and integration, we have

$$\sum_{n=1}^\infty t^n h_n(nx) = \frac{2}{\pi} \int_0^{\pi/2} \frac{t\sin(x\tan\theta - \theta)}{1 - 2t\cos(x\tan\theta - \theta) + t^2} d\theta; \quad t^2 < 1. \tag{61}$$

In a similar way it is possible to obtain for series of the Bateman functions

$$\sum_{n=1}^\infty t^n k_n(nx) = \frac{2}{\pi} \int_0^{\pi/2} \frac{1 - t\cos(x\tan\theta - \theta)}{1 - 2t\cos(x\tan\theta - \theta) + t^2} d\theta; \quad t^2 < 1. \tag{62}$$

By this procedure, using various finite and infinite trigonometric series from [42], many sums of the Bateman $k_n(nx)$ and Havelock $h_n(nx)$ series with different coefficients, can be expressed by corresponding integrals.

4. The Generalized Bateman and Havelock Functions with Integer Orders

In order to solve dual, triple or multi series equations, a number of generalized Bateman and Havelock functions were introduced [25,26,29–35]. From the generalized functions only two considered in 1972 by Srivastava [31] are presented here. There is no agreed uniform notation of the generalized Bateman and Havelock functions. They are defined by using different letters, with upper and lower indexes. Here these functions are presented with an additional lower index with $k > -1$ as

$$\begin{aligned} k_{n,k}(x) &= \frac{2}{\pi} \int_0^{\pi/2} (\cos\theta)^k \cos(x\tan\theta - n\theta)\, d\theta, \\ h_{n,k}(x) &= \frac{2}{\pi} \int_0^{\pi/2} (\cos\theta)^k \sin(x\tan\theta - n\theta)\, d\theta. \end{aligned} \tag{63}$$

It is suggested that if powers of cosine and sine functions appear also in (63), then the third lower index m is included

$$\begin{aligned} k_{n,k,m}(x) &= \frac{2}{\pi} \int_0^{\pi/2} (\cos\theta)^k (\sin\theta)^m \cos(x\tan\theta - n\theta)\, d\theta, \\ h_{n,k,m}(x) &= \frac{2}{\pi} \int_0^{\pi/2} (\cos\theta)^k (\sin\theta)^m \sin(x\tan\theta - n\theta)\, d\theta, \end{aligned} \tag{64}$$

where this notation differs from that used in (7).

Values of three such integrals having $n = 0$ and $k = 0, 1, 2$ are known

$$\begin{aligned} \int_0^{\pi/2} (\cos\theta)^2 \cos(x\tan\theta - n\theta)\, d\theta &= \frac{\pi(1+x)e^{-x}}{4} = \frac{\pi}{2} k_{0,2}(x) \\ \int_0^{\pi/2} (\sin\theta)^2 \cos(x\tan\theta - n\theta)\, d\theta &= \frac{\pi(1-x)e^{-x}}{4} = \frac{\pi}{2} k_{0,0,2}(x) \\ \int_0^{\pi/2} \cos\theta \sin\theta \sin(x\tan\theta - n\theta)\, d\theta &= \frac{\pi x e^{-x}}{4} = \frac{\pi}{2} h_{0,1,1}(x). \end{aligned} \tag{65}$$

The recurrence and differential expressions for the generalized Havelock functions are [31]

$$\begin{aligned} &[(n-k-2)h_{n-2,k}(x) + (n+k+2)h_{n+2,k}(x) + (2n-x)h_{n,k}(x)] = -\frac{8}{\pi} \\ &4xh'_{n,k}(x) = [(n-k-2)h_{n-2,k}(x) - (n+k+2)h_{n+2,k}(x) + 2k\,h_{n,k}(x)] \\ &2xh'_{n,k}(x) - \frac{4}{\pi} = [(n-k-2)h_{n-2,k}(x) + (n+k-2x)h_{n+2,k}(x)] \\ &2h'_{0,2k}(x) = [2h_{0,2k+2}(x) - h_{0,2k}(x) - h_{2,2k+2}(x)] \\ &xh''_{n,k}(x) - kh'_{n,k}(x) + (n-x)h_{n,k}(x) = -\frac{2}{\pi}, \end{aligned} \tag{66}$$

and for the generalized Bateman functions

$$2k'_{0,2k}(x) = [2k_{0,2k+2}(x) - k_{0,2k}(x) - k_{2,2k+2}(x)]. \tag{67}$$

In 1972 Srivastava [31] was able to show that

$$\begin{aligned} k_{0,2k}(x) &= \frac{2}{\pi} \int_0^{\pi/2} (\cos\theta)^{2k} \cos(x\tan\theta)\, d\theta = \\ &\quad \frac{2}{\sqrt{\pi}\,\Gamma(k+1)} \left(\frac{x}{2}\right)^{k+1/2} K_{k+1/2}(x) \\ h_{0,2k}(x) &= \frac{2}{\pi} \int_0^{\pi/2} (\cos\theta)^{2k} \sin(x\tan\theta)\, d\theta = \\ &\quad \frac{2\Gamma(-k)}{\sqrt{\pi}} \left(\frac{x}{2}\right)^{k+1/2} [I_{k+1/2}(x) - L_{-k-1/2}(x)], \end{aligned} \tag{68}$$

and in the explicit form for the generalized Havelock function

$$h_{2n,2k}(x) = \frac{1}{\pi}[k_{2n}(x)\,li(e^x) - 2S_{n-k-1,k}(x)]; \quad n \geq k+1, \tag{69}$$

where he determined the following polynomials for the expression in (69)

$$\begin{aligned}
S_{2,1}(x) &= \frac{1}{6}\left(2 + x + x^2\right) \\
S_{3,1}(x) &= \frac{1}{12}\left(2 - x^2 + x^3\right) \\
S_{4,1}(x) &= \frac{1}{30}\left(4 + x + 2x^2 - 4x^3 + x^4\right) \\
S_{5,1}(x) &= \frac{1}{180}\left(18 - 9x^2 + 31x^3 - 16x^4 + 2x^5\right) \\
S_{5,1}(x) &= \frac{1}{180}\left(18 - 9x^2 + 31x^3 - 16x^4 + 2x^5\right),
\end{aligned} \tag{70}$$

and

$$\begin{aligned}
S_{3,2}(x) &= \frac{1}{48}\left(16 + 7x + 3x^2 + x^3\right) \\
S_{4,2}(x) &= \frac{1}{120}\left(24 + 6x + 2x^2 + x^3 + x^4\right) \\
S_{5,2}(x) &= \frac{1}{360}\left(48 + 6x - x^3 - 2x^4 + x^5\right) \\
S_{6,2}(x) &= \frac{1}{2520}\left(268 + 30x + 6x^2 + 5x^3 + 11x^4 - 44x^5 + 2x^6\right).
\end{aligned} \tag{71}$$

Besides, in 1972 H.M. Srivastava [31] evaluated four Laplace transforms of the generalized Bateman and Havelock functions. Two are presented here, long but complex expressions for the functions $k_{2,2k}(x)$ and $h_{2,2k}(x)$ are omitted here:

$$\begin{aligned}
L\{k_{0,2k}(x)\} &= \left[\frac{(1-s)}{(1-s^2)^{k+1}} - \frac{s}{\sqrt{\pi}}\sum_{m=1}^{k}\frac{\Gamma(k-m+3/2)}{\Gamma(k-m+2)(1-s^2)^m}\right] \\
L\{h_{0,2k}(x)\} &= \frac{1}{\pi}\left[\frac{2\ln(s)}{(1-s^2)^{k+1}} + \sum_{m=1}^{k}\frac{1}{(k-m+1)(1-s^2)^m}\right].
\end{aligned} \tag{72}$$

For the solution of pairs of dual equations, other researchers called Srivastava [28,29] reported a few more properties of the generalized Bateman functions, but these functions are slightly modified in their definitions.

5. The Bateman and Havelock Functions with Unrestricted Orders

General case of the Bateman and Havelock with any order

$$\begin{aligned}
k_\nu(x) &= \frac{2}{\pi}\int_0^{\pi/2} \cos(x\tan\theta - \nu\theta)\,d\theta \\
h_\nu(x) &= \frac{2}{\pi}\int_0^{\pi/2} \sin(x\tan\theta - \nu\theta)\,d\theta
\end{aligned} \tag{73}$$

is practically unknown in the literature, with only one exception, the definition of the Bateman function in terms of the Whittaker function $W_{k,\mu}(z)$ or Tricomi function $U(a,b,z)$ (particular cases of the confluent hypergeometric function) [7]

$$\begin{aligned}
k_{2\nu}(x) &= \frac{1}{\Gamma(\nu+1)}W_{\nu,1/2}(2x) = \frac{e^{-x}}{\Gamma(\nu+1)}U(-\nu,0;2x) \\
U(-\nu,0;2x) &= 2x\,U(1-\nu,2;2x) \\
k_{2n+2}(x) &= 2xe^{-x}\,_1F_1(-2n;2;2x); \quad n = 0,1,2,3,\ldots
\end{aligned} \tag{74}$$

Evidently, the corresponding generalized functions are

$$k_{\nu,\alpha,\beta}(x) = \frac{2}{\pi}\int_0^{\pi/2}(\cos\theta)^\alpha(\sin\theta)^\beta \cos(x\tan\theta - \nu\theta)\,d\theta$$
$$h_{\nu,\alpha,\beta}(x) = \frac{2}{\pi}\int_0^{\pi/2}(\cos\theta)^\alpha(\sin\theta)^\beta \sin(x\tan\theta - \nu\theta)\,d\theta,$$
(75)

where α, β and ν have any real value. By changing the integration variable in (73) and (75), $t = \tan(\theta)$, these functions can be expressed by infinite integrals

$$k_\nu(x) = \frac{2}{\pi}\int_0^\infty \frac{[\cos(xt)\cos[\nu\tan^{-1}(t)] + \sin(xt)\sin[\nu\tan^{-1}(t)]]}{1+t^2}\,dt$$
$$k_{\nu,\alpha,\beta}(x) = \frac{2}{\pi}\int_0^\infty \frac{t^\beta[\cos(xt)\cos[\nu\tan^{-1}(t)] + \sin(xt)\sin[\nu\tan^{-1}(t)]]}{(1+t^2)^{\alpha/2+\beta/2+1}}\,dt$$
$$h_\nu(x) = \frac{2}{\pi}\int_0^\infty \frac{[\sin(xt)\cos[\nu\tan^{-1}(t)] - \cos(xt)\sin[\nu\tan^{-1}(t)]]}{1+t^2}\,dt$$
$$h_{\nu,\alpha,\beta}(x) = \frac{2}{\pi}\int_0^\infty \frac{t^\beta[\sin(xt)\cos[\nu\tan^{-1}(t)] - \cos(xt)\sin[\nu\tan^{-1}(t)]]}{(1+t^2)^{\alpha/2+\beta/2+1}}\,dt.$$
(76)

In Figures 5 and 6 we illustrate the behavior and the symmetries with respect to the order of the Bateman functions with fractional positive and negative values: $\nu = n + 1/2$ and $\nu = -(n+1/2)$, with $n = 0, 1, 2, 3, 4, 5$. The same is demonstrated in Figures 7 and 8 for the Havelock-functions. Similarly as in (28), differentiation of the Bateman functions with respect to the argument x is for $k = 0, 1, 2, 3, \ldots$

$$\frac{\partial^{2k} k_\nu(x)}{\partial x^{2k}} = (-1)^k \frac{2}{\pi}\int_0^{\pi/2}(\tan\theta)^{2k}\cos(x\tan\theta - \nu\theta)\,d\theta$$
$$\frac{\partial^{2k+1} k_\nu(x)}{\partial x^{2k+1}} = (-1)^k \frac{2}{\pi}\int_0^{\pi/2}(\tan\theta)^{2k+1}\sin(x\tan\theta - \nu\theta)\,d\theta.$$
(77)

and in the case of the Havelock functions

$$\frac{\partial^{2k} h_\nu(x)}{\partial x^{2k}} = (-1)^k \frac{2}{\pi}\int_0^{\pi/2}(\tan\theta)^{2k}\sin(x\tan\theta - \nu\theta)\,d\theta$$
$$\frac{\partial^{2k+1} h_\nu(x)}{\partial x^{2k+1}} = (-1)^k \frac{2}{\pi}\int_0^{\pi/2}(\tan\theta)^{2k+1}\cos(x\tan\theta - \nu\theta)\,d\theta$$
(78)

$k = 0, 1, 2, 3, \ldots$

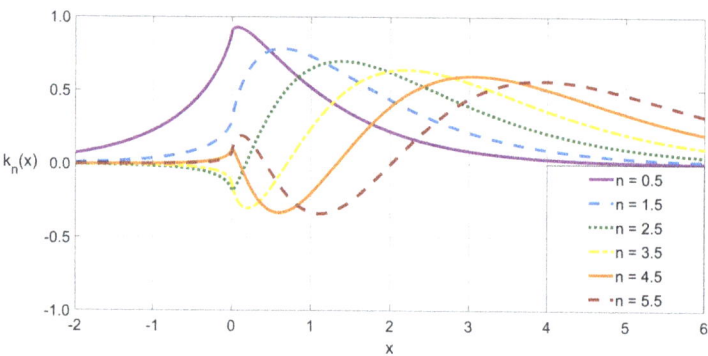

Figure 5. Bateman functions with positive $n + 1/2$ orders as a function of argument x.

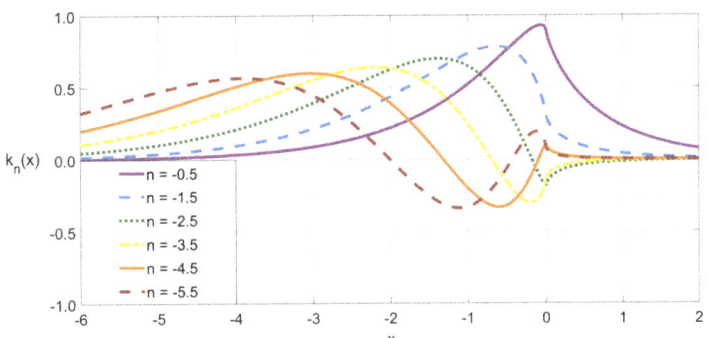

Figure 6. Bateman functions with negative $n + 1/2$ orders as a function of argument x.

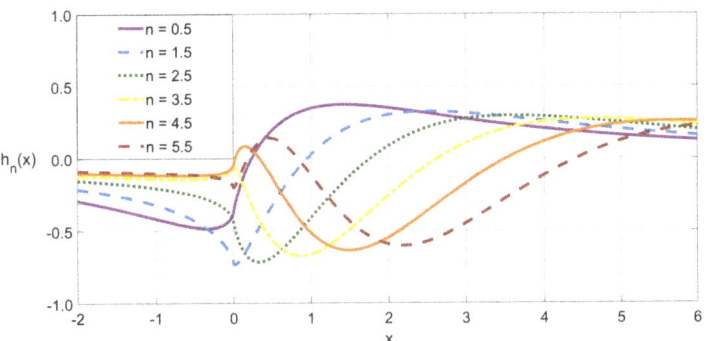

Figure 7. Havelock functions with positive $n + 1/2$ orders as a function of argument x.

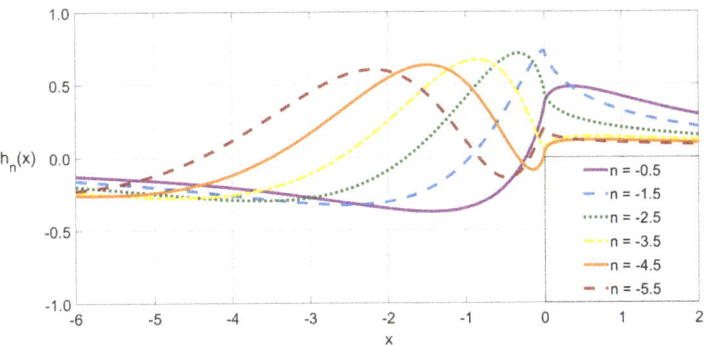

Figure 8. Havelock functions with negative $n + 1/2$ orders as a function of argument x.

Using the definition of these function from (73), it is possible to consider the Bateman and Havelock functions as functions of two variables x and ν. Thus, it is possible also to perform differentiation with respect to ν

$$\frac{\partial^{2k} k_\nu(x)}{\partial \nu^{2k}} = (-1)^k \frac{2}{\pi} \int_0^{\pi/2} \theta^{2k} \cos(x \tan\theta - \nu\theta) \, d\theta$$
$$\frac{\partial^{2k+1} k_\nu(x)}{\partial \nu^{2k+1}} = (-1)^k \frac{2}{\pi} \int_0^{\pi/2} \theta^{2k+1} \sin(x \tan\theta - \nu\theta) \, d\theta \qquad (79)$$
$$k = 0, 1, 2, 3, \ldots$$

and

$$\frac{\partial^{2k} h_\nu(x)}{\partial \nu^{2k}} = (-1)^k \frac{2}{\pi} \int_0^{\pi/2} \theta^{2k} \sin(x \tan \theta - \nu \theta) \, d\theta$$
$$\frac{\partial^{2k+1} h_\nu(x)}{\partial \nu^{2k+1}} = (-1)^k \frac{2}{\pi} \int_0^{\pi/2} \theta^{2k+1} \cos(x \tan \theta - \nu \theta) \, d\theta \quad (80)$$
$$k = 0, 1, 2, 3, \ldots$$

The first derivatives with respect to the order at fixed positive and negative values of argument x of the Bateman functions are plotted in Figures 9 and 10, and the same for the Havelock functions in Figures 11 and 12. As can be observed, these functions are symmetrical in both cases.

If orders are pure imaginary numbers $\nu = i\alpha$ then the Bateman and Havelock functions become complex functions which are expressed by integrals with integrands having products of trigonometric and hyperbolic functions.

As pointed out above, the Bateman and Havelock functions were introduced to the mathematical literature as solutions of particular problems in fluid mechanics [20,36].

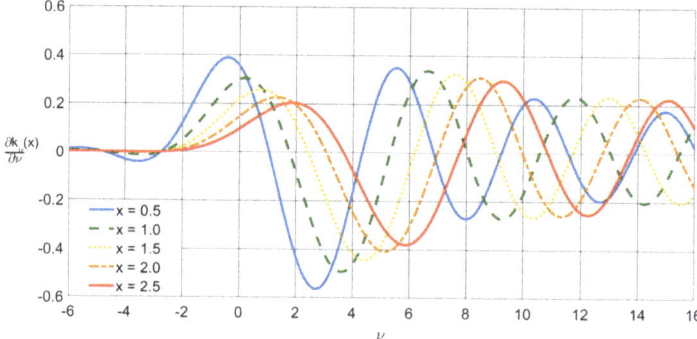

Figure 9. First derivatives of the Bateman functions with respect to the order at fixed positive values of argument x.

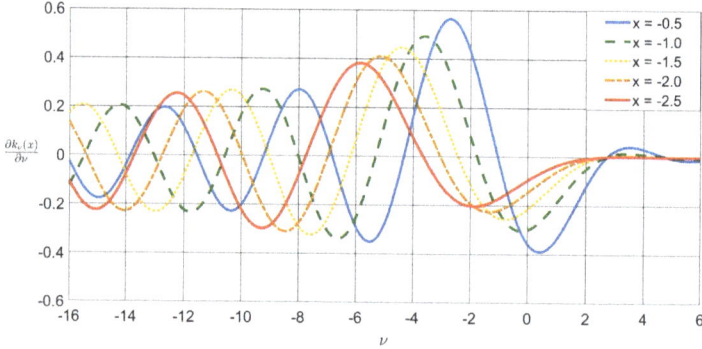

Figure 10. First derivatives of the Bateman functions with respect to the order at fixed negative values of argument x.

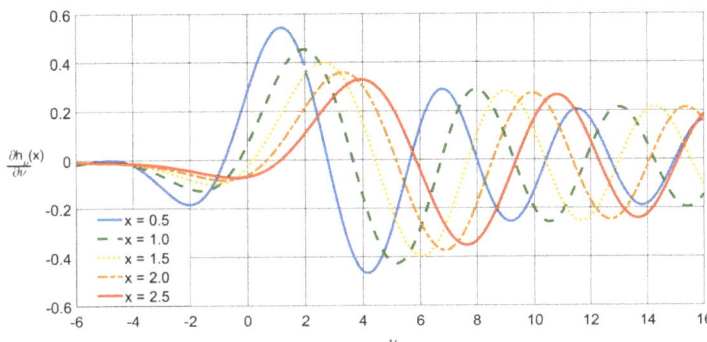

Figure 11. First derivatives of the Havelock functions with respect to the order at fixed positive values of argument x.

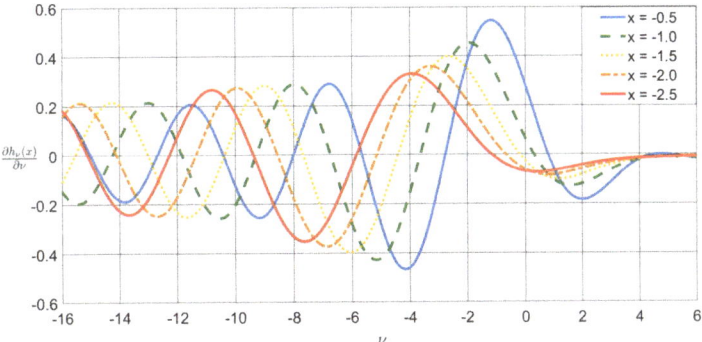

Figure 12. First derivatives of the Havelock functions with respect to the order at fixed negative values of argument x.

Years later, these functions were generalized to the form given in (64) and (75) [25,26,29–35]. It should be mentioned however, that historically, these proposed generalizations are not new, and they were already discussed much earlier by Giuliani in 1888 [43] and by Bateman in 1931 [44]. They also introduced similar trigonometric integrals, but in the context of particular cases of the Kummer confluent hypergeometric functions. It is rather strange, that in the later investigations [25,26,29–35], when the generalized Bateman and Havelock functions were proposed, previous studies on this subject were completely ignored. Considering that the trigonometric integrals and associated with them differential equations presented in the Giuliani and Bateman papers are of particular importance and interest, it was decided to summarize their results separately, in Appendix B.

6. The Bateman-Integral Functions

Analogous to the sine-integral, cosine integral and the Bessel-integral functions

$$\begin{aligned} si(x) &= -\int_x^\infty \frac{\sin t}{t}\,dt \\ Ci(x) &= -\int_x^\infty \frac{\cos t}{t}\,dt \\ Ji_\nu(x) &= -\int_x^\infty \frac{J_\nu(t)}{t}\,dt. \end{aligned} \tag{81}$$

Chaudhuri [26] has introduced the Bateman-integral function

$$ki_{2n}(x) = -\int_x^\infty \frac{k_{2n}(t)}{t} dt; \quad t > 0, \tag{82}$$

and mainly using operational calculus he has discussed its properties.

$$\begin{aligned} ki_{2n}(x) &= \int_0^x \frac{k_{2n}(t)}{t} dt + ki_{2n}(0) \\ ki_{2n}(0) &= 0 \quad ; \quad n = 2k; \quad k = 0, 1, 2, 3, \ldots \\ ki_{2n}(0) &= -\frac{2}{n}; \quad n = 2k+1 \end{aligned} \tag{83}$$

Using similarity with the Laguerre polynomials, Chaudhuri [26] derived the following series expressions for the Bateman-integral functions

$$\begin{aligned} ki_{2n}(x) &= \frac{e^{-x}}{n} \sum_{k=1}^n (-2)^k \binom{n}{k} L_{k-1}(x) \\ ki_{2n}(x) &= \frac{1}{nx}\left[n\, k_{2n}(x) - 2 \sum_{m=1}^n (-1)^k \binom{n}{m} [m\, k_{2m}(2x) + (m+1)k_{2m+2}(2x) - 2k_0(2x)]\right] \\ ki_{2n}(x) &= \frac{(-1)^{n-1} e^x}{2^{n+1}} \left[\sum_{m=1}^n m\, k_{2k}(x)\right] \\ L_{n-1}(x) &= \frac{e^x}{2^n} \sum_{m=1}^n (-1)^m \binom{n}{m} m\, ki_{2m}(x) \\ ki_2(x) &= -2\, k_0(x), \end{aligned} \tag{84}$$

and the recurrence and differential expressions

$$\begin{aligned} k_{2n}(x) &= \frac{(n-1)ki_{2n-2}(x) - (n+1)ki_{2n+2}(x)}{2} \\ n\, ki_{2n}(x) + (n+1)\, ki_{2n+2}(x) &= -2 \sum_{k=0}^n ki_{2k}(x) \\ x\, ki'_{2n}(x) &= \frac{(n-1)ki_{2n-2}(x) - (n+1)ki_{2n+2}(x)}{2} \\ x\, ki'_{2n}(x) &= k_{2n}(x). \end{aligned} \tag{85}$$

He was also able to relate the Bateman-integral functions with the Bessel and the Bessel integral functions

$$\begin{aligned} (n+1)\left[Ji_{n+1}(x)ki_{2n-2}(x) - Ji_{n-1}(x)ki_{2n+2}(x)\right] &= \\ 2xJi_{n-1}(x)ki'_{2n}(x) - 2nJi'_n(x)ki_{2n-2}(x) \\ \sum_{m=1}^\infty (-1)^m m\, ki_{2m}(x)\, ki_{2m}(y) &= J_0(2\sqrt{xy}). \end{aligned} \tag{86}$$

From integral expressions, the Laplace transform are presented here, when indefinite, definite and infinite integrals related to of the Bateman-integral functions are given in Appendix A:

$$L\{ki_{2n}(x)\} = \frac{1}{ns}\left[\left(\frac{1-s}{s+1}\right)^n - 1\right] = \frac{1}{ns}\sum_{k=1}^{n}(-1)^k\binom{n}{k}\left(\frac{2s}{s+1}\right)^k$$

$$L\{ki_{2n}(2x)\} = \frac{1}{ns}\left[\left(\frac{2-s}{s+2}\right)^n - 1\right] \quad (87)$$

$$L\{ki_0(x)\} = -\frac{\ln(s)}{s}$$

$$L\{ki_2(x)\} = -\frac{2}{s+1}.$$

It is also worthwhile to mention that Srivastava [25] expressed the Bateman-integral function in the following way

$$ki_{2n}(x) = \frac{\pi}{2}\left[k'_{2n}(x)h_{2n}(x) - h'_{2n}(x)k_{2n}(x)\right] =$$
$$\frac{\pi}{8x}[(2n+2)[k_{2n}(x)h_{2n+2}(x) - k_{2n+2}(x)h_{2n}(x)] - \quad (88)$$
$$(2n-2)[k_{2n}(x)h_{2n-2}(x) - k_{2n-2}(x)h_{2n}(x)]],$$

by including products of the Bateman and Havelock functions.

7. Conclusions

As solutions of fluid mechanics problems, more than ninety years ago, Havelock in 1925 and Bateman in 1931 introduced new functions which are expressed in terms of finite trigonometric integrals and discussed their properties. Initially, these functions found attention of a number of mathematicians who further developed this subject and proposed some generalizations. However, unfortunately, after a rather short period, the Havelock and Bateman functions were practically abandoned. Today, only the Bateman function is listed in mathematical handbooks as a particular case of the confluent hypergeometric function, thus as a minor special function. However, as is clearly showed in this survey, these functions have interesting properties and a rather large mathematical material was devoted and associated with them. This leads to conclusion that they should be treated as independent special functions. Since at present, in reference books, our knowledge about these functions is very limited, we decided to prepare this survey where basic properties of the Havelock and Bateman functions are presented. We have found useful for the reader's convenience to add two Appendixes: Appendix A is devoted to integrals associated with the Bateman and Bateman-integral functions whereas Appendix B is devoted to trigonometric integrals and differential equations associated with the Kummer Confluent Hypergeometric Functions according to the almost unknown papers by Giuliani [43] and by Bateman himself [44].

In Appendix C we have added the integral representations of the special functions used in this survey.

It is worth to note that the Bateman Manuscript is currently under revision with the name *Encyclopedia of Special Functions: the Askey-Bateman Project*, see [45]. However the volume dealing with the confluent hypergeometric functions is not yet available.

Funding: This research received no external funding.

Institutional Review Board Statement: Not applicable.

Informed Consent Statement: Not applicable.

Data Availability Statement: Not applicable.

Acknowledgments: The research of F.M. and A.C. has been carried out in the framework of the activities of the National Group of Mathematical Physics (GNFM, INdAM). All the the authors like to acknowledge the librarians of the Department of Physics and Astronomy of the University of Bologna to have found the pdf of several articles cited in the bibliography. The reader is kindly requested to accept the authors' somewhat informal style and to contact the corresponding author for pointing out possible misprints and mathematical errors.

Conflicts of Interest: The authors declare no conflict of interest.

Appendix A. Integrals Associated with the Bateman and Bateman-Integral Functions

The integrals presented here are compiled from the literature and they have a definite form. Their number can be enlarged by applying interconnections between the Bateman, Bateman-integral and other special functions and using operational calculus. Besides, there are many integrals which are expressed in term of infinite series, but they are omitted from this tabulation.

$$\int_0^1 (1-t)^{\beta-1} e^{\alpha t} k_{2n}(\alpha t)\, dt = \frac{(-1)^{n-1}(n-1)!\,\Gamma(\beta)}{\Gamma(\beta+n+1)} L_{n-1}^{(\beta+1)}(2\alpha);\quad \beta>0 \tag{A1}$$

$$\int_0^x k_{2m}(t) k_{2n}(x-t)\, dt = \int_0^x k_{2n}(t) k_{2m}(x-t)\, dt = \frac{1}{2}[k_{2m+2n-2}(x) + 2k_{2m+2n}(x) + k_{2m+2n+2}(x)] \tag{A2}$$

$$\int_0^x \frac{J_0(t) - k_0(t)}{t}\, dt = Ji_0(x) - ki_0(x) + \ln 2$$
$$\int_0^x \frac{J_n(t) - k_{2n}(t)}{t}\, dt = Ji_n(x) - ki_{2n}(x) + \frac{(-1)^n}{n} \tag{A3}$$

$$\int_0^\infty J_0(2\sqrt{at})\, k_{2n}(t)\, dt =$$
$$\frac{(-1)^{n-1}}{2}[(n-1)ki_{2n-2}(a) - 2nki_{2n}(a) + (n+1)ki_{2n+2}(a)] \tag{A4}$$

$$\int_0^\infty J_0(2\sqrt{at})\, k_{2n}(t)\, \frac{dt}{t} = (-1)^n ki_{2n}(a) \tag{A5}$$

$$\int_0^\infty e^{-t} J_1(2^{3/2}\sqrt{xt})\, k_{2n}(t)\, \frac{dt}{t} = \frac{(-1)^{n-1} x^{n-1/2} e^{-x}}{\sqrt{2}\, n!} \tag{A6}$$

$$\int_0^\infty e^{-at} t^{n+1/2} J_1(2\sqrt{xt})\, dt = \frac{(-1)^n \Gamma(n+2)\, e^{-x/2a}}{a^{n+1}\sqrt{x}} k_{2n+2}\left(\frac{x}{2a}\right);\quad a>0 \tag{A7}$$

$$\int_0^x \frac{J_n(t) - k_{2n}(t)}{t}\, dt = Ji_n(x) - ki_{2n}(x) + \frac{(-1)^n}{n} \tag{A8}$$

$$\int_0^\infty t^{n/2-1} e^{-t} J_{2-n}(4\sqrt{xt})\, k_{2n}(t)\, \frac{dt}{t} = \frac{x^{n/2-1} e^{-x}}{2} k_{2n}(x) \tag{A9}$$

$$\int_0^\infty \frac{e^{-bt^2} J_\lambda(a\sqrt{t^2+x^2})\, J_\nu(a\sqrt{t^2+x^2})}{t(t^2+x^2)^{(\lambda+\nu)/2}} k_{2n+2}(bt^2)\, dt = \frac{(-1)^n J_\lambda(ax) J_\nu(ax)}{(2n+2)\, x^{\lambda+\nu}} \tag{A10}$$
$$Re(\lambda+\nu) > -3/2$$

$$\int_0^\pi U_{2n}(\sqrt{x}\cos\theta)(\sin\theta)^2\, d\theta = \frac{\pi(2n)!\, e^{x/2}}{2xn!} k_{2n+2}\left(\frac{x}{2}\right) \tag{A11}$$
$$U_n(x) = (-1)^n e^{x^2} \frac{d^n}{dx^n}\{e^{-x^2}\}$$

$$\int_0^x \sin(x-t)\, ki_2(t)\, dt = \cos x - \sin x - e^{-x} \tag{A12}$$

$$\int_0^x \cos(x-t)\, ki_2(t)\, dt = \cos x - \sin x + e^{-x} \tag{A13}$$

$$\int_0^x \sinh(x-t)\,ki_2(t)\,dt = e^{-x}(1+x) - \cosh x \tag{A14}$$

$$\int_0^x \cosh(x-t)\,ki_2(t)\,dt = -xe^{-x} - \sinh x \tag{A15}$$

$$\int_0^x e^{x-t}\,ki_2(t)\,dt = -2\sinh x \tag{A16}$$

$$\int_0^x (x-t)\,e^{x-t}\,ki_4(t)\,dt = \sinh x - x\cosh x \tag{A17}$$

$$\int_0^\infty e^{-at}\,ki_0(bt)\,dt = \frac{1}{a}\ln\left(\frac{b}{a+b}\right); \quad a,b > 0 \tag{A18}$$

$$\int_0^\infty \frac{ki_2(at) - ki_2(bt)}{t}\,dt = 2\ln\left(\frac{a}{b}\right); \quad a,b > 0 \tag{A19}$$

$$\int_0^\infty J_0(2\sqrt{at})\,ki_{2n}(t)\,dt = (-1)^n \frac{k_{2n}(a)}{a} \tag{A20}$$

Appendix B. Trigonometric Integrals and Differential Equations Associated with the Kummer Confluent Hypergeometric Functions

In a paper devoted to the note of Kummer, where he introduced into mathematics the confluent hypergeometric function defined by the following polynomial series

$$_1F_1(a;b;x) = M(a,b,x) = 1 + \frac{a}{b}\frac{x}{1!} + \frac{a(a+1)}{b(b+1)}\frac{x^2}{2!} + \frac{a(a+1)(a+2)}{b(b+1)(b+2)}\frac{x^3}{3!} + \cdots \tag{B1}$$

$Re(b) > Re(a) > 0$.

The Italian mathematician Giulio Giuliani [43] in 1888 considered the trigonometric integral (the original notation is replaced here by that used in this survey)

$$I(x) = \int_0^{\pi/2} (\cos\theta)^{\alpha-1} \cos\left(\frac{x}{2}\tan\theta + n\theta\right) d\theta = \frac{\pi}{2} k_{-n,\alpha-1}\left(\frac{x}{2}\right), \quad \alpha > 1. \tag{B2}$$

We note that this integral is one of particular solutions of the following differential equation

$$4x\frac{d^2 I(x)}{dx^2} - 4(\alpha-1)\frac{dI(x)}{dx} - (x+2n)I(x) = 0. \tag{B3}$$

Besides, Giuliani introduced two integrals coming from (B2)

$$U_n(\alpha, x) = \int_0^{\pi/2} (\cos\theta)^{\alpha-1} \cos\left(\frac{x}{2}\tan\theta\right)\cos(n\theta)\,d\theta,$$

$$V_n(\alpha, x) = \int_0^{\pi/2} (\cos\theta)^{\alpha-1} \sin\left(\frac{x}{2}\tan\theta\right)\sin(n\theta)\,d\theta, \tag{B4}$$

when

$$\int_0^{\pi/2} (\cos\theta)^{\alpha-1} \cos\left(\frac{x}{2}\tan\theta + n\theta\right) d\theta = U_n(\alpha, x) - V_n(\alpha, x). \tag{B5}$$

He showed that these integrals are solutions of the set of differential equations of the first order

$$2(\alpha-1)\frac{dU_n(\alpha, x)}{dx} + \frac{x}{2}U_n(\alpha-2, x) - nV_n(\alpha, x) = 0,$$

$$2(\alpha-1)\frac{dV_n(\alpha, x)}{dx} + \frac{x}{2}V_n(\alpha-2, x) - nU_n(\alpha, x) = 0, \tag{B6}$$

and of the second order

$$2x\frac{d^2U_n(\alpha,x)}{dx^2} - 2(\alpha-1)\frac{dU_n(\alpha,x)}{dx} - \frac{x}{2}U_n(\alpha,x) + nV_n(\alpha,x) = 0,$$
$$2x\frac{d^2V_n(\alpha,x)}{dx^2} - 2(\alpha-1)\frac{dV_n(\alpha,x)}{dx} - \frac{x}{2}V_n(\alpha,x) + nU_n(\alpha,x) = 0.$$
(B7)

From (B6) and (B7) it is possible to obtain a differential equation of the fourth order

$$4x^2\frac{d^4U_n(\alpha,x)}{dx^4} - 8(\alpha-2)x\frac{d^3U_n(\alpha,x)}{dx^3} -$$
$$2\left[x^2 - 2(\alpha-1)(\alpha-2)\right]\frac{d^2U_n(\alpha,x)}{dx^2} + 2x(\alpha-2)\frac{dU_n(\alpha,x)}{dx} -$$
$$\left(\frac{x^2}{4} + n^2 + 1 - \alpha\right)U_n(\alpha,x) = 0,$$
$$V_n(\alpha,x) = \frac{1}{n}\left(-2x\frac{d^2U_n(\alpha,x)}{dx^2} + 2(\alpha-1)\frac{dU_n(\alpha,x)}{dx} + \frac{x}{2}U_n(\alpha,x)\right).$$
(B8)

In terms of the Kummer confluent hypergeometric functions Giuliani was able to obtain that

$$\int_0^{\pi/2} (\cos\theta)^{\alpha-1}\cos(\frac{x}{2}\tan\theta + n\theta)\,d\theta = U_n(\alpha,x) - V_n(\alpha,x) =$$
$$\left[\frac{\pi\Gamma(\alpha-1)e^{-x/2}}{2^\alpha\Gamma\left(\frac{\alpha-n+1}{2}\right)\Gamma\left(\frac{\alpha+n+1}{2}\right)}\,{}_1F_1(\frac{\alpha-n+1}{2};1-\alpha;x) - \right.$$
$$\left.\frac{\pi^2\cos\left(\frac{\alpha-n}{2}\right)x^\alpha e^{-x/2}}{2^\alpha\sin(\pi\alpha)\Gamma(\alpha)}\,{}_1F_1(\frac{\alpha+n+1}{2};\alpha+1;x)\right],$$
(B9)

and

$$U_n(\alpha,x) + V_n(\alpha,x) = \left[\frac{\pi\Gamma(\alpha-1)e^{-x/2}}{2^\alpha\Gamma\left(\frac{\alpha+n+1}{2}\right)\Gamma\left(\frac{\alpha-n+1}{2}\right)}\,{}_1F_1(\frac{1-\alpha-n}{2};1-\alpha;x),\right.$$
$$\left.-\frac{\pi^2\cos\left(\frac{\alpha+n}{2}\right)x^\alpha e^{-x/2}}{2^\alpha\sin(\pi\alpha)\Gamma(\alpha)}\,{}_1F_1(\frac{\alpha-n+1}{2};\alpha+1;x)\right].$$
(B10)

These expressions can be presented in terms of the generalized Bateman functions defined in (75)

$$k_{-\nu,\alpha,0}(x) =$$
$$\left[\frac{\Gamma(\alpha)e^{-x}}{2^\alpha\Gamma\left(\frac{\alpha-\nu}{2}+1\right)\Gamma\left(\frac{\alpha+\nu}{2}\right)}\,{}_1F_1(\frac{\alpha-\nu}{2}+1;-\alpha;2x) - \right.$$
$$\left.\frac{\pi\cos\left(\frac{\alpha-\nu+1}{2}\right)x^{\alpha+1}e^{-x}}{2^\alpha\sin[\pi(\alpha+1)]\Gamma(\alpha+1)}\,{}_1F_1(\frac{\alpha+\nu}{2}+1;\alpha+2;2x)\right],$$
(B11)

and

$$k_{\nu,\alpha,0}(x) = \left[\frac{\Gamma(\alpha)e^{-x}}{2^\alpha\Gamma\left(\frac{\alpha+\nu}{2}+1\right)\Gamma\left(\frac{\alpha-\nu}{2}+1\right)}\,{}_1F_1(\frac{-\alpha-\nu}{2};-\alpha;2x)\right.$$
$$\left.-\frac{\pi\cos\left(\frac{\alpha+\nu+1}{2}\right)x^{\alpha+1}e^{-x}}{2^\alpha\sin[\pi(\alpha+1)]\Gamma(\alpha+1)}\,{}_1F_1(\frac{\alpha-\nu}{2}+1;\alpha+2;2x)\right].$$
(B12)

As shown by Giuliani, by changing the integration variable, the finite trigonometric integrals can be presented as the infinite integrals, for example

$$\int_0^{\pi/2} (\cos\theta)^\alpha \cos(\frac{x}{2}\tan\theta)\, d\theta = \int_0^\infty \frac{\cos(\frac{xt}{2})}{(1+t^2)^{\alpha/2+1}}\, dt. \tag{B13}$$

Considering the case $\alpha = 1$ in (B2), Bateman [44] in 1931 noted the link that exists between the investigated by Giuliani integral and the k-Bateman function with negative order. He also found that the solution of the following third order differential equation

$$x\frac{d^3 I(x)}{dx^3} - (\alpha-1)\frac{d^2 I(x)}{dx^2} - (x+n)\frac{dI(x)}{dx} - \beta I(x) = 0, \tag{B14}$$

is given by the following trigonometric integral

$$I(x) = \int_0^{\pi/2} (\cos\theta)^\alpha (\sin\theta)^{\beta-1} \cos(x\tan\theta + n\theta)\, d\theta = \frac{\pi}{2} k_{n,\alpha,\beta-1}(x). \tag{B15}$$

Besides, Bateman showed that for $x > 0$:

$$\int_0^{\pi/2} (\cos\theta)^m \cos[x\tan\theta + (m+2n)\theta]\, d\theta = \frac{e^x \sin(\pi n)}{2^{k+1}} \int_0^1 t^k (1-t)^{n-1} e^{-2x/t}\, dt,$$
$$\int_0^{\pi/2} \cos[x\tan\theta + (m+2n)\theta]\, d\theta = \frac{e^x \sin(\pi n)}{2} \int_0^1 (1-t)^{n-1} e^{-2x/t}\, dt = \frac{\pi}{2} k_{-2n}(x). \tag{B16}$$

As can be observed, the included material from the 1888 paper by Giuliani and from the 1931 paper by Bateman is important from the historical and mathematical points of view.

Appendix C. Integral Representations of Special Functions Used in This Survey

Hypergeometric Function

$$_2F_1(a,b;c;x) = \frac{\Gamma(c)}{\Gamma(a)\Gamma(b)} \int_0^1 \frac{t^{b-1}(1-t)^{c-b-1}}{(1-xt)^a}\, dt \quad Re(c) > Re(a) > 0. \tag{C1}$$

Kummer Confluent Hypergeometric Function

$$_1F_1(a,b;x) = M(a,b,x) = \frac{\Gamma(b)}{\Gamma(a)\Gamma(b-a)} \int_0^1 t^{a-1} e^{xt} (1-t)^{b-a-1}\, dt \quad Re(b) > Re(a) > 0. \tag{C2}$$

Tricomi Confluent Hypergeometric Function

$$U(a,b,x) = \frac{1}{\Gamma(a)} \int_0^\infty t^{a-1} e^{-xt} (1+t)^{b-a-1}\, dt, \quad Re(b) > Re(a) > 0. \tag{C3}$$

Whittaker Functions

$$M_{\kappa,\mu}(x) = \frac{\Gamma(1+2\mu)\, x^{\mu+1/2}\, e^{-x/2}}{\Gamma(\mu+\kappa+1/2)\,\Gamma(\mu-\kappa+1/2)} \int_0^1 t^{\mu-\kappa-1/2} e^{xt} (1-t)^{\mu+\kappa-1/2}\, dt,$$
$$M_{\kappa,\mu}(x) = x^{\mu+1/2} e^{-x/2} M(\mu-\kappa+1/2, 1+2\mu, x)$$
$$Re(\mu \pm \kappa + 1/2) > 0. \tag{C4}$$

$$W_{\kappa,\mu}(x) = \frac{x^{\mu+1/2}e^{-x/2}}{\Gamma(\mu-\kappa+1/2)} \int_0^\infty t^{\mu-\kappa-1/2} e^{-xt}(1+t)^{\mu+\kappa-1/2} dt,$$

$$W_{\kappa,\mu}(x) = x^{\mu+1/2}e^{-x/2} U(\mu-\kappa+1/2, 1+2\mu, x) \qquad \text{(C5)}$$

$$Re(\mu - \kappa + 12) > 0.$$

Bessel Functions

$$J_\nu(x) = \frac{1}{\pi}\int_0^\pi \cos(x\sin\theta - \nu\theta)\, d\theta - \frac{\sin(\pi\nu)}{\pi}\int_0^\infty e^{-x\sinh t - \nu t}\, dt. \qquad \text{(C6)}$$

$$Y_\nu(x) = \frac{1}{\pi}\int_0^\pi \sin(x\sin\theta - \nu\theta)\, d\theta - \frac{\sin(\pi\nu)}{\pi}\int_0^\infty e^{-x\sinh t - \nu t}\left[e^{\nu t} + e^{-\nu t}\cos(\pi\nu)\right] dt. \qquad \text{(C7)}$$

$$I_\nu(x) = \frac{1}{\pi}\int_0^\pi e^{x\cos\theta}\cos(\nu\theta)\, d\theta - \frac{\sin(\pi\nu)}{\pi}\int_0^\infty e^{-x\cosh t - \nu t}\, dt. \qquad \text{(C8)}$$

$$K_\nu(x) = \frac{\Gamma(\nu+1/2)(2x)^\nu}{\sqrt{\pi}}\int_0^\infty \frac{\cos(xt)}{(1+t^2)^{\nu+1/2}}\, dt = \int_0^\infty e^{-x\cosh t}\cosh(\nu t)\, dt. \qquad \text{(C9)}$$

Struve Functions

$$H_\nu(x) = \frac{2(x/2)^\nu}{\Gamma(\nu+1/2)\sqrt{\pi}}\int_0^1 (1-t^2)^{\nu-1/2}\sin(xt)\, dt, \quad Re(\nu) > -1/2. \qquad \text{(C10)}$$

$$L_\nu(x) = \frac{2(x/2)^\nu}{\Gamma(\nu+1/2)\sqrt{\pi}}\int_0^{\pi/2} (\sin t)^{2\nu}\sinh(x\cos t)\, dt, \quad Re(\nu) > -1/2. \qquad \text{(C11)}$$

Lommel Functions

$$S_{\mu,\nu}(x)x^\mu \int_0^\infty e^{-xt}\, {}_2F_1\left(\frac{1-\mu+|n\nu}{2}, \frac{1-\mu-\nu}{2}; \frac{1}{2}; -t^2\right) dt, \quad Re(x) > 0. \qquad \text{(C12)}$$

References

1. Erdélyi, A.; Magnus, W.; Oberhettinger, F.; Tricomi, F.G. *Higher Transcendental Functions*; McGraw-Hill: New York, NY, USA, 1953; Volume 1.
2. Martin Amima, P.A. Harry Bateman: From Manchester to Manuscript Project. *Math. Today* **2010**, *46*, 82–85.
3. Bateman, H. *Partial Differential Equations of Mathematical Physics*; Cambridge University Press: Cambridge, UK, 1932.
4. Roanes-Lozano, E.; Gonzáles-Bermejo, A.; Roanes-Macias, E.; Cabezas, J. Application of computer algebra to pharmacokinetics: The Bateman equation. *SIAM Rev.* **2006**, *48*, 133–146. [CrossRef]
5. Merle, U.; Laßmann, A.; Dressel, A.R.; Braun, P. Evaluation of the COVID-19 pandemic using an algorithm based on the Bateman function: Prediction of disease progression using observational data for the city of Heidelberg, Germany. *Int. J. Pharm. Therap.* **2020**, *58*, 366–374. [CrossRef] [PubMed]
6. Erdélyi, A.; Magnus, W.; Oberhettinger, F.; Tricomi, F.G. *Tables of Integral Transforms*; McGraw-Hill: New York, NY, USA, 1954.
7. Abramowitz, M.; Stegun, I.A. *Handbook of Mathematical Functions with Formulas, Graphs, and Mathematical Tables*; Applied Mathematics Series; U.S. National Bureau of Standards: Washington, DC, USA, 1964; Volume 55.
8. Magnus, W.; Oberhettinger, F.; Soni, R.P. *Formulas and Theorems for the Special Functions of Mathematical Physics*, 3rd ed.; Springer: Berlin, Germeny, 1966.
9. Roberts, G.E.; Kaufman, H. *Table of Laplace Transforms*; W.B. Saunders Co.: Philadelphia, PA, USA, 1966.
10. Apelblat, A. *Table of Definite and Infinite Integrals*; Elsevier Sci. Publ. Co.: Amsterdam, The Netherlands, 1983.
11. Oberhettinger, F.; Badii, L. *Laplace Transforms*; Springer: Berlin, Germany, 1970.
12. Oberhettinger, F.; Badii, L. *Table of Bessel Transforms*; Springer: Berlin, Germeny, 1972.
13. Oberhettinger, F.; Badii, L. *Table of Mellin Transforms*; Springer: Berlin, Germany, 1974.
14. Prudnikov, A.P.; Brychkov, Y.A.; Marichev, D.I. *Special Functions*; Integrals and Series; Gordon and Breach: New York, NY, USA, 1986; Volumes 1 and 2.

15. Prudnikov, A.P.; Brychkov, Y.A.; Marichev, D.I. *More Special Functions*; Integrals and Series; Gordon and Breach: New York, NY, USA, 1990; Volume 3.
16. Apelblat, A. *Tables of Integrals and Series*; Verlag Harri Deutsch: Frankfurt am Main, Germany, 1996
17. Oldham, K.B.; Myl, J.M.; Spanier, J. *An Atlas of Functions*, 2nd ed.; Springer: New York, NY, USA, 2008.
18. Brychkov, Y.A. *Handbook of Special Functions*; Derivatives, Integrals, Series and Other Formulas; CRC Press: Boca Raton, FL, USA, 2008.
19. Olver, F.W.J.; Lozier, D.W.; Boisvert, R.F.; Clark, C.W. (Eds.) *NIST Handbook of Mathematical Functions*; US Department of Commerce, National Institute of Standards and Technology: Washington, DC, USA, 2010.
20. Bateman, H. The k-function, a particular case of the confluent hypergeometric function. *Trans. Am. Math. Soc.* **1931**, *33*, 817–831.
21. Shastri, N.A. On some properties of the k-function. *Phil. Mag.* **1935**, *20*, 468–478. [CrossRef]
22. Shabde, N.G. On some series and integrals involving k-functions. *J. Indian Math. Soc.* **1939**, *3*, 307–311.
23. Shabde, N.G. On some results involving confluent hypergeometric functions. *J. Indian Math. Soc.* **1940**, *4*, 151–157.
24. Shastri, N.A. Some results involving Bateman's polynomials. *Bull. Calcutta Math. Soc.* **1940**, *32*, 89–94.
25. Srivastava, H.M. On Bateman's function and an allied function. *Bull. Calcutta Math. Soc.* **1950**, *42*, 82–88.
26. Charkrabarty, N.K. On generalization of Bateman K-function. *Bull. Calcutta Math. Soc.* **1953**, *45*, 1–7.
27. Sarkar, G.K. On integral representations of the generalized k-functions of Bateman and its connection with Legendre and parabolic cylinder functions. *Bull. Calcutta Math. Soc.* **1954**, *46*, 89–94.
28. Srivastava, H.M. On certain relations involving the generalized K-function of Bateman. *Ganita* **1954**, *5*, 183–189.
29. Srivastava, K.N. On dual series relations involving series of generalized Bateman K-functions. *Proc. Am. Math. Soc.* **1966**, *17*, 796–802.
30. Srivastava, H.M. A pair of dual series equations involving generalized Bateman k-functions. *Proc. Ind. Math.* **1972**, *75*, 53–61. [CrossRef]
31. Srivastava, H.M. On a generalization of a function allied to Bateman's function. *Mat. Vestnik* **1972**, *9*, 197–204.
32. Srivastava, T.N. Some theorems on generalized Bateman K-functions. *Panjab Univ. J. Math. (Lahore)* **1973**, *8*, 239–249.
33. Joshi, B.K. On an inversion integral involving generalized Bateman function. *Math. Student* **1974**, *42*, 183–184.
34. Narain, K.; Singh, V.B.; Lal, M. Triple series equations involving generalized Bateman k-functions. *Indian J. Pure Appl. Math.* **1984**, *15*, 435–440.
35. Dwivedi, A.P.; Chandel, J. n-series equations involving generalized Bateman K functions. *Acta Cienc. Indica Math.* **1996**, *22*, 236–240.
36. Havelock, T.H. The method of images in some problems of surface waves. *Proc. R. Soc. Lond. A* **1925**, *108*, 582–591.
37. Bateman, H. Some definite integrals occurring in Havelock's work on the wave resistance of ships. *Math. Mag.* **1949**, *23*, 1–4. [CrossRef]
38. Chaudhur, J. On Bateman-integer function. *Math. Zeitschr.* **1968**, *78*, 25–32. [CrossRef]
39. Koepf, W.; Schmersau, D. Bounded nonvanishing functions and Bateman functions. *Complex Var.* **1994**, *25*, 237–259. [CrossRef]
40. Koepf, W. Identities for families of orthogonal polynomials and special functions. *ITSP* **1997**, *5*, 69–102. [CrossRef]
41. Koepf, W. *Hypergeometric Summations. An Algorithmic Approach to Summation and Special Function Identities*; F. Vieweg and Sohn Verlag: Braunschweig, Germany, 1998.
42. Jolley, L.B.W. *Summation of Series*; Collection of Series; Dover Publ. Inc.: New York, NY, USA, 1961.
43. Giuliani, G. Aggiunte ad una memoria del Sig. Kummer. Battaglini. *Giornale Matematiche* **1888**, *26*, 234–250. (In Italian)
44. Bateman, H. Solution of a certain partial differential equation. *Proc. Natl. Acad. Sci. USA* **1931**, *17*, 562–567. [CrossRef]
45. Ismail, M.E.H.; Van Assch, W. (Eds.) *Encyclopedia of Special Functions: The Askey-Bateman Project: Vol. I: Orthogonal Polynomials (Published), Vol. 2: Multivariable Special Functions (Published), Vol. 3: Hypergeometric and Basic Hypergeometric Functions (in Preparation)*; Cambridge University Press: Cambridge, UK, 2020–2022.

Review

Series in Le Roy Type Functions: A Set of Results in the Complex Plane—A Survey

Jordanka Paneva-Konovska

Institute of Mathematics and Informatics, Bulgarian Academy of Sciences, 1113 Sofia, Bulgaria;
jpanevakonovska@gmail.com

Abstract: This study is based on a part of the results obtained in the author's publications. An enumerable family of the Le Roy type functions is considered herein. The asymptotic formula for these special functions in the cases of 'large' values of indices, that has been previously obtained, is provided. Further, series defined by means of the Le Roy type functions are considered. These series are studied in the complex plane. Their domains of convergence are given and their behaviour is investigated 'near' the boundaries of the domains of convergence. The discussed asymptotic formula is used in the proofs of the convergence theorems for the considered series. A theorem of the Cauchy–Hadamard type is provided. Results of Abel, Tauber and Littlewood type, which are analogues to the corresponding theorems for the classical power series, are also proved. At last, various interesting particular cases of the discussed special functions are considered.

Keywords: Le Roy functions and series in them; inequalities; asymptotic formula; convergence of power and functional series in complex plane; Cauchy–Hadamard, Abel, Tauber and Littlewood type theorems

MSC: 30D20; 33E12; 30A10; 40E10; 30D15; 40A30; 40G10; 40E05

Citation: Paneva-Konovska, J. Series in Le Roy Type Functions: A Set of Results in the Complex Plane—A Survey. *Mathematics* **2021**, *9*, 1361. https://doi.org/10.3390/math9121361

Academic Editor: Francesco Mainardi

Received: 26 May 2021
Accepted: 9 June 2021
Published: 12 June 2021

Publisher's Note: MDPI stays neutral with regard to jurisdictional claims in published maps and institutional affiliations.

Copyright: © 2021 by the authors. Licensee MDPI, Basel, Switzerland. This article is an open access article distributed under the terms and conditions of the Creative Commons Attribution (CC BY) license (https://creativecommons.org/licenses/by/4.0/).

1. Introduction

In two recent papers, S. Gerhold [1] and, independently, R. Garra and F. Polito [2] introduced the new special function

$$F_{\alpha,\beta}^{(\gamma)}(z) = \sum_{k=0}^{\infty} \frac{z^k}{[\Gamma(\alpha k + \beta)]^\gamma}, \quad z \in \mathbb{C}, \tag{1}$$

for complex values of the variable z and values of parameters $\alpha > 0$, $\beta > 0$, $\gamma > 0$. On a later stage its definition is extended by R. Garrappa, S. Rogosin and F. Mainardi [3] under more general conditions for the parameters. However, making sure that the coefficients $[\Gamma(\alpha k + \beta)]^{-\gamma}$ in the Expansion (1) exist, the values of the parameters have to be restricted. A natural restriction in this direction would be the following:

$$\alpha, \beta \in \mathbb{C}, \quad \gamma > 0. \tag{2}$$

As is established in [3], this function turns out to be an entire function of the complex variable z for all values of the parameters such that

$$\Re(\alpha) > 0, \ \beta \in \mathbb{C}, \ \gamma > 0. \tag{3}$$

Actually, this function has been recently considered in [1–6] from various points of view. Some of its important properties can be seen therein. For example, different asymptotic formulae can be found in S. Gerhold [1] and R. Garrappa, S. Rogosin, F. Mainardi [3], for complete monotonicity see K. Gorska, A. Horzela, R. Garrappa [4] and T. Simon [5]. For studying its properties in relation to some integro-differential operators involving

Hadamard fractional derivatives or hyper-Bessel-type operators see Garra-Polito [2], different integral representations can be seen in [3] and Pogány [6].

The function (1) is a natural generalization of the so-called Le Roy function

$$F^{(\gamma)}(z) = \sum_{k=0}^{\infty} \frac{z^k}{[\Gamma(k+1)]^\gamma} = \sum_{k=0}^{\infty} \frac{z^k}{[k!]^\gamma}, \quad z \in \mathbb{C}, \gamma \in \mathbb{C}, \quad (4)$$

which was named after the great French mathematician Édouard Louis Emmanuel Julien Le Roy (1870–1954), and probably for that reason the authors of [3] use the name Le Roy type function for the function $F^{(\gamma)}_{\alpha,\beta}$.

Keeping with this, and for the sake of brevity, we often use in this paper the name Le Roy type function for the function $F^{(\gamma)}_{\alpha,\beta}$, defined by (1). In this paper, considering the Le Roy type functions (1), we discuss various earlier results which are needed here. These are results related to inequalities in the complex plane $\mathbb{C}$ and on its compact subsets and asymptotic formula for 'large' values of indices of the functions (1). Further, considering series in such a kind of functions, we provide results for their domains of convergence and investigate their behaviour 'near' the boundaries of their domains of convergence.

In the series of papers [7–10], as well as in the recent book [11], we studied series in systems of some representatives of the special functions of fractional calculus, which are fractional index analogues of the Bessel functions and also multi-index Mittag–Leffler functions (in the sense of [12–15]), and we have proved various results connected with their convergence in the complex domains.

2. Inequalities and an Asymptotic Formula

For our purpose we consider the family

$$F^{(\gamma)}_{\alpha,n}(z) = \sum_{k=0}^{\infty} \frac{z^k}{[\Gamma(\alpha k + n)]^\gamma}, \quad z \in \mathbb{C}; n \in \mathbb{N}, \alpha > 0, \gamma > 0, \quad (5)$$

where $\mathbb{N}$ means the set of positive integers.

We are going to deal with some analytical transformations of the function (5) for each value of the parameter n. The following results hold true (for the formulation and proof see Paneva-Konovska [16]).

Lemma 1. *Let $z \in \mathbb{C}$, $\alpha > 0$, $\gamma > 0$, $n \in \mathbb{N}$ and let $K \subset \mathbb{C}$ be a nonempty compact set. Then there exists an entire function $\vartheta^{(\gamma)}_{\alpha,n}$ such that*

$$F^{(\gamma)}_{\alpha,n}(z) = \frac{1}{[\Gamma(n)]^\gamma}\left(1 + \vartheta^{(\gamma)}_{\alpha,n}(z)\right). \quad (6)$$

The entire function $\vartheta^{\gamma}_{\alpha,n}$ satisfies the following inequality

$$\left|\vartheta^{(\gamma)}_{\alpha,n}(z)\right| \leq \frac{[\Gamma(\alpha+1)]^\gamma [\Gamma(n)]^\gamma}{[\Gamma(\alpha+n)]^\gamma}\left(F^{(\gamma)}_{\alpha,1}(|z|) - 1\right), \quad z \in \mathbb{C}, \quad (7)$$

Moreover there exists a positive constant $C = C(K)$, such that

$$\max_{z \in K}\left|\vartheta^{(\gamma)}_{\alpha,n}(z)\right| \leq C \frac{[\Gamma(n)]^\gamma}{[\Gamma(\alpha+n)]^\gamma}, \quad (8)$$

for all the positive integers n.

Theorem 1. *Let $z \in \mathbb{C}$; $n \in \mathbb{N}$, $\alpha > 0$, $\gamma > 0$. Then the Le Roy type functions $F_{\alpha,n}^{(\gamma)}$ have the following asymptotic formula*

$$F_{\alpha,n}^{(\gamma)}(z) = \frac{1}{[\Gamma(n)]^\gamma} \left(1 + \vartheta_{\alpha,n}^{(\gamma)}(z)\right), \quad \vartheta_{\alpha,n}^{(\gamma)}(z) \to 0 \text{ as } n \to \infty. \tag{9}$$

The convergence is uniform in the nonempty compact subsets of the complex plane.

The results above allow us to write the next two remarks.

Remark 1. *According to the asymptotic Formula (9), it follows that there exists a natural number M such that the functions $[\Gamma(n)]^\gamma F_{\alpha,n}^{(\gamma)}(z)$ do not vanish for any n great enough (say $n > M$).*

Remark 2. *Note that each function $F_{\alpha,n}^{(\gamma)}(z)$ ($n \in \mathbb{N}$), being an entire function, no identically zero, has at most finite number of zeros in the closed and bounded set $|z| \leq R$ ([17], p. 305). Moreover, because of Remark 1, at most finite number of these functions have some zeros.*

3. Series in Le Roy Type Functions

For the sake of simplicity, we introduce an auxiliary family of functions, related to the Le Roy type functions, adding $\widetilde{F}_{\alpha,0}^{(\gamma)}(z)$ just for completeness, namely:

$$\widetilde{F}_{\alpha,0}^{(\gamma)}(z) = 1, \quad \widetilde{F}_{\alpha,n}^{(\gamma)}(z) = z^n \, [\Gamma(n)]^\gamma \, F_{\alpha,n}^{(\gamma)}(z), \quad n \in \mathbb{N}; \, \alpha > 0, \, \gamma > 0, \tag{10}$$

and we study the series with complex coefficients a_n ($n \in \mathbb{N}_0$, i.e., $n = 0, 1, 2, ...$) in these functions for $z \in \mathbb{C}$, namely:

$$\sum_{n=0}^{\infty} a_n \, \widetilde{F}_{\alpha,n}^{(\gamma)}(z). \tag{11}$$

Our major goal is to study the convergence of the series (11) in the complex plane. We give results, corresponding to the classical Cauchy–Hadamard theorem and Abel lemma for the power series and more precise results, giving the behaviour of the series 'near' the boundary of the domain of convergence, as well. Such kind of results may be useful for studying the solutions of some fractional order differential and integral equations, expressed in terms of series (or series of integrals) in special functions of the type (10) and their special cases (as for example in Kiryakova et al., in [18]—for the Mittag–Leffler functions; in [19]—for the hyper-Bessel functions; in [14,20]—for the multi-index Mittag–Leffler functions). Convergence theorems are also obtained for series in other special functions, for example, for series in Laguerre and Hermite polynomials (the results are obtained in a number of publications and they can be seen in Rusev [21]), and respectively by the author for series in Bessel and Mittag–Leffler types functions in the previous papers [7–10] and the book [11].

4. Cauchy–Hadamard Type Theorem and Corollaries

Let us denote by $D(0; R)$ the open disk with the radius R and centred at the origin, and let the circle $C(0; R)$ be its boundary, i.e.

$$D(0; R) = \{z : |z| < R\} \quad \text{and} \quad C(0; R) = \{z : |z| = R\} \quad (z \in \mathbb{C}).$$

In the beginning, we give a theorem of the Cauchy–Hadamard type for the series (11).

Theorem 2 (of Cauchy–Hadamard type). *Let $z \in \mathbb{C}$, $n \in \mathbb{N}_0$, $\alpha > 0$, $\gamma > 0$. Then the domain of convergence of the series (11) with complex coefficients a_n is the disk $D(0; R)$ with a radius of convergence*

$$R = 1 / \limsup_{n \to \infty} (\,|a_n|\,)^{1/n}. \tag{12}$$

The cases $R = 0$ and $R = \infty$ are included in the general case.

Let us note that the series (11) absolutely converges in the open disk $D(0; R)$ with the radius R, given by (12), and it diverges in its outside (i.e., for $z \in \mathbb{C}$ with $|z| > R$), like in the classical theory of the power series. These facts are established in the process of proving this basic theorem. Further, three corollaries are formulated. First of them is analogical to the classical Abel lemma.

Corollary 1. *Let $z \in \mathbb{C}$, $n \in \mathbb{N}_0$, $\alpha > 0$, $\gamma > 0$, and let the series (11) converge at the point $z_0 \neq 0$. Then it is absolutely convergent in the disk $D(0; |z_0|)$.*

Additionally, it turns out that the convergence of the discussed series is uniform inside the disk $D(0; R)$, i.e., on each closed disk $|z| \leq r < R$.

Corollary 2. *Let $z \in \mathbb{C}$, $n \in \mathbb{N}_0$, $\alpha > 0$, $\gamma > 0$. Then the convergence of the series (11) is uniform inside the disk $D(0; R)$, with R defined by (12), i.e., on each closed disk $[D(0;r)] = \{z \in \mathbb{C} : |z| \leq r < R\}$.*

The third corollary considers the behaviour of the series (11) outside the disk $D(0; |z_0|)$, described in Corollary 1.

Corollary 3. *Let $z \in \mathbb{C}$, $n \in \mathbb{N}_0$, $\alpha > 0$, $\gamma > 0$, and let the series (11) diverge at the point $z_0 \neq 0$. Then it is divergent for each z with $|z| > |z_0|$.*

Theorem 2 and Corollaries 1 and 2 are formulated and proved in [16]. The formulation and proof of Corollary 3 can be found in author's paper [22].

Thus, the series (11) absolutely converges in the open disk $D(0; R)$ and it diverges in the region $\{z \in \mathbb{C} : |z| > R\}$. Inside the open disk $D(0; R)$, i.e., in each closed disk $|z| \leq r$ which is a subset of $D(0; R)$, the convergence of the discussed series is uniform. However, the very disk of convergence is not obligatorily a domain of uniform convergence and at the points on its boundary divergence cannot be excluded. More precise results, connected with the behaviour of the series (11) 'near' the boundary $C(0; R)$, are obtained and discussed in the next sections.

5. Abel Type Theorem

Let $z_0 \in \mathbb{C}$, $0 < R < \infty$, $|z_0| = R$ and g_φ be an arbitrary angular region with size $2\varphi < \pi$ and with a vertex at the point $z = z_0$. Let additionally this region be symmetric with respect to the straight line passing through the points 0 and z_0 and d_φ be its part, bounded by the arms of the angle g_φ and the arc of the circle centred at the point 0 and touching the arms of g_φ. The following inequality can be verified for $z \in d_\varphi$ [11] (p. 21):

$$|z - z_0| \cos \varphi < 2(|z_0| - |z|). \tag{13}$$

The next theorem refers to the uniform convergence of the series (11) in the set d_φ and the existence of the limit of its sum at the point z_0, provided $z \in D(0; R) \cap g_\varphi$.

Theorem 3 (of Abel type). *Let $\{a_n\}_{n=0}^{\infty}$ be a sequence of complex numbers, R be the real number defined by (12) and $0 < R < \infty$. If $f(z; \alpha, \gamma)$ is the sum of the series (11) in the open disk $D(0; R)$, i.e.,*

$$f(z; \alpha, \gamma) = \sum_{n=0}^{\infty} a_n \widetilde{F}_{\alpha, n}^{(\gamma)}(z), \quad z \in D(0; R),$$

and this series converges at the point z_0 of the boundary $C(0; R)$, then:

(i) The following relation holds

$$\lim_{z \to z_0} f(z; \alpha, \gamma) = \sum_{n=0}^{\infty} a_n \widetilde{F}_{\alpha,n}^{(\gamma)}(z_0), \qquad (14)$$

provided $z \in D(0; R) \cap g_\varphi$.

(ii) The series (11) is uniformly convergent in the region d_φ.

Proof. The proofs of the two assertions (i) and (ii) are separately given.

(i) Beginning with (i) we only note that the detailed idea of its proof is given in [22] and that is why the proof is omitted here.

(ii) In order to prove (ii), we use the inequality (13) which is a key point of the proof. So, letting $z \in d_\varphi$ and setting for convenience

$$S_k(z) = \sum_{n=0}^{k} a_n \widetilde{F}_{\alpha,n}^{(\gamma)}(z), \quad S_k(z_0) = \sum_{n=0}^{k} a_n \widetilde{F}_{\alpha,n}^{(\gamma)}(z_0), \quad \lim_{k \to \infty} S_k(z_0) = s, \qquad (15)$$

we obtain

$$S_{k+p}(z) - S_k(z) = \sum_{n=0}^{k+p} a_n \widetilde{F}_{\alpha,n}^{(\gamma)}(z) - \sum_{n=0}^{k} a_n \widetilde{F}_{\alpha,n}^{(\gamma)}(z) = \sum_{n=k+1}^{k+p} a_n \widetilde{F}_{\alpha,n}^{(\gamma)}(z).$$

According to Remark 2, there exists a natural number N_0 such that $\widetilde{F}_{\alpha,n}^{(\gamma)}(z_0) \neq 0$ when $n > N_0$. Let $k > N_0$ and $p > 0$. Then, using the denotation

$$\gamma_n(z; z_0) = \widetilde{F}_{\alpha,n}^{(\gamma)}(z) / \widetilde{F}_{\alpha,n}^{(\gamma)}(z_0),$$

the difference $S_{k+p}(z) - S_k(z)$ can be written as follows:

$$S_{k+p}(z) - S_k(z) = \sum_{n=k+1}^{k+p} a_n \widetilde{F}_{\alpha,n}^{(\gamma)}(z_0) \frac{\widetilde{F}_{\alpha,n}^{(\gamma)}(z)}{\widetilde{F}_{\alpha,n}^{(\gamma)}(z_0)} = \sum_{n=k+1}^{k+p} a_n \widetilde{F}_{\alpha,n}^{(\gamma)}(z_0) \gamma_n(z; z_0).$$

Now, by the Abel transformation (see in [17]),

$$\sum_{n=k+1}^{k+p} (\beta_n - \beta_{n-1}) \gamma_n = \beta_{k+p} \gamma_{k+p} - \beta_k \gamma_{k+1} - \sum_{n=k+1}^{k+p-1} \beta_n (\gamma_{n+1} - \gamma_n),$$

and additionally denoting $\beta_n = S_n(z_0) - s$, we obtain consecutively:

$$S_{k+p}(z) - S_k(z) = \sum_{n=k+1}^{k+p} (\beta_n - \beta_{n-1}) \gamma_n(z; z_0)$$

$$= \beta_{k+p} \gamma_{k+p}(z; z_0) - \beta_k \gamma_{k+1}(z; z_0) - \sum_{n=k+1}^{k+p-1} \beta_n (\gamma_{n+1}(z; z_0) - \gamma_n(z; z_0)),$$

and

$$|S_{k+p}(z) - S_k(z)| \leq |S_{k+p}(z_0) - s||\gamma_{k+p}(z; z_0)| + |S_k(z_0) - s||\gamma_{k+1}(z; z_0)|$$

$$+ \sum_{n=k+1}^{k+p-1} |S_n(z_0) - s| \times \left| \frac{\widetilde{F}_{\alpha,n}^{(\gamma)}(z)}{\widetilde{F}_{\alpha,n}^{(\gamma)}(z_0)} - \frac{\widetilde{F}_{\alpha,n+1}^{(\gamma)}(z)}{\widetilde{F}_{\alpha,n+1}^{(\gamma)}(z_0)} \right|. \qquad (16)$$

Then, using the inequality (16), we intend to estimate the module of the difference $S_{k+p}(z) - S_k(z)$. Due to (8) and (9), along with the Γ-functions quotient property (see e.g., [11] (p. 101)) and the equalities $\lim_{n\to\infty} \frac{1}{n^{\alpha\gamma}} = 0$, $\lim_{n\to\infty} (1 + \theta_n(z_0))^{-1} = 1$, we conclude that there exist numbers $A > 0$ and $N_1 > N_0$ such that $|1 + \theta_n(z)| \leq A/2$ for all the positive integers n and $|1 + \theta_n(z_0)|^{-1} < 2$ for $n > N_1$, whence

$$|\gamma_n(z; z_0)| \leq A \quad \text{for } n > N_1. \tag{17}$$

Further, denoting

$$f_n(z; z_0) = \gamma_n(z; z_0) - \gamma_{n+1}(z; z_0),$$

which is the same as

$$f_n(z; z_0) = \frac{\widetilde{F}_{\alpha,n}^{(\gamma)}(z)}{\widetilde{F}_{\alpha,n}^{(\gamma)}(z_0)} - \frac{\widetilde{F}_{\alpha,n+1}^{(\gamma)}(z)}{\widetilde{F}_{\alpha,n+1}^{(\gamma)}(z_0)},$$

and observing that $f_n(z_0; z_0) = 0$, we apply the Schwartz lemma for the function $f_n(z; z_0)$. So, we obtain that there exists a positive constant C such that:

$$|f_n(z; z_0)| = \left| \frac{\widetilde{F}_{\alpha,n}^{(\gamma)}(z)}{\widetilde{F}_{\alpha,n}^{(\gamma)}(z_0)} - \frac{\widetilde{F}_{\alpha,n+1}^{(\gamma)}(z)}{\widetilde{F}_{\alpha,n+1}^{(\gamma)}(z_0)} \right| \leq C|z - z_0||z/z_0|^n,$$

whence, and according to (13) as well, we have:

$$\sum_{n=k+1}^{k+p+1} |f_n(z; z_0)| \leq \sum_{n=0}^{\infty} C|z - z_0||z/z_0|^n = C|z_0| \times \frac{|z - z_0|}{|z_0| - |z|} < \frac{2C|z_0|}{\cos \varphi}. \tag{18}$$

Let ε be an arbitrary positive number. Taking in view the third relation (15), we deduce that there exists a positive number $N_2 > N_0$ so large that

$$|S_n(z_0) - s| < \min\left(\frac{\varepsilon}{3A}, \frac{\varepsilon \cos \varphi}{6C|z_0|}\right) \quad \text{for } n > N_2. \tag{19}$$

Now, let us take $N = N(\varepsilon) = \max(N_1, N_2)$ and $k > N$. Therefore the inequalities (16)–(19) give

$$|S_{k+p}(z) - S_k(z)| < \frac{2\varepsilon}{3} + \frac{\varepsilon \cos \varphi}{6C|z_0|} \sum_{n=k+1}^{k+p+1} |f_n(z; z_0)| < \frac{2\varepsilon}{3} + \frac{\varepsilon \cos \varphi}{6C|z_0|} \frac{2C|z_0|}{\cos \varphi} = \varepsilon,$$

that completes the proof of (ii).

Thus, the theorem is completely proved. □

6. Tauber Type Theorem

It is established in Section 5 that the convergence of the considered series in Le Roy type functions at the point z_0 from the boundary of $D(0; R)$ implies the existing of the limit of its sum when z tends to z_0, provided $z \in D(0; R) \cap g_\varphi$. It turns out that under additional conditions on the coefficients of the considered series, the inverse proposition is also valid.

Now, let $z_0 \in \mathbb{C}$, $|z_0| = R$, $0 < R < \infty$, and let $\widetilde{F}_{\alpha,n}^{(\gamma)}(z_0) \neq 0$ for $n = 0, 1, 2, \ldots$. Note that, the last condition is fulfilled due to Remark 2, since each function $\widetilde{F}_{\alpha,n}^{(\gamma)}(z)$ ($n \in \mathbb{N}$), being an entire function, no identically zero, has at most a finite number of zeros in the closed and bounded set $|z| \leq R$, and moreover, no more than a finite number of these functions have some zeros.

For the sake of brevity, denote

$$F_{n,\alpha,\gamma}^*(z; z_0) = \frac{\widetilde{F}_{\alpha,n}^{(\gamma)}(z)}{\widetilde{F}_{\alpha,n}^{(\gamma)}(z_0)}.$$

Let the series $\sum_{n=0}^{\infty} a_n F_{n,\alpha,\gamma}^*(z; z_0)$, with $a_n \in \mathbb{C}$, be convergent for $|z| < R$, and

$$F(z) = \sum_{n=0}^{\infty} a_n F_{n,\alpha,\gamma}^*(z; z_0), \quad |z| < R. \tag{20}$$

Then the following theorem can be formulated.

Theorem 4 (of Tauber type). *If $\{a_n\}_{n=0}^{\infty}$ is a sequence of complex numbers with*

$$\lim\{n a_n\} = 0, \tag{21}$$

and there exists

$$\lim_{z \to z_0} F(z) = S \quad (|z| < R, z \to z_0 \text{ radially}), \tag{22}$$

then the numerical series $\sum_{n=0}^{\infty} a_n$ is convergent and $\sum_{n=0}^{\infty} a_n = S$.

Proof. Let z belong to the segment $[0, z_0]$. By using the asymptotic Formula (9) for the Le Roy type functions, we obtain:

$$a_n F_{n,\alpha,\gamma}^*(z; z_0) = a_n \left(\frac{z}{z_0}\right)^n \frac{1 + \vartheta_{\alpha,n}^{(\gamma)}(z)}{1 + \vartheta_{\alpha,n}^{(\gamma)}(z_0)} = a_n \left(\frac{z}{z_0}\right)^n \left(1 + \widetilde{\vartheta}_{\alpha,n}^{(\gamma)}(z; z_0)\right), \tag{23}$$

where $\widetilde{\vartheta}_{\alpha,n}^{(\gamma)}(z; z_0) = \dfrac{\vartheta_{\alpha,n}^{(\gamma)}(z) - \vartheta_{\alpha,n}^{(\gamma)}(z_0)}{1 + \vartheta_{\alpha,n}^{(\gamma)}(z_0)}$. Then, due to (8) and the Γ-functions quotient property, $\widetilde{\vartheta}_{\alpha,n}^{(\gamma)}(z; z_0)$ satisfies the following relation

$$\widetilde{\vartheta}_{\alpha,n}^{(\gamma)}(z; z_0) = O\left(\frac{1}{n^{\alpha\gamma}}\right). \tag{24}$$

Writing $\sum_{n=0}^{\infty} a_n F_{n,\alpha,\gamma}^*(z; z_0)$ in the form

$$\sum_{n=0}^{\infty} a_n F_{n,\alpha,\gamma}^*(z; z_0) = \sum_{n=0}^{\infty} a_n \left(\frac{z}{z_0}\right)^n \frac{1 + \vartheta_{\alpha,n}^{(\gamma)}(z)}{1 + \vartheta_{\alpha,n}^{(\gamma)}(z_0)} \tag{25}$$

$$= \sum_{n=0}^{\infty} a_n \left(\frac{z}{z_0}\right)^n \left(1 + \widetilde{\vartheta}_{\alpha,n}^{(\gamma)}(z; z_0)\right),$$

and denoting $w_n(z) = a_n \left(\dfrac{z}{z_0}\right)^n \widetilde{\vartheta}_{\alpha,n}^{(\gamma)}(z; z_0)$, we consider the series $\sum_{n=0}^{\infty} w_n(z)$.

According to condition (21), the numerical sequence $\{n a_n\}_{n=0}^{\infty}$, being a convergent sequence, is bounded. Then, since $|w_n(z)| \leq |a_n| |\widetilde{\vartheta}_{\alpha,n}^{(\gamma)}(z; z_0)|$ and having in view (8), there exists a constant C, such that $|w_n(z)| \leq C/n^{1+\alpha\gamma}$ for all the positive integers n. Since $\sum_{n=1}^{\infty} 1/n^{1+\alpha\gamma}$ converges, the series $\sum_{n=0}^{\infty} w_n(z)$ also converges, even absolutely and uniformly on the segment $[0, z_0]$. Therefore, changing the order of the limit and summation, in view of the equality $\lim_{z \to z_0} w_n(z) = 0$, we deduce that

$$\lim_{z \to z_0} \sum_{n=0}^{\infty} w_n(z) = \sum_{n=0}^{\infty} \lim_{z \to z_0} w_n(z) = 0. \tag{26}$$

Then, bearing in mind that (20) can be written in the form

$$F(z) = \sum_{n=0}^{\infty} a_n F_{n,\alpha,\gamma}^*(z;z_0) = \sum_{n=0}^{\infty} a_n \left(\frac{z}{z_0}\right)^n + \sum_{n=0}^{\infty} w_n(z),$$

along with the assumption (22), we conclude that the limit

$$\lim_{z \to z_0} \sum_{n=0}^{\infty} a_n \left(\frac{z}{z_0}\right)^n \qquad (27)$$

also exists and, moreover, in view of (26),

$$\lim_{z \to z_0} F(z) = \lim_{z \to z_0} \sum_{n=0}^{\infty} a_n F_{n,\alpha,\gamma}^*(z;z_0) = S = \lim_{z \to z_0} \sum_{n=0}^{\infty} a_n \left(\frac{z}{z_0}\right)^n. \qquad (28)$$

Now, from (28) and the existence of the limit (27), by the classical Tauber theorem for the power series, it follows that the series $\sum_{n=0}^{\infty} a_n$ converges and $\sum_{n=0}^{\infty} a_n = S$. □

The conclusion of the above theorem is still valid even if the condition imposed on the coefficients a_n is weakened. Namely, the following theorem holds true.

Theorem 5 (of Littlewood type). *If $\{a_n\}_{n=0}^{\infty}$ is a sequence of complex numbers with*

$$a_n = O(1/n), \qquad (29)$$

$F(z)$ is the function defined by (20), and if there exists

$$\lim_{z \to z_0} F(z) = S \quad (|z| < R, z \to z_0 \text{ radially}), \qquad (30)$$

then the numerical series $\sum_{n=0}^{\infty} a_n$ is convergent and $\sum_{n=0}^{\infty} a_n = S$.

Proof. Let z belong to the segment $[0, z_0]$. The proof goes in the same way as the proof of Theorem 4, and using the same denotations. The only difference is in proving the estimation for $|w_n(z)|$. More especially, according to the relation (24) and the condition (29), it follows that there exists a constant C, such that $|w_n(z)| \leq C/n^{1+\alpha\gamma}$ for all the positive integers n. Finally, the proof ends applying in the last step Littlewood's classical theorem instead of Tauber's theorem. The details are omitted. □

7. $(F_{\alpha,\gamma}, Z_0)$—Summation and (J, Z_0)—Summation

The theorems in the previous section can be formulated in alternative forms. For this purpose, two additional definitions are firstly given.

Let us consider the numerical series

$$\sum_{n=0}^{\infty} a_n, \quad a_n \in \mathbb{C}, \quad n = 0, 1, 2, \ldots \qquad (31)$$

To define its Abel summability ([23], p. 20), we consider also the power series $\sum_{n=0}^{\infty} a_n z^n$.

Definition 1. *The series (31) is called A—summable if the series $\sum_{n=0}^{\infty} a_n z^n$ converges in the open unit disk $D(0;1)$ and moreover there exists*

$$\lim_{z \to 1-0} \sum_{n=0}^{\infty} a_n z^n = S.$$

The complex number S is called A-sum of the series (31) and the usual notation of that is

$$\sum_{n=0}^{\infty} a_n = S \quad (A).$$

Remark 3. *The A-summation is regular. It means that if the series (31) converges, then it is A-summable, and its A-sum is equal to its usual sum.*

Remark 4. *It is well known that in general, the A-summability of the series (31) does not imply its convergence. However, with additional conditions imposed on the growth of the general term of the series (31), the convergence can be provided.*

Let $z_0 \in \mathbb{C}$, $|z_0| = R$, $0 < R < \infty$ and $\widetilde{F}_{\alpha,n}^{(\gamma)}(z_0) \neq 0$ (note that, the last condition is again fulfilled due to Remark 2). For the sake of convenience, denote

$$F_{n,\alpha,\gamma}^*(z; z_0) = \frac{\widetilde{F}_{\alpha,n}^{(\gamma)}(z)}{\widetilde{F}_{\alpha,n}^{(\gamma)}(z_0)}. \tag{32}$$

Further, by analogy with the A-summability of the series (31), another definition is introduced, where the power series $\sum_{n=0}^{\infty} a_n z^n$ is replaced by the series in the Le Roy type functions (32) with the same coefficients.

Definition 2. *The numerical series (31) is said to be $(F_{\alpha,\gamma}, z_0)$—summable if the series*

$$\sum_{n=0}^{\infty} a_n F_{n,\alpha,\gamma}^*(z; z_0), \tag{33}$$

converges in the open disk $D(0; R)$ and, moreover, there exists the limit

$$\lim_{z \to z_0} \sum_{n=0}^{\infty} a_n F_{n,\alpha,\gamma}^*(z; z_0), \tag{34}$$

provided z remains on the segment $[0, z_0)$ (i.e., z radially tends to z_0).

Remark 5. *The $(F_{\alpha,\gamma}, z_0)$—summation is regular, and this property is merely a particular case of Theorem 3.*

Taking into account the latest definitions and remarks, Theorems 4 and 5 can be formulated in the following alternative ways.

Theorem 6 (of Tauber type). *If the numerical series (31) is $(F_{\alpha,\gamma}, z_0)$—summable and*

$$\lim \{n a_n\} = 0, \tag{35}$$

then it is convergent.

Theorem 7 (of Littlewood type). *If the numerical series (31) is $(F_{\alpha,\gamma}, z_0)$—summable and*

$$a_n = O(1/n), \tag{36}$$

then it is convergent.

Remark 6. *We observe that all the functions of the family*

$$(F_{\alpha,\gamma}; z_0) = \{F_{n,\alpha,\gamma}^*(z; z_0), \quad n = 0, 1, \dots\} \tag{37}$$

are entire functions satisfying the condition $F^*_{n,\alpha,\gamma}(z_0;z_0) = 1$.

For convenience, in order to make Definition 2 more universal and usable for various considerations, we intend to paraphrase it in the way, given in [11] (p. 35). For this purpose, we firstly introduce one more denotation.

Let $z_0 \in \mathbb{C}$, $z_0 \neq 0$, $|z_0| = R$, $0 < R < \infty$ and let $(J;z_0)$ be the following family of functions

$$(J;z_0) := \{j_n : j_n - \text{entire function}, \ j_n(z_0) = 1\}_{n \in \mathbb{N}_0}. \tag{38}$$

Now, considering the series given below

$$\sum_{n=0}^{\infty} a_n j_n(z), \quad j_n \in (J;z_0), \tag{39}$$

Definition 2 can be expanded as follows.

Definition 3. *The numerical series (31) is said to be (J, z_0)-summable, if the series (39) converges in the disk $D(0; R)$, and moreover, there exists the limit*

$$\lim_{z \to z_0} \sum_{n=0}^{\infty} a_n j_n(z), \tag{40}$$

provided z remains on the segment $[0, z_0]$.

Remark 7. *Let us note that using this definition must necessarily take into account of the regularity of the summation.*

Ending this section we are going to make one more remark.

Remark 8. *Taking $j_n(z) = F^*_{n,\alpha,\gamma}(z;z_0)$, the family (38) of entire functions reduces to the family (37). Therefore, in this case the (J, z_0)—summation and $(F_{\alpha,\gamma}, z_0)$—summation are the same. Thus Theorems 6 and 7 can be written in equivalent ways, using the notion (J, z_0)—summation (with $j_n(z) = F^*_{n,\alpha,\gamma}(z;z_0)$), instead of $(F_{\alpha,\gamma}, z_0)$—summation. That means that the theorems of Tauber and Littlewood type are statements relating the (J, z_0)—summability and the usual convergence of a numerical series by means of some assumptions imposed on the general term of the numerical series under consideration.*

8. Special Cases

In this section we consider some interesting special cases of the Le Roy type function $F^{(\gamma)}_{\alpha,\beta}$, given by (1), taking the parameters

$$\alpha, \beta \in \mathbb{C}, \ \Re(\alpha) > 0 \text{ and } \gamma > 0,$$

when (1) is an entire function.

Case 1. If γ is an arbitrary positive number, $\alpha = 1$ and $\beta = 1$, then the function (1) coincides with the Le Roy function (confer with (4)), i.e.,

$$F^{(\gamma)}(z) = F^{(\gamma)}_{1,1}(z) = \sum_{k=0}^{\infty} \frac{z^k}{[\Gamma(k+1)]^\gamma}, \quad z \in \mathbb{C}. \tag{41}$$

We have to note that, studying the asymptotics of the analytic continuation of the sum of power series, Le Roy himself used it in [24]. This reason for the origin of (41) sounds somehow similar to the Mittag–Leffler's idea to introduce the function $E_\alpha(z)$ for the aims of analytic continuation (it have to be noted that Mittag–Leffler and Le Roy were working on this idea in competition). The Le Roy function is involved in the solution of various problems; in particular it has been recently used in the construction of a Conway–Maxwell–

Poisson distribution [25] which is important due to its ability to model count data with different degrees of over- and under-dispersion [26,27].

Case 2. If $\gamma = 1$, then the function (1) gives the Mittag–Leffler function $E_{\alpha,\beta}$, namely

$$E_{\alpha,\beta}(z) = F^{(1)}_{\alpha,\beta}(z) = \sum_{k=0}^{\infty} \frac{z^k}{\Gamma(\alpha k + \beta)}, \quad z \in \mathbb{C}. \qquad (42)$$

In addition, when $\beta = 1$, the function (1) reduces to E_α, and to the exponential function, if $\alpha = \beta = 1$, i.e.,

$$E_\alpha(z) = F^{(1)}_{\alpha,1}(z) = \sum_{k=0}^{\infty} \frac{z^k}{\Gamma(\alpha k + 1)}, \quad \exp z = F^{(1)}_{1,1}(z) = \sum_{k=0}^{\infty} \frac{z^k}{k!}; \quad z \in \mathbb{C}. \qquad (43)$$

The functions (42) and (43) are named after the great Swedish mathematician Gösta Magnus Mittag–Leffler (1846–1927) who defined the 1-parametric function $E_\alpha(z)$ by a power series (given by (43)) and he studied its properties in 1902–1905 (detailed description can be seen in [28]). Actually, Mittag–Leffler introduced the function $E_\alpha(z)$ for the purposes of his method for summation of divergent series. Later, the function (43) was recognized as the 'Queen function of fractional calculus' [29–31], see also [11], for its basic role for analytic solutions of fractional order integral and differential equations and systems. In the recent decades successful applications of the Mittag–Leffler function and its generalizations in problems of physics, biology, chemistry, engineering and other applied sciences made it better known among scientists. A considerable literature is devoted to the investigation of the analytical properties of these functions; among the references of [11,28,32], where are quoted several authors who, after Mittag–Leffler, have investigated such kinds of functions from a pure mathematical, applied and numerical oriented point of view as well.

Case 3. If $\gamma = 1/2$ and $\alpha = \beta = 1$, then the function (1) becomes the function $R(z)$, given by the series (see Kolokoltsov [33] Formula (50))

$$R(z) = F^{(1/2)}_{1,1}(z) = \sum_{k=0}^{\infty} \frac{z^k}{\sqrt{k!}}, \quad z \in \mathbb{C}. \qquad (44)$$

The function (44) is used by Kolokoltsov in [33] to estimate the solution of initial stochastic differential equations. As he comments in his paper, the function $R(z)$ plays the same role for stochastic equations as the exponential and the Mittag–Leffler functions for deterministic equations.

Case 4. If the parameter $\gamma = 2$ and $\alpha = \beta = 1$, then the function (1) can be presented as Bessel function of the first kind and related to it, and as 2-parametric Bessel–Maitland function, as well. Namely, the function (1) can be written in the following alternative forms:

$$F^{(2)}_{1,1}(z) = J_0(2i\sqrt{z}) = I_0(2\sqrt{z}) = C_0(-z) = J^1_0(-z) = \sum_{k=0}^{\infty} \frac{z^k}{(k!)^2}, \quad z \in \mathbb{C}. \qquad (45)$$

In this relation J_0 and I_0 are respectively the classical Bessel function of the first kind J_ν and its modified function I_ν with an index $\nu = 0$, C_0 is the Bessel–Clifford function C_ν with an index $\nu = 0$, and J^1_0 is its 2-parametric Bessel–Maitland generalization J^μ_ν (named after Sir Edward Maitland Wright and also known as Bessel–Wright function) with indices $\nu = 0$ and $\mu = 1$.

Case 5. If the number m is a positive integer, $\gamma = m+1$, $\beta = \lambda + 1$ ($\lambda \neq 0$), and $\alpha = 1$, then the function (1) can be expressed with 3–index generalization, as well as by the 4–index generalization of the Bessel function of the first kind. More especially if $m = 1$, then the special function (1) turns, with an exactness to a power function, into the generalized Bessel–Maitland (or Wright's) function $J_{\nu,\lambda}^{\mu}$ (with $\nu = 0$ and $\mu = 1$) of the Bessel function $J_\nu(z)$, introduced by Pathak (for details see [14]):

$$J_{\nu,\lambda}^{\mu}(z) = (z/2)^{\nu+2\lambda}\, \widetilde{J}_{\nu,\lambda}^{\mu}(z) = (z/2)^{\nu+2\lambda} \sum_{k=0}^{\infty} \frac{(-1)^k (z/2)^{2k}}{\Gamma(k+\lambda+1)\Gamma(\mu k + \nu + \lambda + 1)}. \quad (46)$$

More precisely,

$$\widetilde{J}_{0,\lambda}^{1}(2i\sqrt{z}) = F_{1,\lambda+1}^{(2)}(z) = \sum_{k=0}^{\infty} \frac{z^k}{[\Gamma(k+\lambda+1)]^2}. \quad (47)$$

The special case (for $m \geq 2$) is expressed by the generalized Lommel–Wright function $J_{\nu,\lambda}^{\mu,m}$ with 4 indices (with $\nu = 0$ and $\mu = 1$), introduced by de Oteiza, Kalla and Conde (for details see [14]):

$$J_{\nu,\lambda}^{\mu,m}(z) = (z/2)^{\nu+2\lambda}\, \widetilde{J}_{\nu,\lambda}^{\mu,m}(z) = (z/2)^{\nu+2\lambda} \sum_{k=0}^{\infty} \frac{(-1)^k (z/2)^{2k}}{(\Gamma(k+\lambda+1))^m \Gamma(\mu k + \nu + \lambda + 1)}. \quad (48)$$

Especially,

$$\widetilde{J}_{0,\lambda}^{1,m}(2i\sqrt{z}) = F_{1,\lambda+1}^{(m+1)}(z) = \sum_{k=0}^{\infty} \frac{z^k}{[\Gamma(k+\lambda+1)]^{m+1}}. \quad (49)$$

Just to mention that $J_{\nu,\lambda}^{\mu,1} = J_{\nu,\lambda}^{\mu}$, as well as $\widetilde{J}_{\nu,\lambda}^{\mu,1} = \widetilde{J}_{\nu,\lambda}^{\mu}$.

Case 6. If the number m is a positive integer $m \geq 2$, then the function (1) can be presented as the multi-index extensions of (42) (with $2m$ and $3m$ parameters, $m = 1, 2, \ldots$, [11,13,34–36]), i.e., the so-called multi-index Mittag–Leffler functions. The first one was introduced by Yakubovich and Luchko [37] and studied in details by Kiryakova [12,34]. It is defined by the formula

$$E_{(\alpha_i),(\beta_i)}(z) = E_{(\alpha_i),(\beta_i)}^m(z) = \sum_{k=0}^{\infty} \frac{z^k}{\Gamma(\alpha_1 k + \beta_1)\ldots\Gamma(\alpha_m k + \beta_m)}, \quad (50)$$

for $z \in \mathbb{C}$ and $m > 1$. The parameters α_i, β_i are all complex for $i = 1, 2, \ldots m$ and $\Re(\alpha_i) > 0$. The second one has m additional complex parameters γ_i. It was introduced and studied in details by Paneva-Konovska (for its properties see e.g., [11]). It is defined by the formula

$$E_{(\alpha_i),(\beta_i)}^{(\gamma_i),m}(z) = \sum_{k=0}^{\infty} \frac{(\gamma_1)_k \ldots (\gamma_m)_k}{\Gamma(\alpha_1 k + \beta_1)\ldots\Gamma(\alpha_m k + \beta_m)} \frac{z^k}{(k!)^m}, \quad (51)$$

where $(\gamma)_k$ is the Pochhammer symbol: $(\gamma)_k = \gamma(\gamma+1)\ldots(\gamma+k-1)$, $k = 1, 2, \ldots$, $(\gamma)_0 = 1$. More precisely, in this case the function (1) turns into the above multi-index Mittag–Leffler functions, with indices $\alpha_i = \alpha$, $\beta_i = \beta$ and $\gamma_i = 1$ ($i = 1, 2, \ldots m$), namely

$$E_{(\alpha),(\beta)}(z) = E_{(\alpha),(\beta)}^{(1),m}(z) = F_{\alpha,\beta}^{(m)}(z) = \sum_{k=0}^{\infty} \frac{z^k}{[\Gamma(\alpha k + \beta)]^m}. \quad (52)$$

Case 7. If the number m is a positive integer $m \geq 2$, $\alpha = 1$ and $\beta = 1$ then the function (1) is the hyper-Bessel function

$$J_{\nu_1,\ldots,\nu_{m-1}}^{(m-1)}(z) = \left(\frac{z}{m}\right)^{\sum_{i=1}^{m-1}\nu_i} \sum_{k=0}^{\infty} \frac{(-1)^k \left(\frac{z}{m}\right)^{km}}{\Gamma(k+\nu_1+1)\ldots\Gamma(k+\nu_{m-1}+1)}\frac{1}{k!}.$$

introduced by Delerue in 1953 [38]. It is a generalization of the Bessel function of the first type J_ν with vector indices $\nu = (\nu_1, \nu_2, \ldots, \nu_{m-1})$. The hyper-Bessel function of Delerue is closely related to the hyper-Bessel differential operators of arbitrary order $m > 1$, introduced by Dimovski [39]. The function (1) is represented as the hyper-Bessel function with parameters $\nu_i = 0$ ($i = 1, 2, \ldots m - 1$), i.e.,

$$J_{0,\ldots,0}^{(m-1)}\left(m(-z)^{1/m}\right) = F_{\alpha,\beta}^{(m)}(z) = \sum_{k=0}^{\infty} \frac{z^k}{[\Gamma(k+1)]^m}. \tag{53}$$

At last, let us note that if $\gamma = m$ is a positive integer, then the Le Roy function $F_{\alpha,\beta}^{(m)}$ is the Wright generalized hypergeometric function with $2 \times m$ indices $\alpha_i = \alpha$, $\beta_i = \beta$ ($i = 1, \ldots, m$), namely

$$F_{\alpha,\beta}^{(m)}(z) = \sum_{k=0}^{\infty} \frac{z^k}{[\Gamma(\alpha k + \beta)]^m} = {}_1\Psi_m\left[\begin{array}{c}(1, 1)\\(\beta_i, \alpha_i)_1^m\end{array}\bigg| z\right] = {}_1\Psi_m\left[\begin{array}{c}(1, 1)\\(\beta, \alpha)_1^m\end{array}\bigg| z\right],$$

and it is a particular case of the Wright generalized hypergeometric function with $2 \times (p + q)$ indices a_i, A_i ($i = 1, \ldots, p$), and b_j, B_j ($j = 1, \ldots, q$), defined by the formula

$$_p\Psi_q\left[\begin{array}{c}(a_1, A_1) \ldots (a_p, A_p)\\(b_1, B_1) \ldots (b_q, B_q)\end{array}\bigg| \sigma\right] = \sum_{k=0}^{\infty} \frac{\Gamma(a_1 + kA_1) \ldots \Gamma(a_p + kA_p)}{\Gamma(b_1 + kB_1) \ldots \Gamma(b_q + kB_q)} \frac{\sigma^k}{k!}.$$

9. Conclusions

Letting the parameter β in the condition (2) be a positive integer, we consider the family of Le Roy type functions (5) with parameters as follows:

$$\alpha > 0, \ \gamma > 0, \text{ and } \beta = n \in \mathbb{N}.$$

In Section 2 we provide an asymptotic formula for these functions for large values of the parameter n (Theorem 1). We also give upper estimates for the moduli of their remainder terms in the nonempty compact subsets of the complex plane and in the whole complex plane as well (Lemma 1). Further, in order to summarize the results obtained here, we consider the family of the type

$$\left\{\widetilde{j}_n(z)\right\}_{n \in \mathbb{N}}, \tag{54}$$

with the functions $\widetilde{j}_n$ as in (10), and the series

$$\sum_{n=0}^{\infty} a_n \widetilde{j}_n(z), \tag{55}$$

in this case coinciding with the series (11) in Le Roy type functions with complex coefficients a_n ($n = 0, 1, 2, \ldots$) and for $z \in \mathbb{C}$.

It turns out that the series (55) absolutely converges in the open disk $D(0; R)$ with the corresponding radius R, given by the Formula (12) and it diverges in its outside, i.e., for $z \in \mathbb{C}$ with $|z| > R$. Moreover, inside the disk $D(0; R)$, i.e., in each closed disk $[D(0; r)] = \{z : z \in \mathbb{C}, |z| \leq r\}$ with $r < R$, the convergence is uniform. Near the boundary $C(0; R)$ the series (55) satisfies Theorem 3 of Abel type. At last, the series fulfills the theorem of Tauber and Littlewood types, which are inverse of the Abel type theorem.

Now, let us consider the functions from the Section 8 with the same types of parameters. Since in this case they are of the types (5), then all of them satisfy Lemma 1 and the inequalities therein. Further, paying attention to the fact, that the functions (42), (52), (47) and (49) can be considered as representatives of different families of the types (5), we have to note that the functions of each family, discussed above, have the asymptotic Formula (9) with the corresponding values of the parameters α and γ. Further, taking the family of the type (54) with the functions $\widetilde{j}_n$ as follows:

$$\widetilde{j}_n(z) = z^n \left[\Gamma(n)\right]^m E_{((\alpha),(n))}(z) = z^n \left[\Gamma(n)\right]^m F_{\alpha,n}^{(m)}(z), \quad m, n \in \mathbb{N}, \tag{56}$$

in the case (52) (in particular $m = 1$ in the case (42)), and respectively

$$\widetilde{j}_n(z) = z^n \left[\Gamma(n)\right]^{m+1} \widetilde{j}_{0,n-1}^{1,m}(2i\sqrt{z}) = z^n \left[\Gamma(n)\right]^{m+1} F_{1,n}^{(m+1)}(z), \quad m, n \in \mathbb{N}, \tag{57}$$

in the case (49) ($m = 1$ in the case (47)), and adding, just for completeness $\widetilde{j}_0(z) = 1$, we consider the corresponding series (55) with complex coefficients a_n ($n = 0, 1, 2, \dots$) for $z \in \mathbb{C}$, namely the series $\sum_{n=0}^{\infty} a_n \widetilde{j}_n(z)$.

Taking into account that the series (55) is of the type (11) (however with special values of the parameters), it has the same behaviour. That means that the series (55) absolutely converges in the open disk $D(0; R)$ with the corresponding radius R, and it diverges in its outside, i.e., for $z \in \mathbb{C}$ with $|z| > R$. Moreover, inside the disk $D(0; R)$, i.e., in each closed disk $[D(0; r)]$ with $r < R$, the convergence is uniform. Replacing the parameter γ with the corresponding value in Theorem 3, it is reduced to the Abel type theorem for the series (55), referring to the behaviour of (55) near the boundary $C(0; R)$. At last, the series (55) fulfills the theorem of Tauber and Littlewood types, which are inverse of the Abel type theorem.

Thus, generally speaking, the described behaviour of the series (11) in Le Roy type functions, as well as in particular the behaviour of the corresponding series (55) (in the functions of the families (56), respectively (57)), and that of the classical power series are the same. Moreover, the results discussed here are analogues to the Cauchy–Hadamard, Abel, Tauber and Littlewood theorems for the widely used power series.

Author Contributions: The ideas and results in this survey-paper reflect the author's own contributions, resulting from more than 25 years' research on the topic. The author has read and agreed to the published version of the manuscript.

Funding: This research received no financial funding.

Institutional Review Board Statement: Not applicable.

Informed Consent Statement: Not applicable.

Data Availability Statement: Not applicable.

Acknowledgments: This paper is done under the working programs on bilateral collaboration contracts of Bulgarian Academy of Sciences with Serbian and Macedonian Academies of Sciences and Arts, and under the COST program, COST Action CA15225 'Fractional'.

Conflicts of Interest: The author declares no conflict of interest.

References

1. Gerhold, S. Asymptotics for a variant of the Mittag–Leffler function. *Integr. Trans. Spec. Funct.* **2012**, *23*, 397–403. [CrossRef]
2. Garra, R.; Polito, F. On some operators involving Hadamard derivatives. *Integr. Trans. Spec. Func.* **2013**, *24*, 773–782. [CrossRef]
3. Garrappa, R.; Rogosin, S.; Mainardi, F. On a generalized three-parameter Wright function of Le Roy type. *Fract. Calc. Appl. Anal.* **2017**, *20*, 1196–1215. [CrossRef]
4. Gorska, K.; Horzela, A.; Garrappa, R. Some results on the complete monotonicity of Mittag-Leffler functions of le Roy type. *Fract. Calc. Appl. Anal.* **2019**, *22*, 1284–1306. [CrossRef]
5. Simon, T. Remark on a Mittag–Leffler function of Le Roy type. *Integr. Trans. Spec. Func.* **2021**. [CrossRef]
6. Pogány, T.K. Integral form of Le Roy-type hypergeometric function. *Integr. Trans. Spec. Func.* **2018**, *29*, 580–584. [CrossRef]
7. Paneva-Konovska, J. A family of hyper-Bessel functions and convergent series in them. *Fract. Calc. Appl. Anal.* **2014**, *17*, 1001–1015. [CrossRef]
8. Paneva-Konovska, J. Periphery behaviour of series in Mittag-Leffler type functions, I. *Int. J. Appl. Math.* **2016**, *29*, 69–78. [CrossRef]
9. Paneva-Konovska, J. Periphery behaviour of series in Mittag-Leffler type functions, II. *Int. J. Appl. Math.* **2016**, *29*, 175–186. [CrossRef]

10. Paneva-Konovska, J. Overconvergence of series in generalized Mittag–Leffler functions. *Fract. Calc. Appl. Anal.* **2017**, *20*, 506–520. [CrossRef]
11. Paneva-Konovska, J. *From Bessel to Multi-Index Mittag Leffler Functions: Enumerable Families, Series in them and Convergence*, 1st ed.; World Scientific Publishing: London, UK, 2016. [CrossRef]
12. Kiryakova, V. Multiple (multiindex) Mittag–Leffler functions and relations to generalized fractional calculus. *J. Comput. Appl. Math.* **2000**, *118*, 241–259. [CrossRef]
13. Kiryakova, V. The multi-index Mittag–Leffler functions as important class of special functions of fractional calculus. *Comput. Math. Appl.* **2010**, *59*, 1885–1895. [CrossRef]
14. Kiryakova, V. The special functions of fractional calculus as generalized fractional calculus operators of some basic functions. *Comput. Math. Appl.* **2010**, *59*, 1128–1141. [CrossRef]
15. Kiryakova, V. A guide to special functions in fractional calculus. *Mathematics* **2021**, *9*, 106. [CrossRef]
16. Paneva-Konovska, J. Series in Le Roy type functions: Inequalities and convergence theorems. *Int. J. Appl. Math.* **2020**, *33*, 995–1007. [CrossRef]
17. Markushevich, A. *A Theory of Analytic Functions. 1, 2*; Nauka: Moscow, Russia, 1967. (In Russian)
18. Kiryakova, V. Fractional order differential and integral equations with Erdélyi-Kober operators: Explicit solutions by means of the transmutation method. *AIP Conf. Proc.* **2011**, *1410*, 247–258. [CrossRef]
19. Kiryakova, V. Transmutation method for solving hyper-Bessel differential equations based on the Poisson-Dimovski transformation. *Fract. Calc. Appl. Anal.* **2008**, *11*, 299–316.
20. Ali, I.; Kiryakova, V.; Kalla, S. Solutions of fractional multi-order integral and differential equations using a Poisson-type transform. *J. Math. Anal. Appl.* **2002**, *269*, 172–199. [CrossRef]
21. Rusev, P. *Classical Orthogonal Polynomials and their Associated Functions in Complex Domain*; Publ. House Bulg. Acad. Sci.: Sofia, Bulgaria, 2005.
22. Paneva-Konovska, J. Series in Le Roy type functions: Theorems in the complex plane. *C. R. Acad. Bulg. Sci.* **2021**, *74*, 315–323. [CrossRef]
23. Hardy, G. *Divergent Series*, 1st ed.; Clarendon Press: Oxford, UK, 1949.
24. Le Roy, É. Valéurs asymptotiques de certaines séries procédant suivant les puissances entères et positives d'une variable réelle. *Darboux Bull.* **1899**, *24*, 245–268. (In French)
25. Conway, R.W.; Maxwell, W.L. A queueing model with state dependent service rate. *J. Industr. Eng.* **1962**, *12*, 132–136.
26. Pogány, T. Integral form of the COM-Poisson renormalization constant. *Stat. Probab. Lett.* **2016**, *119*, 144–145. [CrossRef]
27. Santarelli, M.F.; Della Latta, D.; Scipioni, M.; Positano, V.; Landini, L. A Conway-Maxwell-Poisson (CMP) model to address data dispersion on positron emission tomography. *Comput. Biol. Med.* **2016**, *77*, 90–101. [CrossRef]
28. Gorenflo, R.; Kilbas, A.A.; Mainardi, F.; Rogosin, S.V. *Mittag–Leffler Functions, Related Topics and Applications*, 2nd ed.; Springer: New York, NY, USA, 2020. [CrossRef]
29. Gorenflo, R.; Mainardi, F. Fractional calculus: Integral and differential equations of fractional order. In *Fractals and Fractional Calculus in Continuum Mechanics*; Carpinteri, A., Mainardi, F., Eds.; Springer: Wien, Ausria, 1997; pp. 223–276.
30. Mainardi, F.; Gorenflo, R. Time-fractional derivatives in relaxation processes: A tutorial survey. *Fract. Calc. Appl. Anal.* **2007**, *10*, 269–308.
31. Rogosin, S. The role of the Mittag–Leffler function in fractional modeling. *Mathematics* **2015**, *3*, 368–381. [CrossRef]
32. Bazhlekova, E. Completely monotone multinomial Mittag–Leffler type functions and diffusion equations with multiple time-derivatives. *Fract. Calc. Appl. Anal.* **2021**, *24*, 88–111. [CrossRef]
33. Kolokoltsov, V.N. The law of large numbers for quantum stochastic filtering and control of many particle systems. *arXiv* **2020**, arXiv:2008.07375.
34. Kiryakova, V. Multiindex Mittag–Leffler functions, related Gelfond-Leontiev operators and Laplace type integral transforms. *Fract. Calc. Appl. Anal.* **1999**, *2*, 445–462.
35. Al-Bassam, M.A.; Luchko, Y.F. On generalized fractional calculus and it application to the solution of integro-differential equations. *J. Fract. Calc.* **1995**, *7*, 69–88.
36. Kilbas, A.A.; Koroleva, A.A.; Rogosin, S.V. Multi-parametric Mittag–Leffler functions and their extension. *Fract. Calc. Appl. Anal.* **2013**, *16*, 378–404. [CrossRef]
37. Yakubovich, S.; Luchko, Y. *The Hypergeometric Approach to Integral Transforms and Convolutions*; Kluwer Academic Publishers: Dordrecht, The Netherlands; Boston, MA, USA; London, UK, 1994.
38. Delerue, P. Sur le calcul symbolic à *n* variables et fonctions hyperbesséliennes (II). *Ann. Soc. Sci. Bruxelle Ser. 1* **1953**, *3*, 229–274.
39. Dimovski, I. Operational calculus for a class of differential operators. *C. R. Acad. Bulg. Sci.* **1966**, *19*, 1111–1114.

Article

The Asymptotic Expansion of a Function Introduced by L.L. Karasheva

Richard Paris

Division of Computing and Mathematics, Abertay University, Dundee DD1 1HG, UK; r.paris@abertay.ac.uk

Abstract: The asymptotic expansion for $x \to \pm\infty$ of the entire function $F_{n,\sigma}(x;\mu) = \sum_{k=0}^{\infty} \frac{\sin(n\gamma_k)}{\sin \gamma_k} \frac{x^k}{k!\Gamma(\mu-\sigma k)}$, $\gamma_k = \frac{(k+1)\pi}{2n}$ for $\mu > 0$, $0 < \sigma < 1$ and $n = 1, 2, \ldots$ is considered. In the special case $\sigma = \alpha/(2n)$, with $0 < \alpha < 1$, this function was recently introduced by L.L. Karasheva (*J. Math. Sciences*, **250** (2020) 753–759) as a solution of a fractional-order partial differential equation. By expressing $F_{n,\sigma}(x;\mu)$ as a finite sum of Wright functions, we employ the standard asymptotics of integral functions of hypergeometric type to determine its asymptotic expansion. This was found to depend critically on the parameter σ (and to a lesser extent on the integer n). Numerical results are presented to illustrate the accuracy of the different expansions obtained.

Keywords: wright function; asymptotic expansions; Stokes phenomenon

MSC: 33C70; 34E05; 41A30; 41A60

1. Introduction

In a recent paper, L.L. Karasheva [1] introduced the entire function

$$\Theta_{n,\alpha}(x;\mu) := \sum_{k=0}^{\infty} \frac{\sin(n\gamma_k)}{\sin \gamma_k} \frac{x^k}{k!\Gamma(\mu - \frac{\alpha k}{2n})}, \qquad \gamma_k := \frac{(k+1)\pi}{2n}, \tag{1}$$

where $\mu > 0$, $0 < \alpha < 1$ and $n = 1, 2, \ldots$ and, throughout, x is a real variable. This function is of interest as it is involved in the fundamental solution of the differential equation

$$\frac{\partial^\alpha u}{\partial t^\alpha} + (-1)^n \frac{\partial^{2n} u}{\partial x^{2n}} = f(x,t)$$

for positive integer n, where the derivative with respect to t is the fractional derivative of the order α. In the simplest case $n = 1$, we have $\Theta_{1,\alpha}(x;\mu) = \phi(-\sigma, \mu; x)$, $\sigma := \alpha/(2n)$, where $\phi(-\sigma, \mu; x)$ is the Wright function

$$\phi(-\sigma, \mu; x) := \sum_{k=0}^{\infty} \frac{x^k}{k!\Gamma(\mu - \sigma k)} \qquad (\sigma < 1), \tag{2}$$

which finds application as a fundamental solution of the diffusion-wave equation [2]. Under the above assumptions on n and α it follows that the parameter σ associated with (1) satisfies $0 < \sigma < \frac{1}{2}$.

In this study, however, we shall allow the parameter σ to satisfy $0 < \sigma < 1$ and consider the function

$$F_{n,\sigma}(x;\mu) := \sum_{k=0}^{\infty} \frac{\sin(n\gamma_k)}{\sin \gamma_k} \frac{x^k}{k!\Gamma(\mu - \sigma k)} \qquad (0 < \sigma < 1), \tag{3}$$

which coincides with $\Theta_{n,\alpha}(x;\mu)$ when $\sigma = \alpha/(2n)$. From the well-known expansion

$$\frac{\sin(n\gamma_k)}{\sin\gamma_k} = \sum_{r=0}^{n-1} e^{i\gamma_k(2r-n+1)} = \sum_{r=0}^{n-1} e^{-i(k+1)\omega_r},$$

where

$$\omega_r := \frac{(n-2r-1)\pi}{2n} \qquad (0 \leq r \leq n-1), \tag{4}$$

it follows that (3) can be expressed as a finite sum of Wright functions defined in (2) with rotated arguments (compare [1], Equation (4))

$$F_{n,\sigma}(x;\mu) = \sum_{r=0}^{n-1} e^{-i\omega_r} \phi(-\sigma, \mu; xe^{-i\omega_r}). \tag{5}$$

We note that the extreme values of ω_r satisfy $\omega_0 = -\omega_{n-1} = (n-1)\pi/(2n)$, whence $|\omega_r| < \frac{1}{2}\pi$ for $0 \leq r \leq n-1$.

We use the representation in (5), with the values of ω_r in (4), to determine the asymptotic expansion of $F_{n,\sigma}(x;\mu)$ for $x \to \pm\infty$ by application of the asymptotic theory of the Wright function. A summary of the expansion of $\phi(-\sigma, \mu; z)$ for large $|z|$ is given in Section 3. The expansions of $F_{n,\sigma}(x;\mu)$ for $x \to \pm\infty$ are given in Sections 4 and 5, where they are shown to depend critically on the parameter σ (and to a lesser extent on the integer n). A concluding section presents our numerical results confirming the accuracy of the different expansions obtained.

2. An Alternative Representation of $F_{n,\sigma}(x;\mu)$

The Wright function appearing in (2) can be written alternatively as

$$\phi(-\sigma,\mu;x) = \frac{1}{\pi} \sum_{k=0}^{\infty} \frac{x^k}{k!} \Gamma(1-\mu+\sigma k) \sin\pi(\mu-\sigma k)$$

$$= \frac{1}{2\pi}\left\{ e^{\pi i\vartheta}\Psi(xe^{\pi i\sigma}) + e^{-\pi i\vartheta}\Psi(xe^{-\pi i\sigma}) \right\}$$

upon use of the reflection formula for the gamma function, where $\vartheta := \frac{1}{2} - \mu$. The associated Wright function $\Psi(z)$ is defined by

$$\Psi(z) := \sum_{k=0}^{\infty} \frac{z^k}{k!} \Gamma(\sigma k + \delta) \qquad (0 < \sigma < 1, \delta = 1-\mu), \tag{6}$$

which is valid for $|z| < \infty$. Hence, we obtain the representation

$$F_{n,\sigma}(x;\mu) = \frac{1}{2\pi} \sum_{r=0}^{n-1} e^{-i\omega_r} Y_r(\sigma;x),$$

where

$$Y_r(\sigma;x) := e^{\pi i\vartheta}\Psi(xe^{\pi i\sigma - i\omega_r}) + e^{-\pi i\vartheta}\Psi(xe^{-\pi i\sigma - i\omega_r}).$$

If we now exploit the symmetry of the ω_r in (4) (and the fact that x is a real variable), we observe that the values of ω_r for $0 \leq r \leq N-1$, where $N = \lfloor n/2 \rfloor$, satisfy

$$\{\omega_0, \omega_1, \ldots, \omega_{N-1}\} = \left\{ \frac{(n-1)\pi}{2n}, \frac{(n-3)\pi}{2n}, \ldots, \frac{\pi}{2n}\epsilon_n \right\}, \qquad \epsilon_n = \begin{cases} 1 & (n \text{ even}) \\ 2 & (n \text{ odd}). \end{cases} \tag{7}$$

Then, we can write

$$F_{n,\sigma}(x;\mu) = \frac{1}{\pi}\Re\left\{\sum_{r=0}^{N-1} e^{-i\omega_r}Y_r(\sigma;x) + \Delta_n e^{\pi i\vartheta}\Psi(xe^{\pi i\sigma})\right\}, \qquad (8)$$

where

$$\Delta_n = \begin{cases} 0 & (n \text{ even}) \\ 1 & (n \text{ odd}). \end{cases}$$

The form (8) involves half the number of Wright functions $\Psi(z)$ and will be used to determine the asymptotic expansion of $F_{n,\sigma}(x;\mu)$ as $x \to \pm\infty$ in Sections 4 and 5.

3. The Asymptotic Expansion of $\Psi(z)$ for $|z| \to \infty$

We first present the large-$|z|$ asymptotics of the function $\Psi(z)$ in (6) based on the presentation described in ([3], Section 4); see also ([4], Section 4.2), ([5], §2.3). We introduce the following parameters:

$$\kappa = 1-\sigma, \quad h = \sigma^\sigma, \quad \vartheta = \delta - \tfrac{1}{2}, \quad \delta = 1-\mu, \qquad (9)$$

together with the associated (formal) exponential and algebraic expansions

$$E(z) := Z^\vartheta e^Z \sum_{j=0}^\infty A_j(\sigma) Z^{-j}, \quad H(z) := \frac{1}{\sigma}\sum_{k=0}^\infty \frac{(-1)^k}{k!}\Gamma\left(\frac{k+\delta}{\sigma}\right) z^{-(k+\delta)/\sigma}, \qquad (10)$$

where (The dependence of the coefficients $A_j(\sigma)$ on the parameter δ is not indicated.)

$$Z := \kappa(hz)^{1/\kappa}, \qquad A_0(\sigma) = (2\pi/\kappa)^{1/2}(\sigma/\kappa)^\vartheta. \qquad (11)$$

Then, since $0 < \kappa < 1$, we obtain from ([5], p. 57) the large-z expansion

$$\Psi(z) \sim \begin{cases} E(z) + H(ze^{\mp\pi i}) & (|\arg z| \leq \tfrac{1}{2}\pi\kappa) \\ H(ze^{\mp\pi i}) & (\tfrac{1}{2}\pi\kappa < |\arg z| \leq \pi), \end{cases} \qquad (12)$$

where the upper or lower signs are chosen according as $\arg z > 0$ or $\arg z < 0$, respectively.

The expansion $E(z)$ is exponentially large as $|z| \to \infty$ in the sector $|\arg z| < \tfrac{1}{2}\pi\kappa$, and oscillatory (multiplied by the algebraic factor $z^{\vartheta/\kappa}$) on the anti-Stokes lines $\arg z = \pm\tfrac{1}{2}\pi\kappa$. In the adjacent sectors $\tfrac{1}{2}\pi\kappa < |\arg z| < \pi\kappa$, the expansion $E(z)$ continues to be present, but is exponentially small reaching maximal subdominance relative to the algebraic expansion on the Stokes lines (On these rays, $E(z)$ undergoes a Stokes phenomenon where it switches off in a smooth manner (see [6], p. 67).) $\arg z = \pm\pi\kappa$. In our treatment of $F_{n,\sigma}(x;\mu)$, we will not be concerned with exponentially small contributions, except in one special case when $x \to -\infty$ where the expansion of $F_{n,\sigma}(x;\mu)$ is exponentially small.

The first few normalised coefficients $c_j = A_j(\sigma)/A_0(\sigma)$ are [3,4]:

$$c_0 = 1, \qquad c_1 = \frac{1}{24\sigma}\{2 + 7\sigma + 2\sigma^2 - 12\delta(1+\sigma) + 12\delta^2\},$$

$$c_2 = \frac{1}{1152\sigma^2}\{4 + 172\sigma + 417\sigma^2 + 172\sigma^3 + 4\sigma^4 - 24\delta(6 + 41\sigma + 41\sigma^2 + 6\sigma^3)$$
$$+ 120\delta^2(4 + 11\sigma + 4\sigma^2) - 480\delta^3(1+\sigma) + 144\delta^4\},$$

$$c_3 = \frac{1}{414{,}720\sigma^3}\{(-1112 + 9636\sigma + 163{,}734\sigma^2 + 336{,}347\sigma^3 + 163{,}734\sigma^4 + 9636\sigma^5$$
$$-1112\sigma^6) - \delta(3600 + 220{,}320\sigma + 929{,}700\sigma^2 + 929{,}700\sigma^3 + 220{,}320\sigma^4 + 3600\sigma^5)$$
$$+\delta^2(65{,}520 + 715{,}680\sigma + 1{,}440{,}180\sigma^2 + 715{,}680\sigma^3 + 65{,}520\sigma^4)$$

$$-\delta^3(161{,}280 + 816{,}480\sigma + 816{,}480\sigma^2 + 161{,}280\sigma^3)$$
$$+ \delta^4(151{,}200 + 378{,}000\sigma + 151{,}200\sigma^2) - 60{,}480\delta^5(1+\sigma) + 8640\delta^6\}. \qquad (13)$$

In addition to the Stokes lines $\arg z = \pm \pi\kappa$, where $E(z)$ is maximally subdominant relative to the algebraic expansion, the positive real axis is also a Stokes line. Here, the algebraic expansion is maximally subdominant relative to $E(z)$. As the positive real axis is crossed from the upper to the lower half plane the factor $e^{-\pi i}$ appearing in $H(ze^{-\pi i})$ changes to $e^{\pi i}$, and vice versa. The details of this transition will not be considered here; see ([5], p. 248) for the case of the confluent hypergeometric function $_1F_1(a;b;z)$.

4. The Asymptotic Expansion of $F_{n,\sigma}(x;\mu)$ for $x \to +\infty$

4.1. Asymptotic Character as a Function of σ

Let us denote the arguments of the Ψ functions appearing in (8) by

$$z_r^\pm = x \exp[i\phi_r^\pm], \qquad \phi_r^\pm = \pm\pi\sigma - \omega_r.$$

The representation of the asymptotic structure of the functions $\Psi(z_r^\pm)$ is illustrated in Figure 1 for different values of σ. The figures show the rays $\arg z = \pm\pi\sigma$ and the anti-Stokes lines (dashed lines) $\arg z = \pm\frac{1}{2}\pi\kappa$. In the case $\sigma = \frac{2}{3}$, the exponentially large sector is $|\arg z| < \frac{1}{6}\pi$, and it is seen from Figure 1a that the arguments $z_r^\pm$ for $0 \leq r \leq N-1$ and $xe^{\pm\pi i\sigma}$ all lie in the domain where $\Psi(z)$ has an algebraic expansion; this conclusion applies a fortiori when $\frac{2}{3} < \sigma < 1$. When $\sigma = \frac{1}{2}$, the exponentially large sector is $|\arg z| < \frac{1}{4}\pi$; when $n = 2$, we have $\omega_0 = \frac{1}{4}\pi$ so that z_0^+ is situated on the boundary of the exponentially large sector.

Other values of $n \geq 3$ will have some z_r^+ inside this sector, whereas the z_r^- are in the algebraic sector for $n \geq 2$. Similarly, the case $\sigma = \frac{1}{3}$, where the rays $\arg z = \pm\pi\sigma$ and $\arg z = \pm\frac{1}{2}\pi\kappa$ coincide, has all the z_r^+ situated in the exponentially large sector, with the z_r^- situated in the algebraic domain. Finally, when $\sigma = \frac{1}{6}$, the exponentially large sector $|\arg z| < \frac{5}{12}\pi$ encloses the rays $\arg z = \pm\pi\sigma$ with the result that all the z_r^+ lie in the exponentially large sector, whereas the z_r^- lie in the algebraic domain (except when $n = 2$ when z_0^- lies on the lower boundary of the exponentially large sector).

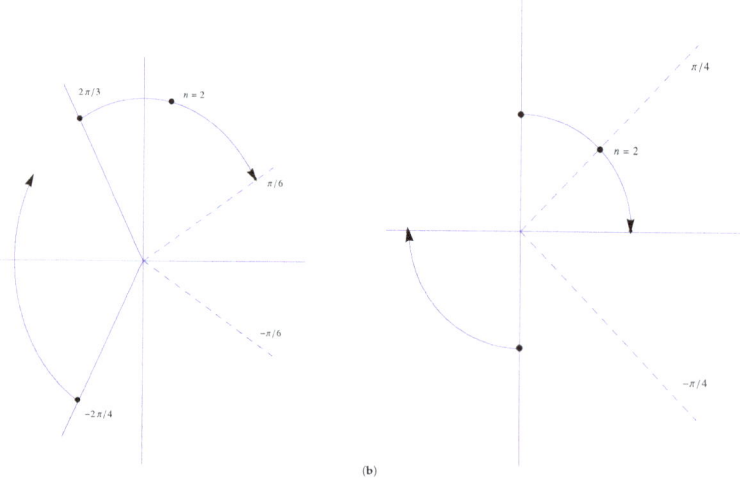

Figure 1. Cont.

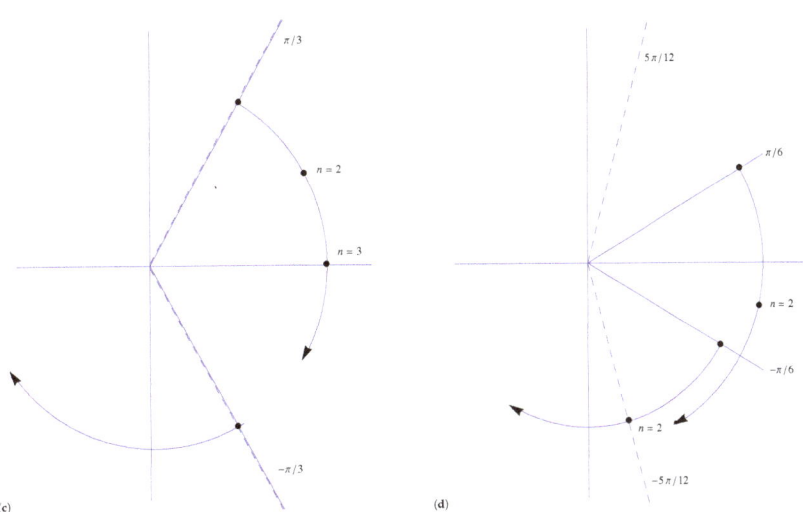

Figure 1. Diagrams representing the rays $\arg z = \pm \pi \sigma$ and the boundaries of the exponentially large sector (shown by dashed rays) $|\arg z| < \frac{1}{2}\pi\kappa$, $\kappa = 1 - \sigma$ for (**a**) $\sigma = 2/3$, (**b**) $\sigma = 1/2$, (**c**) $\sigma = 1/3$, and (**d**) $\sigma = 1/6$. Outside the exponentially large sector, the expansion of $\Psi(z)$ is algebraic in character. The circular quadrants represent the range of the arguments $\arg z = \pm \pi \sigma - \omega_r$ for $0 \leq r \leq \lfloor n/2 \rfloor - 1$, with $n \geq 2$ and the arrow-head corresponds to $n = \infty$. When $\sigma = 1/3$, the rays $\arg z = \pm \pi \sigma$ and $\arg z = \pm \frac{1}{2}\pi\kappa$ coincide.

To summarise, we have the following asymptotic character of $F_{n,\sigma}(x;\mu)$ when $x \to +\infty$ as a function of the parameter σ:

$$\left. \begin{array}{ll} 0 < \sigma < \tfrac{1}{2} & \text{Exp. large} + \text{Algebraic (for } n \geq 2) \\ \tfrac{1}{2} \leq \sigma < \tfrac{2}{3} & \text{Exp, large (dependent on } n) + \text{Algebraic} \\ \tfrac{2}{3} \leq \sigma < 1 & \text{Algebraic (for } n \geq 2). \end{array} \right\} \quad (14)$$

4.2. Asymptotic Expansion

From (8) and (10), we have the algebraic expansion associated with $F_{n,\sigma}(x;\mu)$ given by

$$\mathbf{H}(x) = \frac{1}{\sigma} \sum_{k=0}^{\infty} \frac{x^{-K}}{k!\Gamma(1-K)} \theta_{n,k}, \qquad K := \frac{k+\delta}{\sigma}, \quad (15)$$

where, with appropriate choices of the factors $e^{\pm \pi i}$ in $H(z)$,

$$\begin{aligned} \theta_{n,k} &= \frac{(-1)^k}{\sin \pi K} \Re \left\{ \sum_{r=0}^{N-1} e^{\pi i \vartheta - i \omega_r} \left(e^{\pi i \sigma - i \omega_r} \cdot e^{-\pi i} \right)^{-K} + e^{-\pi i \vartheta - i \omega_r} \left(e^{-\pi i \sigma - i \omega_r} \cdot e^{\pi i} \right)^{-K} \right. \\ &\qquad\qquad \left. + \Delta_n e^{\pi i \vartheta} \left(e^{\pi i \sigma} \cdot e^{-\pi i} \right)^{-K} \right\} \\ &= \frac{(-1)^k}{\sin \pi K} \Re \left\{ \sum_{r=0}^{N-1} e^{(K-1)i\omega_r} \left(e^{\pi i (\vartheta + \kappa K)} + e^{-\pi i (\vartheta + \kappa K)} \right) + \Delta_n e^{\pi i (\vartheta + \kappa K)} \right\} \\ &= \Re \left\{ 2 \sum_{r=0}^{N-1} e^{(K-1)i\omega_r} + \Delta_n \right\}, \quad (16) \end{aligned}$$

as $\cos \pi(\vartheta + \kappa K) = \cos \pi(K - k - \tfrac{1}{2}) = (-1)^k \sin \pi K$.

For the exponential component, we introduce the quantities

$$X = \kappa(hx)^{1/\kappa}, \qquad \Phi_r^\pm = \pm\frac{\pi\vartheta}{\kappa} - \omega_r\left(1 + \frac{\vartheta}{\kappa}\right) \tag{17}$$

and the formal asymptotic sum

$$S(Xe^{i\Omega}) := \sum_{j=0}^{\infty} A_j(\sigma)(Xe^{i\Omega/\kappa})^{-j}. \tag{18}$$

Then, from (8) and (10), we have the exponential expansion in the form

$$\mathbf{E}(x) = \frac{X^\vartheta}{\pi} \Re\left\{ \sum_{r=0}^{N-1} \left(\exp[Xe^{i\Phi_r^+/\kappa} + i\Phi_r^+] S(Xe^{i\Phi_r^+}) + \exp[Xe^{i\Phi_r^-/\kappa} + i\Phi_r^-] S(Xe^{i\Phi_r^-}) \right) \right.$$

$$\left. + \Delta_n \exp[Xe^{\pi i\sigma/\kappa} + \pi i\vartheta/\kappa] S(Xe^{\pi i\sigma}) \right\}. \tag{19}$$

It is important to stress that only the exponential terms with $|\Phi_r^\pm| \leq \frac{1}{2}\pi\kappa$, that is those with

$$|\pm\pi\sigma - \omega_r| \leq \tfrac{1}{2}\pi\kappa,$$

are to be retained in $\mathbf{E}(x)$ in (19). In addition, it is seen by inspection of Figure 1 that the second term involving $S(Xe^{i\Phi_r^-})$ does not contribute to $\mathbf{E}(x)$ when $\frac{1}{3} \leq \sigma < 1$, since, for this range of σ, the ray $\arg z = -\pi\sigma$ lies outside (or, when $\sigma = \frac{1}{3}$, on the lower boundary of) the exponentially large sector $|\arg z| < \frac{1}{2}\pi\kappa$. Thus, when $\frac{1}{2} \leq \sigma < \frac{2}{3}$, the exponential expansion is significant if $\pi\sigma - \omega_0 \leq \frac{1}{2}\pi\kappa$; that is, if $n \geq n_0 = 1/(2 - 3\sigma)$.

In summary, we have the following theorem.

Theorem 1. *The following expansion holds for* $x \to +\infty$:

$$F_{n,\sigma}(x; \mu) \sim \begin{cases} \mathbf{E}(x) + \mathbf{H}(x) & (0 < \sigma < \tfrac{1}{2}; \ n \geq 2) \\ \mathbf{E}(x) + \mathbf{H}(x) & (\tfrac{1}{2} \leq \sigma < \tfrac{2}{3}; \ n \geq n_0) \\ \mathbf{H}(x) & (\tfrac{1}{2} \leq \sigma < \tfrac{2}{3}; \ n < n_0) \\ \mathbf{H}(x) & (\tfrac{2}{3} \leq \sigma < 1; \ n \geq 2), \end{cases}$$

where $n_0 = 1/(2 - 3\sigma)$ *and the exponential and algebraic expansions* $\mathbf{E}(x)$ *and* $\mathbf{H}(x)$ *are defined in* (15) *and* (19).

4.3. Karasheva's Estimate for $|\Theta_{n,\alpha}(x; \mu)|$

When $\sigma = \alpha/(2n) < \frac{1}{2}$, we see from Theorem 1 that the dominant exponential expansion as $x \to +\infty$ corresponds to $r = 0$, yielding

$$\Theta_{n,\alpha}(x; \mu) \sim \frac{A_0(\sigma) X^\vartheta}{\pi} \Re \exp[Xe^{i(\pi\sigma - \omega_0)/\kappa} + i\Phi_0^+]$$

$$= \frac{A_0(\sigma) X^\vartheta}{\pi} \exp[X\cos(\pi\sigma - \omega_0)/\kappa] \cos[X\sin(\pi\sigma - \omega_0)/\kappa) + \Phi_0^+],$$

where

$$\frac{\pi\sigma - \omega_0}{\kappa} = \frac{2n\pi\sigma - (n-1)\pi}{2n - \alpha} = \frac{(\alpha + 1 - n)\pi}{2n - \alpha}.$$

Thus, we have the leading order estimate

$$\Theta_{n,\alpha}(x;\mu) \sim \frac{A_0(\sigma)X^{\vartheta}}{\pi} \exp\left[X\cos\left(\frac{(n-1-\alpha)\pi}{2n-\alpha}\right)\right] \cos\left[X\sin\left(\frac{(n-1-\alpha)\pi}{2n-\alpha}\right) - \Phi_0^{\pm}\right] \quad (20)$$

as $x \to +\infty$. When expressed in our notation, Karasheva's estimate for $|\Theta_{n,\alpha}(x;\mu)|$ in ([1], §8) agrees with (20) (when the second cosine term is replaced by 1), except that she did not give the value of the multiplicative constant $A_0(\sigma)/\pi$ given in (11). However, the presentation of her result as an upper bound is not evident due to the presence of possibly less dominant exponential expansions and also the subdominant algebraic expansion.

5. The Expansion of $F_{n,\sigma}(x;\mu)$ for $x \to -\infty$

To examine the case of negative x, we replace x by $e^{\mp \pi i}x$, with $x > 0$, and use the fact that $\Psi(ze^{2\pi i}) = \Psi(z)$ to find, from (8), that

$$F_{n,\sigma}(-x;\mu) = \frac{1}{\pi}\Re\left\{\sum_{r=0}^{N-1} e^{-i\omega_r}Y_r(-\kappa;x) + \Delta_n\, e^{\pi i\vartheta}\Psi(xe^{-\pi i\kappa})\right\}. \quad (21)$$

The rays $\arg z = \pm\pi\sigma$ in Figure 1 are now replaced by the Stokes lines $\arg z = \pm\pi\kappa$. The Stokes and anti-Stokes lines $\arg z = \pm\frac{1}{2}\pi\kappa$ are illustrated in Figure 2 when $0 < \sigma < \frac{1}{2}$ and $\frac{1}{2} < \sigma < 1$. In the sectors $\frac{1}{2}\pi\kappa < |\arg z| < \pi\kappa$, we recall that the exponential expansion $E(z)$ is still present but is exponentially small as $|z| \to \infty$.

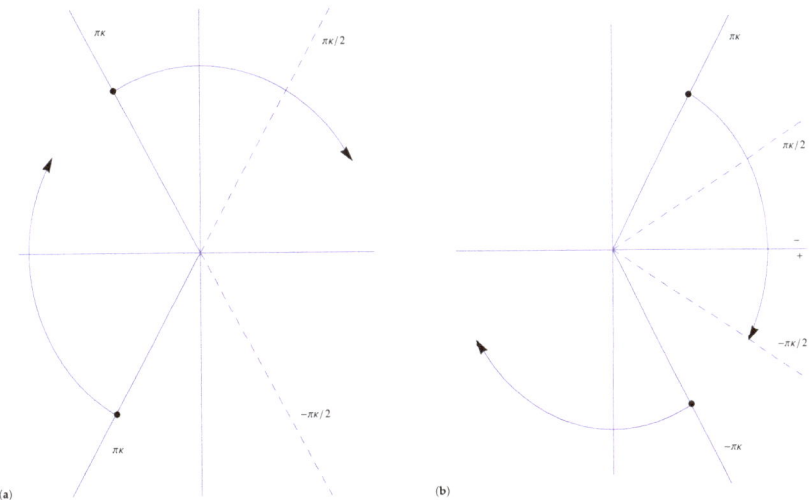

Figure 2. Diagrams representing the rays $\arg z = \pm\pi\kappa$ and the boundaries of the exponentially large sector (shown by dashed rays) $|\arg z| < \frac{1}{2}\pi\kappa$, $\kappa = 1 - \sigma$ for (**a**) $0 < \sigma < \frac{1}{2}$ and (**b**) $\frac{1}{2} < \sigma < 1$. The circular quadrants represent the range of the arguments $\arg z = \pm\pi\kappa - \omega_r$ for $0 \leq r \leq N-1$ with the arrow-head corresponding to $n = \infty$. The $\pm$ signs in (**b**) denote the signs to be chosen in $H(z)$ on either side of the Stokes line $\arg z = 0$.

For the algebraic component of the expansion two cases arise when the argument $\pi\kappa - \omega_r$ of the second Ψ function in $Y_r(-\kappa;x)$ is either (i) positive or (ii) negative. In case (i), the algebraic expansion $H(z)$ does not encounter a Stokes phenomenon as its argument does not cross $\arg z = 0$, whereas in case (ii), a Stokes phenomenon arises for those values of r that make $\pi\kappa - \omega_r < 0$. In case (i), the algebraic component contains the factor inside the sum over r in (21)

$$e^{\pi i\vartheta}\left(e^{-\pi i\kappa - i\omega_r} \cdot e^{\pi i}\right)^{-K} + e^{-\pi i\vartheta}\left(e^{\pi i\kappa - i\omega_r} \cdot e^{-\pi i}\right)^{-K}$$
$$= e^{i\omega_r K}\left(e^{\pi i(\vartheta - \sigma K)} + e^{-\pi i(\vartheta - \sigma K)}\right) = 2e^{i\omega_r K}\cos\pi(k + \tfrac{1}{2}) \equiv 0$$

upon recalling the definition of K in (15) and noting that $\delta - \vartheta = \tfrac{1}{2}$. Similarly, the final term involves the factor $\Re e^{\pi i\vartheta}(e^{-\pi i\kappa} \cdot e^{\pi i})^{-K} = \cos\pi(\vartheta - \sigma K) = 0$. Thus, the algebraic contribution to $F_{n,\sigma}(-x;\mu)$ vanishes in case (i).

For case (ii) to apply, we require that $\pi\kappa - \omega_0 < 0$; that is, $n > n^* = 1/(2\sigma - 1)$. Suppose that $\pi\kappa - \omega_r < 0$ for $0 \leq r \leq r_0$. Then, the algebraic component resulting from the terms with $r \leq r_0$ becomes

$$\frac{1}{\pi\sigma}\Re\left\{\sum_{k=0}^{\infty}\frac{(-1)^k\Gamma(K)}{k!}x^{-K}\sum_{r=0}^{r_0}e^{(K-1)i\omega_r}\left(e^{\pi i\vartheta}(e^{-\pi i\kappa} \cdot e^{\pi i})^{-K} + e^{-\pi i\vartheta}(e^{\pi i\kappa} \cdot e^{\pi i})^{-K}\right)\right\}$$
$$= \frac{2}{\pi\sigma}\Re\left\{\sum_{k=0}^{\infty}\frac{(-1)^k\Gamma(K)}{k!}x^{-K}\sum_{r=0}^{r_0}e^{(K-1)i\omega_r - \pi iK}\cos\pi(\vartheta - \sigma K + \pi K)\right\},$$

where, in the second term in round braces, we have taken account of the Stokes phenomenon (the first term and that multiplied by Δ_n are unaffected). Some routine algebra then produces the algebraic contribution

$$\hat{\mathbf{H}}(x) := \frac{2}{\sigma}\sum_{k=0}^{\infty}\frac{x^{-K}}{k!\Gamma(1-K)}\hat{\theta}_{n,k}, \qquad \hat{\theta}_{n,k} := \sum_{r=0}^{r_0}\cos\left\{\pi K - (K-1)\omega_r\right\} \tag{22}$$

when $n > n^*$ and $\hat{\mathbf{H}}(x) \equiv 0$ when $n < n^*$. (We avoid here consideration of the algebraic contribution when $\pi\kappa - \omega_r = 0$, that is, on the Stokes line $\arg z = 0$.)

Reference to Figure 2 shows that there is no exponential contribution to $F_{n,\sigma}(-x;\mu)$ from the terms $\Psi(xe^{-\pi i\kappa})$ and $\Psi(xe^{-\pi i\kappa - i\omega_r})$. From (10) and (21), we find the exponential expansion results from the terms $\Psi(xe^{\pi i\kappa - i\omega_r})$, which is given by

$$\hat{\mathbf{E}}(x) := \frac{X^\vartheta}{\pi}\Re\sum_{r=0}^{N-1}\exp\left[-Xe^{-i\omega_r/\kappa} - i\Phi\right]S(-Xe^{-i\omega_r/\kappa}), \tag{23}$$

where X and the asymptotic sum S are defined in (17) and (18) with $\Phi := \omega_r(1 + \vartheta/\kappa)$. For $\sigma < \tfrac{1}{2}$ (when the algebraic expansion vanishes), the expansion of $F_{n,\sigma}(-x;\mu)$ will be exponentially small provided $\pi\kappa - \omega_0 > \tfrac{1}{2}\pi\kappa$; that is, when $n < 1/\sigma$. If $n = 1/\sigma$, there is an exponentially oscillatory contribution, and when $n > 1/\sigma$, the expansion is exponentially large.

To summarise, we have the theorem:

Theorem 2. *The following expansion holds for $x \to +\infty$:*

$$F_{n,\sigma}(-x;\mu) \sim \begin{cases} \hat{\mathbf{E}}(x) & (0 < \sigma \leq \tfrac{1}{2}) \\ \hat{\mathbf{E}}(x) + \hat{\mathbf{H}}(x) & (\tfrac{1}{2} < \sigma < 1), \end{cases} \tag{24}$$

where the exponential expansion $\hat{\mathbf{E}}(x)$ is defined in (23). This last expansion is exponentially small as $x \to -\infty$ when $0 < \sigma < \tfrac{1}{2}$ and $n < 1/\sigma$. The algebraic expansion $\hat{\mathbf{H}}(x)$ is given by

$$\hat{\mathbf{H}}(x) := \frac{2}{\sigma}\sum_{k=0}^{\infty}\frac{x^{-K}}{k!\Gamma(1-K)}\hat{\theta}_{n,k} \quad (n > n^*), \qquad 0 \quad (n < n^*),$$

where $n^* = 1/(2\sigma - 1)$ and K, $\hat{\theta}_{n,k}$ are specified in (15) and (22).

6. Numerical Results

In this section, we describe numerical calculations that support the expansions given in Theorems 1 and 2. The function $F_{n,\sigma}(x;\mu)$ was evaluated using the expression in terms of Wright functions (valid for real x)

$$F_{n,\sigma}(x;\mu) = 2\Re \sum_{r=0}^{N-1} e^{i\omega_r}\phi(-\sigma,\mu;xe^{i\omega_r}) + \Delta_n\phi(-\sigma,\mu;x), \qquad N = \lfloor n/2 \rfloor, \qquad (25)$$

which follows from (5) and the symmetry of ω_r.

In Table 1, we present the results of numerical calculations for $x \to +\infty$ compared with the expansions given in Theorem 1. We choose four representative values of σ that focus on the different cases of Theorem 1 and $n = 2, 3$ and 4. The numerical value of $F_{n,\sigma}(x;\mu)$ was obtained by high-precision evaluation of (25). The exponential expansion $\mathbf{E}(x)$ was computed with the truncation index $j = 3$ and the algebraic expansion $\mathbf{H}(x)$ was optimally truncated (that is, at or near its smallest term).

The first case $\sigma = \frac{1}{3}$ has an exponentially large expansion with a subdominant algebraic contribution for all three values of n. The second case $\sigma = \frac{1}{2}$ corresponds to $n_0 = 2$; when $n = 2$, $\mathbf{E}(x)$ is oscillatory and makes a similar contribution as $\mathbf{H}(x)$, whereas when $n = 3$ and 4, $\mathbf{E}(x)$ is exponentially large. The third case $\sigma = \frac{5}{9}$ corresponds to $n_0 = 3$; when $n = 2$, there is no exponential contribution, whereas when $n = 3$, $\mathbf{E}(x)$ is oscillatory and thus makes a similar contribution as $\mathbf{H}(x)$; when $n = 4$, $\mathbf{E}(x)$ is exponentially large. Finally, when $\sigma = \frac{2}{3}$, the expansion of $F_{n,\sigma}(x;\mu)$ is purely algebraic in character.

Table 1. The values of the exponential and algebraic expansions compared with $F_{n,\sigma}(x;\mu)$ for large $x > 0$ for different values of σ and n when $\mu = 3/4$ and $x = 8$.

σ		$n = 2$	$n = 3$	$n = 4$
1/3	$\mathbf{E}(x)$	-1.81418881×10^2	-1.08294258×10^3	-3.08231679×10^3
	$\mathbf{H}(x)$	$+0.34241316$	$+0.17280892$	$+0.34497729$
	$\mathbf{E}(x) + \mathbf{H}(x)$	-1.81076468×10^2	-1.08276977×10^3	-3.08197181×10^3
	$F_{n,\sigma}(x;\mu)$	-1.80709370×10^2	-1.08284759×10^3	-3.08254767×10^3
1/2	$\mathbf{E}(x)$	$+0.06317153$	$+1.15957937 \times 10^3$	-4.47945373×10^4
	$\mathbf{H}(x)$	$+0.74012019$	$+1.09449277$	$+1.45169481$
	$\mathbf{E}(x) + \mathbf{H}(x)$	$+0.80329172$	$+1.16067387 \times 10^3$	-4.47930856×10^4
	$F_{n,\sigma}(x;\mu)$	$+0.80329527$	$+1.16069221 \times 10^3$	-4.47921506×10^4
5/9	$\mathbf{E}(x)$	—	-0.14805870	$+2.77243091 \times 10^2$
	$\mathbf{H}(x)$	$+0.79825166$	$+1.17615555$	$+1.55857242$
	$\mathbf{E}(x) + \mathbf{H}(x)$	$+0.79825166$	$+1.02809685$	$+2.78801663 \times 10^2$
	$F_{n,\sigma}(x;\mu)$	$+0.79825119$	$+1.02809649$	$+2.78801134 \times 10^2$
2/3	$\mathbf{H}(x)$	$+0.84046066$	$+1.23266920$	$+1.63072031$
	$F_{n,\sigma}(x;\mu)$	$+0.84046066$	$+1.23266920$	$+1.63072031$

In Table 2, we present illustrative examples of Theorem 2 when $x \to -\infty$. The first case, $\sigma = \frac{1}{4}$ ($\kappa = \frac{3}{4}$), has an expansion that is exponential in character; for $n < 1/\sigma = 4$, $\hat{\mathbf{E}}(x)$ is exponentially small, whereas for $n = 4$, the argument $\pi\kappa - \omega_0 = \frac{3}{8}\pi$ lies on the upper boundary of the exponentially large sector $|\arg z| < \frac{3}{8}\pi$, and thus $\hat{\mathbf{E}}(x)$ is oscillatory. For $n \geq 5$, $\hat{\mathbf{E}}(x)$ becomes exponentially large as $x \to -\infty$. In the second case, $\sigma = \frac{2}{5}$ ($\kappa = \frac{3}{5}$), $\hat{\mathbf{E}}(x)$ is exponentially small for $n = 2$ and exponentially large for $n \geq 3$.

In the third case, $\sigma = \kappa = \frac{1}{2}$, $\hat{\mathbf{E}}(x)$ is oscillatory for $n = 2$ and exponentially large for $n \geq 3$. Finally, when $\sigma = \frac{3}{4}$ ($\kappa = \frac{1}{4}$), the function $F_{n,\sigma}(x;\mu)$ is exponentially large for

$n = 2, 3$ and $n \geq 5$. However, for $n = 4$, the two values $\omega_0 = \frac{3}{8}\pi$ and $\omega_1 = \frac{1}{8}\pi$ yield arguments $\pi\kappa - \omega_r$ ($r = 0, 1$) situated on *both* boundaries of the exponentially large sector $|\arg z| < \frac{1}{8}\pi$. In this case $\hat{\mathbf{E}}(x)$ is oscillatory and, since $n^* = 2$, there is, in addition, an algebraic contribution $\hat{\mathbf{H}}(x)$.

Table 2. The values of the exponential and algebraic expansions compared with $F_{n,\sigma}(x;\mu)$ for large $x < 0$ for different values of σ and n when $\mu = 3/4$ and $|x| = 8$ (for $\sigma = 1/4, 1/2, 2/5$), $|x| = 5$ (for $\sigma = 3/4$).

σ		$n = 2$	$n = 3$	$n = 4$
1/4	$\hat{\mathbf{E}}(x)$	$+1.59003829 \times 10^{-2}$	$+1.77442984 \times 10^{-1}$	$+6.49578248 \times 10^{-1}$
	$F_{n,\sigma}(-x;\mu)$	$+1.59003416 \times 10^{-2}$	$+1.77011100 \times 10^{-1}$	$+6.49580223 \times 10^{-1}$
2/5	$\hat{\mathbf{E}}(x)$	$-4.18901636 \times 10^{-2}$	-3.79446870×10^0	-3.02428770×10^1
	$F_{n,\sigma}(-x;\mu)$	$-4.18889220 \times 10^{-2}$	-3.79475882×10^0	-3.02402120×10^1
1/2	$\hat{\mathbf{E}}(x)$	-0.56022532	$+1.23070020 \times 10^3$	-1.28808653×10^4
	$F_{n,\sigma}(-x;\mu)$	-0.56023534	$+1.23066913 \times 10^3$	-1.28803505×10^4
3/4	$\hat{\mathbf{E}}(x)$	$+1.81213632 \times 10^{28}$	$+7.55354383 \times 10^{13}$	-0.84956415
	$\hat{\mathbf{H}}(x)$	—	$-1.93112636 \times 10^{-1}$	-0.28756658
	$\hat{\mathbf{E}}(x) + \hat{\mathbf{H}}(x)$	$+1.81213632 \times 10^{28}$	$+7.55354383 \times 10^{13}$	-1.13713072
	$F_{n,\sigma}(-x;\mu)$	$+1.81213650 \times 10^{28}$	$+7.55354314 \times 10^{13}$	-1.13713081

7. Concluding Remarks

We employed the standard asymptotics of the Wright function $\Psi(z)$ defined in (6) to determine the asymptotic expansion of $F_{n,\sigma}(x;\mu)$ for $x \to \pm\infty$. We found that this behaviour depended critically on the parameter σ. The numerical results presented in Tables 1 and 2 demonstrate that the asymptotic forms of $F_{n,\sigma}(x;\mu)$ stated in Theorems 1 and 2 agreed well with the numerically computed values of $F_{n,\sigma}(\pm x;\mu)$. In particular, we showed that, when $\sigma < \frac{1}{2}$, the expansion of $F_{n,\sigma}(x;\mu)$ exponentially decays as $x \to -\infty$.

Funding: This research received no external funding.

Institutional Review Board Statement: Not applicable.

Informed Consent Statement: Not applicable.

Data Availability Statement: Not applicable.

Conflicts of Interest: The author declares no conflict of interest.

References

1. Karasheva, L.L. On properties of an entire function that is a generalization of the Wright function. *J. Math. Sci.* **2020**, *250*, 753–759. [CrossRef]
2. Mainardi, F. *Fractional Calculus and Waves in Linear Viscoelasticity*; Imperial College Press: London, UK, 2010.
3. Paris, R.B. The asymptotics of the generalised Bessel function. *Math. Aeterna* **2017**, *7*, 381–406.
4. Paris, R.B. Asymptotics of the special functions of fractional calculus. In *Handbook of Fractional Calculus with Applications*; Kochubei, A., Luchko, Y., Eds.; De Gruyter: Berlin, Germany, 2019; Volume 1, pp. 297–325.
5. Paris, R.B.; Kaminski, D. *Asymptotics and Mellin-Barnes Integrals*; Cambridge University Press: Cambridge, UK, 2001.
6. Olver, F.W.J.; Lozier, D.W.; Boisvert, R.F.; Clark, C.W. (Eds.) *NIST Handbook of Mathematical Functions*; Cambridge University Press: Cambridge, UK, 2010.

Article

Fractional Calculus in Russia at the End of XIX Century

Sergei Rogosin * and Maryna Dubatovskaya

Department of Economics, Belarusian State University, 4, Nezavisimosti Ave, 220030 Minsk, Belarus; dubatovska@bsu.by
* Correspondence: rogosin@bsu.by; Tel.: +375-17-282-22-84

Abstract: In this survey paper, we analyze the development of Fractional Calculus in Russia at the end of the XIX century, in particular, the results by A. V. Letnikov, N. Ya. Sonine, and P. A. Nekrasov. Some of the discussed results are either unknown or inaccessible.

Keywords: fractional integrals and derivatives; Grünwald-Letnikov approach; Sonine kernel; Nekrasov fractional derivative

MSC: primary 26A33; secondary 34A08; 34K37; 35R11; 39A70

1. Introduction

The year of the birth of Fractional Calculus is considered 1695, when Leibniz discussed the possibility of introducing the derivative of an arbitrary order in his letters to Wallis and Bernoulli. Several attempts were made to give a precise meaning to this new notion. A comprehensive detailed analysis of the history of Fractional Calculus is given in Reference [1]. One of most productive periods in this history was the middle-end of the XIX century. Here, we can mention works by Legendre, Fourier, Peacock, Kelland, Tardi, Roberts, and others. The most advanced approach to the determination of the fractional derivative of an arbitrary order was proposed by Liouville. A deep analysis of the results on this subject was given in the article [2] by Letnikov. In particular, he recognized the basic role of Liouville's approach. Letnikov said ([2], p. 92): "... we give a survey of the results by Liouville whom we ought to consider as the first scientist paid a serious attention to clarifying the question on the derivative of an arbitrary order. In 1832, he started to publish a series of articles devoted to the foundation and the application of his theory of general differentiation which is the first complete discussion of this topic. Before his work, only few very important but not completely clear remarks were made on this subject".

It should be noted that the works by A. V. Letnikov constitute the first rigorous and comprehensive construction of the theory of the fractional integro-differentiation. An extended description of the results by Letnikov is presented in the articles of References [3,4] and in the book of Reference [5], written in Russian.

In the middle-end of the XIX century, an interest to Fractional Calculus in Russia grew significantly [6–10]. One of the reasons for it was a high standard in the research in Real and Complex Analysis in Russia in this period. Russian Universities took care of the level of the education of young scientists. Many applicants for a professorship had been given the opportunity to spend 1–2 years at the leading research centers and to attend the lectures of known mathematicians.

Letnikov's results attracted people to this branch of the Science, at least in Russia. Nevertheless, these works remained unknown abroad and, for a long time, were unaccessible. After contribution by Letnikov, the serious works on Fractional Calculus in Russia in the second part of the XIX century were published by N. Ya. Sonine and P. A. Nekrasov. They introduced the complex-analytic technique into the study and application of derivatives and integrals of an arbitrary order. It should be noted that Complex Analysis was

Citation: Rogosin, S.; Dubatovskaya, M. Fractional Calculus in Russia at the End of XIX Century. *Mathematics* **2021**, *9*, 1736. https://doi.org/10.3390/math9151736

Academic Editor: Clemente Cesarano

Received: 28 June 2021
Accepted: 19 July 2021
Published: 22 July 2021

Publisher's Note: MDPI stays neutral with regard to jurisdictional claims in published maps and institutional affiliations.

Copyright: © 2021 by the authors. Licensee MDPI, Basel, Switzerland. This article is an open access article distributed under the terms and conditions of the Creative Commons Attribution (CC BY) license (https://creativecommons.org/licenses/by/4.0/).

traditionally highly developed discipline starting from Leonard Euler, who worked for a long period in Russia (1726–1741 and 1776–1783). This part of Mathematical Analysis was essentially developed in the XIX century by M. V. Ostrogradsky, V. Ya. Bunyakovsky, P. L. Chebyshev, A. M. Lyapunov, and many others. In particular, Sonine and Nekrasov found a fractional analog of the classical Cauchy integral formula for analytic functions.

In our article, we describe the contribution of Alexey Vasil'evich Letnikov (1837–1888), Nikolai Yakovlevich Sonine (1849–1915), and Pavel Alekseevich Nekrasov (1853–1924) to Fractional Calculus and the role of these results in the modern Fractional Calculus and its Applications.

2. Liouville's Approach and Its Analysis by Letnikov

As it was already said, A. V. Letnikov considered (see, e.g., Reference [2,6,7]) that the Liouville's theory constitutes the only complete treatment of differentiation of an arbitrary order. Realizing the great importance of this theory, Lentnikov had seen that its certain parts did not receive a proper justification and led to some misunderstanding in the works of Liouville's followers.

Let us present here Letnikov's description of the elements of the Lioville's theory following Reference [2]. Letnikov started his analysis with the definitions given by Liouville.

Definition 1. *Let the function $y(x)$ be represented in the form of the following series of exponents:*

$$y(x) = A_1 e^{m_1 x} + A_2 e^{m_2 x} + \ldots, \tag{1}$$

which is denoted for shortness as $\sum A_m e^{mx}$.

Fractional derivative of the order p is defined by multiplying each term of the series by p-th power of the index m:

$$\frac{d^p y}{dx^p} = \sum A_m m^p e^{mx}. \tag{2}$$

If p is negative, then Formula (2) determined the fractional integral of order $-p$.

Fractional integral of order $-p$ is denoted by Liouville as $\int^{-p} y dx^{-p}$. Liouville considered this definition as the only possible way to generalize the usual derivative. Evaluating its role, Letnikov stressed that Definition 1 contains a key ideas to establish a deep analogy with differences and powers and, thus, could lead to a more simple construction.

Nevertheless, the above definition had a very important restriction. It cannot be applied to an arbitrary function since not all of them possess representations in series of exponents. Liouville himself understood this difficulty. He proposed a way to overcome it. By performing the change of variable $z = e^x$ for the function $y = F(x)$, one can expand the composite function $y = F(\ln z)$ (with $x = \ln z$) in a converging power series:

$$F(\ln z) = \sum A_m z^m. \tag{3}$$

Thus, the initial function $y = F(x)$ admits representation via series of exponents

$$F(x) = \sum A_m e^{mx}. \tag{4}$$

But the possibility to represent $y = F(\ln z)$ in form (3) met several restrictions. For instance, if we suppose to get representation of $y = F(\ln z)$ in a form of series in positive powers of z, then all derivatives of $y = F(x)$ at $x = \infty$ should be equal to zero since

$$(F(\ln z))'_z = \frac{F'(\ln z)}{z}, (F(\ln z))''_z = \frac{F''(\ln z) - F'(\ln z)}{z^2}, \ldots.$$

Similar restriction appears if we suppose to represent $y = F(\ln z)$ in a form of series in negative powers of z since we deal in this case with the function $y = F(-\ln z)$. Such conditions look fairly strong. Moreover, they are neither necessary nor sufficient for the representation of the type (4).

Liouville met such a restriction trying to calculate the derivative of a fractional order of the power function. He started with the Euler formula

$$\frac{1}{x^m} = \frac{\int_0^\infty e^{-xz} z^{m-1} dz}{\Gamma(m)}.$$

Liouville supposed that the above integral can be represented in a form of the exponential sum $\sum A_n e^{-nx}$. Here, all coefficients in such representation should be infinitely small. Then, using his main definition, Liouville arrived at the formula of the derivative of this function:

$$\frac{d^p}{dx^p} \frac{1}{x^m} = \frac{\int_0^\infty e^{-xz}(-z)^p z^{m-1} dz}{\Gamma(m)}. \tag{5}$$

Thus, by definition of Γ-function, we get, after substituting $xz = t$, the following formula:

$$\frac{d^p}{dx^p} \frac{1}{x^m} = \frac{(-1)^p \Gamma(m+p)}{\Gamma(m) x^{m+p}}. \tag{6}$$

In his first articles, Liouville used the only definition of the Γ-function of positive variable (later, he noted that he was not familiar with the general definition of the Γ-function by Legendre and Gauss). Therefore, he supposed that, due to assumptions $m > 0$, $m + p > 0$, one needs to use in the above definition so-called auxiliary functions (more detailed discussion of the role of auxiliary function is presented below in Section 4; in fact, such notion appeared in the works by Liouville since he used indefinite integral for fractional integration). Being very important, the use of auxiliary functions did not lead to a general definition of the fractional derivative. Liouville showed that, if one supposed an existence of auxiliary functions, then these necessarily had to be entire functions.

Letnikov claimed and proved that it follows from his analysis that Liouville's formulas were so general that they had no need of any auxiliary function. Later on, several attempts to correct Liouville's approach were made. In particular, Letnikov analyzed in Reference [2] the works by Kelland, Tardi, and Roberts. But the really rigorous approach which transformed Liouville's formulas to the general definition of the fractional derivative was proposed by Letnikov. We have to note that Letnikov used a definite integral in his construction (see Section 3.1). For such construction, the notion of auxiliary function becomes needless.

3. Letnikov's Contribution to Fractional Calculus

3.1. Letnikov or Grünwald-Letnikov Derivative

Starting his work on determination of the derivative of an arbitrary order, Letnikov posed this problem [6,7] as interpolation in form of the elements of two sequences consisting of successive derivatives of the function $f(x)$

$$(a) \quad f(x), f'(x), f''(x), \ldots, f^{(n)}(x), \ldots \tag{7}$$

and of successive n-fold integrals of this function:

$$(b) \quad f(x), \int f(x) dx, \int^2 f(x) dx^2, \ldots, \int^n f(x) dx^n, \ldots \tag{8}$$

In other words, he tried to find such a formula of the derivative of an arbitrary order α which, for nonnegative integer, $\alpha = 0, 1, 2, \ldots$ coincides with the corresponding elements of the sequence (a), and, for nonpositive integer, $\alpha = 0, -1, -2, \ldots$ coincides with the corresponding elements of the sequence (b). Denoting this formula by

$$D^\alpha f(x) \quad \text{or} \quad \frac{d^\alpha f(x)}{dx^\alpha},$$

he expected to get this new object to have (whenever it is possible) that same properties as elements of sequence (a) or (b) when α is an integer.

The next idea by Letnikov was to restrict the generality of the above question and to consider, instead of the sequence (b) (of indefinite n-fold integrals), the sequence of definite integrals, supposing that $f(x)$ is continuous on certain interval $[a, x]$, i.e., to interpolate in form elements of the double sequence

$$(A) \quad \ldots \int_a^x \int_a^x f(x)dx^2, \int_a^x f(x)dx, f'(x), f''(x), \ldots, \tag{9}$$

in which any element is the derivative of the previous one.

The corresponding interpolating object he denoted as

$$[D^\alpha f(x)]_a^x.$$

In order to get such interpolation, Letnikov proposed to examine the following formula:

$$\frac{\sum_{k=0}^{n}(-1)^k \binom{\alpha}{k} y(x-kh)}{h^\alpha}, \tag{10}$$

where $h = \dfrac{x-a}{n}$, and $\binom{\alpha}{k}$ denotes the binomial coefficient. This approach was independently used by Grünwald [11] and by Letnikov [6]. When Letnikov found the paper by Grünwald, he decided to decline publication of his work, but later changed his mind. Letnikov developed in Reference [6] more rigorously than in Reference [11] the theory of the derivative of an arbitrary order and found its relationship with many results known in this area.

Elementary algebra yields that, for $\alpha = m$ being positive integer number, the derivatives of the corresponding order can be defined as a limit of the above expression

$$f^{(m)}(x) = \lim_{\delta \to 0} \frac{f(x) - \binom{m}{1}f(x-\delta) + \binom{m}{2}f(x-2\delta) + \ldots + (-1)^n \binom{m}{n}f(x-n\delta)}{\delta^m}. \tag{11}$$

Here, $\delta \to 0$ is equivalent to $n \to \infty$, but the sum in the numerator remains finite since all binomial coefficients with $n > m$ vanishing. Thus, Formula (11) can be taken as the definition of the derivative of order $m \in \mathbb{N}$.

Vice versa, for $\alpha = -m$ being negative integer, the expression under the limit sign in the right-hand side of (11) equals to

$$\frac{f(x) + \binom{m}{1}f(x-\delta) + \binom{m}{2}f(x-2\delta) + \ldots + \binom{m}{n}f(x-n\delta)}{\delta^{-m}}. \tag{12}$$

Letnikov showed [6] (pp. 5–12) that the limit of this expression as $\delta \to 0$, or equivalently as $n \to \infty$, is equal to the multiple integral, i.e., (in his notation)

$$[D^{-m}f(x)]_a^x = \lim_{\delta \to 0} \frac{f(x) + \binom{m}{1}f(x-\delta) + \binom{m}{2}f(x-2\delta) + \ldots + \binom{m}{n}f(x-n\delta)}{\delta^{-m}} \\ = \int_a^x dx_1 \int_a^{x_1} dx_2 \ldots \int_a^{x_{m-1}} f(x_m)dx_m. \tag{13}$$

This magnitude $[D^{-m}f(x)]_a^x$ satisfies certain properties. First of all, if we apply to it similar operation of order $-p$, $p > 0$, then we will have

$$[D^{-p}D^{-m}f(x)]_a^x = [D^{-m-p}f(x)]_a^x.$$

Next, if we take the derivative $\frac{d^p}{dx^p}$ of order $p > 0$, then we will have

$$\frac{d^p}{dx^p}[D^{-m}f(x)]_a^x = [D^{-m+p}f(x)]_a^x, \quad if \quad m > p,$$

and

$$\frac{d^p}{dx^p}[D^{-m}f(x)]_a^x = \frac{d^{p-m}f(x)}{dx^{p-m}}, \quad if \quad m < p.$$

Thus, in particular, the symbol $[D^{-m}f(x)]_a^x$ means m-times differentiable function whose all derivatives up to m-th order are vanishing at $x = a$.

Formulas (13) and (11) coincide with the corresponding elements of the double sequence (A). Therefore, it led Letnikov to the conclusion that the limit

$$[D^\alpha f(x)]_a^x := \lim_{\delta \to 0} \frac{f(x) - \binom{\alpha}{1}f(x-\delta) + \binom{\alpha}{2}f(x-2\delta) + \ldots + (-1)^n \binom{\alpha}{n}f(x-n\delta)}{\delta^\alpha} \quad (14)$$

is a good candidate to solve the interpolation problem for the sequence (A), i.e., to be the derivative of arbitrary order.

The relations of this new object to known formulas of the fractional derivatives (integrals) were described by Letnikov [6] (p. 15) using the following elementary:

Lemma 1. *Let (α_k) be a sequence of (real or complex) numbers such that*

$$\lim_{k \to \infty} \alpha_k = 0 \quad and \quad \lim_{k \to \infty}(\alpha_1 + \alpha_2 + \ldots + \alpha_k) = C,$$

and let (β_k) be a sequence of (real or complex) numbers such that

$$\lim_{k \to \infty} \beta_k = 1.$$

Then, the sequence of their products has the limits equal to C, i.e.,

$$\lim_{k \to \infty}(\alpha_1 \beta_1 + \alpha_2 \beta_2 + \ldots + \alpha_k \beta_k) = \lim_{k \to \infty}(\alpha_1 + \alpha_2 + \ldots + \alpha_k) = C.$$

The above formula is valid for $[D^\alpha f(x)]_a^x$ with $\alpha < 0$ (i.e., for fractional integral of order $-\alpha$ in modern language).

The corresponding justification of the formula of $[D^\alpha f(x)]_a^x$ with $\alpha > 0$ (i.e., representation of fractional derivative) Letnikov supposed additionally that the function $f(x)$ is $(n+1)$-times continuously differentiable on the interval (a, x), where n is a largest integer smaller than α, i.e., $n < \alpha < n + 1$. Then, using quite cumbersome transformation of the binomial coefficients [6] (pp. 21–26), he has got that the limit in (14) is equal:

$$[D^\alpha f(x)]_a^x = \frac{f(a)(x-a)^{-\alpha}}{\Gamma(-\alpha+1)} + \frac{f'(a)(x-a)^{-\alpha+1}}{\Gamma(-\alpha+2)} + \ldots + \frac{f^{(n)}(a)(x-a)^{-\alpha+n}}{\Gamma(-\alpha+n+1)} + \quad (15)$$

$$+ \frac{1}{\Gamma(-\alpha+n+1)} \int_a^x (x-\tau)^{-\alpha+n} f^{(n+1)}(\tau) d\tau.$$

Note that the same result is true if $\alpha \in \mathbb{C}$, $\text{Re}\,\alpha > 0$. Integration by parts showed that (15) can be taken as the definition of fractional derivative of an arbitrary order $\alpha > 0$. A slightly more general form can be written for any $s \in \mathbb{Z}$, $s \geq n$, $n < \text{Re}\,\alpha < n + 1$ (of course, under additional smoothness conditions):

$$[D^\alpha f(x)]_a^x = \sum_{k=0}^{s} \frac{f^{(k)}(a)(x-a)^{-\alpha+k}}{\Gamma(-\alpha+k+1)} + \frac{1}{\Gamma(-\alpha+s+1)} \int_a^x (x-\tau)^{-\alpha+s} f^{(s+1)}(\tau) d\tau. \quad (16)$$

In Reference [6], Letnikov paid attention to relationship of his formulas with known constructions. In particular, he showed that, if the function $f(x)$ is defined, infinitely differentiable on $[x, \infty)$, and vanishes together with any derivative when x is tending to ∞, then the following formula hold for any $\alpha, \operatorname{Re}\alpha < 0$,:

$$[D^\alpha f(x)]^x_{+\infty} = \frac{1}{\Gamma(-\alpha)} \int_{+\infty}^{x} (x-\tau)^{-\alpha-1} f(\tau) d\tau = \frac{1}{(-1)^\alpha \Gamma(-\alpha)} \int_0^{+\infty} z^{-\alpha-1} f(x+z) dz,$$

i.e., coincides with the corresponding integral defined by Liouville. Similarly, for any $\alpha, 0 \le n < \operatorname{Re}\alpha < n+1$ Letnikov discovered that

$$[D^\alpha f(x)]^x_{+\infty} = \frac{1}{\Gamma(-\alpha+n+1)} \int_{+\infty}^{x} \frac{f^{(n+1)}(\tau) d\tau}{(x-\tau)^{\alpha-n}} = \frac{1}{(-1)^{\alpha-n-1}\Gamma(-\alpha+n+1)} \int_0^{+\infty} \frac{f^{(n+1)}(x+z) dz}{z^{\alpha-n}}. \tag{17}$$

He also noted that the considered class of functions is not empty, it contains, in particular, all functions of the form $x^m e^{-x}$.

In Reference [6], Letnikov also presented a series of formulas for the values of his derivative of an arbitrary order of elementary functions, such as power function $(x-a)^\beta$, exponential function e^{mx}, logarithmic function $\log x$, exponential-trigonometric functions $e^{\beta x} \sin \gamma x$, $e^{\beta x} \cos \gamma x$, and rational functions $\frac{P(x)}{Q(x)}$. These formulas coincide with nowadays known formulas (see, e.g., Reference [1,12]). Composition formulas for fractional derivatives and integrals were found in Reference [6], too. The last result, presented in Reference [6], was the so-called Leibniz rule for the fractional derivative/integral of the product of functions. Note, that after the death of A. V. Letnikov, it was created a committee examined some of his manuscripts [13]. A few results were then published, but not all were found. In particular, the members of the committee reported that they did not find any results on Abel integrals, as it was expected by some researchers.

3.2. Solution to Certain Differential Equations

In Reference [14], Liouville made a background for further development of Fractional Calculus. In order to show an importance of the new branch of Science, he solved in Reference [15] a number of problems (mainly from geometry, classical mechanics, and mathematical physics) by using his constructions of integral and derivatives of an arbitrary order. Later, in Reference [16], he also discussed the tautochrone problem and usage of fractional derivatives to its solution.

In his master thesis, Letnikov carefully examined these results by Liouville and came to the conclusion that Liouville's solutions of the problems can be obtained by using more traditional methods, too. He also remarked that incorrect usage of Liouville construction by his followers led to certain misunderstandings, and even mistakes. Note that the master thesis by Letnikov was reprinted in Russian recently in Reference [4,5].

Nevertheless, Letnikov believed that newly created technique could find proper applications. One of these applications was presented in his article, Reference [17], devoted to use of the fractional derivative to the solution of the differential equation

$$(a_n + b_n x) \frac{d^n y}{dx^n} + (a_{n-1} + b_{n-1} x) \frac{d^{n-1} y}{dx^{n-1}} + \ldots + (a_0 + b_0 x) y = 0. \tag{18}$$

These results were lectured by Letnikov at the meeting of Mathematical Society on 16 April 1876, and at the meeting of the Warsaw Congress of naturalists on 3 September 1876. They were reprinted by P. A. Nekrasov, who parsed the Letnikov's archive after his death.

Denoting

$$\varphi(\rho) := a_n \rho^n + a_{n-1} \rho^{n-1} + \ldots + a_1 \rho + a_0, \quad \psi(\rho) := b_n \rho^n + b_{n-1} \rho^{n-1} + \ldots + b_1 \rho + b_0,$$

Equation (18) can be rewritten in the following symbolic form:

$$\varphi\left(\frac{d}{dx}\right)y + x\psi\left(\frac{d}{dx}\right)y = 0. \qquad (19)$$

Suppose that equation

$$\psi(\lambda) = 0 \qquad (20)$$

has different zeroes $\lambda_1, \lambda_2, \ldots, \lambda_n$. Denoting for each $j = 1, 2, \ldots, n$

$$y := e^{\lambda_j x} Y, \qquad (21)$$

one can rewrite (19)

$$\varphi\left(\lambda_j + \frac{d}{dx}\right)Y + x\psi\left(\lambda_j + \frac{d}{dx}\right)Y = 0. \qquad (22)$$

A crucial idea by Letnikov was to look for the solution to Equation (22) in the form:

$$Y = [D^p y_j]_a^x, \qquad (23)$$

where y_j is a new unknown function, and $[D^p]_a^x$ is an inter-limit (Letnikov-type) derivative whose order p is to be defined later.

Let there exist the function y_j, vanishing at $x = a$, together with all derivatives up to order $n - 2$, satisfying the following equation:

$$\varphi_j\left(\frac{d}{dx}\right)y_j + x\psi_j\left(\frac{d}{dx}\right)y_j = 0, \qquad (24)$$

where

$$\psi_j(\rho) = \frac{\psi(\lambda_j + \rho)}{\rho}, \quad p + 1 = \frac{\varphi(\lambda_j)}{\psi_j(0)} = \frac{\varphi(\lambda_j)}{\psi'(\lambda_j)} = A_j, \quad \varphi_j(\rho) = \frac{\varphi(\lambda_j + \rho) - A_j \psi_j(\rho)}{\rho}.$$

Then, there exists the solution to Equation (19), which is represented in the form:

$$y = e^{\lambda_j x}\left[D^{A_j - 1} y_j\right]_a^x. \qquad (25)$$

Thus, by this transformation, we reduce Equation (18) of order n to Equation (24) of order $n - 1$. By applying this method, one can reduce the order of the equation up to 1 and get the possible solution via successive application of the inter-limit derivative.

4. Sonine-Letnikov Discussion

In 1868, A. V. Letnikov published the main part of his master thesis as an article in Mathematical Sbornik [6], supplemented by the historical survey on the development of the theory of differentiation of an arbitrary order [2]. This article, and Grünwald's article [11], was criticized by N. Ya. Sonine in Reference [9], who also presented in Reference [9] his own approach to determine the derivatives of an arbitrary order. Sonine's article started with the discussion of Liouville's definition of the derivative of an arbitrary order (not necessarily positive). The latter definition is based on the derivative of an arbitrary order $p \in \mathbb{R}$ of exponential function

$$\frac{d^p}{dx^p} e^{mx} = m^p e^{mx}$$

and on the possibility to expand a differentiable function into exponential series (the Dirichlet series in modern language):

$$f(x) = \sum_{k=-\infty}^{+\infty} A_k e^{m_k x}.$$

Sonine made two important remarks concerning Liouville's definition. First, he showed that the derivative of negative order (i.e., the fractional integral) cannot be considered as an inverse to the fractional derivative. The second remark by Sonine is related to the problem discovered by Liouvile himself. If one applies Liouville's definition of the derivative of arbitrary order to power function, then it leads immediately to a kind of contradiction. Liouville founded this phenomena using the following representation of x:

$$x = \lim_{\beta \to 0} \frac{e^{\beta x} - e^{-\beta x}}{2\beta}.$$

If we suppose that the limit and the fractional derivative are interchangeable, then the half-derivative of x becomes infinite. Moreover, the derivative of an infinitesimal quantity could be finite. From these facts, Liouville concluded the existence of additional functions that are the derivative of zero function and coincide with an entire function with arbitrary coefficients. Contrariwise, Sonine has shown that such contradiction follows from a not completely rigorous way of expansion of the function into the series in exponents. He also remarked that Liouville's proof of existence of additional functions does not have a proper rigor.

In the second part of his article [9], Sonine criticized an approach by Grünwald and Letnikov, in which the fractional derivative is defined by the following limit:

$$D^\alpha [f(x)]_{x=a}^{x=x} := \lim_{\delta \to 0} \frac{f(x) - \binom{\alpha}{1} f(x-h) + \binom{\alpha}{2} f(x-2h) + \ldots + (-1)^n \binom{\alpha}{n} f(x-nh)}{h^\alpha}, \quad (26)$$

where $nh = x - a$.

Sonine had two main objections. First, he noted that the series in the numerator of (26) is converging. Hence, the fraction should be infinite. The second remark by Sonine was that, if we apply to the fractional integral $D^{-\alpha}$ the fractional derivative D^β ($\alpha, \beta > 0$), then, by Leibnitz rule, the result should coincide with $D^{\beta-\alpha}$. It leads to contradiction, even for the function $f(x) = 1$, since it should exist for any β, but it is so only for $\operatorname{Re} \alpha > [\operatorname{Re} \beta]$, where $[\cdot]$ means the integer part of the number.

Concerning the first remark, Letnikov noted in Reference [7] that, in the case $\operatorname{Re} \alpha > 0$, Formula (26) determines the fractional integral (coinciding with m-times repeated integral if $\alpha = m \in \mathbb{N}$) if the series in the numerator of (26) is converging, and its sum is equal to zero. Moreover, Letnikov gave sufficient conditions for existence of the limit in (26).

The second question by Sonine appeared due to his incorrect application of the Leibnitz rule. Letnikov noted that fractional integral and fractional derivative are defined by different formulas:

$$D^\alpha [f(x)]_{x=a}^{x=x} = \frac{1}{\Gamma(-\alpha)} \int_a^x (x-t)^{-\alpha-1} f(t) dt, \quad \alpha < 0, \quad (27)$$

$$D^\alpha [f(x)]_{x=a}^{x=x} = \sum_{k=0}^m \frac{f^{(k)}(a)(x-a)^{-\alpha+k}}{\Gamma(-\alpha+k+1)} + \frac{1}{\Gamma(-\alpha+m+1)} \int_a^x (x-t)^{-\alpha+m} f^{(m+1)}(t) dt, \quad (28)$$

$$\alpha > 0, m = [\alpha].$$

Both formulas are applied under certain conditions. Thus, successive application of these two formulas can lead to certain contradiction if we do not take into account the above conditions.

5. Sonine's Contribution to Fractional Calculus

5.1. Sonine's Fractional Derivative and Integral

In his polemical article [9], N. Ya. Sonine not only criticized Grúnwald-Letnikov approach but also proposed another form of "general" fractional derivative. For his formula, Sonine used generalization of the Cauchy integral (or, better to say, the Cauchy-type integral, since the below integral is defined for any continuous function f):

$$\frac{d^\alpha f(x)}{dx^\alpha} = \frac{\Gamma(p+1)}{2\pi i} \int_\gamma \frac{f(\tau)d\tau}{(\tau-x)^{\alpha+1}}, \qquad (29)$$

where γ is a closed simple smooth curve on the complex plane surrounding the point x (without loss of generality, one can assume that γ is the circle of radius r centered at the point x). This formula is really a good candidate for generalization of usual derivatives since, for $\alpha = p$, positive integer (29) gives the value of p-th derivative at the point x, assuming that the function f is p-time differentiable at x.

This formula was analyzed by Letnikov in his answer [7] on the remarks by Sonine. Letnikov proved that, under assumption that the function f is $(m+1) = ([\alpha]+1)$-times continuously differentiable inside the circle γ, Formula (29) coincides with Letnikov's Formula (28). Note that Letnikov discussed Sonine's Formula (29) under the stronger assumptions on the function f.

$$\frac{d^\alpha f(x)}{dx^\alpha} = \frac{(-r)^{-\alpha}f(x+r)}{\Gamma(-\alpha+1)} + \frac{(-r)^{-\alpha+1}f'(x+r)}{\Gamma(-\alpha+2)} + \ldots + \frac{(-r)^{-\alpha+m}f^{(m)}(x+r)}{\Gamma(-\alpha+m+1)} \qquad (30)$$

$$+ \frac{\Gamma(\alpha-m)}{2\pi r^{\alpha-m-1}} \int_0^{2\pi} f^{(m+1)}(x+re^{i\theta})e^{-(\alpha-m-1)i\theta}d\theta.$$

Since the last integral can be transformed to the form

$$\frac{\Gamma(p-m)}{2\pi r^{\alpha-m-1}} \int_0^{2\pi} f^{(m+1)}(x+re^{i\theta})e^{-(\alpha-m-1)i\theta}d\theta =$$

$$\frac{(-1)^{-\alpha+m}}{\Gamma(-\alpha+m+1)} \int_{x+r}^x f^{(m+1)}(\tau)(\tau-x)^{-\alpha+m}d\tau,$$

then Formula (29) coincides with the definition of the fractional derivative given by Letnikov (28). In Reference [9], Sonine concluded that his formula cannot coincide with Grünwald-Letnikov formula for $\alpha > 0$ without adding auxiliary function.

Sonine's definition of a fractional derivative of negative order (i.e., fractional integral) has been criticized by Letnikov in Reference [7]. Sonine used Leibniz's rule (composition formula) for fractional derivatives (which is generally not valid; see Reference [1]). A fractional integral by Sonine is defined as inverse operation for fractional derivative:

$$\frac{d^\alpha}{dx^\alpha} \frac{d^{-\alpha}f(x)}{dx^{-\alpha}} = f(x). \qquad (31)$$

From this formula, Sonine concluded that it should be exist as a so-called auxiliary function $\psi(x)$, satisfying the following relation:

$$\frac{d^\alpha \psi(x)}{dx^\alpha} = 0. \qquad (32)$$

Sonine took the function $\psi(x)$ in the form

$$\psi(x) = A_1(x-a)^{\alpha-1} + A_2(x-a)^{\alpha-2} + \ldots + A_k(x-a)^{\alpha-k},$$

where $A_j, j = 1, 2, \ldots, k$, are arbitrary constants, $k = [p]+1$, but a is not defined by Sonine. If Cauchy formula is taken as the definition of the derivative of an arbitrary order, then the auxiliary function has to satisfy the relation

$$\int_\gamma \frac{\psi(\tau)d\tau}{(\tau-x)^{\alpha+1}} = 0.$$

Since it was shown that by integration by parts Formula (29) is reduced to the definition of the fractional derivative (28) without any auxiliary function, then Letnikov concluded that the following alternative holds: either (1) there exists no auxiliary function of the above form, or (2) formula (29) cannot be taken as the general definition of fractional derivative of an arbitrary order. Instead, he said that his definition is free of necessity to add an auxiliary function (in spite of the fact that this definition is reduced to different a form than (27) and (28) whenever α is negative or positive, respectively).

5.2. Sonine Kernel and Sonine Integral Equations

In one of his pioneering articles [18], Abel presented solution to the integral equation

$$\int_0^x \frac{\varphi(\tau)d\tau}{(x-\tau)^{1-\alpha}} = F(x),\ 0 < \alpha < 1. \tag{33}$$

The main component of Abel's method was the following identity:

$$\int_0^x f(t)dt = \frac{\sin \pi \alpha}{\pi} \int_0^x \frac{dt}{(x-t)^{1-\alpha}} \int_0^t \frac{f(\tau)d\tau}{(t-\tau)^\alpha}, \tag{34}$$

where $f(x) = F'(x)$.

Sonine tried to generalize Abel's identity (34) in order to solve more general equation than integral Equation (33). He looked for a pair of functions $\sigma(x), \psi(x)$ satisfying the identity

$$\int_0^x f(t)dt = \int_0^x \psi(x-t)dt \int_0^t f(\tau)\sigma(t-\tau)d\tau, \tag{35}$$

or, i.e., a pair of functions generating integral representation of unity

$$1 = \int_0^x \psi(x-t)dt \int_0^t \sigma(t-\tau)d\tau. \tag{36}$$

Sonine described in Reference [19] a possible form of the functions $\sigma(x), \psi(x)$,

$$\sigma(t) = \frac{t^{-p}}{\Gamma(1-p)} \sum_{k=0}^\infty a_k t^k,\quad \psi(t) = \frac{t^{-q}}{\Gamma(1-q)} \sum_{k=0}^\infty b_k t^k,$$

where $p + q = 1$, and coefficients a_k, b_k are defined by the following relations

$$a_0 b_0 = 1,\ \sum_{k=0}^n \Gamma(k+p)\Gamma(q+n-k)a_{n-k}b_k = 0,\ n = 1, 2, \ldots.$$

He also applied relation (36) for representation of the solution to the first kind of integral equations with one of these functions as the kernel:

$$\int_0^x \sigma(x-\tau)\varphi(\tau)d\tau = f(x). \tag{37}$$

Both functions $\sigma(x), \psi(x)$ are known as Sonine kernels, and integral Equation (37), generalizing Abel integral Equation (33), is called a Sonine integral equation. In modern language (see, e.g., Reference [20]), a locally integrable function $\sigma(x)$ is called the Sonine kernel if there exists another locally integrable function $\psi(x)$ such that the following identity holds:

$$\int_0^x \sigma(x-\tau)\psi(\tau)d\tau = 1,\ x > 0. \tag{38}$$

In fact, the function $\psi(x)$ is also called the Sonine kernel (sometimes, these functions are called the associated Sonine kernels).

Several special examples of Sonine kernel are presented in Reference [21]. We can mention also Reference [20], in which the properties of the Sonine kernel are discussed in modern setting. Several difficulties which one has to overcome by formal application of Sonine's approach to the solution of the corresponding integral equations are discovered. Possible ways to overcome these difficulties are shown.

In Reference [22], the general fractional integrals and derivatives of arbitrary order are introduced, along with study some of their basic properties and particular cases. First, a suitable generalization of the Sonine condition is presented, and some important classes of the kernels that satisfy this condition are introduced. In the introduction of the general fractional integrals and derivatives, the author follows a recent approach by Kochubei [23]. The general fractional integrals and derivatives with Sonine kernel are defined in the Riemann-Liouville form (see Reference [21,22] and references therein)

$$(\mathbb{I}_\sigma f)(x) = \int_0^x \sigma(x-t)f(t)dt, \ x > 0, \tag{39}$$

$$(\mathbb{D}_\psi f)(x) = \frac{d}{dx}\int_0^x \psi(x-t)f(t)dt, \ x > 0, \tag{40}$$

where the functions $\sigma(x), \psi(x)$ are associated Sonine kernels. Operators (39) and (40) are discussed in Reference [21–23] under different conditions on Sonine kernels, and the constructions are not only similar to Riemann-Liouville-type fractional integrals and derivatives but also to Dzhrbashian-Caputo-type and to Marshaud-type.

5.3. Higher Order Hypergeometric Functions

The main research interest by Sonine was to study the properties of several classes of special functions. His results served as an impetus for the development of the theory of cylindrical functions (or Bessel-type functions) in the second half of the XIX century. These results are based on the achievements by C. Neumann, O. Schlömilch, E. Lommel, H. Hankel, N. Nielsen, L. Schlafli, L. Gegenbauer, and others (see, e.g., Reference [24,25]). Sonine defined in Reference [26] the cylindrical functions $S_\nu(z)$ as a partial solutions to the following system of functional-differential equations:

$$\begin{cases} S_{\nu+1}(z) + 2S'_\nu(z) - S_{\nu-1}(z) = 0, \\ 2\nu S_\nu(z) = z[S_{\nu-1}(z) + S_{\nu+1}(z)], \\ S_1(z) = -S'_0(z), \end{cases} \tag{41}$$

where z is the complex variable, and ν is an arbitrary complex parameter. Sonine proved that these partial solutions admit an integral representation:

$$S_\nu(z) = \frac{1}{2\pi i}\int_a^b \exp\left\{\frac{z}{2}\left(t - \frac{1}{t}\right)\right\}\frac{dt}{t^{\nu+1}}. \tag{42}$$

He found four possible cases for the limits of integration, namely: (1) $\infty \cdot \alpha, \infty \cdot \beta$; (2) $-\frac{0}{\alpha}, -\frac{0}{\beta}$; (3) $-\frac{0}{\alpha}, \infty \cdot \beta$; (4) Im$(za) = \pm\infty$, Im$(zb) = \pm\infty$, where, in cases (1)–(3), Re $(za) < 0$, Re $(zb) < 0$, but, in case (4) Re $(\nu) > 0$. Sonine denoted the functions obtained in these four cases by $S_\nu^{(k)}(z)$ and showed that

$$S_\nu^{(1)}(z) = J_\nu(z), \ S_\nu^{(2)}(z) = e^{-\nu\pi i}J_{-\nu}(z), \ S_\nu^{(3)}(z) = \frac{1}{2}H_\nu^{(1)}(z), \ S_\nu^{(4)}(z) = J_\nu(z).$$

The above integral representation (42) is called Sonine integral representation. It is a source for obtaining new representations for cylindrical functions (see Reference [25]), as well as for calculation of certain definite integrals. Among these integrals are those known as the first and the second Sonine integrals, respectively (or classical Sonine formulas); see, e.g., Reference [27]:

$$J_{\nu+\mu+1}(aq) = \frac{q^\nu}{2^\nu \Gamma(\nu+1) a^{\nu+\mu+1}} \int_0^a J_\mu(qx)(a^2-x^2)^\nu x^{\mu+1} dx, \tag{43}$$

$$\int_0^a J_\mu(qx) J_\nu[z\sqrt{a^2-x^2}](a^2-x^2)^{\frac{\nu}{2}} x^{\mu+1} dx = a^{\nu+\mu+1} q^\mu z^\nu \frac{J_{\nu+\mu+1}(a\sqrt{q^2+z^2})}{(\sqrt{q^2+z^2})^{\nu+\mu+1}}, \tag{44}$$

where $\operatorname{Re}\nu, \operatorname{Re}\mu > -1$. Sonine formulas find interest in different questions of analysis (e.g., in Dunkl theory, as in Reference [28], or in the study of Levy processes [29]).

There exist several multivariate extensions of the classical Sonine integral representation for Bessel functions of some index $\mu + \nu$ in terms of such functions of lower index μ (see, e.g., Reference [30]). For Bessel functions on matrix cones, Sonine formulas involve beta densities $\beta_{\mu,nu}$ on the cone and go already back to Herz.

Several important results dealing with properties of Γ-function were obtained by Sonine during his career. They are based on the study of the solution to the difference equation

$$F(x+1) - F(x) = f(x). \tag{45}$$

In these works, Sonine followed the idea by Binet (1838), who examined the relation

$$\log \Gamma(x+1) - \log \Gamma(x) = \log x.$$

Sonine found [31], in particular, the form of the remainder factor in the product representation of Γ-function

$$\Gamma(x+1) = \frac{n!(n+1)^x}{(x+1)(x+2)\cdots(x+n)} \frac{\left(1+\frac{x}{n+1}\right)^{x+n+\theta}}{\left(1+\frac{1}{n+1}\right)^{x(1+n+\theta)}}, \ x \in \mathbb{R}, \ 0 < \theta < 1. \tag{46}$$

In his article on Bernoulli polynomials, Sonine obtained one more representation, related to Γ-functions (this formula was rediscovered by Ch. Hermite in 1895)

$$\log \frac{\Gamma(x+y)}{\Gamma(y)} = x \log y + \sum_{k=2}^n \frac{(-1)^k \varphi_k(x)}{(k-1)k y^{k-1}} + R_n(x,y), \tag{47}$$

where $\varphi_k(x)$ are Bernoulli polynomials defined by Sonine using difference equation

$$\varphi_k(x+1) - \varphi_k(x) = k x^{k-1}, \ \varphi_k(0) = 0, k = 1, 2, \ldots.$$

Reference [32] contains a number of the most important articles by N. Ya. Sonine, as well as a survey on his other research.

6. Nekrasov's Contribution to Fractional Calculus

In Reference [8], Nekrasov proposed a new definition of the general differentiation. In fact, this definition includes Letnikov's definition as a special case. The main idea by Nekrasov was to define the derivative by using integration along closed contour L crossing the point x and encircling a group of singular points of the differentiable function $f(x)$. This definition gives, in fact, the differentiation with respect to a doubly connected domain, which is free of the singular points of $f(x)$, and contains the above said contour L. Therefore, Nekrasov used the ideas by Sonine (to take into account the properties of the analytic continuation of a given function and to apply the properties of functions in complex domains). The main aim of Nekrasov's construction is to extend the class of functions to which the general differentiation can be applied.

It should be noted that the construction proposed by Nekrasov is fairly complicated and needs to use properties of the functions on Riemann surfaces. It follows from the properties of functions to which Nekrasov tried to apply his definition. The starting point of his construction is the notion of classes (q, μ) of function. Let L be a closed contour encircled a group of singular points of the function $f(z)$. Let the function $f(z)$ have the following property: if the point z makes a complete detour along L in counter clockwise direction, then the function $f(z)$, continuously changing, gains the multiplier $e^{2\pi q i}$. Then, this function is of class $(q, 0)$. Thus, any function of the class $(q, 0)$ can be represented in the form $f(z) = (z - a)^q \phi(z)$, where a lies inside L, and $\phi(z)$ is of the class $(0, 0)$. The function of the form $f(z) = (z - a)^q \log^\mu z \phi(z)$, with $\phi(z)$ being of the class $(0, 0)$, is said to belong to the class (q, μ) (with q being the power index, and μ being the logarithmic index which is supposed to be nonnegative integer number). It is clear that, if the function $f(z)$ belongs to the class (q, μ), then it belongs to any class $(q \pm m, \mu)$, $m \in \mathbb{N}$. Clearly, this definition depends on the choice of the contour L.

The function $f(z)$, which can be represented in form of a sum of finite number n of functions belonging to different classes with respect to the contour L, is said to be reducible to n classes (or simply reducible).

Let the function $f(z)$ be reducible to n classes with zero logarithmic indices, i.e.,

$$f(z) = f_0(z) + f_1(z) + \ldots + f_{n-1}(z), \tag{48}$$

where $f_j(z)$ is of class $(q_j, 0)$. Then, we have the following representation:

$$f(z) + \int\limits_{(L^k)}^{(z)} \frac{df(t)}{dt} dt = \alpha_0^k f_0(z) + \alpha_1^k f_1(z) + \ldots + \alpha_{n-1}^k f_{n-1}(z), \tag{49}$$

where integration is performed along the contour L, traversable k-times in counter clockwise direction starting from the point z, $\alpha_j = e^{2\pi q_j i}$. By assigning the values $0, 1, \ldots, n-1$ to the parameter k, we obtain the system of equations sufficient for determination of functions $f_0(z), f_1(z), \ldots, f_{n-1}(z)$.

Let the function $f(z)$ be reducible to n classes with non-zero logarithmic indices, i.e.,

$$f(z) = \sum_{s=0}^{n_0-1}(z-a)^{q_{s,0}}\phi_{s,0}(z) + \\ \sum_{s=0}^{n_1-1}(z-a)^{q_{s,1}}\log(z-a)\phi_{s,1}(z) + \ldots + \sum_{s=0}^{n_\mu-1}(z-a)^{q_{s,\mu}}\log^\mu(z-a)\phi_{s,\mu}(z), \tag{50}$$

where $n_0 + n_1 + \ldots + n_\mu = n$, and all functions $\phi_{s,j}(z)$ are of class $(0,0)$. Then, we have the following representation:

$$f(z) + \int\limits_{(L^k)}^{(z)} \frac{df(t)}{dt} dt = \sum_{s=0}^{n_0-1} \alpha_{s,0}^k (z-a)^{q_{s,0}} \phi_{s,0}(z) + \\ \sum_{s=0}^{n_1-1} \alpha_{s,1}^k (z-a)^{q_{s,1}} \{2k\pi i + \log(z-a)\} \phi_{s,1}(z) + \ldots \\ + \sum_{s=0}^{n_\mu-1} \alpha_{s,\mu}^k (z-a)^{q_{s,\mu}} \{2k\pi i + \log(z-a)\}^\mu \phi_{s,\mu}(z), \tag{51}$$

where integration is performed along the contour L, traversable k-times in counter clockwise direction starting from the point z, $\alpha_{s,j} = e^{2\pi q_{s,j} i}$. By assigning the values $0, 1, \ldots, n-1$ to the parameter k, we obtain the system of equations sufficient for determination of functions $\phi_{s,0}(z), \phi_{s,1}(z), \ldots, \phi_{s,n-1}(z)$.

Therefore, in both cases, we have a finite sum representation of the function $f(z)$ of the considered form. Now, the question is to define the integral/derivative of arbitrary order of each components of the representation (48) or (50). Moreover, any function $f(z)$ of the class (q, μ) can be determined as the following limit:

$$f(z) = \lim_{h \to 0} \left[(z-a)^q \left(\frac{(z-a)^h - 1}{h} \right)^\mu \right]. \tag{52}$$

Thus, we have the limit of the finite sum of functions belonging to the classes $(q, 0), (q+h, 0), \ldots, (q + \mu h, 0)$. Therefore, the definition of the integral/derivative of arbitrary order of any reducible function can be completely described if one can define the definition of a function of class $(q, 0)$. Nekrasov noted that his construction of his derivative generally speaking cannot be rigorously defined in the case when $f(z)$ is reducible to infinite number of classes.

For this definition, Nekrasov used Letnikov's formulas. The only difference is that the contour of integration is now specially deformed curve on the Riemann surfaces (which depend on the order of the derivative, i.e., can be either finite-sheeted or infinite-sheeted).

In Reference [33], Nekrasov applied his construction to determinate the solution of the following differential equation:

$$\sum_{s=0}^{n}(a_s + b_s x)x^s \frac{d^s y}{dx^s} = 0, \tag{53}$$

which is highly related to the generalized hypergeometric function $_pF_q(z)$.

7. Conclusions

In this article, we analyzed some important results by Russian mathematicians of the end of the XIX century, namely A. V. Letnikov, N. Ya. Sonine, and P. A. Nekrasov. The main attention is paid to their contribution related to Fractional Calculus and Special Functions Theory. Some of these results are presented for the first time in English.

Our article serves to clarify the beginning steps of the development of Fractional Calculus. We believe that it would be useful and interesting for members of fractional society.

Short Biographies

Letnikov Alexey Vasil'evich (1837–1888), Russian mathematician. A short biography.

A. V. Letnikov was born on 1 January 1837, in Moscow, Russia. When Alexey was 8 years old, his father died. His mother tried to give education to Alexey and his sister. The mother sent Alexey to a grammar school in 1847. In spite of his evident abilities, he was not too successful in education. Therefore, he was moved to Konstantin's land-surveyors institute (full-time provisional military-type institute). That was a second rank educational establishment. Its director discovered high interest of Alexey to mathematics and supported his growth in the subject. The director decided to prepare him to the career of a teacher in mathematics in Konstantin's land-surveyors institute. To get the corresponding position, Letnikov was sent to Moscow University and studied mathematics there for two years (1856–1858) as an extern student.

After graduation, he was sent to Paris in order to extend his knowledge in the most known mathematical center for two years and to study the structure and the content of the technical education in France. In Paris, Letnikov attended lectures of many well-known mathematicians (Liouville among them) in the Ecole Polytechnique, College de France and Sorbonne.

Returning from Paris in December 1860, he was appointed as a teacher in the engineering class of the Konstantin's land-surveyors institute and started to teach Probability Theory. Letnikov actively participated in mathematical life in Moscow. In particular, he was among the founders of Moscow Mathematical Society in 1864. In 1863, it was approved a new Statute of Higher Education. Among other regulations, it was supposed to enlarge a number of chairs at universities and to recruit new university teachers. To get a position at university, one needed either to pass graduation gymnasium's exams or to receive the degree at a foreign university. Letnikov decided to use the second option. In 1867, he defended his PhD, "Über die Bedingungen der Integrabilität einiger Differential-Gleichungen", at Leipzig University. In 1868, he got a position at recently reopened Imperial Technical College (now Bauman's Technical University). Letnikov was working in this College up to 1883, when he moved to Alexandrov's Commercial College, sharing this job with a part-time teacher in Konstantin's land-surveyors institute and in the Imperial Technical College.

It was active time for him, and he was awarded the degree of a state councillor, got the order of Saint Stanislav, and was appointed in 1884 as a corresponding member of St.-Petersburg Academy of Sciences (by recommendation of V. Imshenetsky, V. Bunyakovsky, and O. Backlund). At the end of 1880s, Letnikov should have received a state pension and was supposed to leave teaching and to concentrate on the research. He was dreaming of getting a position at the Moscow University. It was not to happen since, at the opening ceremony of a new building of Alexandrov's Commercial College, he caught a cold. He had no serious illness before and continued to deliver lectures. But, this time, the illness was strong enough, and he died on 27 February 1888.

Sonine Nikolai Yakovlevich (1849–1915), Russian mathematician. A short biography.

N.Ya.Sonine was born on February 10th, 1849, in Tula, Russia.

He studied at Physical-mathematical Faculty of Moscow University (1865–1869). After graduation, he continued research in Moscow University for two years and, in 1871, defended his Master Thesis, "On expansion of functions into infinite series". In June 1871, he became Associate Professor (dozent) of Warsaw University.

In 1873, he was sent to Paris to continue research study. In Paris, he attended lectures by Liouville, Hermite, Bertrand, Serre and Darboux. In September 1874, in Moscow University, N. Ya. Sonine defended his PhD Thesis, "On integration of partial differential equations of the second order". In 1877, he became extra-ordinary Professor of Warsaw University and, in 1879, becamoe an ordinary Professor of Warsaw University. In 1891, he resigned from his position at Warsaw University, but he still continued his research. In 1891, N. Ya. Sonine was elected a corresponding member of Academy of Sciences, and, in 1893, he became an academician of St.-Petersburg Academy of Sciences (by recommendation of P. L. Chebyshev). In 1890, he was awarded by V. Ya. Bunyakovsky Prize for the Best Results in Mathematics.

Starting in 1899, N. Ya. Sonine occupied different administrative positions, mostly in the education. He died on 18 February 1915, in St.-Petersburg.

Nekrasov Pavel Alekseevich (1853–1924), Russian mathematician and philosopher. A short biography.

P. A. Nekrasov was born on 1st (13th) February 1853, in Ryazan region, Russia.

After graduation at Ryazan Orthodox seminary in 1874, he entered Physical-mathematical faculty of the Moscow University. In 1878, he graduated from the Moscow University with degree of the candidate of sciences and was left at the department of pure mathematics for preparation to the professorship. From August 1879, P. A. Nekrasov shared his research with teaching mathematics at the private Voskresensky's real school. In 1883, he defended his master thesis, "Study of the equation $u^m - pu^n - q = 0$". For this work, he was awarded by V. Ya. Bunyakovsky Prize for the Best Results in Mathematics.

In 1985, P. A. Nekrasov became a Privatdozent in the Moscow University (having defended his Russian PhD "On Lagrange series" in 1886), and, in 1886, he got the position of an associate professor (extraordinary professor) at Moscow University. In 1890, he received a full professorship. In 1893, he became the rector of Moscow University. After his term as the rector, he actually wanted to retire, but he was not allowed to. He also taught, from 1885–1891, the Probability Theory and the Higher Mathematics in the Land-surveyors institute.

Starting in 1898, he performed only administrative duties for the Ministry of Education (he was curator of the Moscow University and responsible for the schools in Moscow and the surrounding area) and moved in 1905 to Saint Petersburg as a member of the Council of the Ministry of Education. After the Russian Revolution, he tried to adapt himself to the new realities, dealt with mathematical economics (which he lectured in 1918–1919), and studied Marxism. He died of pneumonia on 20 December 1924, in Moscow.

Author Contributions: Conceptualization, S.R. and M.D. Writing—review and editing, S.R. and M.D. The authors made equal contributions to this article. Both authors have read and agreed to the published version of the manuscript.

Funding: The work has been supported by Belarusian Fund for Fundamental Scientific Research through the grant F20R-083.

Data Availability Statement: The study did not report any data.

Conflicts of Interest: The authors declare no conflict of interest.

References

1. Samko, S.G.; Kilbas, A.A.; Marichev, O.I. *Fractional Integrals and Derivatives. Theory and Applications*; Gordon and Breach Science Publishers: Yverdon, Switzerland, 1993; Revised in 1987
2. Letnikov, A.V. On historical development of the theory of differentiation of an arbitrary order. *Mat. Sb.* **1868**, *3*, 85–112. (In Russian)
3. Potapov, A.A. A short essay on the origin and formation of the theory of fractional integro-differentiation. *Nonlinear World* **2003**, *1*, 69–81. (In Russian)
4. Potapov, A.A. Essays on the development of fractional calculus in the A.V. Letnikov's works. To 175th anniversary of A. V. Letnikov, Radioelectronics, Nanosystems. *Inf. Technol. (Radioelektron. Nanosistemy Inf. Tehnol.)* **2012**, *4*, 3–102. (In Russian)
5. Letnikov, A.V.; Chernykh, V.A. *The Foundation of Fractional Calculus (with Applications to the Theory of Oil and Gas Production, Underground Hydrodynamics and Dynamics of Biological Systems)*; Neftegaz: Moscow, Russia, 2011; 429p. (In Russian)
6. Letnikov, A.V. Theory of differentiation of an arbitrary order. *Mat. Sb.* **1868**, *3*, 1–68. (In Russian)
7. Letnikov, A.V. To explanation of the main statements of the theory of differentiation of an arbitrary order. *Mat. Sb.* **1873**, *6*, 413–445. (In Russian)
8. Nekrasov, P.A. General differentiation. *Mat. Sb.* **1888**, *14*, 45–166. (In Russian)
9. Sonine, N.Y. On differentiation of an arbitrary order. *Mat. Sb.* **1872**, *6*, 1–38. (In Russian)
10. Vashchenko-Zakharchenko, M.E. On fractional differentiation. *Q. J. Pure Appl. Math. Ser. 1* **1861**, *IV*, 237–243.
11. Grünwald, A.K. Über "begrentze" Derivationen und deren Anwendung. *Zeitschrift für Angew. Math. Phys.* **1867**, *12*, 441–480.
12. Gorenflo, R.; Kilbas, A.A.; Mainardi, F.; Rogosin, S. *Mittag-Leffler Functions: Related Topics and Applications*, 2nd ed.; Springer: Berlin, Germany; New York, NY, USA, 2020; 540p.
13. Nekrasov, P.A.; Pokrovskii, P.M. On examination of manuscripts by A. V. Letnikov, presented after his death to the Moscow Mathematical Society. *Mat. Sb.* **1889**, *14*, 202–204. (In Russian)
14. Liouville, J. Memoire sur quelques Questions de Geometrie et de Mecanique, et sur un nouveau genre de Calcul pour resoudre ces Questions. *J. Ecole Polytech.* **1832**, *13*, 1–69.
15. Liouville, J. Memoire sur le Calcul des differentielles a indices quelconques. *J. Ecole Polytech.* **1832**, *13*, 71–162.
16. Liouville, J. Memoire sur une formule d analyse. *J. Reine Angew. Math. (Grelle's J.)* **1834**, *12*, 273–287.
17. Letnikov, A.V. On integration of the equation $(a_n + b_n x)\frac{d^n y}{dx^n} + (a_{n-1} + b_{n-1}x)\frac{d^{n-1}y}{dx^{n-1}} + \ldots + (a_0 + b_0 x)y = 0$. *Mat. Sb.* **1889**, *14*, 205–215. (In Russian)
18. Abel, N. H. Auflösung einer mechanischen Aufgabe. *J. Für Die Reine und Angew. Math.* **1826**, *1*, 153–157.
19. Sonine, N.Y. Sur la généralisation d'une formule d'Abel. *Acta Math.* **1884**, *4*, 171–176. (In French) [CrossRef]
20. Samko, S.G.; Cardoso, R.P. Integral equations of the first kind of Sonine type. *Int. J. Math. Math. Sci.* **2003**, *57*, 3609–3632. [CrossRef]
21. Luchko, Y. General Fractional Integrals and Derivatives with the Sonine Kernels. *Mathematics* **2021**, *9*, 594. [CrossRef]
22. Luchko, Y. General Fractional Integrals and Derivatives of Arbitrary Order. *Symmetry* **2021**, *13*, 755 [CrossRef]
23. Kochubei, A.N. General fractional calculus, evolution equations, and renewal processes. *Integral Equ. Oper. Theory* **2011**, *71*, 583–600. [CrossRef]
24. Kropotov, A.I. *Nikolai Yakovlevich Sonine*; Nauka: Leningrad, USSR, 1967; 136p. (In Russian)
25. Olver, F.J.; Lozier, D.W.; Boisvert, R.F.; Clark, C.W. (Eds.) *NIST Handbook of Mathematical Functions*; National Institute of Standards and Technology: Gaithersburg, MD, USA; Cambridge University Press: New York, NY, USA, 2010; 951 + xv pages and a CD.
26. Sonine, N.Y. Recherches sur les fonctions celindriques et le développment des fonctions continues en series. *Math. Ann.* **1880**, *16*, 1–80. [CrossRef]
27. Grandits, P. Some notes on Sonine-Gegenbauer integrals. *Int. Transf. Spec. Funct.* **2019**, *30*, 128–137. [CrossRef]
28. Rösler, M.; Voit, M. Sonine Formulas and Intertwining Operators in Dunkl Theory. *Int. Math. Res. Not.* **2020**. [CrossRef]
29. Barndorff-Nielsen, O.E.; Mikosch, T.; Resnick, S.I. (Eds.) *Levy Processes. Theory and Applications*; Birkhäuser: Boston, MA, USA, 2001; 418p.
30. Rösler, M.; Voit, M. Beta distributions and Sonine integrals for Bessel functions on symmetric cones. *Stud. Appl. Math.* **2018**, *141*, 474–500. [CrossRef]
31. Sonine, N.Y. Note sur une formule de Gauss. *Bull. Soc. Math. Fr.* **1881**, *9*, 162–166. (In French) [CrossRef]
32. Sonine, N.Y. *Research on Cylindric Functions and Orthogonal Polynomials*; Akhieser, N.I., Ed.; GITTL: Moscow, Russia, 1954; 244p. (In Russian)
33. Nekrasov, P.A. Application of the general differentiation to the integration of the equation of the type $\sum(a_s + b_s x)x^s D^s y = 0$. *Mat. Sb.* **1889**, *14*, 344–393. (In Russian)

Article

Special Functions of Fractional Calculus in the Form of Convolution Series and Their Applications

Yuri Luchko

Department of Mathematics, Physics, and Chemistry, Beuth Technical University of Applied Sciences Berlin, Luxemburger Str. 10, 13353 Berlin, Germany; luchko@beuth-hochschule.de

Abstract: In this paper, we first discuss the convolution series that are generated by Sonine kernels from a class of functions continuous on a real positive semi-axis that have an integrable singularity of power function type at point zero. These convolution series are closely related to the general fractional integrals and derivatives with Sonine kernels and represent a new class of special functions of fractional calculus. The Mittag-Leffler functions as solutions to the fractional differential equations with the fractional derivatives of both Riemann-Liouville and Caputo types are particular cases of the convolution series generated by the Sonine kernel $\kappa(t) = t^{\alpha-1}/\Gamma(\alpha)$, $0 < \alpha < 1$. The main result of the paper is the derivation of analytic solutions to the single- and multi-term fractional differential equations with the general fractional derivatives of the Riemann-Liouville type that have not yet been studied in the fractional calculus literature.

Keywords: Sonine kernel; Sonine condition; general fractional derivative; general fractional integral; convolution series; fundamental theorems of fractional calculus; fractional differential equations

MSC: 26A33; 26B30; 44A10; 45E10

1. Introduction

Special functions of mathematical physics are usually defined in the form of a power series, or as solutions to some differential equations, or via integral representations. Of course, for a given function, these three (and possibly other) forms coincide for all arguments and parameter values for which they exist. However, the validity domains of different representations can be unequal. Very often, the series representations of the special functions hold valid only on some restricted domains. To define the corresponding functions for other values of their arguments and parameters, analytical continuation of the series in the form of integral representations is usually employed.

For special functions of fractional calculus (FC), the situation is very similar to the one described above. For instance, one of the most important FC special functions – the two-parameter Mittag-Leffler function – is usually defined in the form of a power series:

$$E_{\alpha,\beta}(z) = \sum_{k=0}^{+\infty} \frac{z^k}{\Gamma(\alpha k + \beta)}, \ \alpha > 0, \ \beta, z \in \mathbb{C}. \tag{1}$$

Because the series is convergent for all $z \in \mathbb{C}$, this definition can be used for all $z \in \mathbb{C}$ without any analytical continuation. Still, the integral representations of the Mittag-Leffler function are very important, say, for derivation of its asymptotic behavior [1] and for its numerical calculation [2]. For $0 < \alpha < 2$ and $\Re(\beta) > 0$, the following integral representations of the Mittag-Leffler function in terms of the integrals over the Hankel-type contours were presented in [1]:

$$E_{\alpha,\beta}(z) = \frac{1}{2\pi\alpha i} \int_{\gamma(\epsilon;\delta)} \frac{e^{\zeta^{1/\alpha}} \zeta^{(1-\beta)/\alpha}}{\zeta - z} d\zeta, \ z \in G^{(-)}(\epsilon;\delta),$$

$$E_{\alpha,\beta}(z) = \frac{1}{\alpha} z^{(1-\beta)/\alpha} e^{z^{1/\alpha}} + \frac{1}{2\pi\alpha i} \int_{\gamma(\epsilon;\delta)} \frac{e^{\zeta^{1/\alpha}} \zeta^{(1-\beta)/\alpha}}{\zeta - z} d\zeta, \ z \in G^{(+)}(\epsilon;\delta),$$

where the integration contour $\gamma(\epsilon;\delta)$ ($\epsilon > 0$, $0 < \delta \leq \pi$) with non-decreasing $\arg \zeta$ consists of the following parts:

(1) the ray $\arg \zeta = -\delta$, $|\zeta| \geq \epsilon$;
(2) the arc $-\delta \leq \arg \zeta \leq \delta$ of the circumference $|\zeta| = \epsilon$;
(3) the ray $\arg \zeta = \delta$, $|\zeta| \geq \epsilon$.

For $0 < \delta < \pi$, the domain $G^{(-)}(\epsilon;\delta)$ is to the left of the contour $\gamma(\epsilon;\delta)$ and the domain $G^{(+)}(\epsilon;\delta)$ is to the right of this contour. If $\delta = \pi$, the contour $\gamma(\epsilon;\delta)$ consists of the circumference $|\zeta| = \epsilon$ and of the cut $-\infty < \zeta \leq -\epsilon$. In this case, the domain $G^{(-)}(\epsilon;\delta)$ is the circle $|\zeta| < \epsilon$ and $G^{(+)}(\epsilon;\alpha) = \{\zeta : |\arg \zeta| < \pi, |\zeta| > \epsilon\}$.

For some parameter values, the Mittag-Leffler function can be also introduced in terms of solutions to the fractional differential equations with the Riemann-Liouville or Caputo fractional derivatives. For instance, for $0 < \alpha \leq 1$, the equation

$$(D_{0+}^{\alpha} y)(t) = \lambda y(t) \tag{2}$$

has the general solution [3]

$$y(t) = C t^{\alpha-1} E_{\alpha,\alpha}(\lambda t^{\alpha}), \ C \in \mathbb{R}. \tag{3}$$

In Equation (2), the Riemann-Liouville fractional derivative D_{0+}^{α} is defined by

$$(D_{0+}^{\alpha} f)(t) = \frac{d}{dt}(I_{0+}^{1-\alpha} f)(t), \ t > 0, \tag{4}$$

where I_{0+}^{α} is the Riemann-Liouville fractional integral of order α ($\alpha > 0$):

$$(I_{0+}^{\alpha} f)(t) = \frac{1}{\Gamma(\alpha)} \int_0^t (t - \tau)^{\alpha-1} f(\tau) d\tau, \ t > 0. \tag{5}$$

The general solution to the equation

$$(_*D_{0+}^{\alpha} y)(t) = \lambda y(t) \tag{6}$$

with the Caputo fractional derivative

$$(_*D_{0+}^{\alpha} f)(t) = (D_{0+}^{\alpha} f)(t) - f(0) \frac{t^{-\alpha}}{\Gamma(1-\alpha)}, \ t > 0 \tag{7}$$

has the form [4]

$$y(t) = C E_{\alpha,1}(\lambda t^{\alpha}), \ C \in \mathbb{R}. \tag{8}$$

As we can see, the solutions to the fractional differential Equations (2) and (6) are expressed in terms of the Mittag-Leffler functions. However, the arguments of these functions are λt^{α} and not just λt. Thus, these solutions are represented in the form of power series with the fractional and not integer exponents. For more advanced properties and applications of the Mittag-Leffler type functions, see [1] and the recent book [5].

In [6], the single- and multi-term fractional differential equations with the general fractional derivatives of the Caputo type have been studied. By definition, their solutions belong to the class of the FC special functions (as the ones represented in form of solutions to the fractional differential equations). Moreover, in [6], another representation of these new FC special functions was derived, namely in terms of the convolution series generated by the Sonine kernels.

The convolution series are a far-reaching generalization of the conventional power series and the power series with the fractional exponents including the Mittag-Leffler Functions (3) and (8). They represent a new class of the FC special functions worth for investigation. In [7], the convolution series were employed for derivation of two different forms of the generalized convolution Taylor formula for representation of a function as a convolution polynomial with a remainder in the form of a composition of the n-fold general fractional integral and the n-fold general sequential fractional derivative of the Riemann-Liouville and the Caputo types, respectively. In [7], the generalized Taylor series in form of convolution series were also discussed. In this paper, we employ the convolution series for derivation of analytical solutions to the single- and multi-terms fractional differential equations with the general fractional derivatives in the Riemann-Liouville sense. This type of the fractional differential equations has not yet been studied in the FC literature.

One of the main reasons for this situation is that until recently, it was not clear at all what type of initial conditions is required while dealing with fractional differential equations with general fractional derivatives of the Riemann-Liouville type. A solution to this problem was provided in a very recent publication [7], where an explicit form of the projector operator of the n-fold sequential general fractional derivative in the Riemann-Liouville sense has been derived for the first time. Another challenge for treatment of fractional differential equations with general fractional derivatives in the Riemann-Liouville sense is an absence of methods for derivation of their analytical solutions. In [6], fractional differential equations with general fractional derivatives of the Caputo type have been studied by means of an operational calculus developed for these derivatives. An operational calculus for general fractional derivatives of the Riemann-Liouville has not yet been constructed. Thus, in this paper, we employ another method for analytical treatment of fractional differential equations with general fractional derivatives of the Riemann-Liouville type, namely the method of convolution series. This method is introduced and applied to fractional differential equations for the first time in the FC literature.

The rest of this paper is organized as follows. In the next section, we introduce general fractional derivatives of the Riemann-Liouville and Caputo types with Sonine kernels from a special class of functions and discuss some of their properties needed for further discussion. In the third section, we first provide some results regarding the convolution series generated by Sonine kernels. Then, convolution series are applied for derivation of analytical solutions to single- and multi-term fractional differential equations with general fractional derivatives in the Riemann-Liouville sense. For a treatment of single- and multi-term fractional differential equations with general fractional derivatives in the Caputo sense, we refer interested readers to [6].

2. General Fractional Integrals and Derivatives

General fractional derivatives (GFDs) with kernel k in the Riemann-Liouville and in the Caputo sense, respectively, are defined as follows [8–13]:

$$(\mathbb{D}_{(k)} f)(t) = \frac{d}{dt}(k * f)(t) = \frac{d}{dt} \int_0^t k(t-\tau)f(\tau)\,d\tau, \tag{9}$$

$$({}_*\mathbb{D}_{(k)} f)(t) = (\mathbb{D}_{(k)} f)(t) - f(0)k(t), \tag{10}$$

where by $*$ the Laplace convolution is denoted:

$$(f * g)(t) = \int_0^t f(t-\tau)g(\tau)\,d\tau. \tag{11}$$

The Riemann-Liouville and the Caputo fractional derivatives of order α, $0 < \alpha < 1$, defined by (4) and (7), respectively, are particular cases of the GFDs (9) and (10) with the kernel

$$k(t) = h_{1-\alpha}(t),\ 0 < \alpha < 1,\ h_\beta(t) := \frac{t^{\beta-1}}{\Gamma(\beta)},\ \beta > 0. \tag{12}$$

The multi-term fractional derivatives and fractional derivatives of distributed order are also particular cases of the GFDs (9) and (10) with the kernels

$$k(t) = \sum_{k=1}^{n} a_k h_{1-\alpha_k}(t), \ 0 < \alpha_1 < \cdots < \alpha_n < 1, \ a_k \in \mathbb{R}, \ k = 1, \ldots, n, \tag{13}$$

$$k(t) = \int_0^1 h_{1-\alpha}(t) \, d\rho(\alpha), \tag{14}$$

respectively, where ρ is a Borel measure defined on the interval $[0, 1]$.

Several useful properties of the Riemann-Liouville fractional integral and the Riemann-Liouville and Caputo fractional derivatives are based on the formula

$$(h_\alpha * h_\beta)(t) = h_{\alpha+\beta}(t), \ \alpha, \beta > 0, \ t > 0 \tag{15}$$

that immediately follows from the well-known representation of the Euler Beta-function in terms of the Gamma-function:

$$B(\alpha, \beta) := \int_0^1 (1-\tau)^{\alpha-1} \tau^{\beta-1} \, d\tau = \frac{\Gamma(\alpha)\Gamma(\beta)}{\Gamma(\alpha+\beta)}, \ \alpha, \beta > 0.$$

In Formula (15) and in what follows, the power function h_α is defined as in (12).

In our discussions, we employ the integer order convolution powers that for a function $f = f(t)$, $t > 0$ are defined by the expression

$$f^{<n>}(t) := \begin{cases} 1, & n = 0, \\ f(t), & n = 1, \\ \underbrace{(f * \ldots * f)}_{n \text{ times}}(t), & n = 2, 3, \ldots. \end{cases} \tag{16}$$

For the kernel $\kappa(t) = h_\alpha(t)$ of the Riemann-Liouville fractional integral, we apply Formula (15) and arrive at the important representation

$$h_\alpha^{<n>}(t) = h_{n\alpha}(t), \ n \in \mathbb{N}. \tag{17}$$

A well-known particular case of (17) is the formula

$$\{1\}^n(t) = h_1^n(t) = h_n(t) = \frac{t^{n-1}}{\Gamma(n)} = \frac{t^{n-1}}{(n-1)!}, \ n \in \mathbb{N}, \tag{18}$$

where by $\{1\}$ we denoted the function that is identically equal to 1 for $t > 0$.

Now let us write down Formula (15) for $\beta = 1 - \alpha$, $0 < \alpha < 1$:

$$(h_\alpha * h_{1-\alpha})(t) = h_1(t) = \{1\}, \ 0 < \alpha < 1, \ t > 0. \tag{19}$$

In [14,15], Abel employed the relation (19) to derive an inversion formula for the operator that is presently referred to as the Caputo fractional derivative and obtained it in form of the Riemann-Liouville fractional integral (solution to the Abel model for the tautochrone problem).

By an attempt to extend the Abel solution method to more general integral equations of convolution type, Sonine introduced in [16] the relation

$$(\kappa * k)(t) = \{1\}, \ t > 0 \tag{20}$$

that is presently referred to as the Sonine condition. The functions that satisfy the Sonine condition are called Sonine kernels. For a Sonine kernel κ, the kernel k that satisfies the

Sonine condition (20) is called an associated kernel to κ. Of course, κ is then an associated kernel to k. In what follows, we denote the set of the Sonine kernels by $\mathcal{S}$.

In [16], Sonine introduced a class of Sonine kernels in the form

$$\kappa(t) = h_\alpha(t) \cdot \kappa_1(t), \ \kappa_1(t) = \sum_{k=0}^{+\infty} a_k t^k, \ a_0 \neq 0, \ 0 < \alpha < 1, \tag{21}$$

$$k(t) = h_{1-\alpha}(t) \cdot k_1(t), \ k_1(t) = \sum_{k=0}^{+\infty} b_k t^k, \tag{22}$$

where $\kappa_1 = \kappa_1(t)$ and $k_1 = k_1(t)$ are analytical functions and the coefficients a_k, b_k, $k \in \mathbb{N}_0$ satisfy the following triangular system of linear equations:

$$a_0 b_0 = 1, \ \sum_{k=0}^{n} \Gamma(k+1-\alpha)\Gamma(\alpha+n-k) a_{n-k} b_k = 0, \ n \geq 1. \tag{23}$$

An important example of the kernels from $\mathcal{S}$ in the form (21), (22) was derived in [16] in terms of the Bessel function J_ν and the modified Bessel function I_ν:

$$\kappa(t) = (\sqrt{t})^{\alpha-1} J_{\alpha-1}(2\sqrt{t}), \ k(t) = (\sqrt{t})^{-\alpha} I_{-\alpha}(2\sqrt{t}), \ 0 < \alpha < 1, \tag{24}$$

where

$$J_\nu(t) = \sum_{k=0}^{+\infty} \frac{(-1)^k (t/2)^{2k+\nu}}{k! \Gamma(k+\nu+1)}, \ I_\nu(t) = \sum_{k=0}^{+\infty} \frac{(t/2)^{2k+\nu}}{k! \Gamma(k+\nu+1)}.$$

For other examples of Sonine kernels we refer readers to [8,12,13,17].

In this paper, we deal with general fractional integrals (GFIs) with kernels $\kappa \in \mathcal{S}$ defined by the formula

$$(\mathbb{I}_{(\kappa)} f)(t) := (\kappa * f)(t) = \int_0^t \kappa(t-\tau) f(\tau) \, d\tau, \ t > 0 \tag{25}$$

and with GFDs with associated Sonine kernels k in the Riemann-Liouville and Caputo senses defined by (9) and (10), respectively.

In our discussions, we restrict ourselves to a class of the Sonine kernels from space $C_{-1,0}(0,+\infty)$ that is an important particular case of the following two-parameter family of spaces [6,12,13]:

$$C_{\alpha,\beta}(0,+\infty) = \{f : f(t) = t^p f_1(t), \ t > 0, \ \alpha < p < \beta, \ f_1 \in C[0,+\infty)\}. \tag{26}$$

By $C_{-1}(0,+\infty)$ we mean the space $C_{-1,+\infty}(0,+\infty)$.

The set of such Sonine kernels will be denoted by $\mathcal{L}_1$ [13]:

$$(\kappa, k \in \mathcal{L}_1) \Leftrightarrow (\kappa, k \in C_{-1,0}(0,+\infty)) \wedge ((\kappa * k)(t) = \{1\}). \tag{27}$$

In the rest of this section, we present some important results for GFIs and GFDs with Sonine kernels from $\mathcal{L}_1$ on space $C_{-1}(0,+\infty)$ and its sub-spaces.

The basic properties of the GFI (25) on space $C_{-1}(0,+\infty)$ easily follow from the known properties of the Laplace convolution:

$$\mathbb{I}_{(\kappa)} : C_{-1}(0,+\infty) \to C_{-1}(0,+\infty), \tag{28}$$

$$\mathbb{I}_{(\kappa_1)} \mathbb{I}_{(\kappa_2)} = \mathbb{I}_{(\kappa_2)} \mathbb{I}_{(\kappa_1)}, \ \kappa_1, \kappa_2 \in \mathcal{L}_1, \tag{29}$$

$$\mathbb{I}_{(\kappa_1)} \mathbb{I}_{(\kappa_2)} = \mathbb{I}_{(\kappa_1 * \kappa_2)}, \ \kappa_1, \kappa_2 \in \mathcal{L}_1. \tag{30}$$

For functions $f \in C^1_{-1}(0, +\infty) := \{f : f' \in C_{-1}(0, +\infty)\}$, GFDs of the Riemann-Liouville type can be represented as follows [12]:

$$(\mathbb{D}_{(k)} f)(t) = (k * f')(t) + f(0)k(t), \ t > 0. \tag{31}$$

Thus, for $f \in C^1_{-1}(0, +\infty)$, GFD (10) of the Caputo type takes the form

$$({}_*\mathbb{D}_{(k)} f)(t) = (k * f')(t), \ t > 0. \tag{32}$$

It is worth mentioning that in FC publications, the Caputo fractional derivative (7) is often defined as in Formula (32):

$$({}_*D^\alpha_{0+} f)(t) = (h_{1-\alpha} * f')(t) = (I^{1-\alpha}_{0+} f')(t), \ t > 0. \tag{33}$$

Now, following [7,12], we define the n-fold GFI and the n-fold sequential GFDs in the Riemann-Liouville and Caputo senses.

Definition 1 ([12]). *Let $\kappa \in \mathcal{L}_1$. The n-fold GFI ($n \in \mathbb{N}$) is a composition of n GFIs with the kernel κ:*

$$(\mathbb{I}^{<n>}_{(\kappa)} f)(t) := (\underbrace{\mathbb{I}_{(\kappa)} \ldots \mathbb{I}_{(\kappa)}}_{n \ times} f)(t), \ t > 0. \tag{34}$$

It is worth mentioning that the index law (30) leads to a representation of the n-fold GFI (34) in the form of GFI with kernel $\kappa^{<n>}$:

$$(\mathbb{I}^{<n>}_{(\kappa)} f)(t) = (\kappa^{<n>} * f)(t) = (\mathbb{I}_{(\kappa)^{<n>}} f)(t), \ t > 0. \tag{35}$$

Kernel $\kappa^{<n>}$, $n \in \mathbb{N}$ belongs to space $C_{-1}(0, +\infty)$, but it is not always a Sonine kernel.

Definition 2 ([7]). *Let $\kappa \in \mathcal{L}_1$ and k be its associated Sonine kernel. The n-fold sequential GFDs in the Riemann-Liouville and in the Caputo sense, respectively, are defined as follows:*

$$(\mathbb{D}^{<n>}_{(k)} f)(t) := (\underbrace{\mathbb{D}_{(k)} \ldots \mathbb{D}_{(k)}}_{n \ times} f)(t), \ t > 0, \tag{36}$$

$$({}_*\mathbb{D}^{<n>}_{(k)} f)(t) := (\underbrace{{}_*\mathbb{D}_{(k)} \ldots {}_*\mathbb{D}_{(k)}}_{n \ times} f)(t), \ t > 0. \tag{37}$$

It is worth mentioning that in [6,12], the n-fold GFDs ($n \in \mathbb{N}$) were defined in a different form:

$$(\mathbb{D}^n_{(k)} f)(t) := \frac{d^n}{dt^n}(k^{<n>} * f)(t), \ t > 0, \tag{38}$$

$$({}_*\mathbb{D}^n_{(k)} f)(t) := (k^{<n>} * f^{(n)})(t), \ t > 0. \tag{39}$$

The n-fold sequential GFDs (36) and (37) are a far-reaching generalization of the Riemann-Liouville and the Caputo sequential fractional derivatives to the case of Sonine kernels from $\mathcal{L}_1$.

Some important connections between n-fold GFI (34) and n-fold sequential GFDs (36) and (37) in the Riemann-Liouville and Caputo senses are provided in the so-called first and second fundamental theorems of FC ([18]) formulated below.

Theorem 1 ([7]). *Let $\kappa \in \mathcal{L}_1$ and k be its associated Sonine kernel.*

Then, the n-fold sequential GFD (36) in the Riemann-Liouville sense is a left inverse operator to the n-fold GFI (34) on the space $C_{-1}(0,+\infty)$:

$$(\mathbb{D}_{(k)}^{<n>} \mathbb{I}_{(\kappa)}^{<n>} f)(t) = f(t),\ f \in C_{-1}(0,+\infty),\ t > 0, \qquad (40)$$

and the n-fold sequential GFD (37) in the Caputo sense is a left inverse operator to the n-fold GFI (34) on the space $C_{-1,(k)}^n(0,+\infty)$:

$$({}_*\mathbb{D}_{(k)}^{<n>} \mathbb{I}_{(\kappa)}^{<n>} f)(t) = f(t),\ f \in C_{-1,(k)}^n(0,+\infty),\ t > 0, \qquad (41)$$

where $C_{-1,(k)}^n(0,+\infty) := \{f:\ f(t) = (\mathbb{I}_{(k)}^{<n>} \phi)(t),\ \phi \in C_{-1}(0,+\infty)\}$.

Theorem 2 ([7]). *Let $\kappa \in \mathcal{L}_1$ and k be its associated Sonine kernel.*

For a function $f \in C_{-1,(k)}^{(n)}(0,+\infty) = \{f \in C_{-1}(0,+\infty):\ (\mathbb{D}_{(k)}^{<j>} f) \in C_{-1}(0,+\infty),\ j = 1,\ldots,n\}$, the formula

$$(\mathbb{I}_{(\kappa)}^{<n>} \mathbb{D}_{(k)}^{<n>} f)(t) = f(t) - \sum_{j=0}^{n-1} \left(k * \mathbb{D}_{(k)}^{<j>} f\right)(0) \kappa^{<j+1>}(t) = \qquad (42)$$

$$f(t) - \sum_{j=0}^{n-1} \left(\mathbb{I}_{(k)} \mathbb{D}_{(k)}^{<j>} f\right)(0) \kappa^{<j+1>}(t),\ t > 0$$

holds valid, where $\mathbb{I}_{(\kappa)}^{<n>}$ is the n-fold GFI (34) and $\mathbb{D}_{(k)}^{<n>}$ is the n-fold sequential GFD (36) in the Riemann-Liouville sense.

For a function $f \in C_{-1}^n(0,+\infty) := \{f:\ f^{(n)} \in C_{-1}(0,+\infty)\}$, the formula

$$(\mathbb{I}_{(\kappa)}^{<n>} {}_*\mathbb{D}_{(k)}^{<n>} f)(t) = f(t) - f(0) - \sum_{j=1}^{n-1} \left({}_*\mathbb{D}_{(k)}^{<j>} f\right)(0) \left(\{1\} * \kappa^{<j>}\right)(t) \qquad (43)$$

holds valid, where $\mathbb{I}_{(\kappa)}^{<n>}$ is the n-fold GFI (34) and ${}_*\mathbb{D}_{(k)}^{<n>}$ is the n-fold sequential GFD (37).

For proofs of Theorems 1 and 2 and their particular cases we refer interested readers to [7].

3. Solutions to Fractional Differential Equations with GFDs in the Riemann-Liouville Sense in Terms of the Convolution Series

First, we introduce the convolution series and treat some of their properties needed for the further discussions.

Definition 3. *For a function $\kappa \in C_{-1}(0,+\infty)$, the series in form*

$$\Sigma_\kappa(t) = \sum_{j=0}^{+\infty} a_j \kappa^{<j+1>}(t),\ a_j \in \mathbb{R}\ (a_j \in \mathbb{C}) \qquad (44)$$

is called convolution series generated by κ.

Convolution series generated by Sonine kernels $\kappa \in \mathcal{L}_1$ were introduced in [13] for analytical treatment of fractional differential equations with n-fold GFDs of the Caputo type by means of an operational calculus developed for these GFDs. In [7], some of the results presented in [13] were extended to convolution series in the form (44) generated by any function $\kappa \in C_{-1}(0,+\infty)$ (i.e., not necessarily a Sonine kernel).

A very important question regarding convergence of the convolution series (44) was answered in [6,7].

Theorem 3 ([7]). *Let a function $\kappa \in C_{-1}(0, +\infty)$ be represented in the form*

$$\kappa(t) = h_p(t)\kappa_1(t), \; t > 0, \; p > 0, \; \kappa_1 \in C[0, +\infty) \tag{45}$$

and the convergence radius of the power series

$$\Sigma(z) = \sum_{j=0}^{+\infty} a_j z^j, \; a_j \in \mathbb{C}, \; z \in \mathbb{C} \tag{46}$$

be non-zero. Then the convolution series (44) *is convergent for all $t > 0$ and defines a function from the space $C_{-1}(0, +\infty)$. Moreover, the series*

$$t^{1-\alpha} \Sigma_\kappa(t) = \sum_{j=0}^{+\infty} a_j t^{1-\alpha} \kappa^{<j+1>}(t), \; \alpha = \min\{p, 1\} \tag{47}$$

is uniformly convergent for $t \in [0, T]$ for any $T > 0$.

In what follows, we always assume that the coefficients of the convolution series satisfy the condition that the convergence radius of the corresponding power series is non-zero and thus Theorem 3 is applicable for these convolution series.

As an example, let us consider the geometric series

$$\Sigma(z) = \sum_{j=0}^{+\infty} \lambda^j z^j, \; \lambda \in \mathbb{C}, \; z \in \mathbb{C}. \tag{48}$$

For $\lambda \neq 0$, the convergence radius r of this series is equal to $1/|\lambda|$ and thus we can apply Theorem 3 that says that the convolution series generated by a function $\kappa \in C_{-1}(0, +\infty)$ in form

$$l_{\kappa,\lambda}(t) = \sum_{j=0}^{+\infty} \lambda^j \kappa^{<j+1>}(t), \; \lambda \in \mathbb{C} \tag{49}$$

is convergent for all $t > 0$ and defines a function from the space $C_{-1}(0, +\infty)$.

The convolution series $l_{\kappa,\lambda}$ defined by (49) plays a very important role in the operational calculus for GFD of Caputo type developed in [6]. It provides a far-reaching generalization of both the exponential function and the two-parameter Mittag-Leffler function in form (3).

Indeed, let us consider the convolution series (49) in the case of the kernel function $\kappa = \{1\}$. Due to the formula $\kappa^{<j+1>}(t) = \{1\}^{<j+1>}(t) = h_{j+1}(t)$ (see (17)), the convolution series (49) is reduced to the power series for the exponential function:

$$l_{\kappa,\lambda}(t) = \sum_{j=0}^{+\infty} \lambda^j h_{j+1}(t) = \sum_{j=0}^{+\infty} \frac{(\lambda t)^j}{j!} = e^{\lambda t}. \tag{50}$$

For the kernel $\kappa(t) = h_\alpha(t)$ of the Riemann-Liouville fractional integral, the formula $\kappa^{<j+1>}(t) = h_\alpha^{<j+1>}(t) = h_{(j+1)\alpha}(t)$ (see (17)) holds valid. Thus, the convolution series (49) takes the form

$$l_{\kappa,\lambda}(t) = \sum_{j=0}^{+\infty} \lambda^j h_{(j+1)\alpha}(t) = t^{\alpha-1} \sum_{j=0}^{+\infty} \frac{\lambda^j t^{j\alpha}}{\Gamma(j\alpha + \alpha)} = t^{\alpha-1} E_{\alpha,\alpha}(\lambda t^\alpha) \tag{51}$$

that is the same as the two-parameter Mittag-Leffler Function (3).

For $\kappa \in \mathcal{L}_1$, another important convolution series was introduced in [6] as follows:

$$L_{\kappa,\lambda}(t) = (k * l_{\kappa,\lambda})(t) = 1 + \left(\{1\} * \sum_{j=1}^{+\infty} \lambda^j \kappa^{<j>}(\cdot)\right)(t), \ \lambda \in \mathbb{C}, \tag{52}$$

where k is Sonine kernel associated with the kernel κ. It is easy to see that in the case $\kappa = \{1\}$, the convolution series (52) coincides with the exponential function:

$$L_{\kappa,\lambda}(t) = 1 + \left(\{1\} * \sum_{j=1}^{+\infty} \lambda^j h_j(\cdot)\right)(t) = 1 + \sum_{j=1}^{+\infty} \lambda^j h_{j+1}(t) = e^{\lambda t}. \tag{53}$$

In the case of the kernel $\kappa(t) = h_\alpha(t)$, $t > 0$, $0 < \alpha < 1$, the convolution series $L_{\kappa,\lambda}$ is reduced to the two-parameter Mittag-Leffler Function (8):

$$L_{\kappa,\lambda}(t) = 1 + \left(\{1\} * \sum_{j=1}^{+\infty} \lambda^j h_{j\alpha}(\cdot)\right)(t) = 1 + \sum_{j=1}^{+\infty} \lambda^j h_{j\alpha+1}(t) = E_{\alpha,1}(\lambda t^\alpha). \tag{54}$$

Analytical solutions to single- and multi-term fractional differential equations with n-fold GFDs of the Caputo type were presented in [6] in terms of the convolution series $l_{\kappa,\lambda}$ and $L_{\kappa,\lambda}$. In the rest of this section, we treat linear single- and multi-term fractional differential equations with n-fold GFDs in the Riemann-Liouville sense.

We start with the following auxiliary result:

Theorem 4. *Two convolution series generated by the same Sonine kernel $\kappa \in \mathcal{L}_1$ coincide for all $t > 0$, i.e.,*

$$\sum_{j=0}^{+\infty} b_j \kappa^{<j+1>}(t) \equiv \sum_{j=0}^{+\infty} c_j \kappa^{<j+1>}(t), \ t > 0 \tag{55}$$

if and only if the corresponding coefficients of these series are equal:

$$a_j = b_j, \ j = 0, 1, 2, \ldots. \tag{56}$$

Proof. If the corresponding coefficients of two convolution series generated by the same Sonine kernel $\kappa \in \mathcal{L}_1$ are equal, then we have just one series and evidently the identity (55) holds valid.

The idea of the proof of the second part of this theorem is the same as the one for the proof of the analogous calculus result for the power series, i.e., under the condition that the identity (55) holds valid we first show that $b_0 = c_0$ and then apply the same arguments to prove that $b_1 = c_1$, $b_2 = c_2$, etc.

According to Theorem 3, the convolution series in the form (44) is uniformly convergent on any interval $[\epsilon, T]$, and thus we can apply the GFI $\mathbb{I}_{(k)}$ to this series term by term:

$$\left(\mathbb{I}_{(k)} \sum_{j=0}^{+\infty} a_j \kappa^{<j+1>}(\cdot)\right)(t) = \sum_{j=0}^{+\infty} \left(\mathbb{I}_{(k)} a_j \kappa^{<j+1>}(\cdot)\right)(t) = \sum_{j=0}^{+\infty} \left(a_j (k(\cdot) * \kappa^{<j+1>}(\cdot)\right)(t) =$$

$$a_0 + \sum_{j=1}^{+\infty} a_j \left(\{1\} * \kappa^{<j>}(\cdot)\right)(t) = a_0 + \left(\{1\} * \sum_{j=1}^{+\infty} a_j \kappa^{<j>}(\cdot)\right)(t) = a_0 + (\{1\} * f_1)(t),$$

where f_1 is the following convolution series:

$$f_1(t) = \sum_{j=1}^{+\infty} a_j \kappa^{<j>}(t) = \sum_{j=0}^{+\infty} a_{j+1} \kappa^{<j+1>}(t). \tag{57}$$

Summarizing the calculations from above, for the convolution series in form (44), the formula

$$\left(\mathbb{I}_{(k)} \sum_{j=0}^{+\infty} a_j \kappa^{<j+1>}(\cdot) \right)(t) = a_0 + \left(\{1\} * \sum_{j=0}^{+\infty} a_{j+1} \kappa^{<j+1>}(\cdot) \right)(t) \qquad (58)$$

holds valid.

Because the convergence radius of the power series $\Sigma_1(t) = \sum_{j=0}^{+\infty} a_{j+1} z^j$ is the same as the convergence radius of the power series $\Sigma(t) = \sum_{j=0}^{+\infty} a_j z^j$, Theorem 3 ensures the inclusion $f_1 \in C_{-1}(0, +\infty)$, where f_1 is defined by Formula (57). As has been shown in [4], the definite integral of a function from $C_{-1}(0, +\infty)$ is a continuous function on the whole interval $[0, +\infty)$ that takes the value zero at the point zero:

$$(\{1\} * f_1)(t) = (I_{0+}^1 f_1)(t) \in C[0, +\infty), \quad (I_{0+}^1 f_1)(0) = 0. \qquad (59)$$

Now we act with the GFI $\mathbb{I}_{(k)}$ on the equality (55) and apply Formula (58) to obtain the relationship

$$b_0 + \left(\{1\} * \sum_{j=0}^{+\infty} b_{j+1} \kappa^{<j+1>}(\cdot) \right)(t) \equiv c_0 + \left(\{1\} * \sum_{j=0}^{+\infty} c_{j+1} \kappa^{<j+1>}(\cdot) \right)(t), \ t > 0. \qquad (60)$$

Substituting point $t = 0$ into equality (60) and using Formula (59), we deduce that $b_0 = c_0$. Now we differentiate equality (60) and obtain the following identity:

$$\sum_{j=0}^{+\infty} b_{j+1} \kappa^{<j+1>}(t) \equiv \sum_{j=0}^{+\infty} c_{j+1} \kappa^{<j+1>}(t), \ t > 0. \qquad (61)$$

This identity has exactly same structure as identity (55) from Theorem 4. Thus, we can apply the same arguments as above and derive the relationhship $b_1 = c_1$. By repeating the same reasoning repeatedly, we arrive at Formula (56) that we wanted to prove. □

Now we are ready to apply the method of convolution series for derivation of solutions to the fractional differential equations with GFDs, and start with the fractional relaxation equation with the GFD of the Riemann-Liouville type:

$$(\mathbb{D}_{(k)} y)(t) = \lambda y(t), \ \lambda \in \mathbb{R}, \ t > 0. \qquad (62)$$

As in the case of the power series, we look for a general solution to this equation in the form of a convolution series generated by the Sonine kernel κ that is an associated kernel to the kernel k of the GFD from Equation (62):

$$y(t) = \sum_{j=0}^{+\infty} b_j \kappa^{<j+1>}(t), \ b_j \in \mathbb{R}. \qquad (63)$$

To proceed, let us first calculate the image of the convolution series (63) by action of the GFD $\mathbb{D}_{(k)}$:

$$(\mathbb{D}_{(k)} y)(t) = \left(\mathbb{D}_{(k)} \sum_{j=0}^{+\infty} b_j \kappa^{<j+1>}(\cdot) \right)(t) = \frac{d}{dt} \left(\mathbb{I}_{(k)} \sum_{j=0}^{+\infty} b_j \kappa^{<j+1>}(\cdot) \right)(t).$$

In the proof of Theorem 4 we already calculated the image of the convolution series (63) by action of the GFI $\mathbb{I}_{(k)}$ (Formula (58)). Applying this formula, we arrive at the representation

$$(\mathbb{D}_{(k)} y)(t) = \frac{d}{dt}\left(b_0 + \left(\{1\} * \sum_{j=0}^{+\infty} b_{j+1} \kappa^{<j+1>}(\cdot)\right)(t)\right) = \sum_{j=0}^{+\infty} b_{j+1} \kappa^{<j+1>}(t). \quad (64)$$

In the next step, we substitute the right-hand side of (64) into Equation (62) and obtain an equality of two convolution series generated by the same kernel κ:

$$\sum_{j=0}^{+\infty} b_{j+1} \kappa^{<j+1>}(t) = \sum_{j=0}^{+\infty} \lambda\, b_j \kappa^{<j+1>}(t), \ t > 0.$$

Application of Theorem 4 to the above identity leads to the following relationships for the coefficients of the convolution series (63):

$$b_{j+1} = \lambda\, b_j,\ j = 0, 1, 2, \dots. \quad (65)$$

The infinite system (65) of linear equations can be easily solved step by step and we arrive at the explicit solution in form

$$b_j = b_0 \lambda^j,\ j = 1, 2, \dots, \quad (66)$$

where $b_0 \in \mathbb{R}$ is an arbitrary constant. Summarizing the arguments presented above, we proved the following theorem:

Theorem 5. *The general solution to the fractional relaxation Equation (62) with GFD (9) in the Riemann-Liouville sense can be represented as follows:*

$$y(t) = \sum_{j=0}^{+\infty} b_0 \lambda^j \kappa^{<j+1>}(t) = b_0\, l_{\kappa,\lambda}(t),\ b_0 \in \mathbb{R}, \quad (67)$$

where $l_{\kappa,\lambda}$ is the convolution series (49).

Remark 1. *The constant b_0 in the general solution (67) to Equation (62) can be determined from a suitably posed initial condition. The form of this initial condition is prescribed by Theorem 2 (see also Formula (58)). Indeed, setting $n = 1$ in the relation (42), we obtain the following representation of the projector operator of the GFD (9) in the Riemann-Liouville sense:*

$$(Pf)(t) = f(t) - (\mathbb{I}_{(\kappa)}\, \mathbb{D}_{(k)} f)(t) = \left(\mathbb{I}_{(k)} f\right)(0)\kappa(t),\ f \in C^{(1)}_{-1,(k)}(0, +\infty). \quad (68)$$

Thus, the initial-value problem

$$\begin{cases} (\mathbb{D}_{(k)} y)(t) = \lambda y(t),\ \lambda \in \mathbb{R},\ t > 0, \\ \left(\mathbb{I}_{(k)} y\right)(0) = b_0 \end{cases} \quad (69)$$

has a unique solution given by Formula (67).

In the case of the Sonine kernel $k(t) = h_{1-\alpha}(t)$, $0 < \alpha < 1$, the Equation (62) is reduced to Equation (2) with the Riemann-Liouville fractional derivative and its solution

(67) is exactly the solution (3) of Equation (2) in terms of the two-parameter Mittag-Leffler function (see Formula (51)). The initial-value problem (69) takes the well-known form

$$\begin{cases} (D_{0+}^\alpha y)(t) = \lambda y(t), \ \lambda \in \mathbb{R}, \ t > 0, \\ \left(I_{0+}^{1-\alpha} y\right)(0) = b_0. \end{cases} \tag{70}$$

Its unique solution is given by the formula $y(t) = b_0 \, t^{\alpha-1} E_{\alpha,\alpha}(\lambda \, t^\alpha)$.

Now we proceed with the inhomogeneous equation of type (62)

$$(\mathbb{D}_{(k)} y)(t) = \lambda y(t) + f(t), \ \lambda \in \mathbb{R}, \ t > 0, \tag{71}$$

where the source function f is represented in form of a convolution series

$$f(t) = \sum_{j=0}^{+\infty} a_j \kappa^{<j+1>}(t), \ a_j \in \mathbb{R}. \tag{72}$$

Again, we look for solutions to Equation (71) in the form of the convolution series (63). Applying exactly the same reasoning as above, we arrive at the following infinite system of linear equations for the coefficients of the convolution series (63):

$$b_{j+1} = \lambda \, b_j + a_j, \ j = 0, 1, 2, \ldots. \tag{73}$$

The explicit form of solutions to this system of equations is as follows:

$$b_j = b_0 \lambda^j + \sum_{i=0}^{j-1} a_i \lambda^{j-i-1}, \ j = 1, 2, \ldots, \tag{74}$$

where $b_0 \in \mathbb{R}$ is an arbitrary constant. Then the general solution to Equation (71) can be written in form of the following convolution series:

$$y(t) = b_0 \kappa(t) + \sum_{j=1}^{+\infty} \left(b_0 \lambda^j + \sum_{i=0}^{j-1} a_i \lambda^{j-i-1} \right) \kappa^{<j+1>}(t) = b_0 \sum_{j=0}^{+\infty} \lambda^j \kappa^{<j+1>}(t) + \sum_{j=1}^{+\infty} \sum_{i=0}^{j-1} a_i \lambda^{j-i-1} \kappa^{<j+1>}(t).$$

By direct calculation, we verify that the second sum in the last formula can be written in a more compact form:

$$\sum_{j=1}^{+\infty} \sum_{i=0}^{j-1} a_i \lambda^{j-i-1} \kappa^{<j+1>}(t) = \sum_{i=0}^{+\infty} a_i \sum_{j=1}^{+\infty} \lambda^{j-1} \kappa^{<j+i+1>}(t) = (f * l_{\kappa,\lambda})(t),$$

where the convolution series $l_{\kappa,\lambda}$ is defined by (49). We thus have proved the following result:

Theorem 6. *The general solution to the inhomogeneous Equation (71) has the form*

$$y(t) = b_0 \, l_{\kappa,\lambda}(t) + (f * l_{\kappa,\lambda})(t), \ b_0 \in \mathbb{R}, \tag{75}$$

where the convolution series $l_{\kappa,\lambda}$ is defined by (49).

The constant b_0 is uniquely determined by the initial condition

$$\left(\mathbb{I}_{(k)} y\right)(0) = b_0. \tag{76}$$

Applying Theorem 6 to the case of the Riemann-Liouville fractional derivative (kernel $k(t) = h_{1-\alpha}(t)$, $0 < \alpha < 1$), we obtain the well-known result ([3]):

The unique solution to the initial-value problem

$$\begin{cases} (D_{0+}^{\alpha} y)(t) = \lambda y(t) + f(t), \ \lambda \in \mathbb{R}, \ t > 0, \\ \left(I_{0+}^{1-\alpha} y \right)(0) = b_0 \end{cases}$$

is given by the formula

$$y(t) = b_0 \, t^{\alpha-1} E_{\alpha,\alpha}(\lambda \, t^\alpha) + \int_0^t \tau^{\alpha-1} E_{\alpha,\alpha}(\lambda \, \tau^\alpha) f(t-\tau) \, d\tau.$$

Remark 2. *In [6], single- and multi-term fractional differential equations with general fractional derivatives of the Caputo type have been studied. In particular, the unique solution to the initial-value problem*

$$\begin{cases} (_*\mathbb{D}_{(k)} y)(t) = \lambda y(t) + f(t), & \lambda \in \mathbb{R}, \ t > 0, \\ y(0) = b_0, & b_0 \in \mathbb{R} \end{cases} \quad (77)$$

with the GFD of the Caputo type defined by (10) *was derived in the form*

$$y(t) = (f * l_{\kappa,\lambda})(t) + b_0 \, L_{\kappa,\lambda}(t), \quad (78)$$

where $\kappa \in \mathcal{L}_1$ *is the Sonine kernel associated with the kernel k and* $l_{\kappa,\lambda}$, $L_{\kappa,\lambda}$ *are the convolution series* (49) *and* (52), *respectively.*

In the case of the homogeneous initial condition ($y(0) = b_0 = 0$), Formula (10) says that GFDs of the Riemann–Liouville and Caputo types coincide. As we see, the solutions to the initial-value problems with the homogeneous initial conditions for Equations (71) and (77) are also identical.

Let us now consider a linear inhomogeneous multi-term fractional differential equation with the sequential GFDs (36) of the Riemann–Liouville type and with the constant coefficients:

$$\sum_{i=0}^{n} \lambda_i (\mathbb{D}_{(k)}^{<i>} y)(t) = f(t), \ \lambda_i \in \mathbb{R}, \ i = 0, 1, \ldots, n, \ \lambda_n \neq 0, \ t > 0, \quad (79)$$

where the source function f is represented in form of the convolution series (72).

As in the case of the single-term Equation (71), we look for solutions to the multi-term Equation (79) in the form of the convolution series (63). First, we determine the images of the convolution series (63) by action of the sequential GFDs $\mathbb{D}_{(k)}^{<i>}$, $i = 1, 2, \ldots, n$. For $i = 1$, the image is provided by Formula (64). For $i = 2, \ldots, n$, Formula (64) is applied iteratively and we arrive at the following result:

$$(\mathbb{D}_{(k)}^{<i>} y)(t) = \sum_{j=0}^{+\infty} b_{j+i} \kappa^{<j+1>}(t), \ i = 1, 2, \ldots, n. \quad (80)$$

Now we substitute the convolution series (63), its images by action of the sequential GFDs $\mathbb{D}_{(k)}^{<i>}$, $i = 1, 2, \ldots, n$ provided by Formula (80), and the convolution series (72) for the source function into Equation (79) and arrive at the following identity:

$$\sum_{i=0}^{n} \lambda_i \left(\sum_{j=0}^{+\infty} b_{j+i} \kappa^{<j+1>}(t) \right) = \sum_{j=0}^{+\infty} a_j \kappa^{<j+1>}(t), \ t > 0.$$

Application of Theorem 4 to the above identity leads to the following infinite triangular system of linear equations for the coefficients of the convolution series (63):

$$\begin{cases} \lambda_0 b_0 + \lambda_1 b_1 + \cdots + \lambda_n b_n = a_0, \\ \lambda_0 b_1 + \lambda_1 b_2 + \cdots + \lambda_n b_{n+1} = a_1, \\ \cdots \\ \lambda_0 b_n + \lambda_1 b_{n+1} + \cdots + \lambda_n b_{2n} = a_n, \\ \lambda_0 b_{n+1} + \lambda_1 b_{n+2} + \cdots + \lambda_n b_{2n+1} = a_{n+1} \\ \cdots \end{cases} \qquad (81)$$

In this system, the first n coefficients $(b_0, b_1, \ldots, b_{n-1})$ can be chosen arbitrarily and all other coefficients are determined step by step as solutions to the infinite triangular system (81) of linear equations:

$$b_{n+l} = (a_l - \lambda_0 b_l - \cdots - \lambda_{n-1} b_{n+l-1})/\lambda_n, \ l = 0, 1, 2, \ldots \qquad (82)$$

We thus proved the following theorem:

Theorem 7. *The general solution to the inhomogeneous multi-term fractional differential Equation (79) can be represented as the convolution series (63), where the first n coefficients $(b_0, b_1, \ldots, b_{n-1})$ are arbitrary real constants and other coefficients are calculated according to Formula (82).*

The constants $b_0, b_1, \ldots, b_{n-1}$ in the general solution to Equation (79) presented in Theorem (7) can be determined based on the suitably posed initial conditions. The form of these initial conditions is prescribed by Theorem 2. Indeed, Formula (42) can be rewritten as follows:

$$(Pf)(t) = f(t) - (\mathbb{I}_{(\kappa)}^{<n>} \mathbb{D}_{(k)}^{<n>} f)(t) = \sum_{j=0}^{n-1} \left(\mathbb{I}_{(k)} \mathbb{D}_{(k)}^{<j>} f \right)(0) \kappa^{<j+1>}(t), \ t > 0, \ f \in C_{-1,(k)}^{(n)}(0, +\infty), \qquad (83)$$

where P is the projector operator of the n-fold sequential GFD of the Riemann-Liouville type. Thus, to uniquely determine the constants $b_0, b_1, \ldots, b_{n-1}$ in the general solution, Equation (79) has to be equipped with the initial conditions in the form

$$\left(\mathbb{I}_{(k)} \mathbb{D}_{(k)}^{<j>} y \right)(0) = b_j, \ j = 0, 1, \ldots, n-1. \qquad (84)$$

Finally, we mention that the inhomogeneous multi-term fractional differential equation of type (79) with sequential Riemann-Liouville fractional derivatives (the case of the kernel $k(t) = h_{1-\alpha}(t)$ in Equation (79)) was treated in [3,19] using operational calculus of the Mikusiński type for the Riemann-Liouville fractional derivative.

Funding: This research received no external funding.

Institutional Review Board Statement: Not applicable.

Informed Consent Statement: Not applicable.

Data Availability Statement: Not applicable.

Conflicts of Interest: The author declares no conflict of interest.

References

1. Dzherbashyan, M.M. *Integral Transforms and Representations of Functions in the Complex Plane*; Nauka: Moscow, Russia, 1966. (In Russian)
2. Gorenflo, R.; Loutchko, J.; Luchko, Yu, Computation of the Mittag-Leffler function and its derivatives. *Fract. Calc. Appl. Anal.* **2002**, *5*, 491–518.

3. Luchko, Y.; Srivastava, H.M. The exact solution of certain differential equations of fractional order by using operational calculus. *Comput. Math. Appl.* **1995**, *29*, 73–85. [CrossRef]
4. Luchko, Y.; Gorenflo, R. An operational method for solving fractional differential equations. *Acta Math. Vietnam.* **1999**, *24*, 207–234.
5. Gorenflo, R.; Kilbas, A.A.; Mainardi, F.; Rogosin, S. *Mittag-Leffler Functions, Related Topics and Applications*, 2nd ed.; Springer: Berlin/Heidelberg, Germany, 2020.
6. Luchko, Y. Operational Calculus for the general fractional derivatives with the Sonine kernels. *Fract. Calc. Appl. Anal.* **2021**, *24*, 338–375. [CrossRef]
7. Luchko, Y. Convolution series and the generalized convolution Taylor formula. *arXiv* **2021**, arXiv:2107.10198v2.
8. Kochubei, A.N. General fractional calculus, evolution equations, and renewal processes. *Integr. Equa. Operator Theory* **2011**, *71*, 583–600. [CrossRef]
9. Luchko, Y.; Yamamoto, M. General time-fractional diffusion equation: some uniqueness and existence results for the initial-boundary-value problems. *Fract. Calc. Appl. Anal.* **2016**, *19*, 675–695. [CrossRef]
10. Kochubei, A.N. General fractional calculus. In *Handbook of Fractional Calculus with Applications. Volume 1: Basic Theory*; Kochubei, A., Luchko, Y., Eds.; De Gruyter: Berlin/Heidelberg, Germany, 2019; pp. 111–126.
11. Luchko, Y.; Yamamoto, M. The General Fractional Derivative and Related Fractional Differential Equations. *Mathematics* **2020**, *8*, 2115. [CrossRef]
12. Luchko, Y. General Fractional Integrals and Derivatives with the Sonine Kernels. *Mathematics* **2021**, *9*, 594. [CrossRef]
13. Luchko, Y. General Fractional Integrals and Derivatives of Arbitrary Order. *Symmetry* **2021**, *13*, 755. [CrossRef]
14. Abel, N.H. Oplösning af et par opgaver ved hjelp af bestemte integraler. *Mag. Naturvidenskaberne* **1823**, *2*, 2.
15. Abel, N.H. Auflösung einer mechanischen Aufgabe. *J. Die Reine Angew. Math.* **1826**, *1*, 153–157.
16. Sonine, N. Sur la généralisation d'une formule d'Abel. *Acta Math.* **1884**, *4*, 171–176. [CrossRef]
17. Samko, S.G.; Cardoso, R.P. Integral equations of the first kind of Sonine type. *Intern. J. Math. Sci.* **2003**, *57*, 3609–3632. [CrossRef]
18. Luchko, Y. Fractional derivatives and the fundamental theorem of Fractional Calculus. *Fract. Calc. Appl. Anal.* **2020**, *23*, 939–966. [CrossRef]
19. Luchko, Y. Operational method in fractional calculus. *Fract. Calc. Appl. Anal.* **1999**, *2*, 463–489.

Article

Basic Fundamental Formulas for Wiener Transforms Associated with a Pair of Operators on Hilbert Space

Hyun Soo Chung

Department of Mathematics, Dankook University, Cheonan 31116, Korea; hschung@dankook.ac.kr

Abstract: Segal introduce the Fourier–Wiener transform for the class of polynomial cylinder functions on Hilbert space, and Hida then develop this concept. Negrin define the extended Wiener transform with Hayker et al. In recent papers, Hayker et al. establish the existence, the composition formula, the inversion formula, and the Parseval relation for the Wiener transform. But, they do not establish homomorphism properties for the Wiener transform. In this paper, the author establishes some basic fundamental formulas for the Wiener transform via some concepts and motivations introduced by Segal and used by Hayker et al. We then state the usefulness of basic fundamental formulas as some applications.

Keywords: Hilbert space; convolution product; first variation; integration by parts formula; translation theorem

MSC: 60J65; 28C20

Citation: Chung, H.S. Basic Fundamental Formulas for Wiener Transforms Associated with a Pair of Operators on Hilbert Space. *Mathematics* 2021, 9, 2738. https://doi.org/10.3390/math9212738

Academic Editor: Francesco Mainardi

Received: 23 September 2021
Accepted: 25 October 2021
Published: 28 October 2021

Publisher's Note: MDPI stays neutral with regard to jurisdictional claims in published maps and institutional affiliations.

Copyright: © 2021 by the authors. Licensee MDPI, Basel, Switzerland. This article is an open access article distributed under the terms and conditions of the Creative Commons Attribution (CC BY) license (https://creativecommons.org/licenses/by/4.0/).

1. Introduction

Let X be a normed space and let T be a operator on X. In functional analysis theory and algebraic structures, the homomorphism properties

$$T(f * g) = T(f)T(g) \qquad (1)$$

and

$$(T(f) * T(g)) = T(fg) \qquad (2)$$

are very important subjects to various fields of mathematics for $f, g \in X$, where $*$ denotes a corresponding convolution product of T.

In [1–3], Segal introduce the Fourier–Wiener transform for the class of polynomial cylinder functions on Hilbert space. Hida then develop this concept via the Fourier analysis on the dual space of nuclear spaces [4,5]. In addition, Negrin obtain an explicit integral representation of the second quantization by use of an integral operator and hence the Wiener transform [6] is extended. Later, Hayker et al. analyze and study some results and formulas of them via the matrix expressions [7].

In [8,9], the authors establish the existence, the composition formula, the inversion formula and the parseval relationship for the Wiener transform. But, they do not establish homomorphism properties (1) and (2) for the Wiener transform.

In this paper, we shall establish homomorphism properties for the Wiener transform. In addition, we obtain an integration by parts formula, and give some applications of it with respect to the Wiener transform. Our integration by parts formula takes a different form than in the Euclidean space. The reason is that the measure used in this paper is a probability measure, unlike the Lebesgue measure.

2. Definitions and Preliminaries

In this section, we first state some definitions and notations to understand the paper.

Let **H'** be a real Hilbert space and **H** be a complexification of **H'**. The inner product on **H** is given by the formula

$$\langle x+iy, x'+iy' \rangle_{\mathbf{H}} = \langle x, x' \rangle_{\mathbf{H'}} + \langle y, y' \rangle_{\mathbf{H'}} + i \langle y, x' \rangle_{\mathbf{H'}} - i \langle x, y' \rangle_{\mathbf{H'}}.$$

Let A and B be operators defined on **H** such that there exists an orthonormal basis $\mathcal{B} = \{e_\alpha\}_{\alpha \in \mathcal{A}}$ of **H** ($\mathcal{A}$ being some index set) consisting of elements of **H** with

$$Ae_\alpha = \mu_\alpha e_\alpha, \qquad Be_\alpha = \lambda_\alpha e_\alpha \qquad (3)$$

for some complex numbers μ_α and λ_α. Then we note that for each $x \in \mathbf{H}$,

$$x = \sum_{\alpha \in \mathcal{A}} \langle x, e_\alpha \rangle_{\mathbf{H}} e_\alpha$$

and so

$$Ax = \sum_{\alpha \in \mathcal{A}} \langle x, e_\alpha \rangle_{\mathbf{H}} \mu_\alpha e_\alpha$$

and

$$Bx = \sum_{\alpha \in \mathcal{A}} \langle x, e_\alpha \rangle_{\mathbf{H}} \lambda_\alpha e_\alpha.$$

We now state a class of functions used in this paper.

Definition 1. *Let f be a polynomial function on $\mathbf{H'}$ defined by the formula*

$$f(x) = \langle x, e_{\alpha_1} \rangle_{\mathbf{H}}^{n_1} \langle x, e_{\alpha_2} \rangle_{\mathbf{H}}^{n_2} \cdots \langle x, e_{\alpha_r} \rangle_{\mathbf{H}}^{n_r} \qquad (4)$$

where $n_1, \cdots, n_r \in \mathbb{N} \cup \{0\}$. Let $\mathcal{P}$ be the space of all complex-valued polynomial on $\mathbf{H'}$.

We are ready to state definitions of the Wiener transform, the convolution product and the first variation for functions in $\mathcal{P}$.

Definition 2. *For each pair of operators A and B on $\mathbf{H}$, we define the Wiener transform $\mathcal{F}_{c,A,B}(f)$ of f by the formula*

$$\mathcal{F}_{c,A,B}(f)(y) = \int_{\mathbf{H'}} f(Ax + By) dg_c(x) \qquad (5)$$

where f is in $\mathcal{P}$ and the integration on $\mathbf{H'}$ is performed with respect to the normalized distribution g_c of the variance parameter $c > 0$. In addition, we define the convolution product $(f_1 * f_2)_A$ of f_1 and f_2 by the formula

$$(f_1 * f_2)_A(y) = \int_{\mathbf{H'}} f_1\left(\frac{y + Ax}{\sqrt{2}}\right) f_2\left(\frac{y - Ax}{\sqrt{2}}\right) dg_c(x) \qquad (6)$$

and the first variation $\delta_B f$ of f is defined by the formula

$$\delta_B f(x|u) = \frac{\partial}{\partial k} f(x + kBu) \bigg|_{k=0} \qquad (7)$$

where $f, f_1, f_2 \in \mathcal{P}$ if they exist.

3. Existence

In this section, we establish the existence of the convolution product and the first variation for function f of the form (4). Before doing this, we give a theorem for some formulas with respect to the Wiener transform $\mathcal{F}_{c,A,B}$ which are established by Hayker et al. [9].

Theorem 1. Let A, B, A', B', A'' and B'' be operators on $\mathbf{H}$ given by

$$Ae_\alpha = \mu_\alpha e_\alpha, \quad Be_\alpha = \lambda_\alpha e_\alpha, \quad A'e_\alpha = \mu'_\alpha e_\alpha, \quad B'e_\alpha = \lambda'_\alpha e_\alpha$$
$$A''e_\alpha = \mu''_\alpha e_\alpha, \quad B''e_\alpha = \lambda''_\alpha e_\alpha$$

where $\mu_\alpha, \mu'_\alpha, \mu''_\alpha, \lambda_\alpha, \lambda'_\alpha$ and λ''_α are complex numbers. Then we have the following assertions.

(a) (Existence): for any $f \in \mathcal{P}$,

$$\mathcal{F}_{c,A,B}(f)(y) = \prod_{j=1}^{r}\left(\sum_{p=0}^{[n_j/2]} {}_{n_j}C_p \mu_{\alpha_j}^{2p} \lambda_{\alpha_j}^{n_j-2p} \langle y, e_{\alpha_j}\rangle_{\mathbf{H}}^{n_j-2p} \frac{(2p)!}{p!}\left(\frac{c}{2}\right)^p\right) \tag{8}$$

and $\mathcal{F}_{c,A,B}(f) \in \mathcal{P}$.

(b) (Composition formula [9], Theorem 1):

$$\mathcal{F}_{c,A',B'}(\mathcal{F}_{c,A,B}(f))(y) = \mathcal{F}_{c,A'',B''}(f)(y)$$

if and only if

$$\mu_\alpha^2 + (\mu'_\alpha \lambda_\alpha)^2 = (\mu''_\alpha)^2 \text{ and } \lambda_\alpha \lambda'_\alpha = \lambda''_\alpha$$

for $\alpha \in \mathcal{A}$.

(c) (Inversion formula [9], Corollary 2):

$$\mathcal{F}_{c,A',B'}(\mathcal{F}_{c,A,B}(f))(y) = f(y) \tag{9}$$

if and only if

$$\mu_\alpha^2 + (\mu'_\alpha \lambda_\alpha)^2 = 0 \text{ and } \lambda_\alpha \lambda'_\alpha = 1$$

for $\alpha \in \mathcal{A}$.

(d) (Parseval relation [9], Theorem 2):

$$\int_{\mathbf{H}'} \mathcal{F}_{c,A,B}(f_1)(y) f_2(y) dg_c(y) = \int_{\mathbf{H}'} \mathcal{F}_{c,A,B}(f_2)(y) f_1(y) dg_c(y)$$

if and only if

$$\mu_\alpha^2 + \lambda_\alpha^2 = 1$$

for $\alpha \in \mathcal{A}$. Furthermore, they show that it can be extended to the Unitary extension.

We shall obtain the existence of the convolution product and the first variation. To do this, we need an observation as below.

Remark 1. For any f_1 and f_2 in $\mathcal{P}$, we can always express f_1 by Equation (4) and f_2 by

$$f_2(x) = \langle x, e_{\alpha_1}\rangle_{\mathbf{H}}^{m_1} \langle x, e_{\alpha_2}\rangle_{\mathbf{H}}^{m_2} \cdots \langle x, e_{\alpha_r}\rangle_{\mathbf{H}}^{m_r} \tag{10}$$

using the same nonnegative integer r and α_j's. Because, if $f_1(x) = \langle x, e_{\alpha_1}\rangle_{\mathbf{H}}^{n_1} \langle x, e_{\alpha_3}\rangle_{\mathbf{H}}^{n_3}$ and $f_2(x) = \langle x, e_{\alpha_1}\rangle_{\mathbf{H}}^{n_1} \langle x, e_{\alpha_2}\rangle_{\mathbf{H}}^{n_2}$, then we can set

$$f_1(x) = \langle x, e_{\alpha_1}\rangle_{\mathbf{H}}^{n_1} \langle x, e_{\alpha_2}\rangle_{\mathbf{H}}^{0} \langle x, e_{\alpha_3}\rangle_{\mathbf{H}}^{n_3}$$

and

$$f_2(x) = \langle x, e_{\alpha_1}\rangle_{\mathbf{H}}^{m_1} \langle x, e_{\alpha_2}\rangle_{\mathbf{H}}^{m_2} \langle x, e_{\alpha_3}\rangle_{\mathbf{H}}^{0}.$$

In addition, if $f_1(x) = \langle x, e_\alpha\rangle_{\mathbf{H}}^n$ and $f_2(x) = \langle x, e_\beta\rangle_{\mathbf{H}}^m$ for $n \neq m$, then we can set

$$f_1(x) = \langle x, e_{\gamma_1}\rangle_{\mathbf{H}}^{n_1} \langle x, e_{\gamma_2}\rangle_{\mathbf{H}}^{0}$$

and
$$f_2(x) = \langle x, e_{\gamma_1} \rangle_H^{m_1} \langle x, e_{\gamma_2} \rangle_H^0$$
where $\gamma_1 = \alpha, \gamma_2 = \beta, n_1 = n, n_2 = 0, m_1 = 0$ and $m_2 = m$.

In Theorem 1, we obtain the existence of the convolution product and the first variation for functions in $\mathcal{P}$.

Theorem 2. *Let f_1 and f_2 be elements of $\mathcal{P}$ and A as in Theorem 1. Then the convolution product $(f_1 * f_2)_A$ of f_1 and f_2 exists, belongs to $\mathcal{P}$ and is given by the formula*

$$(f_1 * f_2)_A(y) = \left(\frac{1}{2\pi c}\right)^{\frac{r}{2}} \prod_{j=1}^{r} \left[\int_{\mathbb{R}} \left(\frac{1}{\sqrt{2}} \langle y, e_{\alpha_j} \rangle_H + \frac{\lambda_{\alpha_j}}{\sqrt{2}} u_j \right)^{n_j} \right. \tag{11}$$

$$\left. \times \left(\frac{1}{\sqrt{2}} \langle y, e_{\alpha_j} \rangle_H - \frac{\lambda_{\alpha_j}}{\sqrt{2}} u_j \right)^{m_j} \exp\left\{ -\frac{u_j^2}{2c} \right\} du_j \right].$$

Furthermore, the first variation $\delta_A f$ of f exists, belongs to $\mathcal{P}$ and is given by the formula

$$\delta_A f(x|u) = \sum_{j=1}^{r} n_j \lambda_{\alpha_j} \langle u, e_{\alpha_j} \rangle_H f_j(x) \tag{12}$$

where

$$f_j(x) = \langle x, e_{\alpha_1} \rangle_H^{n_1} \times \cdots \times \langle x, e_{\alpha_j} \rangle_H^{n_j-1} \times \cdots \times \langle x, e_{\alpha_r} \rangle_H^{n_r}. \tag{13}$$

Proof. Using Equations (5) and (6), we have

$$(f_1 * f_2)_A(y)$$
$$= \int_{H'} \prod_{j=1}^{r} \left(\frac{1}{\sqrt{2}} \langle y, e_{\alpha_j} \rangle_H + \frac{\lambda_{\alpha_j}}{\sqrt{2}} \langle x, e_{\alpha_j} \rangle_H \right)^{n_j} \left(\frac{1}{\sqrt{2}} \langle y, e_{\alpha_j} \rangle_H - \frac{\lambda_{\alpha_j}}{\sqrt{2}} \langle x, e_{\alpha_j} \rangle_H \right)^{m_j} dg_c(x)$$
$$= \left(\frac{1}{2\pi c}\right)^{\frac{r}{2}} \prod_{j=1}^{r} \left[\int_{\mathbb{R}} \left(\frac{1}{\sqrt{2}} v_j + \frac{\lambda_{\alpha_j}}{\sqrt{2}} u_j \right)^{n_j} \left(\frac{1}{\sqrt{2}} v_j - \frac{\lambda_{\alpha_j}}{\sqrt{2}} u_j \right)^{m_j} \exp\left\{ -\frac{u_j^2}{2c} \right\} du_j \right]$$

where $v_j = \langle y, e_{\alpha_j} \rangle_H$ for $j = 1, 2, \cdots, r$. The last integral always exists because

$$\int_{\mathbb{R}} p(u) \exp\left\{ -\frac{u_j^2}{2c} \right\} du < \infty$$

for any polynomial function p. In addition, it is a polynomial in the variables

$$\langle y, e_{\alpha_1} \rangle_H, \cdots, \langle y, e_{\alpha_r} \rangle_H.$$

We next establish Equation (12). From Equation (7), we have

$$\delta_A f(x|u) = \frac{\partial}{\partial k} \prod_{j=1}^{r} (\langle x, e_{\alpha_j} \rangle_H + k \lambda_{\alpha_j} \langle u, e_{\alpha_j} \rangle_H)^{n_j} \bigg|_{k=0}$$
$$= \sum_{j=1}^{r} n_j \lambda_{\alpha_j} \langle u, e_{\alpha_j} \rangle_H f_j(x).$$

Finally, $\delta_A f$ is in $\mathcal{P}$ since $f_j \in \mathcal{P}$ for all $j = 1, 2 \cdots, r$. □

4. Homomorphism Properties and Basic Relationships

In this section, we establish some basic relationships among the Wiener transform, the convolution product and the first variation.

Theorem 3 tells us that the Wiener transform of the convolution product is the product of their Wiener transforms.

Theorem 3. *Let f_1, f_2, A, B and A' be as in Theorem 1. Then*

$$\mathcal{F}_{c,A,B}(f_1 * f_2)_A(y) = \mathcal{F}_{c,A,B}(f_1)\left(\frac{y}{\sqrt{2}}\right)\mathcal{F}_{c,A,B}(f_2)\left(\frac{y}{\sqrt{2}}\right). \tag{14}$$

Furthermore, under the hypothesis of Theorem 1, we have

$$(\mathcal{F}_{c,A,B}(f_1) * \mathcal{F}_{c,A,B}(f_2))_{A'}(y) = \mathcal{F}_{c,A,B}\left(f_1(\frac{\cdot}{\sqrt{2}})f_2(\frac{\cdot}{\sqrt{2}})\right)(y). \tag{15}$$

Proof. Using Equations (2), (6) and (11), we have

$$\mathcal{F}_{c,A,B}(f_1 * f_2)_A(y)$$

$$= \int_{\mathbf{H}'}\int_{\mathbf{H}'} f_1\left(\frac{Ax + By + Az}{\sqrt{2}}\right) f_2\left(\frac{Ax + By - Az}{\sqrt{2}}\right) dg_c(x) dg_c(z)$$

$$= \int_{\mathbf{H}'}\int_{\mathbf{H}'} \prod_{j=1}^{r}\left(\frac{\lambda_{\alpha_j}}{\sqrt{2}}\langle x, e_{\alpha_j}\rangle_\mathbf{H} + \frac{\mu_{\alpha_j}}{\sqrt{2}}\langle y, e_{\alpha_j}\rangle_\mathbf{H} + \frac{\lambda_{\alpha_j}}{\sqrt{2}}\langle z, e_{\alpha_j}\rangle_\mathbf{H}\right)^{n_j}$$

$$\times \prod_{j=1}^{r}\left(\frac{\lambda_{\alpha_j}}{\sqrt{2}}\langle x, e_{\alpha_j}\rangle_\mathbf{H} + \frac{\mu_{\alpha_j}}{\sqrt{2}}\langle y, e_{\alpha_j}\rangle_\mathbf{H} - \frac{\lambda_{\alpha_j}}{\sqrt{2}}\langle z, e_{\alpha_j}\rangle_\mathbf{H}\right)^{m_j} dg_c(x) dg_c(z)$$

$$= \left(\frac{1}{2\pi c}\right)^r \int_{\mathbb{R}^r}\int_{\mathbb{R}^r} \prod_{j=1}^{r}\left(\frac{\lambda_{\alpha_j}}{\sqrt{2}}u_j + \frac{\mu_{\alpha_j}}{\sqrt{2}}v_j + \frac{\lambda_{\alpha_j}}{\sqrt{2}}w_j\right)^{n_j}$$

$$\times \prod_{j=1}^{r}\left(\frac{\lambda_{\alpha_j}}{\sqrt{2}}u_j + \frac{\mu_{\alpha_j}}{\sqrt{2}}v_j - \frac{\lambda_{\alpha_j}}{\sqrt{2}}w_j\right)^{m_j} \exp\left\{-\sum_{j=1}^{r}\frac{u_j^2 + w_j^2}{2c}\right\} d\vec{u}\,d\vec{w}$$

where $v_j = \langle y, e_{\alpha_j}\rangle_\mathbf{H}$ for $j = 1, 2, \cdots, r$. Now let $u'_j = \frac{u_j + w_j}{\sqrt{2}}$ and $w'_j = \frac{u_j - w_j}{\sqrt{2}}$ for $j = 1, 2, \cdots, r$. Then we have

$$\mathcal{F}_{c,A,B}(f_1 * f_2)_A(y)$$

$$= \left(\frac{1}{2\pi c}\right)^r \int_{\mathbb{R}^r}\int_{\mathbb{R}^r} \prod_{j=1}^{r}\left(\lambda_{\alpha_j}u'_j + \frac{\mu_{\alpha_j}}{\sqrt{2}}v_j\right)^{n_j}$$

$$\times \prod_{j=1}^{r}\left(\lambda_{\alpha_j}w'_j + \frac{\mu_{\alpha_j}}{\sqrt{2}}v_j\right)^{m_j} \exp\left\{-\sum_{j=1}^{r}\frac{(u'_j)^2 + (w'_j)^2}{2c}\right\} d\vec{u}'\,d\vec{w}'$$

$$= \left(\frac{1}{2\pi c}\right)^{\frac{r}{2}} \int_{\mathbb{R}^r} \prod_{j=1}^{r}\left(\lambda_{\alpha_j}u'_j + \frac{\mu_{\alpha_j}}{\sqrt{2}}v_j\right)^{n_j} \exp\left\{-\sum_{j=1}^{r}\frac{(u'_j)^2}{2c}\right\} d\vec{u}'$$

$$\times \left(\frac{1}{2\pi c}\right)^{\frac{r}{2}} \int_{\mathbb{R}^r} \prod_{j=1}^{r}\left(\lambda_{\alpha_j}w'_j + \frac{\mu_{\alpha_j}}{\sqrt{2}}v_j\right)^{m_j} \exp\left\{-\sum_{j=1}^{r}\frac{(w'_j)^2}{2c}\right\} d\vec{w}'$$

where $v_j = \langle y, e_{\alpha_j}\rangle_\mathbf{H}$ for $j = 1, 2, \cdots, r$. Hence, using Equation (8), we can conclude that

$$\mathcal{F}_{c,A,B}(f_1 * f_2)_A(y) = \mathcal{F}_{c,A,B}(f_1)\left(\frac{y}{\sqrt{2}}\right)\mathcal{F}_{c,A,B}(f_2)\left(\frac{y}{\sqrt{2}}\right).$$

In addition, using Equation (9), we have

$$\mathcal{F}_{c,A',B'}(\mathcal{F}_{c,A,B}(f_1) * \mathcal{F}_{c,A,B}(f_2))_{A'}(y)$$
$$= \mathcal{F}_{c,A',B'}(\mathcal{F}_{c,A,B}(f_1))\left(\frac{y}{\sqrt{2}}\right)\mathcal{F}_{c,A',B'}(\mathcal{F}_{c,A,B}(f_2))\left(\frac{y}{\sqrt{2}}\right)$$
$$= f_1\left(\frac{y}{\sqrt{2}}\right) f_2\left(\frac{y}{\sqrt{2}}\right),$$

which yields Equation (15) as desired, where $\mathcal{F}_{c,A',B'}$ is as in Theorem 1. □

In our next theorem, we show that the Wiener transform and the first variation are commutable.

Theorem 4. *Let f be as in Theorem 1 and let A and B be as in Theorem 1. Let S be an operator on $\mathbf{H}$ with $Se_\alpha = \gamma_\alpha e_\alpha$ for $\alpha \in \mathcal{A}$. Then*

$$\delta_S \mathcal{F}_{c,A,B}(f)(y|u) = \mathcal{F}_{c,A,B}(\delta_{BS} f(\cdot|u))(y). \tag{16}$$

Proof. Using Equations (5) and (7), we have

$$\delta_S \mathcal{F}_{c,A,B}(f)(y|u)$$
$$= \frac{\partial}{\partial k}\mathcal{F}_{c,A,B}(f)(u+kSu)\Big|_{k=0}$$
$$= \frac{\partial}{\partial k}\int_{\mathbf{H}'} f(Ax+By+kBSu)dg_c(x)\Big|_{k=0}$$
$$= \frac{\partial}{\partial k}\int_{\mathbf{H}'} \prod_{j=1}^{r}(\lambda_{\alpha_j}\langle x, e_{\alpha_j}\rangle_{\mathbf{H}} + \mu_{\alpha_j}\langle y, e_{\alpha_j}\rangle_{\mathbf{H}} + k\mu_{\alpha_j}\gamma_{\alpha_j}\langle u, e_{\alpha_j}\rangle_{\mathbf{H}})^{n_j} dg_c(x)\Big|_{k=0}$$
$$= \sum_{j=1}^{r} n_j \mu_{\alpha_j} \gamma_{\alpha_j} \langle u, e_{\alpha_j}\rangle_{\mathbf{H}} \mathcal{F}_{c,A,B}(f_j)(y)$$

where f_j is as in Equation (13). We next use Equations (5) and (7) again to get

$$\mathcal{F}_{c,A,B}(\delta_S f(\cdot|u))(y)$$
$$= \int_{\mathbf{H}'} \frac{\partial}{\partial k} f(Ax+By+kSu)\Big|_{k=0} dg_c(x)$$
$$= \frac{\partial}{\partial k}\int_{\mathbf{H}'} \prod_{j=1}^{r}(\lambda_{\alpha_j}\langle x, e_{\alpha_j}\rangle_{\mathbf{H}} + \mu_{\alpha_j}\langle y, e_{\alpha_j}\rangle_{\mathbf{H}} + k\gamma_{\alpha_j}\langle u, e_{\alpha_j}\rangle_{\mathbf{H}})^{n_j} dg_c(x)\Big|_{k=0}$$
$$= \sum_{j=1}^{r} n_j \gamma_{\alpha_j} \langle u, e_{\alpha_j}\rangle_{\mathbf{H}} \mathcal{F}_{c,A,B}(f_j)(y)$$

where f_j is as in Equation (13). Comparing two expressions, we obtain Equation (16) as desired. □

From Equations (14) and (16) in Theorems 3 and 4, we have the following basic relationships.

Theorem 5. *Let f_1 and f_2 be as in Theorem 3. Let A and B as in Theorem 1 and let S as in Theorem 4. Then we have*

$$\delta(f_1 * f_2)_S(y|u) = (\delta f_1(\cdot|u/\sqrt{2}) * f_2)_S(y) + (f_1 * \delta f_2(\cdot|u/\sqrt{2}))_S(y), \tag{17}$$

$$\begin{aligned}&\mathcal{F}_{c,A,B}(\delta_{BS}f_1(\cdot|u) * \delta_{BS}f_2(\cdot|u))_A(y)\\&= \delta_S\mathcal{F}_{c,A,B}f_1(y/\sqrt{2}|u)\delta_S\mathcal{F}_{c,A,B}f_2(y/\sqrt{2}|u),\end{aligned} \quad (18)$$

$$\begin{aligned}&\mathcal{F}_{c,A,B}(\delta_{BS}(f_1 * f_2)_A(\cdot|u))(y)\\&= \delta_S(\mathcal{F}_{c,A,B}f_1(\cdot/\sqrt{2})\mathcal{F}_{c,A,B}f_2(\cdot/\sqrt{2}))(y|u)\\&= \delta_S\mathcal{F}_{c,A,B}(f_1 * f_2)_A(y|u)\end{aligned} \quad (19)$$

and

$$(\mathcal{F}_{c,A,B}\delta_{BS}f_1(\cdot|u) * \mathcal{F}_{c,A,B}\delta_{BS}f_2(\cdot|u))_A(z) = (\delta_S\mathcal{F}_{c,A,B}f_1(\cdot|u) * \delta_S\mathcal{F}_{c,A,B}f_2(\cdot|u))_A(y). \quad (20)$$

Proof. We first note that Equation (17) follows directly from the definition of the first variation given by (7). Next we note that Equations (18) and (19) follow from Equations (14)–(16). Finally we note that Equation (20) follows immediately from Equations (14) and (16). □

5. Integration by Parts Formula with an Application

In this section, we obtain an integration by part formula, and give an application with respect to the Wiener transform.

Since the Lebesgue measure m_L on $\mathbb{R}^r$ is an uniform measure and so we see that

$$\int_{\mathbb{R}^r} h(\vec{u} + \vec{v}) dm_L(\vec{u}) = \int_{\mathbb{R}^r} h(\vec{w}) dm_L(\vec{w})$$

by substitution for $w_j = u_j + v_j$ for $j = 1, 2, \cdots, r$ if the integrals exist. It is called the translation theorem for the Lebesgue integrals. However, the distribution measure g_c used in this paper is the Gaussian measure and hence, in generally,

$$\int_{\mathbf{H}'} h(x+y) dg_c(x) \neq \int_{\mathbf{H}'} h(z) dg_c(z)$$

even if the integrals exist, see [10–14]. For this reason, a different form of formula is obtained in this paper.

Lemma 1. *Let s be a non-negative integer and let p be a function on $\mathbf{H}$ defined by the formula*

$$p(x) = \langle x, e_\alpha \rangle_\mathbf{H}^s \quad (21)$$

for some $e_\alpha \in \mathcal{B}$. Then for all $x_0 \in \mathbf{H}'$,

$$\begin{aligned}&\int_{\mathbf{H}'} p(x + x_0) dg_c(x)\\&= \exp\left\{-\frac{1}{2c}\langle x_0, e_\alpha\rangle_\mathbf{H}^2\right\} \int_\mathbf{H} p(x) \exp\left\{\frac{1}{c}\langle x, e_\alpha\rangle_\mathbf{H}\langle x_0, e_\alpha\rangle_{\mathbf{H}'}\right\} dg_c(x).\end{aligned} \quad (22)$$

Proof. We set $v = \langle x_0, e_\alpha \rangle_{\mathbf{H}}$. Using Equations (8) and (21), we have

$$\int_{\mathbf{H}'} p(x+x_0) dg_c(x)$$

$$= \left(\frac{1}{2\pi c}\right)^{\frac{1}{2}} \int_{\mathbb{R}} (u+v)^s \exp\left\{-\frac{u^2}{2c}\right\} du$$

$$= \left(\frac{1}{2\pi c}\right)^{\frac{1}{2}} \int_{\mathbb{R}} w^s \exp\left\{-\frac{(w-v)^2}{2c}\right\} dw$$

$$= \exp\left\{-\frac{1}{2c}v^2\right\} \left(\frac{1}{2\pi c}\right)^{\frac{1}{2}} \int_{\mathbb{R}} w^s \exp\left\{-\frac{w^2}{2c} + \frac{1}{c}vw\right\} dw$$

$$= \exp\left\{-\frac{1}{2c}\langle x_0, e_\alpha \rangle_{\mathbf{H}}^2\right\} \int_{\mathbf{H}'} p(x) \exp\left\{\frac{1}{c}\langle x, e_\alpha \rangle_{\mathbf{H}} \langle x_0, e_\alpha \rangle_{\mathbf{H}}\right\} dg_c(x).$$

Hence, we have the desired result. □

In Theorem 6, we obtain a translation theorem for **H**-integrals.

Theorem 6 (Translation theorem for **H**-integrals). *Let f be as in Equation (4) and let $x_0 \in \mathbf{H}'$. Then*

$$\int_{\mathbf{H}'} f(x+x_0) dg_c(x)$$
$$= \exp\left\{-\frac{1}{2c} \sum_{j=1}^r \langle x_0, e_\alpha \rangle_{\mathbf{H}}^2\right\} \int_{\mathbf{H}'} f(x) \exp\left\{\frac{1}{c} \sum_{j=1}^r \langle x, e_{\alpha_j} \rangle_{\mathbf{H}} \langle x_0, e_{\alpha_j} \rangle_{\mathbf{H}}\right\} dg_c(x). \quad (23)$$

Proof. First, by using fact that

$$\int_{\mathbf{H}'} f(x) dg_c(x) = \int_{\mathbf{H}'} \langle x, e_{\alpha_1} \rangle_{\mathbf{H}}^{n_1} dg_c(x) \cdots \int_{\mathbf{H}'} \langle x, e_{\alpha_r} \rangle_{\mathbf{H}}^{n_r} dg_c(x),$$

and Equation (22) in Lemma 1 we can establish Equation (23) as desired. □

The following theorem is one of main results in this paper.

Theorem 7 (Integration by parts formula). *Let f be as in Theorem 6 and let S be as in Theorem 4. Then*

$$c \int_{\mathbf{H}'} \delta_S f(x|u) dg_c(x)$$
$$= c \int_{\mathbf{H}'} f(x) dg_c(x) + \int_{\mathbf{H}'} f(x) \sum_{j=1}^r \gamma_{\alpha_j} \langle x, e_{\alpha_j} \rangle_{\mathbf{H}} \langle u, e_{\alpha_j} \rangle_{\mathbf{H}} dg_c(x). \quad (24)$$

Proof. Using Equations (1) and (7), we have

$$\int_{\mathbf{H}'} \delta_S f(x|u) dg_c(x)$$

$$= \frac{\partial}{\partial k} \int_{\mathbf{H}'} f(x+kSu) dg_c(x) \Big|_{k=0}$$

$$= \frac{\partial}{\partial k} \left[\exp\left\{-\frac{k^2}{2c} \sum_{j=1}^r \gamma_{\alpha_j}^2 \langle u, e_\alpha \rangle_{\mathbf{H}}^2\right\}\right.$$

$$\left. \times \int_{\mathbf{H}'} f(x) \exp\left\{\frac{k}{c} \sum_{j=1}^r \gamma_{\alpha_j} \langle x, e_{\alpha_j} \rangle_{\mathbf{H}} \langle u, e_{\alpha_j} \rangle_{\mathbf{H}}\right\} dg_c(x)\right]\Big|_{k=0}$$

$$= \int_{\mathbf{H}'} f(x) dg_c(x) + \frac{1}{c} \int_{\mathbf{H}'} f(x) \sum_{j=1}^r \gamma_{\alpha_j} \langle x, e_{\alpha_j} \rangle_{\mathbf{H}} \langle u, e_{\alpha_j} \rangle_{\mathbf{H}} dg_c(x),$$

which yields Equation (24) as desired. □

Finally, we give an application of Theorem 7.

Theorem 8 (Application of Theorem 7). *Let f and S be as in Theorem 7. Let A and B as in Theorem 5. Then*

$$c\mathcal{F}_{c,A,B}(\delta_A f(\cdot|u))(y) = c\mathcal{F}_{c,A,B}(f)(y) + \int_{\mathbf{H}'} f(Ax + By) \sum_{j=1}^{r} \gamma_{\alpha_j} \langle x, e_{\alpha_j} \rangle_{\mathbf{H}} \langle u, e_{\alpha_j} \rangle_{\mathbf{H}} dg_c(x). \qquad (25)$$

Proof. Using Equations (5) and (7), we have

$$\mathcal{F}_{c,A,B}(\delta_A f(\cdot|u))(y) = \frac{\partial}{\partial k} \int_{\mathbf{H}'} f(Ax + By + kAu) dg_c(x) \bigg|_{k=0}.$$

Now, let $f_y(x) = f(x + y)$ and $f^A(x) = f(Ax)$. Then

$$f(Ax + By + kAu) = (f_{By})^A(x + ku)$$

and, hence, using Equation (24) by replacing f with $(f_{By})^A$, we have

$$\mathcal{F}_{c,A,B}(\delta_A f(\cdot|u))(y)$$
$$= \frac{\partial}{\partial k} \int_{\mathbf{H}'} (f_{By})^A(x + ku) dg_c(x) \bigg|_{k=0}$$
$$= \int_{\mathbf{H}'} (f_{By})^A(x) dg_c(x) + \frac{1}{c} \int_{\mathbf{H}'} (f_{By})^A(x) \sum_{j=1}^{r} \gamma_{\alpha_j} \langle x, e_{\alpha_j} \rangle_{\mathbf{H}} \langle u, e_{\alpha_j} \rangle_{\mathbf{H}} dg_c(x)$$
$$= \mathcal{F}_{c,A,B}(f)(y) + \frac{1}{c} \int_{\mathbf{H}'} f(Ax + By) \sum_{j=1}^{r} \gamma_{\alpha_j} \langle x, e_{\alpha_j} \rangle_{\mathbf{H}} \langle u, e_{\alpha_j} \rangle_{\mathbf{H}} dg_c(x).$$

Hence, we have the desired results. □

6. Applications

In this section, we give some applications to apply our fundamental formulas obtained in previous sections.

6.1. Application of Theorem 3

We first give an application to illustrate the usefulness of Equations (14) and (15) in Theorem 3.

Example 1. *Let $r = 2$. Let $f_1(x) = \langle x, e_{\alpha_2} \rangle^2$ and let $f_2(x) = \langle x, e_{\alpha_1} \rangle^2 \langle x, e_{\alpha_2} \rangle$. Let A and B be as in Theorem 3. From Equation (8) we have*

$$\mathcal{F}_{c,A,B}(f_1)(y) = \lambda_{\alpha_2}^2 \langle y, e_{\alpha_2} \rangle_{\mathbf{H}}^2 + 2c\mu_{\alpha_2}^2$$

and

$$\mathcal{F}_{c,A,B}(f_2)(y) = [\lambda_{\alpha_1}^2 \langle y, e_{\alpha_1} \rangle_{\mathbf{H}}^2 + 2c\mu_{\alpha_1}^2][\mu_{\alpha_2} \langle y, e_{\alpha_2} \rangle_{\mathbf{H}}].$$

Hence, using Equation (14), we have

$$\mathcal{F}_{c,A,B}(f_1 * f_2)_A(y)$$
$$= \left[\frac{\lambda_{\alpha_2}^2}{2} \langle y, e_{\alpha_2} \rangle_{\mathbf{H}}^2 + 2c\mu_{\alpha_2}^2\right]\left[\frac{\lambda_{\alpha_1}^2}{2} \langle y, e_{\alpha_1} \rangle_{\mathbf{H}}^2 + 2c\mu_{\alpha_1}^2\right]\left[\frac{\mu_{\alpha_2}}{\sqrt{2}} \langle y, e_{\alpha_2} \rangle_{\mathbf{H}}\right]. \qquad (26)$$

Furthermore, we note that
$$f_1(x)f_2(x) = \langle x, e_{\alpha_1}\rangle^2 \langle x, e_{\alpha_2}\rangle^3$$

and so

$$\mathcal{F}_{c,A,B}\left(f_1(\frac{\cdot}{\sqrt{2}})f_2(\frac{\cdot}{\sqrt{2}})\right)(y)$$
$$= \left[\frac{\lambda_{\alpha_1}^2}{2}\langle y, e_{\alpha_1}\rangle_{\mathbf{H}}^2 + 2c\mu_{\alpha_1}^2\right]\left[\frac{\lambda_{\alpha_2}^3}{2}\langle y, e_{\alpha_2}\rangle_{\mathbf{H}}^3 + \frac{3c(\mu_{\alpha_2}^2 + \lambda_{\alpha_2})}{\sqrt{2}}\langle y, e_{\alpha_2}\rangle_{\mathbf{H}}\right].$$

Hence, using Equation (15), we have

$$(\mathcal{F}_{c,A,B}(f_1) * \mathcal{F}_{c,A,B}(f_2))_{A'}(y)$$
$$= \left[\frac{\lambda_{\alpha_1}^2}{2}\langle y, e_{\alpha_1}\rangle_{\mathbf{H}}^2 + 2c\mu_{\alpha_1}^2\right]\left[\frac{\lambda_{\alpha_2}^3}{2}\langle y, e_{\alpha_2}\rangle_{\mathbf{H}}^3 + \frac{3c(\mu_{\alpha_2}^2 + \lambda_{\alpha_2})}{\sqrt{2}}\langle y, e_{\alpha_2}\rangle_{\mathbf{H}}\right].$$

These tell us that the Wiener transform of convolution product and the convolution product of Wiener transforms can be calculated without concept of convolution product very easily.

6.2. Application of Theorem 5

We next give an application of Equation (19) in Theorem 5.

Example 2. *Let f_1, f_2, A and B be as in Example 1. Using Equation (26), we have*

$$\delta_S \mathcal{F}_{c,A,B}(f_1 * f_2)_A(y|u)$$
$$= \frac{\partial}{\partial k}\mathcal{F}_{c,A,B}(f_1 * f_2)_A(y + kSu)\bigg|_{k=0}$$
$$= \frac{\partial}{\partial k}\left(\left[\frac{\lambda_{\alpha_2}^2}{2}(\langle y, e_{\alpha_2}\rangle_{\mathbf{H}} + k\gamma_{\alpha_2}\langle u, e_{\alpha_2}\rangle_{\mathbf{H}})^2 + 2c\mu_{\alpha_2}^2\right]\right.$$
$$\times \left[\frac{\lambda_{\alpha_1}^2}{2}(\langle y, e_{\alpha_1}\rangle_{\mathbf{H}} + k\gamma_{\alpha_1}\langle u, e_{\alpha_1}\rangle_{\mathbf{H}})^2 + 2c\mu_{\alpha_1}^2\right]$$
$$\left.\times \left[\frac{\mu_{\alpha_2}}{\sqrt{2}}(\langle y, e_{\alpha_2}\rangle_{\mathbf{H}} + k\gamma_{\alpha_2}\langle u, e_{\alpha_2}\rangle_{\mathbf{H}})\right]\right)\bigg|_{k=0}$$
$$= \lambda_{\alpha_2}^2\langle u, e_{\alpha_2}\rangle_{\mathbf{H}}\left[\frac{\lambda_{\alpha_2}^2}{2}\langle y, e_{\alpha_2}\rangle_{\mathbf{H}}^2 + 2c\mu_{\alpha_2}^2\right]\left[\frac{\lambda_{\alpha_1}^2}{2}\langle y, e_{\alpha_1}\rangle_{\mathbf{H}}^2 + 2c\mu_{\alpha_1}^2\right]\left[\frac{\mu_{\alpha_2}}{\sqrt{2}}\langle y, e_{\alpha_2}\rangle_{\mathbf{H}}\right]$$
$$+ \lambda_{\alpha_1}^2\langle u, e_{\alpha_1}\rangle_{\mathbf{H}}\left[\frac{\lambda_{\alpha_2}^2}{2}\langle y, e_{\alpha_2}\rangle_{\mathbf{H}}^2 + 2c\mu_{\alpha_2}^2\right]\left[\frac{\lambda_{\alpha_1}^2}{2}\langle y, e_{\alpha_1}\rangle_{\mathbf{H}}^2 + 2c\mu_{\alpha_1}^2\right]\left[\frac{\mu_{\alpha_2}}{\sqrt{2}}\langle y, e_{\alpha_2}\rangle_{\mathbf{H}}\right]$$
$$+ \frac{\mu_{\alpha_2}}{\sqrt{2}}\langle u, e_{\alpha_1}\rangle_{\mathbf{H}}\left[\frac{\lambda_{\alpha_2}^2}{2}\langle y, e_{\alpha_2}\rangle_{\mathbf{H}}^2 + 2c\mu_{\alpha_2}^2\right]\left[\frac{\lambda_{\alpha_1}^2}{2}\langle y, e_{\alpha_1}\rangle_{\mathbf{H}}^2 + 2c\mu_{\alpha_1}^2\right]\left[\frac{\mu_{\alpha_2}}{\sqrt{2}}\langle y, e_{\alpha_2}\rangle_{\mathbf{H}}\right].$$

Using this, we obtain that

$$\mathcal{F}_{c,A,B}(\delta_{BS}(f_1 * f_2)_A(\cdot|u))(y)$$
$$= \lambda_{\alpha_2}^2\langle u, e_{\alpha_2}\rangle_{\mathbf{H}}\left[\frac{\lambda_{\alpha_2}^2}{2}\langle y, e_{\alpha_2}\rangle_{\mathbf{H}}^2 + 2c\mu_{\alpha_2}^2\right]\left[\frac{\lambda_{\alpha_1}^2}{2}\langle y, e_{\alpha_1}\rangle_{\mathbf{H}}^2 + 2c\mu_{\alpha_1}^2\right]\left[\frac{\mu_{\alpha_2}}{\sqrt{2}}\langle y, e_{\alpha_2}\rangle_{\mathbf{H}}\right]$$
$$+ \lambda_{\alpha_1}^2\langle u, e_{\alpha_1}\rangle_{\mathbf{H}}\left[\frac{\lambda_{\alpha_2}^2}{2}\langle y, e_{\alpha_2}\rangle_{\mathbf{H}}^2 + 2c\mu_{\alpha_2}^2\right]\left[\frac{\lambda_{\alpha_1}^2}{2}\langle y, e_{\alpha_1}\rangle_{\mathbf{H}}^2 + 2c\mu_{\alpha_1}^2\right]\left[\frac{\mu_{\alpha_2}}{\sqrt{2}}\langle y, e_{\alpha_2}\rangle_{\mathbf{H}}\right]$$
$$+ \frac{\mu_{\alpha_2}}{\sqrt{2}}\langle u, e_{\alpha_1}\rangle_{\mathbf{H}}\left[\frac{\lambda_{\alpha_2}^2}{2}\langle y, e_{\alpha_2}\rangle_{\mathbf{H}}^2 + 2c\mu_{\alpha_2}^2\right]\left[\frac{\lambda_{\alpha_1}^2}{2}\langle y, e_{\alpha_1}\rangle_{\mathbf{H}}^2 + 2c\mu_{\alpha_1}^2\right]\left[\frac{\mu_{\alpha_2}}{\sqrt{2}}\langle y, e_{\alpha_2}\rangle_{\mathbf{H}}\right]$$
$$= \left(\lambda_{\alpha_2}^2\langle u, e_{\alpha_2}\rangle_{\mathbf{H}} + \lambda_{\alpha_1}^2\langle u, e_{\alpha_1}\rangle_{\mathbf{H}} + \frac{\mu_{\alpha_2}}{\sqrt{2}}\langle u, e_{\alpha_1}\rangle_{\mathbf{H}}\right)\mathcal{F}_{c,A,B}(f_1 * f_2)_A(y).$$

6.3. Application of Theorem 7

We finish this paper by giving an application of Equation (25) in Theorem 7. Equation (25) tells us that

$$\int_{\mathbf{H}'} f(Ax + By) \sum_{j=1}^{r} \gamma_{\alpha_j} \langle x, e_{\alpha_j} \rangle_{\mathbf{H}} \langle u, e_{\alpha_j} \rangle_{\mathbf{H}} dg_c(x) \qquad (27)$$
$$= c\mathcal{F}_{c,A,B}(\delta_A f(\cdot|u))(y) - c\mathcal{F}_{c,A,B}(f)(y).$$

The left-hand side of Equation (27) contains some polynomial-weight and so it is not easy to calculate. However, by using Equation (27), we can calculate it very easy via the Wiener transform and the first variation. We shall explain this as example.

Example 3. *Let f_1, f_2, A and B be as in Example 1. Then we have*

$$\mathcal{F}_{c,A,B}(\delta_A f_1(\cdot|u))(y) = 2\mu_{\alpha_2} \lambda_{\alpha_2} \langle y, e_{\alpha_2} \rangle_{\mathbf{H}}$$

and

$$\mathcal{F}_{c,A,B}(f_1)(y) = \lambda_{\alpha_2}^2 \langle y, e_{\alpha_2} \rangle_{\mathbf{H}}^2 + 2c\mu_{\alpha_2}^2.$$

Hence, using Equation (27), we obtain that

$$\int_{\mathbf{H}'} [\mu_{\alpha_2} \langle x, e_{\alpha_2} \rangle_{\mathbf{H}} + \lambda_{\alpha_2} \langle y, e_{\alpha_2} \rangle_{\mathbf{H}}]^2 \mu_{\alpha_2} \langle x, e_{\alpha_2} \rangle_{\mathbf{H}} \langle u, e_{\alpha_2} \rangle_{\mathbf{H}} dg_c(x)$$
$$= 2c\mu_{\alpha_2} \lambda_{\alpha_2} \langle y, e_{\alpha_2} \rangle_{\mathbf{H}} - c\lambda_{\alpha_2}^2 \langle y, e_{\alpha_2} \rangle_{\mathbf{H}}^2 + 2c^2 \mu_{\alpha_2}^2.$$

In addition, we have

$$\mathcal{F}_{c,A,B}(\delta_A f_2(\cdot|u))(y) = 2c\mu_{\alpha_1}^3 \lambda_{\alpha_2} \langle u, e_{\alpha_1} \rangle_{\mathbf{H}} \langle y, e_{\alpha_2} \rangle_{\mathbf{H}} + 2\mu_{\alpha_1}^3 \lambda_{\alpha_1}^2 \langle u, e_{\alpha_1} \rangle_{\mathbf{H}} \langle y, e_{\alpha_1} \rangle_{\mathbf{H}}$$
$$+ \mu_{\alpha_2} \langle u, e_{\alpha_2} \rangle_{\mathbf{H}} (c\mu_{\alpha_1}^2 + \lambda_{\alpha_1}^2 \langle y, e_{\alpha_1} \rangle_{\mathbf{H}}^2).$$

and

$$\mathcal{F}_{c,A,B}(f_2)(y) = [\lambda_{\alpha_1}^2 \langle y, e_{\alpha_1} \rangle_{\mathbf{H}}^2 + 2c\mu_{\alpha_1}^2][\mu_{\alpha_2} \langle y, e_{\alpha_2} \rangle_{\mathbf{H}}].$$

Thus, from Equation (27) we conclude that

$$\int_{\mathbf{H}'} [\mu_{\alpha_1} \langle x, e_{\alpha_1} \rangle_{\mathbf{H}} + \lambda_{\alpha_1} \langle y, e_{\alpha_1} \rangle_{\mathbf{H}}]^2$$
$$\times [\mu_{\alpha_2} \langle x, e_{\alpha_2} \rangle_{\mathbf{H}} + \lambda_{\alpha_2} \langle y, e_{\alpha_2} \rangle_{\mathbf{H}}] \sum_{j=1}^{2} \gamma_{\alpha_j} \langle x, e_{\alpha_j} \rangle_{\mathbf{H}} \langle u, e_{\alpha_j} \rangle_{\mathbf{H}} dg_c(x)$$
$$= 2c^2 \mu_{\alpha_1}^3 \lambda_{\alpha_2} \langle u, e_{\alpha_1} \rangle_{\mathbf{H}} \langle y, e_{\alpha_2} \rangle_{\mathbf{H}} + 2c\mu_{\alpha_1}^3 \lambda_{\alpha_1}^2 \langle u, e_{\alpha_1} \rangle_{\mathbf{H}} \langle y, e_{\alpha_1} \rangle_{\mathbf{H}}$$
$$+ c\mu_{\alpha_2} \langle u, e_{\alpha_2} \rangle_{\mathbf{H}} (c\mu_{\alpha_1}^2 + \lambda_{\alpha_1}^2 \langle y, e_{\alpha_1} \rangle_{\mathbf{H}}^2)$$
$$- c[\lambda_{\alpha_1}^2 \langle y, e_{\alpha_1} \rangle_{\mathbf{H}}^2 + 2c\mu_{\alpha_1}^2][\mu_{\alpha_2} \langle y, e_{\alpha_2} \rangle_{\mathbf{H}}].$$

7. Conclusions

According to some results and formula in previous papers [1–3,7–9,15] and our results and formulas in previous Sections 3–5, we note that all results can be explained by the eigenvalue of operators on Hilbert space. As you can see from the results of the previous Sections 3–5, we are able to obtain various relationships that are not found in the previous research results. We also see in Section 6 that our results can be applied to various functions in the application of various fields. Therefore, it can be seen that the results in this paper are structured in a generalized form.

Funding: This research received no external funding.

Institutional Review Board Statement: Not applicable.

Informed Consent Statement: Not applicable.

Data Availability Statement: The study did not report any data.

Acknowledgments: The author would like to express gratitude to the referees for their valuable comments and suggestions, which have improved the original paper.

Conflicts of Interest: The author declares that there is no conflict of interests regarding the publication of this article.

References

1. Segal, I.E. Tensor algebra over Hilbert spaces I. *Tran. Am. Math. Soc.* **1956**, *81*, 106–134. [CrossRef]
2. Segal, I.E. Tensor algebra over Hilbert spaces II. *Ann. Math.* **1956**, *63*, 160–175. [CrossRef]
3. Segal, I.E. Distributions in Hilbert space and canonical systems of operators. *Trans. Am. Math. Soc.* **1958**, *88*, 12–41. [CrossRef]
4. Hida, T. *Stationary Stochastic Process*; Series on Mathematical Notes; Princeton University Press: Princeton, NJ, USA; University of Tokyo Press: Tokyo, Japan, 1970.
5. Hida, T. *Brownian Motion*; Series on Applications of Mathematics; Springer: Berlin/Heidelberg, Germany, 1980.
6. Negrin, E.R. Integral representation of the second quantization via Segal duality transform. *J. Funct. Anal.* **1996**, *141*, 37–44. [CrossRef]
7. Hayker, N.; Gonzalez, B.J.; Negrin, E.R. Matrix Wiener transform. *Appl. Math. Comput.* **2011**, *218*, 773–776.
8. Hayker, N.; Gonzalez, B.J.; Negrin, E.R. The second quantization and its general integral finite-dimensional representation. *Integ. Trans. Spec. Funct.* **2002**, *13*, 373–378.
9. Hayker, N.; Srivastava, H.M.; Gonzalez, B.J.; Negrin, E.R. A family of Wiener transforms associated with a pair of operators on Hilbert space. *Integral Trans. Spec. Funct.* **2012**, *24*, 1–8.
10. Chang, S.J.; Choi, J.G. Analytic Fourier-Feynman transforms and convolution products associated with Gaussian processes on Wiener space. *Banach J. Math.* **2017**, *11*, 785–807. [CrossRef]
11. Chung, H.S.; Chang, S.J. Some Applications of the Spectral Theory for the Integral Transform Involving the Spectral Representation. *J. Funct. Spaces* **2012**, *2012*, 573602. [CrossRef]
12. Chung, H.S. Generalized integral transforms via the series expressions. *Mathematics* **2020**, *8*, 539. [CrossRef]
13. Lee, Y.J. Integral transforms of analytic functions on abstract Wiener spaces. *J. Funct. Anal.* **1982**, *47*, 153–164. [CrossRef]
14. Lee, Y.J. Unitary Operators on the Space of L^2-Functions over Abstract Wiener Spaces. *Soochow J. Math.* **1987**, *13*, 165–174.
15. Chung, H.S.; Lee, I.Y. Relationships between the $*_w$-product and the generalized integral transforms. *Integ. Trans. Spec. Funct.* **2021**, 1–14. [CrossRef]

Article

Mathematical Description and Laboratory Study of Electrophysical Methods of Localization of Geodeformational Changes during the Control of the Railway Roadbed

Artem Bykov [1,*], Anastasia Grecheneva [2], Oleg Kuzichkin [3], Dmitry Surzhik [3], Gleb Vasilyev [3] and Yerbol Yerbayev [4]

1. Department of Data Analysis and Machine Learning, Financial University under the Government of the Russian Federation, Leningrad Avenue 49, 125993 Moscow, Russia
2. Department of Applied Mathematics, Russian State Agrarian University—Moscow Timiryazev Agricultural Academy, Timiryazevskaya Street 49, 127550 Moscow, Russia; grechenevaav@yandex.com
3. Department of Informational and Robototechnical Systems, Belgorod National Research University, Pobedy Street 85, 308015 Belgorod, Russia; oldkuz@yandex.ru (O.K.); arzerum@mail.ru (D.S.); vasilievgleb@yandex.ru (G.V.)
4. Higher School of Mechanical Engineering, Zhangir Khan West Kazakhstan Agrarian-Technical University, Zhangir Khan Street 51, Uralsk 090009, Kazakhstan; Erbol.Erbaev@mail.ru
* Correspondence: ArAbykov@fa.ru

Citation: Bykov, A.; Grecheneva, A.; Kuzichkin, O.; Surzhik, D.; Vasilyev, G.; Yerbayev, Y. Mathematical Description and Laboratory Study of Electrophysical Methods of Localization of Geodeformational Changes during the Control of the Railway Roadbed. *Mathematics* **2021**, *9*, 3164. https://doi.org/10.3390/math9243164

Academic Editor: Francesco Mainardi

Received: 19 October 2021
Accepted: 6 December 2021
Published: 8 December 2021

Publisher's Note: MDPI stays neutral with regard to jurisdictional claims in published maps and institutional affiliations.

Copyright: © 2021 by the authors. Licensee MDPI, Basel, Switzerland. This article is an open access article distributed under the terms and conditions of the Creative Commons Attribution (CC BY) license (https://creativecommons.org/licenses/by/4.0/).

Abstract: Currently, the load on railway tracks is increasing due to the increase in freight traffic. Accordingly, more and more serious requirements are being imposed on the reliability of the roadbed, which means that studies of methods for monitoring the integrity of the railway roadbed are relevant. The article provides a mathematical substantiation of the possibility of using seismoelectric and phasemetric methods of geoelectric control of the roadbed of railway tracks in order to identify defects and deformations at an early stage of their occurrence. The methods of laboratory modeling of the natural–technical system "railway track" are considered in order to assess the prospects of using the presented methods. The results of laboratory studies are presented, which have shown their high efficiency in registering a weak useful electrical signal caused by seismoacoustic effects against the background of high-level external industrial and natural interference. In the course of laboratory modeling, it was found that on the amplitude spectra of the output electrical signals of the investigated geological medium in the presence of an elastic harmonic action with a frequency of 70 Hz, the frequency of a harmonic electrical signal with a frequency of 40 Hz is observed. In laboratory modeling, phase images were obtained for the receiving line when simulating the process of sinking of the soil base of the railway bed, confirming the presence of a transient process that causes a shift in the initial phase of the signal $\Delta\varphi = 40°$ by ~45° ($\Delta\varphi' = 85°$), which allows detection of the initial stage of failure formation.

Keywords: railway transport; roadbed; geodynamic processes; seismoelectric method; stress–strain process; transfer functions; frequency characteristics; phasometric method; laboratory modeling

1. Introduction

The railway is one of the most important cargo transportation systems in the world due to the rapid development of this class of heavy transport, as well as its efficiency in comparison with other transport systems. At the same time, the increase in railway maintenance costs is directly related to the increase in the volume of cargo transportation [1]. Accordingly, in this regard, the requirements for the reliability and safety of the functioning of railway tracks and railway infrastructure in general are significantly increasing, the provision of which through technical monitoring is extremely important and relevant, and, as a result, is the focus of plentiful scientific and applied research [2–4].

This task is complicated by the fact that special requirements for the construction and operation of railways are imposed in areas of activation of geodynamic processes, where there is a significant risk of landslides and the formation of karst craters, which are associated with intensive cyclic impact on the ground base of passing trains [4]. According to the Richter scale, the efficiency of such transport vibration is equivalent to an earthquake of 3–6 points [5]. Moreover, it is known that landslide deformations affect natural and artificial geological environments and can have a varied nature of development, and the activation of karst processes in the area of railway tracks can lead to a gradual destruction of the subgrade (Figure 1) [4]. In addition, the reasons for the deformation of the roadbed may be the discrepancy between the power of the upper structure of the track to the intensive dynamic loads of railway transport, as well as the unfavorable effects of climatic and engineering–geological factors.

It should be noted that the main undesirable activation of near-surface geodynamics arises not at the stage of engineering surveys and construction works, but in the process of direct operation of railway tracks and in most cases is of a sudden, spontaneous nature.

Figure 1. Sudden destruction of the railway roadbed as a result of natural factors.

As a result, the indicated defects and deformations of the subgrade lead to the transition of the "railway track-subsoil" system into an unstable state, the moment of occurrence of which is not predicted by modern control systems. Geodeformational changes in the soil lead to a significant deterioration of the railway track, which, in turn, leads to the need to reduce the negative impact of geodeformational changes on the railway track, since the quality and condition of the track directly affects the safety of the movement of trains [3] and the efficiency of its interaction with infrastructure [4]. To detect the early stage of the development of deformation processes of the roadbed of a railway track and predict the dynamics of their development in time, it is important to have current and model information about its state and possible undesirable changes.

Thus, the purpose of the research is to develop methods for monitoring geodynamic changes in the railway roadbed, to detect the appearance of various kinds of anomalies and inhomogeneities, indicating the development of destructive processes in the soil, to create and test a laboratory model of the natural–technical system "railway track" in order to assess the prospects of using the presented methods.

2. Literature Review

The most reliable method of monitoring the condition of the railway roadbed at the moment is surveying and geodetic observations on the reference points of profile lines [6]. However, due to the large length of railway tracks, the use of this method is very difficult, and the control of the roadbed and the adjacent territory by engineering–

geological methods are inappropriate. In this case, it is relevant to attract geophysical methods that are widely used in exploration and engineering geology.

Currently, to obtain information about the structure of the upper layers of various geological environments, ground-penetrating radar sounding [7], vibroseismic methods [8–11], electrometric [9–12] and some other methods are traditionally used. But in some cases, such as when karst cavities have a developed structure, analysis by groups of these methods is very difficult [13,14].

According to the results of numerous studies, it has been established that when organizing automated control of geodynamic objects, the most promising is the use of geoelectric methods of media sounding. They provide effective observations of geological objects, as well as assessment of the state and forecast of their development, which is determined by their high technology [8,10–12,15].

However, as practice shows, the application in geodynamic monitoring of any single method chosen from the ones considered above is not effective enough. As a result, it is necessary to choose the most preferred research method for each specific task [16]. For instance, some research has been done on dynamic prediction models for tunnels [17,18], railway tracks [19], oil sludge straits [20], as well as various underground structures [21] and geotechnical applications in general [22]. But the usual problem of such methods is the ambiguity of the assessment of geophysical data.

The joint use of geoelectric and seismic methods, that is, the use of the seismoelectric control method [23–29], allows increasing the efficiency of geological media studies by reducing the ambiguity of the assessment of geophysical data.

In the course of laboratory modeling, it is planned to study the amplitude spectra of the output electrical signals of the monitored geological environment in the presence of elastic harmonic action, which will allow using this effect to obtain more detailed information about the structure of the soil during sounding by electric fields. In this case, the registration of the phase component of the signal of the receiving lines will presumably increase the noise immunity of measurements. The presence of a transient process causing a shift in the initial phase of the signal of the receiving lines of the installation will indicate the initial stage of the dip formation.

In this regard, the purpose of the work is to substantiate and study an integrated approach to solving problems of monitoring the subsurface of railway tracks and the adjacent territory to identify the initial stage of defects and deformations in it based on the use of the seismoelectric method.

3. Mathematical Description of the Seismoelectric Method of Monitoring the Ground Bed of Railway Tracks

The seismoelectric control method is based on secondary seismic effects. It consists of the interpretation of signals of an electrical or seismic nature, recorded by the geodynamic monitoring system, which are received when vibrations of these types are simultaneously excited in the studied medium. The method is based on the assumption that the real geological environment is a porous polyphase complex structure in an energetically unstable state. The combined effect of physical fields of various nature (electromagnetic and elastic) on the studied geological environment can lead to a change in its physical properties. The seismoelectric effect of the first kind is the phenomenon of changing the electrical resistance of the geological medium under the influence of elastic vibrations; the second kind is the phenomenon of excitation of an electromagnetic field that occurs under similar conditions. These effects determine, first of all, the nature of the impact on the results of electrical measurements of vibrational seismic–acoustic noise caused by the movement of railway transport. In addition, the nature and degree of their manifestation depend on a number of additional factors, which include the mineral composition of the solid skeleton of the geological environment and its structure, porosity, permeability and structure of pore channels, composition and volume of mineral cement, composition and mineralization of the liquid saturating the pores, etc.

The most important advantage of the seismoelectric method in comparison with traditional methods of applied geophysics is the unambiguity of solving inverse problems. In addition, the role of this method increases significantly with increasing depth and resolution of studies.

In the literature [8,9] it has been demonstrated that with the practical use of the seismoelectric method, the most informative is the study of the recorded electrical signals. This is due to two factors. Firstly, structural changes in various media, first of all, affect their average conductivity, which is characterized by a seismoelectric effect of the first kind. Secondly, there is always a double electric layer with a mobile diffuse part at the boundaries of solid media and pore fluid. Elastic action on such a medium often leads to a relative movement of the porous fluid, which, in turn, generates an external electric current that creates an electromagnetic field—a seismoelectric effect of the second type. In addition, various rock-forming minerals have different types of conductivity: ionic, electron, hole. The contact of such minerals with different types of conductivity can also lead to the emergence of new electric fields. Thus, it is proposed to consider the electrical component as the observed component of the method under study.

The principal possibility of monitoring the roadbed with the use of natural or artificially created geophysical fields is determined by the fact that the selected objects (inhomogeneities) differ in properties from the host medium and as a result create anomalous geophysical fields. In our case, the source of elastic vibrations can be railway trains passing through the studied area (Figure 2). These effects have high energy, and the characteristics of this effect are a priori known [3,4,7,12,20,29]. Geodynamic variations of these fields are a consequence of the action of both natural factors and man-made impacts on the environment, and this makes it possible to distinguish them based on the processing of geodynamic data. At the same time, the seismic impact will highlight the characteristic features in the analyzed signal, adding auxiliary information about the structure of the medium to the controlled parameters, manifested at the combination frequencies of the impact.

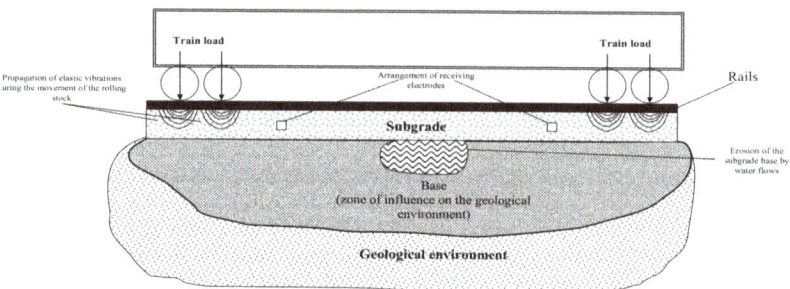

Figure 2. The principle of application of seismoelectric control of the railway trackbed.

Since the ground base of a railway track can be represented as a base formed by solid particles and a liquid pore filler, the type of deformation processes developing in it will be determined primarily by the physical causes and properties of the specified components and described by the dependence of the stress–strain state of the soil depending on the applied load containing four phases (Figure 3).

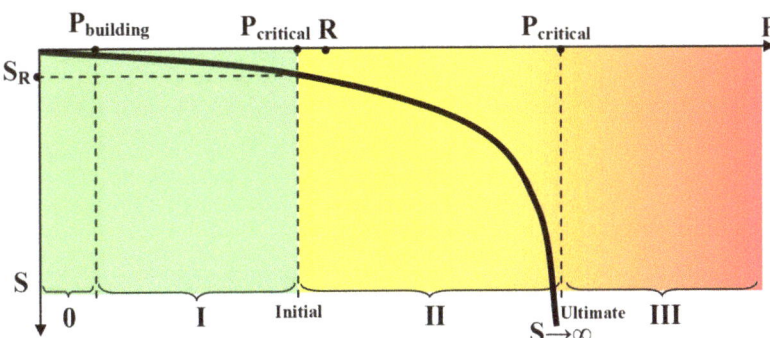

Figure 3. Model of the stress–strain process of the ground base of the railway track.

According to this model, the elastic deformation phase (0) is up to 10% of the permissible load on the soil and is limited by the value of structural strength, assuming that no structural changes occur in the soil under these loads. The compaction phase (I) is determined by the linear relationship between the applied load and the total deformation of the soil base. The shift phase (II) characterizes the processes of destruction determined by significant shear deformations due to exceeding the limit. The extrusion phase (III) characterizes irreversible deformations, in which the soil is squeezed out from under the railway track.

A similar character, in accordance with the seismoelectric effect of the first kind, is the dependence of the electrical resistance of the ground base on the applied mechanical load (due to the movement of railway transport along the track), which can be considered its amplitude characteristics, and the amplitude and phase spectra of the response to this seismic action are indicators of the development of deformation processes.

The advantage of this approach is that the railway transport generates an intense and long-lasting seismic acoustic signal, the parameters of which change slightly over time relative to the passing train, which eliminates the need to use additional sources of seismic signals. Moreover, it is known that the level of noise generated by a rolling stock consists of the following components: drive noise, aerodynamic noise and wheel rolling noise [30,31]. The first two types of noise can be considered stationary background noise, and the third one arises due to the contact of wheels with the rail and is associated with the high rolling pressure of steel on steel, characteristic of the "wheel-rail" system [32]. In the frequency domain, these noises are sources of low-frequency vibrations, the propagation speed of which coincides with the speed of the train movement. As a result, the spectrum of the resulting seismic signal can be divided into four frequency ranges from 3 to 80 Hz.

In accordance with the above, when implementing the seismoelectric method, as an informative parameter, it is necessary to use the complex transfer function of the investigated section of the geological environment [33]:

$$\dot{N}(j\omega, \Delta u) = \frac{\dot{E}(j\omega)}{\dot{I}(j\omega)} = \dot{Z}_A(j\omega) + \dot{Z}_B(j\omega) + \sum_{i=1}^{n} \dot{Z}_i(j\omega, \Delta u) \qquad (1)$$

where $\dot{Z}_A(j\omega), \dot{Z}_B(j\omega)$ are complex resistances of grounding; $E(j\omega), I(j\omega)$ are complex parameters of the electric field; ω is frequency of the probing signal; $\dot{Z}_i(j\omega, \Delta u)$ are complex resistances of the i-th elements of the studied section of the geological environment under seismoacoustic influence Δu.

The expression of the transfer function of the studied geological medium (1) is a geoelectric model of complex resistances connected in series. Such a representation allows us to use the pattern of an N-layer imperfect dielectric. This model contains N elements, and each element has thickness of d and the following electrical parameters of the i-th element—dielectric permissivity ε_i and electrical resistivity ρ_i. The transfer function of the area under investigation can be expressed as the form of RC circuits connected in series, while each circuit has the following parameters [34]:

$$C_i = \varepsilon_i S(j\omega, \Delta u_i)/d(\Delta u_i), \ R_i = \rho_i d(\Delta u_i)/S(j\omega, \Delta u_i) \tag{2}$$

where the effective area $S(j\omega)$ is determined by the skin effect.

If we omit the the grounding parameters, then we might simply represent the transfer function of the geoelectric section from the dielectric parameters (2):

$$\dot{H}(j\omega, \Delta u) = \sum_{i=1}^{N} \frac{R_i}{1+x_i^2} - j\sum_{i=1}^{N} \frac{R_i x_i}{1+x_i^2}, \tag{3}$$

where $x_i = \omega R_i C_i = \omega \varepsilon_i \rho_i$.

At the same time, when an elastic seismic wave propagates in the geological environment, each of its i-th element undergoes a mechanical influence described by the deformation tensor $\Delta u = \{\Delta u_x, \Delta u_y, \Delta u_z\}$.

The transfer function (3) can be written in exponential form as

$$H(p, \Delta u) = A(p, \Delta u)e^{j\phi(p,\Delta u)} \tag{4}$$

where p is the Laplace operator, $A(p, \Delta u) = |H(p, \Delta u)| = \sqrt{\text{Re}^2[H(p, \Delta u)] + \text{Im}^2[H(p, \Delta u)]}$ is the module of the transfer function, Re and Im are real, and the imaginary part of the complex function, accordingly, $\phi(p, \Delta u) = \arg[H(p, \Delta u)] = \arctg\left[\frac{\text{Im}[H(p,\Delta u)]}{\text{Re}[H(p,\Delta u)]}\right]$ is the argument of the transfer function.

The module and phase of this transfer function can be calculated based on the input–output model from the ratio

$$H(p, \Delta u) = \frac{Y(p, \Delta u)}{X(p, \Delta u)} \tag{5}$$

where $X(p,\Delta u)$ is the probing electrical signal in operator form, $Y(p, \Delta u)$ is the recorded electrical signal in operator form.

In this case, the ground base of the railway track can be represented as a dynamic link—Figure 4.

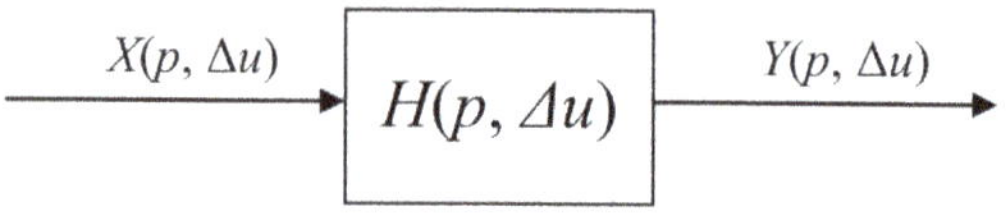

Figure 4. Representation of the ground base of the railway track in the form of a dynamic link.

By converting (5) taking into account (4), we obtain

$$H(p, \Delta u) = \frac{\left|Y(p, \Delta u)e^{j\phi_Y(p,\Delta u)}\right|}{\left|X(p, \Delta u)e^{j\phi_X(p,\Delta u)}\right|} = \frac{|Y(p, \Delta u)|}{|X(p, \Delta u)|}e^{j\Delta\phi(p,\Delta u)} \tag{6}$$

where $\Delta\phi(p, \Delta u) = \phi_Y(p, \Delta u) - \phi_X(p, \Delta u)$.

It follows from the last expression that the control of the ground base of the railway track can be carried out by tracking both the module and the argument (phase) of the recorded geoelectric signals (since the parameters of the probing signals are constant and are a priori known).

4. Methodology of Experimental Research on the Model of the Natural–Technical System "Railway Track"

To assess the prospects of the seismoelectric method in the railway roadbed monitoring, we have done laboratory modeling of the railway track. The laboratory installation is shown in the Figure 5. It allows us to simulate the seismic influence of a train passing by, as well as the natural processes (changes in soil moisture, suffosion, karst, landslide processes, sinkholes).

The installation consists of the following parts:
- a small-scale model of a geodynamic object—a tank with sandy soil, in which it is possible to simulate its full or partial mudflow by means of extracting certain sections of the bottom of the tank; the seismic effect is a vibration source (black box on the left);
- current signal sources, signal registration—metal rods used to generate and register an electric field;
- geodynamic data processing system, which is an analog-to-digital converter, seismic station, personal computer with developed specialized software.

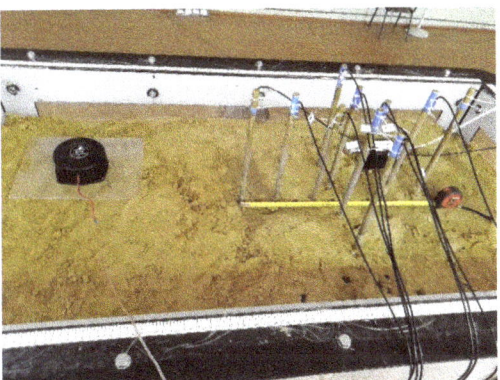

Figure 5. Geodynamic object model.

Primary to the current study, the noise of a railway train was generally registered by a microphone located in the ground base of the railway track. With direct modeling, the seismic impact was simulated by reproducing the recorded noise using a seismic signal source—a vibration loudspeaker.

During the study, we were modeling the soil collapse while registering changes in the characteristics of seismic and electrical signals. Thus, we obtained information about the primary stage of the soil collapse. The proposed approach allows detection of the primary phase of destruction of the railway roadbed, as well as prevention of the man-made catastrophes in the natural–technical system "railway track".

5. Results and Discussion

Figure 6 represents amplitude spectra of the output electrical signals of the investigated geological environment in the absence (a) and presence (b) of elastic action (if applied, an elastic harmonic impact had a frequency of 70 Hz). At the same time, the harmonic electrical signal had a frequency of 40 Hz.

It can be seen from the obtained spectrograms that the presence of an elastic seismoacoustic impact source in the simulated ground base of the railway track leads to the formation of a spectral component with a frequency of seismoacoustic impact in the spectrum of the recorded electrical signal (seismoelectric effect of the second type). This indicates that in this case, the geological environment has a normal specific resistivity (seismoelectric effect of the first type), and there is no deformation of the soil compaction. Depending on the parameters of external elastic influences applied to the investigated area, the electrical parameters of the medium change their characteristics accordingly. This makes it possible, during further processing of the received electrical signal, to determine the presence of heterogeneity in the medium, its depth and deformation state. In this case, there is no need to place the sensors along the entire monitored area.

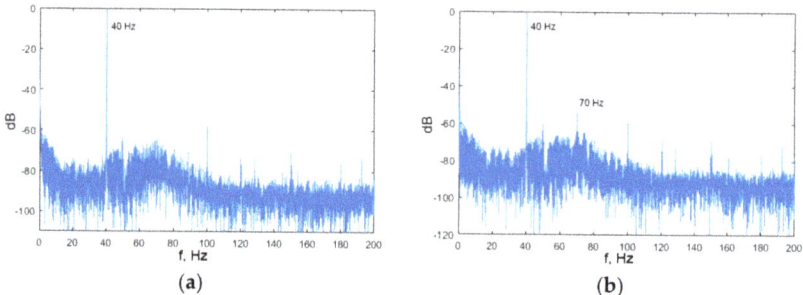

Figure 6. Amplitude spectra of the output electrical signals of the investigated geological environment in the absence (**a**) and presence (**b**) of elastic action.

Further studies have shown that the analysis of changes in the phase characteristics of the transfer function (6) has a number of significant advantages over the amplitude method; in particular, it is characterized by increased sensitivity and noise immunity [30] and also detects and localizes geodynamic processes in geological environments [30–32]. In Figure 7 as an example, the phase spectrum of the output electrical signals of the geological medium under study was measured while applying the 70 Hz elastic action, because at such frequency a phase shift of seismic acoustic action is clearly traced.

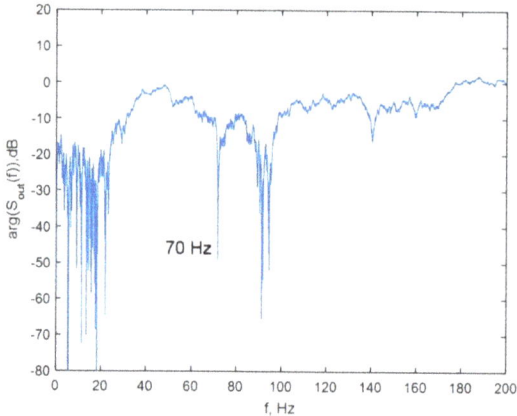

Figure 7. Phase spectrum of the output electrical signals of the studied geological medium in the presence of elastic action.

Studies [25–27] explain the modified phase-measuring method of geoelectric control, namely the use of several current signal sources placed closely to the test control object and an array of vector sensors for measuring the electric field. At the same time, the registration of phase characteristics at a fixed position of the sources and the measuring basis with the possibility of controlling the parameters of the probing signals is based on the fact that the primary and secondary electric fields are vector values.

Figure 8 shows a diagram of the laboratory experiment to control the process of occurrence and subsequent growth of a cavity in the ground foundation of a railway track by the phasometric method. The research was carried out on a physical model of a railway track using a specially developed measuring phasometric system. In this case, the current sources designated as A and B (Figure 8) form quadrature harmonic signals with a phase shift of 90 degrees. Sources A and B generate an electric field signal at point O, described as follows:

$$\vec{E}_{AX} = \vec{E}^0_{AX} + \Delta\vec{\overline{E}}_{AX}, \vec{E}_{BX} = \vec{E}^0_{BX} + \Delta\vec{\overline{E}}_{BX} \tag{7}$$

where $\vec{E}^0$ is the electrical signal recorded before the formation of deformation processes in the ground; $\Delta\overline{E}$ is an abnormal component of the electric field caused by the presence of deformation processes in the ground.

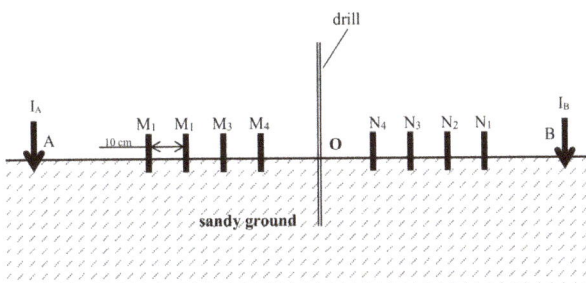

Figure 8. Diagram of the laboratory experiment.

When using multipolar geodynamic control systems at registration points (M_1–M_4, N_1–N_4), we have to deal with an elliptically polarized geoelectric field. Moreover, in this case, vector sensors of electric field measurement with the same indices form pairs; the signals from each pair are sent to the measuring system for processing. In this case, data processing of recorded geoelectric signals presupposes the formation of their difference signal (for filtering in-phase interference), its amplification, phase detection (in relation to the reference signal) and low-pass filtering. The principle of registration for the phase of the geoelectric field at an arbitrary receiving point is illustrated in Figure 9.

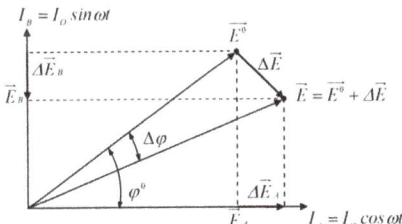

Figure 9. The registration principle for the phase of the geoelectric field at an arbitrary receiving point.

Using this method, simulation of the process of sinkhole formation of the ground base of the railway track in the presence of train noise was carried out (the seismogram taken from the short-period seismic meter ZET 7156 is shown in Figure 10). The electrodes A and B were used as current sources, while the electrodes M and N with corresponding numbers were receivers. The following distances between electrodes were used: between A and B—80 cm, between M1 and N1—70 cm, between M2 and N2—60 cm, between M3 and N3—50 cm, between M4 and N4—40 cm. The frequency of probing electrical signals was 90 Hz, the amplitude of each probing electrical signal was 200 mV, and the analog-to-digital converter was set at a sampling rate of 10,101 Hz.

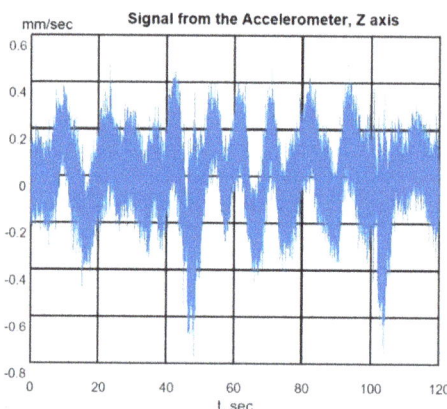

Figure 10. Seismogram of the railway track in the presence of train noise.

Figure 11 shows an example of a phase image obtained as a result of processing in the time domain for the receiving line M3N3.

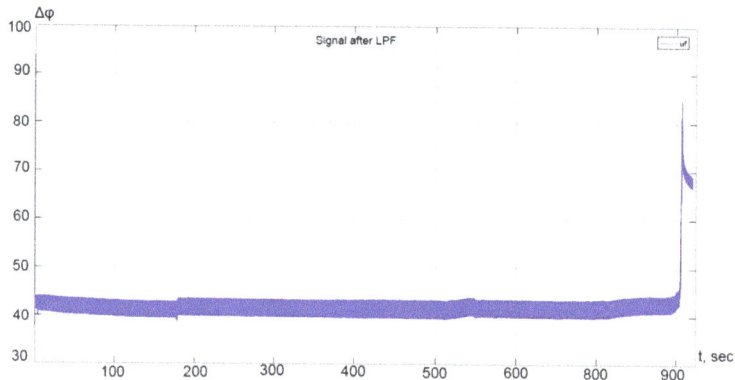

Figure 11. The example of a phase image for the receiving line M3N3, obtained by simulating the process of sinkhole formation of the ground base of the railway track.

The analysis of this process in the time domain makes it possible at an early stage to unambiguously determine the time interval corresponding to the active development of the sinkhole formation process and the direction of its change in time—to localize the source of the deformation process and predict its future dynamics.

For a comparison, Figure 12 shows a variant of solving a similar problem by the classical seismoacoustic method using a highly sensitive seismometer.

A comparison of the two approaches to the control of the underground base of the railway track allows us to conclude that the traditional method allows only the registration of certain geodynamic events in the geological environment without the possibility of their qualitative interpretation, which is eliminated when using the seismoelectric method and monitoring the phase characteristics of the recorded geoelectric signals.

Subsequently, from the recorded signal, it is possible to identify features of soil areas with impaired integrity [31] using Scale Invariant Feature Transform (SIFT)-Support Vector Machine (SVM) methods [35–40]. In the presence of high noise level, it is possible to use a neural back propagation network, which will allow achieving 97% accuracy when processing monitoring data [36,37,39].

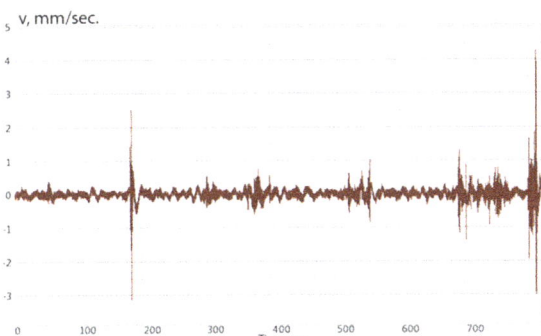

Figure 12. Seismogram of the process of sinkhole formation of the soil base of the railway roadbed.

6. Conclusions

The paper demonstrates that for the diagnostics of the railway roadbed, a number of features not found in traditional engineering geology should be taken into account. The article also theoretically and practically substantiates an integrated approach to solving problems of monitoring the roadbed of a railway on the basis of combining two geophysical methods for monitoring the natural environment: geoelectric and seismoacoustic methods.

From the presented research results, it has been established that the proposed methods for monitoring the railway trackbed have high sensitivity to the primary phase of the soil destruction.

During laboratory modeling, it was found that in the amplitude spectra of the output electrical signals of the investigated geological environment in the presence of an elastic harmonic effect with a frequency of 70 Hz, the frequency of a harmonic electrical signal with a frequency of 40 Hz was observed, which makes it possible to use this effect to obtain more detailed information about the structure of the soil when sounding with electric fields. In laboratory modeling, the obtained phase images for the receiving line of the installation when simulating the process of sinking the soil base of the railway bed confirmed the presence of a transient process that causes a shift in the initial phase of the signal $\Delta\varphi = 40°$ by ~45° ($\Delta\varphi' = 85°$), that precedes the initial stage of failure.

Moreover, the joint processing of geoelectric and seismoacoustic signals makes it possible to detect heterogeneity in the environment, as well as its depth and its geodynamic variations. The phasometric ground control method allows increasing the efficiency of the railway roadbed monitoring systems, in particular, through the use of vertical electrotomography methods with the possible localization of heterogeneity. In this case, the metrological stability of geodynamic measurements, the insensitivity of the technique to seismic interference, the simplicity of varying the installation size and the organization of three-dimensional soil monitoring is ensured.

The information obtained during the monitoring of the roadbed will improve the reliability and safety of the functioning of railway transport, especially in areas with intensive geodynamic processes.

Author Contributions: Data curation, A.B.; Formal analysis, D.S.; Investigation, A.G.; Supervision, O.K.; Writing—original draft, O.K.; Writing—review & editing, G.V. and Y.Y. All authors have read and agreed to the published version of the manuscript.

Funding: This research was funded by the Ministry of Science and Higher Education of the Russian Federation in accordance with agreement No. 075-15-2020-905 date 16 November 2020 on providing a grant in the form of subsidies from the Federal budget of the Russian Federation. The grant was provided for state support for the establishing and development of a world-class Scientific Center "Agrotechnologies for the Future".

Institutional Review Board Statement: Not applicable.

Informed Consent Statement: Not applicable.

Acknowledgments: Anastasia Grecheneva has been supported by the Ministry of Science and Higher Education of the Russian Federation in accordance with agreement No. 075-15-2020-905 date 16 November 2020 on providing a grant in the form of subsidies from the Federal budget of the Russian Federation. The grant was provided for state support for the establishing and development of a world-class Scientific Center "Agrotechnologies for the Future".

Conflicts of Interest: The authors declare no conflict of interest.

References

1. Zhai, W.; Wang, K.; Cai, C. Fundamentals of vehicle-track coupled dynamics. *Veh. Syst. Dyn.* **2009**, *47*, 1349–1376. [CrossRef]
2. Song, Y.; Wang, Z.; Liu, Z.; Wang, R. A spatial coupling model to study dynamic performance of pantograph-catenary with vehicle-track excitation. *Mech. Syst. Signal Process.* **2021**, *151*, 107336. [CrossRef]
3. Yan, H.; Gao, C.; Elzarka, H.; Mostafa, K.; Tang, W. Risk assessment for construction of urban rail transit projects. *Saf. Sci.* **2019**, *118*, 583–594. [CrossRef]
4. Arcos, R.; Soares, P.J.; Costa, P.A.; Godinho, L. An experimental/numerical hybrid methodology for the prediction of railway-induced ground-borne vibration on buildings to be constructed close to existing railway infrastructures: Numerical validation and parametric study. *Soil Dyn. Earthq. Eng.* **2021**, *150*, 106888. [CrossRef]
5. Draft Environmental Impact Report, Preston Property Residential Project State Clearinghouse No: 2012022075; City of Milpitas, CA, USA, 15 November 2012. Available online: http://www.ci.milpitas.ca.gov/_pdfs/plan_eir_PrestonPropertyAppendices-Part1.pdf (accessed on 15 November 2012).
6. Cheverikin, A.V. Modern methods of monitoring geodynamic processes in oil and gas production areas, XXI century. *Technosphere Secur.* **2019**, *13*, 122–133.
7. Brough, M.; Stirling, A.; Ghataora, G.; Madelin, K. Evaluation of railway trackbed and formation: A case study. *NDT E Int.* **2003**, *36*, 145–156. [CrossRef]
8. Fisseha, S.; Mewa, G.; Haile, T. Refraction seismic complementing electrical method in subsurface characterization for tunneling in soft pyroclastic, (a case study). *Heliyon* **2021**, *7*, e07680. [CrossRef]
9. De Pasquale, G.; Linde, N.; Greenwood, A. Joint probabilistic inversion of DC resistivity and seismic refraction data applied to bedrock/regolith interface delineation. *J. Appl. Geophys.* **2019**, *170*, 103839. [CrossRef]
10. Lee, S.C.H.; Noh, K.A.M.; Zakariah, M.N.A. High-resolution electrical resistivity tomography and seismic refraction for groundwater exploration in fracture hard rocks: A case study in Kanthan, Perak, Malaysia. *J. Asian Earth Sci.* **2021**, *218*, 104880. [CrossRef]
11. An, L.; Hao, Y.; Yeh, T.C.J.; Liu, Y.; Liu, W.; Zhang, B. Simulation of karst spring discharge using a combination of time–frequency analysis methods and long short-term memory neural networks. *J. Hydrol.* **2020**, *589*, 125320. [CrossRef]
12. Zheng, Y.; He, S.; Yu, Y.; Zheng, J.; Zhu, Y.; Liu, T. Characteristics, challenges and countermeasures of giant karst cave: A case study of Yujingshan tunnel in high-speed railway. *Tunn. Undergr. Space Technol.* **2021**, *114*, 103988. [CrossRef]
13. Shangxin, F.; Yufei, Z.; Yujie, W.; Shanyong, W.; Ruilang, C. A comprehensive approach to karst identification and groutability evaluation—A case study of the Dehou reservoir, SW China. *Eng. Geol.* **2020**, *269*, 105529. [CrossRef]
14. Drahor, M.G. Identification of gypsum karstification using an electrical resistivity tomography technique: The case-study of the Sivas gypsum karst area (Turkey). *Eng. Geol.* **2019**, *252*, 78–98. [CrossRef]
15. Kaufmann, G.; Romanov, D. Modelling long-term and short-term evolution of karst in vicinity of tunnels. *J. Hydrol.* **2020**, *581*, 124282. [CrossRef]
16. Pazzi, V.; Morelli, S.; Fanti, R. A review of the advantages and limitations of geophysical investigations in Landslide studies. *Int. J. Geophys.* **2019**, *2019*, 2983087. [CrossRef]
17. Mahmoodzadeh, A.; Mohammadi, M.; Ali, H.F.H.; Abdulhamid, S.N.; Ibrahim, H.H.; Noori, K.M.G. Dynamic prediction models of rock quality designation in tunneling projects. *Transp. Geotech.* **2021**, *27*, 100497. [CrossRef]
18. Zhu, H.; Yan, J.; Liang, W. Challenges and Development Prospects of Ultra-Long and Ultra-Deep Mountain Tunnels. *Engineering* **2019**, *5*, 384–392. [CrossRef]
19. Chawla, S.; Shahu, J.T.; Kumar, S. Analysis of cyclic deformation and post-cyclic strength of reinforced railway tracks on soft subgrade. *Transp. Geotech.* **2021**, *28*, 100535. [CrossRef]
20. Baknin, M.D.; Surzhik, D.I.; Vasilyev, G.S.; Dorofeev, N.V. The modeling of the Phase-Metric Method of the Geoelectrical Control of Oil Sludge Straits. *IOP Conf. Ser. Earth Environ. Sci.* **2020**, *459*, 042085. [CrossRef]
21. Yun, B. Design of Underground Structures. In *Underground Engineering*; Yun, B., Ed.; Academic Press: Cambridge, MA, USA, 2019; pp. 47–115; ISBN 9780128127025.
22. Patel, A. Case examples of some geotechnical applications. In *Geotechnical Investigations and Improvement of Ground Conditions*; Woodhead Publishing Series in Civil and Structural Engineering; Patel, A., Ed.; Woodhead Publishing: Cambridge, UK, 2019; pp. 167–191; ISBN 9780128170489.

23. Maxim, B.; Artem, B.; Dmitry, S.; Oleg, K. Geotechnical monitoring of the foundations of structures based on integrated seismoelectric measurements in conditions of karst hazard. In Proceedings of the 20th International Multidisciplinary Scientific GeoConference Proceedings SGEM 2020, Albena, Bulgaria, 16–25 August 2020; Volume 20, pp. 559–566.
24. Kuzichkin, O.; Grecheneva, A.; Bykov, A.; Dorofeev, N.; Surzhik, D. Methods and algorithms of joint processing of geoelectric and seismoacoustic signals in real time. In Proceedings of the SGEM2018 Conference, Albena, Bulgaria, 2–8 July 2018; Volume 18, pp. 877–884.
25. Kuzichkin, O.R.; Vasilyev, G.S.; Grecheneva, A.V.; Mikhaleva, E.V.; Baknin, M.D.; Surzhik, D.I. Application of phase-metric compensation method for geoelectric control of near-surface geodynamic processes. *Bull. Electr. Eng. Inform.* **2020**, *9*, 898–905. [CrossRef]
26. Kuzichkin, O.; Vasilyev, S.; Grecheneva, A.; Mikhaleva, V.; Maxim, B. Application of phase-metric measuring system for geodynamic control of karst processes. *J. Eng. Appl. Sci.* **2017**, *12*, 6858–6863.
27. Kuzichkin, O.R.; Vasilyev, G.S.; Baknin, M.D.; Surzhik, D.I. The phase-metric method of isolating the information component in the distributed processing of geoelectric signals in geoecological monitoring systems. *J. Adv. Res. Dyn. Control Syst.* **2020**, *12*, 463–471.
28. Kuzichkin, O.; Dorofeev, N.; Grecheneva, A.; Bykov, A.; Romanov, R. The use of vertical electrical sounding by the method of two components for allocation of an initial phase of a landslide. In Proceedings of the SGEM2017 Conference, Albena, Bulgaria, 29 June–5 July 2017; Volume 17, pp. 1025–1032.
29. Kovalenko, A.O.; Bykov, A.A.; Kuzichkin, O.R. Railway Traffic Monitoring System by Seismic Methods. *Helix* **2020**, *10*, 218–224. [CrossRef]
30. Aliouache, M.; Wang, X.; Jourde, H.; Huang, Z.; Yao, J. Incipient karst formation in carbonate rocks: Influence of fracture network topology. *J. Hydrol.* **2019**, *575*, 824–837. [CrossRef]
31. Zhang, N.; Zhou, A.; Pan, Y.; Shen, S.L. Measurement and prediction of tunnelling-induced ground settlement in karst region by using expanding deep learning method. *Measurement* **2021**, *183*, 109700. [CrossRef]
32. Eppelbaum, L. Environmental Geophysics. In *Computational Geophysics, Geophysical Potential Fields*; Eppelbaum, L.V., Ed.; Elsevier: Amsterdam, The Netherlands, 2019; Volume 2, pp. 311–364.
33. Jeannin, P.Y.; Artigue, G.; Butscher, C.; Chang, Y.; Charlier, J.; Duran, L.; Gill, L.; Hartmann, A.; Johannet, A.; Jourde, H.; et al. Karst modelling challenge 1: Results of hydrological modelling. *J. Hydrol.* **2021**, *600*, 126508. [CrossRef]
34. Wang, X.; Aliouache, M.; Wang, Y.; Lei, Q.; Jourde, H. The role of aperture heterogeneity in incipient karst evolution in natural fracture networks: Insights from numerical simulations. *Adv. Water Resour.* **2021**, *156*, 104036. [CrossRef]
35. Soloviev, V. Fintech Ecosystem in Russia. In Proceedings of the 2018 11th International Conference; Management of Large-Scale System Development, MLSD, Moscow, Russia, 1–3 October 2018. [CrossRef]
36. Sebyakin, A.; Soloviev, V.; Zolotaryuk, A. Spatio-Temporal Deepfake Detection with Deep Neural Networks. In *Proceedings of the 16th International Conference on Information: iConference 2021: Diversity, Divergence, Dialogue, Beijing, China, 17–31 March 2021*; Lecture Notes in Computer Science (including Subseries Lecture Notes in Artificial Intelligence and Lecture Notes in Bioinformatics); Springer: Cham, Switzerland; Volume 12645, pp. 78–94. [CrossRef]
37. Gataullin, T.M.; Gataullin, S.T.; Ivanova, K.V. Modeling an Electronic Auction. In Proceedings of the Institute of Scientific Communications Conference: ISC 2020: "Smart Technologies" for Society, State and Economy, Volgograd, Russia, 19–20 March 2020; 155, pp. 1108–1117. [CrossRef]
38. Gataullin, T.M.; Gataullin, S.T. Best Economic Approaches under Conditions of Uncertainty. In Proceedings of the 2018 Eleventh International Conference "Management of large-scale system development" (MLSD), Moscow, Russia, 1–3 October 2018. [CrossRef]
39. Dogadina, E.P.; Smirnov, M.V.; Osipov, A.V.; Suvorov, S.V. Evaluation of the forms of education of high school students using a hybrid model based on various optimization methods and a neural network. *Informatics* **2021**, *8*, 46. [CrossRef]
40. Tatarintsev, M.; Korchagin, S.; Nikitin, P.; Gorokhova, R.; Bystrenina, I.; Serdechnyy, D. Analysis of the forecast price as a factor of sustainable development of agriculture. *Agronomy* **2021**, *11*, 1235. [CrossRef]

Article

The Integral Mittag-Leffler, Whittaker and Wright Functions

Alexander Apelblat [1] and Juan Luis González-Santander [2,*]

[1] Department of Chemical Engineering, Ben Gurion University of the Negev, Beer Sheva 84105, Israel; apelblat@bgu.ac.il
[2] Department of Mathematics, Universidad de Oviedo, 33007 Oviedo, Spain
* Correspondence: gonzalezmarjuan@uniovi.es

Abstract: Integral Mittag-Leffler, Whittaker and Wright functions with integrands similar to those which already exist in mathematical literature are introduced for the first time. For particular values of parameters, they can be presented in closed-form. In most reported cases, these new integral functions are expressed as generalized hypergeometric functions but also in terms of elementary and special functions. The behavior of some of the new integral functions is presented in graphical form. By using the MATHEMATICA program to obtain infinite sums that define the Mittag-Leffler, Whittaker, and Wright functions and also their corresponding integral functions, these functions and many new Laplace transforms of them are also reported in the Appendices for integral and fractional values of parameters.

Keywords: integral Mittag-Leffler functions; integral Whittaker functions; integral Wright functions; Laplace transforms

1. Introduction

The appearance of special functions of mathematical physics was associated with solutions of particular ordinary differential equations, while the integral special functions arrived much later in mathematical literature after properties of these functions were investigated. Integral special functions were introduced as new special functions, which can be applied in many circumstances, especially in operational calculus, where they are frequently serving as direct and inverse integral transforms. The form of an integrand is identical for all integral functions, but limits of integration are different in order to assure the convergence of defined integrals. There are two types of integral special functions: those with elementary functions in their integrands and those with special functions. To the first group belong the exponential integral $-\mathrm{Ei}(-x)$, the sine and cosine integrals, $\mathrm{si}(x)$, $\mathrm{Si}(x)$, $\mathrm{ci}(x)$ and $\mathrm{Ci}(x)$, and the corresponding integrals of hyperbolic trigonometric functions, $\mathrm{Shi}(x)$ and $\mathrm{Chi}(x)$. These functions are defined in the following way [1–5]

$$\begin{aligned}
\mathrm{E}_1(x) &= -\mathrm{Ei}(-x) = \int_x^\infty \frac{e^{-t}}{t}dt, \quad x > 0, \\
\mathrm{Si}(x) &= \int_0^x \frac{\sin t}{t}dt, \\
\mathrm{si}(x) &= -\int_x^\infty \frac{\sin t}{t}dt = \mathrm{Si}(x) - \frac{\pi}{2}, \\
\mathrm{Ci}(x) &= -\int_x^\infty \frac{\cos t}{t}dt = \gamma + \ln x - \int_0^x \frac{1-\cos t}{t}dt = -\mathrm{ci}(x), \\
\mathrm{Shi}(x) &= \int_0^x \frac{\sinh t}{t}dt, \\
\mathrm{Chi}(x) &= \gamma + \ln x - \int_0^x \frac{1-\cosh t}{t}dt,
\end{aligned} \quad (1)$$

where γ is the Euler–Mascheroni constant. As can be observed in (1), the integral special functions have integrands in the form, $f(t)/t$, and the intervals of integrations are $0 < t < x$

or $x < t < \infty$. Few direct and inverse integral transforms are presented below to illustrate their applications, for example, in the Laplace transformation [6–8],

$$F(s) := \mathcal{L}[f(t)] := \int_0^\infty e^{-st} f(t) dt, \tag{2}$$

we have

$$\begin{aligned}
\mathcal{L}\left[\frac{1}{\sqrt{t}}\mathrm{Ei}(-t)\right] &= -2\sqrt{\frac{\pi}{s}}\ln\left(\sqrt{s}+\sqrt{s+1}\right), \quad \mathrm{Re}\, s > 0, \\
\mathcal{L}[\mathrm{Si}(t)] &= \frac{\cot^{-1} s}{s}, \quad \mathrm{Re}\, s > 0, \\
\mathcal{L}[\mathrm{si}(t)] &= \frac{\tan^{-1} s}{s}, \\
\mathcal{L}[\mathrm{Ci}(t)] &= -\frac{\ln(1+s^2)}{2s}, \\
\mathcal{L}^{-1}\left[\frac{\ln(s+b)}{s+a}\right] &= e^{-at}[\ln(b-a) - \mathrm{Ei}((a-b)t)], \quad \mathrm{Re}\,(s-a) > 0, \\
\mathcal{L}^{-1}\left[\frac{\ln s}{s^2+1}\right] &= \cos t\,\mathrm{Si}(t) - \sin t\,\mathrm{Ci}(t), \quad \mathrm{Re}\, s > 0, \\
\mathcal{L}^{-1}\left[\frac{s\ln s}{s^2+1}\right] &= -\sin t\,\mathrm{Si}(t) - \cos t\,\mathrm{Ci}(t).
\end{aligned} \tag{3}$$

Integrands in the second group of integral special functions include special functions, the most well-known and applied of which are the integral Bessel functions (see, e.g., [3,7,9–13])

$$\begin{aligned}
\mathrm{Ji}_\nu(x) &= -\int_x^\infty \frac{J_\nu(t)}{t} dt, \\
\mathrm{Yi}_\nu(x) &= -\int_x^\infty \frac{Y_\nu(t)}{t} dt, \\
\mathrm{Ii}_\nu(x) &= -\int_0^x \frac{I_\nu(t)}{t} dt, \\
\mathrm{Ki}_\nu(x) &= -\int_x^\infty \frac{K_\nu(t)}{t} dt.
\end{aligned} \tag{4}$$

Already in 1929, van der Pol [9] showed that it is possible to express the differentiation with respect to the order of the Bessel function of the first kind as a convolution integral, which includes the integral Bessel function of the zero-order:

$$\frac{\partial J_\nu(t)}{\partial \nu} = \frac{1}{2}\int_0^t \mathrm{Ji}_0(t-x)[J_{\nu-1}(x) - J_{\nu+1}(x)] dx. \tag{5}$$

The integral Bessel functions of the zero-order are inverse transforms of the following Laplace transforms [7]

$$\begin{aligned}
\mathcal{L}^{-1}\left[\frac{\sinh^{-1} s}{s}\right] &= \mathrm{Ji}_0(t), \\
\mathcal{L}^{-1}\left[\frac{\left(\sinh^{-1} s\right)^2}{s}\right] &= \mathrm{Yi}_0(t), \\
\mathcal{L}^{-1}\left[\ln\left(s+\sqrt{s^2+1}\right) - \frac{\pi i}{2}\right] &= \mathrm{Ii}_0(t), \\
\mathcal{L}^{-1}\left[\frac{\left(\cosh^{-1} s\right)^2}{2s} + \frac{\pi^2}{8s}\right] &= \mathrm{Ki}_0(t).
\end{aligned} \tag{6}$$

In analogy to the integral Bessel functions and with the possibility of extension to other special functions, this work introduces three new integral functions. Furthermore, these integral functions guide us toward the establishment of integrals and series. Section 2 explores the integral Mittag-Leffler functions. Sections 3 and 4 discuss the integral Whittaker and Wright functions, respectively. Section 5 contains concluding remarks.

In order to preserve the applied form of notation, the following two integral functions are introduced:

$$\text{Fi}(x) = \int_0^x \frac{f(t) - f(0)}{t} dt, \tag{7}$$

and

$$\text{fi}(x) = \int_x^\infty \frac{f(t)}{t} dt. \tag{8}$$

To ensure convergence of integrals in (7) or in (8), which depends on the behavior of $f(t)/t$ integrands at the origin and at infinity, the forms of integral functions $\text{Fi}(x)$ or $\text{fi}(x)$ are chosen. Since the explicit expressions for $f(t)$ functions are sometimes given in the form of $f(t^\alpha)$ where $\alpha = \pm\frac{1}{2}, \pm 1, 2, 3, \ldots$ the corresponding change of integration variables for these equations is desired.

In the case of Mittag-Leffler, Whittaker and Wright functions, for some values of parameters, by using the MATHEMATICA program, it was possible to obtain these integral functions in a closed-form. Derived integral functions are tabulated and also in some cases graphically presented (see [3]).

2. The Integral Mittag-Leffler Functions

The classical one-parameter and the two-parameter Mittag-Leffler functions are defined by [14]:

$$\begin{aligned} E_\alpha(x) &= \sum_{k=0}^\infty \frac{x^k}{\Gamma(\alpha k + 1)}, \quad \text{Re}\,\alpha > 0, \\ E_{\alpha,\beta}(x) &= \sum_{k=0}^\infty \frac{x^k}{\Gamma(\alpha k + \beta)}, \quad \text{Re}\,\alpha > 0,\ \text{Re}\,\beta > 0. \end{aligned} \tag{9}$$

In this investigation, they are only considered for positive real values of the argument, i.e., $x > 0$. In the particular case of positive rational α with $\alpha = p/q$ and p and q positive coprimes, Mittag-Leffler functions are given as a finite sum of generalized hypergeometric functions (see (A3) in Appendix A).

The Laplace transforms of the Mittag-Leffler functions are derived directly from (2) and (9), and we have:

$$\begin{aligned} \mathcal{L}[E_\alpha(t)] &= \int_0^\infty e^{-st} \left[\sum_{k=0}^\infty \frac{t^k}{\Gamma(\alpha k + 1)}\right] dt = \sum_{k=0}^\infty \frac{k!}{\Gamma(\alpha k + 1)} \left(\frac{1}{s}\right)^{k+1}, \\ \mathcal{L}[E_{\alpha,\beta}(t)] &= \int_0^\infty e^{-st} \left[\sum_{k=0}^\infty \frac{t^k}{\Gamma(\alpha k + \beta)}\right] dt = \sum_{k=0}^\infty \frac{k!}{\Gamma(\alpha k + \beta)} \left(\frac{1}{s}\right)^{k+1}, \end{aligned} \tag{10}$$

$\text{Re}\,s > 1$.

For particular values of parameters α and β, the explicit form of the Mittag-Leffler functions can be obtained by applying the MATHEMATICA program to sums of infinite series in (9), and these results are presented in Appendix A. Using Equation (10), many new Laplace transforms of the Mittag-Leffler functions were evaluated, and they are also reported in Appendix A. Similarly as in the case when α is positive rational, the Laplace transforms of the Mittag-Leffler functions can be expressed by the finite sum of products of generalized hypergeometric functions (see (A4) in Appendix A).

The integral Mittag-Leffler functions are introduced by considering their exponential behavior as a function of real, positive variable x (see Appendix A).

$$\begin{aligned} \mathrm{Ei}_\alpha(x) &= \int_0^x \frac{E_\alpha(t) - 1}{t} dt, \\ \mathrm{Ei}_{\alpha,\beta}(x) &= \int_0^x \frac{E_{\alpha,\beta}(t) - 1/\Gamma(\beta)}{t} dt. \end{aligned} \quad (11)$$

Formally, by introducing (9) into (11) we have

$$\begin{aligned} \mathrm{Ei}_\alpha(x) &= \sum_{k=1}^\infty \frac{x^k}{k\,\Gamma(\alpha k + 1)}, \\ \mathrm{Ei}_{\alpha,\beta}(x) &= \sum_{k=1}^\infty \frac{x^k}{k\,\Gamma(\alpha k + \beta)}. \end{aligned} \quad (12)$$

For several values of parameters α and β, it is possible to derive the integral Mittag-Leffler functions in a closed-form by applying the MATHEMATICA program to the sums of infinite series in (12). These functions are presented in Tables 1 and 2. As it is observable, most of these integral functions are expressed as generalized hypergeometric series. Typical behavior of one-parameter and two-parameter integral Mittag-Leffler functions is illustrated in Figures 1 and 2.

Evidently, also direct integration, by using (11), leads to the integral Mittag-Leffler functions. For example, for $E_2(x) = \cosh\sqrt{x}$, according to (1), we have

$$\mathrm{Ei}_2(x) = \int_0^x \frac{\cosh\sqrt{t} - 1}{t} dt = -2\gamma - \ln x + 2\,\mathrm{Chi}\sqrt{x}, \quad (13)$$

and as expected, this result is identical to that derived from (12) (see Tables 1 and 2).

Applying the formulas (A1) and (A2) given in Appendix A, the integral Mittag-Leffler function for positive rational values of parameter α with $\alpha = p/q$ and p,q positive coprimes is

$$\begin{aligned} &\mathrm{Ei}_{p/q,\beta}(x) \\ &= \sum_{k=1}^q \frac{x^k}{k\,\Gamma(k/q + \beta)}\, {}_2F_q\!\left(\begin{matrix} 1, k/q \\ b_0, \ldots, b_{p-1}, k/q+1 \end{matrix} \bigg| \frac{x^q}{p^p}\right). \end{aligned} \quad (14)$$

where

$$b_j = \frac{k}{q} + \frac{\beta + j}{p}.$$

In addition, using the sums in (12), it is possible to derive the Laplace transforms of the integral Mittag-Leffler functions:

$$\begin{aligned} \mathcal{L}[\mathrm{Ei}_\alpha(t)] &= \int_0^\infty e^{-st}\left[\sum_{k=0}^\infty \frac{t^{k+1}}{(k+1)\Gamma(\alpha(k+1)+1)}\right] dt \\ &= \sum_{k=0}^\infty \frac{(k+1)!}{(k+1)\Gamma(\alpha(k+1)+1)}\left(\frac{1}{s}\right)^{k+2}, \quad \mathrm{Re}\,s > 1. \\ \mathcal{L}[\mathrm{Ei}_{\alpha,\beta}(t)] &= \int_0^\infty e^{-st}\left[\sum_{k=0}^\infty \frac{t^{k+1}}{(k+1)\Gamma(\alpha(k+\beta)+1)}\right] dt \\ &= \sum_{k=0}^\infty \frac{(k+1)!}{(k+1)\Gamma(\alpha(k+1)+\beta)}\left(\frac{1}{s}\right)^{k+2}, \quad \mathrm{Re}\,s > 1. \end{aligned} \quad (15)$$

The evaluated Laplace transforms of the integral Mittag-Leffler functions are presented in Tables 3 and 4.

Table 1. The integral Mittag-Leffler functions derived for some values of parameters α and β by using (12).

α	β	$\text{Ei}_{\alpha,\beta}(x)$
$\frac{1}{3}$	$\frac{1}{5}$	$\frac{x}{\Gamma(\frac{8}{15})}\,_2F_2\left(\begin{array}{c}1,\frac{1}{3}\\ \frac{8}{15},\frac{4}{3}\end{array}\bigg\vert x^3\right) + \frac{x^2}{2\Gamma(\frac{13}{15})}\,_2F_2\left(\begin{array}{c}1,\frac{2}{3}\\ \frac{13}{15},\frac{5}{3}\end{array}\bigg\vert x^3\right) + \frac{5x^3}{3\Gamma(\frac{1}{5})}\,_2F_2\left(\begin{array}{c}1,1\\ \frac{6}{5},2\end{array}\bigg\vert x^3\right)$
$\frac{1}{3}$	$\frac{1}{4}$	$\frac{x}{\Gamma(\frac{7}{12})}\,_2F_2\left(\begin{array}{c}1,\frac{1}{3}\\ \frac{7}{12},\frac{4}{3}\end{array}\bigg\vert x^3\right) + \frac{x^2}{2\Gamma(\frac{11}{12})}\,_2F_2\left(\begin{array}{c}1,\frac{2}{3}\\ \frac{11}{12},\frac{5}{3}\end{array}\bigg\vert x^3\right) + \frac{4x^3}{3\Gamma(\frac{1}{4})}\,_2F_2\left(\begin{array}{c}1,1\\ \frac{5}{4},2\end{array}\bigg\vert x^3\right)$
$\frac{1}{3}$	$\frac{1}{2}$	$\frac{2x^3}{3\sqrt{\pi}}\,_2F_2\left(\begin{array}{c}1,1\\ \frac{3}{2},2\end{array}\bigg\vert x^3\right) + \frac{x}{\Gamma(\frac{5}{6})}\,_2F_2\left(\begin{array}{c}1,\frac{1}{3}\\ \frac{5}{6},\frac{4}{3}\end{array}\bigg\vert x^3\right) + \frac{3x^2}{\Gamma(\frac{1}{6})}\,_2F_2\left(\begin{array}{c}1,\frac{2}{3}\\ \frac{7}{6},\frac{5}{3}\end{array}\bigg\vert x^3\right)$
$\frac{1}{3}$	$\frac{3}{2}$	$\frac{x}{\Gamma(\frac{11}{6})}\,_2F_2\left(\begin{array}{c}1,\frac{1}{3}\\ \frac{4}{3},\frac{11}{6}\end{array}\bigg\vert x^3\right) + \frac{18x^2}{7\Gamma(\frac{1}{6})}\,_2F_2\left(\begin{array}{c}1,\frac{2}{3}\\ \frac{5}{3},\frac{13}{6}\end{array}\bigg\vert x^3\right) + \frac{4x^3}{9\sqrt{\pi}}\,_2F_2\left(\begin{array}{c}1,1\\ 2,\frac{5}{2}\end{array}\bigg\vert x^3\right)$
$\frac{1}{2}$	$\frac{1}{2}$	$\frac{x^2}{\sqrt{\pi}}\,_2F_2\left(\begin{array}{c}1,1\\ \frac{3}{2},2\end{array}\bigg\vert x^2\right) + e^{x^2}F(x),\quad F(x)=e^{-x^2}\int_0^x e^{t^2}\,dt$
$\frac{1}{2}$	1	$-\frac{\gamma}{2}-\ln x + \frac{\text{Ei}(x^2)}{2} + \frac{2x}{\sqrt{\pi}}\,_2F_2\left(\begin{array}{c}\frac{1}{2},1\\ \frac{3}{2},\frac{3}{2}\end{array}\bigg\vert x^2\right)$
$\frac{1}{2}$	2	$\frac{1}{2}\left(1-\gamma+\text{Ei}(x^2)+\frac{1-e^{x^2}}{x^2}\right) - \ln x + \frac{4x}{3\sqrt{\pi}}\,_2F_2\left(\begin{array}{c}\frac{1}{2},1\\ \frac{3}{2},\frac{3}{2}\end{array}\bigg\vert x^2\right)$
$\frac{1}{2}$	3	$\frac{2+4x^2+(3-2\gamma)x^4-2e^{x^2}(1+x^2)+2x^4(\text{Ei}(x^2)-2\ln x)}{8x^4} + \frac{8x}{15\sqrt{\pi}}\,_2F_2\left(\begin{array}{c}\frac{1}{2},1\\ \frac{3}{2},\frac{7}{2}\end{array}\bigg\vert x^2\right)$
$\frac{1}{2}$	4	$\frac{12+18x^2(1+x^2)+(11-6\gamma)x^6-6e^{x^2}(2+x^2+x^4)+3x^6(2\text{Ei}(x^2)-4\ln x)}{72x^6} + \frac{16x}{105\sqrt{\pi}}\,_2F_2\left(\begin{array}{c}\frac{1}{2},1\\ \frac{3}{2},\frac{9}{2}\end{array}\bigg\vert x^2\right)$
$\frac{1}{2}$	β	$\frac{x^2}{2\Gamma(\beta+1)}\,_2F_2\left(\begin{array}{c}1,1\\ 2,\beta+1\end{array}\bigg\vert x^2\right) + \frac{x}{\Gamma(\beta+\frac{1}{2})}\,_2F_2\left(\begin{array}{c}\frac{1}{2},1\\ \frac{3}{2},\beta+\frac{1}{2}\end{array}\bigg\vert x^2\right)$
1	$\frac{1}{4}$	$\frac{x}{\Gamma(\frac{5}{4})}\,_2F_2\left(\begin{array}{c}1,1\\ \frac{5}{4},2\end{array}\bigg\vert x\right)$
1	$\frac{1}{3}$	$\frac{x}{\Gamma(\frac{4}{3})}\,_2F_2\left(\begin{array}{c}1,1\\ \frac{4}{3},2\end{array}\bigg\vert x\right)$
1	$\frac{1}{2}$	$\frac{2x}{\sqrt{\pi}}\,_2F_2\left(\begin{array}{c}1,1\\ \frac{3}{2},2\end{array}\bigg\vert x\right)$
1	1	$-\gamma-\ln x+\text{Chi}(x)+\text{Shi}(x)$
1	$\frac{3}{2}$	$\frac{4x}{3\sqrt{\pi}}\,_2F_2\left(\begin{array}{c}1,1\\ \frac{5}{2},2\end{array}\bigg\vert x\right)$
1	β	$\frac{x}{\Gamma(\beta+1)}\,_2F_2\left(\begin{array}{c}1,1\\ 2,1+\beta\end{array}\bigg\vert x\right)$

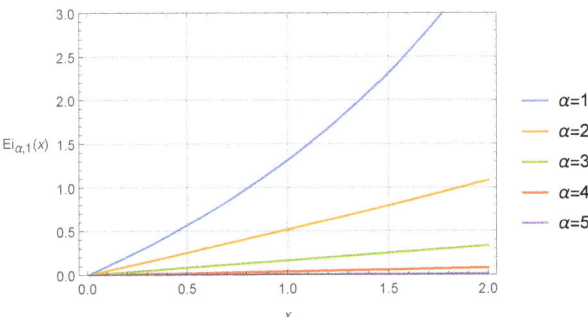

Figure 1. The integral one-parameter Mittag-Leffler function $\text{Ei}_{\alpha,1}(x)$ as a function of variable x and parameters α.

Table 2. The integral Mittag-Leffler functions derived for some values of parameters α and β by using (12).

α	β	$\mathrm{Ei}_{\alpha,\beta}(x)$
$\frac{3}{2}$	$\frac{1}{2}$	$\frac{4x^2}{15\sqrt{\pi}}\,{}_2F_4\left(\begin{array}{c}1,1\\ \frac{7}{6},\frac{3}{2},\frac{11}{6},2\end{array}\Big|\frac{x^2}{27}\right)+\frac{x}{48}\,{}_1F_3\left(\begin{array}{c}\frac{1}{2}\\ 2,\frac{4}{3},\frac{3}{2}\end{array}\Big|\frac{x^2}{27}\right)$
$\frac{3}{2}$	1	$\frac{4x}{3\sqrt{\pi}}\,{}_2F_4\left(\begin{array}{c}\frac{1}{2},1\\ \frac{5}{6},\frac{7}{6},\frac{3}{2},\frac{3}{2}\end{array}\Big|\frac{x^2}{27}\right)+\frac{x^2}{12}\,{}_2F_4\left(\begin{array}{c}1,1\\ \frac{4}{3},\frac{5}{3},2,2\end{array}\Big|\frac{x^2}{27}\right)$
$\frac{3}{2}$	$\frac{3}{2}$	$\frac{x}{2}\,{}_1F_3\left(\begin{array}{c}\frac{1}{2}\\ \frac{4}{3},\frac{3}{2},\frac{5}{3}\end{array}\Big|\frac{x^2}{27}\right)+\frac{8x^2}{105\sqrt{\pi}}\,{}_2F_4\left(\begin{array}{c}1,1\\ \frac{3}{2},\frac{11}{6},2,\frac{13}{6}\end{array}\Big|\frac{x^2}{27}\right)$
$\frac{3}{2}$	2	$\frac{8x}{15\sqrt{\pi}}\,{}_2F_4\left(\begin{array}{c}\frac{1}{2},1\\ \frac{7}{6},\frac{3}{2},\frac{3}{2},\frac{11}{6}\end{array}\Big|\frac{x^2}{27}\right)+\frac{x^2}{48}\,{}_2F_4\left(\begin{array}{c}1,1\\ \frac{5}{3},2,2,\frac{7}{3}\end{array}\Big|\frac{x^2}{27}\right)$
2	$\frac{1}{4}$	$\frac{16x}{5\Gamma(\frac{1}{4})}\,{}_2F_3\left(\begin{array}{c}1,1\\ \frac{9}{8},\frac{11}{8},2\end{array}\Big|\frac{x}{4}\right)$
2	$\frac{1}{3}$	$\frac{9x}{4\Gamma(\frac{1}{3})}\,{}_2F_3\left(\begin{array}{c}1,1\\ \frac{7}{6},\frac{5}{3},2\end{array}\Big|\frac{x}{4}\right)$
2	$\frac{1}{2}$	$\frac{9x}{3\sqrt{\pi}}\,{}_2F_3\left(\begin{array}{c}1,1\\ \frac{5}{4},\frac{7}{4},2\end{array}\Big|\frac{x}{4}\right)$
2	1	$-2\gamma-\ln x+2\mathrm{Chi}(\sqrt{x})$
2	2	$2-2\gamma-\ln x-\frac{2\sinh\sqrt{x}}{\sqrt{x}}+2\mathrm{Chi}(\sqrt{x})$
2	3	$\frac{x}{24}\,{}_2F_3\left(\begin{array}{c}1,1\\ 2,\frac{5}{2},3\end{array}\Big|\frac{x}{4}\right)$
2	4	$\frac{x}{120}\,{}_2F_3\left(\begin{array}{c}1,1\\ 2,\frac{7}{2},3\end{array}\Big|\frac{x}{4}\right)$
2	β	$\frac{x}{\Gamma(\beta+2)}\,{}_2F_3\left(\begin{array}{c}1,1\\ 2,\frac{\beta}{2}+1,\frac{\beta+3}{2}\end{array}\Big|\frac{x}{4}\right)$
3	1	$\frac{x}{6}\,{}_2F_4\left(\begin{array}{c}1,1\\ \frac{4}{3},\frac{5}{3},2,2\end{array}\Big|\frac{x}{27}\right)$
3	β	$\frac{x}{\Gamma(\beta+3)}\,{}_2F_4\left(\begin{array}{c}1,1\\ 2,\frac{\beta}{3}+1,\frac{\beta+4}{3},\frac{\beta+5}{3}\end{array}\Big|\frac{x}{27}\right)$
4	1	$\frac{x}{24}\,{}_2F_5\left(\begin{array}{c}1,1\\ \frac{5}{4},\frac{3}{2},\frac{7}{4},2,2\end{array}\Big|\frac{x}{256}\right)$
4	β	$\frac{x}{\Gamma(\beta+4)}\,{}_2F_5\left(\begin{array}{c}1,1\\ 2,\frac{\beta}{4}+1,\frac{\beta+5}{4},\frac{\beta}{2}+\frac{3}{4},\frac{\beta+7}{4}\end{array}\Big|\frac{x}{256}\right)$
5	1	$\frac{x}{120}\,{}_2F_6\left(\begin{array}{c}1,1\\ \frac{6}{5},\frac{7}{5},\frac{8}{5},\frac{9}{5},2,2\end{array}\Big|\frac{x}{3125}\right)$
5	β	$\frac{x}{\Gamma(\beta+5)}\,{}_2F_6\left(\begin{array}{c}1,1\\ 2,\frac{\beta}{5}+1,\frac{\beta+6}{5},\frac{\beta+7}{5},\frac{\beta+8}{5},\frac{\beta+9}{5}\end{array}\Big|\frac{x}{3125}\right)$

The Laplace transforms of the integral Mittag-Leffler functions with positive rational parameter α with $\alpha = p/q$ and p,q positive coprimes can be evaluated from:

$$\mathcal{L}\left[\mathrm{Ei}_{p/q,\beta}(t)\right] \tag{16}$$

$$= \frac{1}{s^2}\sum_{k=0}^{q-1}\frac{k!\,s^{-k}}{\Gamma\left(\frac{p}{q}(k+1)+\beta\right)}\,{}_{q+1}F_p\left(\begin{array}{c}1,a_0,\ldots,a_{q-1}\\ b_0,\ldots,b_{p-1}\end{array}\Big|\frac{(q/s)^q}{p^p}\right).$$

where

$$a_j = \frac{k+1+j}{q},$$
$$b_j = \frac{k}{q} + \frac{\beta+j}{p}.$$

Furthermore, the following relation is satisfied:

$$\mathcal{L}\left[\text{Ei}_{p/q,\beta}(t)\right] = \frac{1}{p^{p/q}s}\mathcal{L}\left[\text{E}_{p/q,\beta}(t)\right]. \qquad (17)$$

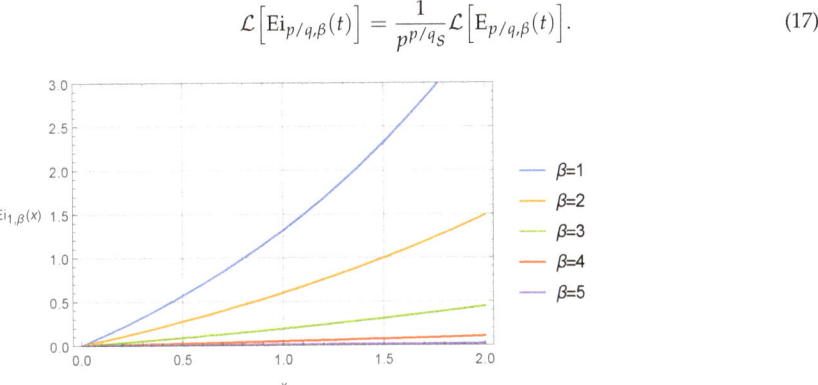

Figure 2. The integral two-parameter Mittag-Leffler function $\text{Ei}_{1,\beta}(x)$ as a function of variable x and parameters β.

Table 3. The Laplace transforms of the integral Mittag-Leffler functions $\text{Ei}_{\alpha,\beta}$ derived for some values of parameters α and β by using (15).

α	β	$\mathcal{L}\left[\text{Ei}_{\alpha,\beta}(t)\right]$
1	$\frac{1}{5}$	$\frac{5}{\Gamma(\frac{1}{5})s^2}\,{}_2F_1\left(\begin{array}{c}1,1\\ \frac{6}{5}\end{array}\bigg|\frac{1}{s}\right)$
1	$\frac{1}{4}$	$\frac{4}{\Gamma(\frac{1}{4})s^2}\,{}_2F_1\left(\begin{array}{c}1,1\\ \frac{5}{4}\end{array}\bigg|\frac{1}{s}\right)$
1	$\frac{1}{3}$	$\frac{3}{\Gamma(\frac{1}{6})s^2}\,{}_2F_1\left(\begin{array}{c}1,1\\ \frac{4}{3}\end{array}\bigg|\frac{1}{s}\right)$
1	$\frac{1}{2}$	$\frac{2\csc^{-1}(\sqrt{s})}{\sqrt{\pi}s\sqrt{s-1}}$
1	1	$-\frac{1}{s}\ln\left(1-\frac{1}{s}\right)$
1	$\frac{3}{2}$	$\frac{4}{\sqrt{\pi}s}\left[1-\sqrt{s-1}\csc^{-1}(\sqrt{s})\right]$
1	β	$\frac{1}{s^2\Gamma(\beta+1)}\,{}_2F_1\left(\begin{array}{c}1,1\\ \beta+1\end{array}\bigg|\frac{1}{s}\right)$
$\frac{3}{2}$	$\frac{1}{2}$	$\frac{8}{15\sqrt{\pi}s^3}\,{}_2F_2\left(\begin{array}{c}1,1\\ \frac{7}{6},\frac{11}{6}\end{array}\bigg|\frac{4}{27s^2}\right)+\frac{1}{s^2}\,{}_2F_2\left(\begin{array}{c}\frac{1}{2},1\\ \frac{2}{3},\frac{4}{3}\end{array}\bigg|\frac{4}{27s^2}\right)$
$\frac{3}{2}$	1	$\frac{4}{3\sqrt{\pi}s^2}\,{}_3F_3\left(\begin{array}{c}\frac{1}{2},1,1\\ \frac{5}{6},\frac{7}{6},\frac{3}{2}\end{array}\bigg|\frac{4}{27s^2}\right)+\frac{1}{6s^3}\,{}_3F_3\left(\begin{array}{c}1,1,\frac{3}{2}\\ \frac{4}{3},\frac{5}{6},2\end{array}\bigg|\frac{4}{27s^2}\right)$

Table 3. Cont.

α	β	$\mathcal{L}\left[\mathrm{Ei}_{\alpha,\beta}(t)\right]$
$\frac{3}{2}$	$\frac{3}{2}$	$\frac{1}{2s^2}\,{}_2F_2\!\left(\begin{array}{c}\frac{1}{2},1\\ \frac{4}{3},\frac{5}{3}\end{array}\bigg|\frac{4}{27s^2}\right)+\frac{16}{105\sqrt{\pi}s^3}\,{}_2F_2\!\left(\begin{array}{c}1,1\\ \frac{11}{6},\frac{13}{6}\end{array}\bigg|\frac{4}{27s^2}\right)$
$\frac{3}{2}$	2	$\frac{8}{15\sqrt{\pi}s^2}\,{}_3F_3\!\left(\begin{array}{c}\frac{1}{2},1,1\\ \frac{7}{6},\frac{3}{2},\frac{11}{6}\end{array}\bigg|\frac{4}{27s^2}\right)+\frac{1}{24s^3}\,{}_3F_3\!\left(\begin{array}{c}1,1,\frac{3}{2}\\ \frac{5}{3},2,\frac{7}{3}\end{array}\bigg|\frac{4}{27s^2}\right)$
2	$\frac{1}{5}$	$\frac{25}{6\Gamma(\frac{1}{5})s^2}\,{}_2F_2\!\left(\begin{array}{c}1,1\\ \frac{11}{10},\frac{8}{5}\end{array}\bigg|\frac{1}{4s}\right)$
2	$\frac{1}{4}$	$\frac{16}{5\Gamma(\frac{1}{4})s^2}\,{}_2F_2\!\left(\begin{array}{c}1,1\\ \frac{9}{8},\frac{13}{8}\end{array}\bigg|\frac{1}{4s}\right)$
2	$\frac{1}{3}$	$\frac{9}{4\Gamma(\frac{1}{3})s^2}\,{}_2F_2\!\left(\begin{array}{c}1,1\\ \frac{7}{6},\frac{5}{3}\end{array}\bigg|\frac{1}{4s}\right)$
2	$\frac{1}{2}$	$\frac{4}{3\sqrt{\pi}s^2}\,{}_2F_2\!\left(\begin{array}{c}1,1\\ \frac{5}{4},\frac{7}{4}\end{array}\bigg|\frac{1}{4s}\right)$
2	1	$\frac{1}{2s^2}\,{}_2F_2\!\left(\begin{array}{c}1,1\\ \frac{3}{2},2\end{array}\bigg|\frac{1}{4s}\right)$
2	2	$\frac{1}{6s^2}\,{}_2F_2\!\left(\begin{array}{c}1,1\\ \frac{5}{2},2\end{array}\bigg|\frac{1}{4s}\right)$
2	3	$\frac{1}{24s^2}\,{}_2F_2\!\left(\begin{array}{c}1,1\\ \frac{5}{2},3\end{array}\bigg|\frac{1}{4s}\right)$
2	4	$\frac{1}{120s^2}\,{}_2F_2\!\left(\begin{array}{c}1,1\\ \frac{7}{2},3\end{array}\bigg|\frac{1}{4s}\right)$
2	β	$\frac{1}{\Gamma(\beta+2)s^2}\,{}_2F_2\!\left(\begin{array}{c}1,1\\ 1+\frac{\beta}{2},\frac{\beta+3}{2}\end{array}\bigg|\frac{1}{4s}\right)$

Table 4. The Laplace transforms of the integral Mittag-Leffler functions $\mathrm{Ei}_{\alpha,\beta}$ derived for some values of parameters α and β by using (15).

α	β	$\mathcal{L}\left[\mathrm{Ei}_{\alpha,\beta}(t)\right]$
3	$\frac{1}{5}$	$\frac{125}{66\Gamma(\frac{1}{5})s^2}\,{}_2F_3\!\left(\begin{array}{c}1,1\\ \frac{16}{15},\frac{7}{5},\frac{26}{15}\end{array}\bigg|\frac{1}{27s}\right)$
3	$\frac{1}{4}$	$\frac{64}{25\Gamma(\frac{1}{4})s^2}\,{}_2F_3\!\left(\begin{array}{c}1,1\\ \frac{13}{12},\frac{17}{12},\frac{7}{4}\end{array}\bigg|\frac{1}{27s}\right)$
3	$\frac{1}{3}$	$\frac{27}{28\Gamma(\frac{1}{3})s^2}\,{}_2F_3\!\left(\begin{array}{c}1,1\\ \frac{10}{9},\frac{13}{9},\frac{10}{9}\end{array}\bigg|\frac{1}{27s}\right)$
3	$\frac{1}{2}$	$\frac{27}{15\sqrt{\pi}s^2}\,{}_2F_3\!\left(\begin{array}{c}1,1\\ \frac{7}{6},\frac{3}{2},\frac{11}{6}\end{array}\bigg|\frac{1}{27s}\right)$
3	1	$\frac{1}{6s^2}\,{}_2F_3\!\left(\begin{array}{c}1,1\\ \frac{4}{3},\frac{5}{3},2\end{array}\bigg|\frac{1}{27s}\right)$
3	3	$\frac{1}{120s^2}\,{}_2F_3\!\left(\begin{array}{c}1,1\\ 2,\frac{7}{3},\frac{8}{3}\end{array}\bigg|\frac{1}{27s}\right)$
3	β	$\frac{1}{\Gamma(\beta+3)s^2}\,{}_2F_3\!\left(\begin{array}{c}1,1\\ 1+\frac{\beta}{3},\frac{\beta+4}{3},\frac{\beta+5}{3}\end{array}\bigg|\frac{1}{27s}\right)$

Table 4. *Cont.*

α	β	$\mathcal{L}[\text{Ei}_{\alpha,\beta}(t)]$
4	1	$\frac{1}{24s^2}\,_2F_4\left(\begin{array}{c}1,1\\ \frac{5}{4},\frac{3}{2},\frac{7}{4},2\end{array}\Big|\frac{1}{256s}\right)$
4	4	$\frac{1}{5040s^2}\,_2F_4\left(\begin{array}{c}1,1\\ 2,\frac{9}{4},\frac{5}{2},\frac{11}{4}\end{array}\Big|\frac{1}{256s}\right)$
4	β	$\frac{1}{\Gamma(\beta+4)s^2}\,_2F_4\left(\begin{array}{c}1,1\\ 1+\frac{\beta}{4},\frac{\beta+5}{3},\frac{3}{2}+\frac{\beta}{4},\frac{\beta+7}{4}\end{array}\Big|\frac{1}{256s}\right)$
5	1	$\frac{1}{120s^2}\,_2F_5\left(\begin{array}{c}1,1\\ \frac{6}{5},\frac{7}{5},\frac{8}{5},\frac{9}{5}\end{array}\Big|\frac{1}{3125s}\right)$
5	β	$\frac{1}{\Gamma(\beta+5)s^2}\,_2F_5\left(\begin{array}{c}1,1\\ 1+\frac{\beta}{5},\frac{\beta+6}{5},\frac{\beta+7}{5},\frac{\beta+8}{5},\frac{\beta+9}{5}\end{array}\Big|\frac{1}{3125s}\right)$

3. The Integral Whittaker Functions

In 1903, Whittaker [15] showed that it is possible to express some special functions such as Bessel functions, parabolic cylinder functions, error functions, incomplete gamma functions, and logarithm and cosine integrals in terms of a new function suggested by him, i.e., the Whittaker function. Two Whittaker functions are applied today, and they are defined by using the Kummer confluent hypergeometric function [3,4]:

$$M_{\kappa,\mu}(x) = x^{\mu-1/2}e^{-x/2}\,_1F_1\left(\begin{array}{c}\mu-\kappa+\frac{1}{2}\\ 1+2\mu\end{array}\Big|x\right),$$
$$W_{\kappa,\mu}(x) = \frac{\Gamma(-2\mu)}{\Gamma\left(\frac{1}{2}-k-\mu\right)}M_{\kappa,\mu}(x) + \frac{\Gamma(2\mu)}{\Gamma\left(\frac{1}{2}-k+\mu\right)}M_{\kappa,-\mu}(x). \quad (18)$$

This permits us to introduce four integral Whittaker functions:

$$\text{Mi}_{\kappa,\mu}(x) = \int_0^x \frac{M_{\kappa,\mu}(t)}{t}\,dt,$$
$$\text{mi}_{\kappa,\mu}(x) = \int_x^\infty \frac{M_{\kappa,\mu}(t)}{t}\,dt. \quad (19)$$

and

$$\text{Wi}_{\kappa,\mu}(x) = \int_0^x \frac{W_{\kappa,\mu}(t)}{t}\,dt,$$
$$\text{wi}_{\kappa,\mu}(x) = \int_x^\infty \frac{W_{\kappa,\mu}(t)}{t}\,dt. \quad (20)$$

The integral Whittaker functions with particular values of parameters κ and μ can be expressed in terms of elementary and special functions. These cases, derived using the MATHEMATICA program, are presented in Tables 5–9. Several integral Whittaker functions $\text{Mi}_{\kappa,\mu}(x)$, $\text{mi}_{\kappa,\mu}(x)$, $\text{Wi}_{\kappa,\mu}(x)$ and $\text{wi}_{\kappa,\mu}(x)$ as a function of variable x at fixed values of parameters κ and μ are plotted in Figures 3–6. Similarly, a long list of the Whittaker functions $M_{\kappa,\mu}(x)$ and $W_{\kappa,\mu}(x)$ with integer and fractional parameters was prepared (see Appendix B). In some cases, it was possible to obtain for them their Laplace transforms, and they are also reported in Appendix B.

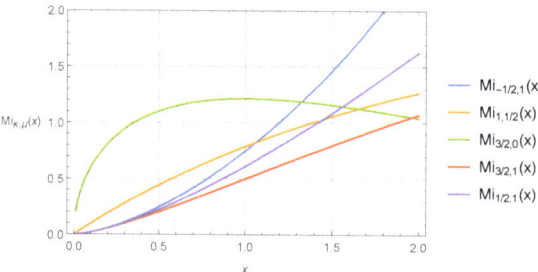

Figure 3. The integral Whittaker functions $\mathrm{Mi}_{\kappa,\mu}(x)$ as a function of variable x at fixed values of parameters κ and μ.

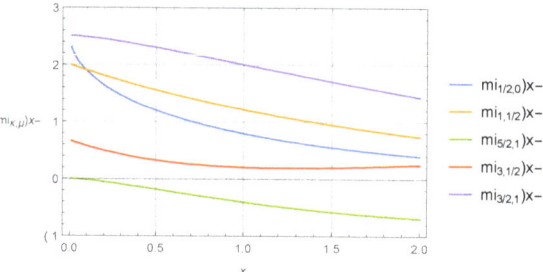

Figure 4. The integral Whittaker functions $\mathrm{mi}_{\kappa,\mu}(x)$ as a function of variable x at fixed values of parameters κ and μ.

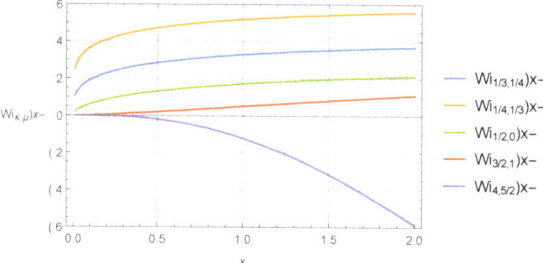

Figure 5. The integral Whittaker functions $\mathrm{Wi}_{\kappa,\mu}(x)$ as a function of variable x at fixed values of parameters κ and μ.

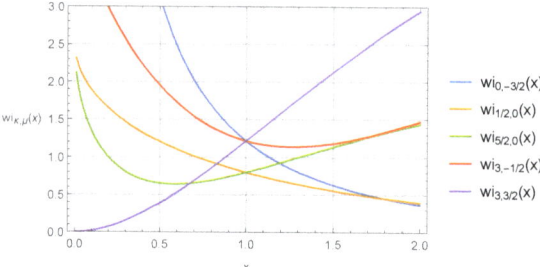

Figure 6. The integral Whittaker functions $\mathrm{wi}_{\kappa,\mu}(x)$ as a function of variable x at fixed values of parameters κ and μ.

Table 5. The integral Whittaker functions $\text{Mi}_{\kappa,\mu}$ derived for some values of parameters κ and μ by using (18) and (19).

κ	μ	$\text{Mi}_{\kappa,\mu}(x)$
$-\frac{5}{2}$	0	$e^{x/2}\left[\sqrt{x}(x+1) + 2F\left(\sqrt{\frac{x}{2}}\right)\right]$, $F(x) = e^{-x^2}\int_0^x e^{t^2} dt$
$-\frac{3}{2}$	0	$2\sqrt{x}e^{x/2}$
$-\frac{3}{2}$	$\frac{1}{2}$	$\frac{1}{6}\left[(8x + 3\pi x\, \mathbf{L}_0(\frac{x}{2}))I_1(\frac{x}{2}) + (-\pi x\, \mathbf{L}_1(\frac{x}{2}) + 6x + 4)I_0(\frac{x}{2}) - 4\right]$
$-\frac{3}{2}$	1	$2e^{x/2}\left[\sqrt{x} - \sqrt{2}F\left(\sqrt{\frac{x}{2}}\right)\right]$
$-\frac{3}{2}$	$\frac{3}{2}$	$\frac{2}{5}\left[(-8 + 8x + 3\pi x\, \mathbf{L}_0(\frac{x}{2}))I_1(\frac{x}{2}) + (-3\pi x\, \mathbf{L}_1(\frac{x}{2}) + 2x - 12)I_0(\frac{x}{2}) + 12\right]$
$-\frac{3}{2}$	2	$8\left[\sqrt{2\pi}\,\text{erf}\left(\sqrt{\frac{x}{2}}\right) + \frac{e^{-x/2}}{x^{-3/2}}(2(x-1) + e^x(x(x-4) + 2))\right]$
$-\frac{3}{2}$	$\frac{5}{2}$	$\frac{16}{7}\left[20 + (5\pi x\, \mathbf{L}_1(\frac{x}{2}) + 18x - 44)I_0(\frac{x}{2}) + \frac{1}{x}(-5\pi x^2\, \mathbf{L}_0(\frac{x}{2}) + 8x(x-9) + 96)I_1(\frac{x}{2})\right]$
$-\frac{3}{2}$	3	$30\left[x^{-5/2}e^{-x/2}(48 + 2e^x(x^3 - 12x^2 + 24x - 24)) + 3\sqrt{2\pi}\,\text{erfi}\left(\sqrt{\frac{x}{2}}\right)\right]$
$-\frac{1}{2}$	0	$\sqrt{2\pi}\,\text{erfi}\left(\sqrt{\frac{x}{2}}\right)$
$-\frac{1}{2}$	$\frac{1}{2}$	$\frac{1}{2}\left[-\pi x\, \mathbf{L}_0(\frac{x}{2})I_1(\frac{x}{2}) + (\pi x\, \mathbf{L}_1(\frac{x}{2}) + 2x + 4)I_0(\frac{x}{2}) - 4\right]$
$-\frac{1}{2}$	1	$8x^{-1/2}\sinh(\frac{x}{2}) - 2\sqrt{2\pi}\,\text{erf}\left(\sqrt{\frac{x}{2}}\right)$
$-\frac{1}{2}$	$\frac{3}{2}$	$2(\pi x\, \mathbf{L}_0(\frac{x}{2}) + 8)I_1(\frac{x}{2}) - 2(\pi x\, \mathbf{L}_1(\frac{x}{2}) + 2x - 4)I_0(\frac{x}{2}) - 8$
$-\frac{1}{2}$	2	$48x^{-3/2}e^{-x/2}[(x-1)e^x + 1] - 12\sqrt{2\pi}\,\text{erfi}\left(\sqrt{\frac{x}{2}}\right)$
0	$\frac{1}{8}$	$\frac{8}{5}x^{5/8}\,{}_1F_2\left(\begin{array}{c}\frac{5}{16}\\ \frac{9}{8},\frac{21}{16}\end{array}\middle|\frac{x^2}{16}\right)$
0	$\frac{1}{7}$	$\frac{14}{9}x^{9/14}\,{}_1F_2\left(\begin{array}{c}\frac{9}{28}\\ \frac{8}{7},\frac{37}{28}\end{array}\middle|\frac{x^2}{16}\right)$
0	$\frac{1}{6}$	$\frac{3}{2}x^{2/3}\,{}_1F_2\left(\begin{array}{c}\frac{1}{3}\\ \frac{7}{6},\frac{4}{3}\end{array}\middle|\frac{x^2}{16}\right)$
0	$\frac{1}{5}$	$\frac{10}{7}x^{7/10}\,{}_1F_2\left(\begin{array}{c}\frac{7}{20}\\ \frac{6}{5},\frac{27}{20}\end{array}\middle|\frac{x^2}{16}\right)$
0	$\frac{1}{4}$	$\frac{4}{3}x^{3/4}\,{}_1F_2\left(\begin{array}{c}\frac{3}{8}\\ \frac{5}{4},\frac{11}{8}\end{array}\middle|\frac{x^2}{16}\right)$
0	$\frac{1}{3}$	$\frac{6}{5}x^{5/6}\,{}_1F_2\left(\begin{array}{c}\frac{5}{12}\\ \frac{4}{3},\frac{17}{12}\end{array}\middle|\frac{x^2}{16}\right)$
0	$\frac{1}{2}$	$2\,\text{Shi}\left(\frac{x}{2}\right)$
0	1	$\frac{2}{3}x^{3/2}\,{}_1F_2\left(\begin{array}{c}\frac{3}{4}\\ \frac{7}{4},2\end{array}\middle|\frac{x^2}{16}\right)$
0	$\frac{3}{2}$	$\frac{24}{x}\sinh\left(\frac{x}{2}\right) - 12$
0	2	$\frac{2}{5}x^{5/2}\,{}_1F_2\left(\begin{array}{c}\frac{5}{4}\\ \frac{9}{4},3\end{array}\middle|\frac{x^2}{16}\right)$
0	$\frac{5}{2}$	$60\left[x^{-2}(6x\cosh(\frac{x}{2}) - 2\sinh(\frac{x}{2})) - \text{Shi}(\frac{x}{2})\right]$

Table 6. The integral Whittaker functions $\text{Mi}_{\kappa,\mu}$ derived for some values of parameters κ and μ by using (18) and (19).

κ	μ	$\text{Mi}_{\kappa,\mu}(x)$
$\frac{1}{2}$	0	$\sqrt{2\pi}\,\text{erf}\left(\sqrt{\frac{x}{2}}\right)$
$\frac{1}{2}$	$\frac{1}{2}$	$\frac{1}{2}\left[-\pi x\,\mathbf{L}_0\left(\frac{x}{2}\right)I_1\left(\frac{x}{2}\right)+\left(\pi x\,\mathbf{L}_1\left(\frac{x}{2}\right)+2x-4\right)I_0\left(\frac{x}{2}\right)+4\right]$
$\frac{1}{2}$	1	$2\sqrt{2\pi}\,\text{erfi}\left(\sqrt{\frac{x}{2}}\right)-8x^{-1/2}\sinh\left(\frac{x}{2}\right)$
$\frac{1}{2}$	$\frac{3}{2}$	$-2\left(\pi x\,\mathbf{L}_0\left(\frac{x}{2}\right)+8\right)I_1\left(\frac{x}{2}\right)+2\left(\pi x\,\mathbf{L}_1\left(\frac{x}{2}\right)+2x+4\right)I_0\left(\frac{x}{2}\right)+8$
$\frac{1}{2}$	$\frac{5}{2}$	$16\left[\left(\pi x\,\mathbf{L}_0\left(\frac{x}{2}\right)+8\right)I_1\left(\frac{x}{2}\right)-\left(\pi x\,\mathbf{L}_1\left(\frac{x}{2}\right)+2x+4\right)I_0\left(\frac{x}{2}\right)+4\right]$ $+8x^2\,{}_2F_3\left(\begin{array}{c}1,1\\2,2,2\end{array}\Big|\frac{x^2}{16}\right)-4x^2\,{}_2F_3\left(\begin{array}{c}1,1\\2,2,3\end{array}\Big|\frac{x^2}{16}\right)$
1	0	$\frac{\sqrt{x}}{30}\left[60\,{}_1F_2\left(\begin{array}{c}\frac{1}{4}\\1,\frac{5}{4}\end{array}\Big|\frac{x^2}{16}\right)+3x^2\,{}_1F_2\left(\begin{array}{c}\frac{5}{4}\\2,\frac{9}{4}\end{array}\Big|\frac{x^2}{16}\right)-20x\,{}_1F_2\left(\begin{array}{c}\frac{3}{4}\\1,\frac{7}{4}\end{array}\Big|\frac{x^2}{16}\right)\right]$
1	$\frac{1}{2}$	$2\left(1-e^{-x/2}\right)$
1	1	$-\frac{2x^{3/2}}{45}\left[-20\,{}_1F_2\left(\begin{array}{c}\frac{3}{4}\\1,\frac{7}{4}\end{array}\Big|\frac{x^2}{16}\right)+5x^2\,{}_1F_2\left(\begin{array}{c}\frac{3}{4}\\\frac{7}{4},\frac{3}{4}\end{array}\Big|\frac{x^2}{16}\right)+3x\,{}_1F_2\left(\begin{array}{c}\frac{5}{4}\\2,\frac{9}{4}\end{array}\Big|\frac{x^2}{16}\right)\right]$
$\frac{3}{2}$	0	$2\sqrt{x}e^{-x/2}$
$\frac{3}{2}$	$\frac{1}{2}$	$\frac{1}{6}\left[x\left(\pi\,\mathbf{L}_0\left(\frac{x}{2}\right)-8\right)I_1\left(\frac{x}{2}\right)+\left(\pi x\,\mathbf{L}_1\left(\frac{x}{2}\right)-6x+4\right)I_0\left(\frac{x}{2}\right)+4\right]$
$\frac{3}{2}$	1	$\sqrt{2\pi}\,\text{erf}\left(\sqrt{\frac{x}{2}}\right)-2\sqrt{x}e^{-x/2}$
$\frac{3}{2}$	$\frac{3}{2}$	$\frac{1}{5}\left[2\left(-3\pi\,\mathbf{L}_0\left(\frac{x}{2}\right)+8x+8\right)I_1\left(\frac{x}{2}\right)+\left(6\pi x\,\mathbf{L}_1\left(\frac{x}{2}\right)-4x+24\right)I_0\left(\frac{x}{2}\right)+24\right]$
2	$\frac{1}{2}$	$xe^{-x/2}$
2	$\frac{3}{2}$	$4-2(2+x)e^{-x/2}$
2	$\frac{5}{2}$	$-5x^{-2}\left(6(2+x)e^x-2x(3+x)^2-12\right)+30\,\text{Shi}\left(\frac{x}{2}\right)$

Table 7. The integral Whittaker functions $\text{mi}_{\kappa,\mu}$ derived for some values of parameters κ and μ by using (18) and (19).

κ	μ	$\text{mi}_{\kappa,\mu}(x)$
$\frac{1}{2}$	0	$\sqrt{2\pi}\,\text{erfc}\left(\sqrt{\frac{x}{2}}\right)$
1	$\frac{1}{2}$	$2e^{-x/2}$
$\frac{3}{2}$	1	$2\sqrt{x}e^{-x/2}+\sqrt{2\pi}\,\text{erfc}\left(\sqrt{\frac{x}{2}}\right)$
2	$\frac{1}{2}$	$-e^{-x/2}$
2	$\frac{3}{2}$	$2(2+x)e^{-x/2}$
$\frac{5}{2}$	1	$-\frac{2}{3}x^{3/2}e^{-x/2}$
$\frac{5}{2}$	2	$2\sqrt{x}(3+x)e^{-x/2}+3\sqrt{2\pi}\,\text{erfc}\left(\sqrt{\frac{x}{2}}\right)$
3	$\frac{1}{2}$	$\frac{1}{3}[2+(x-2)x]e^{-x/2}$
3	$\frac{3}{2}$	$-\frac{1}{2}x^2e^{-x/2}$
4	$\frac{1}{2}$	$-\frac{x}{12}[12+(x-6)x]e^{-x/2}$
4	$\frac{3}{2}$	$\frac{1}{10}\left[8+(x-2)^2x\right]e^{-x/2}$

Table 8. The integral Whittaker functions $\text{Wi}_{\kappa,\mu}$ derived for some values of parameters κ and μ by using (18) and (20).

κ	μ	$\text{Wi}_{\kappa,\mu}(x)$
$-\frac{3}{2}$	0	$-2\sqrt{x}e^{-x/2}\text{Ei}(-x) + \sqrt{2\pi}\,\text{erfc}\left(\sqrt{\frac{x}{2}}\right)$
$\frac{1}{4}$	$\frac{1}{4}$	$2^{1/4}\gamma\left(\frac{1}{4},\frac{x}{2}\right)$
$\frac{1}{2}$	0	$\sqrt{2\pi}\,\text{erf}\left(\sqrt{\frac{x}{2}}\right)$
1	$-\frac{1}{2}$	$2\left(1-e^{-x/2}\right)$
1	$\frac{1}{2}$	$2\left(1-e^{-x/2}\right)$
$\frac{3}{2}$	0	$-2\sqrt{x}e^{-x/2}$
$\frac{3}{2}$	1	$\sqrt{2\pi}\,\text{erf}\left(\sqrt{\frac{x}{2}}\right) - 2\sqrt{x}e^{-x/2}$
2	$-\frac{1}{2}$	$-2xe^{-x/2}$
2	$\frac{1}{2}$	$-2xe^{-x/2}$
2	$\frac{3}{2}$	$4 - 2(2+x)e^{-x/2}$
$\frac{5}{2}$	0	$\sqrt{2\pi}\,\text{erf}\left(\sqrt{\frac{x}{2}}\right) - 2\sqrt{x}(x-1)e^{-x/2}$
3	$\frac{3}{2}$	$-2x^2 e^{-x/2}$
3	$\frac{5}{2}$	$16 - 2[8 + (4+x)x]e^{-x/2}$
4	$-\frac{1}{2}$	$-2x[12 + (x-6)x]e^{-x/2}$
4	$\frac{1}{2}$	$-2x[12 + (x-6)x]e^{-x/2}$
4	$\frac{3}{2}$	$16 - 2\left[8 + x(x-2)^2\right]e^{-x/2}$
4	$\frac{5}{2}$	$-2x^3 e^{-x/2}$

Table 9. The integral Whittaker functions $\text{wi}_{\kappa,\mu}$ derived for some values of parameters κ and μ by using (18) and (20).

κ	μ	$\text{wi}_{\kappa,\mu}(x)$
$-\frac{1}{2}$	1	$2x^{-1/2}e^{-x/2} - \sqrt{2\pi}\,\text{erfc}\left(\sqrt{\frac{x}{2}}\right)$
$-\frac{1}{2}$	2	$2x^{-3/2}e^{-x/2}$
$-\frac{1}{2}$	3	$\frac{1}{3}\left[-2(x-6)(x+2)x^{-5/2}e^{-x/2} + \sqrt{2\pi}\,\text{erfc}\left(\sqrt{\frac{x}{2}}\right)\right]$
0	$-\frac{5}{2}$	$\frac{1}{2}\text{Ei}\left(-\frac{x}{2}\right) + 3x^{-2}(x+2)e^{-x/2}$
0	$-\frac{3}{2}$	$2x^{-1}e^{-x/2}$
0	$-\frac{1}{2}$	$-\text{Ei}\left(-\frac{x}{2}\right)$
$\frac{1}{2}$	0	$\sqrt{2\pi}\,\text{erfc}\left(\sqrt{\frac{x}{2}}\right)$
1	$-\frac{1}{2}$	$2e^{-x/2}$
1	$\frac{1}{2}$	$2e^{-x/2}$
1	$\frac{5}{2}$	$2x^{-2}[6 + x(6+x)]e^{-x/2}$
$\frac{3}{2}$	0	$2\sqrt{x}e^{-x/2}$

Table 9. Cont.

κ	μ	$\mathrm{wi}_{\kappa,\mu}(x)$
$\frac{3}{2}$	1	$\sqrt{2\pi}\,\mathrm{erfc}\left(\sqrt{\frac{x}{2}}\right) + 2\sqrt{x}e^{-x/2}$
2	$\frac{1}{2}$	$2xe^{-x/2}$
2	$\frac{3}{2}$	$2(2+x)e^{-x/2}$
$\frac{5}{2}$	0	$\sqrt{2\pi}\,\mathrm{erfc}\left(\sqrt{\frac{x}{2}}\right) + 2\sqrt{x}(x-1)e^{-x/2}$
$\frac{5}{2}$	1	$2x^{3/2}e^{-x/2}$
3	$-\frac{1}{2}$	$2[2 + x(x-2)]e^{-x/2}$
3	$\frac{1}{2}$	$2[2 + x(x-2)]e^{-x/2}$
3	$\frac{3}{2}$	$2x^2 e^{-x/2}$
3	$\frac{5}{2}$	$2[8 + x(4+x)]e^{-x/2}$
4	$-\frac{1}{2}$	$2x[12 + (x-6)x]e^{-x/2}$
4	$\frac{1}{2}$	$2x[12 + (x-6)x]e^{-x/2}$
4	$\frac{3}{2}$	$2\left[8 + (x-2)^2 x\right]e^{-x/2}$

There is a number of recurrence relations between the Whittaker functions, for example [3,4]

$$2\mu\left[M_{\kappa-1/2,\mu-1/2}(t) - M_{\kappa+1/2,\mu-1/2}(t)\right] = t^{1/2}M_{\kappa,\mu}(t),$$
$$(\kappa+\mu)W_{\kappa-1/2,\mu}(t) + W_{\kappa+1/2,\mu}(t) = t^{1/2}W_{\kappa,\mu+1/2}(t), \tag{21}$$

and this leads to integrals that are expressed in terms of the integral Whittaker functions

$$\int_0^x \frac{M_{\kappa,\mu}(t)}{t^{1/2}}dt = 2\mu\left[\mathrm{Mi}_{\kappa-1/2,\mu-1/2}(t) - \mathrm{Mi}_{\kappa+1/2,\mu-1/2}(t)\right],$$
$$\int_0^x \frac{W_{\kappa,\mu+1/2}(t)}{t^{1/2}}dt = (\kappa+\mu)\mathrm{Wi}_{\kappa-1/2,\mu}(t) + \mathrm{Wi}_{\kappa+1/2,\mu}(t). \tag{22}$$

Using the following representation of the Whittaker functions [5]

$$M_{\kappa,\mu}(t) = t^{\mu+1/2}\sum_{n=0}^{\infty} {}_2F_1\left(\begin{array}{c}-n,\mu-\kappa+\frac{1}{2}\\ 1+2\mu\end{array}\bigg|2\right)\frac{(-t/2)^n}{n!}, \tag{23}$$

it is possible to obtain the integral Whittaker functions in terms of a rapidly convergent alternating series as follows:

$$\mathrm{Mi}_{\kappa,\mu}(x) = x^{\mu+1/2}\sum_{n=0}^{\infty} {}_2F_1\left(\begin{array}{c}-n,\mu-\kappa+\frac{1}{2}\\ 1+2\mu\end{array}\bigg|2\right)\frac{(-x/2)^n}{n!\left(\frac{1}{2}+\mu+n\right)}. \tag{24}$$

There is a number of particular cases where the integral Whittaker functions can be written in a closed-form, for example, from [5]

$$M_{\kappa,\kappa-1/2}(x) = x^\kappa e^{-x/2}, \tag{25}$$

we have

$$\mathrm{Mi}_{\kappa,\kappa-1/2}(x) = 2^\kappa \gamma\left(\kappa, \frac{x}{2}\right), \tag{26}$$

but [5]
$$M_{\kappa,\kappa-1/2}(x) = W_{\kappa,\kappa-1/2}(x) = W_{\kappa,-\kappa+1/2}(x) = e^{-x/2}x^{\kappa}, \tag{27}$$
and therefore
$$\mathrm{Mi}_{\kappa,\kappa-1/2}(x) = \mathrm{Wi}_{\kappa,\kappa-1/2}(x) = \mathrm{Wi}_{\kappa,-\kappa+1/2}(x) = 2^{\kappa}\,\gamma\!\left(\kappa,\frac{x}{2}\right). \tag{28}$$

Furthermore, from [5]
$$M_{0,\mu}(x) = 2^{2\mu+1/2}\Gamma(\mu+1)\sqrt{\frac{t}{2}}\,I_{\mu}\!\left(\frac{x}{2}\right), \tag{29}$$
follows that
$$\mathrm{Mi}_{0,\mu}(x) = \frac{x^{\mu+1/2}}{\mu+1/2}\,{}_1F_2\!\left(\begin{array}{c}\frac{2\mu+1}{4}\\ \mu+1,\frac{2\mu+5}{4}\end{array}\bigg|\,\frac{x^2}{16}\right). \tag{30}$$

Similary from
$$W_{0,\mu}(x) = \sqrt{\frac{x}{\pi}}\,K_{\mu}\!\left(\frac{x}{2}\right), \tag{31}$$
we have
$$\mathrm{Wi}_{0,\mu}(x) = \frac{\sqrt{\pi}}{2\sin\pi\mu}\left[\frac{4^{\mu}\,\mathrm{Mi}_{0,-\mu}(x)}{\Gamma(1-\mu)} - \frac{4^{-\mu}\,\mathrm{Mi}_{0,\mu}(x)}{\Gamma(1+\mu)}\right], \tag{32}$$
and in a general case
$$\mathrm{Wi}_{\kappa,\mu}(x) = \frac{\Gamma(-2\mu)\mathrm{Mi}_{\kappa,\mu}(x)}{\Gamma\!\left(\frac{1}{2}-\kappa-\mu\right)} + \frac{\Gamma(2\mu)\mathrm{Mi}_{\kappa,-\mu}(x)}{\Gamma\!\left(\frac{1}{2}-\kappa+\mu\right)}. \tag{33}$$

For $\kappa = \pm 1/2$, it is possible to obtain
$$\mathrm{Wi}_{\pm\frac{1}{2},\mu}(x) = F_{\mu}^{\pm}(x) + F_{-\mu}^{\pm}(x), \tag{34}$$
where we have set
$$F_{\mu}^{\pm}(x) = \frac{2x^{1/2+\mu}\Gamma(-2\mu)}{(1+2\mu)\Gamma\!\left(\frac{1}{2}\mp\frac{1}{2}-\mu\right)} \\ \left[{}_1F_2\!\left(\begin{array}{c}\frac{1}{4}+\frac{\mu}{2}\\ \frac{1}{2}+\mu,\frac{3}{4}+\frac{\mu}{2}\end{array}\bigg|\,\frac{x^2}{16}\right) \mp \frac{x/2}{3+2\mu}\,{}_1F_2\!\left(\begin{array}{c}\frac{3}{4}+\frac{\mu}{2}\\ \frac{3}{2}+\mu,\frac{7}{4}+\frac{\mu}{2}\end{array}\bigg|\,\frac{x^2}{16}\right)\right]. \tag{35}$$

Since [16]
$${}_2F_1\!\left(\begin{array}{c}-n,\lambda\\ 2\lambda+1\end{array}\bigg|\,2\right) \tag{36}$$
$$= \frac{\Gamma\!\left(\lambda+\frac{1}{2}\right)}{\sqrt{\pi}}\left[\left(\frac{1+(-1)^n}{2}\right)\frac{\Gamma\!\left(\frac{n+1}{2}\right)}{\Gamma\!\left(\lambda+\frac{n+1}{2}\right)} + \left(\frac{1-(-1)^n}{2}\right)\frac{\Gamma\!\left(\frac{n}{2}+1\right)}{\Gamma\!\left(\lambda+\frac{n}{2}+1\right)}\right],$$
and
$${}_2F_1\!\left(\begin{array}{c}-n,\lambda\\ 2\lambda-1\end{array}\bigg|\,2\right) \tag{37}$$
$$= \frac{\Gamma\!\left(\lambda-\frac{1}{2}\right)}{\sqrt{\pi}}\left[\left(\frac{1+(-1)^n}{2}\right)\frac{\Gamma\!\left(\frac{n+1}{2}\right)}{\Gamma\!\left(\lambda+\frac{n-1}{2}\right)} - \left(\frac{1-(-1)^n}{2}\right)\frac{\Gamma\!\left(\frac{n}{2}+1\right)}{\Gamma\!\left(\lambda+\frac{n}{2}\right)}\right],$$

by introducing $\lambda = \mu$ and $\lambda = \mu + 1$, after some steps, it leads to

$$\mathrm{Mi}_{\pm\frac{1}{2},\mu}(x) \tag{38}$$
$$= \frac{x^{\mu+1/2}}{\mu+1/2}\left[{}_1F_2\left(\begin{array}{c}\frac{\mu}{2}+\frac{1}{4}\\ \mu+\frac{1}{2},\frac{\mu}{2}+\frac{3}{4}\end{array}\bigg|\frac{x^2}{16}\right) \mp \frac{x/2}{2\mu+3}{}_1F_2\left(\begin{array}{c}\frac{\mu}{2}+\frac{3}{4}\\ \mu+\frac{3}{2},\frac{\mu}{2}+\frac{7}{4}\end{array}\bigg|\frac{x^2}{16}\right)\right].$$

4. The Integral Wright Functions

In 1933 [17] and in 1940 [18], Wright introduced new special functions that were considered as a kind of generalization of the Bessel functions. However, today they play a significant independent role in mathematics and in solutions of physical problems by modeling space diffusion, stochastic processes, probability distributions and other diverse natural phenomena [19,20]. The Wright functions are defined by the following series

$$W_{\alpha,\beta}(x) = \sum_{k=0}^{\infty} \frac{x^k}{k!\,\Gamma(\alpha k + \beta)}. \tag{39}$$

If the parameter α is a positive real number, they are called the Wright functions of the first kind, and when $-1 < \alpha < 0$, the Wright functions of the second kind.

Furthermore, consider the following functions:

$$\begin{aligned} F_\alpha(x) &= W_{-\alpha,0}(-x), \quad 0 < \alpha < 1, \\ M_\alpha(x) &= W_{-\alpha,1-\alpha}(-x), \quad 0 < \alpha < 1, \\ F_\alpha(x) &= \alpha\, x\, M_\alpha(x). \end{aligned} \tag{40}$$

These functions with negative arguments x and with particular values of parameters are frequently named as the Mainardi functions and are denoted as $F_\alpha(x)$ and $M_\alpha(x)$ [19,20].

Their explicit form is

$$\begin{aligned} F_\alpha(x) &= \sum_{k=1}^{\infty} \frac{(-x)^k}{k!\,\Gamma(-\alpha k)} \\ &= -\frac{1}{\pi}\sum_{k=1}^{\infty} \frac{(-x)^k}{k!}\Gamma(\alpha k + 1)\sin(\pi\alpha k), \\ M_\alpha(x) &= \sum_{k=0}^{\infty} \frac{(-x)^k}{k!\,\Gamma(-\alpha(k+1)+1)} \\ &= \frac{1}{\pi}\sum_{k=0}^{\infty} \frac{(-x)^k}{k!}\Gamma(\alpha(k+1))\sin(\pi\alpha(k+1)). \end{aligned} \tag{41}$$

For positive rational $\alpha = p/q$, where p, q are positive coprimes, we have obtained reduction formulas for $F_{p/q}(x)$ and $M_{p/q}(x)$ in Appendix C. Furthermore, by applying the MATHEMATICA program to sums of infinite series in (39), it is possible to obtain the Wright functions of the first and second kinds for particular values of parameters α and β in an explicit form (Appendix C). The Laplace transforms of these functions are expressed in terms of the Mittag-Leffler functions, so they are omitted here [19–21].

The two-parameter $E_{\alpha,\beta}(t)$ Mittag-Leffler functions defined in (9) differ only by the absence of factorials from the Wright functions and, therefore, the form of series in (39) leads to the integral Wright function, which is similar to that introduced in (11) and (12).

$$\mathrm{Wi}_{\alpha,\beta}(x) = \int_0^x \frac{W_{\alpha,\beta}(t) - 1/\Gamma(\beta)}{t}\,dt. \tag{42}$$

Unfortunately, the notation is the same as the integral Whittaker functions. In an explicit form from (39), we have

$$\text{Wi}_{\alpha,\beta}(x) = \sum_{k=1}^{\infty} \frac{x^k}{k\,k!\,\Gamma(\alpha k + \beta)}. \tag{43}$$

For p and q positive coprimes, applying (A1) and (A2), the corresponding expression to (14) is

$$\text{Wi}_{p/q,\beta}(x) \tag{44}$$
$$= \sum_{k=1}^{q-1} \frac{x^k}{k\,k!\,\Gamma\left(\frac{p}{q}k + \beta\right)}\,{}_2F_{p+q}\left(\begin{array}{c} 1, k/q \\ b_0,\ldots,b_{p-1}, c_0,\ldots,c_{q-1} \end{array} \Big| \frac{x^q}{p^p q^q}\right),$$
$$b_j = \frac{k}{q} + \frac{\beta + j}{p}, \quad c_j = \frac{k+1+j}{q}.$$

In the case of the Mainardi functions, we have

$$\begin{aligned}
\text{Fi}_{p/q}(x) &= -\frac{1}{\pi}\sum_{k=1}^{q}\frac{(-x)^k}{k\,k!}\Gamma\left(\frac{p}{q}k+1\right)\sin\left(\pi\frac{p}{q}k\right)S_k(x), \\
\text{Mi}_{p/q}(x) &= \frac{1}{\pi}\sum_{k=1}^{q}\frac{(-x)^k}{k\,k!}\sin\left(\pi\frac{p}{q}(k+1)\right)\Gamma\left(\frac{p}{q}(k+1)\right)S_k(x),
\end{aligned} \tag{45}$$

where

$$S_k(x) = {}_{p+2}F_{q+1}\left(\begin{array}{c} 1, \frac{k}{q}, a_0,\ldots,a_{p-1} \\ \frac{k}{q}+1, b_0,\ldots,b_{q-1} \end{array} \Big| \frac{(-1)^{p+q}x^q p^p}{q^q}\right), \tag{46}$$
$$a_j = \frac{k}{q} + \frac{j+1}{p}, \quad b_j = \frac{k+1+j}{q}.$$

In Tables 10 and 11, the integral Wright functions derived with the help of MATHEMATICA program for some values of parameters α and β are derived. There are many other expressions for these functions, which are available using this program, but being long and complex, they were omitted. The integral Mainardi fuctions $\text{Fi}_\alpha(x)$ and $\text{Mi}_\alpha(x)$ for $0 < \alpha < 1$, are presented in Tables 12 and 13. As can be expected, most of these integral functions are expressed in terms of generalized hypergeometric functions.

Table 10. The integral Wright functions $\text{Wi}_{\alpha,\beta}$ derived for some values of parameters α and β by using (43).

α	β	$\text{Wi}_{\alpha,\beta}(x)$
-1	$\frac{1}{2}$	$\frac{1}{\sqrt{\pi}}\left[\ln 4 - 2\ln(\sqrt{1+x}+1)\right]$
-1	$\frac{3}{2}$	$\frac{x}{\sqrt{\pi}}\,{}_3F_2\left(\begin{array}{c}\frac{1}{2},1,1\\2,2\end{array}\Big|-x\right)$
-1	β	$\frac{x}{\Gamma(\beta-1)}\,{}_3F_2\left(\begin{array}{c}1,1,2-\beta\\2,2\end{array}\Big|-x\right)$
0	$-\frac{4}{3}$	$\frac{-\gamma-\ln x+\text{Chi}(x)+\text{Shi}(x)}{\Gamma(-4/3)}$
0	β	$\frac{-\gamma-\ln x+\text{Chi}(x)+\text{Shi}(x)}{\Gamma(\beta)}$
$\frac{1}{2}$	0	$-\frac{1}{2}+\frac{1}{2}\,{}_0F_2\left(\begin{array}{c}-\\\frac{1}{2},1\end{array}\Big|\frac{x^2}{4}\right)+\frac{x}{\sqrt{\pi}}\,{}_0F_2\left(\begin{array}{c}-\\\frac{1}{2},\frac{3}{2}\end{array}\Big|\frac{x^2}{4}\right)$
$\frac{1}{2}$	$\frac{1}{2}$	$\frac{x^2}{2\sqrt{\pi}}\,{}_2F_4\left(\begin{array}{c}1,1\\\frac{3}{2},\frac{3}{2},2,2\end{array}\Big|\frac{x^2}{4}\right)+x\,{}_1F_3\left(\begin{array}{c}\frac{1}{2}\\1,\frac{3}{2},\frac{3}{2}\end{array}\Big|\frac{x^2}{4}\right)$

Table 10. Cont.

α	β	$\text{Wi}_{\alpha,\beta}(x)$
$\frac{1}{2}$	1	$\frac{x}{4}\left[\frac{8}{\sqrt{\pi}}\,{}_1F_3\!\left(\begin{array}{c}\frac{1}{2}\\\frac{3}{2},\frac{3}{2},\frac{3}{2}\end{array}\Big|\frac{x^2}{4}\right)+x\,{}_2F_4\!\left(\begin{array}{c}1,1\\\frac{3}{2},\frac{3}{2},2,2\end{array}\Big|\frac{x^2}{4}\right)\right]$
$\frac{1}{2}$	2	$\frac{x^2}{8}\,{}_2F_4\!\left(\begin{array}{c}1,1\\\frac{3}{2},2,2,3\end{array}\Big|\frac{x^2}{4}\right)+\frac{4x}{3\sqrt{\pi}}\,{}_1F_3\!\left(\begin{array}{c}\frac{1}{2}\\\frac{3}{2},\frac{3}{2},\frac{5}{2}\end{array}\Big|\frac{x^2}{4}\right)$
$\frac{1}{2}$	β	$\frac{x^2}{4\Gamma(1+\beta)}\,{}_2F_4\!\left(\begin{array}{c}1,1\\\frac{3}{2},2,2,\beta+1\end{array}\Big|\frac{x^2}{4}\right)+\frac{x}{\Gamma(\frac{1}{2}+\beta)}\,{}_1F_3\!\left(\begin{array}{c}\frac{1}{2}\\\frac{3}{2},\frac{3}{2},\beta+\frac{1}{2}\end{array}\Big|\frac{x^2}{4}\right)$
1	$-\frac{3}{2}$	$-\frac{x}{2\sqrt{\pi}}\,{}_2F_3\!\left(\begin{array}{c}1,1\\-\frac{1}{2},2,2\end{array}\Big|x\right)$
3	$\frac{3}{2}$	$\frac{x}{\sqrt{\pi}}\,{}_2F_3\!\left(\begin{array}{c}1,1\\\frac{1}{2},2,2\end{array}\Big|x\right)$
1	0	$-1+I_0(2\sqrt{x})$
1	$\frac{1}{4}$	$\frac{x}{\Gamma(5/4)}\,{}_2F_3\!\left(\begin{array}{c}1,1\\\frac{5}{2},2,2\end{array}\Big|x\right)$
1	$\frac{1}{2}$	$-\frac{2\gamma+\ln 4+\ln x-2\,\text{Chi}(2\sqrt{x})}{\sqrt{\pi}}$
1	1	$x\,{}_2F_3\!\left(\begin{array}{c}1,1\\2,2,2\end{array}\Big|x\right)$
1	$\frac{3}{2}$	$-\frac{1}{\sqrt{\pi x}}\left[2\sinh(2\sqrt{x})-2\sqrt{x}(2\gamma-2+\ln 4-2\,\text{Chi}(2\sqrt{x}))\right]$
1	β	$x\,{}_2F_3\!\left(\begin{array}{c}1,1\\2,2,\beta+1\end{array}\Big|x\right)$

Table 11. The integral Wright functions $\text{Wi}_{\alpha,\beta}$ derived for some values of parameters α and β by using (43).

α	β	$\text{Wi}_{\alpha,\beta}(x)$
$\frac{3}{2}$	$\frac{1}{2}$	$\frac{2x^2}{15\sqrt{\pi}}\,{}_2F_6\!\left(\begin{array}{c}1,1\\\frac{7}{2},\frac{3}{2},\frac{3}{2},\frac{11}{6},2,2\end{array}\Big|\frac{x^2}{108}\right)+x\,{}_1F_5\!\left(\begin{array}{c}\frac{1}{2}\\\frac{2}{3},1,\frac{4}{3},\frac{3}{2},\frac{3}{2}\end{array}\Big|\frac{x^2}{108}\right)$
2	$\frac{1}{4}$	$\frac{16x}{5\Gamma(1/4)}\,{}_2F_4\!\left(\begin{array}{c}1,1\\\frac{9}{8},\frac{13}{8},2,2\end{array}\Big|\frac{x}{4}\right)$
2	$\frac{1}{3}$	$\frac{9x}{4\Gamma(1/3)}\,{}_2F_4\!\left(\begin{array}{c}1,1\\\frac{7}{6},\frac{5}{3},2,2\end{array}\Big|\frac{x}{4}\right)$
2	$\frac{1}{2}$	$\frac{4x}{3\sqrt{\pi}}\,{}_2F_4\!\left(\begin{array}{c}1,1\\\frac{5}{4},\frac{7}{4},2,2\end{array}\Big|\frac{x}{4}\right)$
2	1	$\frac{x}{2}\,{}_2F_4\!\left(\begin{array}{c}1,1\\\frac{3}{2},2,2,2\end{array}\Big|\frac{x}{4}\right)$
2	2	$\frac{x}{6}\,{}_2F_4\!\left(\begin{array}{c}1,1\\\frac{5}{2},2,2,2\end{array}\Big|\frac{x}{4}\right)$
2	β	$\frac{x}{\Gamma(\beta+2)}\,{}_2F_4\!\left(\begin{array}{c}1,1\\2,2,\frac{\beta}{2}+1,\frac{\beta+3}{2}\end{array}\Big|\frac{x}{4}\right)$
3	1	$\frac{x}{6}\,{}_2F_4\!\left(\begin{array}{c}1,1\\\frac{4}{3},\frac{5}{3},2,2\end{array}\Big|\frac{x}{27}\right)$
3	β	$\frac{x}{\Gamma(\beta+3)}\,{}_2F_5\!\left(\begin{array}{c}1,1\\2,2,\frac{\beta}{3}+1,\frac{\beta+4}{3},\frac{\beta+5}{3}\end{array}\Big|\frac{x}{27}\right)$
4	β	$\frac{x}{\Gamma(\beta+4)}\,{}_2F_6\!\left(\begin{array}{c}1,1\\2,2,\frac{\beta}{4}+1,\frac{\beta+5}{4},\frac{\beta+6}{4},\frac{\beta+7}{4}\end{array}\Big|\frac{x}{256}\right)$
5	β	$\frac{x}{\Gamma(\beta+5)}\,{}_2F_7\!\left(\begin{array}{c}1,1\\2,2,\frac{\beta}{4}+1,\frac{\beta+6}{5},\frac{\beta+7}{5},\frac{\beta+8}{5},\frac{\beta+9}{5}\end{array}\Big|\frac{x}{3125}\right)$

Table 12. The integral Mainardi function Fi_α derived for some values of parameter α by using (45).

α	$\text{Fi}_\alpha(x)$
$\frac{3}{4}$	$-x\left[\frac{1}{\Gamma(-\frac{3}{4})}\,_3F_3\left(\begin{array}{c}\frac{1}{4},\frac{7}{12},\frac{11}{12}\\\frac{1}{2},\frac{3}{4},\frac{5}{4}\end{array}\bigg| -\frac{27x^4}{256}\right)+\frac{x}{144}\right.$ $\left.\left(\frac{8x}{\Gamma(-\frac{9}{4})}\,_3F_3\left(\begin{array}{c}\frac{3}{4},\frac{13}{12},\frac{17}{12}\\\frac{5}{4},\frac{3}{2},\frac{7}{4}\end{array}\bigg| -\frac{27x^4}{256}\right)-\frac{27}{\sqrt{\pi}}\,_3F_4\left(\begin{array}{c}\frac{1}{2},\frac{5}{6},\frac{7}{6}\\\frac{3}{4},\frac{5}{4},\frac{3}{2}\end{array}\bigg| -\frac{27x^4}{256}\right)\right)\right]$
$\frac{2}{3}$	$\frac{x}{4}\left[\frac{x}{\Gamma(-\frac{4}{3})}\,_2F_2\left(\begin{array}{c}\frac{2}{3},\frac{7}{6}\\\frac{4}{3},\frac{5}{3}\end{array}\bigg| -\frac{4x^3}{27}\right)-\frac{4}{\Gamma(-\frac{2}{3})}\,_2F_2\left(\begin{array}{c}\frac{1}{3},\frac{5}{6}\\\frac{2}{3},\frac{4}{3}\end{array}\bigg| -\frac{4x^3}{27}\right)\right]$
$\frac{1}{2}$	$\frac{1}{2}\,\text{erf}\left(\frac{x}{2}\right)$
$\frac{1}{3}$	$\frac{x}{4}\left[\frac{x}{\Gamma(-\frac{2}{3})}\,_1F_2\left(\begin{array}{c}\frac{2}{3}\\\frac{4}{3},\frac{5}{3}\end{array}\bigg| \frac{x^3}{27}\right)-\frac{4}{\Gamma(-\frac{1}{3})}\,_1F_2\left(\begin{array}{c}\frac{1}{3}\\\frac{2}{3},\frac{4}{3}\end{array}\bigg| \frac{x^3}{27}\right)\right]$
$\frac{1}{4}$	$-x\left[\frac{1}{\Gamma(-\frac{1}{4})}\,_1F_3\left(\begin{array}{c}\frac{1}{4}\\\frac{1}{2},\frac{3}{4},\frac{5}{4}\end{array}\bigg| -\frac{x^4}{256}\right)+\frac{x}{72}\right.$ $\left.\left(\frac{9}{\sqrt{\pi}}\,_1F_3\left(\begin{array}{c}\frac{1}{2}\\\frac{3}{4},\frac{5}{4},\frac{3}{2}\end{array}\bigg| -\frac{x^4}{256}\right)+\frac{4x}{\Gamma(-\frac{3}{4})}\,_3F_4\left(\begin{array}{c}\frac{3}{4}\\\frac{5}{4},\frac{3}{2},\frac{7}{4}\end{array}\bigg| -\frac{x^4}{256}\right)\right)\right]$

Table 13. The integral Mainardi function Mi_α derived for some values of parameter α by using (45).

α	$\text{Mi}_\alpha(x)$
$\frac{3}{4}$	$\frac{x}{96}\left[\frac{48}{\sqrt{\pi}}\,_3F_3\left(\begin{array}{c}\frac{1}{4},\frac{5}{6},\frac{7}{6}\\\frac{3}{4},\frac{5}{4},\frac{3}{2}\end{array}\bigg| -\frac{27x^4}{256}\right)\right.$ $\left.+\frac{24x}{\Gamma(-\frac{5}{4})}\,_3F_3\left(\begin{array}{c}\frac{1}{2},\frac{13}{12},\frac{17}{12}\\\frac{5}{4},\frac{3}{2},\frac{3}{2}\end{array}\bigg| -\frac{27x^4}{256}\right)+\frac{x^3}{\Gamma(-\frac{11}{4})}\,_4F_4\left(\begin{array}{c}1,1,\frac{19}{12},\frac{23}{12}\\\frac{3}{2},\frac{7}{4},2,2\end{array}\bigg| -\frac{27x^4}{256}\right)\right]$
$\frac{2}{3}$	$-\frac{x^3}{18\Gamma(-\frac{5}{3})}\,_3F_3\left(\begin{array}{c}1,1,\frac{11}{6}\\\frac{5}{3},2,2\end{array}\bigg| -\frac{4x^3}{27}\right)-\frac{x}{\Gamma(-\frac{1}{3})}\,_2F_2\left(\begin{array}{c}\frac{1}{3},\frac{7}{6}\\\frac{4}{3},\frac{4}{3}\end{array}\bigg| -\frac{4x^3}{27}\right)$
$\frac{1}{2}$	$\frac{1}{2\sqrt{\pi}}\left[\text{Chi}\left(\frac{x^2}{4}\right)-\text{Shi}\left(\frac{x^2}{4}\right)-\ln\left(\frac{x^2}{4}\right)-\gamma\right]$
$\frac{1}{3}$	$-\frac{x^3}{18\Gamma(-\frac{1}{3})}\,_2F_3\left(\begin{array}{c}1,1\\\frac{5}{3},2,2\end{array}\bigg| \frac{x^3}{27}\right)-\frac{x}{\Gamma(\frac{1}{3})}\,_1F_2\left(\begin{array}{c}\frac{1}{3}\\\frac{4}{3},\frac{4}{3}\end{array}\bigg| \frac{x^3}{27}\right)$
$\frac{1}{4}$	$-\frac{x}{\sqrt{\pi}}\,_1F_3\left(\begin{array}{c}\frac{1}{4}\\\frac{3}{4},\frac{5}{4},\frac{5}{4}\end{array}\bigg| -\frac{x^4}{256}\right)$ $+\frac{x^2}{4\Gamma(\frac{1}{4})}\,_1F_3\left(\begin{array}{c}\frac{1}{2}\\\frac{5}{4},\frac{3}{2},\frac{3}{2}\end{array}\bigg| -\frac{x^4}{256}\right)+\frac{x^4}{96\Gamma(-\frac{1}{4})}\,_2F_4\left(\begin{array}{c}1,1\\\frac{3}{2},\frac{7}{4},2,2\end{array}\bigg| -\frac{x^4}{256}\right)$

5. Conclusions

For the first time, three new special functions are presented in this investigation: the integral Mittag-Leffler functions, the integral Whittaker functions, and the integral Wright functions. These functions are defined in the mathematical literature in the same manner as other elementary and special integral functions. It is feasible to generate these functions in an explicit form for certain parameters values using the MATHEMATICA application. These integral functions are often represented in terms of generalized hypergeometric functions. The behavior of some of them is shown graphically. In the Appendices, a large number of Mittag-Leffler, Whittaker, and Wright functions with integral and fractional parameters, as well as their Laplace transforms, are presented in tabular form.

It may be observed that, generally, it is highly possible to make general integral functions such as (19) and (20) by using generalized hypergeometric $_pF_p(t)$, because they converge in the whole complex t-plane, or, for every real number t.

Author Contributions: Conceptualization, A.A. and J.L.G.-S.; Methodology, A.A. and J.L.G.-S.; Resources, A.A.; Writing—original draft, A.A. and J.L.G.-S.; Writing—review & editing, A.A. and J.L.G.-S. All authors have read and agreed to the published version of the manuscript.

Funding: This research received no external funding.

Institutional Review Board Statement: Not applicable.

Informed Consent Statement: Not applicable.

Data Availability Statement: Not applicable.

Acknowledgments: We are grateful to Armando Consiglio from Institut für Theoretische Physik und Astrophysik of University of Würzburg, Germany, who performed numerical evaluations of the integral functions presented in Figures 1 and 2, and to Francesco Mainardi from the Department of Physics and Astronomy, University of Bologna, Bologna, Italy, for his kind encouragement and interest in our work.

Conflicts of Interest: The authors declare no conflict of interest.

Appendix A. Representations of the One- and Two-Parameter Mittag-Leffler Functions and Their Laplace Transforms

The Mittag-Leffler functions are defined by the sums of infinite series presented in (9) and their Laplace transforms in (10). For positive variable x and some values of parameters α and β, these sums can be expressed in terms of elementary and special functions, especially in terms of generalized hypergeometric functions. They were derived by using the MATHEMATICA program and presented in Tables A1 and A2 for the Mittag-Leffler functions, as well as Tables A3 and A4 for the Laplace transforms. These results, given in terms of infinite series, are mostly new, and only they are only partly known in the mathematical literature. Knowing that any infinite sum can be split as

$$\sum_{k=0}^{\infty} a(k) = \sum_{j=0}^{q-1} \sum_{k=0}^{\infty} a(qk+j), \tag{A1}$$

and applying the multiplication formula of the gamma function ([5] (Eqn. 5.5.6)), for $nt \neq 0, -1, -2, \ldots$

$$\Gamma(nt) = (2\pi)^{(1-n)/2} n^{nt-1/2} \prod_{j=0}^{n-1} \Gamma\left(t + \frac{j}{n}\right), \tag{A2}$$

it is possible to express from (9) the Mittag-Leffler function in the case of positive rational $\alpha = p/q$ with p and q positive coprimes,

$$E_{p/q,\beta}(x) = \sum_{k=0}^{q-1} \frac{x^k}{\Gamma\left(\frac{p}{q}k + \beta\right)} {}_1F_p\left(\begin{array}{c} 1 \\ b_0, \ldots, b_{p-1} \end{array} \bigg| \frac{x^q}{p^p}\right), \tag{A3}$$

where

$$b_j = \frac{k}{q} + \frac{\beta + j}{p}.$$

The corresponding Laplace transforms are

$$\mathcal{L}\left[E_{p/q,\beta}(t)\right] \tag{A4}$$

$$= \sum_{k=0}^{q-1} \frac{s^{-k-1}}{\Gamma\left(\frac{p}{q}k + \beta\right)} {}_{q+1}F_p\left(\begin{array}{c} 1, a_0, \ldots, a_{q-1} \\ b_0, \ldots, b_{p-1} \end{array} \bigg| \frac{(q/s)^q}{p^p}\right),$$

where

$$a_j = \frac{k+1+j}{q},$$
$$b_j = \frac{k}{q} + \frac{\beta+j}{p}.$$

Table A1. The Mittag-Leffler functions derived for some values of parameters α and β by using (9).

α	β	$E_{\alpha,\beta}(x)$
$\frac{1}{2}$	$\frac{1}{2}$	$\frac{1}{\sqrt{\pi}} + x e^{x^2}[\text{erf}(x) + 1]$
$\frac{1}{2}$	1	$e^{x^2}[\text{erf}(x) + 1]$
$\frac{1}{2}$	$\frac{3}{2}$	$\frac{e^{x^2}[\text{erf}(x)+1]-1}{x}$
$\frac{1}{2}$	2	$\frac{1}{x^2}\left[e^{x^2}[\text{erf}(x)+1] - 1 - \frac{2x}{\sqrt{\pi}}\right]$
$\frac{1}{2}$	3	$-\frac{1}{3x^4}\left[3e^{x^2}[\text{erfc}(x) - 2] + \frac{4x^3+6x}{\sqrt{\pi}} + 3(1+x^2)\right]$
$\frac{1}{2}$	4	$\frac{1}{30x^4}\left[30 e^{x^2}[\text{erf}(x)+1] - \frac{4x(4x^4+10x^2+15)}{\sqrt{\pi}} - 15(x^4+2x^2+2)\right]$
$\frac{1}{2}$	β	$e^{x^2} x^{2(1-\beta)}\left[2 - \frac{\Gamma(\beta-1,x^2)}{\Gamma(\beta-1)} - \frac{\Gamma(\beta-\frac{1}{2},x^2)}{\Gamma(\beta-\frac{1}{2})}\right]$
1	$\frac{1}{2}$	$\frac{1}{\sqrt{\pi}} + \sqrt{x} e^x \text{erf}(\sqrt{x})$
1	1	e^x
1	$\frac{3}{2}$	$\frac{e^x \text{erf}(\sqrt{x})}{\sqrt{x}}$
1	2	$\frac{e^x - 1}{x}$
$\frac{3}{2}$	$\frac{1}{2}$	$\frac{1}{\sqrt{\pi}} {}_1F_3\left(\begin{array}{c}1\\ \frac{1}{6},\frac{1}{2},\frac{5}{6}\end{array}\middle\vert \frac{x^2}{27}\right) + \frac{x^{1/3}}{3}\left[e^{x^{2/3}} - 2e^{-x^{2/3}/2}\sin\left(\frac{\pi - 3\sqrt{3}x^{2/3}}{6}\right)\right]$
$\frac{3}{2}$	1	$\frac{1}{3}\left[\frac{4x}{\sqrt{\pi}} {}_1F_3\left(\begin{array}{c}1\\ \frac{5}{6},\frac{7}{6},\frac{3}{2}\end{array}\middle\vert \frac{x^2}{27}\right) + e^{x^{2/3}} + 2e^{-x^{2/3}/2}\cos\left(\frac{\sqrt{3}}{2}x^{2/3}\right)\right]$
$\frac{3}{2}$	$\frac{3}{2}$	$\frac{2}{\sqrt{\pi}} {}_1F_3\left(\begin{array}{c}1\\ \frac{1}{2},\frac{5}{6},\frac{7}{6}\end{array}\middle\vert \frac{x^2}{27}\right) + \frac{x^{-1/3}}{3}\left[e^{x^{2/3}} - 2e^{-x^{2/3}/2}\sin\left(\frac{\pi + 3\sqrt{3}x^{2/3}}{6}\right)\right]$
$\frac{3}{2}$	2	$\frac{8x}{15\sqrt{\pi}} {}_1F_3\left(\begin{array}{c}1\\ \frac{7}{6},\frac{3}{2},\frac{11}{6}\end{array}\middle\vert \frac{x^2}{27}\right) + \frac{x^{-2/3}}{3}\left[e^{x^{2/3}} - 2e^{-x^{2/3}/2}\sin\left(\frac{\pi - 3\sqrt{3}x^{2/3}}{6}\right)\right]$
$\frac{3}{2}$	β	$\frac{x}{\Gamma(\beta+\frac{1}{2})} {}_1F_3\left(\begin{array}{c}1\\ \frac{2\beta+3}{6},\frac{2\beta+5}{6},\frac{2\beta+7}{6}\end{array}\middle\vert \frac{x^2}{27}\right) + \frac{1}{\Gamma(\beta)} {}_1F_3\left(\begin{array}{c}1\\ \frac{\beta+1}{3},\frac{\beta+2}{3},\frac{\beta}{3}\end{array}\middle\vert \frac{x^2}{27}\right)$
2	$\frac{1}{2}$	$\frac{1}{\sqrt{\pi}} {}_1F_2\left(\begin{array}{c}1\\ \frac{1}{4},\frac{3}{4}\end{array}\middle\vert \frac{x}{4}\right)$
2	1	$\cosh(\sqrt{x})$
2	2	$\frac{\sinh(\sqrt{x})}{\sqrt{x}}$
2	3	$\frac{\cosh(\sqrt{x})-1}{x}$
2	4	$\frac{\sinh(\sqrt{x})}{x^{3/2}} - \frac{1}{x}$
2	β	$\frac{1}{\Gamma(\beta)} {}_1F_2\left(\begin{array}{c}1\\ \frac{\beta+1}{2},\frac{\beta}{2}\end{array}\middle\vert \frac{x}{4}\right)$

Table A2. The Mittag-Leffler functions derived for some values of parameters α and β by using (9).

α	β	$E_{\alpha,\beta}(x)$
3	1	$\frac{1}{3}\left[e^{x^{1/3}} + 2e^{-x^{1/3}/2}\cos\left(\frac{\sqrt{3}}{2}x^{1/3}\right)\right]$
3	2	$\frac{x^{-1/3}}{3}\left[e^{x^{1/3}} - 2e^{-x^{1/3}/2}\sin\left(\frac{\pi - 3\sqrt{3}x^{1/3}}{6}\right)\right]$
3	3	$\frac{x^{-2/3}}{3}\left[e^{x^{1/3}} - 2e^{-x^{1/3}/2}\sin\left(\frac{\pi + 3\sqrt{3}x^{1/3}}{6}\right)\right]$
3	β	$\frac{1}{\Gamma(\beta)}{}_1F_3\left(\begin{array}{c}1\\ \frac{\beta+1}{3},\frac{\beta+2}{3},\frac{\beta}{3}\end{array}\bigg|\frac{x}{27}\right)$
4	1	$\frac{1}{2}\left[\cos\left(x^{1/4}\right) + \cosh\left(x^{1/4}\right)\right]$
4	2	$\frac{\sin(x^{1/4}) + \sinh(x^{1/4})}{2x^{1/4}}$
4	3	$\frac{\cosh(x^{1/4}) - \cos(x^{1/4})}{2\sqrt{x}}$
4	4	$\frac{\sinh(x^{1/4}) - \sin(x^{1/4})}{2x^{3/4}}$
4	β	$\frac{1}{\Gamma(\beta)}{}_1F_4\left(\begin{array}{c}1\\ \frac{\beta+1}{4},\frac{\beta+2}{4},\frac{\beta+3}{4},\frac{\beta}{4}\end{array}\bigg|\frac{x}{256}\right)$
5	1	${}_0F_4\left(\begin{array}{c}-\\ \frac{1}{5},\frac{2}{5},\frac{3}{5},\frac{4}{5}\end{array}\bigg|\frac{x}{3125}\right)$
5	2	${}_0F_4\left(\begin{array}{c}-\\ \frac{2}{5},\frac{3}{5},\frac{4}{5},\frac{6}{5}\end{array}\bigg|\frac{x}{3125}\right)$
5	3	$\frac{1}{2}{}_0F_4\left(\begin{array}{c}-\\ \frac{3}{5},\frac{4}{5},\frac{6}{5},\frac{7}{5}\end{array}\bigg|\frac{x}{3125}\right)$
5	4	$\frac{1}{6}{}_0F_4\left(\begin{array}{c}-\\ \frac{4}{5},\frac{6}{5},\frac{7}{5},\frac{8}{5}\end{array}\bigg|\frac{x}{3125}\right)$
5	5	$\frac{1}{24}{}_0F_4\left(\begin{array}{c}-\\ \frac{6}{5},\frac{7}{5},\frac{8}{5},\frac{9}{5}\end{array}\bigg|\frac{x}{3125}\right)$
5	β	$\frac{1}{\Gamma(\beta)}{}_1F_5\left(\begin{array}{c}1\\ \frac{\beta+1}{5},\frac{\beta+2}{5},\frac{\beta+3}{5},\frac{\beta+4}{5},\frac{\beta}{5}\end{array}\bigg|\frac{x}{3125}\right)$

Table A3. The Laplace transforms Mittag-Leffler functions derived for some values of parameters α and β by using (10).

α	β	$\mathcal{L}[E_{\alpha,\beta}(t)]$
1	$\frac{1}{2}$	$\frac{\sqrt{s-1} + \csc^{-1}(\sqrt{s})}{\sqrt{\pi}(s-1)^{3/2}}$
$\frac{1}{2}$	1	$\frac{1}{s-1}$
1	$\frac{3}{2}$	$\frac{2\csc^{-1}(\sqrt{s})}{\sqrt{\pi}\sqrt{s-1}}$
1	2	$\ln\left(\frac{s}{s-1}\right)$
1	β	$\frac{1}{s\Gamma(\beta)}{}_2F_1\left(\begin{array}{c}1,1\\ \beta\end{array}\bigg|\frac{1}{s}\right)$
$\frac{3}{2}$	$\frac{1}{2}$	$\frac{1}{\sqrt{\pi s}}{}_2F_2\left(\begin{array}{c}1,1\\ \frac{1}{6},\frac{5}{6}\end{array}\bigg|\frac{4}{27s^2}\right) + \frac{1}{s^2}{}_2F_2\left(\begin{array}{c}1,\frac{3}{2}\\ \frac{2}{3},\frac{4}{3}\end{array}\bigg|\frac{4}{27s^2}\right)$

Table A3. Cont.

α	β	$\mathcal{L}[E_{\alpha,\beta}(t)]$
$\frac{3}{2}$	1	$\frac{4}{3\sqrt{\pi}s^2}\,_2F_2\!\left(\begin{array}{c}1,1\\ \frac{5}{6},\frac{7}{6}\end{array}\bigg|\frac{4}{27s^2}\right)+\frac{1}{s}\,_2F_2\!\left(\begin{array}{c}\frac{1}{2},1\\ \frac{1}{3},\frac{2}{3}\end{array}\bigg|\frac{4}{27s^2}\right)$
$\frac{3}{2}$	$\frac{3}{2}$	$\frac{2}{\sqrt{\pi}s}\,_2F_2\!\left(\begin{array}{c}1,1\\ \frac{5}{6},\frac{7}{6}\end{array}\bigg|\frac{4}{27s^2}\right)+\frac{1}{2s^2}\,_2F_2\!\left(\begin{array}{c}1,\frac{3}{2}\\ \frac{4}{3},\frac{5}{3}\end{array}\bigg|\frac{4}{27s^2}\right)$
$\frac{3}{2}$	2	$\frac{8}{15\sqrt{\pi}s^2}\,_2F_2\!\left(\begin{array}{c}1,1\\ \frac{7}{6},\frac{11}{6}\end{array}\bigg|\frac{4}{27s^2}\right)+\frac{1}{s}\,_2F_2\!\left(\begin{array}{c}\frac{1}{2},1\\ \frac{2}{3},\frac{4}{3}\end{array}\bigg|\frac{4}{27s^2}\right)$
$\frac{3}{2}$	β	$\frac{1}{\Gamma(\beta)s}\,_3F_3\!\left(\begin{array}{c}\frac{1}{2},1,1\\ \frac{\beta+1}{3},\frac{\beta+2}{3},\frac{\beta}{3}\end{array}\bigg|\frac{4}{27s^2}\right)+$ $\frac{1}{\Gamma(\beta+\frac{3}{2})s^2}\,_3F_3\!\left(\begin{array}{c}1,1,\frac{3}{2}\\ \frac{2\beta+3}{6},\frac{2\beta+5}{6},\frac{2\beta+7}{6}\end{array}\bigg|\frac{4}{27s^2}\right)$
2	$\frac{1}{2}$	$\frac{1}{\sqrt{\pi}s}\,_2F_2\!\left(\begin{array}{c}1,1\\ \frac{1}{4},\frac{3}{4}\end{array}\bigg|\frac{1}{4s}\right)$
2	1	$\frac{1}{s}+\frac{\sqrt{\pi}}{2}s^{-3/2}e^{1/(4s)}\operatorname{erf}\!\left(\frac{1}{2\sqrt{s}}\right)$
2	2	$\sqrt{\pi}s^{-1/2}e^{1/(4s)}\operatorname{erf}\!\left(\frac{1}{2\sqrt{s}}\right)$
2	3	$\frac{1}{2s}\,_2F_2\!\left(\begin{array}{c}1,1\\ \frac{3}{2},2\end{array}\bigg|\frac{1}{4s}\right)$
2	4	$\frac{1}{6s}\,_2F_2\!\left(\begin{array}{c}1,1\\ 2,\frac{5}{2}\end{array}\bigg|\frac{1}{4s}\right)$
2	β	$\frac{1}{\Gamma(\beta)s}\,_2F_2\!\left(\begin{array}{c}1,1\\ \frac{\beta+1}{2},\frac{\beta}{2}\end{array}\bigg|\frac{1}{4s}\right)$

Table A4. The Laplace transforms Mittag-Leffler functions derived for some values of parameters α and β by using (10).

α	β	$\mathcal{L}[E_{\alpha,\beta}(t)]$
3	1	$\frac{1}{s}\,_1F_2\!\left(\begin{array}{c}1\\ \frac{1}{3},\frac{2}{3}\end{array}\bigg|\frac{1}{27s}\right)$
3	2	$\frac{1}{s}\,_1F_2\!\left(\begin{array}{c}1\\ \frac{2}{3},\frac{4}{3}\end{array}\bigg|\frac{1}{27s}\right)$
3	3	$\frac{1}{2s}\,_1F_2\!\left(\begin{array}{c}1\\ \frac{4}{3},\frac{5}{3}\end{array}\bigg|\frac{1}{27s}\right)$
3	β	$\frac{1}{\Gamma(\beta)s}\,_2F_3\!\left(\begin{array}{c}1,1\\ \frac{\beta+1}{3},\frac{\beta+2}{3},\frac{\beta}{3}\end{array}\bigg|\frac{1}{27s}\right)$
4	1	$\frac{1}{s}\,_1F_3\!\left(\begin{array}{c}1\\ \frac{1}{4},\frac{1}{2},\frac{3}{4}\end{array}\bigg|\frac{1}{256s}\right)$
4	2	$\frac{1}{s}\,_1F_3\!\left(\begin{array}{c}1\\ \frac{1}{2},\frac{3}{4},\frac{5}{4}\end{array}\bigg|\frac{1}{256s}\right)$
4	3	$\frac{1}{2s}\,_1F_3\!\left(\begin{array}{c}1\\ \frac{3}{4},\frac{5}{4},\frac{3}{2}\end{array}\bigg|\frac{1}{256s}\right)$
4	4	$\frac{1}{6s}\,_1F_3\!\left(\begin{array}{c}1\\ \frac{5}{4},\frac{3}{2},\frac{7}{4}\end{array}\bigg|\frac{1}{256s}\right)$

Table A4. *Cont.*

α	β	$\mathcal{L}[E_{\alpha,\beta}(t)]$
4	β	$\frac{1}{\Gamma(\beta)s}\,{}_2F_4\!\left(\begin{array}{c}1,1\\ \frac{\beta+1}{4},\frac{\beta+2}{4},\frac{\beta+3}{4},\frac{\beta}{4}\end{array}\bigg|\frac{1}{256s}\right)$
5	1	$\frac{1}{s}\,{}_1F_4\!\left(\begin{array}{c}1\\ \frac{1}{5},\frac{2}{5},\frac{3}{5},\frac{4}{5}\end{array}\bigg|\frac{1}{3125s}\right)$
5	2	$\frac{1}{s}\,{}_1F_4\!\left(\begin{array}{c}1\\ \frac{2}{5},\frac{3}{5},\frac{4}{5},\frac{6}{5}\end{array}\bigg|\frac{1}{3125s}\right)$
5	3	$\frac{1}{2s}\,{}_1F_4\!\left(\begin{array}{c}1\\ \frac{3}{5},\frac{4}{5},\frac{6}{5},\frac{7}{5}\end{array}\bigg|\frac{1}{3125s}\right)$
5	4	$\frac{1}{6s}\,{}_1F_4\!\left(\begin{array}{c}1\\ \frac{4}{5},\frac{6}{5},\frac{7}{5},\frac{8}{5}\end{array}\bigg|\frac{1}{3125s}\right)$
5	5	$\frac{1}{24s}\,{}_1F_4\!\left(\begin{array}{c}1\\ \frac{6}{5},\frac{7}{5},\frac{8}{5},\frac{9}{5}\end{array}\bigg|\frac{1}{3125s}\right)$
5	β	$\frac{1}{\Gamma(\beta)s}\,{}_2F_5\!\left(\begin{array}{c}1,1\\ \frac{\beta+1}{5},\frac{\beta+2}{5},\frac{\beta+3}{5},\frac{\beta+4}{5},\frac{\beta}{5}\end{array}\bigg|\frac{1}{3125s}\right)$

Appendix B. Representations of the Whittaker Functions and Their Laplace Transforms

The Whittaker functions $M_{\kappa,\mu}(x)$ and $W_{\kappa,\mu}(x)$ defined in (18) were derived by using the MATHEMATICA program, and they are presented in Tables A5, A6, A10 and A11. The corresponding Laplace transforms are in Tables A7–A9 and A12. Most of the reported results in these tables are unknown in the mathematical reference literature.

Table A5. The Whittaker functions $M_{\kappa,\mu}$ derived for some values of parameters κ and μ by using (18).

κ	μ	$M_{\kappa,\mu}(x)$
$-\frac{5}{2}$	0	$\frac{\sqrt{x}}{2}e^{x/2}[x(x+4)+2]$
$-\frac{3}{2}$	0	$\sqrt{x}e^{x/2}(x+1)$
$-\frac{3}{2}$	$\frac{1}{2}$	$\frac{x}{3}\left[(2x+3)I_0\!\left(\frac{x}{2}\right)+(2x+1)I_1\!\left(\frac{x}{2}\right)\right]$
$-\frac{3}{2}$	1	$e^{x/2}x^{3/2}$
$-\frac{3}{2}$	$\frac{3}{2}$	$\frac{4}{5}\left[x(2x-3)I_0\!\left(\frac{x}{2}\right)+[x(2x-3)+4]I_1\!\left(\frac{x}{2}\right)\right]$
$-\frac{3}{2}$	2	$4x^{-3/2}e^{-x/2}\left[e^x(x^3-3x^2+6x-6)+6\right]$
$-\frac{3}{2}$	$\frac{5}{2}$	$\frac{32}{7x}\left[x(2x^2-9x+24)I_0\!\left(\frac{x}{2}\right)+(2x^3-11x^2+36x-96)I_1\!\left(\frac{x}{2}\right)\right]$
$-\frac{3}{2}$	3	$30x^{-5/2}e^{-x/2}\left[e^x(x^4-8x^3+36x^2-96x+120)-24(x+5)\right]$
$-\frac{1}{6}$	0	$e^{-x/2}\sqrt{x}L_{-2/3}(x)$
$-\frac{1}{4}$	0	$e^{-x/2}\sqrt{x}L_{-3/4}(x)$
$-\frac{1}{4}$	$\frac{1}{4}$	$\frac{\sqrt{\pi}}{2}e^{x/2}x^{1/4}\mathrm{erf}(\sqrt{x})$
$-\frac{1}{3}$	0	$e^{-x/2}\sqrt{x}L_{-5/6}(x)$
$-\frac{1}{2}$	$\frac{1}{2}$	$x\left[I_0\!\left(\frac{x}{2}\right)+I_1\!\left(\frac{x}{2}\right)\right]$
$-\frac{1}{2}$	1	$x^{-1/2}e^{-x/2}[2e^x(x-1)+2]$
$-\frac{1}{2}$	$\frac{3}{2}$	$4\left[xI_0\!\left(\frac{x}{2}\right)+(x-4)I_1\!\left(\frac{x}{2}\right)\right]$
$-\frac{1}{2}$	2	$12x^{-3/2}e^{-x/2}\left[e^x(x^2-4x+6)-2(x+3)\right]$

Table A5. *Cont.*

κ	μ	$M_{\kappa,\mu}(x)$
0	$\frac{1}{8}$	$x^{5/8}\,_0F_1\left(\begin{array}{c}-\\ \frac{9}{8}\end{array}\bigg|\frac{x^2}{16}\right)$
0	$\frac{1}{7}$	$x^{9/14}\,_0F_1\left(\begin{array}{c}-\\ \frac{8}{7}\end{array}\bigg|\frac{x^2}{16}\right)$
0	$\frac{1}{6}$	$x^{2/3}\,_0F_1\left(\begin{array}{c}-\\ \frac{7}{6}\end{array}\bigg|\frac{x^2}{16}\right)$
0	$\frac{1}{5}$	$x^{7/10}\,_0F_1\left(\begin{array}{c}-\\ \frac{6}{5}\end{array}\bigg|\frac{x^2}{16}\right)$
0	$\frac{1}{4}$	$x^{3/4}\,_0F_1\left(\begin{array}{c}-\\ \frac{5}{4}\end{array}\bigg|\frac{x^2}{16}\right)$
0	$\frac{1}{3}$	$x^{5/6}\,_0F_1\left(\begin{array}{c}-\\ \frac{4}{3}\end{array}\bigg|\frac{x^2}{16}\right)$

Table A6. The Whittaker functions $M_{\kappa,\mu}$ derived for some values of parameters κ and μ by using (18).

κ	μ	$M_{\kappa,\mu}(x)$
0	$\frac{1}{2}$	$2\sinh\left(\frac{x}{2}\right)$
0	1	$4\sqrt{x}\,I_1\left(\frac{x}{2}\right)$
0	$\frac{3}{2}$	$12\left[\cosh\left(\frac{x}{2}\right)-\frac{2}{x}\sinh\left(\frac{x}{2}\right)\right]$
0	2	$32\sqrt{x}\,I_2\left(\frac{x}{2}\right)$
0	$\frac{5}{2}$	$\frac{120}{x^2}\left[(x^2+12)\sinh\left(\frac{x}{2}\right)-6x\cosh\left(\frac{x}{2}\right)\right]$
$\frac{1}{6}$	0	$e^{-x/2}\sqrt{x}\,L_{-1/3}(x)$
$\frac{1}{4}$	$-\frac{5}{4}$	$x^{-3/4}e^{-x/2}\left(\frac{2x}{3}+1\right)$
$\frac{1}{4}$	$-\frac{3}{4}$	$x^{-1/4}e^{x/2}$
$\frac{1}{4}$	$-\frac{1}{4}$	$x^{1/4}e^{-x/2}$
$\frac{1}{4}$	0	$e^{-x/2}\sqrt{x}\,L_{-1/4}(x)$
$\frac{1}{3}$	0	$e^{-x/2}\sqrt{x}\,L_{-1/6}(x)$
$\frac{1}{2}$	0	$e^{-x/2}\sqrt{x}$
$\frac{1}{2}$	$\frac{1}{2}$	$x\left[I_0\left(\frac{x}{2}\right)-I_1\left(\frac{x}{2}\right)\right]$
$\frac{1}{2}$	1	$2x^{-1/2}e^{-x/2}(e^x-x-1)$
$\frac{1}{2}$	$\frac{3}{2}$	$4\left[-xI_0\left(\frac{x}{2}\right)+(x+4)I_1\left(\frac{x}{2}\right)\right]$
$\frac{1}{2}$	$\frac{5}{2}$	$32\left[(x+8)I_0\left(\frac{x}{2}\right)-\left(x+4+\frac{32}{x}\right)I_1\left(\frac{x}{2}\right)\right]$
1	0	$\sqrt{x}\left[-(x-1)I_0\left(\frac{x}{2}\right)+xI_1\left(\frac{x}{2}\right)\right]$
1	$\frac{1}{2}$	$x\,e^{-x/2}$
1	1	$\frac{4}{3}\sqrt{x}\left[xI_0\left(\frac{x}{2}\right)-(x+1)I_1\left(\frac{x}{2}\right)\right]$
$\frac{3}{2}$	0	$-\sqrt{x}\,e^{-x/2}(x-1)$
$\frac{3}{2}$	$\frac{1}{2}$	$-\frac{x}{3}\left[(2x-3)I_0\left(\frac{x}{2}\right)+(1-2x)I_1\left(\frac{x}{2}\right)\right]$

Table A6. Cont.

κ	μ	$M_{\kappa,\mu}(x)$
$\frac{3}{2}$	1	$x^{3/2} e^{-x/2}$
$\frac{3}{2}$	$\frac{3}{2}$	$\frac{4}{5}\left[x(2x+1)I_0\left(\frac{x}{2}\right) - (2x^2+3x+4)I_1\left(\frac{x}{2}\right)\right]$
2	0	$\frac{1}{3}\sqrt{x}\left[(2x^2-6x+3)I_0\left(\frac{x}{2}\right) - 2x(x-2)I_1\left(\frac{x}{2}\right)\right]$
2	$\frac{1}{2}$	$-\frac{1}{2}e^{-x/2}x(x-2)$
2	1	$-\frac{4}{15}\sqrt{x}\left[2x(x-2)I_0\left(\frac{x}{2}\right) + (-2x^2+2x+1)I_1\left(\frac{x}{2}\right)\right]$
2	$\frac{3}{2}$	$x^2 e^{-x/2}$
2	2	$\frac{32}{35\sqrt{x}}\left[x(2x^2+2x+3)I_0\left(\frac{x}{2}\right) - (x^3+2x^2+4x+6)I_1\left(\frac{x}{2}\right)\right]$
$\frac{5}{2}$	0	$\frac{1}{2}e^{-x/2}\sqrt{x}(x^2-4x+2)$
$\frac{5}{2}$	$\frac{1}{2}$	$\frac{x}{15}\left[(4x^2-18x+15)I_0\left(\frac{x}{2}\right) + (-4x^2+14x-3)I_1\left(\frac{x}{2}\right)\right]$
$\frac{5}{2}$	1	$-\frac{1}{3}e^{-x/2}x^{3/2}(x-3)$
$\frac{5}{2}$	2	$x^{5/2} e^{-x/2}$
3	$\frac{1}{2}$	$\frac{1}{6}e^{-x/2}x(x^2-6x+6)$
3	1	$\frac{4\sqrt{x}}{105}\left[x(4x^2-24x+27)I_0\left(\frac{x}{2}\right) - (4x^3+20x^2+9x+3)I_1\left(\frac{x}{2}\right)\right]$
3	$\frac{3}{2}$	$-\frac{1}{4}e^{-x/2}x^2(x-4)$
3	$\frac{5}{2}$	$x^3 e^{-x/2}$
$\frac{7}{2}$	0	$-\frac{1}{6}e^{-x/2}\sqrt{x}(x^3-9x^2+18x-6)$
4	$\frac{1}{2}$	$-\frac{1}{24}e^{-x/2}x(x^3-12x^2+36x-24)$
4	$\frac{3}{2}$	$\frac{1}{20}e^{-x/2}x^2(x^2-10x+20)$

Table A7. The Laplace transforms of the Whittaker function $M_{\kappa,\mu}$ derived for some values of parameters κ and μ.

κ	μ	$\mathcal{L}[M_{\kappa,\mu}(t)]$
$-\frac{5}{2}$	0	$\sqrt{\frac{\pi}{2}}\frac{8s^2+16s+5}{(2s-1)^{7/2}}$
$-\frac{3}{2}$	0	$\frac{2\sqrt{2\pi}(s+1)}{(2s-1)^{5/2}}$
$-\frac{3}{2}$	$\frac{1}{2}$	$\begin{cases} \frac{4\sqrt{4s^2-1}}{(2s-1)^3}, & s > \frac{1}{2} \\ 0, & s < \frac{1}{2} \end{cases}$
$-\frac{3}{2}$	1	$\frac{2\sqrt{2\pi}}{(2s-1)^{5/2}}$
$-\frac{3}{2}$	$\frac{3}{2}$	$\begin{cases} \frac{16\{-4s[4s^2-2s(\sqrt{4s^2-1}+3)+3(\sqrt{4s^2-1}+1)]+7\sqrt{4s^2-1}+2\}}{5(2s-1)^3}, & s > \frac{1}{2} \\ -\frac{32}{5}, & s < \frac{1}{2} \end{cases}$
$-\frac{1}{6}$	0	$\begin{cases} \frac{\sqrt{2\pi}\left[(6s+3)\,_2F_1\left(\begin{array}{c}-\frac{1}{2},\frac{2}{3}\\1\end{array}\bigg\|\frac{2}{2s+1}\right)+2\,_2F_1\left(\begin{array}{c}\frac{1}{2},\frac{2}{3}\\1\end{array}\bigg\|\frac{2}{2s+1}\right)\right]}{5(2s-1)^3}, & s > \frac{1}{2} \\ -\frac{\pi\,_2F_1\left(\begin{array}{c}\frac{2}{3},\frac{2}{3}\\\frac{1}{6}\end{array}\bigg\|s+\frac{1}{2}\right)}{(s+\frac{1}{2})^{5/6}\Gamma(\frac{1}{6})\Gamma(\frac{1}{3})}, & s < \frac{1}{2} \end{cases}$

Table A7. Cont.

κ	μ	$\mathcal{L}\left[M_{\kappa,\mu}(t)\right]$
$-\frac{1}{4}$	$\frac{1}{4}$	$\dfrac{2^{15/4}\Gamma\left(\frac{11}{4}\right)\left[(6s+3)\,_2F_1\!\left(\begin{array}{c}-\frac{1}{4},\frac{1}{2}\\ \frac{3}{2}\end{array}\bigg|\frac{2}{1-2s}\right)-4\left(\frac{2}{2s+1}+1\right)^{1/4}\right]}{21(2s-1)^{7/4}(2s+1)}$
$-\frac{1}{3}$	0	$\begin{cases}\dfrac{\sqrt{2\pi}\left[(6s+3)\,_2F_1\!\left(\begin{array}{c}-\frac{1}{2},\frac{5}{6}\\ 1\end{array}\bigg|\frac{2}{2s+1}\right)+4\,_2F_1\!\left(\begin{array}{c}\frac{1}{2},\frac{5}{6}\\ 1\end{array}\bigg|\frac{2}{2s+1}\right)\right]}{3(2s-1)(2s+1)^{3/2}},\ s>\frac{1}{2}\\[2ex] -\dfrac{\pi\,_2F_1\!\left(\begin{array}{c}\frac{5}{6},\frac{5}{6}\\ \frac{1}{3}\end{array}\bigg|s+\frac{1}{2}\right)}{\left(s+\frac{1}{2}\right)^{2/3}\Gamma\left(\frac{1}{6}\right)\Gamma\left(\frac{1}{3}\right)},\ s<\frac{1}{2}\end{cases}$
$-\frac{1}{2}$	0	$\dfrac{2\sqrt{2\pi}}{(2s-1)^{3/2}}$
$-\frac{1}{2}$	$\frac{1}{2}$	$\begin{cases}\dfrac{4}{(2s-1)\sqrt{4s^2-1}},& s>\frac{1}{2}\\ 0,& s<\frac{1}{2}\end{cases}$
$-\frac{1}{2}$	1	$2\sqrt{2\pi}\left[\dfrac{1}{\sqrt{2s+1}}-\dfrac{1}{\sqrt{2s-1}}+\dfrac{1}{(2s-1)^{3/2}}\right]$
0	1	$\begin{cases}\dfrac{2^{3/2}\left[(1-2s)K\!\left(\frac{2}{2s+1}\right)+2s\,E\!\left(\frac{2}{2s+1}\right)\right]}{\sqrt{\pi}(2s-1)\sqrt{2s+1}},& s>\frac{1}{2}\\[1.5ex] \dfrac{8\left[(1-2s)K\!\left(s+\frac{1}{2}\right)+4s\,E\!\left(s+\frac{1}{2}\right)\right]}{\sqrt{\pi}(4s^2-1)},& s<\frac{1}{2}\end{cases}$
0	2	$\begin{cases}\dfrac{64\left[8s(1-2s)K\!\left(\frac{2}{2s+1}\right)+(16s^2-3)\,E\!\left(\frac{2}{2s+1}\right)\right]}{\sqrt{\pi}(2s-1)\sqrt{s+\frac{1}{2}}},& s>\frac{1}{2}\\[1.5ex] \dfrac{64\left[(-16s^2+2s+3)K\!\left(s+\frac{1}{2}\right)+(32s^2-6)\,E\!\left(s+\frac{1}{2}\right)\right]}{\sqrt{\pi}(4s^2-1)},& s<\frac{1}{2}\end{cases}$
$\frac{1}{6}$	0	$\begin{cases}\dfrac{\sqrt{2\pi}\left[(6s+3)\,_2F_1\!\left(\begin{array}{c}-\frac{1}{2},\frac{1}{3}\\ 1\end{array}\bigg|\frac{2}{2s+1}\right)-2\,_2F_1\!\left(\begin{array}{c}\frac{1}{3},\frac{1}{2}\\ 1\end{array}\bigg|\frac{2}{2s+1}\right)\right]}{3(2s-1)(2s+1)^{3/2}},\ s>\frac{1}{2}\\[2ex] \dfrac{\sqrt{2\pi}\,\Gamma\!\left(\frac{7}{6}\right)\,_2F_1\!\left(\begin{array}{c}\frac{1}{3},\frac{1}{3}\\ -\frac{1}{6}\end{array}\bigg|s+\frac{1}{2}\right)}{(2s+1)^{7/6}\Gamma\!\left(\frac{5}{6}\right)\Gamma\!\left(\frac{1}{3}\right)},\ s<\frac{1}{2}\end{cases}$

Table A8. The Laplace transforms of the Whittaker functions $M_{\kappa,\mu}$ derived for some values of parameters κ and μ.

κ	μ	$\mathcal{L}\left[M_{\kappa,\mu}(t)\right]$
$\frac{1}{4}$	$-\frac{5}{4}$	$\dfrac{3\left(s+\frac{1}{2}\right)\Gamma\!\left(\frac{1}{4}\right)+2\,\Gamma\!\left(\frac{5}{4}\right)}{3\left(s+\frac{1}{2}\right)^{5/4}}$
$\frac{1}{4}$	$-\frac{3}{4}$	$\dfrac{\Gamma\!\left(\frac{3}{4}\right)}{\left(s-\frac{1}{2}\right)^{3/4}}$
$\frac{1}{4}$	$-\frac{1}{4}$	$\dfrac{\Gamma\!\left(\frac{5}{4}\right)}{\left(s+\frac{1}{2}\right)^{5/4}}$
0	$\frac{1}{2}$	$\dfrac{4}{4s^2-1}$
$\frac{1}{3}$	0	$\begin{cases}\dfrac{\sqrt{2\pi}\left[(6s+3)\,_2F_1\!\left(\begin{array}{c}-\frac{1}{2},\frac{1}{6}\\ 1\end{array}\bigg|\frac{2}{2s+1}\right)-4\,_2F_1\!\left(\begin{array}{c}\frac{1}{6},\frac{1}{2}\\ 1\end{array}\bigg|\frac{2}{2s+1}\right)\right]}{3(2s-1)(2s+1)^{3/2}},\ s>\frac{1}{2}\\[2ex] \dfrac{\Gamma\!\left(\frac{1}{6}\right)\Gamma\!\left(\frac{1}{3}\right)\,_2F_1\!\left(\begin{array}{c}\frac{1}{6},\frac{1}{2}\\ -\frac{1}{3}\end{array}\bigg|s+\frac{1}{2}\right)}{2^{2/3}\sqrt{3}\pi(2s+1)^{4/3}},\ s<\frac{1}{2}\end{cases}$
$\frac{1}{2}$	$\frac{1}{2}$	$\begin{cases}\dfrac{4}{(2s+1)\sqrt{4s^2-1}},& s>\frac{1}{2}\\ 0,& s<\frac{1}{2}\end{cases}$

Table A8. *Cont.*

κ	μ	$\mathcal{L}[\mathbf{M}_{\kappa,\mu}(t)]$
$\frac{1}{2}$	0	$\dfrac{\sqrt{2\pi}}{(2s+1)^{3/2}}$
$\frac{1}{2}$	1	$2\sqrt{2\pi}\left[-\dfrac{1}{\sqrt{2s+1}} - \dfrac{1}{(2s+1)^{3/2}} + \dfrac{1}{\sqrt{2s-1}}\right]$
1	0	$\begin{cases} \dfrac{2^{3/2}\left[-K\left(\frac{2}{2s+1}\right)+2E\left(\frac{2}{2s+1}\right)\right]}{\sqrt{\pi}(2s+1)^{3/2}}, & s > \frac{1}{2} \\ \dfrac{2(2s-3)K\left(s+\frac{1}{2}\right)+8E\left(s+\frac{1}{2}\right)}{\sqrt{\pi}(2s+1)^{2}}, & s < \frac{1}{2} \end{cases}$
1	$\frac{1}{2}$	$\dfrac{4}{(2s+1)^{2}}$
1	1	$\begin{cases} -\dfrac{2^{7/2}\left[-2(s+1)K\left(\frac{2}{2s+1}\right)+(2s+3)E\left(\frac{2}{2s+1}\right)\right]}{3\sqrt{\pi}(2s+1)^{3/2}}, & s > \frac{1}{2} \\ \dfrac{8(2s+5)K\left(s+\frac{1}{2}\right)-8(4s+6)E\left(s+\frac{1}{2}\right)}{3\sqrt{\pi}(2s+1)^{2}}, & s < \frac{1}{2} \end{cases}$
$\frac{3}{2}$	0	$\dfrac{2\sqrt{2\pi}(s-1)}{(2s+1)^{5/2}}$
$\frac{3}{2}$	$\frac{1}{2}$	$\begin{cases} \dfrac{4\sqrt{4s^2-1}}{(2s+1)^{3}}, & s > \frac{1}{2} \\ 0, & s < \frac{1}{2} \end{cases}$
$\frac{3}{2}$	1	$\dfrac{3\sqrt{2\pi}}{(2s+1)^{5/2}}$
2	0	$\begin{cases} \dfrac{2^{3/2}\left[4(1-2s)K\left(\frac{2}{2s+1}\right)+(14s-9)E\left(\frac{2}{2s+1}\right)\right]}{3\sqrt{\pi}(2s+1)^{5/2}}, & s > \frac{1}{2} \\ \dfrac{2(2s-1)(6s-13)K\left(s+\frac{1}{2}\right)+4(14s-9)E\left(s+\frac{1}{2}\right)}{3\sqrt{\pi}(2s+1)^{3}}, & s < \frac{1}{2} \end{cases}$
2	$\frac{1}{2}$	$\dfrac{8s-4}{(2s+1)^{3}}$
2	1	$\begin{cases} \dfrac{2^{7/2}\left[[-5+4s(s+2)]K\left(\frac{2}{2s+1}\right)-2[(2s+5)-6]E\left(\frac{2}{2s+1}\right)\right]}{15\sqrt{\pi}(2s+1)^{5/2}}, & s > \frac{1}{2} \\ \dfrac{8(2s-1)(2s+17)K\left(s+\frac{1}{2}\right)-32[s(2s+5)-6]E\left(s+\frac{1}{2}\right)}{15\sqrt{\pi}(2s+1)^{3}}, & s < \frac{1}{2} \end{cases}$
2	$\frac{3}{2}$	$\dfrac{16}{(2s+1)^{3}}$

Table A9. The Laplace transforms of the Whittaker functions $M_{\kappa,\mu}$ derived for some values of parameters κ and μ.

κ	μ	$\mathcal{L}[\mathbf{M}_{\kappa,\mu}(t)]$
$\frac{5}{2}$	1	$\dfrac{2\sqrt{2\pi}(3s-1)}{(2s+1)^{7/2}}$
$\frac{5}{2}$	2	$\dfrac{15\sqrt{2\pi}}{(2s+1)^{7/2}}$
3	$\frac{1}{2}$	$\dfrac{4(1-2s)^2}{(2s+1)^{4}}$
3	$\frac{3}{2}$	$\dfrac{8(4s-1)}{(2s+1)^{4}}$
3	$\frac{5}{2}$	$\dfrac{96}{(2s+1)^{4}}$
$\frac{7}{2}$	0	$\dfrac{\sqrt{2\pi}(8s^3-24s^2+15s-3)}{(2s+1)^{9/2}}$
4	$\frac{1}{2}$	$\dfrac{4(2s-1)^3}{(2s+1)^{5}}$
4	$\frac{3}{2}$	$\dfrac{32(10s^2-5s+1)}{5(2s+1)^{5}}$

Table A10. The Whittaker functions $W_{\kappa,\mu}$ derived for some values of parameters κ and μ by using (18).

κ	μ	$W_{\kappa,\mu}(x)$
$-\frac{5}{2}$	0	$\frac{\sqrt{x}}{4}e^{-x/2}\left[e^x(x^2+4x+2)\Gamma(0,x)-x-3\right]$
$-\frac{3}{2}$	0	$\sqrt{x}e^{-x/2}[e^x(x+1)\Gamma(0,x)-1]$
$-\frac{3}{2}$	1	$x^{3/2}e^{x/2}\Gamma(-2,x)$
$-\frac{1}{4}$	$\frac{1}{4}$	$x^{1/4}e^{x/2}\Gamma\left(\frac{1}{2},x\right)$
$-\frac{1}{2}$	0	$x^{1/2}e^{x/2}\Gamma(0,x)$
$-\frac{1}{2}$	1	$x^{-1/2}e^{-x/2}$
$-\frac{1}{2}$	2	$x^{-3/2}e^{-x/2}(x+3)$
$-\frac{1}{2}$	3	$x^{-5/2}e^{-x/2}(x^2+8x+20)$
$-\frac{3}{4}$	$\frac{3}{4}$	$\frac{e^{-x/2}}{2x^{-1/4}}\left[2\sqrt{x}-\sqrt{\pi}e^x(2x-1)\operatorname{erfc}(\sqrt{x})\right]$
0	β	$\sqrt{\frac{x}{\pi}}K_\beta\left(\frac{x}{2}\right)$
0	$\frac{1}{2}$	$e^{-x/2}$
0	$\frac{3}{2}$	$e^{-x/2}\left(1+\frac{2}{x}\right)$
0	$\frac{5}{2}$	$e^{-x/2}\left(1+\frac{6}{x}+\frac{12}{x^2}\right)$
$\frac{1}{4}$	$-\frac{5}{4}$	$x^{-3/4}e^{-x/2}\left(x+\frac{3}{2}\right)$
$\frac{1}{4}$	$-\frac{3}{4}$	$x^{-1/4}e^{x/2}\Gamma\left(\frac{3}{2},x\right)$
$\frac{1}{4}$	$-\frac{1}{4}$	$x^{1/4}e^{-x/2}$
$\frac{1}{2}$	0	$x^{1/2}e^{-x/2}$
$\frac{1}{2}$	1	$x^{-1/2}e^{x/2}\Gamma(2,x)$
$\frac{3}{4}$	$\frac{1}{4}$	$x^{3/4}e^{-x/2}$
$\frac{3}{4}$	$\frac{3}{4}$	$\frac{1}{2}x^{-1/4}e^{-x/2}(2x+1)$

Table A11. The Whittaker functions $W_{\kappa,\mu}$ derived for some values of parameters κ and μ by using (18).

κ	μ	$W_{\kappa,\mu}(x)$
$\frac{3}{4}$	$\frac{5}{4}$	$x^{-3/4}e^{x/2}\Gamma\left(\frac{5}{2},x\right)$
1	$\frac{1}{2}$	$xe^{-x/2}$
$\frac{3}{2}$	0	$\sqrt{x}e^{-x/2}(x-1)$
$\frac{3}{2}$	1	$x^{3/2}e^{-x/2}$
$\frac{3}{2}$	2	$x^{-3/2}e^{x/2}\Gamma(4,x)$
2	$\frac{1}{2}$	$x(x-2)e^{-x/2}$
2	$\frac{3}{2}$	$x^2 e^{-x/2}$
2	$\frac{5}{2}$	$x^{-2}e^{x/2}\Gamma(5,x)$
$\frac{5}{2}$	0	$\sqrt{x}e^{-x/2}(x^2-4x+2)$
$\frac{5}{2}$	1	$x^{3/2}e^{-x/2}(x-3)$
$\frac{5}{2}$	2	$x^{5/2}e^{-x/2}$

Table A11. *Cont.*

κ	μ	$W_{\kappa,\mu}(x)$
$\frac{5}{2}$	3	$x^{-5/2}e^{x/2}\Gamma(6,x)$
3	$\frac{1}{2}$	$e^{-x/2}x(x^2-6x+6)$
3	$\frac{3}{2}$	$e^{-x/2}x^2(x-4)$
3	$\frac{5}{2}$	$x^3 e^{-x/2}$
$\frac{7}{2}$	0	$e^{-x/2}\sqrt{x}(x^3-9x^2+18x-6)$
4	$\frac{1}{2}$	$e^{-x/2}x(x^3-12x^2+36x-24)$
4	$\frac{3}{2}$	$e^{-x/2}x^2(x^2-10x+20)$

Table A12. The Laplace transforms of the Whittaker function $W_{\kappa,\mu}$ derived for some values of parameters κ and μ.

κ	μ	$\mathcal{L}[W_{\kappa,\mu}(t)]$
$-\frac{1}{2}$	0	$\sqrt{2\pi}\left[\dfrac{\ln(\sqrt{4s^2-1}+2s)}{(2s-1)^{3/2}}+\dfrac{2}{(1-2s)\sqrt{2s+1}}\right]$
$-\frac{1}{2}$	1	$\sqrt{\dfrac{2\pi}{2s+1}}$
$\frac{1}{4}$	$-\frac{5}{4}$	$\dfrac{2^{1/4}\Gamma(\frac{1}{4})(3s+1)}{(2s+1)^{5/4}}$
$\frac{1}{4}$	$-\frac{1}{4}$	$\dfrac{\Gamma(\frac{5}{4})}{(s+\frac{1}{2})^{5/4}}$
0	$\frac{1}{2}$	$\dfrac{2}{2s+1}$
0	1	$\dfrac{8s\,E(\frac{1}{2}-s)-2(2s+1)K(\frac{1}{2}-s)}{4s^2-1}$
$\frac{1}{2}$	0	$\dfrac{\sqrt{2\pi}}{(2s+1)^{3/2}}$
$\frac{3}{4}$	$\frac{1}{4}$	$\dfrac{\Gamma(\frac{7}{4})}{(s+\frac{1}{2})^{7/4}}$
$\frac{3}{4}$	$\frac{3}{4}$	$\dfrac{2^{3/4}\Gamma(\frac{3}{4})(s+2)}{(2s+1)^{7/4}}$
$\frac{3}{4}$	$\frac{5}{4}$	$4\Gamma\left(\frac{11}{4}\right){}_2F_1\left(\begin{smallmatrix}\frac{1}{4},\frac{11}{4}\\\frac{5}{4}\end{smallmatrix}\bigg|\frac{1}{2}-s\right)$
1	$\frac{1}{2}$	$\dfrac{4}{(2s+1)^2}$
$\frac{3}{2}$	0	$\dfrac{2\sqrt{2\pi}(1-s)}{(2s+1)^{5/2}}$
$\frac{3}{2}$	1	$\dfrac{3\sqrt{2\pi}}{(2s+1)^{5/2}}$
2	$\frac{1}{2}$	$\dfrac{8-16s}{(2s+1)^3}$
2	$\frac{3}{2}$	$\dfrac{16}{(2s+1)^3}$
$\frac{5}{2}$	0	$\dfrac{\sqrt{2\pi}(8s^2-16s+5)}{(2s+1)^{7/2}}$
$\frac{5}{2}$	1	$\dfrac{6\sqrt{2\pi}(1-3s)}{(2s+1)^{7/2}}$
$\frac{5}{2}$	2	$\dfrac{15\sqrt{2\pi}}{(2s+1)^{7/2}}$

Table A12. *Cont.*

κ	μ	$\mathcal{L}[W_{\kappa,\mu}(t)]$
3	$\frac{1}{2}$	$\frac{24(1-2s)^2}{(2s+1)^4}$
3	$\frac{3}{2}$	$\frac{32(4s-1)}{(2s+1)^4}$
3	$\frac{5}{2}$	$\frac{96}{(2s+1)^4}$
$\frac{7}{2}$	0	$-\frac{6\sqrt{2\pi}(8s^3-24s^2+15s-3)}{(2s+1)^{9/2}}$
4	$\frac{1}{2}$	$\frac{96(1-2s)^3}{(2s+1)^5}$
4	$\frac{3}{2}$	$\frac{128(10s^2-5s+1)}{(2s+1)^5}$

Appendix C. Representations of the Wright Functions

The Wright functions $W_{\alpha,\beta}(x)$, defined in (39), and presented in Tables A13 and A14, as well as the Mainardi functions $F_\alpha(x)$ and $M_\alpha(x)$, defined in (40), and presented in Tables A15 and A16, were derived by using the MATHEMATICA program. Only a small part of these Wright functions is known in the mathematical reference literature.

In the case of positive rational $\alpha = p/q$ with p and q positive coprimes, applying (A1) and (A2), it is possible to express the Wright function by

$$W_{p/q,\beta}(x) = \sum_{k=0}^{q-1} \frac{x^k}{k!\,\Gamma\left(\frac{p}{q}k+\beta\right)} {}_0F_{p+q-1}\left(\begin{array}{c}-\\b_0,\ldots,b_{p-1},c_0^*,\ldots,c_{q-2}^*\end{array}\middle|\frac{x^q}{p^p q^q}\right), \quad (A5)$$

where

$$b_j = \frac{k}{q}+\frac{\beta+j}{p},$$
$$c_j = \frac{k+1+j}{q}, \quad (A6)$$

and the set of numbers $\{c_j^*\} = \{c_j\}\setminus\{1\}$.

For the Mainardi functions, we have the following reduction formulas for positive rational $\alpha = p/q$ with p and q positive coprimes:

$$F_{p/q}(x) \quad (A7)$$
$$= -\frac{1}{\pi}\sum_{k=1}^{q}\frac{(-x)^k}{k!}\Gamma\left(\frac{p}{q}k+1\right)\sin\left(\pi\frac{p}{q}k\right)$$
$$\quad {}_pF_{q-1}\left(\begin{array}{c}a_0,\ldots,a_{p-1}\\b_0^*,\ldots,b_{q-2}^*\end{array}\middle|\frac{(-1)^{p+q}x^q p^p}{q^q}\right),$$

and

$$M_{p/q}(x) = \frac{q}{px}F_{p/q}(x), \quad (A8)$$

where

$$a_j = \frac{k}{q}+\frac{j+1}{p},$$
$$b_j = \frac{k+1+j}{q},$$

and the set of numbers $\left\{b_j^*\right\} = \{b_j\}\setminus\{1\}$.

Table A13. The Wright functions $W_{\alpha,\beta}$ derived for some values of parameters α and β by using (39).

α	β	$W_{\alpha,\beta}(x)$
-1	$\frac{1}{2}$	$\frac{1}{2\sqrt{\pi}(x+1)^{3/2}}$
-1	$\frac{3}{2}$	$\frac{1}{\sqrt{\pi}(x+1)^{1/2}}$
-1	β	$\frac{(x+1)^{\beta-1}}{\Gamma(\beta)}$
$-\frac{1}{2}$	β	$\frac{1}{\Gamma(\beta)}\,_1F_1\left(\begin{array}{c}\frac{1-\beta}{2}\\ \frac{1}{2}\end{array}\Big| -\frac{x^2}{4}\right) + \frac{x}{\Gamma(\beta-\frac{1}{2})}\,_1F_1\left(\begin{array}{c}\frac{3}{2}-\beta\\ \frac{3}{2}\end{array}\Big| -\frac{x^2}{4}\right)$
$-\frac{1}{2}$	-1	$\frac{1}{8\sqrt{\pi}}\left[x(6-x^2)e^{-x^2/4}\right]$
$-\frac{1}{2}$	$-\frac{1}{2}$	$\frac{1}{4\sqrt{\pi}}\left[(x^2-2)e^{-x^2/4}\right]$
$-\frac{1}{2}$	0	$-\frac{xe^{-x^2/4}}{2\sqrt{\pi}}$
$-\frac{1}{2}$	$\frac{1}{2}$	$\frac{e^{-x^2/4}}{\sqrt{\pi}}$
$-\frac{1}{2}$	1	$\operatorname{erf}\left(\frac{x}{2}\right)+1$
$-\frac{1}{2}$	$\frac{3}{2}$	$x\left[\operatorname{erf}\left(\frac{x}{2}\right)+1\right]+\frac{2}{\sqrt{\pi}}e^{-x^2/4}$
0	$-\frac{3}{2}$	$\frac{3e^x}{4\sqrt{\pi}}$
0	$-\frac{1}{2}$	$\frac{e^x}{2\sqrt{\pi}}$
0	1	e^x
0	β	$\frac{e^x}{\Gamma(\beta)}$
$\frac{1}{3}$	β	$\frac{1}{\Gamma(\beta)}\,_0F_3\left(\begin{array}{c}-\\ \frac{1}{3},\frac{2}{3},\beta\end{array}\Big|\frac{x^3}{27}\right)+\frac{x}{\Gamma(\beta+\frac{1}{3})}\,_0F_3\left(\begin{array}{c}-\\ \frac{2}{3},\frac{4}{3},\beta+\frac{1}{3}\end{array}\Big|\frac{x^3}{27}\right)+\frac{x^2}{2\Gamma(\beta+\frac{2}{3})}\,_0F_3\left(\begin{array}{c}-\\ \frac{4}{3},\frac{5}{3},\beta+\frac{2}{3}\end{array}\Big|\frac{x^3}{27}\right)$
$\frac{1}{2}$	β	$\frac{1}{\Gamma(\beta)}\,_0F_2\left(\begin{array}{c}-\\ \frac{1}{2},\beta\end{array}\Big|\frac{x^2}{4}\right)+\frac{x}{\Gamma(\beta+\frac{1}{2})}\,_0F_3\left(\begin{array}{c}-\\ \frac{3}{2},\beta+\frac{1}{2}\end{array}\Big|\frac{x^2}{4}\right)$
1	β	$x^{(1-\beta)/2}I_{\beta-1}(2\sqrt{x})$
1	$-\frac{3}{2}$	$\frac{(4x+3)\cosh(2\sqrt{x})-6\sqrt{x}\sinh(2\sqrt{x})}{4\sqrt{\pi}}$
1	$-\frac{1}{2}$	$\frac{2\sqrt{x}\sinh(2\sqrt{x})-\cosh(2\sqrt{x})}{2\sqrt{\pi}}$
1	0	$\sqrt{x}I_1(2\sqrt{x})$
1	$\frac{1}{2}$	$\frac{\cosh(2\sqrt{x})}{\sqrt{\pi}}$
1	1	$I_0(2\sqrt{x})$
1	$\frac{3}{2}$	$\frac{\sinh(2\sqrt{x})}{\sqrt{\pi x}}$
1	$\frac{5}{2}$	$\frac{2\sqrt{x}\cosh(2\sqrt{x})-\sinh(2\sqrt{x})}{2\sqrt{\pi}x^{3/2}}$

Table A14. The Wright functions $W_{\alpha,\beta}$ derived for some values of parameters α and β by using (39).

α	β	$W_{\alpha,\beta}(x)$
$\frac{3}{2}$	β	$\frac{1}{\Gamma(\beta)} {}_0F_4\left(\begin{array}{c}-\\ \frac{1}{2},\frac{\beta+1}{3},\frac{\beta+2}{3},\frac{\beta}{3}\end{array}\bigg|\frac{x^2}{108}\right) + \frac{x}{\Gamma(\beta+\frac{3}{2})} {}_0F_4\left(\begin{array}{c}-\\ \frac{3}{2},\frac{2\beta+3}{6},\frac{2\beta+5}{6},\frac{2\beta+7}{6}\end{array}\bigg|\frac{x^2}{108}\right)$
2	β	$\frac{1}{\Gamma(\beta)} {}_0F_2\left(\begin{array}{c}-\\ \frac{\beta+1}{2},\frac{\beta}{2}\end{array}\bigg|\frac{x}{4}\right)$
3	β	$\frac{1}{\Gamma(\beta)} {}_0F_3\left(\begin{array}{c}-\\ \frac{\beta+1}{3},\frac{\beta+2}{3},\frac{\beta}{3}\end{array}\bigg|\frac{x}{27}\right)$
4	β	$\frac{1}{\Gamma(\beta)} {}_0F_4\left(\begin{array}{c}-\\ \frac{\beta+1}{4},\frac{\beta+2}{4},\frac{\beta+3}{4},\frac{\beta}{4}\end{array}\bigg|\frac{x}{256}\right)$
5	β	$\frac{1}{\Gamma(\beta)} {}_0F_5\left(\begin{array}{c}-\\ \frac{\beta+1}{5},\frac{\beta+2}{5},\frac{\beta+3}{5},\frac{\beta+4}{5},\frac{\beta}{5}\end{array}\bigg|\frac{x}{3125}\right)$

Table A15. The Mainardi function F_α derived for some values of parameter α by using (A7).

α	$F_\alpha(x)$
$\frac{3}{4}$	$\frac{x\Gamma(\frac{7}{4})}{\sqrt{2\pi}} {}_2F_2\left(\begin{array}{c}\frac{7}{12},\frac{11}{12}\\ \frac{1}{2},\frac{3}{4}\end{array}\bigg|-\frac{27x^4}{256}\right) + \frac{3x^2}{8\sqrt{\pi}} {}_2F_2\left(\begin{array}{c}\frac{5}{8},\frac{7}{8}\\ \frac{3}{4},\frac{5}{4}\end{array}\bigg|-\frac{27x^4}{256}\right) + \frac{x^3\Gamma(\frac{13}{4})}{6\sqrt{2\pi}} {}_2F_2\left(\begin{array}{c}\frac{13}{12},\frac{17}{12}\\ \frac{5}{4},\frac{3}{2}\end{array}\bigg|-\frac{27x^4}{256}\right)$
$\frac{2}{3}$	$\frac{\sqrt{3}x}{4\pi}\left[2\Gamma(\frac{5}{3}) {}_1F_1\left(\begin{array}{c}\frac{5}{6}\\ \frac{2}{3}\end{array}\bigg|-\frac{4x^3}{27}\right) + x\Gamma(\frac{7}{3}) {}_1F_1\left(\begin{array}{c}\frac{7}{6}\\ \frac{4}{3}\end{array}\bigg|-\frac{4x^3}{27}\right)\right]$
$\frac{1}{2}$	$\frac{xe^{-x^2/4}}{2\sqrt{\pi}}$
$\frac{1}{3}$	$3^{-1/3}x\,\mathrm{Ai}(3^{-1/3}x)$
$\frac{1}{4}$	$\frac{x\Gamma(\frac{5}{4})}{\sqrt{2\pi}} {}_0F_2\left(\begin{array}{c}-\\ \frac{1}{2},\frac{3}{4}\end{array}\bigg|-\frac{x^4}{256}\right) - \frac{x^2}{4\sqrt{\pi}} {}_0F_2\left(\begin{array}{c}-\\ \frac{3}{4},\frac{5}{4}\end{array}\bigg|-\frac{x^4}{256}\right) + \frac{x^3\Gamma(\frac{7}{4})}{6\sqrt{2\pi}} {}_0F_2\left(\begin{array}{c}-\\ \frac{5}{4},\frac{3}{2}\end{array}\bigg|-\frac{x^4}{256}\right)$

Table A16. The Mainardi function M_α derived for some values of parameter α by using (A8).

α	$M_\alpha(x)$
$\frac{3}{4}$	$\frac{1}{\Gamma(\frac{1}{4})} {}_2F_2\left(\begin{array}{c}\frac{7}{12},\frac{11}{12}\\ \frac{1}{2},\frac{3}{4}\end{array}\bigg|-\frac{27x^4}{256}\right) + \frac{x}{2\sqrt{\pi}} {}_2F_2\left(\begin{array}{c}\frac{5}{8},\frac{7}{8}\\ \frac{3}{4},\frac{5}{4}\end{array}\bigg|-\frac{27x^4}{256}\right) + \frac{x^2}{2\Gamma(-\frac{5}{4})} {}_2F_2\left(\begin{array}{c}\frac{13}{12},\frac{17}{12}\\ \frac{5}{4},\frac{3}{2}\end{array}\bigg|-\frac{27x^4}{256}\right)$
$\frac{2}{3}$	$3^{-2/3}e^{-2x^3/27}\left[3^{1/3}\mathrm{Ai}(3^{-4/3}x^2) - 3\mathrm{Ai}'(3^{-4/3}x^2)\right]$
$\frac{1}{2}$	$\frac{e^{-x^2/4}}{\sqrt{\pi}}$
$\frac{1}{3}$	$3^{2/3}\,\mathrm{Ai}(3^{-1/3}x)$
$\frac{1}{4}$	$\frac{2\sqrt{2}\Gamma(\frac{5}{4})}{\pi} {}_0F_2\left(\begin{array}{c}-\\ \frac{1}{2},\frac{3}{4}\end{array}\bigg|-\frac{x^4}{256}\right) - \frac{x}{\sqrt{\pi}} {}_0F_2\left(\begin{array}{c}-\\ \frac{3}{4},\frac{5}{4}\end{array}\bigg|-\frac{x^4}{256}\right) + \frac{\sqrt{2}x^2\Gamma(\frac{7}{4})}{3\pi} {}_0F_2\left(\begin{array}{c}-\\ \frac{5}{4},\frac{3}{2}\end{array}\bigg|-\frac{x^4}{256}\right)$

References

1. Erdélyi, A.; Magnus, W.; Oberhettinger, F.; Tricomi, F. *Higher Transcendental Functions*; McGraw-Hill: New York, NY, USA, 1953; Volume 1.
2. Erdélyi, A.; Magnus, W.; Oberhettinger, F.; Tricomi, F. *Tables of Integral Transforms*; McGraw-Hill: New York, NY, USA, 1954.
3. Abramowitz, M.; Stegun, I. *Handbook of Mathematical Functions with Formulas, Graphs, and Mathematical Tables*; U.S. National Bureau of Standards: Washington, DC, USA, 1964; Volume 55.
4. Magnus, W.; Oberhettinger, F.; Soni, R. *Formulas and Theorems for the Special Functions of Mathematical Physics*, 3rd ed.; Springer: Berlin, Germany, 1966.
5. Olver, F.; Lozier, D.; Boisvert, R.; Clark, C. *NIST Handbook of Mathematical Functions*; Cambridge University Press: Cambridge, UK, 2010.
6. Roberts, G.; Kaufman, H. *Table of Laplace Transforms*; WB Saunders Co.: Philadelphia, PA, USA, 1966.
7. Oberhettinger, F.; Badii, L. *Tables of Laplace Transforms*; Springer: Berlin, Germany, 1970.
8. Apelblat, A. *Laplace Transforms and Their Applications*; Nova Science Publishers Inc.: New York, NY, USA, 2012.
9. Van der Pol, B. On the operational solution of linear differential equations and an investigation of the properties of these solutions. *Lond. Edinb. Dublin Philos. Mag. J. Sci.* **1929**, *8*, 861–898. [CrossRef]
10. Van der Pol, B.; Bremmer, H. *Operational Calculus: Based on the Two-Sided Laplace Integral*; Taylor & Francis: Abingdon, UK, 1987; Volume 327.
11. Humbert, P. Bessel-integral functions. *Proc. Edinb. Math. Soc.* **1933**, *3*, 276–285. [CrossRef]
12. Apelblat, A.; Kravitsky, N. Integral representations of derivatives and integrals with respect to the order of the Bessel functions $J_\nu(t)$, $I_\nu(t)$, the Anger function $\mathbf{J}_\nu(t)$ and the integral Bessel function $Ji_\nu(t)$. *IMA J. Appl. Math.* **1985**, *34*, 187–210. [CrossRef]
13. Apelblat, A. *Bessel and Related Functions. Mathematical Operations with Respect to the Order*; Theoretical Aspects; Walter de Gruyter GmbH: Berlin, Germany, 2020; Volume 1.
14. Gorenflo, R.; Kilbas, A.; Mainardi, F.; Rogosin, S. *Mittag-Leffler Functions, Related Topics and Applications*; Springer: Berlin, Germany, 2014; Volume 2.
15. Whittaker, E. An expression of certain known functions as generalized hypergeometric functions. *Bull. Am. Math. Soc.* **1903**, *10*, 125–134. [CrossRef]
16. Prudnikov, A.; Brychkov, Y.; Marichev, O. *Integrals and Series: More Specific Function*; CRC Press: Boca Raton, FL, USA, 1986; Volume 3.
17. Wright, E. On the coefficients of power series having exponential singularities. *J. Lond. Math. Soc.* **1933**, *1*, 71–79. [CrossRef]
18. Wright, E. The generalized Bessel function of order greater than one. *Q. J. Math.* **1940**, *1*, 36–48. [CrossRef]
19. Gorenflo, R.; Luchko, Y.; Mainardi, F. Analytical properties and applications of the Wright function. *Fract. Calc. Appl. Anal.* **1999**, *2*, 383–414.
20. Mainardi, F. *Fractional Calculus and Waves in Linear Viscoelasticity: An Introduction to Mathematical Models*; World Scientific: Singapore, 2010.
21. Apelblat, A. Differentiation of the Mittag-Leffler functions with respect to parameters in the Laplace transform approach. *Mathematics* **2020**, *8*, 657. [CrossRef]

MDPI
St. Alban-Anlage 66
4052 Basel
Switzerland
Tel. +41 61 683 77 34
Fax +41 61 302 89 18
www.mdpi.com

Mathematics Editorial Office
E-mail: mathematics@mdpi.com
www.mdpi.com/journal/mathematics

www.ingramcontent.com/pod-product-compliance
Lightning Source LLC
LaVergne TN
LVHW070125100526
838202LV00016B/2232